T0213342

Texts in Computational Science and Engineering 18

Editors

Timothy J. Barth
Michael Griebel
David E. Keyes
Risto M. Nieminen
Dirk Roose
Tamar Schlick

More information about this series at http://www.springer.com/series/5151

John A. Trangenstein

Scientific Computing

Vol. I – Linear and Nonlinear Equations

Springer

John A. Trangenstein
Professor Emeritus of Mathematics
Department of Mathematics
Duke University
Durham
North Carolina, USA

Additional material to this book can be downloaded from http://extras.springer.com.

ISSN 1611-0994 ISSN 2197-179X (electronic)
Texts in Computational Science and Engineering
ISBN 978-3-030-09870-4 ISBN 978-3-319-69105-3 (eBook)
https://doi.org/10.1007/978-3-319-69105-3

Mathematics Subject Classification (2010): 15, 65

© Springer International Publishing AG, part of Springer Nature 2017
Softcover re-print of the Hardcover 1st edition 2017
This work is subject to copyright. All rights are reserved by the Publisher, whether the whole or part of the material is concerned, specifically the rights of translation, reprinting, reuse of illustrations, recitation, broadcasting, reproduction on microfilms or in any other physical way, and transmission or information storage and retrieval, electronic adaptation, computer software, or by similar or dissimilar methodology now known or hereafter developed.
The use of general descriptive names, registered names, trademarks, service marks, etc. in this publication does not imply, even in the absence of a specific statement, that such names are exempt from the relevant protective laws and regulations and therefore free for general use.
The publisher, the authors and the editors are safe to assume that the advice and information in this book are believed to be true and accurate at the date of publication. Neither the publisher nor the authors or the editors give a warranty, express or implied, with respect to the material contained herein or for any errors or omissions that may have been made. The publisher remains neutral with regard to jurisdictional claims in published maps and institutional affiliations.

Printed on acid-free paper

This Springer imprint is published by the registered company Springer International Publishing AG part of Springer Nature.
The registered company address is: Gewerbestrasse 11, 6330 Cham, Switzerland

To my daughter Christina

Preface

This is the first volume in a three-volume book about scientific computing. The primary goal in these volumes is to present many of the important computational topics and algorithms used in applications such as engineering and physics, together with the theory needed to understand their proper operation. However, a secondary goal in the design of this book is to allow readers to experiment with a number of interactive programs *within the book*, so that readers can improve their understanding of the problems and algorithms. This interactivity is available in the HTML form of the book, through JavaScript programs.

The intended audience for this book are upper level undergraduate students and beginning graduate students. Due to the self-contained and comprehensive treatment of the topics, this book should also serve as a useful reference for practicing numerical scientists. Instructors could use this book for multisemester courses on numerical methods. They could also use individual chapters for specialized courses such as numerical linear algebra, constrained optimization, or numerical solution of ordinary differential equations.

In order to read all volumes of this book, readers should have a basic understanding of both linear algebra and multivariable calculus. However, for this volume it will suffice to be familiar with linear algebra and single variable calculus. Some of the basic ideas for both of these prerequisites are reviewed in this text, but at a level that would be very hard to follow without prior familiarity with those topics. Some experience with computer programming would also be helpful, but not essential. Students should understand the purpose of a computer program, and roughly how it operates on computer memory to generate output.

Many of the computer programming examples will describe the use of a Linux operating system. This is the only publicly available option in our mathematics department, and it is freely available to all. Students who are using proprietary operating systems, such as Microsoft and Apple systems, will need to replace statements specific to Linux with the corresponding statements that are appropriate to their environment.

This book also references a large number of programs available in several programming languages, such as C, C^{++}, Fortran and JavaScript, as well as

MATLAB modules. These programs should provide examples that can train readers to develop their own programs, from existing software whenever possible or from scratch whenever necessary.

Chapters begin with an overview of topics and goals, followed by recommended books and relevant software. Some chapters also contain a case study, in which the techniques of the chapter are used to solve an important scientific computing problem in depth.

Chapter 1 uses a simple example to show the steps in solving a scientific problem through computation and analysis, and how to measure success with the computation. Some simple computer programs for solving this problem are provided in several computer languages.

Chapter 2 examines the fundamentals of computer arithmetic and discusses the pros and cons of various programming languages. It also describes how computer memory is organized and how this affects the proper design of programs. The case study on matrix-matrix multiplication in Sect. 2.11 demonstrates the importance of understanding computer architecture in designing software.

Chapter 3 discusses the numerical solution of systems of linear equations. This chapter begins with a review of the important theory, followed by the solution of a collection of easy linear systems. The key idea in this chapter is to reduce each general problem of solving a general linear system into a small number of easily solved linear systems. This chapter contains two case studies, related to the use of standard software to solve very large systems of linear equations and to estimate the quality of the computed results.

Chapter 4 discusses computer graphics. This topic is seldom discussed in scientific computing texts, but can be very useful in understanding and presenting scientific computing results. Interactive computer graphics can also be very helpful in designing and debugging scientific computer programs. This chapter will show readers how they can build visualization into their scientific computing programs. Readers will find that the discussion in this chapter makes use of some notions in linear algebra, discussed in Chap. 3.

Chapter 5 examines the numerical solution of nonlinear equations involving a single variable. For the most part, this chapter introduces the notion of *iterative* solutions to problems, as opposed to the *sequential* approaches that were used in solving small systems of linear equations in Chap. 3. The mathematical analysis of these problems involves calculus, rather than linear algebra. Although linear systems of equations are hard to visualize, nonlinear equations in a single variable are easy to visualize.

Chapter 6 discusses least squares problems. These are important in statistics, especially for parameter estimation and data fitting. The algorithms developed in this chapter are important in later chapters, especially in Chap. 1 of Volume II, on eigenvalues. The case study in this chapter involves a large least squares problem for parameter estimation. This chapter depends on Chap. 3.

In summary, this volume covers mathematical and numerical analysis, algorithm selection, and software development. The goal is to prepare readers to build programs for solving important problems in their chosen discipline. Furthermore, they should develop enough mathematical sophistication to know the limitations of the pieces of their algorithm, and to recognize when numerical features are due to programming bugs rather than the correct response of their problem.

I am indebted to many teachers and colleagues who have shaped my professional experience. I thank Jim Douglas Jr. for introducing me to numerical analysis as an undergrad. (Indeed, I could also thank a class in category theory for motivating me to look for an alternative field of mathematical study.) John Dennis, James Bunch, and Jorge Moré all provided a firm foundation for my training in numerical analysis, while Todd Dupont, Jim Bramble, and Al Schatz gave me important training in finite element analysis for my PhD thesis. But I did not really learn to program until I met Bill Gragg, who also emphasized the importance of classical analysis in the development of fundamental algorithms. I also learned from my students, particularly Randy LeVeque, who was in the first numerical analysis class I ever taught. Finally, I want to thank Bill Allard for many conversations about the deficiencies in numerical analysis texts. I hope that this book moves the field a bit in the direction that Bill envisions.

Most of all, I want to thank my family for their love and support.

John A. Trangenstein

Durham, NC, USA
July 7, 2017

Contents

Contents for Volume 2

Contents for Volume 3

Chapter 1
Introduction to Scientific Computing

There is nothing more difficult to take in hand, more perilous to conduct, or more uncertain in its success, than to take the lead in the introduction of a new order of things.
Nicolò Machiavelli, The Prince

Abstract This chapter introduces five basic steps in scientific computing applied to an initial value problem.The first step constructs a mathematical model, consisting of an ordinary differential equation and an initial value. The second step examines the mathematical well-posedness of the model problem, to see if the a solution exists, is unique, and depends continuously on the data. In the third step, we construct a simple numerical method for this problem, and in the fourth step we develop computer programs to execute this method. In the final step, we perform a mathematical analysis of the numerical method to determine its stability and convergence properties, even in the presence of computer roundoff errors. This analysis helps us to choose appropriate parameters for the numerical method, such as the time step size, and to compare the relative efficiency of competing methods.

1.1 Overview

Scientific computing involves both problem-solving with a computer, and computing with a scientific purpose. As a reader engages this subject, he or she should always keep in mind a practical problem to be solved. This aspect of scientific computing typically involves interaction of mathematicians or computer scientists with engineers, physicists, chemists, biologists, physicians, economists or social scientists. Of course, the problem needs to be formulated as a mathematical model, and the mathematical properties of this model should be investigated to understand whether the model is **well-posed**. As we will often note, a problem is well-posed

---

Additional Material: The details of the computer programs referred in the text are available in the Springer website (http://extras.springer.com/2018/978-3-319-69105-3) for authorized users.

© Springer International Publishing AG, part of Springer Nature 2017
J.A. Trangenstein, *Scientific Computing*, Texts in Computational Science and Engineering 18, https://doi.org/10.1007/978-3-319-69105-3_1

when a solution is guaranteed to exist, the solution is unique, and perturbations to the solution due to small errors in computation remain small.

Mathematics also comes into play when a practitioner of scientific computing develops numerical methods and numerical approximations to solve problems. Since a computer is employed to perform these numerical calculations, a reader needs to know how to interact with the computer through one or more *programming languages*. Some programming languages make this task easier than others, depending on the problem to be solved and the available computer environment. Finally, since the computer program must be implemented on an actual machine, the *computer architecture* is also an issue. This naturally leads the reader to consider the hierarchical nature of computer memory, the form of floating point operations, and possibly communication between processors.

In order to help the reader to make a clear connection between the mathematical model and its numerical solution on a particular machine, we will examine the use of several computer programming languages. In this book, we will primarily discuss the use of C, C^{++}, Fortran and MATLAB. There are some good reasons for these choices of programming languages. First, there are a tremendous number of high-quality scientific algorithms available in Fortran, and Fortran still provides the most user-friendly and efficient implementation of multi-dimensional arrays. The C programming language is used for many tasks that involve the operating system or references to computer memory. C^{++} allows us to develop simple interfaces to complicated operations such as graphics; if implemented properly, C^{++} can also perform many common scientific computations efficiently. MATLAB is an interactive, interpreted language; its use does not involve compilation or assembly steps, and it provides built-in graphical display. MATLAB is very popular in mathematics and engineering departments at a large number of universities.

In this chapter, we will introduce the reader to some simple scientific computing problems involving ordinary differential equations. Ordinary differential equations provide a nice example of a scientific computing problem because

- they motivate a number of related scientific computing issues (such as systems of linear equations, nonlinear equations, interpolation and approximation),
- they illustrate interesting mathematical issues (such as well-posedness, numerical stability and convergence),
- they lead to interesting programming issues (such as overflow and underflow, and the need for data structures and visualization), and
- they are important in solving even more complicated problems (such as partial differential equations).

Our first goal in this chapter is to describe the basic steps in solving a scientific computing problem. Once we understand those steps, we will present techniques for comparing scientific computing algorithms, by measuring their accuracy and relative efficiency. We will end this chapter with an overview of the major topics in this book.

1.2 Examples of Ordinary Differential Equations

In this section we will present two examples of ordinary differential equations. The first is an example of an initial-value problem, and should be familiar to anyone who has studied either calculus or physics. The second example is a boundary-value problem from solid mechanics.

Example 1.2.1 (Radioactive Decay) Let $u(t)$ represent the amount (measured in mass, moles, etc.) of some radioactive element. Experimentally, it is well-known that the rate of change of the amount of the radioactive isotope is proportional to the current amount. We will let λ represent the decay rate, with units of one over time. In order to quantify the amount of the isotope present at later times, we need to specify the amount u_0 present at the initial time. These two sentences can be written as the mathematical model

$$\frac{\mathrm{d}u}{\mathrm{d}t} = -\lambda u \text{ for } t > 0 \text{ , and } u(0) = u_0 \ .$$

Such a combination of an ordinary differential equation and a value at some prescribed time is called an **initial value problem**. Using techniques from calculus, we know that the analytic solution of this problem is

$$u(t) = u_0 \mathrm{e}^{-\lambda t} \text{ for } t \geq 0 \ .$$

It is easy to check that this function satisfies both the ordinary differential equation and the initial value.

If necessary, the decay rate can be determined indirectly from this solution. Let $t_{1/2} > 0$ be the half-life, namely the time at which the amount of the radioactive isotope is one-half its initial value. In other words, $t_{1/2}$ is defined by the equation $u(t_{1/2}) = u_0/2$. Then the analytical solution of the initial value problem implies that

$$u_0 \mathrm{e}^{-\lambda t_{1/2}} = u\left(t_{1/2}\right) = u_0/2 = u_0 \mathrm{e}^{-\log 2} \ .$$

Thus the decay rate can be determined as

$$\lambda t_{1/2} = \log 2 \Longrightarrow \lambda = \frac{\log 2}{t_{1/2}} \ .$$

Because the radioactive decay problem is so simple to state and solve, it has become an important test problem for numerical methods. We will discuss the merits of a number of methods for solving this problem in Chap. 3 of Volume III.

Example 1.2.2 (Bending of an Elastic Beam) Suppose that an elastic beam is rigidly supported at two ends, which are located at $x = 0$ and $x = L$. Let the *displacement* $u(x)$ of the beam be *infinitesimal*, so that the strain is accurately given by du/dx. We assume that *Hooke's law* applies, so that the stress is proportional to the strain, with proportionality constant k. Suppose that the beam has a constant density ϱ and is subjected to a single body force, namely gravity with acceleration g. At equilibrium, the material *restoring force* per volume must equal the force per volume due to gravity:

$$-\frac{\mathrm{d}}{\mathrm{d}x}\left(k\frac{\mathrm{d}u}{\mathrm{d}x}\right) = \rho g \;,$$

Also, the deflection must satisfy the specified boundary values:

$$u(0) = 0 = u(L) \;.$$

This is an example of a **boundary-value problem**.

It is easy to check that the analytical solution is

$$u(x) = \frac{\rho g}{2k}x(L-x) \;.$$

Readers should convince themselves that this function satisfies both the differential equation and the boundary values.

In practice, the acceleration of gravity g is well-known, and the length L of the beam is easily measured. The density ρ can be determined easily by weighing the beam and dividing by its volume. Next, we will show that the constant k in Hooke's law can be determined indirectly from the analytical solution, by measuring the maximum displacement $u_{\max}$ at the middle of the beam. Since

$$u_{\max} = u(L/2) = \frac{\rho g L^2}{8k} \;,$$

the elastic constant can be computed as

$$k = \frac{\rho g L^2}{8 u_{\max}} \;.$$

Later, in Chap. 4 of Volume III we will see that boundary-value problems typically require numerical methods that are very different from those used to solve initial-value problems.

Exercise 1.2.1 Reformulate the radioactive decay problem in terms of dimensionless time $\tau = t/t_{1/2}$ and the ratio $\upsilon = u/u_0$ of the current amount of radioactive material relative to the initial amount. What is the ordinary differential equation for υ in terms of τ, and how may we evaluate the solution υ in terms of τ?

Exercise 1.2.2 Reformulate the beam-bending problem in terms of dimensionless space $\xi = x/L$ and the dimensionless displacement $\upsilon = u/u_{\max}$.

1.3 Steps in Solving a Problem

We will identify five steps used in scientific computing to solve a problem. First, we must construct a mathematical model. Next, we must determine if the problem is well-posed; in other words, we need to guarantee that the problem has a unique solution that depends continuously on the problem data. Afterward, we will construct one or more numerical methods to solve the problem. These methods might be exact (if the available arithmetic permits it), or approximate. The real benefit from our work is obtained by implementing the numerical methods in a computer program.

However, we typically get the best numerical results when we analyze the numerical method to understand its computational work, memory requirements, numerical convergence and stability. This analysis guides us in selecting program parameters that are appropriate for specific scientific computing problems, and in choosing between competing algorithms.

1.3.1 Construct a Model

Since we began in Sect. 1.2 with ordinary differential equations as examples of problems that can motivate the use of scientific computing, we will continue to use ordinary differential equations for our discussion. We shall assume that we are given an ordinary differential equation in **autonomous form**

$$\frac{\mathrm{d}u}{\mathrm{d}t} = f(u) \; . \tag{1.1}$$

In general, u and f may be vectors of the same length. For simplicity, we will restrict our discussion to scalars in this chapter.

1.3.2 Examine Well-Posedness

The next step in applying scientific computing to a problem is to guarantee that the problem is **well-posed**. Let us remind the reader that a problem is well-posed when the problem has a solution, the solution is unique, and the solution depends continuously on the data. Generally, if the autonomous form (1.1) involves vectors of length n, then somehow we must specify n parameters. For an initial value problem, we might specify the initial value $u(0)$. For boundary value problems, uniquely specifying the solution can be more difficult, as the next example shows.

Example 1.3.1 (Non-uniqueness in Boundary-Value Problem) Consider the system of ordinary differential equations

$$\frac{\mathrm{d}}{\mathrm{d}t}\begin{bmatrix} u \\ w \end{bmatrix} = \begin{bmatrix} w \\ -u \end{bmatrix} = \begin{bmatrix} 0 & 1 \\ -1 & 0 \end{bmatrix}\begin{bmatrix} u \\ w \end{bmatrix} .$$

The reader can easily check that the general analytical solution can be written in the form

$$\begin{bmatrix} u \\ w \end{bmatrix}(t) = \begin{bmatrix} \sin(t+\alpha) \\ \cos(t+\alpha) \end{bmatrix} a .$$

Here the amplitude a and the phase shift α are the two parameters that specify a unique solution. Suppose that instead of specifying the amplitude and phase shift, we specify two boundary values. If we specify $u(0) = 0$, it follows that $\alpha = k\pi$ for some integer k. Then there is no solution if we specify a second boundary value of the form $u(n\pi) = 0$ or $w([n + 1/2]\pi) = 0$.

For general initial-value problems, Theorem 3.2.2 of Chap. 3 in Volume III will guarantee existence of a unique solution for general systems of initial value problems. For our less ambitious purposes in this chapter, the following theorem provides a simpler version of this theorem for a single (scalar) ordinary differential equation.

Theorem 1.3.1 (Existence for Initial-Value Problems) *Suppose that we are given an initial value u_0, and a function $f(u,t)$ that is continuous in both arguments for $|u - u_0| \le \alpha$ and $|t| \le T$. Also assume that f is* ***Lipschitz continuous*** *in u, meaning that there exists $L > 0$ such that for all $|t| \le T$ and for all $u, \tilde{u} \in [u_0 - \alpha, u_0 + \alpha]$ we have*

$$|f(u,t) - f(\tilde{u},t)| \le L|u - \tilde{u}| .$$

Then there is time $T_0 \le T$ and a unique function $u(t)$ so that $u(t)$ continuously differentiable for $0 < t < T_0$ and $u(t)$ solves the initial value problem

$$\frac{\mathrm{d}u}{\mathrm{d}t} = f(u,t) \ , \ u(0) = u_0 .$$

In order to assist the reader, we note that if f is continuously differentiable in u and continuous in t for $|t| \le T$, then it satisfies the Lipschitz continuity condition.

The conclusion of the theorem says that a solution is guaranteed to exist for some time interval, after which the solution could become infinite. We also note that this theorem says nothing about continuous dependence on the data; in other words, We have not yet ruled out the possibility that nearly equal initial values lead to radically different solutions to the initial value problem. Of course, we skip the proof of this theorem because it is too complicated for such an early stage of our development.

The next two examples show that when an initial-value problem is not Lipschitz continuous, the problem may have multiple solutions.

Example 1.3.2 (Non-uniqueness in Initial-Value Problem) Consider the initial-value problem

$$\frac{\mathrm{d}u}{\mathrm{d}t} = -\sqrt{1-u^2}\ ,\ u(0) = 1\ .$$

By inspection, we can see that both $u(t) \equiv 1$ and $u(t) = \cos(t)$ are solutions. In view of Theorem 1.3.1 we conclude that $f(u) = -\sqrt{1-u^2}$ cannot be Lipschitz continuous in u on a closed interval containing time $t = 0$. In fact,

$$\frac{|f(u_0) - f(u_0 - \varepsilon)|}{|u_0 - (u_0 - \varepsilon)|} = \frac{|\sqrt{1-1^2} - \sqrt{1-(1-\varepsilon)^2}|}{|1-(1-\varepsilon)|} = \frac{\sqrt{2\varepsilon}}{\varepsilon} \to \infty \text{ as } \varepsilon \to 0\ .$$

In this case, f is not continuously differentiable in u at the initial value, so Theorem 1.3.1 does not apply to this example.

Example 1.3.3 (Non-uniqueness in Initial-Value Problem) Consider the initial-value problem

$$\frac{\mathrm{d}u}{\mathrm{d}t} = \frac{u}{t}\ ,\ u(0) = 0\ .$$

Note that $u(t) = ct$ is a solution for any value of c. In this case, f is not continuous in t at the initial time, so Theorem 1.3.1 does not apply to this example.

Next, we will provide a theorem that guarantees continuous dependence on the data. Again, we provide a simpler statement than the more general form in Theorem 3.2.3 of Chap. 3 in Volume III.

Theorem 1.3.2 (Continuous Dependence for Initial-Value Problems) *Suppose that we are given an initial value u_0, and a function $f(u,t)$ that is continuous in both arguments for $|u - u_0| \le \alpha$ and $|t| \le T$. Also assume that f is Lipschitz continuous in u. Further, suppose that $u(t)$ solves the initial value problem*

$$\frac{\mathrm{d}u}{\mathrm{d}t} = f(u,t)\ ,\ u(0) = u_0\ ,$$

and $\tilde{u}(t)$ solves the modified initial value problem

$$\frac{\mathrm{d}\tilde{u}}{\mathrm{d}t} = f(\tilde{u},t) + \delta(t)\ ,\ \tilde{u}(0) = \tilde{u}_0 + \varepsilon\ ,$$

Then the difference of the two solutions satisfies

$$\max_{0\le t\le T_0} |\tilde{u}(t) - u(t)| \le \max\left\{|\varepsilon|, \max_{0\le t\le T_0} |\delta(t)|\right\} \frac{(L+1)\mathrm{e}^{Lt} - 1}{L}\ .$$

Although perturbations in the initial value u_0 or in the rate of change f may grow exponentially in time, the size of the perturbation at any fixed later time t is bounded. The conclusion of this theorem describes precisely the sense in which the initial value problem depends continuously on its data.

1.3.3 Construct a Numerical Method

Our next step in scientific computing is to construct a numerical method. For the numerical solution of the initial-value problem

$$\frac{\mathrm{d}u}{\mathrm{d}t} = f(u,t)\ ,\ u(0) = u_0\ , \tag{1.2}$$

the simplest approach begins by approximating the solution $u(t)$ at N distinct times t_n with $0 < t_1 < \ldots < t_N = T$. The approximate solution values will be denoted by $\tilde{u}_n \approx u(t_n)$. There are several sources of errors in this approach. First, we replace $u(t_n)$ by $\tilde{u}_n$. Next, we may also replace $f(\tilde{u}_n, t)$ by a numerical approximation, The largest errors will likely arise from replacing the derivative du/dt by a difference quotient. Finally, we use finite-precision arithmetic on a computer to perform the computations. All of these errors may accumulate, in ways that depend on the problem, numerical method and the computer.

In order to discuss the relative merits of different numerical methods, we will need several new concepts. First, an **explicit one-step method** for approximating the solution of $du/dt = f(u,t)$ can be written in the form

$$\frac{\tilde{u}_{n+1} - \tilde{u}_n}{t_{n+1} - t_n} = \phi(\tilde{u}_n, t_n, t_{n+1})\ . \tag{1.3}$$

Next, we need to what it means for a numerical method to be convergent.

Definition 1.3.1 Suppose that we specify some desired accuracy ε for a numerical approximation. Then a sequence of approximate solutions $\{\tilde{u}_n\}_{n=0}^{\infty}$ is **convergent** to a function $u(t)$ for all times $t \in [0,T]$ if and only if there exists some minimum total number of time steps N_0 so that for all sufficiently large total number of time steps $N \geq N_0$, we have that for all time step indices $1 \leq n \leq N$ with $t_n \in [0,T]$ the error satisfies $|\tilde{u}_n - u(t_n)| \leq \varepsilon$.

Basically, this definition says that for any specified time interval $[0,T]$ and any error tolerance ε, we can take enough small time steps so that the error in the numerical solution is within the desired tolerance during the specified time interval.

The previous definition says absolutely nothing about converging to a solution of a given differential equation. This means that we need another definition.

Definition 1.3.2 The explicit one-step method (1.3) is **consistent** with the initial-value problem (1.2) on the time interval $0 \le t \le T$ if and only if $\tilde{u}_0 = u_0$ and for all $0 \le t \le T$ we have $\phi(u(t), t, t) = f(u(t), t)$.
In simple terms, this definition says that the true solution would satisfy the numerical method if we could take infinitesimally small time steps.

We need one more definition to describe our numerical methods.

Definition 1.3.3 An explicit one-step method is **stable** if and only if there exists some time step bound $h > 0$, and perturbation bound $B > 0$ so that the following is true: if for all $0 \le n \le N$ we have $|t_{n+1} - t_n| \le h$, then any two sequences $\{\tilde{u}_n\}_{n=0}^N$ and $\left\{\tilde{\tilde{u}}_n\right\}_{n=0}^N$ computed by a given explicit one-step method (1.3) are such that for all $1 \le N$

$$\left|\tilde{u}_n - \tilde{\tilde{u}}_n\right| \le B\left|\tilde{u}_0 - \tilde{\tilde{u}}_0\right| .$$

In other words, explicit one-step methods are stable when the perturbations at later times are at worst a bounded multiple of perturbations in the initial values.

The following example shows that a method can be stable without being consistent.

Example 1.3.4 The method $\tilde{u}_n = \tilde{u}_{n-1}$ for all n is stable, but not necessarily consistent for solving the initial-value problem (1.2).

There are also examples of consistent but unconditionally unstable methods, but these are found most commonly among *multistep methods* (see Sect. 3.4 of Chap. 3 in Volume III). In general, we will find that *consistency and stability imply convergence*, but we will have to phrase this claim carefully.

Here is an example of a simple consistent and conditionally stable one-step method.

Example 1.3.5 (Forward-Euler Method) The **forward Euler method** is defined by the finite-difference approximation

$$\frac{\tilde{u}_{n+1} - \tilde{u}_n}{t_{n+1} - t_n} = f(\tilde{u}_n, t_n) \tag{1.4}$$

to the ordinary differential equation (1.2). This method can be rewritten in the form

$$\tilde{u}_{n+1} = \tilde{u}_n + [t_{n+1} - t_n] f(\tilde{u}_n, t_n) \ , \ \tilde{u}_0 = u_0 \ .$$

In the special case $f(u, t) = \lambda u$, the analytical solution to the initial value problem is

$$u(t) = e^{\lambda t} u(0) \ .$$

If we take equally-spaced time steps, then the forward-Euler solution (in the absence of rounding errors) has values

$$\tilde{u}_n = (1 + \lambda h)^n \tilde{u}_0 \ .$$

Readers may remember the result from calculus that this approximation tends to the analytical solution as $n \to \infty$. This limit shows that the forward Euler method is convergent for at least this special case.
We will analyze the consistency, stability and convergence of the forward-Euler method in Sect. 1.3.5.

1.3.4 Implement a Computer Program

Our next step in scientific computing is to construct a computer program to implement the numerical method. The following algorithm a quick description of the forward-Euler method.

Algorithm 1.3.1

$$\tilde{u}_0 = u_0$$

for $0 \le n < N$

$$\tilde{u}_{n+1} = \tilde{u}_n + (t_{n+1} - t_n) f(\tilde{u}_n, t_n)$$

The Fortran program euler.f, the C program euler.c or the C^{++} program euler.C are examples of computer programs implementing the forward-Euler method. Each program begins by declaring the types of variables (both `integer` and `real`). We recommend this programming style for Fortran; it is required for C. Each program also uses *arithmetic statement functions* for the problem description (the function `f`) and the numerical method (the function `euler`). Most of each program is devoted to reading the input parameters (`rate`, `x`, `tmax`, and `nsteps`). The evolution of the solution of the initial-value problem is computed by a loop (`do` in Fortran, `for` in C or C^{++}). Inside the loop, the numerical solution at $t_n = nh$ is represented by the program variable `u`. The computed results are written to a file (`fort.10` in Fortran, `coutput` in C and `cppoutput` in C^{++}).

In MATLAB, the same kind of program organization could be accomplished by three files fode.m, euler.m and m_exec.m. MATLAB functions must appear in separate files with related names, and all parameters for a MATLAB function must appear in the argument list for the function. In other words, MATLAB functions cannot make use of "global variables."

Before running any one of the Fortran, C or C^{++} programs, we have to **compile** the program and **assemble** an executable. The compile step translates our program statements into machine instructions, and the assemble step combines our machine instructions with system utilities for handling input and output. Suppose that we have stored the Fortran program in a file `euler.f`. For the Fortran file, both compilation and assembly steps can be accomplished under Linux by the following command:

```
gfortran -g -o f_exec euler.f
```

This command compiles the program file `euler.f` and constructs the executable `f_exec`. For the C program stored in a file `euler.c`, we would type

```
gcc -g -o c_exec euler.c
```

For the C^{++} program stored in a file `euler.C`, we would type

```
g++ -g -o C_exec euler.C
```

Under operating systems other than Linux, the user will need to replace the names of the Linux compilers (`gfortran`, `gcc` and `g++`) with the appropriate names for the resident compilers. Note that MATLAB is an interpreted language; it does not involve a compiler or an assembler.

If compilation and assembly is successful (i.e., neither the compiler nor the assembler find any errors), then the executable is placed in the specified output file. The Fortran program can be executed by typing the command

```
f_exec
```

If the `gfortran` command is not successful, then the compiler will provide some complaints that should guide the user in fixing their mistakes. Similarly, the C program can be executed by typing the command

```
c_exec
```

and the C^{++} program can be executed by typing the command

```
C_exec
```

To execute the MATLAB program, inside the MATLAB command window just type

```
m_exec
```

When any one of the Fortran, C or C^{++} programs executes, the input parameters are echoed to the screen, and the results of the numerical method are written to a file (`fort.10` for Fortran, `coutput` for C and `cppoutput` for C^{++}). The Fortran results can be viewed graphically by typing the Linux command

```
xmgrace fort.10
```

With other operating systems, it may be necessary to replace the `xmgrace` plotting command with its appropriate substitute. To plot the C or C^{++} output, we would replace the file name following the `xmgrace` command. The MATLAB program `m_exec.m` contains an instruction to plot the results from MATLAB. All of these instructions can be found in the README file.

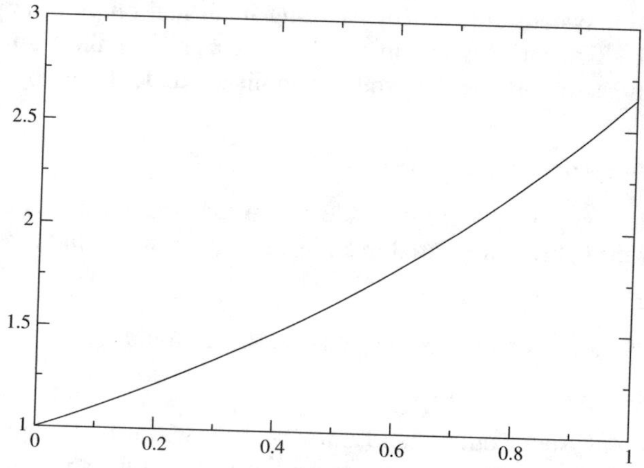

Fig. 1.1 Forward Euler method for $u'(t) = u(t)$, $u(0) = 1$ using 20 time steps

For example, if we choose the `rate` to be 1.0, the `initial u` to be 1.0, `tmax` to be 0.9999 and `number steps` to be 20, then `xmgrace` produces the plot shown in Fig. 1.1.

To solve

$$u'(t) = -10u(t) \ , \ u(0) = 1 \ , \ 0 < t < 1 \ ,$$

readers may also experiment with the following small JavaScript Program for Euler's Method. (Clicking on this link should open your default browser to run the JavaScript program.)

Exercise 1.3.1 Copy euler.f, euler.c or euler.C onto your computer, compile it and form an executable. Then plot the results for `rate = 1.`, initial x = 1., `tmax` = 1. and `nsteps` = $2^k, k = 1, \ldots, 6$.

Exercise 1.3.2 Modify `euler.f`, `euler.c` or `euler.C` to integrate the initial value problem with $f(u,t) = -1/u^2$, $u(0) = 1$. Run the numerical method for $0 \le t \le 1$ and plot the results for `number steps =` $10^k, k = 1, \ldots, 6$.

Exercise 1.3.3 Modify `euler.f`, `euler.c` or `euler.C` to use the **modified Euler method**

$$\tilde{u}_{n+1} = \tilde{u}_n + hf\left(\tilde{u}_n + [h/2]f(\tilde{u}_n, nh), [n + 1/2]\, h\right) \ .$$

Compare this method to the forward euler method for $f(u,t) = u$. Which method is more accurate for the same number of time steps? Which method requires more work per time step? What would you measure to decide which of these two methods is preferable?

Exercise 1.3.4 Consider the ordinary differential equation

$$\frac{\mathrm{d}u}{\mathrm{d}t} = -\sqrt{|u|}(1+u^2) \;,\; u(0) = 3 \;.$$

First, modify `euler.f`, `euler.c` or `euler.C` to compute a numerical solution to this problem. Next, plot the numerical solution for $0 \leq t \leq 3$ using $h = 10^{-k}, k = 1, 2, 3$. Afterward, compute the numerical solution for $0 \leq t \leq 4$. Use the existence and continuous dependence theorems to explain what is going on.

Exercise 1.3.5 Experimentally, we have found that the rate of growth of the enrollment in scientific computing is a quadratic function of the current population. This gives us the following model:

$$\frac{\mathrm{d}u}{\mathrm{d}t} = au(b-u) \;,\; u(0) = u_0 \;.$$

In order to apply this model to data, we need some information:

- Initially, 15 students were registered.
- One-half year later, 16 students were registered.
- The registrar won't let more than 30 students register.

First, find the analytical solution to this problem (usually called the *logistics equation*). Next, use the information above to find values for a, b and u_0. Afterward, modify the code in `euler.f`, `euler.c` or `euler.C` to solve this problem. Also, modify the code to compute the error in the numerical solution. (Here, the error is the absolute value of the numerical solution minus the analytical solution at some specified time.) Suppose that our goal is to determine the enrollment after 5 years. Plot $\log(error)$ versus $\log(1/h)$ for $h = 2^{-k}, k = 1, 2, \ldots, 10$.

1.3.5 Analyze the Method

Previously in this chapter, we have examined some applied problems involving ordinary differential equation, and listed several important steps in applying scientific computing to solve problems. In particular, we presented the forward euler method for solving initial value problems, and showed how to implement this method in Fortran. In this section, we will examine the forward euler method analytically, in order to understand its numerical behavior. This analysis will help us to understand the circumstances under which the method works, as well as the limitations imposed on it by its design and by computer rounding errors. After performing this analysis, we should be able to understand how to use the method most effectively.

1.3.5.1 Stability

Let us begin by discussing the **stability** of the forward Euler method. We have already seen one definition of stability in Sect. 1.3.3, namely that a perturbation in the initial data for the numerical method leads to a bounded perturbation in the numerical solution.

Lemma 1.3.1 *Suppose that the function $f(u,t)$ is Lipschitz continuous in u with Lipschitz constant L, and suppose that $0 = t_0 < t_1 < \ldots < t_N = T$. Let $\{\tilde{u}_n\}_{n=0}^N$ and $\left\{\tilde{\tilde{u}}_n\right\}_{n=0}^N$ be any two sequences computed by the forward Euler method (1.4) for the given time sequence $\{t_n\}_{n=0}^N$. Then the differences in the two forward Euler sequence entries satisfies*

$$\left|\tilde{u}_n - \tilde{\tilde{u}}_n\right| \leq \mathrm{e}^{Lt_n} \left|\tilde{u}_0 - \tilde{\tilde{u}}_0\right| .$$

Proof Suppose $\tilde{u}_n$ and $\tilde{\tilde{u}}_n$ satisfy

$$\tilde{u}_{n+1} = \tilde{u}_n + [t_{n+1} - t_n] f(\tilde{u}_n, t_n) \ , \ \tilde{\tilde{u}}_{n+1} = \tilde{\tilde{u}}_n + [t_{n+1} - t_n] f(\tilde{\tilde{u}}_n, t_n) .$$

Then the difference in the two results satisfies

$$\tilde{\tilde{u}}_{n+1} - \tilde{u}_{n+1} = \left[\tilde{\tilde{u}}_n - \tilde{u}_n\right] + [t_{n+1} - t_n] \left[f\left(\tilde{\tilde{u}}_n, t_n\right) - f\left(\tilde{u}_n, t_n\right)\right] .$$

Then the Lipschitz continuity of f implies that

$$\begin{aligned}\left|\tilde{\tilde{u}}_{n+1} - \tilde{u}_{n+1}\right| &\leq \left|\tilde{\tilde{u}}_n - \tilde{u}_n\right| + L[t_{n+1} - t_n] \left|\tilde{\tilde{u}}_n - \tilde{u}_n\right| \\ &= (1 + L[t_{n+1} - t_n]) \left|\tilde{\tilde{u}}_n - \tilde{u}_n\right| \leq \left|\tilde{\tilde{u}}_n - \tilde{u}_n\right| \mathrm{e}^{L[t_{n+1} - t_n]} .\end{aligned}$$

An argument by induction shows us that

$$\left|\tilde{\tilde{u}}_n - \tilde{u}_n\right| \leq \left|\tilde{\tilde{u}}_0 - \tilde{u}_0\right| \prod_{k=0}^{n-1} \mathrm{e}^{L[t_{k+1} - t_k]} = \left|\tilde{\tilde{u}}_0 - \tilde{u}_0\right| \mathrm{e}^{L \sum_{k=0}^{n-1} [t_{k+1} - t_k]} = \left|\tilde{\tilde{u}}_0 - \tilde{u}_0\right| \mathrm{e}^{Lt_n} .$$

With this notion of stability, the forward euler method for a Lipschitz continuous function f is stable, no matter what size we choose for the time steps. We shall soon see that this meaning of stability is useless for problems with decaying solutions.

1.3.5.2 Absolute Stability

Let us define a notion of stability that is *moderately* useful for ordinary differential equations involving decay.

Definition 1.3.4 The region of **absolute stability** for an explicit one-step method is the set of growth rates λ and time steps h for which a perturbation in the numerical solution to

$$\frac{\mathrm{d}u}{\mathrm{d}t} = \lambda u$$

does not increase from one step to the next.

In the case of the forward Euler method

$$\tilde{u}_{n+1} = \tilde{u}_n + \lambda h \tilde{u}_n = (1 + \lambda h)\tilde{u}_n ,$$

the region of absolute stability is given by those values of λ and h for which

$$|1 + \lambda h| \leq 1 .$$

This is a circle in the complex λh plane with center at -1 and radius 1, as shown by the blue curve in Fig. 1.2. The green curve in that figure is the locus of points where $|1 + \lambda h| = 0.9$, and the red curve is the locus of points where $|1 + \lambda h| = 1.1$.

Note that if the solution of an initial value problem is increasing in time, then it is undesirable to select a method and a time step to obtain absolute stability. On the other hand, if the analytical solution is decreasing, then absolute stability is highly desirable.

Example 1.3.6 (Absolute Stability for Rapid Decay) Suppose that we want to solve

$$\frac{\mathrm{d}u}{\mathrm{d}t} = -100u .$$

Since the solution $u(t) = \mathrm{e}^{-100t}u(0)$ decreases rapidly in time, absolute stability is desirable. If we use the forward euler method, then absolute stability requires

$$|1 + \lambda h| = |1 - 100h| \leq 1 ,$$

or $h \leq 0.02$. If the time steps are not chosen to satisfy this inequality, the forward Euler method will produce a sequence that grows exponentially and oscillates in sign with each step.

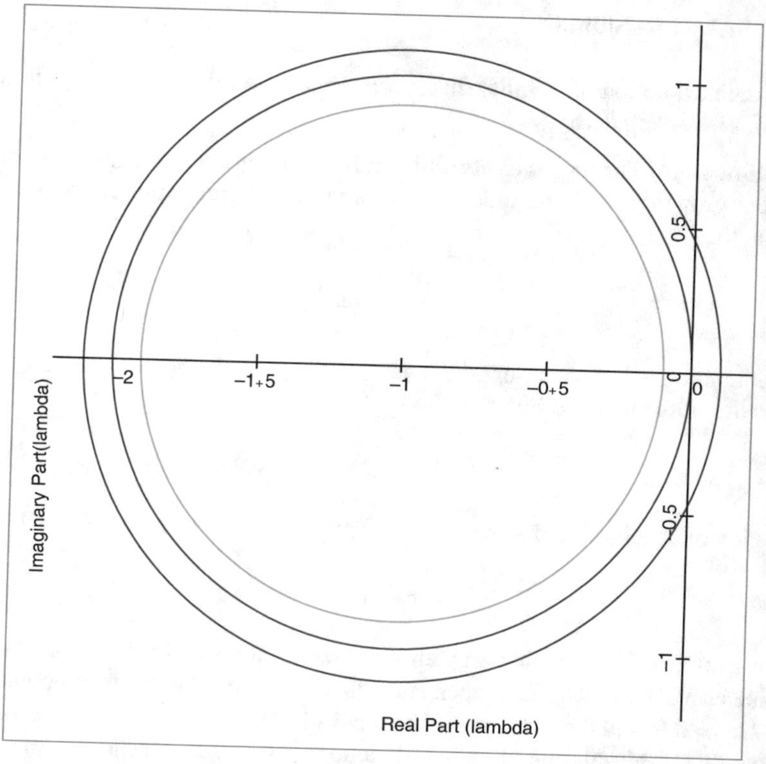

Fig. 1.2 Forward Euler absolute stability region: blue is absolute stability region, red is outside absolute stability region, green is inside absolute stability region

Example 1.3.7 (Absolute Stability and Rapid Growth) Suppose that we want to solve

$$\frac{du}{dt} = 100u \;.$$

Since the solution $u(t) = e^{100t}u(0)$ increases in time, absolute stability is undesirable. If we use the forward euler method, then

$$\tilde{u}_{n+1} = (1 + \lambda h)\tilde{u}_n \;.$$

Note that for any time step h

$$1 + \lambda h = 1 + 100h \geq 1 \;;$$

in other words, any size time step will produce an increasing numerical solution with the forward euler method. There is no choice for the time step h that can lead to absolute stability for this particular problem.

1.3.5.3 Monotonicity

In practice, the absolute stability condition does not necessarily produce acceptable numerical results. In particular, it is possible to choose time steps satisfying the absolute stability condition, and obtain decaying *oscillations* for problems involving monotonic decay. A more useful condition is **monotonicity preservation**, which requires the numerical solution to be monotone whenever the analytical solution is monotone.

Example 1.3.8 (Monotonicity for Decay) Suppose that we want to solve

$$\frac{du}{dt} = -\lambda u \;,\; u(0) > 0 \;,\; \lambda > 0 \;.$$

The analytical solution to this problem is positive and monotonically decreasing. The forward Euler approximation

$$\tilde{u}_{n+1} = (1 - \lambda h)\tilde{u}_n$$

remains positive and decreases if and only if

$$0 < 1 - \lambda h < 1 \Longleftrightarrow 0 < h < \frac{1}{\lambda} \;.$$

As we have seen, absolute stability allows h to be twice as large.

Example 1.3.9 (Monotonicity for Growth) Suppose that we want to solve

$$\frac{du}{dt} = \lambda u \;,\; u(0) > 0 \;,\; \lambda > 0 \;.$$

Then the forward Euler approximation is

$$\tilde{u}_{n+1} = (1 + \lambda h)\tilde{u}_n \;.$$

Here, the analytical solution is positive and increasing, and so is the forward Euler approximation for any time step $h > 0$.

Figure 1.3 shows some computational results with the forward Euler method for the initial value problem

$$\frac{du}{dt} = -20u \;,\; u(0) = 1 \;.$$

The analytical solution of this problem is $u(t) = e^{-20t}$, which involves rapid and monotone decay to zero as t increases. In graph (a) of this figure, we show that the numerical solution grows and oscillates when 6 equally spaced time steps are used; this behavior occurs because this time step is not absolutely stable. In graph (b),

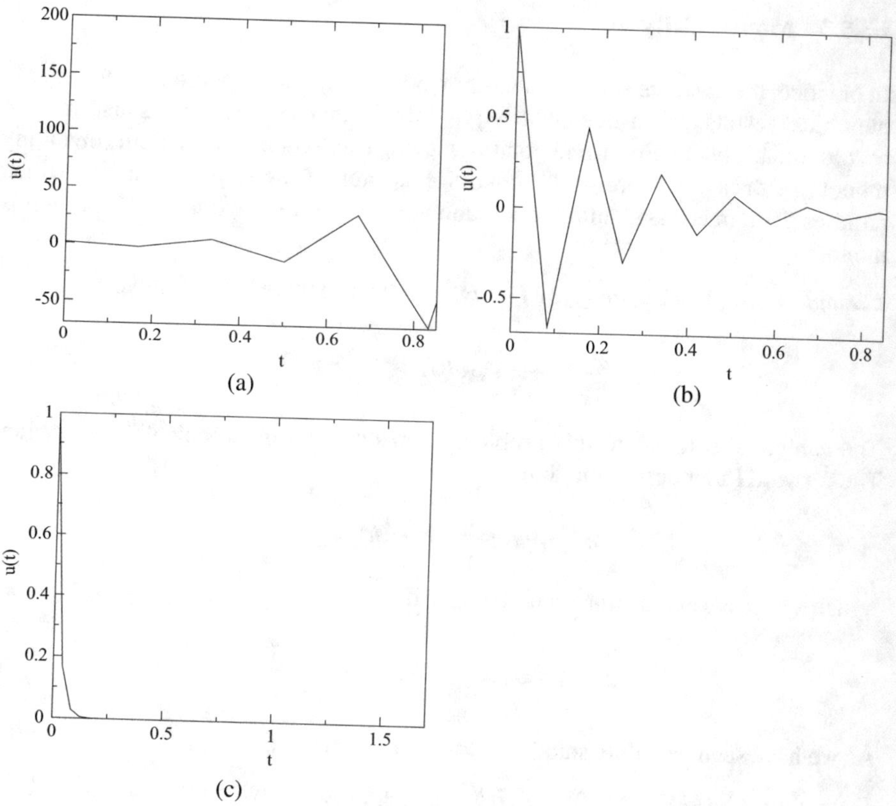

Fig. 1.3 Forward Euler method for $u' = -20u$, $u(0) = 1$: $u(t)$ versus t. (**a**) 6 steps not absolutely stable; (**b**) 12 steps absolutely stable, not monotone; (**c**) 24 steps absolutely stable and monotone

we show that the numerical solution decays and oscillates when 12 equally spaced time steps are used; although this time step is absolutely stable, it is not monotone. In graph (c), we show that the numerical solution decays monotonically when 24 equally spaced time steps are used. These computational results were obtained by running the executable for `euler.f`, and the figures were obtained by running `xmgrace`.

1.3.5.4 Convergence

The following theorem will estimate errors in the forward Euler method with inexact arithmetic.

Theorem 1.3.3 (Error Bound for Forward Euler) *Suppose that $f(u, t)$ is bounded (with bound F), Lipschitz continuous in u (with constant L) and Lipschitz continuous in t (with constant M), for $0 \leq t \leq T$ and all u. Let $u(t)$ solve the* *initial-value problem*

$$\frac{du}{dt} = f(u,t) \ , \ u(0) = u_0 \ .$$

Suppose that the forward Euler method computations for this initial-value problem are subject to rounding errors:

$$\tilde{u}_{n+1} = \tilde{u}_n + [t_{n+1} - t_n] f(\tilde{u}_n, t_n) + r_n \ .$$

Then the error in the forward Euler method satisfies

$$|u(t_n) - \tilde{u}_n| \leq |u(t_0) - \tilde{u}_0| \, e^{Lt_n} + (M + LF) \left[\max_{0 \leq k < n} h_k \right] \frac{e^{Lt_n} - 1}{L} + \frac{\max_{k=0}^{n-1} |r_k|}{\min_{0 \leq k < n} h_k} \frac{e^{Lt_n} - 1}{L} \ . \tag{1.5}$$

We will discuss the implications of this theorem after the following proof.

Proof Let us define the time step

$$h_n = t_{n+1} - t_n \ ,$$

the **local truncation error**

$$d_N \equiv u(t_{n+1}) - u(t_n) - [t_{n+1} - t_n] f(u(t_n), t_n) \ ,$$

and the error in the solution

$$e_n \equiv u(t_n) - \tilde{u}_n \ .$$

Since f is Lipschitz continuous in both of its arguments and the derivative of u is f, the mean-value theorem implies that there is a number $\theta \in (0,1)$ such that

$$u(t_{n+1}) - u(t_n) = [t_{n+1} - t_n] f(u(t_n + \theta h_n), t_n + \theta h_n) \ .$$

This implies that the local truncation error satisfies

$$\begin{aligned} |d_n| &= |h_n f(u(t_n + \theta h_n), t_n + \theta h_n) - h_n f(u(t_n), t_n)| \\ &\leq h_n |f(u(t_n), t_n + \theta h_n) - f(u(t_n), t_n)| \\ &+ h_n |f(u(t_n + \theta h_n), t_n + \theta h_n) - f(u(t_n), t_n + \theta h_n)| \\ &\leq M\theta h_n^2 + L h_n |u(t_n + \theta h_n) - u(t_n)| \end{aligned}$$

Again, the mean value theorem implies that there exists $s \in (t_n, t_n + \theta h_n)$ so that

$$|u(t_n + \theta h_n) - u(t_n)| = \theta h_n |f(u(s), s)| \leq \theta h_n F \ .$$

Since $\theta \in (0,1)$, we conclude that

$$|d_n| \le h_n^2 (M + LF) \ .$$

Next, we note that since the exponential function is monotonically increasing, the Riemann sum approximation satisfies

$$\sum_{k=0}^{n-1} \mathrm{e}^{Lt_k} h_k \le \int_0^{t_n} \mathrm{e}^{Lt} \, dt = \frac{\mathrm{e}^{Lt_n} - 1}{L} \ .$$

As a result,

$$\left[\min_{0 \le k < n} h_k \right] \sum_{k=0}^{n-1} \mathrm{e}^{Lt_k} \le \frac{\mathrm{e}^{Lt_n} - 1}{L} \ ,$$

and

$$\sum_{k=0}^{n-1} \mathrm{e}^{Lt_k} h_k^2 \le \left[\max_{0 \le k < n} h_k \right] \int_0^{t_n} \mathrm{e}^{Lt} \, dt = \left[\max_{0 \le k < n} h_k \right] \frac{\mathrm{e}^{Lt_n} - 1}{L} \ .$$

Finally, let us estimate the error in the numerical solution. The triangle inequality and Lipschitz continuity imply that

$$\begin{aligned} |e_{n+1}| &= |u(t_{n+1}) - \tilde{u}_{n+1}| \\ &= |[u(t_n) + h_n f(u(t_n), t_n) + d_n] - [\tilde{u}_n + h_n f(\tilde{u}_n, t_n) + r_n]| \\ &\le (1 + L h_n) |e_n| + |d_n| + |r_n| \ . \\ &\le \mathrm{e}^{L h_n} |e_n| + |d_n| + |r_n| \ . \end{aligned}$$

An inductive argument shows that

$$|e_n| \le |e_0| \prod_{k=0}^{n-1} \mathrm{e}^{L h_k} + \sum_{k=0}^{n-1} [|d_k| + |r_k|] \prod_{j=0}^{k-1} \mathrm{e}^{L h_j}$$

then our bound on the local truncation error gives us

$$\begin{aligned} &\le |e_0| \, \mathrm{e}^{L \sum_{k=0}^{n-1} h_k} + (M + LF) \sum_{k=0}^{n-1} h_k^2 \mathrm{e}^{L \sum_{j=0}^{k-1} h_j} + \left[\max_{k=0}^{n-1} |r_k| \right] \sum_{k=0}^{n-1} \mathrm{e}^{L \sum_{j=0}^{k-1} h_j} \\ &= |e_0| \, \mathrm{e}^{L t_n} + (M + LF) \sum_{k=0}^{n-1} h_k^2 \mathrm{e}^{L t_k} + \left[\max_{k=0}^{n-1} |r_k| \right] \sum_{k=0}^{n-1} \mathrm{e}^{L t_k} \end{aligned}$$

and finally our Riemann sum bounds give us

$$\leq |e_0| \, e^{Lt_n} + (M + LF) \left[\max_{0 \leq k < n} h_k \right] \frac{e^{Lt_n} - 1}{L} + \frac{\max_{k=0}^{n-1} |r_k|}{\min_{0 \leq k < n} h_k} \frac{e^{Lt_n} - 1}{L} \, .$$

1.3.5.5 Rate of Convergence

The error bound in Theorem 1.3.3 merits some discussion. Let us begin by considering how the forward Euler method behaves when rounding errors are negligibly small. In such a case, we can take the initial value for the forward Euler method to be the true initial value for the initial value problem, and assume that the rounding errors r_n are all zero. This leaves us with the error bound

$$|u(t_n) - \tilde{u}_n| \leq (M + LF) \frac{e^{Lt_n} - 1}{L} \max_{0 \leq k < n} h_k \, ,$$

in which the only source of error is due to **accumulated local truncation error**. We have shown that, if rounding errors are negligible, then the error in the forward Euler method is $O(h)$, where h is the time step size. We typically say that the forward Euler method has **first-order accuracy**.

Given a maximum error tolerance ε, we could choose a sequence of computation times $\{t_n\}_{n=0}^{N}$ so that

$$\max_{0 \leq n < N} [t_{n+1} - t_n] \leq \varepsilon \frac{L}{(M + LF)\left[e^{Lt_N} - 1\right]}$$

and use the error bound theorem to show that the error in the numerical solution at all computation times is within the tolerance ε. Although such an approach would require knowledge of the bounds F, L and M, such time step selection is theoretically possible.

The purpose of an error estimate should be to tell us how to choose the time step in order to obtain a desired accuracy in the solution. One difficulty in using the error estimate (1.5) is that it can produce estimates for the errors that are far larger than the actual error. In practice, we will want to have other ways to estimate the error. The art of time step selection is an advanced topic that we shall discuss later, such as in Sect. 3.4.11 of Chap. 3 in Volume III.

1.3.5.6 Accumulation of Rounding Errors

Next, let us consider what happens when rounding errors are *not* negligible. Suppose that the rounding errors due to entering the initial value $u(t_0)$ into the machine to get $\tilde{u}_0$, and due to computing the new forward Euler solution value $\tilde{u}_{n+1}$ from $\tilde{u}_n$, are both bounded above by some value η, which is small relative to the solution values $\tilde{u}_n$. Under these circumstances, Theorem 1.3.3 produces the error estimate

$$|u(t_n) - \tilde{u}_n| \leq \eta e^{Lt_n} + (M + LF)\left[\max_{0\leq k<n} h_k\right]\frac{e^{Lt_n} - 1}{L} + \frac{\eta}{\min_{0\leq k<n} h_k}\frac{e^{Lt_n} - 1}{L}.$$

As the maximum time step size is reduced, the portion of the error bound due to the accumulated local truncation error decreases (this is the middle term), the portion due to the error in the initial values remains the same (this is the first term), and the portion of the error due to rounding errors in the computation of $\tilde{u}_{n+1}$ increases, inversely proportional to the smallest time step size. Eventually, the rounding errors will dominate all other errors, and *further reduction in the time step size will actually decrease the accuracy of the computations.* This means that convergence of the forward Euler method, in the strict mathematical sense, is impossible on a fixed-precision machine.

Figure 1.4 shows the errors in the forward-Euler numerical solution to $u'(t) = u(t)$ with $u(0) = 1$ at $t = 2$ for various values of the time step h. The base 10 logarithm of the error in u is plotted as a function of the base 10 logarithm of $1/h$. This figure shows that the error initially decreases at a rate proportional to h. However, once the time step becomes sufficiently small, the error begins to grow at a rate proportional to $1/h$. Both behaviors were predicted by Theorem 1.3.3. For those who are interested, the data for Fig. 1.4 was generated by mesh_refinement.f, and the results were plotted with Linux utility `xmgrace`.

For more information regarding the nature of computer rounding errors, please see Sect. 2.3.

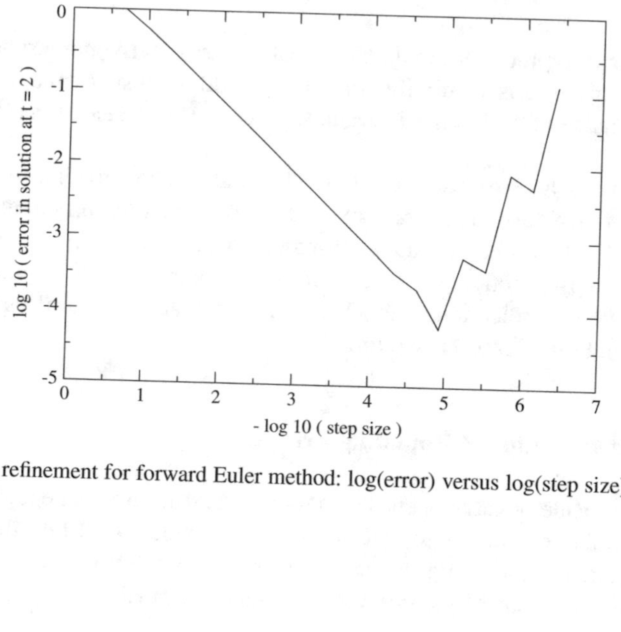

Fig. 1.4 Mesh refinement for forward Euler method: log(error) versus log(step size)

1.3.5.7 Time Step Selection

The practical goal in this discussion is to develop criteria to use in choosing a good time step. It is natural to want to use Theorem 1.3.3 to predict the error in a forward-Euler algorithm, typically by ignoring rounding errors. The nice feature of this theorem is that it gives us an *a priori error estimate*, meaning that the error estimate can be determined before we execute the algorithm. Unfortunately, the Lipschitz constants in that theorem are hard to determine.

As we saw in Sect. 1.3.5.3, monotonicity leads to nice pictures. So, if we are using the forward-Euler method to solve

$$\frac{\mathrm{d}u}{\mathrm{d}t} = f(u) ,$$

and if $\frac{\partial f}{\partial u} < 0$, we might want to choose the time step h so that

$$h \left| \frac{\partial f}{\partial u} \right| < 1 .$$

This condition provides an upper bound on the acceptable time step size.

However, monotonicity is not the only consideration. The time step should be reduced further, if necessary, to produce a desired accuracy. One way to guess the time step size is to make several runs and examine how much the solution varies for different values of h at the same time t. This is called a **convergence study**, and is an essential part of many scientific computing efforts. Oftentimes, the convergence study is combined with **Richardson extrapolation** of the error. Here is a brief introduction to these ideas. Suppose that we want to solve

$$\frac{\mathrm{d}u}{\mathrm{d}t} = f(u,t)$$

by the forward Euler method. If we can ignore rounding errors, and if f is Lipschitz continuous, then the proof of the error bound Theorem 1.3.3 shows that the local truncation error in the method at time t satisfies

$$d(h) = Ch + O(h^2)$$

for some unknown constant C. Suppose that we make two calculations over some time increment using step sizes h and h', in order to obtain numerical solutions $\tilde{u}$ and $\tilde{u}'$ at some later time t. Then

$$\tilde{u} - \tilde{u}' = d(h) - d\left(h'\right) = C\left(h - h'\right) + O(h^2) .$$

If we ignore the higher-order terms, we can use this result to replace C in the error expression above:

$$d(h) = \frac{\tilde{u} - \tilde{u}'}{h - h'} h + O(h^2) .$$

In particular, if we want $|d(h)| < \varepsilon$, this expression suggests that we can choose a new time step $\tilde{h}$ to satisfy

$$\tilde{h} < \varepsilon \frac{|h - h'|}{|\tilde{u} - \tilde{u}'|} .$$

We will examine various strategies for choosing time steps in Sect. 3.4.11 of Chap. 3 in Volume III, and we will develop Richardson extrapolation in Sect. 2.2.4 of Chap. 2 in Volume III.

Exercise 1.3.6 Modify program euler.f, euler.c or euler.C to compute the absolute value of the error. You could write an arithmetic statement function to compute the analytical solution or the error. For $du/dt = u$, $u(0) = 1$, $0 < t < 1$ compute the error in the solution at $t = 1$ for 2^k time steps, $k = 4, \ldots, 20$. Plot the log of the absolute value of the error at $t = 1$ (on the vertical axis) versus minus the log of the time step h (on the horizontal axis). On the same plot, draw (by hand) and label the curves that should represent the accumulated local truncation error, and the accumulated rounding error. (You should be able to estimate these curves from the shape of the computed error curve.) Roughly what value of $h = 2^{-1}$ gives the most accurate solution possible?

Exercise 1.3.7 Modify program `euler.f`, `euler.c` or `euler.C` to solve $\frac{du}{dt} = -128u$, $u(0) = 1$, $0 < t < 1$ using either the forward Euler method. Plot the numerical solutions for $h = 2^{-k}$, $k = 4, \ldots, 10$. At which time steps do each of the methods become stable, absolutely stable and/or monotone?

Exercise 1.3.8 Modify program `euler.f`, `euler.c` or `euler.C` to solve $\frac{du}{dt} = 16u$, $u(0) = 1$, $0 < t < 1$ using either the forward Euler method. Plot the numerical solutions for $h = 2^{-k}$, $k = 2, \ldots, 8$. At which time steps do each of the methods become stable, absolutely stable and/or monotone?

Exercise 1.3.9 Use program `euler.f`, `euler.c` or `euler.C` to solve $\frac{du}{dt} = u$, $u(0) = 1$, $0 < t < 1$ with $h = 0.1$ and $h = 0.05$. Use Richardson extrapolation to estimate a step size h that will produce a numerical approximation to $u(1)$ with error at most 10^{-6}. Then run `euler.f` with that step size and see what error you actually obtain.

1.4 Program Efficiency

How should we decide if our computation is producing an accurate answer? If possible, we should compute an error estimate. Often, error estimates serve the dual purpose of selecting appropriate time steps. In any case, we can always solve the same problem with different numbers of time steps, and compare each to a reasonably well-refined computation. If the errors show the expected order of the method, and if the numerical scheme is known to be convergent on similar model problems, then the accuracy in the results is probably being well-estimated.

Specifically, we could plot the logarithm of the absolute value of the difference between intermediate results and our finest result versus the logarithm of the reciprocal step size. When the asymptotic estimates for the scheme are valid, the slope of the plot should be equal to minus the order of the method. We call these **order of accuracy** plots. As an example of such a program, readers may view meshRefinement.C. Figure 1.5 shows the results of applying `meshRefinement.C` to solve

$$\frac{du}{dt} = -u \ , \ u(0) = 1$$

and taking 2^k time steps for $k = 3, \ldots, 12$ to generate an order of accuracy plot. This plot shows the order of accuracy for both the forward Euler method (upper line) and the modified Euler method

$$\tilde{u}_{n+1} = \tilde{u}_n + hf\left(\tilde{u}_n + (h/2)f(\tilde{u}_n, nh), [n + 1/2]\,h\right) \ .$$

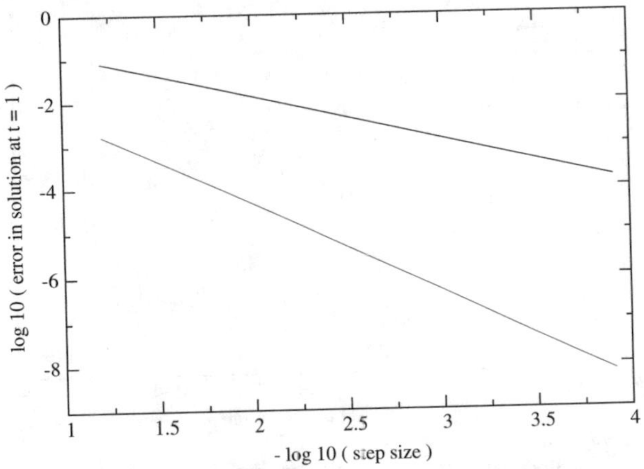

Fig. 1.5 Log(error) versus log(step size); top curve = forward Euler, bottom curve = modified Euler

This plot demonstrates that the forward Euler method is first-order in the time step size, and that the modified Euler method is second-order. The graph was generated by `xmgrace`.

In the scientific computing literature, it is common to find order of accuracy plots in discussions of computational efficiency. Since high-order schemes have faster rates of convergence, this method of comparing methods almost always favors the high-order scheme. To get an accurate measure of program efficiency, however, it is better to plot the logarithm of the error versus the logarithm of the elapsed computer time. We will call these **relative efficiency** plots. Since high-order schemes typically require more computational work per time step, it is possible that the point at which the high-order curve crosses the low-order curve, and therefore becomes relatively more efficient, may occur at an unacceptably large computer time.

We have compared the relative efficiency of the forward-Euler method against the modified Euler method in Fig. 1.6. For the range of time steps considered in the computations, the relative efficiency curves for the forward Euler method and the modified Euler method do not cross. However, we could extrapolate backwards in time step sizes to estimate that they would cross for very coarse time steps. The data for this graph was generated by the same program that generated this figure, and the graph was generated by Linux command `xmgrace`.

In order to measure efficiency we need to have some technique to measure computational time. The program `meshRefinement.C` contains one way to do this in C^{++}. Note that we start the timing after reading the input data, and stop the timing before we write the output.

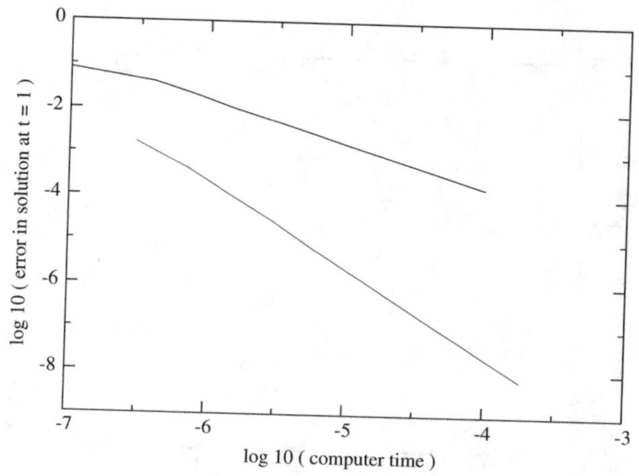

Fig. 1.6 Log(error) versus log(computer time); top curve = forward Euler, bottom curve = modified Euler

Here is an important warning about computer timing. Most of the timing routines are not very accurate. The routine we used in `meshRefinement.C` is not able to measure elapsed times of less than a millisecond. Since it is possible to perform about 100 million floating point operations in a second, this means that it is not possible to time anything that requires fewer than roughly 100 thousand floating point operations. To get meaningful timing information, readers may have to repeat the computations many times, as in `meshRefinement.C`.

Exercise 1.4.1 Modify euler.C so that

- it performs all floating-point computations in double precision (i.e., using type `double`),
- the program computes the logarithm of the error, minus the logarithm of the step size, and minus the logarithm of the elapsed computational time.
- initially you take 100,000 time steps.
- you solve the problem 10 times, doubling the number of time steps with each pass.

Plot the logarithm of the error (on the vertical axis) versus minus the logarithm of the step size (on the horizontal axis), and the logarithm of the error versus the logarithm of the elapsed computer time (on the horizontal axis). Verify that both plots have the same slope, and explain why.

Exercise 1.4.2 Modify `euler.C` so that you can run either the forward Euler method or the modified Euler method in double precision. The modified Euler method can be written in the form

$$\tilde{u}_{n+1/2} = \tilde{u}_n + \frac{t_{n+1} - t_n}{2} f\left(\tilde{u}_n, t_n\right)$$

$$\tilde{u}_{n+1} = \tilde{u}_n + [t_{n+1} - t_n] f\left(\tilde{u}_{n+1/2}, \frac{t_n + t_{n+1}}{2}\right) .$$

Solve both

1. $\frac{du}{dt} = u, u(0) = 1, 0 \le t \le 1,$
2. $\frac{du}{dt} = -\frac{1}{u^2},] \, u(0) = 1, 0 \le t \le 1/2.$

Compare the two for efficiency, by plotting the logarithm of the error versus the logarithm of the elapsed computer time for both methods on the same graph for each of the problems. Which method is more efficient for low accuracy, and which is more for high accuracy? Do both show the expected orders of accuracy in both problems?

Chapter 2
Working with a Computer

To err is human, but to really foul things up requires a computer.

Farmers' Almanac, 1978

"There are only 10 types of people in the world: those who understand binary, and those who don't." Wikipedia webpage on Mathematical Jokes

Part of the inhumanity of the computer is that, once it is competently programmed and working smoothly, it is completely honest.

Isaac Asimov

Abstract This chapter begins with basic information about how a digital computer represents and works with numbers. For floating point numbers, the chapter describes how rounding errors occur in arithmetic operations, and examines the consequences for basic computations such as subtraction, finite differences and quadratic equations. Problems such as overflow, underflow and floating point exceptions are examined. Next, programming languages are introduced, compared and applied to the development of scientific computing algorithms. Techniques for timing the execution of programs are described, for later use in comparing the relative efficiency of competing algorithms. Object-oriented programming is introduced in C++. The organization of computer memory and its relevance to the design of computer programs is examined and applied to matrix operations. The fundamentals of computer input and output are provided for various computer languages. Important programming suggestions and techniques of defensive programming to detect and protect against floating point exceptions and memory faults are provided. The fundamental algorithm libraries LAPACK, STL and Pthread are introduced.

Additional Material: The details of the computer programs referred in the text are available in the Springer website (http://extras.springer.com/2018/978-3-319-69105-3) for authorized users.

© Springer International Publishing AG, part of Springer Nature 2017
J.A. Trangenstein, *Scientific Computing*, Texts in Computational Science and Engineering 18, https://doi.org/10.1007/978-3-319-69105-3_2

2.1 Overview

In this chapter, we will present some basic knowledge about floating point numbers and arithmetic, computer memory and programming languages. The discussion of floating point numbers will explain why their set is finite, implying both that there is a largest floating point number and a smallest positive floating point number. Attempts to generate numbers not in this finite set will lead to errors, most of which are benignly treated as *rounding errors*. However, some of these errors must be treated differently as *floating point exceptions*. By adopting the rules of IEEE arithmetic, programming languages (with the exception of JAVA) can guarantee certain bounds for rounding errors. These bounds become the basis for analyzing numerical methods, such as in Sect. 3.8.

Floating point exceptions, which correspond to exceeding the upper or lower bounds on the range of floating point numbers, do not lend themselves to mathematical analysis. Computing with exceptional floating point numbers is akin to operating on infinity, and the result is another exception. For readers who are new to programming, it can be very difficult to find the root cause of floating point exceptions. In Sect. 2.3.2, we will see that programming languages such as Fortran, C and C^{++} often allow us to terminate execution at the first occurrence of a floating point exception, and to examine the state of the program in a debugger; MATLAB also provides a similar capability, with the command window available as the debugger. We can use such floating point exception traps to catch other kinds of programming errors, by adopting *defensive programming* ideas suggested in Sect. 2.9.3.

In Sect. 2.4.1, we will see that each programming language has advantages and disadvantages. In order to use the best language for distinct computational tasks, we may choose to build programs by using multiple programming languages, as described in Sect. 2.4.3. This process is greatly simplified by developing *makefiles*, which are described in Sect. 2.5.

The design of computer hardware affects our use of computer memory for programs and their data. Various alternatives for accessing computer memory can produce drastically different execution times. By understanding certain principles described in Sect. 2.6, readers should be able to improve their program efficiency.

Some basic algorithms of linear algebra and computer science occur very frequently. These algorithms have been programmed efficiently and collected into a couple of very useful libraries, which are described in Sect. 2.10. We recommend that readers avoid writing their own naive versions of these algorithms, and merely access the more sophisticated forms in these libraries.

Throughout this chapter, we will necessarily make specific references to several programming languages. We will limit our discussion to Fortran, C, C^{++}, MATLAB and JavaScript. The first three are well-respected scientific programming *languages*, while MATLAB is a popular scientific *programming environment*, and the JavaScript programming language is useful for web page displays.

The material in this chapter is drawn from a number of sources. For a detailed discussion of computer architecture, readers might consult Patterson and Hennessy [87]. For an introduction to rounding error analysis, see Wilkinson [110] or Higham [56]. For information regarding the implementation of basic algorithms, such as the multiplication of floating point numbers, see Aho et al. [1], or Knuth [68]. An excellent source for learning the C programming language is the classic book by Kernighan and Ritchie [64]. For instruction in the C^{++} programming language, we recommend Stroustrup [100], or Barton and Nackman [5]. For help with Fortran programming, try Chapman [22], and for help with JavaScript, see Flanagan [39]. Potential Python programmers might use Lutz [79]. For information about LAPACK, see Anderson et al. [2]. Readers can find information about the standard template library in Josuttis [60], or Vandevoorde and Josuttis [107].

For software related to this chapter, we recommend LAPACK BLAS (Basic Linear Algebra Subroutines) and the Standard Template Library (STL). For most users, the best performance for the BLAS routines will be provided by ATLAS. When we discuss how to pass a function as an argument to a computer subroutine in Sect. 2.4.2.5, we will refer to MINPACK.

2.2 Machine Numbers

In Sect. 1.3.5.6, we saw that the accumulation of rounding errors can dominate the numerical solution of initial value problems. This same difficulty also arises in other topics in scientific computing. In order to understand how rounding errors occur, and potentially to control their growth, we first need to understand how machines and programming languages represent numbers. There are several distinct types of numbers, requiring separate discussions.

2.2.1 *Booleans*

In order to represent "false" or "true" in the computer, only a single **bit** is needed. Mathematically, we represent the bit by the symbol $b = 0$ (for false) or 1 (for true). However, most computers do not provide distinct memory addresses for individual bits. As a result, **boolean** numbers are usually given more than a single bit in storage.

In C^{++}, use type `bool` to represent booleans. Under the Linux `g++` compiler, a `bool` typically uses eight bits (i.e., one **byte**). The boolean value `false` is stored as the bit string '00000000' and `true` is stored as the bit string '00000001'. Writing out all these bits can be tedious. Because 4 bits can represent up to $2^4 = 16$ different values, it is common to use **hexadecimal** notation, in which the 16 possible values of a byte are represented by the numbers 0 through 9, followed by the letters A through F. In hexadecimal, `false` is '0x00' and `true` is '0x01'.

Table 2.1 Boolean operations

Operation	C or C++	Fortran	JavaScript	MATLAB
not	`! b`	`.not b`	`! b`	`~ b`
and	`b1 && b2`	`b1 .and. b2`	`b1 && b2`	`b1 & b2`
or	`b1 || b2`	`b1 .or. b2`	`b1 || b2`	`b1 | b2`
xor	`b1 ^ b2`	`b1 .neqv. b2`	`b1 != b2`	`b1 != b2`
iff	`b1 == b2`	`b1 .eqv. b2`	`b1 == b2`	`b1 == b2`

In Fortran, it is common to use type `logical` for booleans. However, this type typically uses four bytes, or four times as many bits as `bool` in C++. In order to get a one-byte boolean in Fortran, use type `logical*1`. In Fortran, the fixed boolean values are `.false.` and `.true.`

In JavaScript, the primitive type `Boolean` represents booleans. These have two possible values, `true` and `false`. In all four of the previously mentioned programming languages, booleans are convertible into other data types.

Both JavaScript and MATLAB are weakly dynamically typed programming languages, because variables can be assigned without declaring a type, and the type can be implicitly converted. The logical class in MATLAB can involve variables that would be considered arrays of Booleans in other programming languages. In MATLAB, the two possible logical values are `0` and `1`.

The legal operations on booleans are described in Table 2.1

None of the languages has an **implies** operator. Instead of writing "`b1 implies b2`" we must write "`b1 or ( not b2 )`". In C and C++, if-tests can be performed on integer and pointer types as well. Thus if *n* is an integer, a zero value of `n` is treated as `false` and a nonzero value as `true`. This can lead to some subtle programming errors in C++. For example, the statement

```
if (i=0)...
```

will assign variable `i` the value zero and return `false`, thereby never executing the code that follows. On the other hand, the statement

```
if (i==0)...
```

will test if `i` is zero, and if so then execute that follows. Java prevents such type conversions, but C, C++ and JavaScript do not.

Exercise 2.2.1 How many bits are used by the boolean type in your favorite programming language?

Exercise 2.2.2 How could you program the condition that `b1` does not imply `b2`?

2.2.2 *Integers*

The C and C^{++} integer data types are `short`, `int` and `long`. all of these can be qualified with the word `unsigned`. On an Intel x86 machine, these use 2, 4 and 8 bytes respectively. The Fortran integer data types are `integer*1`, `integer*2`, `integer*4`, `integer*8` and `integer*16`, but none of these are unsigned. Fortran `integer` is the same as `integer*4`. In Fortran 90, the same integer data types are available as `integer(kind=`n`)` where n is the number of bytes; the integer 1 using n bytes would be written in Fortran 90 as 1_n. In JavaScript, there are no integer types; the only numeric primitive data type is a floating-point number. MATLAB has two signed integer types, namely `int32` and `int64`. It also has four unsigned integer types, namely `unint8`, `uint16`, `uint32` and `uint64`. The numbers in these data types refer to the number of bits used in the data storage. The most common choices are `int` in C or C^{++}, and `integer` in Fortran or Fortran 90.

Each C or C^{++} `int`, Fortran or Fortran 90 `integer`, or MATLAB `int32` has the form

$$\mathtt{i} = \begin{cases} \sum_{j=0}^{30} b_j 2^j \;, & i \geq 0 \\ -1 - \sum_{j=0}^{30} (1 - b_j) 2^j \;, & i < 0 \end{cases}$$

where the bits b_j are stored in the binary computer representation. If the **byte order** is **little endian** (the form used by Intel machines), then the sign bit is bit 31 (the left-most bit) and b_0 is the right-most bit. Macs use **big endian** byte order, in which the sign bit is first (i.e., right-most) and the left-most bit is b_0. On Intel machines, the negative number $-i$ is stored as the **twos complement** of $-i + 1$. Thus in hexadecimal, 1 is '0x00000001' and -1 is '0xFFFFFFFF'. Note that the zero `int` has bit representation via all bits equal to zero, and note that negative zero is equal to zero. Little endian hexadecimal representations of some useful integers can be found in Table 2.2.

Sometimes, it can be helpful to know the largest and smallest integers that can be represented by a given data type. These are listed in Table 2.3. In C^{++}, users can

Table 2.2 Little Endian integer representations

Decimal	Hexadecimal
1	1
(short) -1	FFFF
(short) -2	FFFE
(int) -1	FFFFFFFF
(short) 32767	7FFF
(short) -32768	8000
(int) 2147483647	7FFFFFFF
(int) -2147483648	80000000

Table 2.3 Largest and smallest integers

C++ numeric_limits	C	Fortran	MATLAB	Decimal
		huge(1_1)		127
<short>::max()	SHRT_MAX	huge(1_2)		32767
<short>::min()	SHRT_MIN			-32768
<int>::max()	INT_MAX	huge(1_4)	intmax('int32')	2147483647
<int>::min()	INT_MIN		intmin('int32')	-2147483648
<long>::max()	LONG_MAX	huge(1_8)	intmax('int64')	9223372036854775807
<long>::min()	LONG_MIN		intmin('int64')	-9223372036854775808

easily refer to the largest and smallest integer as follows:

```
int imax=numeric_limits<int>::max(),
imin=numeric_limits<int>::min();
```

For other integer data types, such as `short` or `long`, replace the string '`int`' between the symbols '<' and '>' with the alternative data type. Because it will not fit in the table, we note that in Fortran

```
huge(1_16) = 170141183460469231731687303715884105727
```

which is $2^{127} - 1$.

Next, we show that integer arithmetic can overflow. Consider the following algorithm.

Algorithm 2.2.1 (Integer Overflow)

$$i = 1$$

$$\text{while } i > 0$$

$$i = 2 * i + 1$$

Under the rules of mathematics, this algorithm should not terminate. However, something very different happens inside a computer. The binary representation of one has a single nonzero bit. When we multiply by 2, we shift the non-zero bit one location (to the left, if little endian). So, if we repetitively perform the replacement `i = 2 * i + 1` we begin to fill up all of the bits with ones in the binary representation. Eventually, all of the bits are set to one, *including the sign bit*, and the result is the machine number for -1. Typically, this overflow does not generate a run-time exception.

Readers can examine programs that generate integer overflow in Fortran, integer overflow in Fortran 90, integer overflow in C or integer overflow in C^{++}. Each of these programs shows how to declare the various integer data types, and how to display the integers in both decimal and hexadecimal form.

Integer overflow is prevented in MATLAB. For example, the MATLAB commands

```
k=int32(1);
while k>0
  disp(k);
  k=2*k+1;
end;
disp(k)
```

produce an infinite loop that eventually displays `7FFFFFFF` indefinitely. For more examination of integers in MATLAB, see the exercises at the end of this section.

Exercise 2.2.3 What is the hexadecimal representation of the integer -2 in your favorite programming language on your machine?

Exercise 2.2.4 What is the decimal value of the most negative integer your machine can represent? What happens if we multiply that number by 2? Explain what went on in the bits.

Exercise 2.2.5 Recall that JavaScript uses floating point numbers to represent integers such as loop counters. What will happen if we try to loop over odd integers in JavaScript? For example, describe what will happen in the bit representation of `i` when we do the replacement "`i=2*i+1`" for large values of `i`.

Exercise 2.2.6 Is there a distinction between integers and floating point numbers in MATLAB? For example, does the MATLAB script

$$m = 1; n = 2 * m; while(n < n + 1 \& n + 1 < n + 2)$$
$$m = n + 1; n = 2 * m;$$
$$end; m$$

stop before overflow? Consider the implications: if MATLAB integers are floating point numbers, can large loop counters in MATLAB be odd?

2.2.3 *Floating Point Numbers*

The C or C^{++} floating point data types are `float`, `double` and `long double`. In Linux on an Intel x86 machine, these use 4, 8 and 16 bytes respectively. There are two Fortran or Fortran 90 floating point types, namely `real*4` and `real*8`, which have the alternative declarations `real` and `double precision`, respectively. JavaScript has a single floating point primitive data type, called `Number`. MATLAB has two floating point data types, namely `single` and `double`.

On an Intel x86 machine, a C or C^{++} **normal** `float` has the binary representation

$$\mathtt{f} = (-1)^s 2^{e-bias} \left(\sum_{j=0}^{22} b_j 2^{j-23} + b \right) ,$$

where

- s = sign (the left-most bit)
- e = biased exponent (the next 8 bits)
- $bias$ = 127 (implied, not stored)
- b = leading bit (determined by e, not stored)
- b_j = mantissa bit, $b_j = 0$ or 1 (right-most 22 bits).

Table 2.4 Representations of floats

Decimal	Hexadecimal	Sign bit	Biased exponent	Leading bit	Mantissa bits
1.	3F800000	0	127	1	all 0
2.	40000000	0	128	1	all 0
0.5	3F000000	0	126	1	all 0
0.75	3F400000	0	126	1	all 0 except the first
0.	00000000	0	0	0	all 0
-0.	80000000	1	0	0	all 0
+Infinity	7F800000	0	255	1	all 0
0.34028235e+39	7F7FFFFF	0	254	1	all 1
0.11754944e-37	800000	0	1	1	all 0
0.11920929e-06	34000000	0	104	1	all 0

Similarly, a `double` normally has the binary representation

$$\mathtt{d} = (-1)^s 2^{e-bias} \left(\sum_{j=0}^{51} b_j 2^{j-52} + b \right) ,$$

where $bias = 1023$. For very small numbers with exponent $e = bias$, we can have **sub-normal numbers**, in which the most significant bits in the mantissa are allowed to be zero, so that underflow progresses gradually to zero. Several floating point numbers and their hexadecimal representations can be found in Table 2.4.

Note that floating point zero has a sign, so the floating point number "`0.`" is different from the floating point number "`-0.`" Nevertheless, contemporary compilers will treat negative zero as if it has no sign.

For each floating point data type, there is an infinity, a largest possible number, a smallest positive number, and a smallest positive number x so that the floating point representation of $1 + x$ is not equal to 1. The values of these numbers in various programming languages can be found in Table 2.5.

There are three important problems that are consequences of the floating point number representations. **Overflow** occurs when the exponent needed to represent the number is greater than the largest possible machine representation. Overflow eventually occurs if we compute the sequence

$$f_n = 2. * f_{n-1} \text{ with } f_1 = 1 .$$

Roundoff occurs when the mantissa needed to represent the number is smaller than needed. Roundoff eventually occurs if we compute the sequence

$$f_n = 0.5 * f_{n-1} + 1. \text{ with } f_1 = 1 .$$

Table 2.5 Floating point bounds

C++ numeric_limits	C	Fortran	MATLAB	JavaScript
<float>::infinity()	HUGE_VALF		inf('single')	
<float>::max()	FLT_MAX	huge(r_4)	realmax('single')	
<float>::min()	FLT_MIN	tiny(r_4)	realmin('single')	
<float>::epsilon()	FLT_EPSILON	epsilon(r_4)	eps('single')	
<double>::infinity()	HUGE_VAL		inf('double')	
<double>::max()	DBL_MAX	huge(r_4)	realmax('double')	Number.MAX_VALUE
<double>::min()	DBL_MIN	tiny(r_4)	realmin('double')	Number.MIN_VALUE
<double>::epsilon()	DBL_EPSILON	epsilon(r_4)	eps('double')	
<long double>::infinity()	HUGE_VALL			
<long double>::max()	LDBL_MAX			
<long double>::min()	LDBL_MIN			
<long double>::epsilon()	LDBL_EPSILON			

Underflow occurs when the exponent needed to represent the number is smaller than the smallest possible exponent and resulting mantissa is then smaller than the smallest possible mantissa. As floating point numbers decrease, the smallest possible exponent is eventually reached with a nonzero most-significant bit in the mantissa. During gradual underflow, this smallest exponent remains but more and more of the significant bits of the mantissa become zero. Underflow occurs if we compute the sequence

$$f_n = 0.5 * f_{n-1} \text{ with } f_1 = 1 .$$

Readers can examine programs that generate real overflow in Fortran, real overflow in Fortran 90, float overflow in C, float overflow in C^{++} or floating point number overflow in MATLAB. All of these programs display the numbers in both scientific and hexadecimal form. Furthermore, all of the programs except MATLAB demonstrate how to declare, operate on, and print floating point numbers of various sizes.

Readers can also experiment with a JavaScript **Number Overflow-Roundoff-Underflow** program. (Clicking on this link should open your default browser to run the JavaScript program.) Note that JavaScript does not contain a function to display the hexadecimal form of a `Number`. Instead, we use the function numberToHexadecimal to compute the hexadecimal representation for finite numbers.

In summary, the binary representation of machine numbers has the following implications.

1. There is a finite set of integers and a finite set of floating point numbers available in on digital computers.
2. Entering floating point numbers into the machine usually leads to an error. For example, `0.1` is not a machine number, and `( 10. * 0.1 ) - 1.` will be nonzero on binary computers. (The same computation will produce precisely zero on most hand calculators, because they use inefficient binary-coded decimal number representations.)
3. It may or may not be safe to test whether a floating point number equals zero, since zero can have a sign. It is safe to test if a floating point number is greater than zero or less than zero. To see if a floating point number is nonzero, it is safe to test if its *absolute value* is greater than zero.

Exercise 2.2.7 Determine the binary representation of a `long double`. In particular, what is its bias, and how many bits are in its mantissa?

Exercise 2.2.8 Consider the program roundingErrors in Fortran, roundingErrors in Fortran 90, roundingErrors in C, roundingErrors in C^{++} or roundingErrors in MATLAB. Each of these programs shows that there are different machine numbers ε at which $(1.+\varepsilon)-1. = 0.$, $(1./\varepsilon-1.)*\varepsilon = 1.$ or at which $(1.-1./\varepsilon)+1./\varepsilon = 0.$. Examine the different values of ε to explain the different results. You might conclude that "machine precision" means different things in these different circumstances.

Table 2.6 Special math numbers in C or C++

M_E	e	2.7182818284590451
M_LOG10E	$\log_{10}(e)$	0.43429448190325182
M_LN2	ln(2)	0.69314718055994529
M_PI	π	3.1415926535897931
M_PI_2	$\pi/2$	1.5707963267948966
M_PI_4	$\pi/4$	0.78539816339744828
M_1_PI	$1/\pi$	0.31830988618379069
M_2_PI	$2/\pi$	0.63661977236758138
M_SQRT2	$\sqrt{2}$	1.4142135623730951
M_SQRT1_2	$1/\sqrt{2}$	0.70710678118654757

2.2.4 *Special Numbers*

The C and C++ programming languages define some useful numbers in file `math.h`. Several of these are listed in Table 2.6.

For `long double` computations, there are extended precision versions of these numbers with different names (such as `M_El` instead of `M_E`). These values may also be available in Fortran, if the Fortran compiler is based on a C compiler (as in the GNU system).

JavaScript provides various mathematical constant in its `Math` class. These include `Math.E`, `Math.LN10`, `Math.LN2`, `Math.LOG10E`, `Math.LOG2E`, `Math.PI`, `Math.SQRT1_2` and `Math.SQRT2`.

In MATLAB, the constant `pi` is defined, but `e` is available only through the MATLAB MuPad Symbolic Math Toolbox.

2.2.5 *Floating Point Arithmetic*

In this section, we will examine how the floating point representations of numbers influence the basic arithmetic operations of addition, subtraction, multiplication and division.

2.2.5.1 Addition

Integer addition is exact, unless it overflows. On the other hand, floating point addition almost always introduces an error. To understand why, consider the addition of two floating point numbers:

$$f^{(1)} + f^{(2)} = \left\{(-1)^{s^{(1)}} 2^{e^{(1)}-bias} \sum_{j=0}^{m} b_j^{(1)} 2^{j-m}\right\} + \left\{(-1)^{s^{(2)}} 2^{e^{(2)}-bias} \sum_{j=0}^{m} b_j^{(2)} 2^{j-m}\right\} .$$

To simplify the discussion, we will assume that both numbers have the same sign (i.e., their sign bits satisfy $s^{(1)} = s^{(2)} = s$), both are normal floating point numbers (i.e., their leading bits are $b_m^{(1)} = b_m^{(2)} = 1$, and $|f^{(1)}| \geq |f^{(2)}|$ (i.e., their exponents satisfy $e^{(1)} \geq e^{(2)}$). Then

$$f^{(1)} + f^{(2)} = (-1)^s 2^{e^{(1)}-bias} \left(\sum_{j=0}^{m} b_j^{(1)} 2^{j-m} + \sum_{j=0}^{m} b_j^{(2)} 2^{j-m+e^{(2)}-e^{(1)}} \right)$$

$$= (-1)^s 2^{e^{(1)}-bias} \left(\sum_{j=m+1-(e^{(1)}-e^{(2)})}^{m} b_j^{(1)} 2^{j-m} + \sum_{j=0}^{m-(e^{(1)}-e^{(2)})} \left[b_j^{(1)} + b_{j+e^{(1)}-e^{(2)}}^{(2)} \right] 2^{j-m} \right.$$

$$\left. + \sum_{j=-(e^{(1)}-e^{(2)})}^{-1} b_j^{(2)} 2^{j-m} \right) .$$

$$= (-1)^s 2^{e^{(1)}-bias} \sum_{j=-(e^{(1)}-e^{(2)})}^{m+1} b_j 2^{j-m} \approx (-1)^s 2^{e-bias} \sum_{j=0}^{m} b_j 2^{j-m} .$$

To represent this result as a floating-point number, the bits in the red sum must be added, the full sum must be represented in binary form, the position of the leading bit must be determined, and *the least significant bits in the sum must be appropriately rounded to significance*. The relative error due to rounding the insignificant bits is called the **rounding error**, and satisfies

$$\frac{\left|(-1)^s 2^{e-bias} \sum_{j=-(e^{(1)}-e^{(2)})}^{-2} b_j 2^{j-m}\right|}{\left|(-1)^s 2^{e-bias} \sum_{j=-(e^{(1)}-e^{(2)})}^{m+1} b_j 2^{j-m}\right|} \leq \frac{\sum_{j=-(e^{(1)}-e^{(2)})}^{-2} 1 \cdot 2^{j-m}}{1 + \sum_{j=-(e^{(1)}-e^{(2)})}^{m} 0 \cdot 2^{j-m}} \leq 2^{-m-1} .$$

If $\mathrm{fl}\left(f^{(1)} + f^{(2)}\right)$ is the floating point representation of the sum of two normal floating point numbers with the same sign, then we have shown that

$$\left| \mathrm{fl}\left(f^{(1)} + f^{(2)}\right) - \left(f^{(1)} + f^{(2)}\right)\right| \leq \varepsilon \left|f^{(1)} + f^{(2)}\right| .$$

where $\varepsilon = 2^{-m-1}$ is the **machine precision**.

Example 2.2.1 Suppose that our floating mantissa has four bits (i.e. $m = 4$), and we want to add

$$f^{(1)} + f^{(2)} = \frac{13}{16} + \frac{11}{32} = 2^{-1}\left(1 + 2^{-1} + 0 \cdot 2^{-2} + 2^{-3}\right)$$
$$+ 2^{-2}\left(1 + 0 \cdot 2^{-1} + 2^{-2} + 2^{-3}\right)$$
$$= 2^{-1}\left(1 + 2^{-1} + 0 \cdot 2^{-2} + 2^{-3} + 0 \cdot 2^{-4}\right) + 2^{-1}\left(0 + 2^{-1} + 0 \cdot 2^{-2} + 2^{-3} + 2^{-4}\right)$$

$$= 2^{-1}\left(1 + 2 \cdot 2^{-1} + 0 \cdot 2^{-2} + 2 \cdot 2^{-3} + 2^{-4}\right)$$

$$= 2^{0}\left(1 + 0 \cdot 2^{-1} + 0 \cdot 2^{-2} + 2^{-3} + 0 \cdot 2^{-4} + 2^{-5}\right) = \frac{37}{32}$$

$$\approx 2^{0}\left(1 + 0 \cdot 2^{-1} + 0 \cdot 2^{-2} + 2^{-3}\right) = \frac{9}{8} .$$

In this example, the relative error due to rounding is

$$\frac{|9/8 - 37/32|}{|37/32|} = \frac{1}{37} < \frac{1}{32} = 2^{-5} = 2^{-m-1} .$$

2.2.5.2 Subtraction

In general, the subtraction of two normal floating point numbers involves rounding errors similar to addition. However, it is important to note that **subtraction of two nearly equal floating point numbers is computed exactly**. Suppose that two floating point numbers $f^{(1)}$ and $f^{(2)}$ have the same signs and the same exponent, and satisfy $|f^{(1)}| \geq |f^{(2)}|$. Then

$$f^{(1)} - f^{(2)} = (-1)^s 2^{e-bias}\left(\sum_{j=0}^{m} b_j^{(1)} 2^{j-m} + \sum_{j=0}^{m} b_j^{(2)} 2^{j-m}\right)$$

$$= (-1)^s 2^{e-bias} \sum_{j=0}^{m}\left(b_j^{(1)} - b_j^{(2)}\right) 2^{j-m} = (-1)^s 2^{e-bias} \sum_{j=0}^{m} b_j 2^{j-m} .$$

It is possible that $b_m = 0$ in this sum, so the exponent may need to be adjusted to correspond to the most significant nonzero bit. Such an adjustment would introduce one or more zero bits in the least significant positions, but would not introduce a rounding error.

Example 2.2.2 Suppose that our floating mantissa has four bits (i.e. $m = 4$), and we want to subtract

$$f^{(1)} - f^{(2)} = \frac{9}{8} - \frac{15}{16} = 2^{0}\left(1 + 0 \cdot 2^{-1} + 0 \cdot 2^{-2} + 2^{-3}\right) - 2^{-1}\left(1 + 2^{-1} + 2^{-2} + 2^{-3}\right)$$

$$= 2^{0}\left(1 + 0 \cdot 2^{-1} + 0 \cdot 2^{-2} + 2^{-3}\right) - 2^{0}\left(0 + 2^{-1} + 2^{-2} + 2^{-3} + 2^{-4}\right)$$

$$= 2^{0}\left(0 + 0 \cdot 2^{-1} + 0 \cdot 2^{-2} + 1 \cdot 2^{-3} + 1 \cdot 2^{-4}\right)$$

$$= 2^{-3}\left(1 + 1 \cdot 2^{-1} + 0 \cdot 2^{-2} + 1 \cdot 2^{-3}\right) = \frac{3}{16} .$$

Since $f^{(1)}$ and $f^{(2)}$ are nearly equal, the subtraction was performed exactly, even though $f^{(1)}$ and $f^{(2)}$ have different exponents.

2.2.5.3 Multiplication

As expected, floating point multiplication typically involves rounding errors. For example, suppose that we multiply two normal floating point numbers:

$$f^{(1)}f^{(2)} = \left\{(-1)^{s^{(1)}} 2^{e^{(1)}-bias} \sum_{i=0}^{m} b_i^{(1)} 2^{i-m}\right\} \left\{(-1)^{s^{(2)}} 2^{e^{(2)}-bias} \sum_{j=0}^{m} b_j^{(2)} 2^{j-m}\right\}$$

$$= (-1)^{s^{(1)}+s^{(2)}} 2^{e^{(1)}+e^{(2)}-2\cdot bias} \sum_{i=0}^{m}\sum_{j=0}^{m} b_i^{(1)} b_j^{(2)} 2^{i+j-2m}$$

$$= (-1)^{s^{(1)}+s^{(2)}} 2^{e^{(1)}+e^{(2)}-2\cdot bias} \sum_{i=0}^{m}\sum_{k=i}^{i+m} b_i^{(1)} b_{k-i}^{(2)} 2^{k-2m}$$

$$= (-1)^{s^{(1)}+s^{(2)}} 2^{e^{(1)}+e^{(2)}-2\cdot bias} \left\{\sum_{k=0}^{m-1}\sum_{i=0}^{k} b_i^{(1)} b_{k-i}^{(2)} 2^{k-2m} + \sum_{k=m}^{2m}\sum_{i=k-m}^{m} b_i^{(1)} b_{k-i}^{(2)} 2^{k-2m}\right\}$$

In order to approximate this result by a floating point number, we would need to compute the bit convolutions in the sums over i, represent the result in binary form, adjust the exponent to correspond to the most significant bit, and discard the insignificant bits. If the approximate product is a normal floating point number, then the resulting rounding error will satisfy the same bounds that we found for floating point addition.

However, the bit convolutions involve more work than the work in floating point addition. This fact motivates some clever algorithms. A **divide and conquer** algorithm is described in Aho et al. [1, p. 62f]. Some other algorithms are described on the Wikipedia Floating-point arithmetic web page.

Example 2.2.3 Suppose that our floating mantissa has four bits (i.e. $m = 4$), and we want to multiply

$$f^{(1)} \times f^{(2)} = \frac{13}{16} \times \frac{11}{32} = 2^{-1}\left(1 + 2^{-1} + 0 \cdot 2^{-2} + 2^{-3}\right)$$

$$\times 2^{-2}\left(1 + 0 \cdot 2^{-1} + 2^{-2} + 2^{-3}\right)$$

$$= 2^{-3}\left(1 + 2^{-1} + 0 \cdot 2^{-2} + 2^{-3}\right) \times \left(1 + 0 \cdot 2^{-1} + 2^{-2} + 2^{-3}\right)$$

$$= 2^{-3}\left(2 + 0 \cdot 2^{0} + 0 \cdot 2^{-1} + 0 \cdot 2^{-2} + 2^{-3} + 2^{-4} + 2^{-5}\right) = \frac{143}{512}$$

$$\approx 2^{-2}\left(1 + 0 \cdot 2^{-1} + 0 \cdot 2^{-2} + 1 \cdot 2^{-3}\right) = \frac{9}{32} \cdot$$

In this example, the relative error due to rounding is

$$\frac{|9/32 - 143/512|}{|143/512|} = \frac{1}{143} < \frac{1}{32} = 2^{-5} = 2^{-m-1} .$$

2.2.5.4 Division

Floating point division is even more complicated than multiplication. It is possible to approximate the quotient of two normal floating point numbers as a floating point number, with rounding error satisfying the same relative error bounds as we found for addition or multiplication. Some algorithms for performing a division can be found on the Wikipedia Division algorithm web page. In particular, we note that one of those algorithms involves a floating point short cut that we will discuss in Sect. 2.3.1.4.

2.2.5.5 Summary

In summary, the nature of floating point arithmetic has the following implications.

- Errors are made in essentially all floating point arithmetic operations.
- The difference of two nearly equal floating point numbers is computed exactly.
- Multiplication is likely to be more expensive than addition or subtraction, and division is likely to be more expensive than multiplication.
- Because there are several hardware stages in computer implementations of arithmetic operations, a clever hardware designer could pipeline the multiple stages of the calculations; in this process, similar operations on multiple pairs of numbers are overlapped in the successive stages to perform the collection more rapidly.

Exercise 2.2.9 Write a program to do the following. First set x = 0.1, then compute 1. - 10.*x and print the result. Run the program and explain your results.

Exercise 2.2.10 Write a program to compute the smallest floating machine number ε such that $1 - \varepsilon \neq 1$. How does this number differ from `numeric_limits<double>::epsilon()`?

2.2.6 Complex Numbers

A **complex number** is an ordered pair of real numbers with certain rules for binary arithmetic operations. Typically, we will write a complex number c in the form

$$c = a + bi$$

where a and b are real numbers and $\mathrm{i} \equiv \sqrt{-1}$. The rules for operating on complex numbers are as follows:

$$c_1 \pm c_2 = (a_1 \pm a_2) + (b_1 \pm b_2)\mathrm{i} \,,$$
$$c_1 \times c_2 = (a_1 a_2 - b_1 b_2) + (a_1 b_2 + a_2 b_1)\mathrm{i} \text{ and}$$
$$c_1 / c_2 = \frac{a_1 a_2 + b_1 b_2}{a_2^2 + b_2^2} + \frac{-a_1 b_2 + a_2 b_1}{a_2^2 + b_2^2}\mathrm{i}$$

In practice, some care should be taken to ensure that the quotient of two complex numbers does not involve unnecessary overflow. Interested readers can examine LAPACK routine dladiv to see how complex division can be conducted safely.

Complex arithmetic is part of the Fortran language. The two complex data types are `complex*8` and `complex*16`, which have the shorthand forms `complex` and `double complex`, respectively. In Fortran 90, we can declare these same variable types as `complex(kind=4)` or `complex(kind=8)`. The Fortran function `cmplx(a,b)` returns a complex number with real part `a` and imaginary part `b`. If `a` and `b` are both `double precision`, then `cmplx` returns a `double complex`. All of the arithmetic operations, trigonometric and exponential/logarithm functions are defined for Fortran complex numbers. The real part of a `complex` or `double complex` number `z` is `real(z)`, and the imaginary part is `imag(z)`.

In C^{++}, complex numbers are available as a template class, with template argument either `float`, `double` or `long double`. Most important operations on complex numbers are defined in the header file `complex`, which is found in a sub-directory of `/usr/include/c++` in most Linux systems. This include file provides functions such as `abs`, `arg`, `conj` (for complex conjugation), `polar`, `cos`, `cosh`, `exp`, `log`, `pow`, `sin`, `sinh`, `sqrt`, `tan` and `tanh`. The file also defines unary and binary operations on complex numbers such as addition, subtraction, multiplication and division. The real part of a `complex<T>` number `z` is `z.real()`, and the imaginary part is `z.imag()`.

Complex numbers are not part of the JavaScript standard language at this time. Instead, we have written a complex class for JavaScript.

Complex numbers in MATLAB can be obtained by commands such as `z=1.+2.*i` or `z=complex(1.,2.)` The rules regarding assignment of MATLAB complex numbers are somewhat tricky. For example, if either `x` or `y` is of type `single`, then `z = x + y*i` is of type `single complex`. Also note that `i` and `j` are used to represent $\sqrt{-1}$ in MATLAB; it could become confusing if these variable names are also used for loop indices in MATLAB.

Exercise 2.2.11 What is the size of a `complex<double>` in C^{++}? Is this twice the size of a double?

Exercise 2.2.12 Where does the complex logarithm choose to make its branch cut in your favorite programming language on your machine? For example, if we choose a positive integer n and set $\theta = 2\pi/n$, for what value of k will we find that $\log(\exp(ik\theta)) \neq k\theta$?

2.3 IEEE Arithmetic

In Sect. 2.2, we learned how numbers are represented in computers, and briefly examined how arithmetic operations are performed on floating point numbers. Our next goal is to examine some of the implications of floating point arithmetic. We will study several examples of important floating point computations in Sect. 2.3.1. Afterward, we will learn about floating point exceptions. These can be very unnerving for readers who are new to scientific computing. We will end the section with some important floating point shortcuts.

2.3.1 *Floating Point Computations*

On machines respecting the IEEE 754 standard, the floating point operations add, subtract, multiply and divide, while acting on and producing normal numbers, are designed so that the result satisfies the equation

$$fl(a \; op \; b) = (a \; op \; b)(1 + \varepsilon) \; .$$

Here fl could represent any one of the floating point arithmetic operations, or entry of a number into the computer. Also, ε is a relative error on the order of **machine precision**. We shall define the machine precision ε to be

$$\varepsilon = \min\{machine \; numbers \; a \; : \; fl(1 - a) < 1\} \; .$$

The IEEE standard has some interesting implications, which will be revealed by the following examples.

2.3.1.1 Subtraction

Let us examine the error in subtracting two user numbers. In this computation, we will make errors of two types. First, there are rounding errors in converting to machine numbers (data entry). Second, arithmetic errors can occur in subtracting two machine numbers. Mathematically, we can represent the result as follows:

$$fl(fl(a) - fl(b)) = fl(a(1 + \varepsilon_a) - b(1 + \varepsilon_b)) = (a(1 + \varepsilon_a) - b(1 + \varepsilon_b))(1 + \varepsilon_-)$$
$$\approx a(1 + \varepsilon_a + \varepsilon_-) - b(1 + \varepsilon_b + \varepsilon_-) = (a - b) + a(\varepsilon_a + \varepsilon_-) - b(\varepsilon_b + \varepsilon_-) \; .$$

Here, the approximation was due to ignoring errors involving products of epsilons. If we take

$$\varepsilon = \max\{|\varepsilon_a| \ , \ |\varepsilon_a| \ , \ |\varepsilon_-|\}$$

then we can estimate the *relative error* in the floating point difference as follows:

$$\left|\frac{fl(fl(a) - fl(b)) - (a-b)}{a-b}\right| \approx \left|\frac{a(\varepsilon_a + \varepsilon_-) - b(\varepsilon_b + \varepsilon_-)}{a-b}\right|$$
$$\leq \frac{|a|\left(|\varepsilon_a| + |\varepsilon_-|\right) + |b|\left(|\varepsilon_b| + |\varepsilon_-|\right)}{|a-b|} \leq 2\varepsilon\frac{|a| + |b|}{|a-b|} \ .$$

We claim that if b is close to a, then $\varepsilon_- = 0$. This is true because Sect. 2.2.5 showed that the floating point difference of two nearly equal machine numbers is computed exactly. However, if b is very close to a, then the denominator $|a-b|$ might be on the same order as ε times the size of a or b, and the relative error in the difference could be very large. In this case, the rounding errors due to data entry lead to large relative error in the difference.

2.3.1.2 Finite Differences

A similar situation arises in computing a derivative by finite differences. Suppose that h is a power of 2, so that no error occurs in division by h. Also suppose that x is a machine number, and that $fl(x+h) = x+h$. We assume that rounding errors occur in the evaluation of $f(x)$ and $f(x+h)$, but not in the subtraction of these two function values (since they should be close to one another). Under these assumptions, the error in the numerical approximation of the derivative by a difference quotient is

$$fl\left(\frac{fl(f(x+h)) - fl(f(x))}{h}\right) - f'(x)$$
$$= fl\left(\frac{f(x+h)\left(1+\varepsilon_{f+}\right) - f(x)\left(1+\varepsilon_f\right)}{h}\right) - f'(x)$$
$$= \frac{f(x+h)\left(1+\varepsilon_{f+}\right) - f(x)\left(1+\varepsilon_f\right)}{h} - f'(x)$$
$$= \left[f(x) + f'(x)h + f''(x)\frac{h^2}{2} + O\left(h^3\right)\right]\frac{1+\varepsilon_{f+}}{h} - f(x)\frac{1+\varepsilon_f}{h} - f'(x)$$
$$= f(x)\frac{\varepsilon_{f+} - \varepsilon_f}{h} + f'(x)\varepsilon_{f+} + f''(x)\frac{h\left(1+\varepsilon_{f+}\right)}{2} + O\left(h^2\right)$$
$$= f(x)\frac{\varepsilon_{f+} - \varepsilon_f}{h} + f'(x)\varepsilon_{f+} + f''(x)\frac{h}{2} + O\left(h^2\right) + O\left(\varepsilon h\right) \ .$$

As $h \to 0$, first term, which is due to rounding errors in the function evaluations, becomes large and dominates the other terms in the final expression. This is

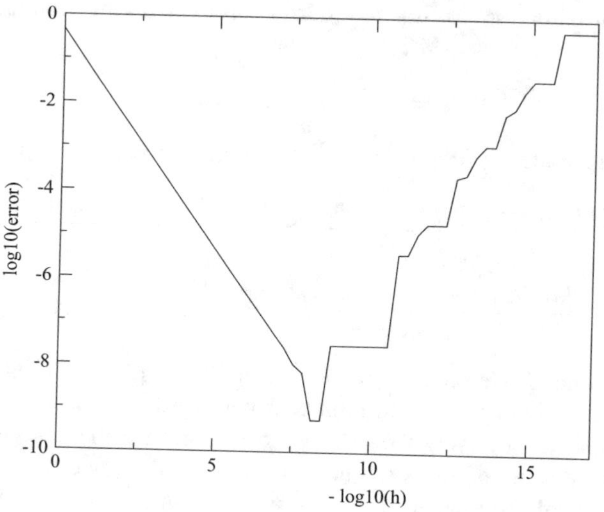

Fig. 2.1 Error in finite difference derivative for $sin(x)$ at $x = 1$

important to understand, because we might want to replace derivatives by difference quotients in solving nonlinear equations by Newton's method (see Sect. 5.4.9). Clearly, we should want to avoid computing finite differences via very small perturbations in the arguments.

To see these computations in action, readers can experiment with programs that compute finite differences in Fortran, finite differences in Fortran 90, finite differences in C, finite differences in C^{++} and finite differences in MATLAB. Figure 2.1 shows a plot of the error in the finite difference computation of the derivative of $sin(x)$ for $x = 1$ and $h = 2^{-n}$. Note that the smallest errors are on the order of the square root of machine precision, and increase as h decreases further to machine precision.

2.3.1.3 Quadratic Equations

Quadratic equations arise commonly enough in applications, and the evaluation of their roots requires some care. Suppose that we want to solve

$$ax^2 + bx + c = 0 .$$

High school mathematics classes teach that the solution can be evaluated in the form

$$x = \frac{-b \pm \sqrt{b^2 - 4ac}}{2a} .$$

This formula has several problems. If the discriminant is real then one of these roots will involve cancellation; for example, if $b > 0$ then cancellation occurs in evaluating $-b + \sqrt{b^2 - 4ac}$. Also, if $4ac \ll b^2$, then this cancellation can destroy the relative precision of one of the roots. A more subtle problem is that unnecessary overflow can occur while evaluating the discriminant $b^2 - 4ac$.

Before developing a better alternative to the quadratic formula, let us make some observations. First, if $| a |= 0$ then there is at most one root, and we should not use the quadratic formula at all. Next, we note that if $| a |> 0$ then the product of the roots is c/a and the sum of the roots is $-b/a$. In particular, if both roots are real then one of the roots is at least as large as $| b/(2a) |$ in absolute value. Furthermore, we note that if $ac > 0$ then cancellation will occur in the discriminant; however, if $4ac \approx b^2$ then the square root of the discriminant will have only a small effect in the quadratic formula, and the relative accuracy of the roots will not be adversely affected by rounding errors in evaluating the discriminant. Finally, we note that the quadratic formula does not involve any cancellation outside the square root when computing the root of larger absolute value.

The following algorithm will carefully solve a quadratic equation.

Algorithm 2.3.1 (Quadratic Equation)

$\sigma = \max\{|a|, |b|, |c|\}$

if $\sigma = 0$ then $r_1 = r_2 =$ undefined /* all scalars are solutions */

else

  $\sigma' = 2^{\lceil \log_2 \sigma \rceil}$ /* scale by power of 2 to avoid rounding error */

  $\alpha = a/\sigma, \beta = b/\sigma, \gamma = c/\sigma$ /* scale to avoid unnecessary overflow */

  if $|\alpha| = 0.$ /* α may underflow; at most one root */

    if $|\beta| = 0.$ then $r_1 = r_2 =$ undefined /* no solutions */

    else $r_1 =$ undefined , $r_2 = -\gamma/\beta$ /* one solution */

  else /* two roots */

    $\beta' = -\beta/(2\alpha)$, $\gamma' = \gamma/\alpha$

    $\delta = \beta'\beta' - \gamma'$ /* discriminant */

    if $\delta \geq 0.$ /* two real roots */

      $r_1 = \beta' + \text{sign}(\sqrt{\delta}\ , \beta')$ /* larger root */

      if $|r_1| > 0$ then $r_2 = \gamma'/r_1$ /* smaller root */

      else $r_2 = 0$

    else /* two complex roots */

      $\delta' = \sqrt{-\delta}$

      $r_1 = \beta' + \mathrm{i}\delta'$, $r_2 = \beta' - \mathrm{i}\delta'$

Although Algorithm 2.3.1 involves more work than the standard quadratic formula, it provides better results. Readers might compare the performance of the quadratic formula and our algorithm for the sets of coefficients in Table 2.7.

Readers can experiment with programs that solve the quadratic equation in Fortran, quadratic equation in Fortran 90, quadratic equation in C and quadratic equation in C++. In MATLAB, readers may run the command

```
r = roots([ a b c ])
```

however, this command fails to find a nonzero value for the smaller root with most of the test problems in Table 2.7.

The reader may also execute the following JavaScript **Quadratic Equation Solver.**

2.3.1.4 Approximate Division

Some machines use Newton iterations to produce fast accurate approximations to important functions. To explain how these approximations work, let us recall Newton's method, which is commonly discussed in elementary calculus. To solve $f(x) = 0$ starting with some initial guess x, **Newton's method** performs the replacement

$$x \leftarrow x - \frac{f(x)}{f'(x)} .$$

If x is close to the zero of a function f with continuous derivative, then this iteration converges rapidly, typically doubling the number of accurate bits with each iteration. See Sect. 5.4 for a discussion of the convergence of Newton's method in the presence of rounding errors.

Some computers perform division iteratively, in order to use pipelined and chained operations for multiplication and addition. Here is the basic idea. Given a number a, the reciprocal $x = 1/a$ solves

$$0 = f(x) \equiv \frac{1}{x} - a .$$

Table 2.7 Test problems for quadratic formula

a	b	c	Larger root	Smaller root
10^{160}	-2×10^{160}	10^{160}	1	1
10^{-20}	-1	1	10^{20}	1
1	-1	10^{-20}	1	10^{-20}
10^{-10}	-1	10^{-10}	10^{10}	1
10^{-5}	-1	10^{-5}	$2 \times 10^5 - 0.5 \times 10^{-5}$	$0.5 \times 10^{-5} + 0.125 \times 10^{-15}$
10^{-220}	-10^{100}	10^{100}	10^{320}	1

The Newton iteration to find a zero of $f(x)$ is

$$x \leftarrow x * (2 - a * x) .$$

Each iteration of this method involves only multiplication and subtraction.

Note that other machines have had their troubles with division, because it is considerably more difficult to implement in binary arithmetic than multiplication. For example, there was a well-publicized Pentium division flaw, and at one time the Dec Alpha chips disabled pipelining in order to perform divisions.

In general, we advise readers to avoid division whenever possible. For example, it is generally better to use the code `y = 0.1 * x` rather than `y = x / 10.`

2.3.1.5 Approximate Square Roots

Given a, many compilers compute $x = \sqrt{a}$ by solving

$$0 = f(x) \equiv x^2 - a .$$

Here they use the Newton iteration

$$x \leftarrow 0.5 * (x + a/x) .$$

This Newton iteration just averages the current guess x with a divided by x. However, this iteration involves a division in each iteration. The next algorithm provides an alternative.

2.3.1.6 Approximate Reciprocal Square Roots

Given a, one might compute $x = 1/\sqrt{a}$ by solving

$$0 = f(x) \equiv \frac{1}{x^2} - a .$$

This leads to the Newton iteration for a reciprocal square root, which might be written simply as

$$x \leftarrow 0.5 * x * (3. - a * x^2) .$$

Since this iteration does not involve division, some machines actually compute fast square roots by first finding the reciprocal square root, and then taking the reciprocal. However, computing the square root of zero on such machines can produce an IEEE floating point exception.

2.3.1.7 Horner's Rule

Some other programming shortcuts involve reorganizing computations in order to reduce the computational cost. Suppose that we want to evaluate the polynomial

$$p_n(x) = \sum_{i=0}^{n} a_{n-i} x^i \; .$$

If we evaluate this as it is written, then each power x^i costs either an exponentiation by an integer, or $i-1$ multiplications of x times itself. Afterward, the term $a_{n-i}x^i$ requires another multiplication. In this way, the evaluation of $p_n(x)$ would cost $\sum_{i=0}^{n} i = n(n+1)/2$ multiplications and n additions.

Alternatively, we could evaluate $p_n(x)$ by Horner's rule:

$$p_n(x) = a_n + x(a_{n-1} + x(a_{n-2} + \ldots)) \; .$$

This can be performed by the following algorithm:

Algorithm 2.3.2 (Horner's Rule)

$$p = a_0$$

for $1 \le i \le n$

$$p = a_i + x * p \; .$$

This algorithm requires n multiplications and n additions. Since this computation is so efficient, polynomials defined by a monomial expansion should always be evaluated by Horner's rule.

2.3.1.8 Synthetic Division

Suppose that $p_n(x)$ is a polynomial of degree n, and r is being investigated as a possible zero of p_n. We would like to evaluate $p_n(r)$ to see if r is in fact a zero of

$p_n(x)$. We would also like to find the quotient polynomial

$$q_{n-1}(x) = \frac{p_n(x) - p_n(r)}{x - r}$$

for use in finding the remaining zeros of the original polynomial $p_n(x)$. The definition of q_{n-1} implies that

$$p_n(x) = (x - r)q_{n-1}(x) + d_0 ,$$

where $d_0 = p_n(r)$. If the monomial expansion for q_{n-1} is

$$q_{n-1}(x) = \sum_{i=0}^{n-1} b_{n-1-i}x^i ,$$

then we can find the b_i and d_0 by the following algorithm:

Algorithm 2.3.3 (Synthetic Division)

$$\begin{aligned}
&b_0 = a_0 \\
&\text{for } 1 \le i \le n \\
&\quad b_i = a_i + r * b_{i-1} \\
&d_0 = b_n
\end{aligned}$$

Note that the computation of $d_0 = p_n(r)$ is identical to Horner's rule. The difference between **synthetic division** and Horner's rule is that the former saves the coefficients b_i of the quotient polynomial.

2.3.1.9 Repeated Synthetic Division

We can continue the synthetic division process $k \le n$ times to obtain

$$\begin{aligned}
p_n(x) &= (x - r)\left\{(x - r)q_{n-2}(x) + d_1\right\} + d_0 = \ldots \\
&= (x - r)^{k+1}q_{n-k-1}(x) + d_k(x - r)^k + \ldots + d_1(x - r) + d_0 .
\end{aligned}$$

If $k = n$, then q_{-1} is understood to be zero. In any case, it can be shown that

$$d_k = \frac{1}{k!}\frac{\mathrm{d}^i p_n}{\mathrm{d}x^k}(r) .$$

For example, if $k = 2$ we can compute $p(r)$ and $p'(r)$ simultaneously with the following algorithm:

Algorithm 2.3.4 (Repeated Synthetic Division)

$$\begin{aligned}
&c_0 = b_0 = a_0 \; ; \\
&\text{for } 1 \leq i < n \\
&\quad b_i = a_i + r * b_{i-1} \\
&\quad c_i = b_i + r * c_{i-1} \\
&b_n = a_n + r * b_{n-1}
\end{aligned}$$

Then $p_n(r) = b_n$, $p_n'(r) = c_{n-1}$ and

$$q_{n-1}(x) = \sum_{i=0}^{n-1} b_{n-1-i}x^i = \frac{p_n(x) - p_n(r) - p_n'(r)(x-r)}{x-r} \; .$$

Note that this algorithm produces $p_n'(r)$ for an additional cost that is essentially the same as that for $p_n(r)$. Thus this algorithm should always be used to evaluate polynomials and their derivatives, as might be necessary when we use Newton's method to find a zero of a polynomial.

2.3.1.10 Binomial Coefficients

In combinatorics or discrete probability, it is well-known that the number of ways to order n distinct items is

$$n \times (n-1) \times \ldots \times 2 \times 1 = n! \; .$$

From this, it follows that the number of ways to choose m items in some order from a group of n distinct items is

$$n \times (n-1) \times \ldots \times (n-m+1) = n!/(n-m)! \; .$$

Finally, the number of ways to choose m items from a group of n in any order is

$$[n!/(n-m)!]/m! = \binom{n}{m} \; .$$

A combinatorial argument can be used to prove the **binomial expansion**

$$(x+y)^n = \sum_{m=0}^{n} \binom{n}{m} x^m y^{n-m} \; . \tag{2.1}$$

Indeed, it is easy to see that in distributing the product

$$(x+y)^n = (x+y) \times \ldots \times (x+y) = \sum_{m=0}^{n} b_m^{(n)} x^m y^{n-m} ,$$

there are $\binom{n}{m}$ ways to choose exactly m x's to form the term involving x^m; since $\binom{n}{m} = \binom{n}{n-m}$, this is exactly the same as the number of ways of choosing $(n-m)$ y's.

The programming issue here is how to evaluate binomial coefficients. One way is to write a recursive function to evaluate the factorial:

if $n = 1$ then

factorial(n) = 1

else if $n > 1$ then

factorial(n) = $n *$ factorial($n - 1$) .

In MATLAB, we could call `factorial(n)`. Using this function, we could program

$$\text{binomialCoefficient}(n, m) = \text{factorial}(n) / (\text{factorial}(m) * \text{factorial}(n-m)) .$$

However, in MATLAB it would be much better to call `nchoosek(n,m)`.

Notice that

$$\binom{n}{0} = 1 = \binom{n}{n} ,$$

so no work is needed to compute these values. To evaluate all binomial coefficients $\binom{n}{m}$ for $1 \leq m \leq n-1$, we could compute the values of $m!$ for $2 \leq m \leq n$ in a total of $n-1$ multiplications, and then compute the binomial coefficients $\binom{n}{m}$ for $1 \leq m \leq n-1$ with one multiplication and one division each, for an additional $n-1$ multiplications and $n-1$ divisions. The total work would be $2(n-1)$ multiplications and $n-1$ divisions.

However, it is far better to evaluate binomial coefficients using **Pascal's triangle**. This computation depends on the fact that for $1 \leq m \leq n-1$

$$\binom{n}{m} = \binom{n-1}{m} + \binom{n-1}{m-1} . \tag{2.2}$$

This can be proved by induction, or by the following combinatorial argument. To choose m items from n, we can mark one of the items before choosing. There are $\binom{n-1}{m-1}$ ways to choose the marked item and $m-1$ others, and there are $\binom{n-1}{m}$ ways to choose m unmarked items. The total number of ways to choose m items, marked or unmarked, is the sum of these two counts.

Note that Pascal's triangle (2.2) uses only addition to compute the binomial coefficients. However, the total number of arithmetic operations needed to compute all values of $\binom{n}{m}$ for $1 \le m \le n-1$ is

$$\sum_{k=2}^{n}\sum_{m=1}^{k-1} 1 = \sum_{k=2}^{n}(k-1) = \frac{n(n-1)}{2}$$

additions. For $n > 6$, this involves more arithmetic operations than using factorials.

However, factorials grow large quite rapidly. On a machine with integers stored in 64 bits, $n!$ will overflow as an integer at $n = 21$, while no values of $\binom{n}{m}$ overflow as integers until $n = 67$. On a machine with floating-point numbers stored in 64 bits, $n!$ will overflow as a floating-point number at $n = 171$, while no values of $\binom{n}{m}$ will overflow until $n = 1030$.

Interested readers may execute the JavaScript **Binomial Coefficient** Program. This program computes the binomial coefficients by factorials, and by Pascal's triangle. Readers can discover values of n for which the factorial approach introduces rounding errors, or leads to overflow.

Exercise 2.3.1 Consider performing a finite difference approximation to the derivative of $f(x) = \sqrt{x}$.

1. Program this computation and test it for $h = 1., 0.5, 0.25, \ldots$. Does the error increase as h approaches machine precision?
2. Repeat the same exercise for $h = 0.1, 0.05, 0.025, \ldots$. Does the error increase as h approaches machine precision?
3. Examine the rounding errors in the finite difference computation for $h = 2^{-n}$. Can you explain why $\varepsilon_{f+} = \varepsilon_f$ in this case?

Exercise 2.3.2 Consider the quadratic equation $10^{-16}x^2 + x - 1 = 0$.

1. If you solve this equation via the quadratic equation using floating point arithmetic, what is the smaller root, and what is the value of the quadratic with x equal to this root?
2. If the leading coefficient were zero instead of $1.e-16$, what would be the root of the equation?
3. What are the exact roots of the original quadratic?

Exercise 2.3.3 Program Newton's method to compute the square root of a number a, by solving $f(x) \equiv x^2 - a = 0$. Run the program to compute $\sqrt{2.}$ using $1.$ as the initial guess. Plot $ln(f(x))$ versus the number of iterations. What is the slope of the curve? What does the slope of the curve tell you about the performance of Newton's method?

Exercise 2.3.4 How would you find a good initial guess for the square root if you had access to the exponent in the binary representation of the argument?

Exercise 2.3.5 Determine a Newton iteration to compute the cube root of a number.

Exercise 2.3.6 Consider the cubic equation $x^3 + px^2 + qx + r = 0$.

1. Show that this equation can be rewritten in the form
$$(x - p/3)^3 + a(x - p/3) + b = 0 ,$$
where
$$a = \left(3q - p^2\right)/3 \text{and } b = \left(2p^3 - 9pq + 27r\right)/27 .$$
2. Explain how repeated synthetic division would compute a and b.
3. Show that if $b^2/4 + a^3/27 < 0$ then there will be three real and unequal values for $x - p/3$; if $b^2/4 + a^3/27 = 0$ then there will be three real roots of which at least two will be equal; and if $b^2/4 + a^3/27 > 0$ then there will be one real root and two conjugate complex roots.
4. If we define
$$A = \sqrt[3]{-\frac{b}{2} + \sqrt{\left[\frac{b}{2}\right]^2 + \left[\frac{a}{3}\right]^3}} \text{ and } B = \sqrt[3]{-\frac{b}{2} - \sqrt{\left[\frac{b}{2}\right]^2 + \left[\frac{a}{3}\right]^3}} ,$$
then show that the three roots of the original cubic are
$$\frac{p}{3} + A + B , \frac{p}{3} - \frac{A+B}{2} + \frac{A-B}{2}\sqrt{-3} \text{ and } \frac{p}{3} - \frac{A+B}{2} - \frac{A-B}{2}\sqrt{-3} .$$
5. Show that if there is only one real root, then it is
$$x = (A + B) + \frac{p}{3} .$$
6. Show that $AB = -a/3$ and
$$A + B = \frac{b}{A(A - B) + B^2} .$$
7. If the cubic has only one real root, explain how you would compute it with minimal roundoff error.
8. Show that if there are three real roots, then they have the form $x = \varrho \cos\theta + p/3$ where θ is one of the three solutions of
$$\cos(3\theta) = \frac{b}{\varrho(g/3)} .$$
9. Explain how to find the largest of three real roots with minimal roundoff error.

10. Explain how to use a Newton iteration to improve the accuracy of a computed root of the cubic.
11. Explain how to use synthetic division and a computed root of the cubic to determine a quadratic whose roots are the remaining roots of the original cubic.
12. How would you find the roots of a cubic equation to minimize the effects of rounding errors, overflow and underflow?

Exercise 2.3.7 Given four real numbers a, b, c and+ d, write a program using real arithmetic to compute real numbers p and q, so that $p + q\mathrm{i} = (a + b\mathrm{i})/(c + d\mathrm{i})$ is the quotient of two complex numbers. When you are done, compare your program to LAPACK routine dladiv.

2.3.2 *Floating Point Exceptions*

In Sect. 2.2.3 we introduced the ideas of floating point overflow and underflow. As we saw, **overflow** occurs when the number we want to represent would have an exponent too large for the available floating point type. For example, an overflow exception should be generated when we perform the C++ statements

```
double d = numeric_limits<double>::max(); d *= 2. ;
```

After executing this statement, the value of `d` will be `infinity`. **Underflow** occurs when the required exponent would be too small. This exception should be generated when we execute the C++ statements

```
double d = numeric_limits<double>::min();
d /= numeric_limits<double>::max();
```

The value of `d` after executing these statements will be `0`.

In addition, there are two other kinds of floating point exceptions that we should understand. The **divide by zero** exception occurs whenever we divide a nonzero number by zero. For example, the following C++ code should generate this exception:

```
double d = 0.; d = 1. / d;
```

Afterward, we will find that `d` is equal to `infinity`. Normally, we should want to know immediately when such an operation occurs, so that we can determine why the denominator was zero, and correct the program.

Finally, the **invalid** exception occurs whenever we try to perform a floating point operation that makes no sense. For example, the following C++ code should generate an invalid exception:

```
double d = 0.; d = 0. / d
```

Th result of this C++ code is that `d` is equal to `NaN`. We could also generate an invalid exception by trying to compute the real logarithm of a non-positive number, or the real square root of a negative number.

By default, IEEE arithmetic will *continue computation in the presence of floating point exceptions.* It does so by making use of the special floating point numbers `Infinity` and `NaN`.

We have seen `Infinity` previously in this section, and in Sect. 2.2.4. This floating point number has all exponent bits equal to one and all mantissa bits equal to zero; the sign of Infinity can be either positive or negative.

The acronym `NaN` stands for "not a number," and these "numbers" come in two types. A **quiet NaN** has exponent bits all one and most significant mantissa bit equal to zero. In C^{++}, such a number is returned by the `quiet_NaN` function, available through file `limits`. Normally, quiet NaN's do not generate floating point exceptions when they are used in arithmetic operations. In contrast, a **signaling NaN** has exponent bits all one and most significant mantissa bit equal to one. In C^{++}, such a number is returned by the `signaling_NaN` function, also available through file `limits`. Normally, when signaling NaN's are used in floating point operations they generate an invalid exception and result in a quiet NaN.

Table 2.8 shows several mathematical operations, their resulting values and the exceptions generated (if any). Since all floating point operations involving `NaNs` produce a `NaN`, We did not show any floating point operations with `NaNs` in this table. The entries in this table can be generated by any one of the Fortran exceptions, Fortran 90 exceptions, C exceptions or C^{++} exceptions programs.

In GNU C, C^{++}, Fortran or Fortran 90, users can detect exceptional numbers by calling the functions isnan or isinf. For C programs, these are defined in `math.h`, and C^{++} users should include `cmath`. However, it is inconvenient to call these functions just to prevent computational problems. An alternative is to trap all dangerous floating point exceptions, as described in Sect. 2.9.3.

MATLAB and JavaScript do not generate floating point exceptions. In either of these languages, the user can call functions `isfinite(x)` or `isnan(x)`, to test whether variable `x` is finite or not a number, respectively. MATLAB also provides a function `isinf(x)`, and in MATLAB exactly one of the functions `isfinite(x)`, `isinf(x)` and `isnan(x)` will be true.

There are some other results in MATLAB that might be surprising. For example, `log(-1.)` produces `0 + 3.1416i` and `acos(2.)` produces `0 + 1.3170i`. In other words, MATLAB uses the complex function of a complex variable for such arguments to avoid an `invalid` exception with the real function.

In the absence of floating point traps, the generation of either an infinity or NaN in any programming language is very unfortunate. For example, in the evolution of some differential equation it is possible for either an `inf` or a `NaN` to be produced. At that point, every future timestep involves operations with that `inf` or `NaN`, and produces additional exceptional numbers.

Example 2.3.1 (Exceptional Values in Initial Value Problems) Suppose that we use Fortran program euler.f, to solve

$$\frac{\mathrm{d}u}{\mathrm{d}t} = \lambda u \;, \; u(0) = 1 \;.$$

Table 2.8 C++ operations and exceptions

Operation	Result	Hexadecimal	Exception
d = `numeric_limits<double>::infinity()`	inf	7FF0000000000000	None
d = infinity()	inf	7FF0000000000000	None
d = quiet_nan()	nan	7FF8000000000000	None
1. / infinity()	0	0	None
infinity() * infinity()	inf	7FF0000000000000	None
infinity() + infinity()	inf	7FF0000000000000	None
d = signaling_nan()	nan	7FFc000000000000	None
0. / 0.	-nan	FFF8000000000000	Invalid
infinity() - infinity()	-nan	FFF8000000000000	Invalid
infinity() / infinity()	-nan	FFF8000000000000	Invalid
infinity() * 0.	-nan	FFF8000000000000	Invalid
ln (-1.)	nan	7FF8000000000000	Invalid
sqrt (-1.)	nan	7FF8000000000000	Invalid
acos (2.)	nan	7FF8000000000000	Invalid
1. / 0.	inf	7FF0000000000000	Invalid
ln (0.)	-inf	FFF0000000000000	Divide by zero
`numeric_limits<double>::max()` * 2.	inf	7FF0000000000000	Divide by zero
			Overflow

If we choose $\lambda = 100.$ and time step = 5 then $u_{20} = \texttt{inf}$, indicating that overflow has occurred. If we choose $\lambda = -10$ and time step = 0.05 then $u_{200} = 0$, indicating that underflow has occurred.

Exercise 2.3.8 Is there a distinction between `+0.` and `-0.` in your favorite programming language on your machine? How should you test a divisor to prevent a division by zero?

Exercise 2.3.9 Examine LAPACK routines disnan and dlaisnan to see how one might test in Fortran whether a double precision floating point number is a `NaN`. How could you make such a test in C or C++?

Exercise 2.3.10 Examine LAPACK routine ieeeck.f to see how one might test in Fortran whether a machine is performing IEEE arithmetic regarding infinity and `NaN`. Run these checks on your machine to verify the arithmetic.

Exercise 2.3.11 Suppose that x and y are floating-point numbers, and you want to compute $\sqrt{x^2 + y^2}$ without causing "unnecessary" overflow.

1. Describe conditions on x and y under which the final value would have to overflow.
2. Under what circumstances would direct evaluation (by squaring x and y, adding the squares and taking the square root) cause overflow?
3. Under what circumstances would direct evaluation lead to taking the square root of underflow?
4. Write a program to compute this square root of a sum of two squares safely. When you are done, compare your program to LAPACK routine dlapy2.

Exercise 2.3.12 Suppose that you want to compute the square root of the sum of squares of a set of numbers:

$$\sigma = \sqrt{\sum_{i=1}^{n} \alpha_i^2} \; .$$

Which floating point exceptions should you try to prevent? After you design your algorithm, compare your program to LAPACK BLAS routine dnrm2.f.

2.4 Programming Languages

In this and the previous chapter, we have provided some simple examples of programs in various languages. Our purpose is to illustrate how a user could instruct the computer to achieve certain computational goals. In this section, we will examine the programming languages themselves in more detail. We will discuss their strengths and weaknesses, and how to use mixed language programming to make best use of individual languages. Modular programs and mixed language programs can be a bit cumbersome to assemble into an executable, so we will describe the use of makefiles to simplify the process. We will also discuss the use of scoping rules.

2.4.1 Advantages and Disadvantages

Computer programs for scientific computing generally involve several important parts. In addition to solving the intended scientific problem, it is necessary to read input parameters, perform data checks, allocate and access memory, start and stop timing, and write the results for post-processing. We want to make these operations as effective as possible. As a result, we will

- use programming languages that are best-suited to each of these tasks,
- use existing software packages or re-use our own programs as much as possible to avoid repeated writing and debugging of common tasks, and
- structure our code to simplify the process of finding programming errors.

Each of C, C^{++}, Fortran, JavaScript, Python and MATLAB has its advantages and disadvantages. C was designed for developing operating systems. Many system commands, such as timing and memory allocation, are written in C. Many graphics programs (e.g., GTK) are also written in C. However, multi-dimensional arrays in C are awkward, and the starting index of arrays is fixed to be zero.

Each of C^{++}, JavaScript and Python is object-oriented. Many new scientific programs are being written in C^{++}, many web-based programs are being written in JavaScript, and Python now has well-developed libraries for scientific computing. Structures to represent multi-dimensional arrays are missing in C^{++}, JavaScript and Python, but easily available in Fortran. Of course, it is possible to write C^{++}, JavaScript or Python classes that can achieve much of the functionality of Fortran arrays. However, the need to develop classes to handle multi-dimensional arrays also increases the code development burden on the programmer. This process also leads to a lack of standardization in the implementation of that functionality, as each developer produces a different version of the same features already available in Fortran.

Fortran 77 was designed for scientific computing, although long before modern programming practices were developed. Fortran has extensive mathematical libraries. Most basic numerical software, such as LAPACK, has been written in Fortran 77. The Fortran 77 compilers are well-developed, so Fortran 77 code that will run on one machine will almost always run on another machine. Some of the Fortran 77 syntax is awkward, such as the use of parentheses for both arrays and function calls. Furthermore, Fortran is **case-insensitive**, so it cannot distinguish between variable names like `UPPER`, `UpPeR` and `upper`. Data structures and pointers are generally unavailable in Fortran 77. For example, if a Fortran 77 array is an argument to a subroutine, the array dimensions are not automatically associated with the array itself.

Fortran 90 was designed to overcome some of the shortcomings of Fortran 77. It allows subroutine array arguments to determine their array dimensions (through the intrinsic functions `SIZE`, `SHAPE`, `LBOUND` and `UBOUND`), but may do so by internally modifying the argument list to the function. This makes it difficult to call some Fortran 90 routines from other languages, such as C or C^{++}. (If a Fortran 90 programmer avoids the use of these intrinsic functions by passing the array bounds as subroutine arguments, then it should be possible to call Fortran 90 from C or C^{++}.) Also note that Fortran 90 does not have good access to operating system functions (unlike C or C^{++}), and does not have good publicly-available graphics.

JavaScript is not really designed for scientific computing. It has limited numerical types, and primitive debugging capabilities via the Firefox Developer Tools debugger. For our purposes, its major advantage is that JavaScript executables can be incorporated into Hypertext Markup Language (HTML) files. We use JavaScript to make the electronic form of this text interactive.

Each of C, C^{++}, Fortran 77 and Fortran 90 involve a **compiler**, which translates the language instructions into machine instructions for faster execution. After the compiler is called, an **assembler** must be called to combine all of the needed program routines into a single **executable**. The compiler makes some useful checks on the soundness of the program, and the assembler will fail if some program pieces are missing. Error messages from the compiler or the assembler can often be mystifying to new programmers.

MATLAB is a programming environment, more than a language. It was originally based on the LinPack package for linear algebra, but has expanded to handle many other computational tasks. MATLAB is an **interpreter** rather than a compiler, meaning that each MATLAB instruction is executed as it is encountered. In other words, readers can see the results of MATLAB instructions without going through the compile and build steps of other programming languages. Another advantage of MATLAB is that 2D and 3D graphics are built into the programming environment, so it is not necessary to write plot files to be read by a separate graphics program. One of the disadvantages of MATLAB is that every piece of data in MATLAB is stored internally as an array, and almost always either a one-dimensional or two-dimensional array. Data type conversions are automatic, may not be the programmer's intention, and probably add to the computational cost. Some automatic type conversions lead to a loss of precision. Debugging of floating point exceptions can be difficult. In many cases, large programs execute significantly faster in C or Fortran than in MATLAB.

Of course, there are a number of other programming languages in common use, and many of them could be useful for scientific computing. Python is an object-oriented language that is gaining popularity. Python implements a single floating-point data type (equivalent to `double` in C) and a single complex number type. Unlike C^{++}, Python is not type-safe (function arguments do not have declared types) and does not have scoping rules (loop arguments are defined outside the loop). Because Python uses an interpreter rather than a compiler, programmer code must be translated into machine instructions as the code is executed, thereby slowing down the execution. However, the PyPy implementation of Python provides a just-in-time compiler for faster execution. Python does have built-in capabilities to generate graphical user interfaces, and the ability to call C or C^{++} code. It is also possible to trap floating point exceptions in Python, and to display 3D graphics. Many programmers find it easier to write Python code than C^{++} code.

Java was developed by Sun Microsystems to run identically on all machine platforms across the web. In order to avoid different number representations on various machines, Java uses its own number machine, and its arithmetic differs from the IEEE standard. Neither Python nor Java use pointers, and both have class inheritance formalities that differ from C^{++}.

The choice of programming language is often a matter of personal preference, although it may be influenced by the availability of software. In this text, we will generally write numerically intensive routines in Fortran, because of the vast scientific computing software available in that language, and because of its robust treatment of multi-dimensional arrays. We will write routines that control input/output, memory allocation, system calls and graphics in C^{++}. We use C^{++} because it is strongly typed and its compiler leads to fast execution. Finally, we will write and use makefiles, in order to avoid repeatedly typing complicated compiler and assembler commands. Programs that are designed for reader interaction with the web browser will be written in JavaScript.

Exercise 2.4.1 Suppose that you have an array of velocity vectors associated with points on a three-dimensional grid. If you have to choose among C, C^{++}, Fortran, JavaScript or MATLAB, which would you choose to work with this array, and why?

Exercise 2.4.2 If you need to examine the file system to check for the existence of a file during program execution, which programming language would you use?

2.4.2 Building Programs

Readers who are new to programming commonly have problems with building an executable from multiple source files. This problem typically arises in calling LAPACK routines. For example, LAPACK routine `dgesv`, which solves a system of linear equations, in turn calls LAPACK routines `dgetrf`, `dgetrs` and `xerbl`; these routines may call other LAPACK routines as well. The advantage of breaking a complicated routine into several subroutines is that the programming done by each subroutine may be shared by the various calling routines. The disadvantage is that users may not easily discover the full list of subroutines that will be called, and may have trouble building an executable.

First, we will suppose that the user knows the full list of routines and the files that contain them, before building the executable. We will consider four cases in this process, depending on whether the files are in Fortran 77, Fortran 90, C^{++} or C. Afterward, we will discuss how to build executables when the list of needed subroutines is not known.

2.4.2.1 Known Fortran Files

Suppose that we have two Fortran 77 files fmain.f and fsub.f. To build an executable from these two files, we begin by compiling each separately to create the object files. Under a Linux operating system, we would type

```
gfortran -c -g fmain.f
gfortran -c -g fsub.f
```

Here `gfortran` is the name of a Linux Fortran compiler. Readers who use other operating systems will need to replace `gfortran` with the name of the appropriate Fortran compiler on their machine. The option "-c" tells the compiler to compile only, and not try to build an executable. The option "-g" tells the compiler to build the symbol table for the object file, so that we can use a debugger (see Sect. 2.8). By default, these two commands will create object files `fmain.o` and `fsub.o`. To assemble these two object files into an executable called `execf`, type

```
gfortran -o execf fmain.o fsub.o
```

Afterward, the executable `execf` can be run by typing

```
execf
```

To make this process easier, we have created a target for `execf` in our GNUmakefile. With this `GNUmakefile`, we could compile and assemble the executable merely by typing

```
make execf
```

To read more about makefiles, see Sect. 2.5.

2.4.2.2 Known Fortran 90 Files

For the two Fortran 90 files f90main.f90 and f90sub.f90, we would begin by using the Linux Fortran 90 compiler `gfortran` to create object files as follows:

```
gfortran -c -g f90main.f90
gfortran -c -g f90sub.f90
```

Readers who use other operating systems will need to replace `gfortran` with the name of the appropriate Fortran 90 compiler on their machine. These commands will create object files `f90main.o` and `f90sub.o`, as well as a Fortran 90 **module file** `f90sub.mod`. The Fortran 90 module allows the subroutine to access the `lbound` and `ubound` functions for dynamically allocated arrays. To assemble these two object files into an executable, type

```
gfortran -o execf90 f90main.o f90sub.o
```

To make this process easier, we have created a target for `execf90` in our GNUmakefile. With this `GNUmakefile`, we could compile the source files and assemble the executable merely by typing

```
make execf90
```

Afterward, the executable `execf90` can be run by typing

```
execf90
```

2.4.2.3 Known C++ Files

Given the two C++ files Cppmain.C and Cppsub.C, we begin by using the Linux `g++` compiler to create object files as follows:

```
g++ -MMD -c -g Cppmain.C
g++ -MMD -c -g Cppsub.C
```

The option "-MMD" tells the compiler to make a dependency ".d" file, which contains a list of header files upon which the object file depends. These two commands will create object files `Cppmain.o` and `Cppsub.o`. To combine these two object files into an executable, type

```
g++ -o execC Cppmain.o Cppsub.o
```

To make this process easier, we have created a target for `execC` in our GNUmakefile. With this `GNUmakefile`, we could compile the source files and assemble the executable merely by typing

```
make execC
```

We can run `execC` by typing

```
execC
```

2.4.2.4 Known C Files

Given two C files cmain.c and csub.c. we use the Linux `gcc` compiler to create the object files as follows:

```
gcc -MMD -g -c -o cmain.o cmain.c
gcc -MMD -g -c -o csub.o csub.c
```

To combine these two object files `cmain.o` and `csub.o` into an executable, type

```
gcc -MMD -o execc cmain.o csub.o
```

To make this process easier, we have created a target for `execc` in our GNUmakefile. With this `GNUmakefile`, we could compile the source files and assemble the executable merely by typing

```
make execc
```

The new executable `execc` can be run by typing

```
execc
```

2.4.2.5 Unknown Files

Typically, when readers call LAPACK or MINPACK routines (or routines from any large package), they can scan the top-level routine to determine easily those other routines that are called by the top-level routine. However, it can be very difficult to discover the full list of routines that are called by these other routines. The LAPACK call stack can be quite deep, especially during routines that compute eigenvalues. It can be painfully slow to try a compilation, get an error message about missing routines, and then download and compile the routines corresponding to those errors, only to get different error messages about other missing routines.

In many cases, LAPACK provides `tar` files that contain an important LAPACK routine and all of its LAPACK dependencies. For example, the LAPACK routine dgesv contains suggestions near the top of the file within the comments describing how to "Download DGESV + dependencies." Users can download such a file, compile the files within, and build an executable. Alternatively, if the LAPACK library of object files is available, readers may link with the library as described in Sect. 2.10.1.

To introduce readers to programming (particularly with LAPACK), we have provided several example programs that execute and time various ways to perform a matrix-vector multiply. These programs also demonstrate how LAPACK BLAS routines (see Sect. 2.10.1) can be called in the different programming languages. We will begin with the Fortran 77 program callingLapackFromFortran.f. The corresponding executable can be built by typing

```
make callingLapackFromFortran
```

Here the Linux command `make` uses our GNUmakefile, which contains instructions on how to compile the source files and assemble the executable. In particular, the `GNUmakefile` assembler instruction links with the LAPACK BLAS library to provide references to the work routines. Afterward, the program can be executed by typing

```
callingLapackFromFortran
```

As another example, the Fortran 77 program callingMinpackFromFortran.f demonstrates how we can call MINPACK to find the zero of a nonlinear system of equations. Note that we define a subroutine `fcn` with arguments as required by MINPACK routine hybrj1. Another key point is that we include the statement

```
external fcn
```

in the main program. The corresponding executable can be built by typing

```
make callingMinpackFromFortran
```

The GNUmakefile compiles our source code and links with our MINPACK library to form the executable.

For more information regarding makefiles, see Sect. 2.5.

2.4.3 *Mixed Language Programming*

Fortran 77 has very useful multi-dimensional arrays and well-tested numerical subroutines (such as LAPACK). However the C, C^{++} and Fortran 90 languages have programming features that may make high-level object functions (like matrix multiplication), operating system calls and graphics more manageable. As a result, it is natural to write programs that mix Fortran 77 with another language.

Suppose that we have a Fortran 77 subroutine named `work` with a **common block** named `hiding`:

```
subroutine work(n,iarray,x,yarray)
integer*4 n,iarray(n)
real*8 x,yarray(n)
...
real*8 r
common/hiding/r
...
end
```

(Mercifully, LAPACK does not use common blocks.) Whenever another Fortran 77 routine calls `work`, it calls by reference, meaning that the variable itself appears in the call, but the language passes the location of the variable. For example, the source file for a routine that calls `work` might look like

```
integer*4 n,iarray(10)
real*8 x,yarray(10)
...
real*8 r
common/hiding/r
...
n=10
x=1.d0
r=0.d0
call work(n,iarray,x,yarray)
...
```

Note that the calling routine must use the correct types for the members of the common block `hiding`, and for the arguments to routine `work`; it is not necessary to use exactly the same variable names in both places.

Our goal is to discover how we might call the Fortran 77 routine `work` from routines written other languages.

2.4.3.1 Fortran 90 Calling Fortran 77

Fortran 90 routines can use Fortran 77 routines and common blocks directly. For example, a Fortran 90 calling routine might look like

```
integer(kind=4) :: n
integer(kind=4), dimension(10) :: iarray
real(kind=8) :: x
```

```
real(kind=8), dimension(10) :: yarray
...
real(kind=8) :: r
common/hiding/r
...
n=10
x=1.d0
r=0.d0
call work(n,iarray,x,yarra)
...
```

Note that the common block name `hiding` must not be the same as any Fortran 90 **module**.

As an example, readers can examine a Fortran 90 program callingLapack-FromFortran90.f90, to compute a matrix-vector multiply by calling LAPACK BLAS. If the reader's operating system is able to use the GNUmakefile, then the corresponding executable can be built by typing

```
make callingLapackFromFortran90
```

Afterward, the program can be executed by typing

```
callingLapackFromFortran90
```

As another example, the Fortran 90 program callingMinpackFromFortran90.f90 demonstrates how we can call MINPACK to find the zero of a nonlinear system of equations. In particular, this program shows how to pass a function as an argument in the call to another routine. The corresponding executable can be built by typing

```
make callingMinpackFromFortran90
```

and the program can be executed by typing

```
callingMinpackFromFortran90
```

2.4.3.2 C++ Calling Fortran 77

In order to call Fortran from C++, we will use an `extern "C"` statement to refer to the Fortran routine `work`, and a `struct` to refer to the common block `hiding`:

```
extern "C" {
  void work_(const int &nsteps,int *iarray,double &x,
  double *yarray);
}
struct hiding_common {
  double r;
};
extern hiding_common hiding_;
```

The `extern "C"` statement prevents C++ from mangling the subroutine name with some encryption of its argument list. In this way, both C++ and Fortran can refer to a procedure with a similar name in the compiled code. Often, the Fortran compiler changes the subroutine name slightly in the object file, typically by appending one or more underscores. The true program name can usually be determine by running the system command nm or objdump -t on the object file as follows:

```
nm work.o
```

or

```
objdump -t work.o
```

Either command will also show the true name of the Fortran common block.

In order to call the Fortran subroutine work from C++, we must remember that Fortran *calls by reference* only, meaning the memory location of each subroutine argument is passed to the called routine. In our extern "C" statement above, we declared that we would pass the first and third arguments to work by reference, but the second and fourth will be passed by location. Consequently, we can call the Fortran subroutine from C++ as follows:

```
...
int n=10;
double x=1.;
int *iarray=new int[10];
double *yarray=new double[10];
hiding_.r=0.;
work_(n,iarray,x,yarray);
...
```

A common mistake is to use **call by value** in C++ for Fortran routines. This mistake would occur with the improper extern "C" statement

```
extern "C" {
  void work_(const int nsteps,int *iarray,double x,
  double *yarray);
}
```

This would tell C++ to pass the *value* of nsteps to the subroutine work, and Fortran 77 will treat this as the *location* of nsteps. If nsteps is reasonably small, this value will point to a location outside of the user program memory, and generate a **segmentation fault**.

Readers interested in more details about calling Fortran from C++ should study the Linux Tutorial Mixing Fortran and C.

As an example, we have provided a C++ program callingLapackFromCpp.C to compute a matrix-vector multiply by calling LAPACK BLAS. If the reader's operating system is able to use the GNUmakefile, then the corresponding executable can often be built by typing

```
make callingLapackFromCpp
```

Afterward, the program can be executed by typing

```
callingLapackFromCpp
```

and so on.

As another example, the C^{++} program callingMinpackFromCpp.C demonstrates how we can call MINPACK to find the zero of a nonlinear system of equations. In particular, this program shows how to pass a C^{++} function as an argument in the call to a Fortran 77 routine. The corresponding executable can be built by typing

```
make callingMinpackFromCpp
```

and the program can be executed by typing

```
callingMinpackFromCpp
```

2.4.3.3 C Calling Fortran 77

C routines can either **call by value** or **call by location**. Since Fortran 77 routines always call by reference, we must call Fortran 77 routines from C by passing the location of all variables. For example, we could call the Fortran 77 routine `work` from C as follows:

```
...
extern void work_(int *n,int *iarray,double *x,double *yarray);
...
struct hiding_common {
  double r;
}
extern hiding_common hiding_;
...
int n;
int *iarray;
double x;
double *yarray;
...
n=10;
x=1.;
hiding_.r=0.;
iarray=(int*) malloc(10*sizeof(int));
yarray=(double*) malloc(10*sizeof(double));
work_(&n,iarray,&x,yarray);
...
```

Furthermore, exactly one C routine should contain the lines

```
struct hiding_common {
  double r;
}
hiding_common hiding_;
```

(i.e., the `extern` word is omitted).

As an example, we have provided a C program callingLapackFromC.c to compute a matrix-vector multiply by calling LAPACK BLAS. If the reader's operating system is able to use the GNUmakefile, then the corresponding executable can be built by typing

```
make callingLapackFromFortranC
```

Afterward, the programs can be executed by typing

```
callingLapackFromC
```

As another example, the C program callingMinpackFromC.c demonstrates how we can call MINPACK to find the zero of a nonlinear system of equations. In particular, this program shows how to pass a C function as an argument in the call to a Fortran 77 routine. The corresponding executable can be built by typing

```
make callingMinpackFromC
```

and the program can be executed by typing

```
callingMinpackFromC
```

2.4.3.4 C or C++ Calling Fortran 90

Fortran 90 prefers **modules** to common blocks, because a module provides a common description of its contents to all routines that use the module. In Fortran 90, our module might look like

```
module hideAndWork
  use iso_c_binding
  real(kind=C_DOUBLE), bind(C) :: r
contains
  subroutine work(n,iarray,x,yarray) bind(C)
  integer(kind=C_INT) :: n
  integer(kind=C_INT), dimension(n) :: iarray
  real(kind=C_DOUBLE) :: x
  real(kind=C_DOUBLE), dimension(n) :: yarray
  ...
  return
  end subroutine
end module
```

We could call work from C as follows:

```
extern void work_(int *n,int *iarray,double *x,double *yarray);
extern double r;
...
int n;
int *iarray;
double x;
double *yarray;
n=10;
x=1.;
r=0.;
iarray=(int*) malloc(10*sizeof(int));
yarray=(double*) malloc(10*sizeof(double));
work(&n,iarray,&x,yarray);
...
free(iarray); iarray=0;
free(yarray); yarray=0;
```

The corresponding C++ code might look like

```
extern "C" {
  void work_(int &n,int *iarray,double &x,double *yarray);
}
extern double r;
...
int n;
int *iarray;
double x;
double *yarray;
int n=10;
double x=1.;
r=0.;
int *iarray=new int[10];
double *yarray=new double[10];
work_(n,iarray,x,yarray);
...
delete [] iarray; iarray=0;
delete [] yarray; yarray=0;
```

It may be necessary to run the system command `nm` or `objdump -t` on the object file as follows

```
nm hideAndWork.o
```

or

```
objdump -t hideAndWork.o
```

Either command should show the true name of the Fortran routine `work` and the hidden variable `r`.

2.4.3.5 Fortran Calling C++

It may also be useful to call some C++ procedures from Fortran. For example, since Fortran 77 does not have any concept of pointers, it may be useful to provide the location of some variable or array entry in Fortran. Here is how we might achieve this goal. In C++, define a procedure `print_loc` inside the scope of `extern "C"` as follows:

```
extern "C" {
  void print_loc__(const void *p) {
    cout << p << endl;
    cout << "\tdouble = " << *static_cast<double*>(p) << endl;
    cout << "\tfloat = " << *static_cast<float*>(p) << endl;
    cout << "\tint = " << *static_cast<int*>(p) << endl;
  }
}
```

Then, in Fortran we can call `print_loc` as follows:

```
subroutine work(n,iarray,x,yarray)
integer n,iarray(n)
```

```
double precision x,yarray(n)
...
call print_loc(yarray(5))
...
end
```

Again, the system command `nm work.o` or `objdump -t work.o` can help us to use the correct symbol table entry for `print_loc__` in C++ or C; the number of appended underscores varies with the compiler.

2.4.3.6 MATLAB Calling C++ or Fortran

The single object type in MATLAB is the MATLAB array. This data structure contains a type, array dimensions, array data, a flag denoting real or complex for numerical data, and sparsity information if appropriate. The MATLAB `mex` compiler can compile C, C++ or Fortran code into a MATLAB binary **MEX** file, which can be called from MATLAB. It is also possible to call the MATLAB engine from C, C++ or Fortran. For more details, see MathWorks Product Support.

2.4.4 Timing

In either Fortran or Fortran 90, you can use cpu_time or dtime to measure elapsed time in some portion of a program. As an example of a program calling `dtime`, see the Fortran 77 subroutine callingLapackFromFortran.f. In either C or C++, you can use times or clock. To examine a program calling `times`, see the C program callingLapackFromC.c. In MATLAB, use the tic and toc commands to time a program segment.

Information about the performance of an executable can be provided by the Linux time or perf stat commands.

2.4.5 C++ Scoping Rules

A name for a C++ variable, function, etc. is defined only within its **scope**. There are four kinds of scope:

local scope is enclosed by braces { } ,
function scope is defined by the limits of the definition of a procedure,
file scope is for those variables declared outside the functions in a file (the global variables; however, new C++ language standards actually place these variables inside a default `Namespace`), and
class scope is for class members, which can be used within any class member function.

We can use C++ scoping rules to simplify our programming, as we will show in the following examples.

2.4.5.1 Tracer Class

In debugging, it is sometimes useful to follow program execution through printed messages. We might want to print a message whenever we enter some scope (e.g., "entering Class::function") and another message when we leave the scope (e.g., "leaving Class::function"). In Fortran, the code might look like

```
print *, "entering function"
...
print *, "leaving function"
```

In both Fortran and C, we have to place tracer messages in two places, and remember to remove both whenever we are no longer interested in tracing the particular code block.

In C++, we could define a `Tracer` as in the files Tracer.H and Tracer.C. Interested readers can examine the file integrateIOMain.C to see the `Tracer` class in use. *Note that a single line of `Tracer`-related code appears inside each program scope.* The `Tracer` constructor is called whenever program execution enters the scope, and the `Tracer` destructor is called when the `Tracer` goes out of scope.

Note that there are multiple `Tracer` constructors, some of which are able to print the file name and line number where the `Tracer` was constructed. The extra information can aid in removing the message when it is no longer needed.

For more information about C++ classes, see Sect. 3.12.

2.4.5.2 Timer Class

Next, suppose that we would like to measure the computational time spent in some program scope. We have to record the time when we enter the scope, then record the time when we leave and subtract to determine the elapsed time. In Fortran we might write

```
real elapsed, start, finish
call cpu_time(start)
...
call cpu_time(finish)
elapsed = finish - start
```

This will compute the `elapsed` time in seconds. Both C and Fortran require two lines of code to be properly placed, in order to perform the timing.

In C++, we can create a `Timer` class as in the files TimedObject.H and TimedObject.C. Interested readers can examine the file integrateIOMain.C to see the `Timer` class in use.

2.5 Makefiles

Programs involving more than one program file, such as mixed-language programs, cannot be compiled and assembled with a single system command. Rather than type in several commands to compile and assemble a mixed language program, we will use the make utility. Simple makefiles consist of lines of the form

```
target : dependencies
        <tab> rule
```

The **target** is the thing (compilation, executable, etc.) to be made. The **dependencies** list those things needed to make the target. If any of the dependencies are newer than the target, then the **rule** is performed to bring the target up to date.

For Chap. 1, we provided (but did not advertize) a very simple **GNUmakefile.** This makefile contained several targets, with corresponding dependencies and rules. Each of the programs in that chapter involved a single source file, so the targets in that makefile had very simple dependencies.

For this chapter, we have a more complicated **GNUmakefile.** This makefile begins by defining several macros for compiler flags. Afterward, it defines the rules

```
%.o: %.C
        $(COMPILE.C++) -c -o $@ $<
%.o: %.c
        $(COMPILE.C) -c -o $@ $<
%.o: %.f
        $(COMPILE.f) -c -o $@ $<
%.o: %.f90
        $(COMPILE.f) -c -O3 -o $@ $<
```

These rules describe how to make an object file `file.o` from either a C++ `file.C`, a C `file.c`, a Fortran 77 `file.f` or a Fortran 90 `file.f90`. The make variable `$@` refers to the file name of the target, and `$<` is the file name of the dependency. An individual program, such as `callingLapackFromC++`, can be assembled by the lines

```
callingLapackFromC++ : callingLapackFromC++.o
        $(C++) -o $@ callingLapackFromC++.o -lblas
```

Here the dependency is `callingLapackFromC++.o`, which could be constructed by `%.o: %.C` first rule above. Once this object file is up-to-date, the code is assembled by the second line (the rule).

Note that we use a number of compiler flags in our compilations. Without `make`, we would have to type these lengthy list of flags every time we call the compiler. These compiler flags include the `-g` option for all languages, to tell the compiler to build a symbol table for the debugger. When compiling Fortran, we use like to use the `-u` option, which declares all variables have undefined type unless they are explicitly declared.

`Make` can be very temperamental. The tab characters are invisible to most editors, but essential to `make`. Since `make` can call itself, it sometimes becomes very difficult to understand just what values the make macros have in each call. Beginners should use make, but keep its use very simple.

For another introduction to makefiles, see Makefiles by example. For more details about make, see Wikibooks or type the command

```
info make
```

on a machine running a Linux operating system.

Exercise 2.5.1 Get a copy of the tarfile and expand its contents in a new directory or folder. Change the compiler names and flags as necessary to correspond to your operating system. Then use the `GNUmakefile` to create all of the targets at once by typing

```
make
```

If one of the targets fails, correct the problem and make that individual target.

2.6 Computer Memory

Many scientific computing algorithms require substantial computer memory. For example, differential equations may require arrays to represent the solution at various times or spatial locations. Also, systems of linear equations can involve large matrices and vectors. In developing scientific computing algorithms, we will have to choose between competing schemes for storing and accessing the required data. The choices will usually have the same numerical reliability, but will differ substantially in computational speed due to their use of computer memory. In this section, we will explain the guiding principles that determine how we can access computer memory efficiently. The case study in Sect. 2.11 will demonstrate the importance of the principles we are about to discuss.

2.6.1 Computer Architecture

There are two basic assumptions in the design of computer memory:

temporal locality: If some datum is referenced, it will tend to be referenced again soon. For example, we often load an old value for some variable and then store a new value for that same variable.

spatial locality: If some datum is referenced, then data near it are likely to be referenced soon. For example, an array entry corresponding to the solution at some time in an initial value problems may be used to compute the solution at the next time.

Essentially all computers now use multiple forms of memory. There are several reasons for this:

- The most recently used data are stored closest to the central processing unit (CPU) for processing; unused data are stored on the disk (or other non-volatile devices) for safe-keeping.
- Speed and access time of memory varies with the type of memory: slow memory (e.g. a disk) might have an access time 10^6 to 10^7 longer than fast memory (e.g. on a central processing unit).
- The fastest memory is more expensive per bit than slower memory; as a result, fast memory is usually small.
- The central processing unit sits idle until the needed memory is available.

As a result, computer manufacturers have adopted the following common strategy. Fast memory is located close to the processor and is generally small, while slow memory is farther away from the processor and is generally larger. This strategy leads to hierarchical memory, which involves a combination of registers, **cache**, various kinds of random-access memory (RAM), disk(s) and a network (or cloud).

Let us examine the implications of hierarchical memory. Consider a simple (idealized) machine with two levels of memory (say register and cache, or cache and RAM). If the needed data is available in fast memory, then the operating system has a "hit"; otherwise it has a "miss." The **hit rate** is the fraction of memory access found in fast memory. The **hit time** is the time required to access fast memory. The **miss penalty** is the time required to replace a block of data in fast memory with a block from slow memory. Because of the differences in the speed of the memory, the hit time is much less than the miss penalty.

In order to find data (whether it is located in fast or slow memory), addresses are mapped, usually modulo some **block** size. When a miss occurs, the operating system fetches multiple adjacent words of memory (a block). Note that the miss rate initially decreases as the block size increases beyond the size of a single word. However, for very large block sizes, fast memory will hold few blocks. As the size of a block increases, the cost of a miss increases because of the amount of data that must be sent. Sometimes, data transfer from slow memory can be improved by increasing the bandwidth of the communication, or by interleaving the memory.

The size of the data blocks increases as the memory hierarchy is traversed from registers to RAM. RAM (often making up what is called "virtual memory") uses large **pages**, consisting of 4–16 KB, to transfer data from disk. Typically, caches use blocks that could consist of anywhere from 4 to 256 bytes, depending on the machine. The trend has been for central processor speed to increase faster than memory speed, so computers are employing more layers of cache with successive new designs.

Under the Linux operating system, the virtual memory page size can be obtained in several ways. The file `limits.h` (or `climits` for the C^{++} compiler) provides the page size in bytes as the value of `PAGESIZE`. Alternatively, the command

```
getconf PAGESIZE
```

will provide this value. For example, the page size is 4096 bytes on my Dell laptop.

It is also possible to get information about the number and sizes of CPU cache. The Linux command

```
getconf -a | grep CACHE_SIZE
```

will return everything known by `getconf` about things involving `CACHE_SIZE`. On my laptop, this command returns

```
LEVEL1_ICACHE_SIZE                  32768
LEVEL1_DCACHE_SIZE                  32768
LEVEL2_CACHE_SIZE                   262144
LEVEL3_CACHE_SIZE                   4194304
LEVEL4_CACHE_SIZE                   0
```

Thus there are three levels of cache on my machine. The largest (and slowest) holds 4 MB, while the smallest and fastest holds 32 KB.

The most important point that readers should learn from this section is the following principle: *always attempt to design programs so that the innermost loops access consecutive array entries*. If extra performance is needed, readers should examine the cache miss rate to see if their algorithm could be designed to make better use of memory pages. The case study in Sect. 2.11 will demonstrate the importance of our programming principle.

Some compilers (especially Fortran 77 and Fortran 90) can look at multidimensional arrays and convert array address offsets into memory access strides. On the other hand, with C or C^{++} pointers for arrays, and especially with matrices being treated as arrays of pointers, address arithmetic is more likely to occur and slow the calculation. We will discuss this issue more carefully in Sect. 2.6.3.

Finally, note that most operating systems provide window-based utilities that provide some idea of how the computer resources are being used. For example, Linux Fedora release 13 contains gnome-system-monitor, which displays information about cpu usage etc. under its `Resources` tab. Also, the Linux ksysguard command provides similar information under the `System Load` tab. The Linux command /usr/bin/time will provide information about page faults during execution of a particular `executable`. Finally, the Linux perf stat command will display a number of hardware statistics involved in running a particular executable. We recommend the version

```
perf stat -B -e cache-references,cache-misses,faults,
migrations <executable>
```

Example 2.6.1 (Matrix Access) In Fortran 77 file stride.f there are three functions that assign the entries in an $m \times n$ matrix $\mathbf{A}$. The first function accesses the entries of **A** at stride one:

```
do j=1,n
  do i=1,m
    A(i,j)=1.d0
  enddo
enddo
```

The second function accesses the entries at stride *m*:

```
do i=1,m
  do j=1,n
    A(i,j)=1.d0
  enddo
enddo
```

Finally, the third function access the array entries randomly:

```
do i=1,m
  do j=1,n
    A(rowperm(i),colperm(j))=1.d0
  enddo
enddo
```

The C++ main program strideMain.C calls the Fortran functions for various array sizes and times their execution. The timing results are shown in Table 2.9. These results demonstrate that array access at stride 1 is the fastest of the three. Furthermore, access at the fixed stride *m* is actually slower than access at a random stride.

Exercise 2.6.1 Consider the three Fortran Subroutines `stride1`, `stridem` and `strider` in Example 2.6.1. Separate the Fortran calls into separate executables and run `/usr/bin/time -v executable` on each `executable`. Then describe how the differences in the observed page faults relates to the observed timings.

Exercise 2.6.2 Examine the LAPACK BLAS routine ddot.f, which computes the inner product of two real vectors.

1. Try to explain the purpose of the loop where `I` ranges from `M+1` to `N`, incremented each time by `5`.
2. Compare the speed of BLAS subroutine `ddot` to a new dot product routine that computes the dot product in the simpler form

```
dtemp = 0.d0
do i = 1,n
  dtemp = dtemp + dx(i)*dy(i)
end do
```

Table 2.9 Array access timings

m = n	stride = 1	stride = *m*	Random stride
512	0.	0.00333333	0.
1024	0.	0.0075	0.005
2048	0.008	0.038	0.03
4096	0.0266667	0.143333	0.105
8192	0.112857	0.478571	0.42

In order to see nonzero run times, you may need to use vectors with at least 10^6 entries.

2.6.2 *Dynamic Memory Allocation*

Sometimes it is useful to choose the size of arrays during execution, rather than before compilation. This process is called **dynamic memory allocation**. This feature is available in C^{++}, C and Fortran 90, but not necessarily in Fortran 77. JavaScript and MATLAB handle their own dynamic memory allocation, automatically resizing arrays as necessary and freeing the memory when it is no longer needed. As a result, a programmer has little control over the memory allocation process in these two languages.

In C, memory is allocated in bytes. An array of 1000 `doubles` might be allocated and deallocated in C as follows:

```
#include <malloc.h>
...
double *array = (double*) malloc(1000 * sizeof(double));
...
free array; array = 0;
```

In C^{++}, the same memory allocation can be accomplished by operators `new` and `delete`:

```
...
double *array=new double[1000];
...
delete [] array; array=0;
```

In Fortran 90, dynamic memory allocation takes the following form:

```
...
real(kind=8), allocatable, dimension(:) :: array
...
allocate( array(1000) )
...
deallocate( array )
```

There are some drawbacks to working with dynamically allocated memory. For example, there is no utility in any of these languages for warning that memory is being addressed out of bounds. Under Linux, some out-of-bounds memory assignments may generate the message glibc detected free(): invalid next size (fast) when the memory is freed. Further, these three programming languages provide no way to guarantee that allocated memory is freed when it is no longer needed. If a program repeatedly requests additional memory without freeing prior obsolete requests, eventually the available resources of the machine will be insufficient to handle a request for new memory, and a zero pointer to the new memory will be returned.

One way to overcome these problems is to use languages like JavaScript, Python, Java or MATLAB. These programming languages perform hidden memory allocation, and automatically check for addressing errors. Alternatively the C^{++} Boost library provides the shared_ptr class template to maintain a reference count, and automatically frees the associated memory when it is no longer needed. The C^{++} Blitz++ library will also use reference counters to free array pointers when the arrays are no longer needed. In the Standard Template Library (STL), which we will discuss in Sect. 2.10.2, memory allocation is encapsulated in the Allocator class. Note, however, that it may be difficult to perform message passing on distributed memory machines when the actual memory storage is hidden from the user, especially if that memory is not guaranteed to be contiguous.

For examples of programs performing dynamic memory allocation, see cmain.c, Cppmain.C and f90main.f90.

2.6.3 Multidimensional Arrays

Matrices in linear algebra are two-dimensional arrays. These are handled easily in Fortran and MATLAB, and not so easily in other languages. Arrays become even more complicated in the numerical solution of partial differential equations. For example, an array representing the stress tensor for a three-dimensional solid might have three subscripts for spatial location on a grid, and either 1 or 2 subscripts for the entries of the stress tensor at each spatial location. The resulting four- or five-dimensional array is easy to handle in Fortran, but much harder to handle in C, C^{++}, MATLAB, Python or Java. We do note that, beginning with version 5, MATLAB can handle multidimensional arrays with more than two subscripts as "extensions" of matrices.

2.6.3.1 Fortran Arrays

Matrices have different forms and data layouts in different languages. Fortran 77 arrays have the same syntax as function calls. As a result, whenever the array declaration is missing, the *Fortran compiler assumes that the statement refers to an undeclared external function*. To avoid this problem, always declare the array. For example, a general two-dimensional array in Fortran 77 could be declared in a subroutine as follows:

```
subroutine sub(beg1,beg2,end1,end2,A)
integer beg1,beg2,end1,end2
double precision A(beg1:end1,beg2:end2)
...
do j=beg2,end2
  do i=beg1,end1
    A(i,j) = ...
  enddo
```

```
      enddo
      ...
      return
      end
```

If the phrase "beg1:" or "beg2:" describing either starting index is omitted, then Fortran assumes that the starting index is 1. In most Fortran programs, array indices begin at one and the second dimension of a matrix is not used in the array declaration. This leads to Fortran code in the form

```
      double precision A(m,*)
```

See, for example, LAPACK BLAS routine dgemv.f

Fortran arrays are stored by columns in computer memory. Thus

```
      double precision A(3,2)
```

is stored in the order

$$\begin{bmatrix} 1 & 4 \\ 2 & 5 \\ 3 & 6 \end{bmatrix} .$$

2.6.3.2 Fortran Subarrays

In some scientific computations, it may be convenient to pass subarrays to subroutines. In the calling routine, we may view the array in the form

$$\begin{array}{ll|ll|}
\alpha_{11} \ldots & & \alpha_{1j} \ldots \alpha_{1n} & \\
\vdots & & \vdots & \\
\hline
\alpha_{i1} \ldots & & \alpha_{ij} \ldots \alpha_{in} & \\
\vdots & & \vdots \quad\; \vdots & \\
\alpha_{m1} \ldots & & \alpha_{mj} \ldots \alpha_{mn} & \\
\hline
\end{array}$$

However, in the called routine, we may only want to refer to the subarray beginning at row i and column j.

Passing subarrays is very common and convenient in Fortran. Generally, there are two crucial pieces of information that are needed to address entries of the subarray, namely the starting location and the distance in memory between row entries. Since Fortran calls by reference, we can use the array entry `A(i,j)` to provide the starting location of the subarray. It is crucial to note that the distance in memory between entries in a row of a subarray is *the same as in the original array*. In order to perform matrix operations with the subarray, we also need to pass the number of rows and columns of the subarray to the called subroutine. So, for an $m \times n$ matrix $\mathbf{A}$, the subarray beginning at entry i, j would have starting location $\mathbf{A}_{ij}$, distance m between row entries, $m + 1 - i$ rows and $n + 1 - j$ columns.

Copies to submatrices are almost always unnecessary in Fortran, because this language can easily refer directly to the entries in a submatrix of an existing matrix. This fact is used to design very effective block algorithms in LAPACK. The block algorithms can be used to optimize the linear algebra routines for the available page size of the current machine, thereby avoiding memory misses. Block algorithms may also assist the implementation of the linear algebra routines on parallel processors. See Sect. 3.7.3.4 for additional discussion of block algorithms for Gaussian factorization.

As an example of a Fortran routine passing subarrays, readers can examine how the LAPACK Gaussian factorization routine dgetrf.f calls the LAPACK BLAS matrix-matrix multiplication routine `dgemm`.

2.6.3.3 C Arrays

Matrices in C and C^{++} are sometimes declared as pointers to pointers. For example, the C^{++} code lines

```
double A[3][5];
proc1(3,A);
proc2(3,5,&A[0][0]);
```

declares `A` to be an array of 3 pointers to 5 doubles. In a called procedure, `A` might be declared as follows:

```
void proc1(int m,double A[][5]) {
  ...
  for (i=0;i<m;i++) {
    for (j=0;j<5;j++) {
      A[i][j] = ...;
    }
  }
  ...
}
```

The problem is that procedure `proc1` has to know the (fixed) number of columns of `A`. Alternatively, Stroustrup [100, p. 838f] recommends a procedure with argument list in the following form:

```
void proc2(int m,int n,double *A) {
  ...
  for (i=0;i<m;i++) {
    for (j=0;j<n;j++) {
      A[i*n+j] = ...;
    }
  }
  ...
}
```

However, the multidimensional nature of the array A is unavailable to the called procedure proc2, and we must perform address computations through arithmetic.

Here is another way in which people sometimes allocate matrices in C or C^{++}. First, they write a procedure allocateMatrix to allocate contiguous memory for the matrix as follows:

```
double** allocateMatrix(int m,int n) {
  int i;
  double *space;
  double **array;
  array=new double*[m];
  space = new double[m*n];
  for (i=0; i<m; i++) {
    array[i] = space;
    space += n;
  }
  return array;
}
```

Next, they write a procedure freeMatrix to deallocate the memory:

```
void freeMatrix(double **array) {
  free(array[0]);
  free array;
}
```

Then the calling routine can use lines of the form

```
double **A=allocateMatrix(3,5);
...
proc3(m,n,A);
...
freeMatrix(A);
```

The called procedure might look like

```
void proc3(int m,int n,double **A) {
  int i,j;
  for (i=0;i<m;i++) {
    for (j=0;j<n;j++) {
      A[i][j] = ...;
    }
  }
}
```

In this way, the called routines would continue to use the C or C^{++} matrix syntax for array entries.

With either array design, the entries in the matrices would be stored sequentially by rows, and array indices would always start at 0. For example,

```
double A[3][2]
```

is stored in the order

$$\begin{bmatrix} 0 & 1 \\ 2 & 3 \\ 4 & 5 \end{bmatrix} .$$

This is inconsistent with Fortran array storage, and would cause serious problems in calling Fortran LAPACK routines.

2.6.3.4 C Arrays for Fortran

In code mixing C or C^{++} and Fortran, we recommend the following approach. First, declare the array in C or C^{++} as a single "vector":

```
extern "C" { void fortran_(int&,int&,double*); }
int main(int /* argc */,char** /* argv */) {
  ...
  int m=3,n=5;
  double *A = new double[m * n];
  fortran_(m,n,A);
  ...
}
```

Next, begin the Fortran array indices at 0 to avoid confusion with C indices:

```
subroutine fortran(m,n,A)
integer m,n
double A(0:m-1,0:n-1)
...
do j=0,n-1
  do i=0,m-1
    A(i,j) = ...
  enddo
enddo
...
return
end
```

This prevents the developer from mixing row and column order between C or C^{++} and Fortran, and avoids confusion between the starting indices in the two languages. The difficulty is that entry `A(i,j)` in Fortran would have to appear as `A[i+j*m]` in C^{++}, which requires some explicit array index arithmetic (and therefore the risk of a programmer error). We recommend avoiding explicit reference to multidimensional array locations in C or C^{++} whenever possible; handle those references in Fortran as much as possible.

2.6.3.5 MATLAB Arrays

Array subscripts in MATLAB are like the Fortran default: subscripts in each dimension always begin at one. Thus

```
>> A = [ 1 4; 2 5; 3 6 ]; A(1,2)
```

produces 4 for the result. This array address notation is similar to Fortran, in that parentheses are used for both array addresses and function calls.

If we ask MATLAB for a location in A that is beyond the original number of rows or columns, the matrix is expanded with zeros to the necessary size before referencing the location. This is an implicit memory allocation, and may not be the intention of the programmer.

In MATLAB, a subarray S beginning at entry i, j of an $m \times n$ matrix A can be defined by

```
S = A(i:m,j:n);
```

In this case, the entries of S are *copied* from A, and *changes to entries of* S *have no effect on entries of* A. Similarly, the MATLAB submatrix command copies entries from one matrix to another.

2.6.4 *Copying Memory*

A common programming task involves copying memory from one location to another. In this section, we will discuss several routines to copy memory and arrays in various ways.

2.6.4.1 Memcpy

In C or C^{++}, if the memory to be copied is contiguous for both the source and the destination, then we recommend using the system routine memcpy.

2.6.4.2 BLAS Copy

If the source and destination locations are not both contiguous, and if the data to be copied are all floating point numbers (either real or complex), the **LAPACK BLAS** routines _copy provide a good solution for C, C^{++} or Fortran. If both arrays copy contiguous memory (that is, in the call to _copy both arrays use stride 1), then the BLAS _copy routines will unroll the loops for better use of the registers. See, for example, dcopy.f. However, even with loop unrolling this BLAS routine will not quite match the speed of memcpy.

2.6.4.3 LAPACK _lacpy

Also note that LAPACK has routines `_lacpy` to copy all or portions of matrices. In particular, these routines can copy the upper or lower triangular portion of a matrix. See, for example, dlacpy.f.

2.6.4.4 STL Copy

A very general copy procedure is available with the C++ Standard Template Library (STL) copy. Additional information can be found by studying Mutating Algorithms. This algorithm is more difficult for beginning readers to use, because it involves the C++ STL class `Iterator`.

2.6.4.5 Fortran 90 or MATLAB

In Fortran 90 or MATLAB, arrays can be copied in whole or in part by a very simple syntax. In Fortran 90, we can copy an entire array `A` into an array `B` by simply writing

```
B = A
```

We can also copy the even-indexed columns of `A` to `C` by the following:

```
real(kind=8), dimension(20,10) :: A
real(kind=8), dimension(20,5)  :: C
C = A(:, 1:10:2)
```

In Fortran 90, the syntax `s:f:i` corresponds to a loop that starts at index `s`, increments by `i`, and runs until index `f`. This is consistent with Fortran `do` loop syntax.

The same syntax is used in MATLAB, but has a different interpretation. MATLAB uses the syntax `s:i:f` to start at index `s`, and increment by `i` until index `f` is reached. Thus the MATLAB command

```
B = A( 3 : -1 : 1, : )
```

will copy the first three rows of `A` into `B` *in reverse order*.

In Table 2.10 we have provided run times for various copy routines. The data for this table was generated by the files copyMain.C and f90copy.f90. In particular, the file `copyMain.C` provides examples of each of the different copy routines mentioned above.

Exercise 2.6.3 Consider the source code for LAPACK BLAS routine dcopy.f.

1. Program a very simple version of this copy routine, in which the loop looks like

```
do i = 1,n
  dy(i) = dx(i)
enddo
```

Table 2.10 Copy times (in seconds for an $n \times n$ matrix and 64 repetitions)

n	memcpy	dcopy	dlacpy	STL copy	F90 copy
512	0.05	0.05	0.06	0.05	0.05
1024	0.21	0.26	0.28	0.61	0.29
2048	0.83	0.87	0.90	0.89	0.87
4096	3.30	3.37	3.53	3.28	3.44
8192	11.32	12.10	14.29	13.65	12.36

Compare the speed of execution for this simpler version with the BLAS routine for copying very large arrays using stride 1.

2. Try unrolling the loop with 6 or 8, instead of 7 copy instances. Do either of these work faster on your machine?
3. Explain how the loop unrolling is making use of the registers on the central processing unit.

2.6.5 *Memory Debugger*

Two kinds of memory management problems occur in scientific computing. The first occurs when we **write out of bounds** that were specified for some data structure, such as an array. Data intended for one array could be written into space reserved for another data structure, which might not be an array of the same data type. Another problem occurs when we repeatedly allocate memory and neglect to free it when it is no longer needed. Eventually, little free memory will remain, and array allocation requests will fail.

Some programming languages attempt to protect users from writing out of bounds. Fortran 77 and Fortran 90 will check array indices against array bounds for all arrays *declared with fixed dimensions*. GNU compilers provide the `-fbounds-check` option to check array subscripts against declared bounds at run time; this provides additional checks for arrays with variable dimensions. By default, the C^{++} library Boost performs range checking, while the C^{++} Blitz++ library performs bounds check in debug mode. However, these checks increase the execution time of the program. Even these software bounds checks could not catch certain kinds of writes out of bounds, such as a call to `memcpy` that tries to copy more data into an array than it can hold.

We have found a way to overcome the memory management problems in C^{++}. We use operator overloading to replace the default operator new and operator delete with procedures that employ a `MemoryDebugger`. The constructor for the `MemoryDebugger` takes an integer that determines the pad width. For every dynamic memory allocation request, the `MemoryDebugger` calls `malloc` to receive space for the requested memory, *plus* the pad width at both the beginning and the end of each request. The `MemoryDebugger` writes a special bit pattern into the padded memory just before the request, and a different special bit pattern just after the request. The user can call a procedure `mem_check` to check all the padded memory for all the active memory allocations to make sure that these special bit patterns have not been changed. When the `MemoryDebugger` destructor is called, if there are any memory allocations that remain on the active list, then the corresponding pointer will be printed (possibly together with the file name and line number where the allocation occurred). This information allows the user to find and fix the memory leaks rapidly.

The code for our `MemoryDebugger` class is contained in the files MemoryDebugger.H and MemoryDebugger.C. The `MemoryDebugger` can be tested by the files testMemoryDebugger.C and testfortroutine.f.

Exercise 2.6.4 Remove the comments from lines in testMemoryDebugger.C that are supposed to write out of bounds. Then use the GNUmakefile and the command `make testMemoryDebugger` to make an executable. Verify that the `MemoryDebugger` does catch errors due to writing out of bounds.

Exercise 2.6.5 Get a copies of the files mixedMain.C, MemoryDebugger.H, MemoryDebugger.C, TimedObject.H, TimedObject.C, Tracer.H, Tracer.C and integrate.f. The C++ main program in `mixedMain.C` will use the `MemoryDebugger`. Write a makefile to compile `integrate.f` using a Fortran compiler, and each of `mixedMain.C`, `MemoryDebugger.C`, `TimedObject.C` and `Tracer.C` using a C++ compiler. Then modify the makefile so that it will build an executable from the object files. Run the executable, and describe the results.

2.7 Input and Output

Computations are useless if they are unobserved, or ignore suggestions for program parameters. Typically, our programs will want to read some user input, act upon it, and write some output. We will want to make these input and output operations easy to understand, and quick to perform. Sometimes, these two goals are in conflict.

If the input and output are alphabetic or numeric, then they are called **formatted**. On the other hand, **unformatted** input and output involves binary or hexadecimal representations of characters or numbers. Computers work most efficiently with binary representations, but these are hard for humans to read, write or understand.

In order to convert a machine number to a decimal, scientific or character representation, a computer must perform a significant amount of work. Oftentimes, this work can greatly exceed the computational cost of performing numerical algorithms. Such work may be unnecessary, if the goal of the output was to act as input to another program, such as a data plotter. Our goal in this section is to describe how different programming languages handle input and output, so that readers can make informed decisions about which form to use.

2.7.1 C

In the C programming language, all input and output is directed to a **stream**. There are three pre-defined streams: **stdin** for standard input, **stdout** for standard output, and **stderr** for standard error output. All three of these streams are normally attached to the console, but can be redirected to a file by the form of the shell command. For example, the Linux **bash shell** command

```
executable > standard_output_file 2> error_output_file
```

will put all `stdout` output in a file named `standard_output_file`, and all `stderr` output in a file named `error_output_file`. Within a C program, we can also create a stream for reading and writing to and from a file by calling `fopen` or `freopen`; we end the stream by calling `fclose`. These streams and functions can be found in file `stdio.h`.

There are a variety of functions for writing to `stdout`. For example, putchar will write a single character, puts will write a character string, and printf will write all kinds of characters, strings and numbers in formatted output. The `printf` web page contains useful information about how to use printf, including examples.

In order to read from `stdin`, you can use getchar, gets and scanf. After calling `gets`, you may need to convert the string to a number. To convert to a `double`, call strtod; to convert to a `float` call strtof; to convert to an `int` call strtol; and to convert to a `long long` call strtoll. These and other string conversions can be found in `stdlib.h`. For additional information, see parsing of floats and parsing of integers.

To create a `stream` for writing to a file or reading from a file, use fopen. When you are finished writing or reading from the file, call fclose. To write to a file `stream`, use fputc, fputs or fprintf. The last of these is the most general, and most often used. Writes to `stderr` can take the form `fprintf(stderr,...)`. To read from a file `stream`, use fgetc fgets or fscanf.

An **unformatted write** in C can be performed on a pointer and some given number of bytes by the command `fwrite`. Similarly, `fread` will read unformatted information from a file.

Readers should note that the C program cio.c shows how to program formatted and unformatted writes and reads for various data types. Also, the C programs intOverflowC.c and floatOverflowC.c demonstrate how to write integers and floating point numbers to formatted files in both decimal and hexadecimal form.

2.7.2 Fortran 77

Fortran 77 provides some simple default options for writing and reading numbers, as well as full control over the form of the output. The Fortran command `print` will write to `stdout` and the Fortran command `read` will read from `stdin`.

Alternatively, we could use the commands `write(6,...)` and `read(5,...)`.[1] This older `write` command provides a way to write to `stderr`, in the form `write(0,...)`. It is convenient that `print *, ...` separates the printed values by a single blank space.

Fortran input is accomplished by means of the `read` command. With the command

```
read *, ...
```

the user can enter the required variables on the console, separated by a comma and/or one or more spaces and/or a carriage return. It is possible to perform formatted read commands in Fortran, but the results may be hard to control.

It is possible to write to files in Fortran. First, the program must call `open` to associate a unit number with a file name. Afterward, the program can `write` to the unit number. You can read more about Fortran Input/Output Statements and a Fortran 77 Tutorial. Readers can also examine the Fortran 77 program fio.f to see formatted and unformatted writes and reads for various Fortran data types. Also, the Fortran programs integerOverflowF.f and realOverflowF.f demonstrate how to write integers and floating point numbers to formatted files in both decimal and hexadecimal form.

We offer one important caution. Fortran input and output is **buffered**, meaning that it is not acted upon until some internally-determined size of data has been requested for input or output. If Fortran output is mixed with C or C++ output, the Fortran output may not appear in the correct position relative to program execution. To remedy this problem, call the Fortran routine flush. For example, to flush the buffer for `stdout`, use the Fortran command `flush(6)`.

2.7.3 C++

C++ also provides simple ways to write and read data. Basic input/output in C++ is performed by the **iostreams** `cin` and `cout`. Both of these are able to display a variety of data types in a user-friendly manner.

We can specify the data display format for C++ output by means of the manipulators. Most beginning C++ users find these more difficult to use than `printf`, which can also be used in C++. However, there are two manipulators that readers may find useful. By default, C++ will print booleans as either 0 or 1, so it is convenient to place the statement

```
cout << boolalpha;
```

early in the main program. This will cause `cout` to write booleans throughout the remainder of the program in the form `false` or `true`. Also, if the user would like

---

[1] Originally, a `read` from unit 5 referred to the card reader; a `write` to unit 6 referred to the paper printer, and a `write` to unit 7 referred to a separate card punch machine.

to see all significant digits of a `double`, the beginning of the main program should contain the line

```
cout << setprecision(16);
```

For additional information, read I/O Manipulators.

One difficulty with manipulators is that they make permanent changes to the behavior of `cout`; all future output in the program will be affected. In contrast, the formatting commands in `printf` affect only the current `printf` statement. For examples of C++ programs using manipulators, see intOverflowCpp.C and floatOverflowCpp.C.

In order to write to a file in C++, the user should use an ofstream. To read from a file in C++, use an ifstream. The C++ program cppio.C shows how to program formatted and unformatted writes and reads for various data types.

2.7.4 MATLAB

MATLAB is very chatty: it normally prints everything it computes. However, the printing can be prevented by placing a semicolon at the end of a command line. On the other hand, its keyboard input only returns character strings. After `input` is called, the information can be converted from strings to other data types.

In MATLAB, unformatted output is handled by the save command. This command can be used to save an individual array, or all arrays in the workspace. The same data can be read by the load command.

2.7.5 JavaScript

The easiest output in JavaScript is through the document.write function. This command is useful for some debugging, but has its problems. For one, it is necessary to include the string `<br>` at the end of a line, in order to obtain a line break; the `document.writeln` function will not produce a line break on the web page. Second, if `document.write` is called after a web page finishes loading, it will create a new page. This may destroy objects that the programmer expects to be available during the debugging process.

An alternative is to open a `textarea` and use it only for debugging messages. Such an approach can be found in the comments within NumberOverflow.html.

2.7.6 Formatted Versus Unformatted

Formatted input and output is much slower than unformatted IO, often by a factor of 1000 or more in execution time. Whenever possible, readers should use unformatted input and output to speed their program execution.

Example 2.7.1 (Compare Formatted and Unformatted Write Times) The program integrateIOMain.C contains three timed segments of code. When this program was called to solve a simple ordinary differential equation by Euler's method on my laptop, the results for 10^7 timesteps were the following:

- integrating the ordinary differential equation took 0.09 s,
- the formatted write took 24.34 s, and
- the unformatted write took less than 0.01 s.

2.8 Program Debuggers

Programmers are rarely able to write programs that execute perfectly, without employing some experimental process that finds and fixes programming errors. Usually, programmers discover these errors by inserting temporary messages that print values of program variables at various stages of the execution. However, it can be painful to insert program lines to produce these debugging messages, and just as painful to remove them.

Debuggers, such as GNU gdb offer a helpful alternative. Under Linux, debuggers can examine a core file that contains information about the state of a program at the point where execution terminated abnormally. The core file contains the **call stack**, which lists the routines being called and the specific program lines at which the execution stopped. The core file also contains the values of program variables at each level of the call stack. Alternatively, programmers can begin execution within a debugger, stop the program at various break points, and examine the state of the program before it crashes. Usually a debugger can debug code written in a mixture of languages, especially C, C^{++}, Fortran and Fortran 90.

Since core files can be large, some institutions limit their size. Under the Linux operating system, you may need to set an environment variable to **unlimit** the **core dump size**. For example, my `.bashrc` file contains the line

```
ulimit -S -c unlimited
```

Other debuggers are incorporated into **integrated development environments**, which can combine editors, compilers, assemblers and debuggers. There are free IDEs, such as eclipse, kdevelop, Code::Blocks and Qt Creator. There are also proprietary IDEs, such as Visual Studio for Windows, or Xcode for Mac OSx and IOS. All of these IDEs make it easy to find, fix, compile, assemble and reexamine programs in one environment. We will leave it to the interested readers to learn more about these IDEs.

Here is a simple example of the use of the Linux gdb debugger. Consider the Fortran program exceptionsf.f. Suppose that we compile this Fortran program via

```
gfortran -g -finit-real=INF -Wunused -Werror -fno-range-check \
  -ffpe-trap=invalid,zero,overflow -o exceptionsf_trap \
  exceptionsf.f
```

This creates an executable called `exceptionsf_trap`, and because of the `-g` option, this executable contains a symbol table for debugging. With the Fedora operating system, we can use the `kdbg` debugger as follows:

```
kdbg exceptionsf_trap
```

Next, we use the left mouse button to pull down on `Execution` to `Run`. Execution stops at line 29, which is

```
d=huge(d)*dtwo
```

In the bottom right-hand corner, the `kdbg` window displays the message

```
Program received signal SIGFPE, Arithmetic exception.
```

Clicking on `Stack` in the bottom left-hand corner will display the call list. On the right in the middle of the display, a window shows that the value of `d` is `0`, and the value of `dtwo` is `2`. Execution has been terminated because the value of `huge(d)` to too large to be multiplied by two. Clicking on the line number for the offending statement will cause the assembler instructions for this statement to be display; for new programmers these will almost always be useless. Click again to remove the assembler instructions.

However, let us click on line 23, which reads

```
d=nanq
```

We suggest clicking on the actual code line, not the line number. Next, use the left mouse to pull down on `Breakpoint` to `Set temporary breakpoint`. Then pull down on `Execution` to `Restart`. Notice that the current value of `d` appears to the right, together with its hexadecimal value. Now you can use the left mouse to click on a later line of code, say

```
d=done/dinfinity
```

Then pull down on `Execution` to `Run to cursor`, and execution will proceed to the chosen line of code. These are some basic debugging features; you can learn more about the debugger by pulling down on `Help` to `KDbg Handbook`. Alternatively, you can view the KDbg-User's Manual online.

JavaScript programmers should be aware of the Firefox Developer Tools debugger for the Firefox browser. This web page gives some simple instructions for installing a JavaScript debugger, and examining html web pages with embedded JavaScript code.

MATLAB programmers should learn about MException and the related `try`, `throw` and `catch` commands. A similar throw and catch mechanism is available for C++ exception handling. Alternatively, MATLAB programmers can insert calls to isfinite, isinf, isnan and/or isreal to catch floating point exceptions when they first occur.

Exercise 2.8.1 Use your debugger to examine floating point errors as they occur in exceptionsCpp.C, exceptionsC.c, exceptionsf.f or exceptionsf90.f90. You will need to compile your choice of the form of the program, and form an executable before running the debugger. You may want to use ideas from Sect. 2.3.2 to help in trapping the exceptions. Describe the steps you had to take to enable the debugger to trace an exception back to a particular line of the program.

Exercise 2.8.2 Get copies of the files mixedMainBad.C, MemoryDebugger.H, MemoryDebugger.C, TimedObject.H, TimedObject.C, Tracer.H, Tracer.C and integrate.f. Form an executable from these files, and use a debugger to fix the program errors. List the bugs, describe their symptoms and how you found and fixed them.

2.9 Programming Suggestions

Programming is an art form. It is a skill that is developed with practice, often by learning from mistakes. Sometimes we can learn from the mistakes of others, so we will attempt to provide some guidance to the reader in the discussion that follows.

2.9.1 Avoid Pitfalls

There are a number of bad habits and unintended consequences available in any programming language. For Fortran programming pitfalls, we recommend reading Sects. 1.7 and 1.8 of USER NOTES ON FORTRAN PROGRAMMING (UNFP). Fortran 90 programmers should read Mistakes in Fortran 90 Programs That Might Surprise You. Some interesting and surprising programming errors in C can be found in the list The Top 10 Ways to get screwed by the "C" programming language. There are also good lists of C^{++} Pitfalls and more C^{++} Pitfalls. There is also a list of Some Common MATLAB Programming Pitfalls and How to Avoid Them. We will provide a short list of problems that readers encounter frequently in scientific computing.

2.9.1.1 Unintended Type Conversion

Many programming languages provide automatic type conversions to make programming easier. Sometimes, these automatic conversions have unintended consequences. A common problem in Fortran is loss of precision due to type conversion. For example, consider the program lines

```
double precision y
y = 0.1
```

Since `y` is double precision and `0.1` is single precision and not a machine number, Fortran will first compute the single precision number closest to 0.1, and then convert that number to double precision. As a result, only about half the bits in the mantissa of y will be accurate to double-precision. Instead, the reader should write

```
double precision y
y = 0.1d0
```

In Fortran 90, the corrected code could also take the form

```
real(kind=8) y
y = 0.1_8
```

C and C++ avoid this problem this problem with floating point numbers, by promoting all values to doubles, before computations and assignments in mixed precision expressions.

However, C and C++ have other problems with type conversions. For example, consider the program lines

```
int i=0;
if (i=2) {
   ...
}
```

Here `i=2` is an assignment, not a boolean operator; it will set `i` equal to `2`, then convert that integer to a bool. This conversion returns `true` for any nonzero integer, and the `if` test will execute the code block that follows. To avoid this problem, write the test as follows:

```
int i=0;
if (i==2) {
   ...
}
```

In some cases, readers perform expensive computations by using the wrong type. For example, the C or C++ code

```
x=pow(y,4.);
```

is slower than

```
x=pow(y,4);
```

The latter uses an integer power, while the former uses a floating point power. The binary representation of the integer `4` would be used to evaluate the latter `pow` via two multiplications, as follows:

```
t=y*y; x=t*t;
```

However, the former would be evaluated as

```
x=exp(4.*log(y));
```

C performs some type conversions through **casts**. For example, array allocation in C often takes the form

```
double *array=(double*) malloc(n*sizeof(double));
```

In C++, no cast is needed for memory allocation:

```
double *array=new double[n];
```

Other type conversions in C++ are handled by constructors, or by the operators `static_cast`, `const_cast` and `reinterpret_cast`. Use the `explicit` qualifier to avoid unintended type conversions by C++ constructors taking a single argument.

2.9.1.2 Repeated Evaluations

Readers often write code in the same way that they write mathematics. This can lead to unnecessary operations. For example, one student wrote the following MATLAB code for a function and Jacobian evaluation:

```
while timup<10e4,
  f1=x1^2+x2^2-2;
  f2=e/x1+x2^3-2;
  f = [f1;f2];
  J = [2*x1 2*x2; -e/x1^2 3*x2^2];
  s = J\(-f);
  timup=timup+1;
end
```

This code evaluates powers of `x1` and `x2` multiple times, thereby increasing the cost of the function evaluations. Even worse, another student's MATLAB code for a conjugate gradient iteration took the form

```
alpha = -(p'*r)/(p'*A*p);
xtilde = xtilde+p.*alpha;
r = A*xtilde-b;
beta = (p'*A*r)/(p'*A*p);
p = -r+p.*beta;
```

This code performs several matrix-vector products and vector dot products more than once. Sadly, the student who wrote this code did not even use the most efficient form of the conjugate gradient algorithm, which requires only one matrix-vector multiplication per iteration.

Whenever possible, speed execution by moving some calculations outside a loop. For example, a student wrote the MATLAB code

```
X=0:0.01:1;
u=zeros(size(X));
i=1;
for loc=X,
  syms x
  theta = solve('sqrt(2)*cosh(x/4)-x');
  u(i) = -2*log(cosh(theta/2*(loc-0.5))/cosh(theta/4));
  i=i+1;
end
```

This code will repeat the computation of `cosh(theta/4)` (and other expensive quantities) with each loop index.

Table 2.11 Loop alternatives: unoptimized (and optimized) times for $n = 2^{24}$

		subroutine fcn(a) real*8 a a = 2.2d0 ** 3.3d0 return ...	function fcn(x) real*8 fcn,x return x ** 3.3d0 ...
t = 2.2d0 ** 3.3d0 do i=1,n a(i) = temp enddo	cubed(x) = x ** 3.3d0 do i=1,n a(i) = cubed(2.2d0) enddo	do i=1,n call fcn(a(i)) enddo	do i=1,n a(i)=fcn(2.2) enddo
.26 (.03) seconds	12.95 (.04) seconds	.68 (.1) seconds	13.16 (6.4) seconds

Fortunately, modern Fortran, C and C++ compilers move the evaluation of many loop constants outside the loop automatically; this is almost sure to happen with compiler optimization. If they are willing to program in compiled languages, this may save some novice programmers from their own devices.

Function and subroutine calls require arguments to be placed on the call stack, which cannot be register addresses. Compiler optimization across the routine calls is generally impossible. It is best to avoid function and subroutine calls inside the innermost loop of a program. Note that in C++, functions can be declared to be inline, so that their instructions are inserted by the compiler at each point of call. In Fortran, readers might try to achieve the same goal with a statement function. However, computational experiments indicate that this is not always a wise choice.

Consider the following four alternative Fortran programs in Table 2.11. The execution times for unoptimized code (and optimized, in parentheses) with loops of length $n = 2^{24}$ appear in the bottom line for each alternative. Note that the timings indicate that it is far more efficient to remove constants from inside loops, and that function calls can prevent full speedup from optimization. The code that generated these timings may be found in fcn.f and fcnMain.C.

It is possible that loops with lots of statements could run slowly because the registers or cache overflow. In such cases, the code might run faster if the computations can be broken into separate loops,

2.9.1.3 Division vs. Multiplication

Floating-point division can be about 25% slower than multiplication, although numerical experiments indicate that division by 2 is about the same speed as multiplication by 0.5. For C++ code that compares the relative speed of multiplication and division, readers can examine multDiv.C.

2.9.1.4 Cut-Offs

It is distressingly common for programmers (particularly engineers) to write lines like

```
if (pressure .lt. 1.e-6) pressure = 1.e-6
```

Typically, their goal is to avoid negative values resulting from some programming error that they do not care to find and fix. Such programming puts an unnecessary scale for the physical value into the code. For example, if at a later time someone wants to switch from Pascals to psi the code might not execute the same way it did previously. Instead, try to make all decisions regarding small values in terms of machine precision, relative to some program parameter that sets the physical scale. Better yet, the code should be programmed to work with dimensionless quantities, and conversions to and from specified units should occur as output and input to the program.

Exercise 2.9.1 Consider the following five Fortran program lines

```
integer i
real x
do i=0,9
  x = i / 10
enddo
```

Verify that x will always be zero in this loop, and explain why. Then correct the code.

Exercise 2.9.2 Consider the following three C or C^{++} program lines

```
int i;
float x;
for (i=0;i<10;i++) x=i/10;
```

Verify that x will always be zero in this loop, and explain why. Then correct the code.

Exercise 2.9.3 Explain why the problem in the previous exercise cannot occur in either MATLAB or JavaScript.

Exercise 2.9.4 Compare the computational cost of the C code

```
#include <math.h>
#include <stdlib.h>
double x,y;
x=rand();
double y=pow(M_E,x);
```

versus the code

```
#include <math.h>
double x,y;
x=rand();
y=exp(x);
```

2.9.2 *Optimized Compilation*

C, C^{++} and Fortran compilers have the ability to reorganize programs in ways that make optimal use of the computer resources, without changing the computed results. This code reorganization process, called compiler optimization, can lead to significant computational speedup. For example, the timings in Table 2.11 show that optimization can speed relatively efficient code by a factor of nearly 10, and inefficient code by a factor of more than 100. On the other hand, compiler optimization may affect the ability of the debugger to trace floating point exceptions to specific lines of code.

Readers can learn how to engage compiler optimization by reading the instructions for their compiler. Under Linux, most C, C^{++} and Fortran compilers will produce good optimization with the `-O3` option. We recommend that readers avoid using compiler optimization until they have successfully debugged their code and are ready to make long and expensive runs.

2.9.3 *Defensive Programming*

We began this chapter with a discussion of floating point numbers and floating point arithmetic, including floating point exceptions. This was followed by a discussion of programming languages, including mixed language programming. There are a number of pitfalls in all of these topics, providing ample opportunity for initial versions of scientific computing programs to fail.

Consider the occurrence of floating point exceptions. To readers who are just being introduced to scientific computing and programming, floating point exceptions occur far too frequently. For example, a reader might choose an unstable timestep for solving an initial value problem, and find output full of infinities or NaNs. In other cases, a reader might perform an order of accuracy computation for a method that works all too well, and take the logarithm of a zero error in the solution. Introductory readers need to understand the possible causes of the floating point exceptions, so that they can begin to find a cure.

We offer the following **defensive programming** suggestions that make productive use available resources, including floating point exceptions.

2.9.3.1 Use Trusted Software

First, we recommend that readers use high-quality software for their basic computations, such as LAPACK or the Standard Template Library. These are discussed in Sect. 2.10. The functions in these packages are designed to protect against unnecessary exceptions. Furthermore, these algorithms are typically very efficient and accurate.

2.9.3.2 Activate Debugger

Second, we suggest that readers compile all Fortran, C and C++ files with the appropriate compiler option to build the symbol table. For example, the GNU compilers all achieve this with the option `-g`.

2.9.3.3 Trap Floating Point Exceptions

We also suggest that readers enable floating point exception traps. In GNU C or C++, we can use the function feenableexcept, which is declared in file `fenv.h`. A typical call might look like

```
feenableexcept(FE_INVALID | FE_DIVBYZERO | FE_OVERFLOW).
```

This line will cause invalid, divide-by-zero and overflow exceptions to stop execution and generate a `core` file. The `core` file can be used in a debugger to examine values in the code at the point in execution when the exception occurred. An example C program that enables floating point traps can be found in C exceptionsC.c. Readers may also examine test-fenv.c, which tests the correctness and compliance of various floating point environment features.

Note that GNU C++ does not throw any floating point exceptions, so the C++ `throw / catch` mechanism is useless here. Also, C++ may call some constructors that generate floating point exceptions before the main program is entered. We recommend that C++ programmers include the lines

```
#define __USE_GNU 1
#include <fenv.h>
static void __attribute__((constructor)) trapfpe () {
  feenableexcept(FE_INVALID | FE_DIVBYZERO | FE_OVERFLOW);
}
```

just *before* the declaration of the `main` program. An examples of a C++ program that enables floating point traps can be found in exceptionsCpp.C.

With the GNU gfortran compiler, floating point traps can be activated by using the

```
-ffpe-trap=invalid,zero,overflow
```

compiler directive. This does not work in exactly the same way as `feenableexcept` works in C or C++. In Fortran, it may also be necessary to use the `-fno-range-check` compiler directive to make assignments to fixed numbers, such as `infinity` and `NaN`. When an exception is detected, a core file will be generated and execution will terminate. Examples of Fortran 77 and Fortran 90 programs that enable floating point traps can be found in exceptionsf.f and exceptionsf90.f90.

2.9.3.4 Initialize Memory

In C^{++}, we recommend that readers initialize all floating point variables and arrays to IEEE infinity before they are used. Recall from Sect. 2.2.3 that the function `numeric_limits<T>::infinity()` is recommended for this purpose. With GNU gfortran, we recommend that readers use the `-finit-real=INF` compiler option to initialize unused floating point variables to IEEE infinity. Note that this idea of initializing variables to undefined values is incorporated into some programming languages, such as JavaScript [39, p. 41].

Suppose that a reader actually follows these recommendations, and initializes memory to infinity. In either of Fortran, C or C^{++}, if we forget to initialize a variable or array properly after the defensive initializations, then any floating point operation involving the improperly initialized data will cause the program to abort immediately. A **core file** will be generated, and an interactive debugger can be used to examine values of variables at the point where the exception occurred.

2.9.3.5 Memory Debugger

In Sect. 2.6.5, we described a C^{++} class and associated functions that can help us to locate various memory access and management problems. These problems include writing out of bounds on arrays, or memory leaks due to forgetting to free memory after it is allocated. The causes of either of these errors can be very difficult to detect without this tool.

2.10 Basic Algorithms

Some problems in scientific computing, such as solving partial differential equations, image processing or constrained optimization, can involve very complicated computer programs. However, these and simpler problems of scientific computing typically use some very fundamental data structures and numerical algorithms at their core. These core functions may be called many times, and should be coded as efficiently as possible.

In this section, we will describe three key software packages for these core data structures and algorithms. LAPACK is designed to handle linear algebra, from solving systems of linear equations to finding eigenvalues and eigenvectors. This software was based on the earlier LINPACK project [34], which was in turn derived from the EISPACK project [112]. LAPACK and LINPACK were written in Fortran, while EISPACK was written in Algol. Within LAPACK, C programmers can find the LAPACK C interface. There was an effort to develop LAPACK for C^{++}, called LAPACK ++. This project has been superseded by the Template Numerical Toolkit (TNT), which contains C^{++} classes to represent arrays with one to three subscripts. Routines to perform many of the algorithms in LAPACK via the TNT are available in the Java Matrix Package (JAMA), which is available in C^{++} on the TNT website. Yet another collection of C^{++} classes providing calls to the LAPACK Fortran routines is available in CPPLAPACK. Many university operating systems provide access to LAPACK object files in a library that can be linked to a user program. We will describe how to link with an existing LAPACK library in Sect. 2.10.1.

At the core of LAPACK are the basic linear algebra subroutines (BLAS). Readers may download the file `blas.tgz` containing the BLAS routines from the BLAS web site. Readers who use Windows operating systems should go to LAPACK for Windows. The LAPACK routines are available in either Fortran or C, with pre-built and optimized libraries for use with `VisualStudio`. We will describe the BLAS in Sect. 2.10.1 below.

We will also discuss the Standard Template Library (STL). This software package was developed by Silicon Graphics Incorporated (SGI) in C^{++}, and is designed to handle some of the fundamental data structures and algorithms of computer science. Readers should also note that boost also has many data structures and algorithms similar to the STL. We will describe the STL in Sect. 2.10.2.

Finally, we will discuss POSIX threads (pthreads). Most modern personal computers now have multiple central processing units (CPUs); such machines are sometimes called **multi-core**. In order to make most effective use of these machines, programmers need to learn how to assign program **threads** to separate CPUs. We will discuss pthreads in Sect. 2.10.3.

2.10.1 BLAS

The Basic Linear Algebra Subroutines (BLAS) are fundamental operations that are routinely performed on vectors and matrices. There are three levels of BLAS:

Level 1: scalar-scalar, vector-scalar and vector-vector operations,
Level 2: matrix-vector operations, and
Level 3: matrix-matrix operations.

For vectors of size n, the Level 1 BLAS all have computational complexity at most $O(n)$. For most systems of linear equations the Level 1 BLAS perform an insignificant amount of work and have little effect on the performance of general linear system solvers [2, p. 36]. For matrices of size $m \times n$, the Level 2 BLAS all have computational complexity at most $O(mn)$. These routines can achieve near peak performance on many vector processors, but their performance on some vector

red-black tree: a `binary tree` with a root; members of the binary tree have a parent (except for the root), one of two colors (e.g., red or black, left or right, 0 or 1), and at most two children,

map: an associative array, mapping unique objects of one type (the **key**) to objects of another type (the **value**), such that the key type provides a weak comparison operator,

multimap: a `map` without the restriction that the `key` values are unique, and

valarray: a `vector` for which the object type is able to perform vector addition, vector subtraction, scalar multiplication.

In addition, there are other containers related to `hashes`, and special consideration given to `vectors` of `bools`.

The STL `vector` class is not useful for linear algebra; rather it is more like a list that allows random access, say for a binary search. The original intention of the `valarray` class was to provide a standard functionality for implementing the BLAS routines, but this intention does not seem to have reached general acceptability. Readers may also wish to read more about valarray versus vector.

2.10.2.2 Iterators

An **iterator** is used to point to elements in a container. Unlike a C pointer, an iterator does not necessarily point to a location in computer memory, although iterators can be constructed from pointers. Iterators provide C++ operators to reference objects and point to class members of the container elements. There are various types of iterators:

input: can be used to read a sequence of objects into a container, such as characters into a `stream`,

output: can be used to write a sequence of objects in to a container, such as characters from a `stream`,

forward: can be used to read, write and move forward in a container,

reverse: can be used to read, write, move forward and move backward in a container,

random_access: have the capabilities of reverse iterators, but can move forward or backward any number of objects in the container.

These iterators are usually defined in file `stl_iterator.h`.

Iterators encapsulate some basic operations common to many containers. For example, an algorithm to reverse a sequence could be programmed once for a general container and general reverse iterator, and then applied to a variety of containers such as lists, vectors and deques. This style of programming is quite different from LAPACK, in which the same functionality is repeatedly programmed for different data types.

In most cases, users are satisfied with the `__normal_iterator`, which is usually the default. This iterator converts a pointer to an object into an STL iterator, and is a random access iterator.

Readers should view copyMain.C, which uses a iterator to copy a vector.

Some words of caution about iterators should be mentioned. Iterators do not automatically check bounds, so it is possible to reference beyond the bounds of a container. Also, different containers may use different iterators, so it is not always possible to change a program to use a different container type without also changing the iterators.

2.10.2.3 Functors

A **functor** is an object that can be called as a function. Functors are useful for keeping and retrieving state information in functions passed to other functions. Regular function pointers can be used as functors. In fact, any object of a class that overloads `operator()` is a functor. Classes that do this are called functors or function objects.

In the standard template library, there are three kinds of functors:

generator: can be called as `f()`,
unary: can be called as `f(x)`,
binary: can be called as `f(x,y)`,

Within the STL, unary functors that return a `bool` are called **Predicate**s, and binary functors that return a `bool` are called **BinaryPredicate**s. For example, a `BinaryPredicate` could be used to compare two objects for use in a sorting algorithm.

Example 2.10.1 Suppose that we want to add all of the entries of a STL `vector`. We could loop over all of the possible indices of a `vector`:

```
int sum=0;
for (int i=0;i<ivector.size();i++) sum+=ivector[i];
```

Alternatively, we could use an iterator:

```
int sum=0;
vector<int>::const_iterator ie=ivector.end();
for (vector<int>::iterator it=ivector.begin();
it!=ie;it++) sum+=*it;
```

Since the STL algorithm `for_each` (see Sect. 2.10.2.4) can be used to loop between two iterators and apply a functor, we could alternatively write an equivalent code as follows:

```
class Adder : public unary_function<int, void> {
  private:
    int total;
  public:
    Adder() : total(0) {}
    void operator()(int x) { total += x; }
    int getTotal() const { return total; }
};
```

```
...
Adder result = for_each(ivector.begin(), ivector.end(),
Adder());
int sum=result.getTotal();
```

We do not necessarily recommend this particular programming approach; we provide it merely as an example of the use of a functor.

2.10.2.4 Algorithms

The standard template library provides a number of algorithms that can be used in scientific computing projects, in order to avoid repetitive code writing. These algorithms fall into three broad categories:

non-mutating: do not change their arguments,
mutating: may change one or more of the arguments without using a comparison,
sorting: change one or more of the arguments by means of a comparison, and
numeric: require arithmetic operators.

Descriptions of these algorithms can be found by viewing GNU.org libstdc++ algorithms and SGI.com Table of Contents for the Standard Template Library.

Non-Mutating Algorithms
Most of the **non-mutating algorithms** involve searching or counting. Here is a list:

adjacent_find: find the first occurrence of two consecutive equal elements in a sequence,
all_of: check if a predicate is true for all elements of a sequence,
any_of: check if a predicate is true for any element of a sequence,
count: determine the number of copies of a value in a sequence,
count_if: determine the number of entries in a sequence for which a predicate is true,
equal: test a range of elements in a sequence for equality,
find: determine the first location of a value in a sequence,
find_end: determine the location of the last matching subsequence of a sequence, using a binary predicate,
find_first_of: given a `set`, find the first location of any element of that `set` withing a given sequence,
find_if: determine the first location of an element in a sequence for which a predicate is true,
find_if_not: determine the first location of an element in a sequence for which a predicate is false,
for_each: apply a unary function to every element of a sequence,
mismatch: determine all pairs of locations in two sequences where the elements do not match,

search: if a given subsequence occurs in a sequence, determine the location of its first occurrence else return the end of the sequence,

search_n: if n consecutive elements of a sequence have binary predicate value equal to true, then return the first such occurrence else return the end of the sequence.

Example 2.10.2 Suppose that we have a `vector<double>`, and we would like to know if it contains any `NaN` or `Infinite` values. Using the STL algorithm `find_if_not`, we could simply write

```
#include <algorithm>
#include <vector>
...
vector<double> dvector(n);
...
bool all_ok = all_of(dvector.begin(),dvector.end(),isfinite);
```

It is not essential to use an STL container in order to use the STL algorithms. For example, we could write

```
#include <algorithm>
...
double *darray = new double[n];
...
bool all_ok = all_of(darray,darray+n,isfinite);
```

Mutating Algorithms

The mutating algorithms are generally related to copying, filling, partitioning, removing and reversing. Here is a list:

copy: copy some range of a sequence into a result,

copy_backward: copy a sequence by starting at the end of its range,

copy_if: copy the elements of a sequence for which a predicate is true,

copy_n: copy n elements of a sequence into the result,

fill: fill some range of a sequence with copies of some value,

fill_n: put n copies of some value into a sequence,

generate: assign the result of some function with no argument to each element of a sequence,

generate_n: assign the result of some function with no argument to n elements of a sequence,

is_partitioned: returns true if and only if some given range of a sequence satisfies a predicate before those element that do not,

iter_swap: swap the values referenced by two iterators,

move: move some range of values from a sequence into a result,

move_backward: move some range of values from a sequence into a result, starting at the end of the source sequence,

partition: move all elements in a sequence for which a predicate is true to the beginning of the sequence (the relative order of the original sequence is not necessarily preserved),

partition_copy: copy elements of a sequence to separate output sequences depending on the truth value of some predicate,
partition_point: find the partition point of a partitioned sequence,
random_shuffle: reorder the elements of a sequence randomly,
remove: remove all elements in a sequence that are equal to a given value, and preserve the original relative order of the remaining elements,
remove_copy: copy all elements of a sequence that are not equal to a given value, and preserve the original relative order,
remove_copy_if: copy all elements of a sequence for which a predicate is false, and preserve the original relative order,
replace: replace each occurrence of one value in a sequence with a second value,
replace_copy_if: copy a sequence, but replace each value for which a predicate is true with another value,
replace_if: replace each element in a sequence for which a predicate is true with another value,
reverse: reverse the order of a sequence in place,
reverse_copy: copy a sequence to another sequence in reverse order,
rotate: given some middle element, swap the range from first to middle with the range from middle to last,
rotate_copy: copy a sequence into a rotated output sequence,
stable_partition: move all elements of a sequence for which a predicate is true to the beginning of the sequence while preserving the original relative ordering,
swap_ranges: swap the elements in some sequence with the elements in a second sequence,
transform: either perform a unary operation on all elements of a sequence, or perform a binary operation on corresponding elements of two sequences,
unique: either remove consecutive duplicate values from a sequence, or remove all but the first element from any subsequence for which a predicate is true,
unique_copy: either copy a sequence while removing consecutive duplicate values, or copy a sequence while removing all but the first element in a subsequence for which a predicate is true.

Example 2.10.3 Suppose that we have a `vector<double>`, and we would like to fill it with IEEE infinity. Using the STL algorithm `fill`, we could simply write

```
  #include <math.h>
  #include <algorithm>
  #include <vector>
  ...
  vector<double> dvector(n);
  ...
fill(dvector.begin(),dvector.end(),
numeric_limits<double>::infinity());
```

On the other hand, if we want to fill the vector with uniformly distributed random numbers in the unit interval, we can write

```
#include <stdlib.h>
#include <algorithm>
#include <vector>
...
vector<double> dvector(n);
...
generate(dvector.begin(),dvector.end(),rand);
```

These STL algorithms allow the user to avoid programming some loops, and possibly gain some efficiency from the STL implementation of the algorithm. As we noted in Example 2.10.2, it is not necessary that these algorithms be applied to an STL container; iterators can be constructed from pointers.

Sorting Algorithms

The sorting algorithms are generally related to imposing some ordering on a sequence, using a comparison operator. Here is a list:

inplace_merge: merge two sorted sequences into one, using a comparison operator,

is_sorted: return true if and only if the elements of a sequence are in sorted order according to some comparison operator,

is_sorted_until: determine the end of a sorted sequence according to some comparison operator,

lexicographical_compare: either perform a "dictionary" comparison of two sequences,

max: return the maximum of two values by means of a comparison operator,

max_element: return the maximum element in a sequence,

merge: merge two sorted sequences into one sorted sequence,

min: return the minimum of two values by means of a comparison operator,

min_element: return the minimum element in a sequence,

minmax: return the `pair` (min,max) of two values,

minmax_element: given two sequences, return a pair of sequences of the element-wise mins and maxes,

next_permutation: permute the current sequence into the next sequence in the set of all sequences,

nth_element: rearrange the elements in a sequence so that the n'th element is in the correct position if the whole sequence had been sorted,

partial_sort: sort some subsequence to put smaller elements first,

partial_sort_copy: copy and sort some range of smaller elements of a sequence,

prev_permutation: permute the current sequence into the previous sequence in the set of all sequences,

sort: sort the elements of a sequence,

stable_sort: sort while preserving the original relative order of equivalent elements.

Example 2.10.4 In order to test if some code is robust, we may need to run it on all possible permutations of the input data. We can use `next_permutation` to do this:

```
#include <algorithm>
#include <vector>
...
vector<int> ordinals(n);
iota( ordinals.begin(), ordinals.end(), 1 );
do {
  ...
} while ( next_permutation( ordinals.begin(),
ordinals.end() ) );
```

The numeric algorithm `iota` is described below. As we noted in Example 2.10.2, it is not necessary that these algorithms be applied to an STL container.

Numeric Algorithms

The numeric algorithms are the following:

iota: assigns values, each sequentially incremented by one, to a sequence,
accumulate: applies a binary function to a given initial value and each element of a sequence,
inner_product: computes the inner product of two sequences of numbers,
partial_sum: returns the sequence of partial sums of some given sequence,
adjacent_difference: returns the sequence of differences of adjacent elements in a given sequence, and
power: raises a given object to a nonnegative integer power.

Example 2.10.5 We can compute the product of all entries in a `vector<double>` as follows:

```
#include <algorithm>
#include <vector>
...
vector<double> dvector(n);
...
double prod=
  accumulate(dvector.begin(),dvector.end(),
  1.,multiplies<double>);
```

Exercise 2.10.3 Determine the largest `int` `n` so that `n` factorial is less than `numeric_limits<int>::max()`. Then form an array of `n+1` or more distinct `int`s and test if `next_permutation` can reorder them without causing an integer overflow.

Exercise 2.10.4 Compare the performance of the STL numeric algorithm `inner_product` with the level 1 BLAS routine `ddot`.

Exercise 2.10.5 Describe how the STL numeric algorithm `power` could use the bit structure of the integer exponent to compute the `power` via multiplications. For example, suppose that we want to compute the *n*th `power` of a square matrix.

2.10.3 *Pthreads*

Most contemporary personal computers use multiple central processing units (CPUs) for greater processing speed. These machines allow for **shared memory** parallel programming, whereby a computer program assigns tasks to multiple processors that have access to the same memory hardware. Unix operating systems have agreed to implement a standard programming interface for shared memory programs through the **POSIX threads** library. These routines are written in C, but they can be called from C^{++} and Fortran. For an introduction to POSIX threads, we recommend the POSIX Threads Programming tutorial.

MATLAB programmers will be interested to note that MATLAB uses threads for linear algebra, fast Fourier transforms and sorting. MATLAB programmers do not have to change their programs in any way to take advantage of multiple shared processors; this is handled automatically.

There are four main groups of pthread routines:

Thread management : create, detach, join and set or query attributes;
Mutexes : create, destroy, lock or unlock mutual exclusion conditions;
Condition variables : communicate between threads that share a mutex; also create, destroy, wait or signal based on variable values; and
Synchronization : manage read or write locks and barriers.

2.10.3.1 Thread Management

Programs begin with a single default thread. Programmers can create additional threads, up to an implementation-dependent limit. With a Linux bash shell, the limit can be determined by the command

```
ulimit -a | grep processes
```

Threads can be terminated when its starting routine returns, the thread calls `pthread_exit`, another thread calls `pthread_cancel` for the given thread, or the entire process is terminated. In order to guarantee that the main program cannot finish before the threads it spawns, call `pthread_exit` from the main program.

The basic thread management routines are

pthread_create : create a new thread that calls a function,
pthread_exit : terminate a thread,
pthread_cancel : request that a thread be cancelled,
pthread_attr_init : initialize a thread attribute object with default values,
pthread_attr_destroy : destroy a thread attributes object,
pthread_join : suspend execution of the calling thread until the target thread terminates,
pthread_detach : indicate to the implementation that storage for the thread can be reclaimed when that thread terminates,

pthread_attr_setdetachstate : control whether a thread is created in a detached state,
pthread_attr_getdetachstate : determine whether a thread is created in a detached state.
pthread_attr_getstacksize : get the thread creation stacksize attribute,
pthread_attr_setstacksize : set the thread creation stacksize attribute,
pthread_attr_getstackaddr : get the thread creation stackaddr attribute,
pthread_attr_setstackaddr : set the thread creation stackaddr attribute,
pthread_self : return the thread ID of the calling thread, and
pthread_equal : compare thread IDs.

The very simple C program hello_arg2.c shows how `pthread_create` and `pthread_exit` might be called. The slightly more complicated C program matmult-dyn.c shows how to use pthreads to perform a simple matrix-matrix multiply.

2.10.3.2 Mutex Variables

Problems arise if multiple threads attempt to write to the same data simultaneously, or if one thread depends on the results from another thread assigned to a different CPU. To prevent these problems, the shared data can be assigned to a `mutex`, and threads can be restricted in their access to the `mutex`. Here are the important pthread mutex functions.

pthread_mutex_init : initialize a mutex with attributes,
pthread_mutex_destroy : destroy a mutex object,
pthread_mutexattr_init : initialize a mutex attributes object with default values,
pthread_mutexattr_destroy : destroy a mutex attributes object.

The C program dotprod_mutex.c shows how to use pthreads and mutexes to perform a simple vector dot product.

2.10.3.3 Condition Variables

Condition variables provide one way for threads to synchronize. Mutexes control thread access to data, while condition variables allow threads to synchronize based on actual data values. Here are the important pthread condition variable functions.

pthread_cond_init : initialize a condition variable with specified attributes,
pthread_cond_destroy : destroy a condition variable,
pthread_condattr_init : initialize a condition variable attributes object with the default values,
pthread_condattr_destroy : destroy a condition variable attributes object,
pthread_cond_wait : automatically release a mutex and cause the calling thread to block on a condition variable,

pthread_cond_signal : unblock at least one of the threads that are blocked on the specified condition variable,
pthread_cond_broadcast : unblock all threads currently blocked on the specified condition variable,

The C program condvar.c shows how to use pthreads and condition variables. The main thread creates three threads. Two of those threads increment a `count` variable, while the third thread watches the value of `count`. When `count` reaches a predefined limit, the waiting thread is signaled by one of the incrementing threads. The waiting thread awakens and then modifies `count`. The program continues until the incrementing threads reach `TCOUNT`. The main program prints the final value of `count`.

2.10.3.4 Synchronization

Here are the important pthread synchronization routines.

pthread_mutex_lock : block the calling thread until the mutex becomes available,
pthread_mutex_trylock : return immediately if the mutex object is currently locked by any thread (including the current thread),
pthread_mutex_unlock : release the lock on a mutex object,
pthread_barrier_init : allocate resources for a barrier and initialize its attributes,
pthread_barrier_wait : synchronize participating threads at a barrier, and
pthread_barrier_destroy : destroy a barrier and release its resources.

The C program dotprod_mutex.c shows how to use mutex locks to perform a simple vector dot product, and the C program pthread_barrier.c demonstrates how to use pthread barriers.

2.11 Case Study: Matrix-Matrix Multiplication

Previously in this chapter, we have learned about machine numbers, floating point arithmetic, programming languages, computer memory and basic algorithms. We are ready to test our knowledge on a common operation in scientific computing, namely the multiplication of two matrices.

Suppose that we are given an $m \times \ell$ matrix $\mathbf{A}$ and an $\ell \times n$ matrix $\mathbf{B}$. Then the $m \times n$ matrix $\mathbf{C} = \mathbf{AB}$ has entries $\mathbf{C}_{ij}$ for all $(i,j) \in [1,m] \times [1,n]$, given by

$$\mathbf{C}_{ij} = \sum_{k=1}^{\ell} \mathbf{A}_{i\ell}\mathbf{B}_{\ell j} \ .$$

There are a variety of ways to compute the entries of $\mathbf{C}$. These methods can differ substantially in their execution time.

For example, if **C** is initialized to have all zero entries, we could compute the entries of **C** by the algorithm

$$\begin{aligned}&\text{for } 1 \leq i \leq m\\ &\quad \text{for } 1 \leq j \leq n\\ &\quad\quad \text{for } 1 \leq k \leq \ell\\ &\quad\quad\quad C_{ij} = C_{ij} + A_{ik} * B_{kj}\end{aligned}$$

This algorithm has been implemented in the Fortran 90 program matrixMatrixMultiply_ijk.f90. Alternatively, we could order the loops over i, j and k in any of $3! = 6$ different ways. Statistics for these computations with $m = \ell = n = 1024$ were obtained by the Linux command

```
perf stat -B -e cache-references,cache-misses,faults, \
  migrations <executable>
```

and appear in Table 2.12. These timings show that looping over i in the innermost loop is substantially faster than the other four alternatives.

To understand the relative speeds of these loop alternatives, let us examine the ways in which the different algorithms access the array data. If the innermost loop is over j, then none of the three arrays is being accessed at stride one, and the memory miss rate is high. If the innermost loop is over k, then **B** is being accessed at stride one, and the memory miss rate is lower. The best results occur when the innermost loop is over i, because both **C** and **A** are being accessed at stride one.

Alternatively, we could perform a matrix-matrix multiply by using the LAPACK BLAS `ddot` routine for an inner product. This leads to Fortran programs of the form

```
C(i,j) = ddot(l,A(i,1),m,B(1,j),1)
```

for $(i,j) \in [1,m] \times [1,n]$. These algorithms have been implemented in the Fortran 90 programs matrixMatrixMultiplyDdot_ij.f90. and matrixMatrixMultiplyDdot_ji.f90. Since `ddot` uses loop unrolling, we might hope that it could make better use of the registers and the CPU. There are two ways to order the loops over i and j, with timings reported in Table 2.13. Loop unrolling in `ddot` does not seem to give any noticeable performance enhancement in comparison to the loops in Table 2.12. This

Table 2.12 Statistics for matrix multiply by loops

Loop order	Time (s)	Cache misses
i / j / k	19.063	0.848×10^9
i / k / j	41.584	2.007×10^9
j / i / k	19.303	0.953×10^9
j / k / i	2.951	0.008×10^9
k / i / j	40.282	1.712×10^9
k / j / i	3.147	0.009×10^9

Table 2.13 Times for matrix multiply by DDOT

Loop order	Time (s)	Cache misses
i / j	19.180	0.897×10^9
j / i	19.418	0.873×10^9

Table 2.14 Times for matrix multiply by DAXPY or DGEMV

BLAS	Time (s)	Cache misses
daxpy, j / k	2.334	6.662×10^6
daxpy, k / j	2.768	8.638×10^6
dgemv	3.833	8.623×10^6

Table 2.15 Times for matrix multiply by DGEMM

Routine (library)	Time (s)	Cache misses
matmul (F90)	2.635	6.935×10^6
dgemm (LAPACK)	3.024	7.871×10^6
dgemm (ATLAS)	1.321	2.250×10^6
dgemm (ptATLAS)	0.435	2.389×10^6

may be due to the fact that the algorithms employing `ddot` involve substantial cache misses, with only array **B** being accessed at stride one.

Next, suppose that we perform matrix-matrix multiplication by calling the LAPACK BLAS routine `daxpy` in Fortran 90 for $(k, j) \in [1, \ell] \times [1, n]$ as follows:

```
daxpy(m,B(k,j),A(:,k),1,C(:,j),1)
```

These algorithms has been implemented in Fortran 90 programs matrixMatrixMultiplyDaxpy_jk.f90. and matrixMatrixMultiplyDaxpy_kj.f90. Since this call to `daxpy` accesses both **A** and **C** at stride one, and uses loop unrolling, it can make good use of the registers, CPU and memory access. Similarly, we could call LAPACK BLAS routine `dgemv` in Fortran 90 to perform matrix-matrix multiplication as follows:

```
dgemv('N',m,n,1.d0,A,m,B(:,j),1,0.d0,1,C(:,j),1)
```

This algorithm has been implemented in the Fortran 90 program matrixMatrixMultiplyDgemv.f90 The statistics for the `daxpy` and `dgemv` calls are reported in Table 2.14. In these timings, the loop unrolling in daxpy appears to provide some improvement over the timing for dgemv.

Fortran 90 has a built-in matrix-matrix multiplication function, called `matmul`. This algorithm has been implemented in the Fortran 90 program matrixMatrixMultiplyMatmul.f90. This is presumably similar to a call to LAPACK routine `daxpy`. We could also call LAPACK BLAS routine `dgemm`, which is implemented like the $j/k/i$ loop above. This algorithm has been implemented in the Fortran 90 program matrixMatrixMultiplyDgemm.f90. Timings for these Fortran 90 matrix-matrix multiplies appear in Table 2.15.

However, there are other alternatives available through ATLAS. This software package implements matrix-matrix multiplication by partitioning the matrices into blocks that fit in memory pages, performing matrix-matrix multiplication on the blocks, and adding up the results. Furthermore, ATLAS has a parallel threaded version that can speed computations on machines with multiple cores. The timing statistics with ATLAS are reported in Table 2.15. Note that the ATLAS version of `dgemm` involved less than one third the number of cache misses of `dgemm` from LAPACK. The threaded version of ATLAS offered substantial speedup by making use of both cores on my dual-core laptop.

To obtain the best timings, ATLAS should be tuned specifically to each machine. The executable for these ATLAS timings was built by loading with `libf77blas.a` and `libatlas.a`, while the executable for the ptATLAS timings was build by loading with `libptf77blas.`, `libatlas.a` and the `pthread` library. See the GNUmakefile for more details regarding how the executables were assembled.

Note that MATLAB is very efficient in its implementation of standard linear algebraic computations. For matrix-matrix multiplication, the MATLAB command

```
tic;A=rand(1024,1024);B=rand(1024,1024);C=A*B;toc
```

reported 0.411 s. This is similar to, and slightly better than, the timing for the parallel threaded ATLAS implementation of `dgemm`.

These examples have shown that the same computations can be implemented in different ways, and lead to substantially different computational speeds. Readers need to understand these programming techniques in order to program efficiently.

Chapter 3
Linear Algebra

It has been estimated that the solution of a linear system of equations enters in at some stage in about 75 percent of all scientific problems.

Germund Dahlquist and Åke Björck, *[26, p. 137]*

It is safe to say that matrix computation has passed well beyond the stage where an amateur is likely to think of computing methods which can compete with the better-known methods. Certainly one cannot learn theoretical linear algebra and an algebraic programming language, and nothing else, and start writing programs which will perform acceptably by today's standards. There is simply too much hard-earned experience behind the better algorithms, and yet this experience is hardly mentioned in mathematical textbooks of linear algebra.

George E. Forsythe, speaking before the ***Society of Industrial and Applied Mathematics*** *(1966)*

Abstract This chapter discusses the theory of linear algebra, and numerical methods for solving systems of linear equations. Easy linear systems are solved first. Next, norms are introduced for an analysis of the sensitivity of linear systems to perturbations. Afterwards, techniques for reducing general linear systems to a small number of easy problems are developed, implemented in computer programs and studied for computational cost. The accumulation of rounding errors is studied and applied to the various steps in solving linear systems. Algorithm refinements such as scaling and iterative improvement are provided. Inverses and determinants are discussed briefly. Object-oriented programming is applied to linear algebra, and appropriate algorithms for special classes of matrices are presented.

Additional Material: The details of the computer programs referred in the text are available in the Springer website (http://extras.springer.com/2018/978-3-319-69105-3) for authorized users.

© Springer International Publishing AG, part of Springer Nature 2017
J.A. Trangenstein, *Scientific Computing*, Texts in Computational Science and Engineering 18, https://doi.org/10.1007/978-3-319-69105-3_3

3.1 Overview

There are good reasons for beginning a serious discussion of scientific computing with linear algebra. First, linear algebra is a fundamental part of other problems, such as solving systems of nonlinear equations, determining a statistical model of data, or solving a partial differential equation Further, linear algebra does not involve calculus and is fairly easy to describe. Linear algebra can be subdivided into simple and basic tasks that can form useful ingredients in other problems. And best of all, there is high quality publicly available software to solve basic linear algebra problems.

Our goals in this chapter are to describe the fundamental concepts of linear algebra, to use those concepts to determine how solutions of linear systems depend on perturbations, and to develop efficient and stable numerical methods for solving linear systems. Readers should be expected to know how to choose an appropriate algorithm for a particular linear system of equations, how to execute that algorithm efficiently on a computer, and to use an *a posteriori* error estimate to judge the quality of their computed solution.

For additional information about the material in this chapter, we recommend books by Demmel [29], Dongarra et al. [34], Golub and van Loan [46], Higham [56], Stewart [97] and [98], Trefethen and Bau [103], Wilkinson [111] and Wilkinson and Reinsch [112].

For linear algebra software, we recommend LAPACK (written in Fortran), CLAPACK (translated from Fortran to C), Automatically Tuned Linear Algebra Software (ATLAS) (written in C, mostly as macros), Template Numerical Toolkit (TNT) (written in C^{++}), numpy.linalg (written in Python), and GNU Scientific Library (GSL) (written in C).

We also recommend MATLAB arithmetic operations *, ' and .' as well as commands lu, ldl, chol, linsolve, mldivide and cond. We do *not* recommend MATLAB commands det, inv or rref.

Scilab is a free and open source alternative to MATLAB that contains a number of routines for solving linear equations and performing matrix factorizations discussed in this chapter.

3.2 Fundamental Concepts

In Sect. 1.3, we listed five steps in solving scientific computing problems. Those steps began with constructing a mathematical model for the problem, and guaranteeing that the model is well-posed. In this section, we will develop basic data structures for systems of linear equations, and present mathematical theory to determine when linear systems have a unique solution.

After this section, a portion of the second step in scientific computing will remain, as well as steps three through five. The numerical stability of linear systems (the remainder of step two) will be studied in Sect. 3.6. To perform step three, we will construct numerical methods for solving linear systems in Sect. 3.4, 3.7, 3.9.2 and 3.13. The computer implementation of those algorithms (step four) will be discussed in Sects. 3.7.2, 3.7.3 and 3.12. Finally, the numerical stability of these algorithms (step five in scientific computing) will be studied in Sect. 3.8.

3.2.1 Scalars

Scalars are numbers, either real or complex. **Real numbers** *in a computer* are those that can be represented as an integer times a power of two, as in Sect. 2.2.3. Discussions of real numbers *in mathematics* typically proceed from rational numbers, to irrational numbers and on to a continuum, as in Hardy [52, pp. 1–16]. It is important to note that real numbers can be ordered: if ξ and η are two real numbers, then exactly one of the three statement "$\xi < \eta$", "$\xi = \eta$" and "$\xi > \eta$" is true. This is also true for floating point numbers treated by contemporary compilers, even if $\xi = 0$ and $\eta = -0$.

A **complex number** is an ordered pair of real numbers. Typically we write a complex number ζ in the form

$$\zeta = \xi + \mathrm{i}\eta$$

where ξ and η are real numbers and $\mathrm{i} \equiv \sqrt{-1}$. When we plot complex numbers, we view them as an ordered pair, as in Fig. 3.1. When we write arithmetic operations with complex numbers, we use i:

$$\begin{aligned}
\zeta_1 + \zeta_2 &= (\xi_1 + \xi_2) + \mathrm{i}(\eta_1 + \eta_2) = \zeta_2 + \zeta_1 \,, \\
\zeta_1\zeta_2 &= (\xi_1\xi_2 - \eta_1\eta_2) + \mathrm{i}(\xi_1\eta_2 + \xi_2\eta_1) = \zeta_2\zeta_1 \text{ and} \\
\frac{\zeta_1}{\zeta_2} &= \frac{\xi_1\xi_2 + \eta_1\eta_2}{\xi_2^2 + \eta_2^2} + \mathrm{i}\frac{-\xi_1\eta_2 + \xi_2\eta_1}{\xi_2^2 + \eta_2^2}
\end{aligned}$$

Here are two useful definitions that relate to complex numbers.

Definition 3.2.1 The **complex conjugate** of the complex scalar $\zeta = \xi + \mathrm{i}\eta$ is $\overline{\zeta} = \xi - \mathrm{i}\eta$.
Note that $\overline{\zeta}$ is the reflection of ζ about the real axis. It is easy to compute

$$\overline{\zeta_1 + \zeta_2} = \overline{\zeta_1} + \overline{\zeta_2} \text{ and} \overline{\zeta_1\zeta_2} = \overline{\zeta_1}\,\overline{\zeta_2} \,.$$

Definition 3.2.2 The **absolute value** or **modulus** of the complex scalar $\zeta = \xi + \mathrm{i}\eta$ is

$$\mid \zeta \mid = \sqrt{\xi^2 + \eta^2} = \sqrt{\zeta\overline{\zeta}} \,.$$

Fig. 3.1 Number in complex plane

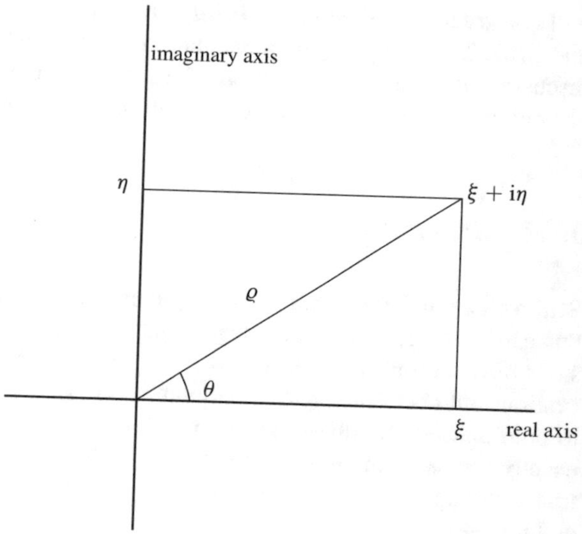

Through Taylor series, it is easy to verify

Theorem 3.2.1 (Euler's Identity) *For all real scalars* θ, $e^{\mathrm{i}\theta} = \cos\theta + \mathrm{i}\sin\theta$.

Since

$$\left|e^{\mathrm{i}\theta}\right| = \sqrt{\cos^2\theta + \sin^2\theta} = 1 ,$$

Euler's identity can be used to provide a polar coordinate representation for complex numbers:

$$\xi + \mathrm{i}\eta = \varrho e^{\mathrm{i}\theta} \Leftrightarrow \varrho = \sqrt{\xi^2 + \eta^2} , \; \cos\theta = \xi/\varrho , \; \sin\theta = \eta/\varrho .$$

The polar form makes it easier to multiply or divide complex numbers:

$$\zeta_1\zeta_2 = (\varrho_1 e^{\mathrm{i}\theta_1})(\varrho_2 e^{\mathrm{i}\theta_2}) = \varrho_1\varrho_2 e^{\mathrm{i}(\theta_1+\theta_2)} \text{ and}$$
$$\zeta_1/\zeta_2 = (\varrho_1 e^{\mathrm{i}\theta_1})/(\varrho_2 e^{\mathrm{i}\theta_2}) = (\varrho_1/\varrho_2) e^{\mathrm{i}(\theta_1-\theta_2)} ,$$

Both real and complex numbers are examples of the following notion from abstract algebra.

Definition 3.2.3 A nonempty set $\mathscr{F}$ is a **field** if and only if there are two operations, + and ·, satisfying

closure: for all α and $\beta \in \mathscr{F}$, $\alpha + \beta \in \mathscr{F}$ and $\alpha \cdot \beta \in \mathscr{F}$;
commutativity: for all α and $\beta \in \mathscr{F}$, $\alpha + \beta = \beta + \alpha$ and $\alpha \cdot \beta = \beta \cdot \alpha$;
associativity: for all α, β and $\gamma \in \mathscr{F}$, $(\alpha+\beta)+\gamma = \alpha+(\beta+\gamma)$ and $(\alpha\cdot\beta)\cdot\gamma = \alpha\cdot(\beta\cdot\gamma)$;
unit: there exists 0 and $1 \in \mathscr{F}$ so that for every $\alpha \in \mathscr{F}$ we have $\alpha + 0 = \alpha$ and $\alpha \cdot 1 = \alpha$.
additive inverse: for every $\alpha \in \mathscr{F}$ there exists $-\alpha \in \mathscr{F}$ so that $\alpha + (-\alpha) = 0$.

multiplicative inverse: for all nonzero $\alpha \in \mathscr{F}$ there exists $1/\alpha \in \mathscr{F}$ so that $\alpha \cdot (1/\alpha) = 1$.
distribution: for all α, β and $\gamma \in \mathscr{F}$, $\alpha \cdot (\beta + \gamma) = \alpha \cdot \beta + \alpha + \gamma$.

3.2.2 Vectors and Matrices

Suppose that we want to solve the following system of m equations in n unknowns:

$$\alpha_{11}\xi_1 + \ldots + \alpha_{1n}\xi_n = \beta_1$$
$$\vdots$$
$$\alpha_{m1}\xi_1 + \ldots + \alpha_{mn}\xi_n = \beta_m\ .$$

We can organize this system as follows. We place the unknowns ξ_j into an n-vector $\mathbf{x}$, the coefficients α_{ij} of the unknowns into an $m \times n$ array $\mathbf{A}$, and the right-hand side entries β_i into an m-vector $\mathbf{b}$. Let us define these ideas a bit more formally.

Definition 3.2.4 Let $\mathscr{F}$ be a field. Then members of $\mathscr{F}$ are called **scalars**. An m-**vector** is a column array of m scalars, and an $m \times n$ **matrix** is a side-by-side array of n m-vectors, called the **columns** of the matrix. If $\mathbf{A}$ is an $m \times n$ matrix, then m is the number of its **rows**. A matrix is **square** if and only if its number of rows is equal to its number of columns.

Sometimes it is useful to consider an m-vector to be an $m \times 1$ matrix; the two are indistinguishable when written as arrays. We also remark that m-vectors and $m \times n$ matrices are examples of the following idea, which is carefully developed by Halmos [51, p. 3].

Definition 3.2.5 Let $\mathscr{F}$ be a field. Then $\mathscr{X}$ is a **linear space** with respect to $\mathscr{F}$ if and only if there are two operations, namely addition and scalar multiplication, such that the following assumptions hold:

Closure under addition: if $\mathbf{x}_1$ and $\mathbf{x}_2 \in \mathscr{X}$, then $\mathbf{x}_1 + \mathbf{x}_2 \in \mathscr{X}$.
Commutativity of addition: if $\mathbf{x}_1$ and $\mathbf{x}_2 \in \mathscr{X}$, then $\mathbf{x}_1 + \mathbf{x}_2 = \mathbf{x}_2 + \mathbf{x}_1$.
Associativity of addition: if $\mathbf{x}_1$, $\mathbf{x}_2$ and $\mathbf{x}_3 \in \mathscr{V}$, then $(\mathbf{x}_1 + \mathbf{x}_2) + \mathbf{x}_3 = \mathbf{x}_1 + (\mathbf{x}_2 + \mathbf{x}_3)$.
Identity element for addition: there is a member $\mathbf{0} \in \mathscr{X}$ so that for all $\mathbf{x} \in \mathscr{X}$ we have $\mathbf{0} + \mathbf{x} = \mathbf{x}$.
Inverse for addition: for each $\mathbf{x} \in \mathscr{X}$ there is a member $-\mathbf{x} \in \mathscr{X}$ such that $\mathbf{x} + (-\mathbf{x}) = \mathbf{0}$.
Closure under scalar multiplication: if $\mathbf{x} \in \mathscr{X}$ and α is a scalar, then $\mathbf{x}\alpha \in \mathscr{X}$.
Identity for scalar multiplication: for every $\mathbf{x} \in \mathscr{X}$, $\mathbf{x} \cdot 1 = \mathbf{x}$.
Distributivity of scalar multiplication over addition: if $\mathbf{x}_1$ and $\mathbf{x}_2 \in \mathscr{X}$ and α is a scalar, then $(\mathbf{x}_1 + \mathbf{x}_2)\alpha = (\mathbf{x}_1\alpha) + (\mathbf{x}_2\alpha)$.
Distributivity of scalar multiplication over scalar addition: if $\mathbf{x} \in \mathscr{X}$ and α_1, α_2 are scalars, then $\mathbf{x}(\alpha_1 + \alpha_2) = \mathbf{x}\alpha_1 + \mathbf{x}\alpha_2$.

Associativity of multiplication: if $\mathbf{x} \in \mathscr{X}$ and α_1, α_2 are scalars, then $\mathbf{x}(\alpha_1\alpha_2) = (\mathbf{x}\alpha_1)\alpha_2$.

Connections to more general linear spaces are possible via **isomorphisms** [51, p. 15].

When discussing linear algebra, we will follow the notation suggested by Stewart [97]. We will use upper case Latin or Greek letters to denote matrices, lower case Latin letters to denote vectors, and lower case Greek letters to denote scalars. So, the columns of the matrix $\mathbf{A}$ will be $\mathbf{a}_j$, and the entries of $\mathbf{A}$ will be α_{ij}. However, we will use different notation for the following important matrix.

Definition 3.2.6 The **identity matrix I** has columns given by the **axis vectors** $\mathbf{e}_j$, and entries given by the **Kronecker delta** δ_{ij}. Specifically,

$$\delta_{ij} = \begin{cases} 1, \; i = j \\ 0, \; i \neq j \end{cases}$$

$$\mathbf{e}_1 = \begin{bmatrix} \delta_{11} \\ \delta_{21} \\ \vdots \\ \delta_{n1} \end{bmatrix} = \begin{bmatrix} 1 \\ 0 \\ \vdots \\ 0 \end{bmatrix}, \; \mathbf{e}_2 = \begin{bmatrix} \delta_{12} \\ \delta_{22} \\ \vdots \\ \delta_{n2} \end{bmatrix} = \begin{bmatrix} 0 \\ 1 \\ \vdots \\ 0 \end{bmatrix}, \ldots, \; \mathbf{e}_n = \begin{bmatrix} \delta_{1n} \\ \delta_{2n} \\ \vdots \\ \delta_{nn} \end{bmatrix} = \begin{bmatrix} 0 \\ 0 \\ \vdots \\ 1 \end{bmatrix} \text{ and}$$

$$\mathbf{I} = \left[\mathbf{e}_1, \mathbf{e}_2, \ldots, \mathbf{e}_n\right] = \begin{bmatrix} \delta_{11} & \delta_{12} & \ldots & \delta_{1n} \\ \delta_{21} & \delta_{22} & \ldots & \delta_{2n} \\ \vdots & \vdots & \ddots & \vdots \\ \delta_{n1} & \delta_{n2} & \ldots & \delta_{nn} \end{bmatrix} = \begin{bmatrix} 1 & 0 & \ldots & 0 \\ 0 & 1 & \ldots & 0 \\ \vdots & \vdots & \ddots & \vdots \\ 0 & 0 & \ldots & 1 \end{bmatrix} .$$

3.2.3 *Linear Combinations*

Vectors are added component-wise, and multiplied by scalars component-wise, according to the following definition.

Definition 3.2.7 If $\mathbf{x}$ and $\mathbf{y}$ are two n-vectors, then their **vector sum** is the n-vector

$$\mathbf{x} + \mathbf{y} = \begin{bmatrix} \xi_1 \\ \vdots \\ \xi_n \end{bmatrix} + \begin{bmatrix} \eta_1 \\ \vdots \\ \eta_n \end{bmatrix} = \begin{bmatrix} \xi_1 + \eta_1 \\ \vdots \\ \xi_n + \eta_n \end{bmatrix} .$$

If α is a scalar, then the **vector-scalar product** $\mathbf{x}\alpha$ is the n-vector

$$\mathbf{x}\alpha = \begin{bmatrix} \xi_1 \\ \vdots \\ \xi_n \end{bmatrix} \alpha = \begin{bmatrix} \xi_1\alpha \\ \vdots \\ \xi_n\alpha \end{bmatrix} .$$

If $\mathbf{a}_1, \ldots, \mathbf{a}_n$ are m-vectors and $\xi_1, \ldots, \xi_n$ are scalars, then the m-vector

$$\mathbf{b} = \sum_{j=1}^{n} \mathbf{a}_j \xi_j$$

is called a **linear combination** of the vectors $\mathbf{a}_1, \ldots, \mathbf{a}_n$. The set of all linear combinations of a set of vectors is called the **span** of the vectors.

We note that vector-scalar products can be computed by LAPACK Level 1 BLAS routines `_scal`; see, for example, dscal.f. Also, vector-vector sums can be computed by `_axpy`; see, for example, daxpy.f. For more information about the BLAS, see Sect. 2.10.1.1. We also note that LAPACK routine dlascl.f. computes matrix-scalar products in a way that avoids overflow and underflow.

Vector addition and vector-scalar multiplication are common operations. As a result, it is useful to develop a mathematical notion of sets that keep the results of such operations within the set.

Definition 3.2.8 A subset $\mathscr{S}$ of a linear space $\mathscr{X}$ is a **subspace** if and only if the following two conditions are satisfied:

- if $s_1, s_2 \in \mathscr{S}$, then $s_1 + s_2 \in \mathscr{S}$, and
- if $s \in \mathscr{S}$ and α is a scalar, then $s\alpha \in \mathscr{S}$.

It is easy to see that the span of a set of m-vectors is a subspace of the set of all m-vectors.

Next, we define matrix-vector and matrix-matrix multiplication.

Definition 3.2.9 Suppose that $\mathbf{A} = [\mathbf{a}_1, \mathbf{a}_2, \ldots, \mathbf{a}_n]$ is an $m \times n$ matrix and $\mathbf{x}$ is an n-vector with entries ξ_j. Then the **matrix-vector product Ax** is the m-vector

$$\mathbf{A}\mathbf{x} = \sum_{j=1}^{n} \mathbf{a}_j \xi_j \; . \tag{3.1}$$

Also, the set of all matrix-vector products **Ax** is called the **range** of the matrix **A**, and is denoted by

$$\mathscr{R}(\mathbf{A}) = \{\mathbf{A}\mathbf{x} : \mathbf{x} \text{ is an } n \text{ vector} \} \; .$$

Finally, if $\mathbf{B} = \left[\mathbf{b}_1, \ldots, \mathbf{b}_k\right]$ is an $n \times k$ matrix, then **AB** is the $m \times k$ matrix

$$\mathbf{A}\mathbf{B} = \mathbf{A}\left[\mathbf{b}_1, \ldots, \mathbf{b}_k\right] = \left[\mathbf{A}\mathbf{b}_1, \ldots, \mathbf{A}\mathbf{b}_k\right] \; .$$

Note that a matrix-vector product can be computed by LAPACK Level 2 BLAS routines `_gemv`. See, for example, dgemv.f. These routines compute the matrix-vector product as described by Eq. (3.1); see Sect. 2.10.1.2 for a discussion of alternative matrix-vector product algorithms. Also, a matrix-matrix product can be computed by LAPACK Level 3 BLAS routine `_gemm`. See, for example, dgemm.f and the discussion of alternative matrix-matrix product programs in Sect. 2.11.

The definition (3.1) of the matrix-vector product shows that $\mathbf{Ax}$ is a linear combination of the columns of $\mathbf{A}$, and that each column of $\mathbf{AB}$ is a linear combination of the columns of $\mathbf{A}$. It follows that $\mathscr{R}(\mathbf{A})$ is a subspace of the set of all m-vectors. Finally, we note the following easy result, which follows immediately from the definition of the range.

Lemma 3.2.1 *The system of linear equations* $\mathbf{Ax} = \mathbf{b}$ *has at least one solution if and only if* $\mathbf{b} \in \mathscr{R}(\mathbf{A})$. *If* $\mathbf{A}$ *is an* $m \times n$ *matrix, then the linear systems* $\mathbf{Ax} = \mathbf{b}$ *each have at least one solution for any right-hand side* $\mathbf{b}$ *if and only if* $\mathscr{R}(\mathbf{A})$ *is the set of all* m*-vectors.*

Exercise 3.2.1 Compare LAPACK Level 1 BLAS routine dscal with LAPACK routine dlascl for computing a vector-scalar product. Under what circumstances would `dlascl` avoid overflow or underflow that `dscal` would encounter? What extra cost does `dlascl` incur to provide this protection? How do the two routines compare in execution speed?

Exercise 3.2.2 Replace the inner loops of the nested loops in LAPACK routine `dgemv` with calls to BLAS routine `daxpy`. Compare the execution speed of the original version of `dgemv` with your modified version. (See Sect. 2.6.1 for computer architecture issues to consider in your comparison.)

3.2.4 Transpose

Next, we will describe a common operation for transforming arrays.

Definition 3.2.10 If $\mathbf{a}$ is an m-vector with entries α_i, then its **transpose** $\mathbf{a}^\top$ is the $1 \times m$ matrix with entries α_i:

$$\mathbf{a}^\top = \begin{bmatrix} \alpha_1 \\ \vdots \\ \alpha_m \end{bmatrix}^\top = \begin{bmatrix} \alpha_1, \ldots, \alpha_m \end{bmatrix} .$$

If $\mathbf{A} = [\mathbf{a}_1, \ldots, \mathbf{a}_n]$ is an $m \times n$ matrix, then the **transpose** of $\mathbf{A}$ is the $n \times m$ matrix

$$\mathbf{A}^\top = \begin{bmatrix} {\mathbf{a}_1}^\top \\ \vdots \\ {\mathbf{a}_n}^\top \end{bmatrix} .$$

The transpose of the jth column of $\mathbf{A}$ is jth **row** of $\mathbf{A}^\top$.

For complex vectors and matrices, we have a related definition.

Definition 3.2.11 If $\mathbf{z}$ is a complex m-vector with entries ζ_i, then the **conjugate transpose** (*a.k.a.* **Hermitian**) $\mathbf{z}^H$ is the $1 \times m$ matrix with entries $\overline{\zeta_i}$:

$$\mathbf{z}^H = \begin{bmatrix} \overline{\zeta_1} \\ \vdots \\ \overline{\zeta_m} \end{bmatrix}^\top = \left[\overline{\zeta_1}, \ldots, \overline{\zeta_m}\right] .$$

If $\mathbf{Z} = [\mathbf{z}_1, \ldots, \mathbf{z}_n]$ is a complex $m \times n$ matrix, then the **conjugate transpose** of $\mathbf{Z}$ is the $n \times m$ matrix

$$\mathbf{Z}^H = \begin{bmatrix} {\mathbf{z}_1}^H \\ \vdots \\ {\mathbf{z}_n}^H \end{bmatrix} .$$

Note that the Hermitian of a real matrix is the same as the transpose. Thus we can use the Hermitian for general purposes.

Example 3.2.1 If

$$\mathbf{A} = \begin{bmatrix} 0 & 3 \\ 1 & 4 \\ 2 & 5 \end{bmatrix}$$

then

$$\mathbf{A}^\top = \begin{bmatrix} 0 & 1 & 2 \\ 3 & 4 & 5 \end{bmatrix} .$$

A matrix transpose is expensive to perform, because it involves a copy from *columns* of the input array to *rows* of the output array. In this copy, the input is accessed at stride one, while the output is accessed at stride n for an $m \times n$ input array. See Sect. 2.6.4 for more discussion about copying memory. We also remark that *LAPACK has no routine to compute a transpose or conjugate transpose*. On the other hand, MATLAB array vs matrix operations provide for the conjugate transpose of a matrix `A` to be computed by `A'`, and the transpose of a matrix to be computed by `A.'`.

Example 3.2.2 Let $\mathbf{A}$ be an $m \times n$ matrix and $\mathbf{b}$ be an m-vector. Suppose that we would like to compute $\mathbf{A}^\top\mathbf{b}$. In MATLAB, the transpose of $\mathbf{A}$ is $\mathbf{A}'$, and the desired result would be given by $\mathbf{A}' * \mathbf{b}$. In LAPACK, the routines `_gemv` can be used to compute either $\mathbf{A} * \mathbf{b}$ or $\mathbf{A}^\top * \mathbf{b}$ without computing the transpose of $\mathbf{A}$, and using only stride one data access. See, for example, zgemv.f.

Some special matrices are unchanged by the transpose.

Definition 3.2.12 A matrix $\mathbf{A}$ is **symmetric** if and only if $\mathbf{A} = \mathbf{A}^\top$. A complex matrix $\mathbf{Z}$ is **Hermitian** if and only if $\mathbf{Z} = \mathbf{Z}^H$.
For example, the identity matrix is both symmetric and Hermitian. Symmetric and Hermitian matrices are necessarily square, and Hermitian matrices necessarily have real diagonal entries.

Exercise 3.2.3 Write a routine to take a given $m \times n$ matrix and return its transpose. Is it possible to access both arrays at stride one simultaneously? Comment on why you think that the designers of LAPACK did not provide a routine to compute a matrix transpose.

Exercise 3.2.4 Show that the set of all $n \times n$ Hermitian matrices is a linear space, defined in 3.2.5.

Exercise 3.2.5 For a large square matrix `A`, is there a significant difference in computational time between the MATLAB matrix-vector products `A * b` and `A' * b`?

3.2.5 Inner Product

The transpose is used to perform a common operation on a pair of vectors.

Definition 3.2.13 If $\mathbf{x}$ and $\mathbf{y}$ are real n-vectors with entries ξ_i and η_i respectively, then the **inner product** of $\mathbf{x}$ and $\mathbf{y}$ is the scalar

$$\mathbf{x} \cdot \mathbf{y} \equiv \mathbf{x}^\top \mathbf{y} = \sum_{i=1}^{n} \xi_i \eta_i \; .$$

The inner product of two complex n-vectors $\mathbf{x}$ and $\mathbf{y}$ is the scalar

$$\mathbf{x} \cdot \mathbf{y} \equiv \mathbf{x}^H \mathbf{y} = \sum_{i=1}^{n} \overline{\xi_i} \eta_i \; .$$

We note that there is a more abstract notion of inner products on general linear spaces; for more details see Sect. 1.7.1 of Chap. 1 in Volume III, or Halmos [51, p. 121].

It is easy to see that the inner product of a vector with itself is the sum of the squares of the absolute values of its entries. It follows that the inner product of a vector with itself is a nonnegative real number. Also, the order of the vectors in a real inner product does not matter, since

$$\mathbf{x}^\top \mathbf{y} = \mathbf{y}^\top \mathbf{x} \; .$$

On the other hand, the order of vectors in a complex inner product does matter, because

$$\mathbf{x}^H\mathbf{y} = \overline{\mathbf{y}^H\mathbf{x}} .$$

Real inner products $\mathbf{x}^\top\mathbf{y}$ can be computed by LAPACK Level 1 BLAS routines `_dot`. See, for example, ddot.f. Complex inner products $\mathbf{x}^H\mathbf{y}$ can be computed by LAPACK Level 1 BLAS routines `_dotc`. See, for example, zdotc.f. The complex product $\mathbf{x}^\top\mathbf{y}$ can be computed by Level 1 BLAS routines `_dotu`. See, for example, zdotu.f. Also see Sect. 2.10.1.1 for a more general discussion of the LAPACK BLAS routines. In MATLAB, dot products can be computed by the command dot. Note that in MATLAB, `dot(x,y)` is the same computation as `x' * y`; in other words, it uses the complex conjugate of x in the dot product.

Sometimes people use inner products to compute matrix-vector products. Suppose that we have an $m \times n$ matrix $\mathbf{A}$. We can view this matrix either in terms of its columns $\mathbf{c}_j$, its rows $\mathbf{r}_i$, or its components α_{ij}:

$$\mathbf{A} = \left[\mathbf{c}_1, \ldots, \mathbf{c}_n\right] = \begin{bmatrix} \mathbf{r}_1{}^H \\ \vdots \\ \mathbf{r}_m{}^H \end{bmatrix} = \begin{bmatrix} \alpha_{11} & \ldots & \alpha_{1n} \\ \vdots & \ddots & \vdots \\ \alpha_{m1} & \ldots & \alpha_{mn} \end{bmatrix} .$$

Suppose that the n-vector $\mathbf{x}$ has components ξ_j. Then Definition 3.2.9 shows that the i-th entry of $\mathbf{Ax}$ is

$$(\mathbf{Ax})_i = \left(\sum_{j=1}^{n} \mathbf{c}_j\xi_j\right)_i = \sum_{j=1}^{n} \alpha_{ij}\xi_j = \overline{\mathbf{r}}_i{}^H\mathbf{x} . \tag{3.2}$$

Thus matrix-vector products could be computed as an array of "inner products" (via BLAS routine `_dot` for real arrays and `_dotu` for complex arrays) of the rows of $\mathbf{A}$ with the vector $\mathbf{x}$.

Example 3.2.3 If

$$\mathbf{A} = \begin{bmatrix} 2 & 1 \\ 1 & 2 \\ 0 & 1 \end{bmatrix}$$

and

$$\mathbf{x} = \begin{bmatrix} 3 \\ 4 \end{bmatrix}$$

then

$$\mathbf{A}\mathbf{x} = \begin{bmatrix} 2 \\ 1 \\ 0 \end{bmatrix} 3 + \begin{bmatrix} 1 \\ 2 \\ 1 \end{bmatrix} 4 = \begin{bmatrix} 6 \\ 3 \\ 0 \end{bmatrix} + \begin{bmatrix} 4 \\ 8 \\ 4 \end{bmatrix} = \begin{bmatrix} 10 \\ 11 \\ 4 \end{bmatrix} .$$

Alternatively,

$$\mathbf{A}\mathbf{x} = \begin{bmatrix} [2, 1] \cdot \mathbf{x} \\ [1, 2] \cdot \mathbf{x} \\ [0, 1] \cdot \mathbf{x} \end{bmatrix} = \begin{bmatrix} 2 \cdot 3 + 1 \cdot 4 \\ 1 \cdot 3 + 2 \cdot 4 \\ 0 \cdot 3 + 1 \cdot 4 \end{bmatrix} = \begin{bmatrix} 10 \\ 11 \\ 4 \end{bmatrix} .$$

Example 3.2.4 If $\mathbf{e}_i$ is the i-th axis vector and $\mathbf{x}$ has entries ξ_i, then

$$\mathbf{e}_i{}^\top \mathbf{x} = \sum_{j=1}^{n} \delta_{ij} \xi_j = \xi_i$$

is the i-th entry of $\mathbf{x}$. Also, $\mathbf{A}\mathbf{e}_j$ is the j-th column of $\mathbf{A}$, and $\mathbf{e}_i{}^\top \mathbf{A}$ is the i-th row of $\mathbf{A}$.

The two ways of computing $\mathbf{A}\mathbf{x}$, namely (3.1) and (3.2), are not equally efficient on a computer. The manner in which the array $\mathbf{A}$ is stored will determine which of these two computational forms should be used. See Sect. 2.6.3 for a discussion of array memory allocation. Note that since Fortran stores matrices by columns, the Level 2 BLAS routine `_gemv` uses equation (3.1) to compute matrix-vector products.

Equation (3.2) has the following easy consequence:

Lemma 3.2.2 *If* $\mathbf{A}$ *and* $\mathbf{B}$ *are matrices,* $\mathbf{x}$ *is a vector, and we can form* $\mathbf{A}\mathbf{x}$ *and* $\mathbf{A}\mathbf{B}$*, then*

$$(\mathbf{A}\mathbf{x})^H = \mathbf{x}^H \mathbf{A}^H \textit{ and } (\mathbf{A}\mathbf{B})^H = \mathbf{B}^H \mathbf{A}^H .$$

Proof The proof uses the equivalence of the two methods (3.1) and (3.2), for computing a matrix-vector product. For more details, see Stewart [97, Theorem 4.20, p. 37].

Exercise 3.2.6 Reverse the order of the nested loops in LAPACK Level 2 BLAS routine dgemv. This will require some modification of the use of temporary variables. Compare the execution speed of the modified routine with the original for computing a matrix-vector product $\mathbf{A}\mathbf{x}$. (See Sect. 2.6.1 for computer architecture issues to consider in your comparison.) Does it make much difference in this computation if the innermost loop of the modified routine is replaced with a call to the Level 1 BLAS inner product routine ddot?

Exercise 3.2.7 Suppose that $\mathbf{A}$ is an $m \times k$ matrix and $\mathbf{B}$ is a $k \times n$ matrix. Consider LAPACK Level 3 BLAS routine dgemm for computing $\mathbf{A}\mathbf{B}$. This routine contains code involving 3 nested loops over loop variables J, L and I. These loops could have been ordered in any of $3! = 6$ ways. Describe mathematically how each of these 6 computations could be written in terms of linear combinations or inner products. Then program each and compare them for execution time. (See Sect. 2.6.1 for computer architecture issues to consider in your comparison.)

3.2.6 *Orthogonality*

The inner product can be used to develop another important concept.

Definition 3.2.14 Two m-vectors $\mathbf{x}$ and $\mathbf{y}$ (either real or complex) are **orthogonal** if and only if $\mathbf{x} \cdot \mathbf{y} = 0$.

If $\mathbf{x}$ is orthogonal to $\mathbf{y}$, then we write $\mathbf{x} \perp \mathbf{y}$. An m-vector $\mathbf{x}$ (either real or complex) is a **unit vector** if and only if $\mathbf{x} \cdot \mathbf{x} = 1$. A set of m-vectors $\{\mathbf{x}_1, \ldots, \mathbf{x}_n\}$ is **orthonormal** if and only if each is orthogonal to all of the others, and each is a unit vector.

If $\mathscr{U}$ and $\mathscr{W}$ are two sets of m-vectors, then $\mathscr{U} \perp \mathscr{W}$ if and only if for all $\mathbf{u} \in \mathscr{U}$ and all $\mathbf{w} \in \mathscr{W}$ we have $\mathbf{u} \perp \mathbf{w}$. If S is a set of m-vectors, then its **orthogonal complement** is

$$S^{\perp} = \left\{ m\text{-vectors } \mathbf{z} : \text{ for all } \mathbf{s} \in S\ ,\ \mathbf{z}^H \mathbf{s} = 0 \right\} .$$

The real $n \times n$ matrix $\mathbf{Q}$ is **orthogonal** if and only if $\mathbf{Q}^\top \mathbf{Q} = \mathbf{I}$. Similarly, the complex $n \times n$ matrix $\mathbf{U}$ is **unitary** if and only if $\mathbf{U}^H \mathbf{U} = \mathbf{I}$.
It is easy to see that the columns of an orthogonal matrix are orthonormal. Also note that the axis m-vectors are orthonormal, and the identity matrix is both orthogonal and unitary.

In addition, we have the following well-known theorem.

Theorem 3.2.2 (Pythagorean) *If* $\mathbf{x} \perp \mathbf{y}$, *then*

$$(\mathbf{x} + \mathbf{y}) \cdot (\mathbf{x} + \mathbf{y}) = \mathbf{x} \cdot \mathbf{x} + \mathbf{y} \cdot \mathbf{y} . \tag{3.3}$$

More generally, if the m-vectors $\mathbf{x}_1, \ldots, \mathbf{x}_n$ *are such that* $\mathbf{x}_i \perp \mathbf{x}_j$ *for all* $i \neq j$, *then*

$$\left(\sum_{j=1}^{n} \mathbf{x}_j \right) \cdot \left(\sum_{j=1}^{n} \mathbf{x}_j \right) = \sum_{j=1}^{n} \mathbf{x}_j \cdot \mathbf{x}_j . \tag{3.4}$$

Proof The first claim is equivalent to the second claim with $n = 2$. To prove the second claim, we note that the pairwise orthogonality of the vectors implies that

$$\left(\sum_{i=1}^{n} \mathbf{x}_i\right) \cdot \left(\sum_{j=1}^{n} \mathbf{x}_j\right) = \sum_{i=1}^{n}\sum_{j=1}^{n} \mathbf{x}_i \cdot \mathbf{x}_j = \sum_{\substack{i=1 \\ i \neq j}}^{n}\sum_{j=1}^{n} \mathbf{x}_i \cdot \mathbf{x}_j + \sum_{j=1}^{n} \mathbf{x}_j \cdot \mathbf{x}_j = \sum_{j=1}^{n} \mathbf{x}_j \cdot \mathbf{x}_j \,.$$

Exercise 3.2.8 Define the 2×2 **Hadamard matrix** by

$$\mathbf{H}_2 = \begin{bmatrix} 1 & 1 \\ 1 & -1 \end{bmatrix} \,.$$

Show that $\mathbf{H}_2$ is orthogonal. Next, if n is a power of 2, define

$$\mathbf{H}_{2n} = \begin{bmatrix} \mathbf{H}_n & \mathbf{H}_n \\ \mathbf{H}_n & -\mathbf{H}_n \end{bmatrix} \,.$$

Show that $\mathbf{H}_n$ is orthogonal for $n \geq 2$ a power of 2. (Hint: write down all of the entries of $\mathbf{H}_4$.)

Exercise 3.2.9 For $1 \leq k \leq n$, define the complex n-vector $\mathbf{z}_k$ to have jth entry equal to $e^{\mathrm{i}2\pi kj/n}$. Show that $\mathbf{z}_k$ is orthogonal to $\mathbf{z}_\ell$ for $k \neq \ell$.

Exercise 3.2.10 If $\mathbf{z}$ is a complex vector with $\mathbf{z}^H\mathbf{z} = 1$, show that $\mathbf{I} - 2\mathbf{z}\mathbf{z}^H$ is a Hermitian unitary matrix.

3.2.7 *Linear Independence*

The next definition will help us to discuss the conditions under which systems of linear equations have unique solutions.

Definition 3.2.15 The m-vectors $\mathbf{a}_1, \ldots, \mathbf{a}_n$ are **linearly independent** if and only if the only possible choice for scalars ξ_j to satisfy the equation $\sum_{j=1}^{n} \mathbf{a}_j\xi_j = \mathbf{0}$ are all $\xi_j = 0$. A related notion is the **nullspace** $\mathscr{N}(\mathbf{A})$ of an $m \times n$ matrix $\mathbf{A}$:

$$\mathscr{N}(\mathbf{A}) \equiv \{n\text{-vectors } \mathbf{z} : \mathbf{A}\mathbf{z} = \mathbf{0}\} \,.$$

It is easy to see that the columns of $\mathbf{A}$ are linearly independent if and only if $\mathscr{N}(\mathbf{A}) = \{\mathbf{0}\}$. We can take the negation of the definition of linear independence to obtain the following definition.

Definition 3.2.16 The vectors $\mathbf{a}_1, \ldots, \mathbf{a}_n$ are **linearly dependent** if and only if there are scalars ξ_j so that $\sum_{j=1}^{n} \mathbf{a}_j\xi_j = \mathbf{0}$, and not all of these scalars are zero.

It is easy to see that the columns of **A** are linearly dependent if and only if there is a nonzero vector $\mathbf{z} \in \mathscr{N}(\mathbf{A})$. It should also be obvious that a set of orthonormal vectors is linearly independent.

The following lemma is an easy consequences of Definition 3.2.15.

Lemma 3.2.3 *If* **A** *is an* $m \times n$ *matrix, then* $\mathscr{N}(\mathbf{A})$ *is a subspace of the set of all n-vectors. If the system of linear equations* $\mathbf{Ax} = \mathbf{b}$ *has a solution, then the set of all solutions is* $\{\mathbf{x} + \mathbf{z} : \mathbf{z} \in \mathscr{N}(\mathbf{A})\}$*. Finally, the system of linear equations* $\mathbf{Ax} = \mathbf{b}$ *has at most one solution if and only if* $\mathscr{N}(\mathbf{A}) = \{\mathbf{0}\}$.

Proof The first claim follows easily from the Definition 3.2.8 of a subspace, and from the definition 3.2.16 of linear dependence. The remainder of the proof merely uses the fact that $\mathbf{Az} = \mathbf{0}$ for all $\mathbf{z} \in \mathscr{N}(\mathbf{Z})$. For more details, see Stewart Stewart [97, Theorem 6.3, p. 56].

3.2.8 Basis

Next, we combine some previous ideas to form an important new concept.

Definition 3.2.17 The n m-vectors $\mathbf{a}_1, \ldots, \mathbf{a}_n$ form a **basis** for the set of all m-vectors if and only if they are linearly independent and span the set of all m-vectors. Similarly, an $m \times n$ matrix **A** is **nonsingular** if and only if $\mathscr{R}(\mathbf{A})$ is the set of all m-vectors, and $\mathscr{N}(\mathbf{A}) = \{\mathbf{0}\}$. A matrix is **singular** if and only if it is not nonsingular. It is easy to see that **A** is nonsingular if and only if its columns form a basis for the set of all m-vectors. An $m \times n$ matrix is singular if and only if is range is smaller than the set of all m-vectors, or its nullspace contains a nonzero vector.

Example 3.2.5 We claim that the axis vectors form a basis for the set of all m-vectors. To see this, we note that any m-vector **b** with entries β_i can be written

$$\mathbf{b} = \sum_{i=1}^{m} \mathbf{e}_i \beta_i \ .$$

This proves that the span of the axis vectors is the set of all m-vectors. It remains to show that the axis vectors are linearly independent. Since $\sum_{i=1}^{m} \mathbf{e}_i \beta_i = \mathbf{b}$ where **b** is the m-vector with entries β_i, if the linear combination is zero then all of the coefficients are zero. Thus the axis vectors are linearly independent and span the set of m-vectors, so they form a basis for the set of all m-vectors.
Another way to state the result in this example is that any identity matrix is nonsingular.

Here are several well-known facts regarding bases, linear independence and spans:

Lemma 3.2.4 *Every nonzero subspace of the set of all m-vectors has a basis consisting of at most m vectors.*

Proof The proof involves selecting successive linearly independent vectors until a spanning set is found. See Halmos [51, Theorem 1, p. 18], or Stewart [97, Corollary 2.18, p. 16].

Lemma 3.2.5 *If the m-vectors* $\mathbf{a}_1, \ldots, \mathbf{a}_n$ *are linearly independent, then* $n \leq m$*; equivalently, if* $\mathbf{A} = [\mathbf{a}_1, \ldots, \mathbf{a}_n]$ *is an* $m \times n$ *matrix and* $\mathscr{N}(\mathbf{A}) = \{\mathbf{0}\}$*, then* $n \leq m$*.*

Proof The proof involves using a basis for the m-vectors, such as the axis vectors, to extend the given set of vectors to a basis for all m-vectors. See Stewart [97, Corollary 2.16, p. 15].

Corollary 3.2.1 *Any linearly independent set of vectors in a subspace of m-vectors can be extended to a basis for that subspace. Any orthonormal set of vectors in a subspace of m-vectors can be extended to an orthonormal basis for that subspace.*

Proof The extension of a linearly independent set to a basis is part of the proof of Lemma 3.2.5. The extension of an orthonormal set to an orthonormal basis can be accomplished by extending to a basis, and then orthonormalizing the extension by the Gram-Schmidt process. This is described in Sect. 6.8.

Lemma 3.2.6 *If the m-vectors* $\mathbf{a}_1, \ldots, \mathbf{a}_n$ *span the set of all m-vectors, then* $n \geq m$*. Equivalently, if* $\mathbf{A}$ *is an* $m \times n$ *matrix and* $\mathscr{R}(\mathbf{A})$ *is the set of all m-vectors, then* $n \geq m$*.*

Proof The proof involves removing linearly dependent vectors to reduce the given set to a basis. See Strang [99, 2L, p. 86].

Lemma 3.2.7 *Any two bases for a subspace must have the same number of vectors. Equivalently, if an* $m \times n$ *matrix is nonsingular, then* $m = n$*. In other words, nonsingular matrices are square.*

Proof This result is a direct consequence of Lemmas 3.2.5 and 3.2.6. See also Halmos [51, Theorem 1, p. 13], Stewart [97, discussion after Corollary 2.18, p. 16] or Strang [99, 2J, p. 85].

Lemma 3.2.8 *Given an* $m \times n$ *matrix* $\mathbf{A}$*, the linear systems* $\mathbf{A}\mathbf{x} = \mathbf{b}$ *have a unique solution for any right-hand side* $\mathbf{b}$ *if and only if* $\mathbf{A}$ *is nonsingular.*

Proof This follows immediately from Lemmas 3.2.1 and 3.2.3.

Lemma 3.2.9 *Suppose that* $\mathbf{A}$ *is a square matrix of size n. Then* $\mathbf{A}$ *is nonsingular if and only if* $\mathscr{N}(\mathbf{A}) = \{\mathbf{0}\}$*. Alternatively, the square matrix* $\mathbf{A}$ *is nonsingular if and only if* $\mathscr{R}(\mathbf{A})$ *is the set of all n-vectors.*

Proof Let us prove the first claim. If $\mathbf{A}$ is nonsingular, then Definition 3.2.17 shows that its nullspace is zero. Let us prove the converse. If the nullspace of $\mathbf{A}$ is zero, then the columns of $\mathbf{A}$ are linearly independent. It follows that the columns of $\mathbf{A}$ must form a basis for the set of all n-vectors, since the existence of another linearly independent n-vector would violate the statement of Lemma 3.2.7.

Let us prove the second claim. If $\mathbf{A}$ is nonsingular, then Definition 3.2.17 shows that its range is the set of all n-vectors. This proves the forward direction of the claim. Let us prove the converse. If the range of $\mathbf{A}$ is the set of all n-vectors, then its n columns span the set of all n-vectors. These columns must be linearly independent, because we cannot remove a linearly dependent column from the set without violating the statement of Lemma 3.2.6.

3.2.9 *Dimension*

As a result of Lemma 3.2.7, the following definitions are possible.

Definition 3.2.18 The **dimension** of a subspace is the number of vectors in any basis for that subspace. The **rank** of a matrix $\mathbf{A}$ is rank $(\mathbf{A}) = \dim(\mathscr{R}(\mathbf{A}))$, and the **nullity** of $\mathbf{A}$ is $\dim(\mathscr{N}(\mathbf{A}))$. The dimension of $\{\mathbf{0}\}$ (i.e., the subspace consisting solely of the zero m-vector) is zero.
Note that not all square matrices are nonsingular: for example, the zero matrix is necessarily singular. Also note that if $\mathbf{u}$ and $\mathbf{v}$ are nonzero vectors, then $\mathbf{u}\mathbf{v}^H$ is a matrix with rank one.

Here is a collection of very important results regarding the range and nullspace of matrices.

Theorem 3.2.3 (Fundamental Theorem of Linear Algebra) *Suppose that* $\mathbf{A}$ *is a nonzero* $m \times n$ *matrix with rank* $(\mathbf{A}) = r$. *Then* $1 \le r \le \min\{m, n\}$, *rank* $\left(\mathbf{A}^H\right) = r$,

$$\dim(\mathscr{R}(\mathbf{A})) + \dim\left(\mathscr{N}\left(\mathbf{A}^H\right)\right) = m \tag{3.5a}$$

and

$$\dim\left(\mathscr{R}\left(\mathbf{A}^H\right)\right) + \dim(\mathscr{N}(\mathbf{A})) = n\ . \tag{3.5b}$$

Furthermore, $\mathscr{R}(\mathbf{A}) \perp \mathscr{N}\left(\mathbf{A}^H\right)$ *and* $\mathscr{R}\left(\mathbf{A}^H\right) \perp \mathscr{N}(\mathbf{A})$. *For every* m*-vector* $\mathbf{b}$ *there exists a unique* m*-vector* $\mathbf{r} \in \mathscr{N}\left(\mathbf{A}^H\right)$ *and an* n*-vector* $\mathbf{x}$ *so that* $\mathbf{b} = \mathbf{A}\mathbf{x} + \mathbf{r}$. *Similarly, for every* n*-vector* $\mathbf{c}$ *there is a unique* n*-vector* $\mathbf{s} \in \mathscr{N}(\mathbf{A})$ *and an* m*-vector* $\mathbf{y}$ *so that* $\mathbf{c} = \mathbf{A}^H\mathbf{y} + \mathbf{s}$.

Proof One proof involves a careful examination of Gaussian elimination as a matrix factorization. See Strang [99, p. 95 and p. 138] for the proof with $\mathbf{A}$ real.
The fundamental theorem of linear algebra has the following useful consequence.

Corollary 3.2.2 *Suppose that* $\mathbf{a}_1, \ldots, \mathbf{a}_n$ *are* n*-vectors. If* $\{\mathbf{a}_1, \ldots, \mathbf{a}_n\}$ *is either linearly independent or spans the set of all* n*-vectors, then it is a basis for all* n*-vectors.*

Proof Let $\mathbf{A} = [\mathbf{a}_1, \ldots, \mathbf{a}_n]$. If $\{\mathbf{a}_1, \ldots, \mathbf{a}_n\}$ is linearly independent, then $\mathscr{N}(\mathbf{A}) = \{\mathbf{0}\}$, so the fundamental theorem of linear algebra 3.2.3 shows that $n = \dim \mathscr{R}(\mathbf{A}^H) = \dim \mathscr{R}(\mathbf{A})$, from which we conclude that $\{\mathbf{a}_1, \ldots, \mathbf{a}_n\}$ spans the set of all n-vectors and is therefore a basis. Alternatively, suppose that $\{\mathbf{a}_1, \ldots, \mathbf{a}_n\}$ spans the set of all n-vectors. Then $n = \dim \mathscr{R}(\mathbf{A})$, and the fundamental theorem of linear algebra implies that $n = \dim \mathscr{R}(\mathbf{A}^H)$, and then that $\mathscr{N}(\mathbf{A}) = \{\mathbf{0}\}$. From this fact, we conclude that $\{\mathbf{a}_1, \ldots, \mathbf{a}_n\}$ is linearly independent, and therefore forms a basis for all n-vectors.

3.2.10 Direct Sums

The fundamental theorem of linear algebra shows that the fundamental subspaces $\mathscr{R}(\mathbf{A})$ and $\mathscr{N}(\mathbf{A}^H)$ of an $m \times n$ matrix $\mathbf{A}$ provide important decompositions of an arbitrary m-vector. In this section, we will generalize this notion. We begin with the following definition.

Definition 3.2.19 If $\mathscr{U}$ and $\mathscr{W}$ are two subspaces of m-vectors such that $\mathscr{U} \cap \mathscr{W} = \{\mathbf{0}\}$, then

$$\mathscr{V} = \mathscr{U} + \mathscr{W} \equiv \{\mathbf{u} + \mathbf{w} : \mathbf{u} \in \mathscr{U} \text{ and } \mathbf{w} \in \mathscr{W}\} \tag{3.6}$$

is the **direct sum** of $\mathscr{U}$ and $\mathscr{W}$. If $\mathscr{V}$ is the direct sum of $\mathscr{U}$ and $\mathscr{W}$, then we will write $\mathscr{V} = \mathscr{U} \oplus \mathscr{W}$.

One important consequence of this definition is contained in the following lemma.

Lemma 3.2.10 *If $\mathscr{U}$ and $\mathscr{W}$ are two subspaces of m-vectors such that $\mathscr{U} \cap \mathscr{W} = \{\mathbf{0}\}$, then for every $\mathbf{v} \in \mathscr{V}$ there are unique m-vectors $\mathbf{u} \in \mathscr{U}$ and $\mathbf{w} \in \mathscr{W}$ so that $\mathbf{v} = \mathbf{u} + \mathbf{w}$.*

Proof By the definition (3.6) of the direct sum, if $\mathbf{v} \in \mathscr{V}$ then there are m-vectors $\mathbf{u} \in \mathscr{U}$ and $\mathbf{w} \in \mathscr{W}$ so that $\mathbf{v} = \mathbf{u} + \mathbf{w}$. We need only show that $\mathbf{u}$ and $\mathbf{w}$ are unique. If $\mathbf{u} + \mathbf{w} = \mathbf{u}' + \mathbf{w}'$, then $\mathbf{u} - \mathbf{u}' = \mathbf{w}' - \mathbf{w} \in \mathscr{U} \cap \mathscr{W}$, so the Definition 3.2.19 of the direct sum implies that $\mathbf{u} - \mathbf{u}' = \mathbf{w}' - \mathbf{w} = \mathbf{0}$.

As an example of direct sums, the fundamental theorem of linear algebra 3.2.3 implies that $\mathscr{R}(\mathbf{A}) \oplus \mathscr{N}(\mathbf{A}^H)$ is the set of all m-vectors, and $\mathscr{R}(\mathbf{A}^H) \oplus \mathscr{N}(\mathbf{A})$ is the set of all n-vectors.

Next, we will use direct sums to provide another useful definition.

Definition 3.2.20 If $\mathscr{V} = \mathscr{U} \oplus \mathscr{W}$ and $\mathbf{v} = \mathbf{u} + \mathbf{w} \in \mathscr{V}$, then $\mathbf{u}$ is the **projection** of $\mathbf{v}$ onto $\mathscr{U}$ along $\mathscr{W}$, and $\mathbf{w}$ is the projection of $\mathbf{v}$ onto $\mathscr{W}$ along $\mathscr{U}$.

For example, the fundamental theorem of linear algebra says that if $\mathbf{A}$ is an $m \times n$ matrix, $\mathbf{b}$ is an m-vector and $\mathbf{b} = \mathbf{Ax} + \mathbf{r}$ where $\mathbf{r} \in \mathscr{N}(\mathbf{A}^H)$, then $\mathbf{Ax}$ is the projection of $\mathbf{b}$ onto $\mathscr{R}(\mathbf{A})$ and $\mathbf{r}$ is the projection of $\mathbf{b}$ onto $\mathscr{N}(\mathbf{A}^H)$. These ideas are illustrated in Fig. 3.2.

Definition 3.2.20 leads to the following useful result.

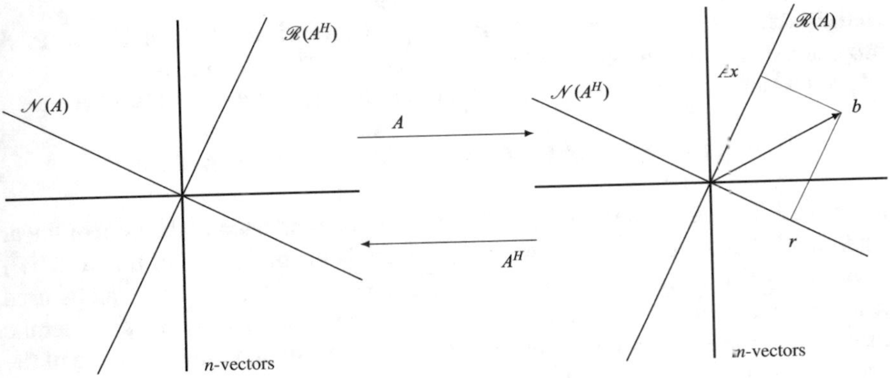

Fig. 3.2 Orthogonality of fundamental subspaces and projections

Lemma 3.2.11 *Suppose that* $\mathbf{P}$ *is a square* $n \times n$ *matrix. If* $\mathbf{P}^2 = \mathbf{P}$*, then for all* n*-vectors* $\mathbf{b}$*,* $\mathbf{Pb}$ *is the projection of* $\mathbf{b}$ *onto* $\mathscr{R}(\mathbf{P})$ *along* $\mathscr{N}(\mathbf{P})$*, and* $\mathbf{b} - \mathbf{Pb}$ *is the projection of* $\mathbf{b}$ *onto* $\mathscr{N}(\mathbf{P})$ *along* $\mathscr{R}(\mathbf{P})$*. Conversely, if for all n-vectors* $\mathbf{b}$ *we have* $\mathbf{b} - \mathbf{Pb} \in \mathscr{N}(\mathbf{P})$*, then* $\mathbf{P}^2 = \mathbf{P}$.

Proof For any n-vector $\mathbf{b}$, we can write

$$\mathbf{b} = \mathbf{Pb} + (\mathbf{I} - \mathbf{P})\mathbf{b} .$$

Note that $\mathbf{Pb} \in \mathscr{R}(\mathbf{P})$, and

$$\mathbf{P}(\mathbf{Pb}) = \mathbf{P}^2\mathbf{b} = \mathbf{Pb} .$$

Also note that

$$\mathbf{P}(\mathbf{I} - \mathbf{P})\mathbf{b} = \mathbf{Pb} - \mathbf{P}^2\mathbf{b} = \mathbf{Pb} - \mathbf{Pb} = \mathbf{0} ,$$

so $(\mathbf{I} - \mathbf{P})\mathbf{b} \in \mathscr{N}(\mathbf{P})$. If $\mathbf{z} \in \mathscr{R}(\mathbf{P}) \cap \mathscr{N}(\mathbf{P})$, then there exists an n-vector $\mathbf{x}$ so that $\mathbf{z} = \mathbf{Px}$, so

$$\mathbf{0} = \mathbf{Pz} = \mathbf{P}^2\mathbf{x} = \mathbf{Px} = \mathbf{z} .$$

Thus $\mathscr{R}(\mathbf{P}) \cap \mathscr{N}(\mathbf{P}) = \{\mathbf{0}\}$. Thus we have proved that $\mathscr{R}(\mathbf{P}) \oplus \mathscr{N}(\mathbf{P})$ is the set of all n-vectors. The first claim in the lemma now follow from Definition 3.2.20.

To prove the second claim, suppose that for all n-vectors $\mathbf{b}$ we have $\mathbf{b} - \mathbf{Pb} \in \mathscr{N}(\mathbf{P})$. Then

$$\mathbf{0} = \mathbf{P}(\mathbf{b} - \mathbf{Pb}) = \mathbf{Pb} - \mathbf{P}^2\mathbf{b} .$$

Since $\mathbf{b}$ was arbitrary, it follows that $\mathbf{P}^2 = \mathbf{P}$.

This result motivates the following definition.

Definition 3.2.21 A square matrix $\mathbf{P}$ is a **projector** if and only if $\mathbf{P}^2 = \mathbf{P}$. A projector $\mathbf{P}$ is an **orthogonal projector** if and only if $\mathbf{P}$ is Hermitian.
We call a Hermitian projector "orthogonal," because for all vectors $\mathbf{b}$ we have

$$[\mathbf{P}\mathbf{b}] \cdot [(\mathbf{I} - \mathbf{P})\mathbf{b}] = \mathbf{b}^H \mathbf{P}^H (\mathbf{I} - \mathbf{P})\mathbf{b} = \mathbf{b}^H \mathbf{P}(\mathbf{I} - \mathbf{P})\mathbf{b} = \mathbf{b}^H \mathbf{0} \mathbf{b} = \mathbf{0} .$$

Exercise 3.2.11 Suppose that $\mathbf{P}^2 = \mathbf{P}$. Note that the fundamental theorem of linear algebra 3.2.3 and Definition 3.2.20 imply that $\mathbf{Pb}$ is the projection of $\mathbf{b}$ onto $\mathscr{R}(\mathbf{P})$ along $\mathscr{N}(\mathbf{P}^H)$. Does this mean that $\mathscr{N}(\mathbf{P}^H) = \mathscr{N}(\mathbf{P})$, since both can be used in a direct sum with $\mathscr{R}(\mathbf{P})$ to represent all vectors? In your discussion, consider the case where $\mathbf{P} = \mathbf{u}\mathbf{w}^H$ and $\mathbf{w}$ is not a scalar multiple of $\mathbf{u}$; and draw a picture of the two projections for $\mathbf{u} = \mathbf{e}_1$ and $\mathbf{w} = \mathbf{e}_1 + \mathbf{e}_2$.

3.2.11 Inverse

In the special case where $\mathbf{A}$ is nonsingular (and therefore square), Theorem 3.2.3 allows us to make the following definition.

Definition 3.2.22 If $\mathbf{A}$ is a nonsingular matrix, then there is a unique matrix $\mathbf{A}^{-1}$, called the **inverse** of $\mathbf{A}$, defined to be the solution of the system of linear equations

$$\mathbf{A}\mathbf{A}^{-1} = \mathbf{I} .$$

The following results are easily proved.

Lemma 3.2.12 *If* $\mathbf{A}$ *and* $\mathbf{B}$ *are nonsingular matrices of the same size, then*

$$(\mathbf{A}\mathbf{B})^{-1} = \mathbf{B}^{-1}\mathbf{A}^{-1} .$$

Proof The proof basically consists of multiplying $\mathbf{AB}$ times $\mathbf{B}^{-1}\mathbf{A}^{-1}$ and showing that the product is equal to the identity. See also Halmos [51, Theorem 3, p. 63], Strang [99, 1L, p. 43] or Stewart [97, Theorem 6.11, p. 60].

Lemma 3.2.13 *If* $\mathbf{A}$ *is nonsingular, then* $\mathbf{A}^{-1}\mathbf{A} = \mathbf{I}$ *and* $\left[\mathbf{A}^{-1}\right]^{-1} = \mathbf{A}$.

Proof The proof depends on the left cancellation property of multiplication by nonsingular matrices. Since $\mathbf{A}$ is nonsingular, $\mathbf{A}\mathbf{A}^{-1} = \mathbf{I}$. We can multiply on the right by $\mathbf{A}$ to get $\mathbf{A}\mathbf{A}^{-1}\mathbf{A} = \mathbf{A}$, and associate terms to get $\mathbf{A}\left(\mathbf{A}^{-1}\mathbf{A}\right) = \mathbf{A}\mathbf{I}$. Since Lemma 3.2.8 shows that the solutions of linear systems involving nonsingular matrices are unique, we must have $\mathbf{A}^{-1}\mathbf{A} = \mathbf{I}$. The Definition 3.2.22 of the inverse now proves that $\left[\mathbf{A}^{-1}\right]^{-1} = \mathbf{A}$.

Lemma 3.2.14 *If* $\mathbf{A}$ *is nonsingular, then* $\left(\mathbf{A}^{-1}\right)^H = \left(\mathbf{A}^H\right)^{-1}$.

Proof Lemmas 3.2.2 and 3.2.13 show that

$$\mathbf{A}^H(\mathbf{A}^{-1})^H = [\mathbf{A}^{-1}\mathbf{A}]^H = \mathbf{I}^H = \mathbf{I}\,.$$

Definition 3.2.22 now proves the claimed result.

In most circumstances it is both unnecessary and inefficient to compute a matrix inverse for use in solving systems of linear equations. Nevertheless, LAPACK contains routines for computing inverses. In particular, routines `_getri` can be used to compute the inverse of a general nonsingular matrix. See, for example, dgetri.f. Routines `_potri`, `_sytri` and `_trtri` compute inverses of symmetric positive, general symmetric and triangular matrices, respectively. In MATLAB, it is possible to use the function inv to compute the inverse of a matrix. MATLAB users should never solve a linear system $\mathbf{Ax} = \mathbf{b}$ by computing `x = inv(A) * b`; instead they should write `x = A \ b`.

3.2.12 Determinant

Elementary linear algebra texts commonly use the determinant of a square matrix to decide if it is nonsingular. This approach is unsuitable for scientific computation, since the computation of determinants of large matrices can easily **underflow** to produce zero values, or **overflow** to produce floating-point infinity. Instead, nonsingularity is usually determined numerically via the singular value decomposition, described in Sects. 6.11 and 1.5 of Chap. 1 in Volume II. We will say more about computing determinants in Sect. 3.11.

Determinants have practical value in measuring volumes of parallelepipeds. Within this context, there are some practical observations about volumes of parallelepipeds with a vertex at the origin. First, we note that if one of the parallelepiped edges emanating from the origin is scaled by some factor (and parallel edges are likewise scaled to maintain a parallelepiped), then the multi-dimensional volume is scaled by that factor. To see an example of this fact, draw a parallelogram in two dimensions and scale two parallel sides, as in Fig. 3.3. Second, if we take two parallelepipeds sharing all edges emanating from the origin except one, and form a third parallelepiped by maintaining the common edges at the origin and adding the other two at the origin, then the new parallelepiped has multidimensional volume equal to the sum of the volumes of the original two parallelepipeds. This idea is illustrated in Fig. 3.4. Finally, if a parallelepiped has two indistinguishable edges emanating from the origin, then the multidimensional volume of the parallelepiped is zero.

To make these statements mathematically consistent, we must allow for the multidimensional volume to have a sign. The correct description of our ideas is contained in the following definition.

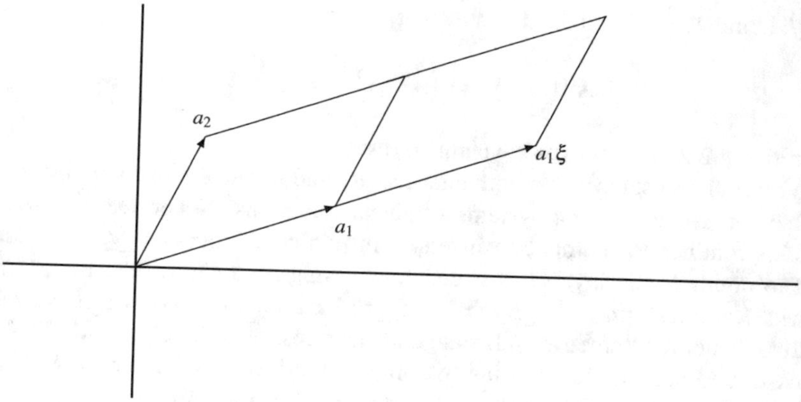

Fig. 3.3 Scaling a parallelogram side

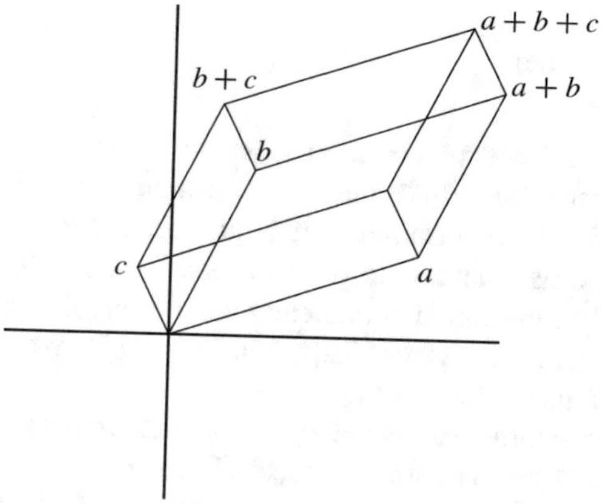

Fig. 3.4 Summing parallelograms

Definition 3.2.23 An **alternating multi-linear functional** on $n \times n$ matrices $\mathbf{A} = [\mathbf{a}_1, \ldots, \mathbf{a}_n]$ is a a function $\phi(\mathbf{A}) = \phi(\mathbf{a}_1, \ldots, \mathbf{a}_n)$ satisfying

- for all scalars ξ and for all $1 \leq j \leq n$

$$\phi(\mathbf{a}_1, \ldots, \mathbf{a}_{j-1}, \mathbf{a}_j\xi, \mathbf{a}_{j+1}, \ldots, \mathbf{a}_n) = \phi(\mathbf{a}_1, \ldots, \mathbf{a}_{j-1}, \mathbf{a}_j, \mathbf{a}_{j+1}, \ldots, \mathbf{a}_n)\xi \ ,$$

- for all n-vectors $\mathbf{b}$ and $\mathbf{c}$ and for all $1 \leq j \leq n$ we have

$$\begin{aligned}&\phi\left(\mathbf{a}_1, \ldots, \mathbf{a}_{j-1}, \mathbf{b}+\mathbf{c}, \mathbf{a}_{j+1}, \ldots, \mathbf{a}_n\right)\\ =&\phi\left(\mathbf{a}_1, \ldots, \mathbf{a}_{j-1}, \mathbf{b}, \mathbf{a}_{j+1}, \ldots, \mathbf{a}_n\right) + \phi\left(\mathbf{a}_1, \ldots, \mathbf{a}_{j-1}, \mathbf{c}, \mathbf{a}_{j+1}, \ldots, \mathbf{a}_n\right) \ , \text{ and}\end{aligned}$$

- for all n-vectors $\mathbf{b}$ and for all $1 \leq i < j \leq n$

$$\phi\left(\mathbf{a}_1, \ldots, \mathbf{a}_{i-1}, \mathbf{b}, \mathbf{a}_{i+1}, \ldots, \mathbf{a}_{j-1}, \mathbf{b}, \mathbf{a}_{j+1}, \ldots, \mathbf{a}_n\right) = 0 \ .$$

In the next lemma, we use Definition 3.2.23 to show that whenever we perform a transposition of the arguments of an alternating multi-linear functional, we change its sign.

Lemma 3.2.15 *If ϕ is an alternating multi-linear functional on $n \times n$ matrices, then for any $n \times n$ matrix $\mathbf{A} = [\mathbf{a}_1, \ldots, \mathbf{a}_n]$ we have*

$$\phi\left(\mathbf{a}_1, \ldots, \mathbf{a}_{i-1}, \mathbf{a}_j, \mathbf{a}_{i+1}, \ldots, \mathbf{a}_{j-1}, \mathbf{a}_i, \mathbf{a}_{j+1}, \ldots, \mathbf{a}_n\right) = -\phi\left(\mathbf{a}_1, \ldots, \mathbf{a}_n\right) \ .$$

Proof We consider the matrix with columns i and j both equal to $\mathbf{a}_i + \mathbf{a}_j$, and recall that $\phi = 0$ whenever two columns are equal. This gives us

$$\begin{aligned}
0 =& \phi\left(\ldots, \mathbf{a}_i + \mathbf{a}_j, \ldots, \mathbf{a}_i + \mathbf{a}_j, \ldots\right) \\
=& \phi\left(\ldots, \mathbf{a}_i, \ldots, \mathbf{a}_i, \ldots\right) + \phi\left(\ldots, \mathbf{a}_j, \ldots, \mathbf{a}_i, \ldots\right) \\
+& \phi\left(\ldots, \mathbf{a}_i, \ldots, \mathbf{a}_j, \ldots\right) + \phi\left(\ldots, \mathbf{a}_j, \ldots, \mathbf{a}_j, \ldots\right) \\
=& \phi\left(\ldots, \mathbf{a}_i, \ldots, \mathbf{a}_j, \ldots\right) + \phi\left(\ldots, \mathbf{a}_j, \ldots, \mathbf{a}_i, \ldots\right) \ ,
\end{aligned}$$

which implies the claimed result.

Next, we remind the reader of the following well-known result regarding permutations.

Lemma 3.2.16 *For every permutation π, there is an inverse permutation π^{-1} so that $\pi \circ \pi^{-1}$ is the identity permutation. Every permutation is a product of transpositions. Further, for any given permutation all such products of transpositions have the same* **parity***: either the products all involve an even number of transpositions, or they all involve an odd number of transpositions.*

Proof See Halmos [51, pp. 43–47].

Lemmas 3.2.15 and 3.2.16 lead to the following theorem.

Theorem 3.2.4 (Laplace Expansion) *If ϕ is an alternating multi-linear functional on $n \times n$ matrices, and $\mathbf{A}$ is an $n \times n$ matrix with entries α_{ij}, then*

$$\phi(\mathbf{A}) = \phi(\mathbf{I}) \sum_{\textit{permutations}\ \pi} (-1)^{N(\pi)} \alpha_{\pi(1),1} \ldots \alpha_{\pi(n),n} \ . \tag{3.7}$$

Here $N(\pi)$ is the parity of the permutation π.

Proof We can write each column of **A** in terms of the axis vectors:

$$\mathbf{a}_j = \sum_{i_j=1}^{n} \mathbf{e}_{i_j} \alpha_{i_j,j} .$$

Then we can use the multi-linearity of ϕ to get

$$\phi(\mathbf{A}) = \phi\left(\sum_{i_1=1}^{n} \mathbf{e}_{i_1}\alpha_{i_1,j}, \ldots, \sum_{i_n=1}^{n} \mathbf{e}_{i_n}\alpha_{i_n,j}\right)$$
$$= \sum_{i_1=1}^{n} \cdots \sum_{i_n=1}^{n} \phi(\mathbf{e}_{i_1}, \ldots, \mathbf{e}_{i_n})\alpha_{i_1,j} \ldots \alpha_{i_n,j} .$$

There are n^n terms in this sum, but many of them are zero because the arguments to ϕ do not involve distinct axis vectors. In other words, there are $n!$ distinct nonzero values of ϕ in the sum, corresponding to cases in which the axis vectors $e_{i_1}, \ldots, e_{i_n}$ are all distinct. Such a case corresponds to $\phi\left(\mathbf{e}_{\pi(1),1}, \ldots, \mathbf{e}_{\pi(n),n}\right)$ for some permutation π. All possible permutations must appear exactly once in the sum. Then Lemmas 3.2.15 and 3.2.16 complete the proof.

The Laplace expansion shows that the following definition makes sense

Definition 3.2.24 The **determinant** of an $n \times n$ matrix **A** is the alternating multi-linear functional det(**A**) such that det(**I**) $= 1$.

Example 3.2.6 For

$$\mathbf{A} = \begin{bmatrix} \alpha_{11} & \alpha_{12} \\ \alpha_{21} & \alpha_{22} \end{bmatrix}$$

there are only two possible permutations of the columns, the permutation that leaves things in place, and a transposition. The Laplace expansion (3.7) shows that

$$\det\left(\begin{bmatrix} \alpha_{11} & \alpha_{12} \\ \alpha_{21} & \alpha_{22} \end{bmatrix}\right) = \alpha_{11}\alpha_{22} - \alpha_{21}\alpha_{12} .$$

The difficulty with the Laplace expansion is that the number of terms in its expression is typically very large. For $n > 3$ there are typically too many terms to make good use of this expression. We need to find simpler ways to evaluate the determinant. Here is a lemma that will help.

Lemma 3.2.17 *If* **A** *and* **B** *are* $n \times n$ *matrices, then* det(**AB**) $=$ det(**A**) det(**B**).

Proof We begin by showing that $\phi_{\mathbf{A}}(\mathbf{B}) \equiv \det(\mathbf{AB})$ is an alternating multi-linear functional. If we multiply a column of $\mathbf{B}$ by the scalar ξ, then

$$\phi_{\mathbf{A}}\left(\ldots, \mathbf{b}_j\xi, \ldots\right) = \det\left(\ldots, \mathbf{A}\mathbf{b}_j\xi, \ldots\right) = \det\left(\ldots, \mathbf{A}\mathbf{b}_j, \ldots\right)\xi = \det(\mathbf{AB})\xi$$
$$= \phi_{\mathbf{A}}(\mathbf{B})\xi$$

If one column of $\mathbf{B}$ is a sum of two vectors, then

$$\phi_{\mathbf{A}}\left(\ldots, \mathbf{c}+\mathbf{d}, \ldots\right) = \det\left(\ldots, \mathbf{Ac}+\mathbf{Ad}, \ldots\right)$$
$$= \det\left(\ldots, \mathbf{Ac}, \ldots\right) + \det\left(\ldots, \mathbf{Ad}, \ldots\right) = \phi_{\mathbf{A}}\left(\ldots, \mathbf{c}, \ldots\right) + \phi_{\mathbf{A}}\left(\ldots, \mathbf{d}, \ldots\right) .$$

If two columns of $\mathbf{B}$ are the same, then

$$\phi_{\mathbf{A}}(\ldots, \mathbf{b}, \ldots, \mathbf{b}, \ldots) = \det(\ldots, \mathbf{Ab}, \ldots, \mathbf{Ab}, \ldots) = 0 .$$

Thus $\phi_{\mathbf{A}}$ is an alternating multi-linear functional. The Laplace expansion (3.7) shows that

$$\phi_{\mathbf{A}}(\mathbf{B}) = \phi_{\mathbf{A}}(\mathbf{I})\det(\mathbf{B}) = \det(\mathbf{A})\det(\mathbf{B}) .$$

We also note the following useful result.

Lemma 3.2.18 *If* $\mathbf{A}$ *is an* $n \times n$ *matrix, then* $\det\left(\mathbf{A}^\top\right) = \det(\mathbf{A})$.

Proof If $\mathbf{A}$ has entries $\alpha_{i,j}$, then

$$\det\left(\mathbf{A}^\top\right) = \sum_{\text{permutations } \pi} (-1)^{N(\pi)}\alpha_{1,\pi(1)} \cdots \alpha_{n,\pi(n)}$$
$$= \sum_{\text{permutations } \pi} (-1)^{N(\pi)}\alpha_{\pi^{-1}(1),1} \cdots \alpha_{\pi^{-1}(n),n} .$$

Since every permutation has an inverse, and since π and its inverse are easily shown to have the same parity, the second sum is the determinant.

The next lemma describes how determinants can be used (in theory) to test nonsingularity.

Lemma 3.2.19 *If* $\mathbf{A}$ *is an* $n \times n$ *matrix, then* $\mathbf{A}$ *is singular if and only if* $\det(\mathbf{A}) = 0$.

Proof $\Rightarrow$ Suppose that $\mathbf{A}$ is singular. Then $\mathscr{N}(\mathbf{A}) \neq \{\mathbf{0}\}$, so there is a nonzero n-vector $\mathbf{z}$ such that $\mathbf{Az} = \mathbf{0}$. Since $\mathbf{z} \neq \mathbf{0}$, there is at least one index j so that the j-th component ζ_j of $\mathbf{z}$ is nonzero. As a result, we can solve the equation $\mathbf{Az} = 0$ for the j-th column $\mathbf{a}_j$ of $\mathbf{A}$ to get

$$\mathbf{a}_j = -\sum_{i \neq j} \mathbf{a}_i \frac{\zeta_i}{\zeta_j} .$$

Then

$$\det(\mathbf{A}) = \det(\mathbf{a}_1, \ldots, \mathbf{a}_{j-1}, -\sum_{i \neq j} \mathbf{a}_i \frac{\zeta_i}{\zeta_j}, \mathbf{a}_{j+1}, \ldots, \mathbf{a}_n)$$

$$= \sum_{i \neq j} \det(\mathbf{a}_1, \ldots, \mathbf{a}_{j-1}, \mathbf{a}_i, \mathbf{a}_{j+1}, \ldots, \mathbf{a}_n) \frac{\zeta_i}{\zeta_j} .$$

Since every determinant in the sum has two columns the same, Definition 3.2.23 shows that $\det(\mathbf{A}) = 0$.

$\Leftarrow$ To prove the other direction of the lemma, we prove the contrapositive. Suppose that $\mathbf{A}$ is nonsingular. Then $\mathbf{A}\mathbf{A}^{-1} = \mathbf{I}$, so $1 = \det(\mathbf{I}) = \det(\mathbf{A}) \det(\mathbf{A}^{-1})$. This implies that $\det(\mathbf{A}) \neq 0$.

The proof of Lemma 3.2.19 also shows the following.

Corollary 3.2.3 *If* $\mathbf{A}$ *is nonsingular, then* $\det\left(\mathbf{A}^{-1}\right) = 1/\det(\mathbf{A})$.

We will also have occasional use for the following three results.

Theorem 3.2.5 (Expansion by Minors) *Let* $\mathbf{A}$ *be an* $n \times n$ *matrix with entries* $\alpha_{i,j}$, *and for all* $1 \leq i,j \leq n$ *let* $\mathbf{A}_{ij}$ *be the* $(n-1) \times (n-1)$ *matrix formed by deleting row* i *and column* j *of* $\mathbf{A}$. *Then*

$$\det \mathbf{A} = \sum_{j=1}^{n} \alpha_{ij} (-1)^{i+j} \det \mathbf{A}_{ij} . \tag{3.8}$$

Proof The Laplace expansion (3.7) shows that

$$\det(\mathbf{A}) = \sum_{\text{permutations } \pi} (-1)^{N(\pi)} \prod_{k=1}^{n} \alpha_{\pi(k),k}$$

$$= \sum_{j=1}^{n} \alpha_{ij} \sum_{\substack{\text{permutations } \pi \\ \pi(j)=i}} (-1)^{N(\pi)} \prod_{\substack{1 \leq k \leq n \\ k \neq j}} \alpha_{\pi(k),k} .$$

Next we note that for each j, if $\pi(j) = i$ then $i - j$ interchanges following π will produce a permutation π' so that $\pi'(i) = i$, which implies that

$$= \sum_{j=1}^{n} \alpha_{ij} (-1)^{i-j} \sum_{\substack{\text{permutations } \pi' \\ \pi'(i)=i}} (-1)^{N(\pi')} \prod_{\substack{1 \leq k \leq n \\ k \neq j}} \alpha_{\pi'(k),k}$$

then using the identity $(-1)^{i-j} = (-1)^{i+j}$ and the Laplace expansion, we obtain

$$= \sum_{j=1}^{n} \alpha_{ij}(-1)^{i+j} \det \mathbf{A}_{ij} \; .$$

Since $\det \mathbf{A} = \det \mathbf{A}^\top$, we can expand by minors using a column of $\mathbf{A}$ as well.

Theorem 3.2.6 (Determinant of Block Right-Triangular) *Suppose that the $n \times n$ matrix can be partitioned in the form*

$$\mathbf{A} = \begin{bmatrix} \mathbf{B} & \mathbf{C} \\ \mathbf{0} & \mathbf{D} \end{bmatrix} \; .$$

Then

$$\det \mathbf{A} = (\det \mathbf{B})(\det \mathbf{D}) \; .$$

Proof Suppose that $\mathbf{B}$ is $m \times m$. The Laplace expansion (3.7) shows that

$$\det(\mathbf{A}) = \sum_{\text{permutations } \pi} (-1)^{N(\pi)} \prod_{k=1}^{n} \alpha_{\pi(k),k}$$

but since $\alpha_{i,j} = 0$ for $i > m$ and $j \le m$, the only useful permutations are of the form $\pi = \pi_1 \circ \pi_2$ where π_1 is a permutation of indices 1 through m and π_2 is a permutation of indices $m+1$ through n; this leads to

$$= \sum_{\text{permutations } \pi_1} \sum_{\text{permutations } \pi_2} (-1)^{N(\pi_1)} \prod_{j=1}^{m} \alpha_{\pi_1(j),j} (-1)^{N(\pi_2)} \prod_{\ell=m+1}^{n} \alpha_{\pi_2(\ell),\ell}$$

$$= \left(\sum_{\text{permutations } \pi_1} (-1)^{N(\pi_1)} \prod_{j=1}^{m} \alpha_{\pi_1(j),j} \right) \left(\sum_{\text{permutations } \pi_2} (-1)^{N(\pi_2)} \prod_{\ell=m+1}^{n} \alpha_{\pi_2(\ell),\ell} \right)$$

$$= (\det \mathbf{B})(\det \mathbf{D}) \; .$$

Theorem 3.2.7 (Sylvester's Determinant Identity) *Let $\mathbf{A}$ be an $n \times n$ matrix with $n \ge 3$. For pairs of distinct integers (i_1, i_2) and (j_1, j_2) that are between 1 and n, let $\mathbf{A}_{i_1,i_2\,;\,j_1,j_2}$ be the $(n-2) \times (n-2)$ matrix formed by deleting rows i_1 and i_2 as well as columns j_1 and j_2 of $\mathbf{A}$. Then*

$$(\det \mathbf{A})(\det \mathbf{A}_{i_1,i_2\,;\,j_1,j_2}) = (\det \mathbf{A}_{i_1,j_1})(\det \mathbf{A}_{i_2,j_2}) - (\det \mathbf{A}_{i_1,j_2})(\det \mathbf{A}_{i_2,j_1}) \; . \tag{3.9}$$

Proof Partition

$$\mathbf{A} = \begin{bmatrix} \mathbf{B} & \mathbf{c}_1 & \mathbf{c}_2 \\ \mathbf{r}_1{}^\top & \sigma_{11} & \sigma_{12} \\ \mathbf{r}_2{}^\top & \sigma_{21} & \sigma_{22} \end{bmatrix}$$

and define

$$\mathbf{M} = \begin{bmatrix} \mathbf{B} & \mathbf{c}_1 & \mathbf{c}_2 & \mathbf{0} \\ \mathbf{r}_1{}^\top & \sigma_{11} & \sigma_{12} & \mathbf{0}^\top \\ \mathbf{r}_2{}^\top & \sigma_{21} & \sigma_{22} & \mathbf{r}_2{}^\top \\ \mathbf{0} & \mathbf{0} & \mathbf{0} & \mathbf{B} \end{bmatrix}$$

Since $\mathbf{M}$ is block right-triangular, Theorem 3.2.6 implies that

$$(\det \mathbf{A})(\det \mathbf{B}) = \det \mathbf{M}$$

then we can add the first block rows to the last to get

$$= \det \begin{bmatrix} \mathbf{B} & \mathbf{c}_1 & \mathbf{c}_2 & \mathbf{0} \\ \mathbf{r}_1{}^\top & \sigma_{11} & \sigma_{12} & \mathbf{0}^\top \\ \mathbf{r}_2{}^\top & \sigma_{21} & \sigma_{22} & \mathbf{r}_2{}^\top \\ \mathbf{B} & \mathbf{c}_1 & \mathbf{c}_2 & \mathbf{B} \end{bmatrix}$$

then we can subtract the last block column from the first to obtain

$$= \det \begin{bmatrix} \mathbf{B} & \mathbf{c}_1 & \mathbf{c}_2 & \mathbf{0} \\ \mathbf{r}_1{}^\top & \sigma_{11} & \sigma_{12} & \mathbf{0}^\top \\ \mathbf{0}^\top & \sigma_{21} & \sigma_{22} & \mathbf{r}_2{}^\top \\ \mathbf{0} & \mathbf{c}_1 & \mathbf{c}_2 & \mathbf{B} \end{bmatrix}$$

then we use the sublinearity of the determinant to see that

$$= \det \begin{bmatrix} \mathbf{B} & \mathbf{c}_1 & \mathbf{c}_2 & \mathbf{0} \\ \mathbf{r}_1{}^\top & \sigma_{11} & \sigma_{12} & \mathbf{0}^\top \\ \mathbf{0}^\top & 0 & \sigma_{22} & \mathbf{r}_2{}^\top \\ \mathbf{0} & \mathbf{0} & \mathbf{c}_2 & \mathbf{B} \end{bmatrix} + \det \begin{bmatrix} \mathbf{B} & \mathbf{0} & \mathbf{c}_2 & \mathbf{0} \\ \mathbf{r}_1{}^\top & 0 & \sigma_{12} & \mathbf{0}^\top \\ \mathbf{0}^\top & \sigma_{21} & \sigma_{22} & \mathbf{r}_2{}^\top \\ \mathbf{0} & \mathbf{c}_1 & \mathbf{c}_2 & \mathbf{B} \end{bmatrix}$$

but the left-hand matrix is block right-triangular, and interchanging the middle two columns produces another block right-triangular matrix, so we get

$$= \det \begin{bmatrix} \mathbf{B} & \mathbf{c}_1 \\ \mathbf{r}_1{}^\top & \sigma_{11} \end{bmatrix} \det \begin{bmatrix} \sigma_{22} & \mathbf{r}_2{}^\top \\ \mathbf{c}_2 & \mathbf{B} \end{bmatrix} - \det \begin{bmatrix} \mathbf{B} & \mathbf{c}_2 \\ \mathbf{r}_1{}^\top & \sigma_{12} \end{bmatrix} \det \begin{bmatrix} \sigma_{21} & \mathbf{r}_2{}^\top \\ \mathbf{c}_1 & \mathbf{B} \end{bmatrix}$$

$$= \det \mathbf{A}_{n+2,n+2} \det \mathbf{A}_{n+1,n+1} - \det \mathbf{A}_{n+2,n+1} \det \mathbf{A}_{n+1,n+2} \,.$$

For general locations of the deleted rows and columns, we can interchange row i_1 with $n+1$, row i_2 with $n+2$, column j_1 with $n+1$ and column j_2 with $n+2$ to produce the desired result.

3.2.13 Summary

A system of n linear equations in n unknowns can be described mathematically as follows. Given an $n \times n$ matrix $\mathbf{A}$ and an n-vector $\mathbf{b}$, find an n-vector $\mathbf{x}$ so that $\mathbf{A}\mathbf{x} = \mathbf{b}$. Given $\mathbf{A}$, the linear system $\mathbf{A}\mathbf{x} = \mathbf{b}$ has a solution if and only if $\mathbf{b} \in \mathscr{R}(\mathbf{A})$. If $\mathbf{A}\mathbf{x} = \mathbf{b}$ has a solution $\mathbf{x}$, then an arbitrary solution can be written as $\mathbf{x}$ plus a solution $\mathbf{z} \in \mathscr{N}(\mathbf{A})$ to the homogeneous system $\mathbf{A}\mathbf{z} = \mathbf{0}$. Linear systems involving the matrix $\mathbf{A}$ always have a solution if and only if $\mathscr{R}(\mathbf{A})$ is the set of all m-vectors. The solution of linear systems involving the matrix $\mathbf{A}$ is unique if and only if $\mathscr{N}(\mathbf{A}) = \{\mathbf{0}\}$. Thus the system of linear equations $\mathbf{A}\mathbf{x} = \mathbf{b}$ has a unique solution for any $\mathbf{b}$ if and only if $\mathbf{A}$ is nonsingular. The matrix $\mathbf{A}$ is nonsingular if and only if it is square and the only solution to the homogeneous system $\mathbf{A}\mathbf{x} = \mathbf{0}$ is $\mathbf{x} = \mathbf{0}$. Equivalently, $\mathbf{A}$ is nonsingular if and only if it is square and $\mathbf{A}\mathbf{x} = \mathbf{b}$ has a solution for all m-vectors $\mathbf{b}$.

Exercise 3.2.12 Suppose that $\mathbf{A}$ is a **diagonal matrix**, meaning that $\mathbf{A}$ is square, and $\mathbf{A}_{ij} = 0$ whenever $i \neq j$. Show that $\mathbf{A}$ is nonsingular if and only if the diagonal entries $\mathbf{A}_{ii}$ are all nonzero.

Exercise 3.2.13 Suppose that $\mathbf{A}$ is a **left trapezoidal matrix**, meaning that $\mathbf{A}_{ij} = 0$ for all $i < j$. Show that $\mathscr{N}(\mathbf{A}) = \{\mathbf{0}\}$ if and only if the diagonal entries are all nonzero.

3.3 Fundamental Problems

There are several important classes of problems in linear algebra, each of which provides an interesting subject for scientific computing:

Systems of Linear Equations: Given a nonsingular $n \times n$ matrix $\mathbf{A}$ and an n-vector $\mathbf{b}$, find an n-vector $\mathbf{x}$ so that $\mathbf{A}\mathbf{x} = \mathbf{b}$.

Least Squares: Given an $m \times n$ matrix $\mathbf{A}$ and an m-vector $\mathbf{b}$, find the smallest n-vector $\mathbf{x}$ that minimizes the $\|\mathbf{b} - \mathbf{A}\mathbf{x}\|_2$.

Eigenvectors: Given an $n \times n$ matrix $\mathbf{A}$, find all scalars λ and associated nonzero n-vectors $\mathbf{x}$ so that $\mathbf{A}\mathbf{x} = \mathbf{x}\lambda$.

Linear Programs: Given a real $m \times n$ matrix $\mathbf{A}$ with rank $(\mathbf{A}) = m$, a real m-vector $\mathbf{b}$ and a real n-vector $\mathbf{c}$, find a real n-vector $\mathbf{x}$ to minimize the objective $\mathbf{c}^\top\mathbf{x}$ subject to the constraints $\mathbf{A}\mathbf{x} = \mathbf{b}$ and $\mathbf{x} \geq \mathbf{0}$.

Systems of linear equations arise in a number of applications, especially physics (e.g., balance of forces), engineering (e.g., finite element analysis) and economics (e.g., determining an input to provide a desired output). Least squares problems arise when there are more equations than unknowns (possibly due to oversampling to reduce the effects of random errors), fewer equations than unknowns (possibly due to uncertainty regarding which independent variables are important), or rank-deficient coefficient arrays. Eigenvalue problems arise in engineering study of vibrations, decay and signal processing, as well as steady-state input-output models in economics. Finally, linear programming problems arise quite often in business and economics, either to maximize profit or minimize cost subject to production and supply constraints, or to determine optimal strategies for two-person zero-sum games.

Since ideas involved in solving systems of linear equations are essential to all of these fundamental problems of linear algebra, we will begin by discussing systems of linear equations. That discussion will consume the remainder of this chapter. Least squares problems will be discussed in Chap. 6, while eigenvalues and eigenvectors will be discussed in Chap. 1 of Volume II. Linear programs will be covered in Chap. 4 of Volume II.

3.4 Easy Linear Systems

The solution of general linear systems involving a large number of unknowns can appear daunting. Fortunately, there are a number of computational approaches that can reduce such general problems to a series of simpler computational problems. Let us examine several of these simpler problems.

3.4.1 *One by One and Diagonal*

When $\alpha \neq 0$, the solution of $\alpha\xi = \beta$ is $\xi = \beta/\alpha$. Although this problem may appear to be too trivial to mention, it is the fundamental step in solving triangular systems, discussed in Sect. 3.4.4 below.

Linear systems for diagonal matrices are similarly easy to solve.

Definition 3.4.1 A diagonal matrix $\boldsymbol{\Sigma}$ is a square matrix with entries σ_{ij} satisfying $\sigma_{ij} = 0$ for $i \neq j$.

It should be obvious that for any diagonal matrix $\boldsymbol{\Sigma}$ the solution $\mathbf{x}$ of $\boldsymbol{\Sigma}\mathbf{x} = \mathbf{b}$ has entries $\xi_j = \beta_j/\sigma_j$.

In LAPACK, diagonal linear systems can be solved by calling the routines `_larscl2`. See, for example, dlarscl2.f. In MATLAB, if `x` and `y` are vectors of the same size, then `x ./ y` will return a vector with entries `x(i)/y(i)`.

3.4.2 *Two by Two*

It is also easy to solve two equations in two unknowns. Whenever the linear system

$$\begin{bmatrix}\alpha_{11} & \alpha_{12}\\ \alpha_{21} & \alpha_{22}\end{bmatrix}\begin{bmatrix}\xi_1\\ \xi_2\end{bmatrix}=\begin{bmatrix}\beta_1\\ \beta_2\end{bmatrix}$$

is such that the determinant

$$\delta \equiv \alpha_{11}\alpha_{22}-\alpha_{12}\alpha_{21}$$

is nonzero, we can compute

$$\begin{bmatrix}\xi_1\\ \xi_2\end{bmatrix}=\begin{bmatrix}\alpha_{22} & -\alpha_{12}\\ -\alpha_{21} & \alpha_{11}\end{bmatrix}\begin{bmatrix}\beta_1\\ \beta_2\end{bmatrix}\frac{1}{\delta}. \tag{3.10}$$

This simple problem is the fundamental task in solving the quasi-triangular systems in Sect. 3.4.6. On the other hand, if the determinant satisfies $\delta = 0$, then the problem has no solution if $\alpha_{11}\beta_2-\alpha_{21}\beta_1 \neq 0$, and infinitely many solutions otherwise.

There is some risk of unnecessary underflow or overflow in computing the determinant δ, or in computing the matrix of minors times the right-hand side. This risk can be reduced by scaling by largest entry of the matrix, as in the following

Algorithm 3.4.1 (2 × 2 **Linear System Solve)**

$$\sigma = \max\{\alpha_{11},\alpha_{21},\alpha_{12},\alpha_{22}\}$$

$$\begin{bmatrix}\mu_{11} & \mu_{12}\\ \mu_{21} & \mu_{22}\end{bmatrix}=\begin{bmatrix}\alpha_{11} & \alpha_{12}\\ \alpha_{21} & \alpha_{22}\end{bmatrix}\frac{1}{\sigma}$$

$$\delta = \mu_{11}\mu_{22}-\mu_{21}\mu_{12}$$

$$\begin{bmatrix}\gamma_1\\ \gamma_2\end{bmatrix}=\begin{bmatrix}\beta_1\\ \beta_2\end{bmatrix}\frac{1}{\sigma}$$

$$\begin{bmatrix}\xi_1\\ \xi_2\end{bmatrix}=\begin{bmatrix}\mu_{22} & -\mu_{12}\\ -\mu_{21} & \mu_{11}\end{bmatrix}\begin{bmatrix}\gamma_1\\ \gamma_2\end{bmatrix}\frac{1}{\delta}.$$

An example of such scaling in solving a 2*x*2 linear system can be found in LAPACK routines `_sytf2`; see, for example, the code near and including loop 30 in dsytf2.f. Of course, there is also risk of cancellation error in forming the determinant δ; this error could affect the accuracy of the computed results. We will need to examine both the problem and the algorithm to see if such errors are unnecessarily large.

Exercise 3.4.1 Determine circumstances under which the solution of a 2 × 2 linear system in Eq. (3.10) will overflow or underflow, but the algorithm in Eqs. (3.4.1) does not. Then program both algorithms for solving a 2×2 linear system and compare the results to see if your predictions were accurate.

3.4.3 Rank-One Modifications of the Identity

We will now consider a slightly more complicated matrix and find its inverse.

Lemma 3.4.1 *If* $\mathbf{u}$ *and* $\mathbf{v}$ *are m-vectors such that* $\mathbf{v}^H\mathbf{u} \neq 1$, *then*

$$\left[\mathbf{I} + \mathbf{u}\mathbf{v}^H\right]^{-1} = \mathbf{I} - \mathbf{u}\frac{1}{1 + \mathbf{v}^H\mathbf{u}}\mathbf{v}^H .$$

Proof Formally, we can multiply

$$\left[\mathbf{I} + \mathbf{u}\mathbf{v}^H\right]\left[\mathbf{I} - \mathbf{u}\frac{1}{1 + \mathbf{v}^H\mathbf{u}}\mathbf{v}^H\right] = \mathbf{I} + \mathbf{u}\mathbf{v}^H - \mathbf{u}\frac{1}{1 + \mathbf{v}^H\mathbf{u}}\mathbf{v}^H - \mathbf{u}\frac{\mathbf{v}^H\mathbf{u}}{1 + \mathbf{v}^H\mathbf{u}}\mathbf{v}^H$$

$$= \mathbf{I} + \mathbf{u}\left[1 - \frac{1 + \mathbf{v}^H\mathbf{u}}{1 + \mathbf{v}^H\mathbf{u}}\right]\mathbf{v}^H = \mathbf{I} .$$

As a result, whenever the scalar

$$\delta \equiv 1 + \mathbf{v}^H\mathbf{u}$$

is nonzero, the solution of

$$\left[\mathbf{I} + \mathbf{u}\mathbf{v}^H\right]\mathbf{x} = \mathbf{b}$$

can be computed by

$$\mathbf{x} = \mathbf{b} - \mathbf{u}\left(\mathbf{v}^H\mathbf{b}/\delta\right) .$$

This computation is fundamental to the development of Gaussian factorization in terms of **elementary multiplier matrices**, as described by Stewart [97, p. 115] or Strang [99, p. 31]. What is more, we will use these rank-one modifications of the identity to solve triangular linear systems in Sect. 3.4.4.

In general, it is possible to reduce scaling difficulties while solving systems involving rank-one modifications of the identity, by pre-processing the vectors $\mathbf{u}$ and $\mathbf{v}$ so that one of them is a unit vector.

Exercise 3.4.2 Write a program to solve $\left[\mathbf{I} + \mathbf{u}\mathbf{v}^H\right]\mathbf{x} = \mathbf{b}$ by calling LAPACK BLAS routines to perform all of the work on vectors.

Exercise 3.4.3 Develop a safer algorithm for solving $\left[\mathbf{I} + \mathbf{u}\mathbf{v}^H\right]\mathbf{x} = \mathbf{b}$, so that unnecessary overflow and underflow is avoided.

3.4.4 *Triangular*

Triangular linear systems have a special form that allows us to solve them as a series of linear equations in a single unknown. Here is their definition.

Definition 3.4.2 A **left triangular matrix L** (also called a **lower triangular matrix**) has entries λ_{ij} satisfying

$$\lambda_{ij} = 0 \text{ for all } i < j \;. \tag{3.11}$$

A **unit left triangular matrix** is a left triangular matrix with diagonal entries satisfying $\lambda_{ii} = 1$ for all i. A **right triangular matrix R** (also called an **upper triangular matrix**) has entries ϱ_{ij} satisfying

$$\varrho_{ij} = 0 \text{ for all } i > j \;.$$

A **unit right triangular matrix** is a right triangular matrix with diagonal entries satisfying $\varrho_{ii} = 1$ for all i.

3.4.4.1 Left-Triangular

Let us begin by studying a unit left triangular system $\mathbf{L}\mathbf{y} = \mathbf{b}$. Using formula (3.2) for the matrix-vector product $\mathbf{L}\mathbf{y}$, and then the left-triangular condition (3.11) and the assumption that the diagonal entries of $\mathbf{L}$ are all one, we see that for all $1 \leq i \leq m$ we have

$$\beta_i = \sum_{j=1}^{m} \lambda_{ij}\eta_j = \sum_{j=1}^{i-1} \lambda_{ij}\eta_j + \eta_i \;.$$

In the case where $i = 1$, the sum is empty and it is easy to see that the first entry of $\mathbf{y}$ is $\eta_1 = \beta_1$. It is also easy to see that η_i depends only on $\eta_1, \ldots, \eta_{i-1}$, as well as entries of $\mathbf{L}$ and $\mathbf{b}$. As a result, we can compute the entries η_i of $\mathbf{y}$ by

Algorithm 3.4.2 (Forward Linear Recurrence)

$$\text{for } 1 \leq i \leq m \;,\; \eta_i = \beta_i - \sum_{j=1}^{i-1} \lambda_{ij}\eta_j \;.$$

In this algorithm, each entry of $\mathbf{y}$ is computed by performing an inner product of a row of $\mathbf{L}$ with the previously-found entries of $\mathbf{y}$. Note that this algorithm costs

- $\sum_{i=1}^{m}(i-1) = (m-1)m/2$ multiplications, and
- $\sum_{i=1}^{m} i = m(m+1)/2$ additions or subtractions.

The total work is on the order of $m^2/2$ multiplications and the same number of additions/subtractions.

If the entries of $\mathbf{L}$ are stored columnwise, then the forward linear recurrence Algorithm 3.4.2 does not access the computer memory for $\mathbf{L}$ at stride one. Alternatively, we can use the following result.

Lemma 3.4.2 *Suppose that $\mathbf{L}$ is an $m \times m$ unit left triangular matrix with strict subdiagonal entries given by the vectors $\boldsymbol{\ell}_1, \ldots, \boldsymbol{\ell}_{m-1}$:*

$$\mathbf{L} = \mathbf{I} + [\boldsymbol{\ell}_1, \ldots, \boldsymbol{\ell}_{m-1}, \mathbf{0}] .$$

Then

$$\mathbf{L} = \left(\mathbf{I} + \boldsymbol{\ell}_1 \mathbf{e}_1{}^\top\right) \ldots \left(\mathbf{I} + \boldsymbol{\ell}_{m-1} \mathbf{e}_{m-1}{}^\top\right) . \tag{3.12}$$

Proof By multiplying each side of (3.12) times an axis vector, it is easy to show that the matrices on both sides have the same columns.

Example 3.4.1 The unit left triangular matrix

$$\mathbf{L} = \begin{bmatrix} 1 & 0 & 0 \\ 2 & 1 & 0 \\ 3 & 2 & 1 \end{bmatrix}$$

can be written as the product

$$\mathbf{L} = \begin{bmatrix} 1 & 0 & 0 \\ 2 & 1 & 0 \\ 3 & 0 & 1 \end{bmatrix} \begin{bmatrix} 1 & 0 & 0 \\ 0 & 1 & 0 \\ 0 & 2 & 1 \end{bmatrix} .$$

It follows from Lemma 3.4.2 and Lemma 3.4.1 that the solution of $\mathbf{L}\mathbf{y} = \mathbf{b}$ is given by

$$\mathbf{y} = \left(\mathbf{I} + \boldsymbol{\ell}_{m-1} \mathbf{e}_{m-1}{}^\top\right)^{-1} \ldots \left(\mathbf{I} + \boldsymbol{\ell}_1 \mathbf{e}_1{}^\top\right)^{-1} \mathbf{b} = \left(\mathbf{I} - \boldsymbol{\ell}_{m-1} \mathbf{e}_{m-1}{}^\top\right) \ldots \left(\mathbf{I} - \boldsymbol{\ell}_1 \mathbf{e}_1{}^\top\right) \mathbf{b} .$$

As a result, we can solve $\mathbf{L}\mathbf{y} = \mathbf{b}$ by the following algorithm:

Algorithm 3.4.3 (Forward Solution by Elementary Multiplier Matrices)

for $1 \leq j \leq m$, $\eta_j = \beta_j$

for $1 \leq j \leq m$

  for $j + 1 \leq i \leq m$, $\eta_i = \eta_i - \lambda_{ij} \eta_j$.

This algorithm accesses the entries of $\mathbf{L}$ columnwise, and stores the solution $\mathbf{y}$ in the vector $\mathbf{b}$. Note that this algorithm costs $\sum_{j=1}^{m}(m-j) = (m-1)m/2$ multiplications and subtractions. The total work is on the order of $m^2/2$ multiplications and the same number of subtractions. Thus Algorithm 3.4.3 costs essentially the same as the forward linear recurrence Algorithm 3.4.2.

It is easy to implement forward solution in a computer program, and even easier to call LAPACK BLAS routines to perform this algorithm. For implementations, readers can view the Fortran 77 program forwardSubF.f, the Fortran 90 program forwardSubF90.f90, the C program forwardSubC.c or the C^{++} program forwardSubCpp.C. Each of these programs calls the general LAPACK Level 2 BLAS routine `_trsv` for solving triangular systems. See, for example, dtrsv.f.

In MATLAB, the command linsolve can be instructed to solve a left triangular linear system. For example,

```
opts.LT=true; opts.TRANS=false; y=linsolve(L,b,opts);
```

will solve $\mathbf{Ly} = \mathbf{b}$ for $\mathbf{y}$ using forward substitution. Note that the MATLAB command `linsolve` *cannot be instructed to assume that the diagonal entries of the matrix are all one.*

3.4.4.2 Right-Triangular

Next, we will consider general right-triangular matrices.

Lemma 3.4.3 *Suppose that* $\mathbf{R}$ *is an* $n \times n$ *right triangular matrix with entries* ϱ_{ij}. *Assume that the diagonal entries of* $\mathbf{R}$ *are given by the diagonal matrix* $\mathbf{D}$ *and the strict superdiagonal entries are given by the vectors* $\mathbf{r}_1, \ldots, \mathbf{r}_n$:

$$\mathbf{R} = \mathbf{D} + [\mathbf{r}_1, \ldots, \mathbf{r}_n] \;.$$

If $\mathbf{r}_j$ *has entries* ϱ_{ij}, *let* $\boldsymbol{\Sigma}^{(j)}$ *be the diagonal matrix with entries*

$$\sigma_{ii}^{(j)} = \begin{cases} \varrho_{jj}, \; i = j \\ 1, \; i \neq j \end{cases}$$

Then

$$\mathbf{R} = \boldsymbol{\Sigma}^{(n)}\left(\mathbf{I} + \mathbf{r}_n \mathbf{e}_n{}^\top\right) \ldots \boldsymbol{\Sigma}^{(2)}\left(\mathbf{I} + \mathbf{r}_2 \mathbf{e}_2{}^\top\right) \boldsymbol{\Sigma}^{(1)} \;. \tag{3.13}$$

Proof The proof is simple, and involves showing that both sides of the claimed equation have the same columns.

The i-th equation in the linear system $\mathbf{Rx} = \mathbf{y}$ is

$$\varrho_{ii}\xi_i + \sum_{j=i}^{n} \varrho_{ij}\xi_j = \eta_i \;.$$

Since the equations in this right-triangular system can be solved for the unknowns in reverse order, we can compute the solution via the

Algorithm 3.4.4 (Backward Linear Recurrence)

$$\text{for } n \geq i \geq 1 \; \xi_i = \left[\eta_i - \sum_{j=i+1}^{n} \varrho_{ij}\xi_j \right] / \varrho_{ii} \; .$$

This algorithm accesses the entries of **R** row-wise. If **R** is stored columnwise, then we can use Lemma 3.4.3 to develop

Algorithm 3.4.5 (Back Substitution by Elementary Multiplier Matrices)

$$\begin{aligned} &\text{for } n \geq j \geq 1 \\ &\quad \xi_j = \eta_j / \varrho_{jj} \\ &\quad \text{for } 1 \leq i < j \; , \; \eta_i = \eta_i - \varrho_{ij}\xi_j \; . \end{aligned}$$

This algorithm destroys the right-hand side **y**, which could share computer storage with the solution **x**.

Note that a backward linear recurrence costs

- $\sum_{i=1}^{n} 1 = n$ divisions,
- $\sum_{i=1}^{n}(n-i) = (n-1)n/2$ multiplications, and
- $\sum_{i=1}^{n}(n-i+1) = n(n+1)/2$ additions or subtractions.

The total work is on the order of $n^2/2$ multiplications and additions. Back substitution by elementary multiplier matrices costs

- $\sum_{j=1}^{n} 1 = n$ divisions,
- $\sum_{j=1}^{n}(j-1) = (n-1)n/2$ multiplications and subtractions.

The total work is on the order of $n^2/2$ multiplications and subtractions, and involves essentially the same work as a backward linear recurrence. recurrence.

It is easy to implement back substitution in a computer program, and even easier to call LAPACK BLAS routines to perform back substitution. For implementations of Algorithm 3.4.5, readers can view the Fortran 77 program backSubF.f, the Fortran 90 program backSubF90.f90, the C program backSubC.c or the C^{++} program backSubCpp.C. Each of these programs calls the general LAPACK Level 2 BLAS routine `_trsv` for solving triangular systems. See, for example, dtrsv.f.

In MATLAB, the command linsolve can be instructed to solve a right triangular linear system. For example,

```
opts.UT=true; opts.TRANS=false; x=linsolve(R,y,opts);
```

will solve $\mathbf{Rx} = \mathbf{y}$ for **x** using back substitution.

Exercise 3.4.4 Program the forward linear recurrence Algorithm 3.4.2, and compare its computational speed to LAPACK Level 2 BLAS routine `dtrsv`. (See

Sect. 2.6.1 for computer architecture issues to consider in your comparison.) Does it help to call LAPACK Level 1 BLAS routine `ddot` to compute the inner product in the forward linear recurrence?

Exercise 3.4.5 Program the backward linear recurrence Algorithm 3.4.4 and compare its computational speed to LAPACK Level 2 BLAS routine `dtrsv`. (See Sect. 2.6.1 for computer architecture issues to consider in your comparison.)

3.4.5 *Trapezoidal*

Our next goal is to develop and analyze numerical methods for solving trapezoidal systems of linear equations. We will see that the mathematical solution of trapezoidal systems of linear equations is complicated by consistency conditions and the possible existence of multiple solutions.

3.4.5.1 Left Trapezoidal

In presenting numerical methods for linear systems, it is often useful to display vectors and matrices as rectangular formations of smaller vectors and matrices. The smaller vectors and matrices are formed by **partitioning** the larger between rows and/or columns. A definition of **matrix partitioning** can be found in Stewart [97, Definition 3.13, p. 27]. Since the idea is intuitive, we will avoid a formal definition. Instead, we will present some illustrative examples.

Example 3.4.2 Consider the left-trapezoidal system

$$\mathbf{L}\mathbf{y} \equiv \begin{bmatrix} 1 & 0 \\ -1 & 1 \\ -1 & -1 \end{bmatrix} \begin{bmatrix} \eta_1 \\ \eta_2 \end{bmatrix} = \begin{bmatrix} 1 \\ 2 \\ 3 \end{bmatrix} \equiv \mathbf{b} \; .$$

Here, it is natural to partition $\mathbf{L}$ into a 2×2 triangular array $\mathbf{L}_B$ on top, and a 1×2 array $\mathbf{L}_N$ on bottom:

$$\mathbf{L} = \left[\begin{array}{cc} 1 & 0 \\ -1 & 1 \\ \hline -1 & -1 \end{array}\right] = \left[\begin{array}{c} \mathbf{L}_B \\ \hline \mathbf{L}_N \end{array}\right] .$$

We partition $\mathbf{b}$ accordingly:

$$\mathbf{b} = \left[\begin{array}{c} 1 \\ 2 \\ \hline 3 \end{array}\right] = \left[\begin{array}{c} \mathbf{b}_B \\ \hline \mathbf{b}_N \end{array}\right] .$$

If $\mathbf{y}$ solves $\mathbf{L}\mathbf{y} = \mathbf{b}$, then $\mathbf{y}$ is determined by solving the **binding equations** $\mathbf{L}_B\mathbf{y} = \mathbf{b}_B$, or

$$\begin{bmatrix} 1 & 0 \\ -1 & 1 \end{bmatrix} \begin{bmatrix} \eta_1 \\ \eta_2 \end{bmatrix} = \begin{bmatrix} 1 \\ 2 \end{bmatrix} \Longrightarrow \mathbf{y} = \begin{bmatrix} \eta_1 \\ \eta_2 \end{bmatrix} = \begin{bmatrix} 1 \\ 3 \end{bmatrix} .$$

Then we see that $\mathbf{L}\mathbf{y} = \mathbf{b}$ has a solution if and only if $\mathbf{b}$ satisfies the **non-binding equation** $\mathbf{b}_N = \mathbf{L}_N\mathbf{y}$, or

$$\beta_3 = [-1 \,,\; -1] \begin{bmatrix} 1 \\ 3 \end{bmatrix} = -4 \,.$$

In general, if $\mathbf{L}$ is an $m \times r$ left trapezoidal matrix and $\mathbf{b}$ is an m-vector, we can solve the linear system $\mathbf{L}\mathbf{y} = \mathbf{b}$ by performing the following steps:

1. Partition

$$\mathbf{L} = \begin{bmatrix} \mathbf{L}_B \\ \mathbf{L}_N \end{bmatrix} \text{ and } \mathbf{b} = \begin{bmatrix} \mathbf{b}_B \\ \mathbf{b}_N \end{bmatrix}$$

   where $\mathbf{L}_B$ is an $r \times r$ left triangular matrix and $\mathbf{b}_B$ is an r-vector;
2. Forward-solve $\mathbf{L}_B\mathbf{y} = \mathbf{b}_B$, using either the Forward Linear Recurrence Algorithm 3.4.2 or the Forward Solution by Elementary Multiplier Matrices Algorithm 3.4.3.
3. Check the **consistency conditions** $\mathbf{L}_N\mathbf{y} = \mathbf{b}_N$ to guarantee that the full linear system $\mathbf{L}\mathbf{y} = \mathbf{b}$ has a solution.

3.4.5.2 Right Trapezoidal

Next, we will consider right-trapezoidal systems of linear equations.

Example 3.4.3 Consider the right-trapezoidal system

$$\mathbf{R}\mathbf{x} \equiv \begin{bmatrix} 1 & 1 & 1 & 1 \\ 0 & 1 & 2 & 3 \end{bmatrix} \begin{bmatrix} \xi_1 \\ \xi_2 \\ \xi_3 \\ \xi_4 \end{bmatrix} = \begin{bmatrix} 4 \\ 6 \end{bmatrix} \equiv \mathbf{y} \,.$$

It is natural to partition $\mathbf{R}$ into a 2×2 right-triangular array $\mathbf{R}_B$, and the remainder $\mathbf{R}_N$:

$$\mathbf{R} = \left[\begin{array}{cc|cc} 1 & 1 & 1 & 1 \\ 0 & 1 & 2 & 3 \end{array}\right] = \left[\, \mathbf{R}_B \,\middle|\, \mathbf{R}_N \,\right] .$$

It is also natural to partition $\mathbf{x}$ into the corresponding **basic variables** $\mathbf{x}_B$ and **non-basic variables** $\mathbf{x}_N$:

$$\mathbf{x} = \left[\begin{array}{c} \xi_1 \\ \xi_2 \\ \hline \xi_3 \\ \xi_4 \end{array}\right] = \left[\begin{array}{c} \mathbf{x}_B \\ \hline \mathbf{x}_N \end{array}\right] .$$

To find the general solution to the right-trapezoidal system, first note that

$$\mathbf{R}\mathbf{x} = [\mathbf{R}_B \mid \mathbf{R}_N] \left[\begin{array}{c} \mathbf{x}_B \\ \hline \mathbf{x}_N \end{array}\right] = \mathbf{R}_B\mathbf{x}_B + \mathbf{R}_N\mathbf{x}_N = \mathbf{y} .$$

Thus, given any values for the non-basic variables $\mathbf{x}_N$ we can back-solve in the linear system

$$\mathbf{R}_B\mathbf{x}_B = \mathbf{y} - \mathbf{R}_N\mathbf{x}_N = \begin{bmatrix} 4 \\ 6 \end{bmatrix} - \begin{bmatrix} 1 & 1 \\ 2 & 3 \end{bmatrix} \begin{bmatrix} \xi_3 \\ \xi_4 \end{bmatrix} = \begin{bmatrix} 4 - \xi_3 - \xi_4 \\ 6 - 2\xi_3 - 3\xi_4 \end{bmatrix}$$

to obtain values for the basic variables $\mathbf{x}_B$:

$$\mathbf{x}_B = \begin{bmatrix} -2 + \xi_3 + 2\xi_4 \\ 6 - 2\xi_3 - 3\xi_4 \end{bmatrix} = \begin{bmatrix} -2 \\ 6 \end{bmatrix} + \begin{bmatrix} 1 & 2 \\ -2 & -3 \end{bmatrix} \begin{bmatrix} \xi_3 \\ \xi_4 \end{bmatrix} .$$

In general if $\mathbf{R}$ is an $r \times n$ right trapezoidal matrix and $\mathbf{y}$ is an r-vector, we can solve the linear system $\mathbf{R}\mathbf{x} = \mathbf{y}$ by performing the following steps:

1. Partition

$$\mathbf{R} = \begin{bmatrix} \mathbf{R}_B & \mathbf{R}_N \end{bmatrix} \text{ and } \mathbf{x} = \begin{bmatrix} \mathbf{x}_B \\ \mathbf{x}_N \end{bmatrix}$$

   where $\mathbf{R}_B$ is an $r \times r$ left triangular matrix and $\mathbf{x}_B$ is an r-vector;
2. Choose values for the non-basic variables in $\mathbf{x}_N$;
3. Back-solve $\mathbf{R}_B\mathbf{x}_B = \mathbf{y} - \mathbf{R}_N\mathbf{x}_N$, using either the Backward Linear Recurrence Algorithm 3.4.4 or the Back Solution by Elementary Multiplier Matrices Algorithm 3.4.5.

3.4.5.3 Partitioned Matrix Products

The reader should note that we multiply partitioned matrices in a natural way. In general, we have

$$\mathbf{AB} = \left[\begin{array}{c|c} \mathbf{A}_{BB} & \mathbf{A}_{BN} \\ \hline \mathbf{A}_{NB} & \mathbf{A}_{NN} \end{array}\right] \left[\begin{array}{c|c} \mathbf{B}_{BB} & \mathbf{B}_{BN} \\ \hline \mathbf{B}_{NB} & \mathbf{B}_{NN} \end{array}\right]$$
$$= \left[\begin{array}{c|c} \mathbf{A}_{BB}\mathbf{B}_{BB} + \mathbf{A}_{BN}\mathbf{B}_{NB} & \mathbf{A}_{BB}\mathbf{B}_{BN} + \mathbf{A}_{BN}\mathbf{B}_{NN} \\ \hline \mathbf{A}_{NB}\mathbf{B}_{BB} + \mathbf{A}_{NN}\mathbf{B}_{NB} & \mathbf{A}_{NB}\mathbf{B}_{BN} + \mathbf{A}_{BN}\mathbf{B}_{NN} \end{array}\right] .$$

Exercise 3.4.6 Given a real scalar α, a real $(m-1)$-vector $\mathbf{b}$, a real $(n-1)$-vector $\mathbf{a}$ and a real $(m-1) \times (n-1)$ matrix $\mathbf{B}$, solve the following equation for the real scalar ϱ, real $(m-1)$-vector $\boldsymbol{\ell}$, real $(n-1)$-vector $\mathbf{r}$ and real $(m-1) \times (n-1)$ matrix $\mathbf{L}$:

$$\begin{bmatrix} \alpha & \mathbf{a}^\top \\ \mathbf{b} & \mathbf{B} \end{bmatrix} = \begin{bmatrix} 1 & 0 \\ \boldsymbol{\ell} & \mathbf{L} \end{bmatrix} \begin{bmatrix} \varrho & \mathbf{r}^\top \\ \mathbf{0} & \mathbf{I} \end{bmatrix} .$$

Exercise 3.4.7 Given a real scalar α, a real $(n-1)$-vector $\mathbf{a}$ and a real $(n-1)\times(n-1)$ matrix $\mathbf{A}$, solve the following equation for the real scalar λ, real $(n-1)$-vector $\boldsymbol{\ell}$ and real $(n-1) \times (n-1)$ matrix $\mathbf{L}$:

$$\begin{bmatrix} \alpha & \mathbf{a}^\top \\ \mathbf{a} & \mathbf{A} \end{bmatrix} = \begin{bmatrix} \lambda & 0 \\ \boldsymbol{\ell} & \mathbf{B} \end{bmatrix} \begin{bmatrix} \lambda & \boldsymbol{\ell}^\top \\ \mathbf{0} & \mathbf{I} \end{bmatrix} .$$

Exercise 3.4.8 Given a real m-vector $\mathbf{a}$ and a real $m \times (n-1)$ matrix $\mathbf{A}$, solve the following equations for the m-vector $\mathbf{q}$, $(n-1)$-vector $\mathbf{r}$ and $m \times (n-1)$ matrix $\mathbf{B}$:

$$\begin{bmatrix} \mathbf{a} & \mathbf{A} \end{bmatrix} = \begin{bmatrix} \mathbf{q} & \mathbf{B} \end{bmatrix} \begin{bmatrix} 1 & \mathbf{r}^\top \\ \mathbf{0} & \mathbf{I} \end{bmatrix} \text{ and } \mathbf{q}^\top \mathbf{B} = 0 .$$

Exercise 3.4.9 Suppose that $\mathbf{A}$ is a symmetric $n \times n$ matrix and $\mathbf{x}$ is an n-vector. It is often convenient to store only the entries on and below the diagonal of $\mathbf{A}$. In order to use these entries to compute $\mathbf{Ax}$, let us partition

$$\mathbf{A} = \begin{bmatrix} \mathbf{A}_1 & \mathbf{a}_1 & \mathbf{A}_2{}^\top \\ \mathbf{a}_1{}^\top & \alpha & \mathbf{a}_2{}^\top \\ \mathbf{A}_2 & \mathbf{a}_2 & \mathbf{A}_3 \end{bmatrix} \text{ and } \mathbf{x} = \begin{bmatrix} \mathbf{x}_1 \\ \xi \\ \mathbf{x}_2 \end{bmatrix} .$$

1. Multiply the partitioned forms for $\mathbf{A}$ and $\mathbf{x}$ together to see how the product $\mathbf{Ax}$ depends on the entries in the partitioned column of $\mathbf{A}$ (i.e., α and $\mathbf{a}_2$).

2. Use this partitioned product to develop an algorithm for computing $\mathbf{Ax}$ by looping over columns of $\mathbf{A}$ from first to last. (See LAPACK Level 2 BLAS subroutine `dsymv` for an implementation of this algorithm).

Exercise 3.4.10 Suppose that $\mathbf{A}$ and $\mathbf{B}$ are symmetric $n \times n$ matrices, whose entries are stored only on and below the diagonal. Partition $\mathbf{A}$ and $\mathbf{B}$ as in the previous exercise, and find the corresponding partitioning of the product $\mathbf{AB}$. Use this partitioned product to develop an algorithm for computing $\mathbf{AB}$ by looping over columns of both matrices.

Exercise 3.4.11 If $\mathbf{L}$ is an $m \times r$ unit left-trapezoidal matrix, find bases for $\mathscr{R}(\mathbf{L})$ and $\mathscr{N}(\mathbf{L})$. Verify the second Sylvester identity (3.5b) for this matrix.

Exercise 3.4.12 If $\mathbf{R}$ is an $r \times n$ right-trapezoidal matrix, find bases for $\mathscr{R}(\mathbf{R})$ and $\mathscr{N}(\mathbf{R})$. Verify the second Sylvester identity (3.5b) for this matrix. Also, use the ideas from the previous exercise to verify the first Sylvester identity for $\mathbf{R}$.

3.4.6 Quasi-Triangular

Occasionally, we will find it convenient to work with the following generalization of right (or upper) triangular matrices. Here is the idea.

Definition 3.4.3 A **right (or upper) quasi-triangular matrix R** has entries ϱ_{ij} satisfying

$\mathbf{R}$ is almost right-triangular: if $i > j+1$ then $\varrho_{ij} = 0$,
diagonal blocks are either 1×1 or 2×2: $\varrho_{j-1,j} \neq 0$ implies that $\varrho_{j-2,j-1} = 0$ and $\varrho_{j,j+1} = 0$,
1×1 diagonal blocks are invertible: if $\varrho_{j-1,j} = 0 = \varrho_{j+1,j}$ then $\varrho_{jj} \neq 0$, and
2×2 diagonal blocks are invertible: if $\varrho_{j,j-1} \neq 0$ then $\varrho_{j-1,j-1}\varrho_{j,j} - \varrho_{j-1,j}\varrho_{j,j-1} \neq 0$.

Right quasi-triangular matrices arise in a Schur decomposition of real matrices (see Sect. 1.4.1.3 of Chap. 1 in Volume II), and quasi-diagonal matrices arise in the solution of linear systems with symmetric matrices that are not necessarily positive(see Sect. 3.13.2).

Example 3.4.4 The matrix

$$\mathbf{R} = \begin{bmatrix} 1 & 3 & 4 & 5 \\ 0 & 2 & 3 & 4 \\ 0 & 1 & 2 & 3 \\ 0 & 0 & 0 & 2 \end{bmatrix}$$

is right quasi-triangular.

It is easy to verify that the following algorithm will solve right quasi-triangular systems $\mathbf{Rx} = \mathbf{y}$ involving n equations with row-wise storage of $\mathbf{R}$:

Algorithm 3.4.6 (Solve Right Quasi-Triangular System)

$i = n$

while $i \geq 1$

if $i > 1$ and $\varrho_{i,i-1} = 0$

$$\xi_i = \left[\eta_i - \sum_{j=i+1}^{n} \varrho_{ij}\xi_j \right] / \varrho_{ii}$$

$i = i - 1$

else

$$\text{solve} \begin{bmatrix} \varrho_{i-1,i-1} & \varrho_{i-1,i} \\ \varrho_{i,i-1} & \varrho_{i,i} \end{bmatrix} \begin{bmatrix} \xi_{i-1} \\ \xi_i \end{bmatrix} = \begin{bmatrix} \eta_{i-1} - \sum_{j=i+1}^{n} \varrho_{i-1,j}\xi_j \\ \eta_i - \sum_{j=i+1}^{n} \varrho_{i,j}\xi_j \end{bmatrix}$$

$i = i - 2$

The 2×2 linear systems can be solved as in Algorithm 3.4.1. On the other hand, if $\mathbf{R}$ is stored columnwise, then the following algorithm is preferable.

Algorithm 3.4.7 (Solve Right Quasi-Triangular System Columnwise)

$j = n$

while $j \geq 1$

if $j > 1$ and $\varrho_{j,j-1} = 0$

$\xi_j = \eta_j / \varrho_{j,j}$

for $1 \leq i < j$, $\eta_i = \eta_i - \varrho_{ij}\xi_j$

$j = j - 1$

else

$$\text{solve} \begin{bmatrix} \varrho_{j-1,j-1} & \varrho_{j-1,j} \\ \varrho_{j,j-1} & \varrho_{j,j} \end{bmatrix} \begin{bmatrix} \xi_{j-1} \\ \xi_j \end{bmatrix} = \begin{bmatrix} \eta_{j-1} \\ \eta_j \end{bmatrix}$$

for $1 \leq i < j - 1$, $\eta_i = \eta_i - \varrho_{i,j-1}\xi_{j-1} - \varrho_{i,j}\xi_j$

$j = j - 2$.

We note that for quasi-diagonal matrices arising in the solution of symmetric indefinite linear systems, the 2×2 diagonal blocks will always have negative determinant. For quasi-triangular matrices arising in the real Schur decomposition, the 2×2 diagonal blocks will always have $\varrho_{j-1,j}\varrho_{j,j-1} < 0$.

In LAPACK, quasi-triangular linear systems can be solved using routines `_sytrs`. See, for example, dsytrs.f. The MATLAB command linsolve does not have an option to solve a quasi-triangular system of linear equations.

3.4.7 Permutations

Whenever it is necessary to reorder entries in an array, the following concept is useful.

Definition 3.4.4 The columns of an $n \times n$ **permutation matrix** are distinct axis vectors in some specified order.
It follows that for every $n \times n$ permutation matrix $\mathbf{P}$ there is a permutation π of the integers $1, \ldots, n$ so that $\mathbf{P} = \left[\mathbf{e}_{\pi(1)}, \ldots, \mathbf{e}_{\pi(n)}\right]$.

Here are some important results involving permutation matrices.

Lemma 3.4.4 *Suppose that $\mathbf{P} = \left[\mathbf{e}_{\pi(1)}, \ldots, \mathbf{e}_{\pi(n)}\right]$ is an $n \times n$ permutation matrix.*

1. *If $\mathbf{A} = [\mathbf{a}_1, \ldots, \mathbf{a}_n]$ is an $m \times n$ matrix, then $\mathbf{AP} = \left[\mathbf{a}_{\pi(1)}, \ldots, \mathbf{a}_{\pi(n)}\right]$.*
2. $\mathbf{P}^\top \mathbf{P} = \mathbf{I}$.
3. *If*

$$\mathbf{B} = \begin{bmatrix} {\mathbf{b}_1}^\top \\ \vdots \\ {\mathbf{b}_n}^\top \end{bmatrix}$$

is an $n \times m$ matrix, then

$$\mathbf{P}^\top \mathbf{B} = \begin{bmatrix} {\mathbf{b}_{\pi(1)}}^\top \\ \vdots \\ {\mathbf{b}_{\pi(n)}}^\top \end{bmatrix} .$$

4. *$\mathbf{P}^\top$ is a permutation matrix.*
5. $\mathbf{P}\mathbf{P}^\top = \mathbf{I}$.
6. *The product of two permutation matrices is a permutation matrix.*

Proof Since $\mathbf{APe}_j = \mathbf{Ae}_{\pi(j)}$, the first claim is trivial.

Let us prove the second claim. Since the permutation π is a one-to-one mapping of the integers, the i,j entry of $\mathbf{P}^\top \mathbf{P}$ is

$${\mathbf{e}_i}^\top \mathbf{P}^\top \mathbf{P}\mathbf{e}_j = [\mathbf{Pe}_i]\left[\mathbf{Pe}_j\right] = \mathbf{e}_{\pi(i)}\mathbf{e}_{\pi(j)} = \delta_{ij} \; .$$

The third claim follows from the first claim, because Lemma 3.2.2: shows that $(\mathbf{P}^\top \mathbf{B})^\top = \mathbf{B}^\top \mathbf{P}$.

Next, we will prove the fourth claim. For any axis vector $\mathbf{e}_{\pi(j)}$,

$$\mathbf{P}^\top \mathbf{e}_{\pi(j)} = \mathbf{P}^\top \mathbf{P} \mathbf{e}_j = \mathbf{I} \mathbf{e}_j = \mathbf{e}_j \ .$$

This shows that the columns of $\mathbf{P}^\top$ are distinct axis vectors, and that $\mathbf{P}^\top$ is associated with the inverse permutation π^{-1}.

The fifth claim follows directly from the second claim and Lemma 3.2.13.

Only the last claim remains to be proved. Suppose that $\mathbf{P}$ and $\mathbf{Q}$ are two $n \times n$ permutation matrices, associated with the permutations π and ω, respectively. Then the j-th column of $\mathbf{PQ}$ is

$$\mathbf{PQ} \mathbf{e}_j = \mathbf{P} \mathbf{e}_{\omega(j)} = \mathbf{e}_{\pi(\omega(j))} \ ,$$

which is an axis n-vector, and distinct from the other columns of $\mathbf{PQ}$.

Lemma 3.4.4 makes it easy to solve linear systems involving a permutation matrix. The solution of $\mathbf{Px} = \mathbf{b}$ is $\mathbf{x} = \mathbf{P}^\top \mathbf{b}$, and the i-th entry of the solution $\mathbf{x}$ is $\xi_i = \beta_{\pi(i)}$.

In Sect. 3.2.12 we recalled the fact that every permutation is a product of transpositions. In terms of matrices, the corresponding statement is that every permutation matrix is a product of interchange matrices, which we define next.

Definition 3.4.5 An **interchange matrix** is a permutation matrix with just two columns in a different order from those in the identity matrix. In other words, if $\mathbf{I}_{ij}$ interchanges distinct entries i and j, then

$$\mathbf{I}_{ij} \mathbf{e}_k = \begin{cases} \mathbf{e}_k, \ k \neq i,j \\ \mathbf{e}_i, \ \ k = j \\ \mathbf{e}_j, \ \ k = i \end{cases}$$

The corresponding permutation is called a **transposition**.

Note that if $\mathbf{I}_{ij}$ is an interchange matrix, then it is symmetric and is equal to its inverse: $\mathbf{I}_{ij}{}^\top = \mathbf{I}_{ij}$ and $\mathbf{I}_{ij} \mathbf{I}_{ij} = \mathbf{I}$.

In LAPACK, permutation matrices are generally represented by a vector of integers, rather than by a matrix of scalars. See, for example, the vector `ipiv` in dgetrf.f. If the ith entry of this vector has value j, this indicates that entries i and j should be interchanged. The direction of the loop over i depends on the algorithm that generated the permutation.

In MATLAB, permutation matrices are stored as full matrices of scalars. If the permutation matrix is generated by a MATLAB matrix factorization, such as the command lu, then the MATLAB permutation matrix is *not sparse*.

Exercise 3.4.13 If I_{ij} is an interchange matrix with $i \neq j$, show that $I_{ij} = \mathbf{I} - (\mathbf{e}_i - \mathbf{e}_j)(\mathbf{e}_i - \mathbf{e}_j)^\top$.

Exercise 3.4.14 Show that a product of interchange matrices is a permutation matrix.

Exercise 3.4.15 Show that every permutation matrix is a product of interchange matrices.

3.4.8 *Orthogonal*

3.4.8.1 Orthogonal Projection

Recall that Definition 3.2.14 described an orthogonal matrix $\mathbf{Q}$ by the equation $\mathbf{Q}^H\mathbf{Q} = \mathbf{I}$. As a result, the solution of the $n \times n$ linear system $\mathbf{Q}\mathbf{y} = \mathbf{b}$ can be computed by $\mathbf{y} = \mathbf{Q}^H\mathbf{b}$. This formula leads to the

Algorithm 3.4.8 (Orthogonal Projection)

$$\text{for } 1 \leq i \leq n \quad \eta_i = \mathbf{q}_i{}^H\mathbf{b}$$

However, there are often good numerical reasons for avoiding this algorithm.

In the Gram-Schmidt orthogonalization process (see Sect. 6.8), the columns of an orthogonal matrix $\mathbf{Q}$ are computed by successive orthogonal projection from some given set of vectors. The projections can become small relative to the original vectors, and may suffer from significant cancellation errors. This means that the columns of a computed orthogonal matrix may not be very orthogonal at all.

An alternative approach is to preserve the successive orthogonal projection process in the solution of linear systems. Suppose that the columns of $\mathbf{Q}$ and entries of $\mathbf{y}$ are given by

$$\mathbf{Q} = [\mathbf{q}_1, \ldots, \mathbf{q}_n] \text{ and } \mathbf{y} = \begin{bmatrix} \eta_1 \\ \vdots \\ \eta_n \end{bmatrix},$$

and assume that $\mathbf{Q}\mathbf{y} = \mathbf{b}$. Then the first unknown is

$$\eta_1 = \mathbf{q}_1{}^H\mathbf{b} .$$

Since the remaining columns of $\mathbf{Q}$ are supposed to be orthogonal to $\mathbf{q}_1$, we can apply the orthogonal projection $\mathbf{I} - \mathbf{q}_1\mathbf{q}_1{}^H$ to the equation $\mathbf{Q}\mathbf{y} = \mathbf{b}$ to get

$$\left[\mathbf{I} - \mathbf{q}_1\mathbf{q}_1{}^H\right][\mathbf{q}_1, \mathbf{q}_2, \ldots, \mathbf{q}_n]\,\mathbf{y} = [0, \mathbf{q}_2, \ldots, \mathbf{q}_n]\,\mathbf{y} = \left[\mathbf{I} - \mathbf{q}_1\mathbf{q}_1{}^H\right]\mathbf{b} = \mathbf{b} - \mathbf{q}_1\eta_1 .$$

As a result, we have a smaller system of linear equations for the remaining unknowns $\eta_2, \ldots, \eta_n$. This process can be continued to produce the following

Algorithm 3.4.9 (Successive Orthogonal Projection)

$$\text{for } 1 \leq i \leq n$$
$$\eta_i = \mathbf{q}_i{}^H \mathbf{b}$$
$$\mathbf{b} = \mathbf{b} - \mathbf{q}_i \eta_i \ .$$

Note that this Successive Orthogonal Projection algorithm destroys the original right-hand side **b**.

3.4.8.2 Elementary Rotations

In other situations, we may not form the orthogonal matrix directly. Instead, we may represent it as a product of elementary rotations or elementary reflections.

Definition 3.4.6 A 2×2 **elementary rotation G** has the form

$$\mathbf{G} = \begin{bmatrix} \gamma & \sigma \\ -\overline{\sigma} & \overline{\gamma} \end{bmatrix}$$

where $|\gamma|^2 + |\sigma|^2 = 1$. For every $n \times n$ **elementary rotation Q** there exists an $n \times n$ permutation matrix **P** and a 2×2 elementary rotation **G** so that we can partition

$$\mathbf{P}^\top \mathbf{Q} \mathbf{P} = \begin{bmatrix} \mathbf{G} & \mathbf{0} \\ \mathbf{0} & \mathbf{I} \end{bmatrix} .$$

It should be easy to see that a real elementary rotation with $\gamma = \cos\theta$ and $\sigma = \sin\theta$ corresponds to rotation clockwise by angle θ.

Example 3.4.5 If the scalars γ and σ satisfy $|\gamma|^2 + |\sigma|^2 = 1$, then the matrix

$$\mathbf{Q} = \begin{bmatrix} 1 & & & & \\ & \gamma & & \sigma & \\ & & 1 & & \\ & -\overline{\sigma} & & \overline{\gamma} & \\ & & & & 1 \end{bmatrix}$$

is an elementary rotation.

The following result is easy to prove.

Lemma 3.4.5 *Every elementary rotation is orthogonal.*

Proof First, let us check 2×2 elementary rotations:

$$\mathbf{G}^H\mathbf{G} = \begin{bmatrix} \overline{\gamma} & -\sigma \\ \overline{\sigma} & \gamma \end{bmatrix} \begin{bmatrix} \gamma & \sigma \\ -\overline{\sigma} & \overline{\gamma} \end{bmatrix} = \begin{bmatrix} |\gamma|^2 + |\sigma|^2 & \overline{\gamma}\sigma - \sigma\overline{\gamma} \\ \overline{\sigma}\gamma - \gamma\overline{\sigma} & |\sigma|^2 + |\gamma|^2 \end{bmatrix} = \begin{bmatrix} 1 & 0 \\ 0 & 1 \end{bmatrix} .$$

Let $\mathbf{P}$ be a permutation that orders first the rows in which the entries of $\mathbf{G}$ appear. Then we use Lemma 3.4.4 to verify the claim for general $n \times n$ elementary rotations:

$$\mathbf{Q}^H\mathbf{Q} = \mathbf{P}\begin{bmatrix} \mathbf{G}^H & \mathbf{0} \\ \mathbf{0} & \mathbf{I} \end{bmatrix}\mathbf{P}^\top\mathbf{P}\begin{bmatrix} \mathbf{G} & \mathbf{0} \\ \mathbf{0} & \mathbf{I} \end{bmatrix}\mathbf{P}^\top = \mathbf{P}\begin{bmatrix} \mathbf{G}^H & \mathbf{0} \\ \mathbf{0} & \mathbf{I} \end{bmatrix}\begin{bmatrix} \mathbf{G} & \mathbf{0} \\ \mathbf{0} & \mathbf{I} \end{bmatrix}\mathbf{P}^\top$$
$$= \mathbf{P}\begin{bmatrix} \mathbf{G}^H\mathbf{G} & \mathbf{0} \\ \mathbf{0} & \mathbf{I} \end{bmatrix}\mathbf{P}^\top = \mathbf{P}\begin{bmatrix} \mathbf{I} & \mathbf{0} \\ \mathbf{0} & \mathbf{I} \end{bmatrix}\mathbf{P}^\top = \mathbf{P}\mathbf{P}^\top = \mathbf{I} .$$

Since elementary rotations are orthogonal, it is easy to solve linear systems involving them. For example, if $\mathbf{Q}$ is an elementary rotation and we want to solve $\mathbf{Q}\mathbf{y} = \mathbf{b}$, we can choose a permutation matrix $\mathbf{P}$ to rewrite the equation in the form

$$\begin{bmatrix} \mathbf{G} & \mathbf{0} \\ \mathbf{0} & \mathbf{I} \end{bmatrix}\mathbf{P}^\top\mathbf{y} = \mathbf{P}^\top\mathbf{b} .$$

If $\mathbf{P}$ reorders entries k and ℓ first and second, respectively, then the first two equations in this linear system are

$$\begin{bmatrix} \gamma & \sigma \\ -\overline{\sigma} & \overline{\gamma} \end{bmatrix}\begin{bmatrix} \eta_k \\ \eta_\ell \end{bmatrix} = \begin{bmatrix} \beta_k \\ \beta_\ell \end{bmatrix} .$$

Since 2×2 elementary rotations are orthogonal, we compute

$$\begin{bmatrix} \eta_k \\ \eta_\ell \end{bmatrix} = \begin{bmatrix} \gamma & -\sigma \\ \sigma & \gamma \end{bmatrix}\begin{bmatrix} \beta_k \\ \beta_\ell \end{bmatrix} .$$

The other entries of $\mathbf{y}$ are equal to the corresponding entries of $\mathbf{b}$.

The LAPACK Level 1 BLAS routines `_rot` implement multiplication of an real elementary rotation times a real $2 \times n$ matrix. See, for example, drot.f. Also, LAPACK Level 1 BLAS routines `csrot` and `zdrot` multiply real elementary rotations times complex $2 \times n$ matrices. Note that the BLAS routines `_rot` have been superseded by LAPACK routines `_lartv`. See, for example, dlartv.f.

3.4.8.3 Elementary Reflectors

The next definition describes another common orthogonal matrix.

Definition 3.4.7 If $\mathbf{v}$ is a nonzero m-vector and $\tau = 2/\mathbf{v}^H\mathbf{v}$, then $\mathbf{H} = \mathbf{I} - \mathbf{v}\tau\mathbf{v}^H$ is an **elementary reflector**.

The following lemma explains why we call $\mathbf{H}$ a reflector, and shows that $\mathbf{H}$ is orthogonal.

Lemma 3.4.6 *If* $\mathbf{H} = \mathbf{I} - \mathbf{v}\tau\mathbf{v}^H$ *is an elementary reflector, then*

- *for all* $\mathbf{x} \perp \mathbf{v}$, $\mathbf{H}\mathbf{x} = \mathbf{x}$;
- $\mathbf{H}\mathbf{v} = -\mathbf{v}$;
- $\mathbf{H}^H = \mathbf{H}$;
- $\mathbf{H}\mathbf{H} = \mathbf{I}$.

Proof Let us prove the first claim. If $\mathbf{x} \perp \mathbf{v}$, then $\mathbf{v}^H\mathbf{x} = 0$ and

$$\mathbf{H}\mathbf{x} = \left[\mathbf{I} - \mathbf{v}\tau\mathbf{v}^H\right]\mathbf{x} = \mathbf{x} - \mathbf{v}\tau\mathbf{v}^H\mathbf{x} = \mathbf{x} \,.$$

The second claim is also easy. We can also compute

$$\mathbf{H}\mathbf{v} = \left[\mathbf{I} - \mathbf{v}\tau\mathbf{v}^H\right]\mathbf{v} = \mathbf{v} - \mathbf{v}\frac{2}{\mathbf{v}^H\mathbf{v}}\mathbf{v}^H\mathbf{v} = \mathbf{v} - \mathbf{v}2 = -\mathbf{v} \,.$$

Thus $\mathbf{H}$ preserves all vectors orthogonal to $\mathbf{v}$, and reflects all vectors in the direction of $\mathbf{v}$. In other words, $\mathbf{H}$ is the reflection about the $(m-1)$-dimensional subspace orthogonal to $\mathbf{v}$.

It is easy to see from the formula for $\mathbf{H}$ that it is Hermitian, proving the third claim.

To see that $\mathbf{H}$ is orthogonal, we compute

$$\begin{aligned}\mathbf{H}\mathbf{H} &= \left[\mathbf{I} - \mathbf{v}\tau\mathbf{v}^H\right]\left[\mathbf{I} - \mathbf{v}\tau\mathbf{v}^H\right] = \mathbf{I} - \mathbf{v}2\tau\mathbf{v}^H + \mathbf{v}\tau\mathbf{v}^H\mathbf{v}\tau\mathbf{v}^H \\ &= \mathbf{I} - \mathbf{v}2\tau\mathbf{v}^H + \mathbf{v}\frac{2}{\mathbf{v}^H\mathbf{v}}\mathbf{v}^H\mathbf{v}\tau\mathbf{v}^H = \mathbf{I} \,.\end{aligned}$$

Lemma 3.4.6 shows us how to solve linear systems involving an elementary reflector. If $\mathbf{H}\mathbf{y} = \mathbf{b}$, then $\mathbf{H}\mathbf{b} = \mathbf{H}^2\mathbf{y} = \mathbf{y}$. To compute $\mathbf{H}\mathbf{b}$, we can perform the following

Algorithm 3.4.10 (Apply Elementary Reflector)

$$\begin{aligned}\alpha &= \tau\left(\mathbf{v}^H\mathbf{b}\right) \\ \mathbf{y} &= \mathbf{b} - \mathbf{v}\alpha \,.\end{aligned}$$

The LAPACK routines `_larz` and `_larf` implement multiplication of an elementary reflector times an $m \times n$ matrix. See, for example, dlarz.f or dlarf.f.

MATLAB does not provide commands to compute or apply an elementary reflector. Instead, use the MATLAB commands

```
alpha = tau * ( v' * b ); y = b - v * alpha
```

Exercise 3.4.16 Implement the Successive Orthogonal Projection Algorithm 3.4.9, using loops with no subroutine calls. Then reprogram the algorithm to make calls to LAPACK BLAS routines `_dot` and `_axpy`. Which version is faster?

3.5 Norms

We have made some progress in pursuing the steps in scientific computing for systems of linear equations. We have formulated the mathematical model. In Sect. 3.2, we determined conditions under which a system of linear equations has a solution and the solution is unique. Section 3.4 presented some simple linear systems and described their solution. Our next goal is to measure the sensitivity of the solution of a linear system to perturbations in the data. In this section, we will develop ways to measure vectors and matrices.

The general idea is contained in the following definition.

Definition 3.5.1 ([97, p. 163]) A **norm** on a linear space $\mathscr{X}$ is a function $\| \cdot \|$ mapping members of X to nonnegative real numbers and satisfying the following hypotheses:

definiteness: for all $\mathbf{x} \in X$, $\mathbf{x} \neq \mathbf{0}$ implies $\|\mathbf{x}\| > 0$,
homogeneity: for all $\mathbf{x} \in X$ and all scalars α, $\|\mathbf{x}\alpha\| = \|\mathbf{x}\| \, |\alpha|$, and
triangle inequality: for all $\mathbf{x}_1, \mathbf{x}_2 \in X$, $\|\mathbf{x}_1 + \mathbf{x}_2\| \leq \|\mathbf{x}_1\| + \|\mathbf{x}_2\|$.

We will provide specific examples of norms in the sections that follow.

3.5.1 *Vector Norms*

In scientific computing, there are three commonly-used vector norms.

Definition 3.5.2 If $\mathbf{x}$ is an n-vector with entries ξ_i, then its **infinity-norm**, **one-norm** and **two-norm** are

$$\|\mathbf{x}\|_\infty = \max_{1 \leq i \leq n} |\xi_i| \ , \tag{3.14a}$$

$$\|\mathbf{x}\|_1 = \sum_{i=1}^{n} |\xi_i| \ , \tag{3.14b}$$

$$\|\mathbf{x}\|_2 = \left[\sum_{i=1}^{n} |\xi_i|^2 \right]^{1/2} = \sqrt{\mathbf{x} \cdot \mathbf{x}} \ . \tag{3.14c}$$

The vector two norm is often called the **Euclidean norm**, and the infinity norm is often called the **max norm**.

The following lemma will justify our use of the word "norm" in Definition 3.5.2.

Lemma 3.5.1 *Each of the three vector norms in Definition 3.5.2 satisfies the conditions in the Definition 3.5.1 of a norm. In addition, for all n-vectors* $\mathbf{x}$ *and* $\mathbf{y}$ *the vector 2-norm satisfies the* ***Cauchy inequality***

$$|\mathbf{y} \cdot \mathbf{x}| \leq \|\mathbf{y}\|_2 \, \|\mathbf{x}\|_2 \,. \tag{3.15}$$

Finally, all n-vectors $\mathbf{x}$ *and* $\mathbf{y}$ *satisfy the following special case of the* ***Hölder inequality***

$$|\mathbf{y} \cdot \mathbf{x}| \leq \|\mathbf{y}\|_\infty \|\mathbf{x}\|_1 \,. \tag{3.16}$$

Proof It is obvious that all three norms are definite and homogeneous. We will verify the triangle inequality in each case. For the infinity-norm,

$$\|\mathbf{x}+\mathbf{y}\|_\infty = \max_{1\leq i\leq n} |\xi_i + \eta_i| \leq \max_{1\leq i\leq n} \{|\xi_i| + |\eta_i|\} \leq \max_{1\leq i\leq n} |\xi_i| + \max_{1\leq i\leq n} |\eta_i|$$
$$= \|\mathbf{x}\|_\infty + \|\mathbf{y}\|_\infty \,.$$

For the one-norm,

$$\|\mathbf{x}+\mathbf{y}\|_1 = \sum_{i=1}^n |\xi_i + \eta_i| \leq \sum_{i=1}^n \{|\xi_i| + |\eta_i|\} = \sum_{i=1}^n |\xi_i| + \sum_{i=1}^n |\eta_i| = \|\mathbf{x}\|_1 + \|\mathbf{y}\|_1 \,.$$

In order to prove the triangle inequality for the 2-norm, we will first verify the Cauchy inequality. Let $\mathbf{w} = \mathbf{y}\,\mathbf{x}\cdot\mathbf{x} - \mathbf{x}\,\mathbf{x}\cdot\mathbf{y}$. Then

$$\mathbf{x}\cdot\mathbf{w} = \mathbf{x}\cdot\mathbf{y}\,\mathbf{x}\cdot\mathbf{x} - \mathbf{x}\cdot\mathbf{x}\,\mathbf{x}\cdot\mathbf{y} = 0 \,,$$

so $\mathbf{w} \perp \mathbf{x}$. Next,

$$0 \leq \mathbf{w}\cdot\mathbf{w} = \mathbf{w}\cdot(\mathbf{y}\,\mathbf{x}\cdot\mathbf{x} - \mathbf{x}\,\mathbf{x}\cdot\mathbf{y}) = (\mathbf{w}\cdot\mathbf{y})(\mathbf{x}\cdot\mathbf{x}) = [\mathbf{y}\,\mathbf{x}\cdot\mathbf{x} - \mathbf{x}\,\mathbf{x}\cdot\mathbf{y}]\cdot\mathbf{y}(\mathbf{x}\cdot\mathbf{x})$$
$$= \|\mathbf{x}\|_2^2 \left\{\|\mathbf{y}\|_2^2\|\mathbf{x}\|_2^2 - |\mathbf{x}\cdot\mathbf{y}|^2\right\} \,.$$

If $\mathbf{x} \neq \mathbf{0}$ then we can cancel $\|\mathbf{x}\|_2^2$ in this equality to prove the Cauchy inequality. On the other had, if $\mathbf{x} = \mathbf{0}$, then the Cauchy inequality is trivial.

Next, we will prove the triangle inequality for the vector 2-norm. Since the real part of a complex number cannot be larger than its modulus, we have

$$\mathbf{x}\cdot\mathbf{y} + \mathbf{y}\cdot\mathbf{x} \leq 2\,|\mathbf{y}\cdot\mathbf{x}| \,.$$

This and the Cauchy inequality imply that

$$\|\mathbf{x}+\mathbf{y}\|_2^2 = (\mathbf{x}+\mathbf{y})\cdot(\mathbf{x}+\mathbf{y}) = \|\mathbf{x}\|_2^2 + \mathbf{x}\cdot\mathbf{y} + \mathbf{y}\cdot\mathbf{x} + \|\mathbf{y}\|_2^2$$
$$\le \|\mathbf{x}\|_2^2 + 2\,|\mathbf{y}\cdot\mathbf{x}| + \|\mathbf{y}\|_2^2 \le \|\mathbf{x}\|_2^2 + 2\|\mathbf{x}\|_2\|\mathbf{y}\|_2 + \|\mathbf{y}\|_2^2 = (\|\mathbf{x}\|_2 + \|\mathbf{y}\|_2)^2 \ .$$

All that remains is to prove the Hölder inequality (3.16):

$$|\mathbf{y}\cdot\mathbf{x}| = \left|\sum_{i=1}^n \overline{\eta_i}\xi_i\right| \le \sum_{i=1}^n |\overline{\eta_i}\xi_i| = \sum_{i=1}^n |\eta_i|\ |\xi_i| \le \max_{1\le i\le n} |\eta_i| \sum_{i=1}^n |\xi_i| = \|\mathbf{y}\|_\infty \|\mathbf{x}\|_1 \ .$$

The Cauchy inequality leads to the following useful characterization of a vector 2-norm.

Corollary 3.5.1 *If* $\mathbf{x}$ *is an n-vector, then*

$$\|\mathbf{x}\|_2 = \max_{\mathbf{y}\neq\mathbf{0}} \frac{|\mathbf{y}\cdot\mathbf{x}|}{\|\mathbf{y}\|_2} \ .$$

Proof If $\mathbf{x}=\mathbf{0}$, the claim is obvious. In the remainder of the proof, we will assume that $\mathbf{x}$ is nonzero. The Cauchy inequality (3.15) implies that

$$\max_{\mathbf{y}\neq\mathbf{0}} \frac{|\mathbf{y}\cdot\mathbf{x}|}{\|\mathbf{y}\|_2} \le \|\mathbf{x}\|_2 \ .$$

Since $\mathbf{x}\neq\mathbf{0}$, we also have

$$\|\mathbf{x}\|_2 = \frac{|\mathbf{x}\cdot\mathbf{x}|}{\|\mathbf{x}\|_2} \le \max_{\mathbf{y}\neq\mathbf{0}} \frac{|\mathbf{y}\cdot\mathbf{x}|}{\|\mathbf{y}\|_2} \ .$$

We will use norms to measure the size of vectors and matrices. As we have seen, we have several choices for vector norms. The following theorem shows that the choice of norm is not that important.

Theorem 3.5.1 (Equivalence of Norms) *If* $\|\cdot\|_\alpha$ *and* $\|\cdot\|_\beta$ *are two norms on a finite-dimensional linear space* $\mathscr{X}$, *then there are positive constants* $\underline{\gamma}$ *and* $\overline{\gamma}$ *so that for all* $\mathbf{x}\in\mathscr{X}$,

$$\underline{\gamma}\|\mathbf{x}\|_\beta \le \|\mathbf{x}\|_\alpha \le \overline{\gamma}\|\mathbf{x}\|_\beta \ .$$

Proof The proof depends on the Bolzano-Weierstrass theorem, which says that a bounded infinite sequence in a finite-dimensional linear space has a convergent subsequence. For a proof of the Bolzano-Weierstrass theorem, see Loomis and Sternberg [76, p. 208]. For a proof of the equivalence of norms, see Kreyszig [69, p. 75] or Loomis and Sternberg [76, p. 209].

For our purposes, it will be sufficient to use the following simplified form of Theorem 3.5.1.

Lemma 3.5.2 *For all n-vectors* $\mathbf{x}$,

$$\|\mathbf{x}\|_1 \leq \sqrt{n}\|\mathbf{x}\|_2 \text{ and } \|\mathbf{x}\|_1 \leq n\|\mathbf{x}\|_\infty ,$$
$$\|\mathbf{x}\|_2 \leq \|\mathbf{x}\|_1 \text{ and } \|\mathbf{x}\|_2 \leq \sqrt{n}\|\mathbf{x}\|_\infty ,$$
$$\|\mathbf{x}\|_\infty \leq \|\mathbf{x}\|_1 \text{ and } \|\mathbf{x}\|_\infty \leq \|\mathbf{x}\|_2 .$$

Proof The inequalities involving the infinity-norm are all easy. Only two inequalities remain. One is a direct consequence of Cauchy's inequality (3.15):

$$\|\mathbf{x}\|_1 = \sum_{j=1}^n |\xi_j| \leq \left(\sum_{j=1}^n 1^2\right)^{1/2} \left(\sum_{j=1}^n |\xi_j|^2\right)^{1/2} = \sqrt{n}\|\mathbf{x}\|_2 .$$

The other inequality involves repeated application of the fact that $(a+b)^2 \geq a^2+b^2$ for all nonnegative a and b:

$$\left(\sum_{j=1}^n |\xi_j|\right)^2 \geq \left(\sum_{j=1}^{n-1} |\xi_j|\right)^2 + |\xi_n|^2 \geq \left(\sum_{j=1}^{n-2} |\xi_j|\right)^2 + |\xi_{n-1}|^2 + |\xi_n|^2 \ldots \geq \sum_{j=1}^n |\xi_j|^2 .$$

In MATLAB, these three norms can be computed by the norm command. In LAPACK, the Euclidean norm of a vector can be computed by Level 1 BLAS subroutine `_nrm2`; see, for example, dnrm2.f. The one-norm can be computed by Level 1 BLAS subroutines `_asum`; see, for example, dasum.f or dzasum.f. There is no LAPACK subroutine to compute the infinity norm of a vector; instead, programmers can call LAPACK Level 1 BLAS subroutine `i_amax` to find the entry of the vector with maximum absolute value, and then evaluate the absolute value of this entry. See, for example, idamax.f.

Exercise 3.5.1 For $p = 1, 2$ and ∞, graph the set of all 2-vectors $\mathbf{x}$ with $\|\mathbf{x}\|_p = 1$.

Exercise 3.5.2 Suppose that $\mathbf{x}$ and $\mathbf{y}$ are real n-vectors. Show that

$$\|\mathbf{x} + \mathbf{y}\|_2 = \|\mathbf{x}\|_2 + \|\mathbf{y}\|_2$$

if and only if $\mathbf{x}$ and $\mathbf{y}$ are linearly *dependent* with $\mathbf{x} \cdot \mathbf{y} \geq 0$. How does this result differ from the Pythagorean Theorem 3.2.2?

Exercise 3.5.3 Define the **angle between two nonzero vectors** $\mathbf{x}$ and $\mathbf{y}$ to be θ where

$$\cos\theta = \frac{|\mathbf{x} \cdot \mathbf{y}|}{\|\mathbf{x}\|_2 \|\mathbf{y}\|_2} .$$

Recall that if $\mathbf{x}$ and $\mathbf{y}$ are 3-vectors, then their **cross product** is

$$\mathbf{x} \times \mathbf{y} = \begin{bmatrix} \mathbf{x}_2\mathbf{y}_3 - \mathbf{x}_3\mathbf{y}_2 \\ \mathbf{x}_3\mathbf{y}_1 - \mathbf{x}_1\mathbf{y}_3 \\ \mathbf{x}_1\mathbf{y}_2 - \mathbf{x}_2\mathbf{y}_1 \end{bmatrix} .$$

Show that

$$\|\mathbf{x} \times \mathbf{y}\|_2 = \|\mathbf{x}\|_2 \|\mathbf{y}\|_2 \sin \theta \; .$$

Exercise 3.5.4 Let $\mathbf{x}$ be an n-vector and $\mathbf{y}$ be an m-vector. Define the **tensor product** to be the mn-vector

$$\mathbf{x} \otimes \mathbf{y} = \begin{bmatrix} \mathbf{x}\mathbf{y}_1 \\ \vdots \\ \mathbf{x}\mathbf{y}_m \end{bmatrix} .$$

Show that

$$\|\mathbf{x} \otimes \mathbf{y}\|_p = \|\mathbf{x}\|_p \|\mathbf{y}\|_p$$

for $p = 1, 2, \infty$.

3.5.2 *Subordinate Matrix Norms*

Systems of linear equations involve both vectors and matrices. In order to bound perturbations in the solution of a linear system by perturbations in its data, we will need to develop norms for both vectors and matrices. Fortunately, we can use vector norms to define some useful matrix norms.

Definition 3.5.3 If $\| \cdot \|$ is a vector norm and $\mathbf{A}$ is a matrix, then the **subordinate matrix norm** of $\mathbf{A}$ is

$$\|\mathbf{A}\| = \max_{\mathbf{x} \neq \mathbf{0}} \frac{\|\mathbf{A}\mathbf{x}\|}{\|\mathbf{x}\|} = \max_{\|\mathbf{x}\|=1} \|\mathbf{A}\mathbf{x}\| \; .$$

For those readers who have studied real analysis, we remark that in order to show that these maximum values are actually attained for some unit vector $\mathbf{x}$, it is common to use the Bolzano-Weierstrass theorem.

The following lemma shows that a subordinate matrix norm is a norm.

Lemma 3.5.3 *If $\| \cdot \|$ is a subordinate matrix norm, then it satisfies the conditions in Definition 3.5.1 to be a norm on matrices. Furthermore, if we can compute the* *matrix-matrix product* **AB** *then*

$$\|\mathbf{AB}\| \leq \|\mathbf{A}\| \; \|\mathbf{B}\| \; . \tag{3.17}$$

Proof First, we will show that $\|\mathbf{A}\|$ is definite. Recall that $\mathbf{e}_j$ is the jth axis vector. If $\mathbf{A} \neq \mathbf{0}$, then there exists an index j so that $\mathbf{Ae}_j \neq \mathbf{0}$. Then $\|\mathbf{A}\| \geq \|\mathbf{Ae}_j\| > 0$.

Next, will show that the matrix norm is homogeneous. Using the homogeneity of the vector norm, we see that for any scalar α

$$\|\mathbf{A}\alpha\| = \max_{\|\mathbf{x}\|=1} \|(\mathbf{A}\alpha)\mathbf{x}\| = \max_{\|\mathbf{x}\|=1} \|\mathbf{Ax}\| \; |\alpha| = \|\mathbf{A}\| \; |\alpha| \; .$$

Let us prove the triangle inequality:

$$\|\mathbf{A}+\mathbf{B}\| = \max_{\|\mathbf{x}\|=1} \|(\mathbf{A}+\mathbf{B})\mathbf{x}\| \leq \max_{\|\mathbf{x}\|=1} (\|\mathbf{Ax}\| + \|\mathbf{Bx}\|) \leq \max_{\|\mathbf{x}\|=1} \|\mathbf{Ax}\| + \max_{\|\mathbf{x}\|=1} \|\mathbf{Bx}\|$$
$$= \|\mathbf{A}\| + \|\mathbf{B}\| \; .$$

Finally, we will prove (3.17). As we mentioned above, the Bolzano-Weierstrass theorem allows us to find a vector $\mathbf{x}$ so that $\|\mathbf{x}\| = 1$ and $\|\mathbf{ABx}\| = \|\mathbf{AB}\|$. Then

$$\|\mathbf{AB}\| = \|\mathbf{ABx}\| \leq \|\mathbf{A}\| \; \|\mathbf{Bx}\| \leq \|\mathbf{A}\| \; \|\mathbf{B}\| \; \|\mathbf{x}\| = \|\mathbf{A}\| \; \|\mathbf{B}\| \; .$$

The following lemma will provide values for the three useful subordinate matrix norms.

Lemma 3.5.4 *If* **A** *is an* $m \times n$ *matrix, then*

$$\|\mathbf{A}\|_1 = \max_{1 \leq j \leq n} \left\{ \sum_{i=1}^{m} |\alpha_{ij}| \right\} \; , \tag{3.18}$$

$$\|\mathbf{A}\|_\infty = \max_{1 \leq i \leq m} \left\{ \sum_{j=1}^{n} |\alpha_{ij}| \right\} \; \textit{and} \tag{3.19}$$

$$\|\mathbf{A}\|_2 = \sqrt{\max\{\lambda : \lambda \textit{ is an eigenvalue of } \mathbf{A}^H\mathbf{A}\}} \; . \tag{3.20}$$

As a result,

$$\left\|\mathbf{A}^H\right\|_1 = \|\mathbf{A}\|_\infty \; \textit{and} \; \left\|\mathbf{A}^H\right\|_\infty = \|\mathbf{A}\|_1 \; .$$

Proof First, we will prove (3.18). If $\mathbf{A}$ has entries α_{ij} and the n-vector $\mathbf{x}$ has entries ξ_j, then

$$\|\mathbf{A}\mathbf{x}\|_1 = \sum_{i=1}^{m}\left|\sum_{j=1}^{n}\alpha_{ij}\xi_j\right| \leq \sum_{i=1}^{m}\sum_{j=1}^{n}|\alpha_{ij}|\,|\xi_j| = \sum_{j=1}^{n}\left[\sum_{i=1}^{m}|\alpha_{ij}|\,|\xi_j|\right]$$
$$\leq \left\{\max_{1\leq j\leq n}\sum_{i=1}^{m}|\alpha_{ij}|\right\}\left\{\sum_{j=1}^{n}|\xi_j|\right\} = \left\{\max_{1\leq j\leq n}\sum_{i=1}^{m}|\alpha_{ij}|\right\}\|\mathbf{x}\|_1 \,.$$

This implies that

$$\|\mathbf{A}\|_1 = \max_{\|\mathbf{x}\|_1=1}\|\mathbf{A}\mathbf{x}\|_1 \leq \max_{1\leq j\leq n}\sum_{i=1}^{m}|\alpha_{ij}| \,.$$

If the max column sum $\sum_{i=1}^{m}|\alpha_{ij}|$ is achieved for column j, then $\|\mathbf{A}\mathbf{e}_j\|_1$ is equal to this maximum column sum. Since $\|\mathbf{e}_j\|_1 = 1$, the claim (3.18) is proved.

Next, we will prove (3.19). First, we compute

$$\|\mathbf{A}\mathbf{x}\|_\infty = \max_{1\leq i\leq m}\left|\sum_{j=1}^{n}\alpha_{ij}\xi_j\right| \leq \max_{1\leq i\leq m}\sum_{j=1}^{n}|\alpha_{ij}|\max_{1\leq j\leq n}|\xi_j| = \max_{1\leq i\leq m}\sum_{j=1}^{n}|\alpha_{ij}|\,\|\mathbf{x}\|_\infty \,.$$

If the max row sum $\sum_{j=1}^{n}|\alpha_{ij}|$ is achieved for row i, then let

$$\xi_j = \begin{cases}\overline{\alpha_{ij}}/|\alpha_{ij}|, & |\alpha_{ij}| \neq 0\\ 1, & |\alpha_{ij}| = 0\end{cases} \,.$$

Then the corresponding vector $\mathbf{x}$ is such that $\|\mathbf{x}\|_\infty = 1$ and $\|\mathbf{A}\mathbf{x}\|_\infty$ is equal to this maximum row sum.

Inequality (3.20) is harder to verify. The spectral Theorem 1.3.1 of Chap. 1 in Volume II will show that there is an orthogonal matrix $\mathbf{Q}$ and a nonnegative diagonal matrix Λ so that $\mathbf{A}^H\mathbf{A} = \mathbf{Q}\Lambda\mathbf{Q}^H$. The diagonal entries of Λ are the eigenvalues of $\mathbf{A}^H\mathbf{A}$. Then

$$\|\mathbf{A}\|_2^2 = \max_{\|\mathbf{x}\|_2=1}\sqrt{\mathbf{x}^H\mathbf{Q}\Lambda\mathbf{Q}^H\mathbf{x}} = \max_{\|\mathbf{y}\|_2=1}\sqrt{\mathbf{y}^H\Lambda\mathbf{y}} \leq \max_{1\leq j\leq n}\sqrt{\lambda_j} \,.$$

In other words, $\|\mathbf{A}\|_2$ is equal to the square root of the the largest eigenvalue of $\mathbf{A}^H\mathbf{A}$.

The final claim is a direct consequence of inequalities (3.18) and (3.19). Alternatively, it is possible to show that $\|\mathbf{A}\|_2$ is equal to the largest singular value of $\mathbf{A}$; see Sect. 1.5 of Chap. 1 in Volume II for a discussion of the singular value decomposition.

We also note the following two important lemmas.

Lemma 3.5.5 *If the* $m \times n$ *matrix* $\mathbf{U}$ *has orthonormal columns, then* $\|\mathbf{U}\mathbf{x}\|_2 = \|\mathbf{x}\|_2$ *for all n-vectors* $\mathbf{x}$.

Proof Since $\mathbf{U}^H\mathbf{U} = \mathbf{I}$, we can compute

$$\|\mathbf{U}\mathbf{x}\|_2^2 = (\mathbf{U}\mathbf{x})^H (\mathbf{U}\mathbf{x}) = \mathbf{x}^H\mathbf{U}^H\mathbf{U}\mathbf{x} = \mathbf{x}^H\mathbf{x} = \|\mathbf{x}\|_2^2 \, .$$

Lemma 3.5.6 *If* $\mathbf{P}$ *is an orthogonal projector, then* $\|\mathbf{P}\|_2 \leq 1$.

Proof For any vector $\mathbf{x}$ of the appropriate size, we can write

$$\mathbf{x} = \mathbf{P}\mathbf{x} + (\mathbf{I} - \mathbf{P})\,\mathbf{x} \, .$$

Since

$$(\mathbf{P}\mathbf{x}) \cdot [(\mathbf{I} - \mathbf{P})\,\mathbf{x}] = \mathbf{x} \cdot \mathbf{P}^H (\mathbf{I} - \mathbf{P})\,\mathbf{x} = \mathbf{x} \cdot \mathbf{P}(\mathbf{I} - \mathbf{P})\,\mathbf{x} = \mathbf{x} \cdot \left(\mathbf{P} - \mathbf{P}^2\right)\mathbf{x} = \mathbf{0} \, ,$$

the Pythagorean Theorem 3.2.2 implies that

$$\|\mathbf{x}\|_2^2 = \|\mathbf{P}\mathbf{x}\|_2^2 + \|(\mathbf{I} - \mathbf{P})\,\mathbf{x}\|_2^2 \, .$$

It follows that for all $\mathbf{x}$ we have

$$\|\mathbf{P}\mathbf{x}\|_2^2 = \|\mathbf{x}\|_2^2 - \|(\mathbf{I} - \mathbf{P})\,\mathbf{x}\|_2^2 \leq \|\mathbf{x}\|_2^2 \, .$$

The claimed result follows.

The next lemma provides constants for the equivalence of these subordinate matrix norms, as required by Theorem 3.5.1.

Lemma 3.5.7 *Let* $\mathbf{A}$ *be an* $m \times n$ *matrix. Then*

$$\|\mathbf{A}\|_1 \leq \sqrt{m}\|\mathbf{A}\|_2 \text{ and } \|\mathbf{A}\|_1 \leq m\|\mathbf{A}\|_\infty \, ,$$
$$\|\mathbf{A}\|_2 \leq \sqrt{n}\|\mathbf{A}\|_1 \text{ and } \|\mathbf{A}\|_2 \leq \sqrt{m}\|\mathbf{A}\|_\infty \, ,$$
$$\|\mathbf{A}\|_\infty \leq n\|\mathbf{A}\|_1 \text{ and } \|\mathbf{A}\|_\infty \leq \sqrt{n}\|\mathbf{A}\|_2 \, .$$

Proof First, we will show that $\|\mathbf{A}\|_1 \leq m\|\mathbf{A}\|_\infty$:

$$\|\mathbf{A}\|_1 = \max_{1 \leq j \leq n} \sum_{i=1}^{m} |\alpha_{ij}| \leq \sum_{j=1}^{n} \sum_{i=1}^{m} |\alpha_{ij}| \leq m \max_{1 \leq i \leq m} \sum_{j=1}^{n} |\alpha_{ij}| = m\|\mathbf{A}\|_\infty \, .$$

Since $\|\mathbf{A}\|_1 = \left\|\mathbf{A}^H\right\|_\infty$, we have also proved the inequality $\|\mathbf{A}\|_\infty \leq n\|\mathbf{A}\|_1$.

Next, let us prove that $\|\mathbf{A}\|_\infty \leq \sqrt{n}\|\mathbf{A}\|_2$. If $\|\mathbf{A}\|_\infty = \sum_{j=1}^n |\alpha_{ij}|$, define the n-vector $\mathbf{x}$ to have components

$$\xi_j = \begin{cases} \overline{\alpha_{ij}}/|\alpha_{ij}|, \; \alpha_{ij} \neq 0 \\ 1, \; \alpha_{ij} = 0 \end{cases} .$$

Then $\|\mathbf{x}\|_\infty = 1$ and

$$\|\mathbf{A}\|_\infty = \sum_{j=1}^n |\alpha_{ij}| = \left|\sum_{j=1}^n \alpha_{ij}\xi_j\right| = \left|\mathbf{e}_i{}^H\mathbf{A}\mathbf{x}\right| \leq \|\mathbf{A}\mathbf{x}\|_\infty \leq \|\mathbf{A}\mathbf{x}\|_2 \leq \|\mathbf{A}\|_2 \, \|\mathbf{x}\|_2$$
$$= \sqrt{n}\|\mathbf{A}\|_2 .$$

Since $\|\mathbf{A}\|_1 = \left\|\mathbf{A}^H\right\|_\infty$, this also proves $\|\mathbf{A}\|_1 \leq \sqrt{m}\|\mathbf{A}\|_2$.

Let us prove that $\|\mathbf{A}\|_2 \leq \sqrt{m}\|\mathbf{A}\|_\infty$. Let $\mathbf{x}$ be such that $\|\mathbf{x}\|_2 = 1$ and $\|\mathbf{A}\|_2 = \|\mathbf{A}\mathbf{x}\|_2$. Here $\mathbf{x}$ is the eigenvector of $\mathbf{A}^H\mathbf{A}$ corresponding to the largest eigenvalue. Then Lemma 3.5.2 implies that

$$\|\mathbf{A}\|_2 = \|\mathbf{A}\mathbf{x}\|_2 \leq \sqrt{m}\|\mathbf{A}\mathbf{x}\|_\infty \leq \sqrt{m}\|\mathbf{A}\|_\infty\|\mathbf{x}\|_\infty \leq \sqrt{m}\|\mathbf{A}\|_\infty\|\mathbf{x}\|_2 = \sqrt{m}\|\mathbf{A}\|_\infty .$$

The proof of $\|\mathbf{A}\|_2 \leq \sqrt{n}\|\mathbf{A}\|_1$ is similar:

$$\|\mathbf{A}\|_2 = \|\mathbf{A}\mathbf{x}\|_2 \leq \|\mathbf{A}\mathbf{x}\|_1 \leq \|\mathbf{A}\|_1\|\mathbf{x}\|_1 \leq \sqrt{n}\|\mathbf{A}\|_1\|\mathbf{x}\|_2 = \sqrt{n}\|\mathbf{A}\|_1 .$$

The following lemma can be used to bound subordinate matrix norms.

Lemma 3.5.8 *Assume that* $\mathbf{A}$ *is an* $m \times n$ *matrix and* $\|\cdot\|$ *is a norm on vectors. Suppose that there is a scalar* $\alpha > 0$ *such that for all n-vectors* $\mathbf{x}$ *we have* $\|\mathbf{A}\mathbf{x}\| \leq \alpha\|\mathbf{x}\|$. *Then the subordinate matrix norm* $\|\mathbf{A}\|$ *satisfies* $\|\mathbf{A}\| \leq \alpha$; *if in addition* $\mathbf{A}$ *is nonsingular, then* $\left\|\mathbf{A}^{-1}\right\| \geq 1/\alpha$. *Next, suppose that* $\mathbf{A}$ *is nonsingular and there is a scalar* $\alpha > 0$ *such that for all n-vectors* $\mathbf{x}$ *we have* $\left\|\mathbf{A}^{-1}\mathbf{x}\right\| \geq \|\mathbf{x}\|/\alpha$. *Then* $\|\mathbf{A}\| \leq \alpha$ *and* $\left\|\mathbf{A}^{-1}\right\| \geq 1/\alpha$.

Proof Let us prove the first claim. By Definition 3.5.3 of the subordinate matrix norm,

$$\|\mathbf{A}\| = \max_{\mathbf{x}\neq\mathbf{0}} \frac{\|\mathbf{A}\mathbf{x}\|}{\|\mathbf{x}\|} \leq \alpha .$$

If $\mathbf{A}$ is nonsingular and $\mathbf{y} = \mathbf{A}\mathbf{x}$, then $\mathbf{x} = \mathbf{A}^{-1}\mathbf{y}$. This allows us to rewrite $\|\mathbf{A}\mathbf{x}\| \leq \alpha\|\mathbf{x}\|$ in the form

$$\|\mathbf{y}\| \leq \alpha \left\|\mathbf{A}^{-1}\mathbf{y}\right\| .$$

Then the definition of the subordinate matrix norm implies that

$$\|\mathbf{A}^{-1}\| = \max_{\mathbf{y}\neq\mathbf{0}} \frac{\|\mathbf{A}^{-1}\mathbf{y}\|}{\|\mathbf{y}\|} \geq \frac{1}{\alpha} .$$

The proof of the second claim is similar.

Exercise 3.5.5 Show that for any n-vector $\mathbf{x}$, $\|\mathbf{x}^H\|_2 = \|\mathbf{x}\|_2$, $\|\mathbf{x}^H\|_\infty = \|\mathbf{x}\|_1$ and $\|\mathbf{x}^H\|_1 = \|\mathbf{x}\|_\infty$.

Exercise 3.5.6 Suppose that $\mathbf{A}$ is an $m \times n$ matrix and $\mathbf{B}$ is a **submatrix** of $\mathbf{A}$, formed by eliminating some number of rows and/or columns of $\mathbf{A}$. Show that $\|\mathbf{B}\|_p \leq \|\mathbf{A}\|_p$ for $p = 1, 2, \infty$.

Exercise 3.5.7 If $\mathbf{A}$ is a nonsingular matrix, show that for $p = 1, 2$ and ∞ we have $\|\mathbf{A}^{-1}\|_p \neq 0$ and

$$\frac{1}{\|\mathbf{A}^{-1}\|_p} = \min_{\mathbf{x}\neq\mathbf{0}} \frac{\|\mathbf{A}\mathbf{x}\|_p}{\|\mathbf{x}\|_p} .$$

Exercise 3.5.8 If $\mathbf{A}$ is an $m \times n$ matrix and $\mathbf{B}$ is an $k \times \ell$ matrix, define their tensor product to be the $mk \times n\ell$ matrix

$$\mathbf{A} \otimes \mathbf{B} = \begin{bmatrix} \mathbf{A}\mathbf{B}_{11} & \dots & \mathbf{A}\mathbf{B}_{1\ell} \\ \vdots & \ddots & \vdots \\ \mathbf{A}\mathbf{B}_{k1} & \dots & \mathbf{A}\mathbf{B}_{k\ell} \end{bmatrix} .$$

Show that for $p = 1$ and ∞ we have

$$\|\mathbf{A} \otimes \mathbf{B}\|_p = \|\mathbf{A}\|_p \|\mathbf{B}\|_p .$$

3.5.3 *Consistent Matrix Norms*

Lemma 3.5.4 shows that the subordinate matrix 2-norm is somewhat difficult to compute. As a result, we would like to find a substitute matrix norm for use with the vector 2-norm. This goal leads to the following definition.

Definition 3.5.4 A matrix norm $\|\cdot\|$ is **consistent** with respect to a given vector norm (also denoted by $\|\cdot\|$) if and only if for all $m \times n$ matrices $\mathbf{A}$ and all n-vectors $\mathbf{x}$,

$$\|\mathbf{A}\mathbf{x}\| \leq \|\mathbf{A}\| \|\mathbf{x}\| .$$

Here is a common example of a consistent matrix norm.

Definition 3.5.5 If $\mathbf{A}$ is an $m \times n$ matrix with entries α_{ij}, the **Frobenius matrix norm** of $\mathbf{A}$ is

$$\|\mathbf{A}\|_F \equiv \sqrt{\sum_{i=1}^{m}\sum_{j=1}^{n}|\alpha_{ij}|^2} \tag{3.21}$$

The following lemma establishes some important results regarding the Frobenius norm.

Lemma 3.5.9 *For any matrix* $\mathbf{A}$, *we have*

$$\|\mathbf{A}^H\|_F = \|\mathbf{A}\|_F \ .$$

Also, for any unitary matrices $\mathbf{U}$ *and* $\mathbf{V}$ *we have*

$$\|\mathbf{U}\mathbf{A}\mathbf{V}^H\|_F = \|\mathbf{A}\|_F \ .$$

Proof The first claim is an easy consequence of the definition of the Frobenius norm. To prove the second claim, we use Lemma 3.5.5 to show that

$$\|\mathbf{U}\mathbf{A}\mathbf{V}^H\|_F^2 = \sum_{j=1}^{n}\|\mathbf{U}\mathbf{A}\mathbf{V}^H\mathbf{e}_j\|_2^2 = \sum_{j=1}^{n}\|\mathbf{A}\mathbf{V}^H\mathbf{e}_j\|_2^2 = \|\mathbf{A}\mathbf{V}^H\|_F^2 = \|\mathbf{V}\mathbf{A}^H\|_F^2$$
$$= \|\mathbf{A}^H\|_F^2 = \|\mathbf{A}\|_F^2 \ .$$

The next lemma proves that the Frobenius norm is a norm, and establishes its equivalence with respect to our subordinate matrix norms.

Lemma 3.5.10 *The Frobenius matrix norm is a matrix norm, and it is consistent with the Euclidean vector norm. Furthermore,*

$$\|\mathbf{A}\|_F \le \sqrt{n}\|\mathbf{A}\|_1 \text{ and } \|\mathbf{A}\|_1 \le \sqrt{m}\|\mathbf{A}\|_F \ ,$$
$$\|\mathbf{A}\|_F \le \sqrt{rank\,(\mathbf{A})}\|\mathbf{A}\|_2 \text{ and } \|\mathbf{A}\|_2 \le \|\mathbf{A}\|_F \ ,$$
$$\|\mathbf{A}\|_F \le \sqrt{m}\|\mathbf{A}\|_\infty \text{ and } \|\mathbf{A}\|_\infty \le \|\mathbf{A}\|_F \ .$$

Proof Given any $m \times n$ matrix $\mathbf{A}$ with entries α_{ij}, we can construct an mn-vector $\mathbf{b}$ with entries $\beta_{i+jm} = \alpha_{ij}$. Then $\|\mathbf{A}\|_F = \|\mathbf{b}\|_2$. We can use this relationship to show that $\|\cdot\|_F$ is definite, homogeneous and satisfies the triangle inequality. Thus $\|\cdot\|_F$ is a norm.

Next, we will prove consistency. If $\mathbf{x}$ has entries ξ_j, then the triangle inequality and homogeneity of the vector 2-norm give us

$$\|\mathbf{A}\mathbf{x}\|_2 = \left\| \sum_{j=1}^{n} \mathbf{A}\mathbf{e}_j \xi_j \right\|_2 \leq \sum_{j=1}^{n} \|\mathbf{A}\mathbf{e}_j\|_2 \, |\xi_j|$$

then the Cauchy inequality (3.15) leads to

$$\leq \left(\sum_{j=1}^{n} \|\mathbf{A}\mathbf{e}_j\|_2^2 \right)^{1/2} \left(\sum_{j=1}^{n} |\xi_j|^2 \right)^{1/2} = \|\mathbf{A}\|_F \|\mathbf{x}\|_2 \, .$$

Our next goal is to establish the norm equivalence inequality $\|\mathbf{A}\|_F \leq \sqrt{n}\|\mathbf{A}\|_1$:

$$\|\mathbf{A}\|_F^2 = \sum_{j=1}^{n} \sum_{i=1}^{m} |\alpha_{ij}|^2 \leq \sum_{j=1}^{n} \left(\sum_{i=1}^{m} |\alpha_{ij}| \right)^2 \leq n \left(\max_{1 \leq j \leq n} \sum_{i=1}^{m} |\alpha_{ij}| \right)^2 = n\|\mathbf{A}\|_1^2 \, .$$

To prove that $\|\mathbf{A}\|_1 \leq \sqrt{m}\|\mathbf{A}\|_F$, we use the Cauchy inequality (3.15) to get

$$\|\mathbf{A}\|_1^2 = \max_{1 \leq j \leq n} \left(\sum_{i=1}^{m} |\alpha_{ij}| \right)^2 \leq \max_{1 \leq j \leq n} \left(\sum_{i=1}^{m} |\alpha_{ij}|^2 \right) \left(\sum_{i=1}^{m} 1^2 \right) = m \max_{1 \leq j \leq n} \sum_{i=1}^{m} |\alpha_{ij}|^2$$
$$\leq m \sum_{j=1}^{n} \sum_{i=1}^{m} |\alpha_{ij}|^2 = \|\mathbf{A}\|_F^2 \, .$$

The proofs that $\|\mathbf{A}\|_F \leq \sqrt{m}\|\mathbf{A}\|_\infty$ and $\|\|\mathbf{A}\|_\infty \leq \sqrt{n}\|\mathbf{A}\|_F$ follow from the previous Frobenius norm equivalences, and the fact that $\|\mathbf{A}\|_\infty = \left\|\mathbf{A}^H\right\|_1$.

To prove that $\|\mathbf{A}\|_F \leq \sqrt{\text{rank}\,(\mathbf{A})}\|\mathbf{A}\|_2$, we will use an argument involving singular values that can be found in Zielke [115]. The spectral Theorem 1.3.1 of Chap. 1 in Volume II allows us to write $\mathbf{A}^H\mathbf{A} = \mathbf{Q}\boldsymbol{\Lambda}\mathbf{Q}^H$, where $\mathbf{Q}$ has orthonormal columns and Λ is a nonnegative diagonal matrix. The number of nonzero diagonal entries of $\boldsymbol{\Lambda}$ is $r = \text{rank}\,(\mathbf{A})$. Then

$$\|\mathbf{A}\|_F^2 = \sum_{j=1}^{n} \sum_{i=1}^{m} |\alpha_{ij}|^2 = \sum_{j=1}^{n} \mathbf{e}_j{}^H \mathbf{A}^H \mathbf{A}\mathbf{e}_j = \sum_{j=1}^{n} (\mathbf{Q}\mathbf{e}_j)^H \lambda_j \, (\mathbf{Q}\mathbf{e}_j) = \sum_{j=1}^{n} \lambda_j$$
$$\leq r \max_{1 \leq j \leq n} \lambda_j = r\|\mathbf{A}\|_2^2 \, .$$

On the other hand,

$$\|\mathbf{A}\|_2^2 = \max_{1\leq j\leq n} \lambda_j \leq \sum_{j=1}^{n} \lambda_j = \operatorname{tr}\left(\mathbf{A}^H\mathbf{A}\right) = \|\mathbf{A}\|_F^2 \,.$$

Here tr refers to the **trace** of a matrix, which will be defined in Definition 1.2.3 of Chap. 1 in Volume II. The connection between the trace and eigenvalues will be determined in Lemma 1.2.8 of Chap. 1 in Volume II.

The Frobenius norm is useful because it is far easier to compute than the Euclidean subordinate matrix norm. We also have the following interesting result about the Frobenius norm of a matrix product.

Lemma 3.5.11 *If* $\mathbf{A}$ *is an* $m \times k$ *matrix and* $\mathbf{B}$ *is a* $k \times n$ *matrix, then*

$$\|\mathbf{AB}\|_F \leq \|\mathbf{A}\|_2\|\mathbf{B}\|_F \leq \|\mathbf{A}\|_F\|\mathbf{B}\|_F \,.$$

Proof The definition (3.21) of the Frobenius norm implies that

$$\|\mathbf{AB}\|_F^2 = \sum_{j=1}^{n} \|\mathbf{ABe}_j\|_2^2 \leq \|\mathbf{A}\|_2^2 \sum_{j=1}^{n} \|\mathbf{Be}_j\|_2^2 = \|\mathbf{A}\|_2^2 \,\|\mathbf{B}\|_F^2$$

then Lemma 3.5.10 implies that

$$\leq \|\mathbf{A}\|_F^2\|\mathbf{B}\|_F^2 \,.$$

In MATLAB, matrix norms can be computed by the norm command. In LAPACK, routines `_lange` can compute the one, infinity, Frobenius or max norm of a matrix. See, for example, dlange.f.

Exercise 3.5.9 Show that for any vector norm, the subordinate matrix norm of the identity matrix is one.

Exercise 3.5.10 If $\mathbf{P}$ is an $m \times m$ permutation matrix, compute $\|\mathbf{P}\|_1$, $\|\mathbf{P}\|_\infty$, $\|\mathbf{P}\|_2$ and $\|\mathbf{P}\|_F$.

Exercise 3.5.11 If $\mathbf{Q}$ is an $m \times m$ orthogonal matrix, compute $\|\mathbf{Q}\|_1$, $\|\mathbf{Q}\|_\infty$, $\|\mathbf{Q}\|_2$ and $\|\mathbf{Q}\|_F$.

Exercise 3.5.12 If $\mathbf{u}$ is an m-vector and $\mathbf{v}$ is an n-vector, show that $\|\mathbf{uv}^H\|_2 = \|\mathbf{u}\|_2 \;\|\mathbf{b}\|_2$. Is this equality true for the subordinate one-norm or infinity norm?

Exercise 3.5.13 Show that

$$\nu(\mathbf{A}) \equiv \max_{i,j} \left|\mathbf{A}_{i,j}\right|$$

is a norm on $m \times n$ matrices. In other words, show that it is definite ($\mathbf{A} \neq \mathbf{0}$ implies that $\nu(\mathbf{A}) > 0$, homogeneous ($\nu(\mathbf{A}\alpha) = \nu(\mathbf{A})|\alpha|$) and satisfies the triangle inequality: $\nu(\mathbf{A} + \mathbf{B}) <= \nu(\mathbf{A}) + \nu(\mathbf{B})$. Also show that $\nu(\mathbf{A})$ is not subordinate to any vector norm, and does not satisfy the consistency condition $\nu(\mathbf{AB}) \leq \nu(\mathbf{A})\nu(\mathbf{B})$.

3.5.4 *Hadamard Inequality*

In order to estimate the growth of Gaussian factorization perturbation errors in Lemma 3.8.10, will make use of the following result.

Lemma 3.5.12 (Hadamard Inequality) *If* $\mathbf{A} = [\mathbf{a}_1, \ldots, \mathbf{a}_n]$ *is an* $n \times n$ *matrix, then*

$$|\det \mathbf{A}| \leq \prod_{j=1}^{n} \|\mathbf{a}_j\|_2 \; . \tag{3.22}$$

Proof See Beckenbach and Bellman [6, p. 64]. The result can also be proved easily from the Gram-Schmidt factorization, which is discussed in Sect. 6.8.

3.6 Perturbation Analysis

In Lemma 3.2.8, we proved that a system of linear equations has a unique solution for any right-hand side if and only if $\mathbf{A}$ is nonsingular. Our goal in this section is to complete our discussion of the well-posedness of a linear system $\mathbf{Ax} = \mathbf{b}$, by examining how perturbations in the data $\mathbf{A}$ and $\mathbf{b}$ affect the solution $\mathbf{x}$. We will use the vector and matrix norms from Sect. 3.5 to measure the perturbations.

Our perturbation bounds will take several forms. We will determine an absolute error bound in Lemma 3.6.2 and a relative error bound in Theorem 3.6.1. These two bounds are called *a priori* because they can be evaluated before the solution of the linear system is computed. We will provide an alternative error estimate in Theorem 3.6.2. This error estimate is called *a posteriori*, because we must compute an approximate value for the solution vector $\mathbf{x}$ in order to estimate its error.

3.6.1 *Absolute Error*

We will begin by bounding perturbations to the identity matrix. Two such bounds appear in the following lemma.

Lemma 3.6.1 *If $\|\cdot\|$ is a subordinate matrix norm and $\|\mathbf{E}\| < 1$, then $\mathbf{I}-\mathbf{E}$ is nonsingular,*

$$\left\|(\mathbf{I}-\mathbf{E})^{-1}\right\| \leq \frac{1}{1-\|\mathbf{E}\|} \tag{3.23}$$

and

$$\left\|\mathbf{I}-(\mathbf{I}-\mathbf{E})^{-1}\right\| \leq \frac{\|\mathbf{E}\|}{1-\|\mathbf{E}\|} . \tag{3.24}$$

Proof Our proof will follow that in Stewart [97, Theorem 3.4, p. 187].

First, we will show that $\mathbf{I}-\mathbf{E}$ is nonsingular. Since the matrix norm is subordinate (and therefore consistent),

$$\|(\mathbf{I}-\mathbf{E})\mathbf{x}\| = \|\mathbf{x}-\mathbf{E}\mathbf{x}\| \geq \|\mathbf{x}\| - \|\mathbf{E}\|\|\mathbf{x}\| = (1-\|\mathbf{E}\|)\|\mathbf{x}\| .$$

Since $\|\mathbf{E}\| < 1$, this inequality implies that $\|(\mathbf{I}-\mathbf{E})\mathbf{x}\| > 0$ whenever $\mathbf{x} \neq \mathbf{0}$. This implies that $\mathscr{N}(\mathbf{I}-\mathbf{E}) = \{\mathbf{0}\}$. As a result, Lemma 3.2.9 proves that $\mathbf{I}-\mathbf{E}$ is nonsingular.

Next, we note that the Definition 3.5.3 of a subordinate matrix norm implies that $\|\mathbf{I}-\mathbf{E}\| \geq 1-\|\mathbf{E}\|$. As a result, Lemma 3.5.8 implies

$$\|(\mathbf{I}-\mathbf{E})^{-1}\| \leq \frac{1}{\|\mathbf{I}-\mathbf{E}\|} \leq \frac{1}{1-\|\mathbf{E}\|} .$$

Let us prove the second claim in the lemma. Since

$$\mathbf{I} = (\mathbf{I}-\mathbf{E})(\mathbf{I}-\mathbf{E})^{-1} = (\mathbf{I}-\mathbf{E})^{-1} - \mathbf{E}(\mathbf{I}-\mathbf{E})^{-1} ,$$

it follows that

$$\mathbf{I}-(\mathbf{I}-\mathbf{E})^{-1} = -\mathbf{E}(\mathbf{I}-\mathbf{E})^{-1} .$$

We can take the norms of the matrices in this equation to obtain

$$\left\|\mathbf{I}-(\mathbf{I}-\mathbf{E})^{-1}\right\| \leq \|\mathbf{E}\| \; \|(\mathbf{I}-\mathbf{E})^{-1}\| \leq \frac{\|\mathbf{E}\|}{1-\|\mathbf{E}\|} .$$

Here is an easy consequence of the previous lemma.

Corollary 3.6.1 *Suppose that $\mathbf{A}$ and $\tilde{\mathbf{A}}$ are two square matrices of the same size, and $\mathbf{A}$ is nonsingular. Let $\|\cdot\|$ be a subordinate matrix norm, and assume that*

$$\left\|\mathbf{A}^{-1}\left(\tilde{\mathbf{A}}-\mathbf{A}\right)\right\| < 1 .$$

Then $\tilde{\mathbf{A}}$ *is nonsingular,*

$$\left\|\tilde{\mathbf{A}}^{-1}\right\| \leq \frac{\|\mathbf{A}\|}{1-\left\|\mathbf{A}^{-1}\left(\tilde{\mathbf{A}}-\mathbf{A}\right)\right\|} . \tag{3.25}$$

and

$$\left\|\tilde{\mathbf{A}}^{-1}-\mathbf{A}^{-1}\right\| \leq \frac{\|\mathbf{A}^{-1}\|\left\|\mathbf{A}^{-1}\left(\tilde{\mathbf{A}}-\mathbf{A}\right)\right\|}{1-\left\|\mathbf{A}^{-1}\left(\tilde{\mathbf{A}}-\mathbf{A}\right)\right\|} . \tag{3.26}$$

Proof Let $\mathbf{E} = \mathbf{A}^{-1}\left(\mathbf{A}-\tilde{\mathbf{A}}\right)$. Then $\|\mathbf{E}\| < 1$, so Lemma 3.6.1 shows that

$$\mathbf{I}-\mathbf{E} = \mathbf{A}^{-1}\tilde{\mathbf{A}}$$

is nonsingular, which implies that $\tilde{\mathbf{A}}$ is nonsingular. As a result,

$$\left\|\tilde{\mathbf{A}}^{-1}\right\| = \left\|\left(\mathbf{A}^{-1}\tilde{\mathbf{A}}\right)^{-1}\mathbf{A}\right\| = \left\|(\mathbf{I}-\mathbf{E})^{-1}\mathbf{A}\right\| \leq \left\|(\mathbf{I}-\mathbf{E})^{-1}\right\| \|\mathbf{A}\| \leq \frac{\|\mathbf{A}\||}{1-\|\mathbf{E}\|}$$
$$= \frac{\|\mathbf{A}\||}{1-\left\|\mathbf{A}^{-1}\left(\mathbf{A}-\tilde{\mathbf{A}}\right)\right\|} .$$

Furthermore,

$$\left\|\tilde{\mathbf{A}}^{-1}-\mathbf{A}^{-1}\right\| = \|(\mathbf{I}-\mathbf{E})^{-1}\mathbf{A}^{-1}-\mathbf{A}^{-1}\| \leq \|(\mathbf{I}-\mathbf{E})^{-1}-\mathbf{I}\| \, \|\mathbf{A}^{-1}\|$$

then inequality (3.24) produces

$$\leq \frac{\|\mathbf{E}\|}{1-\|\mathbf{E}\|}\|\mathbf{A}^{-1}\| = \frac{\|\mathbf{A}^{-1}\|\left\|\mathbf{A}^{-1}\left(\tilde{\mathbf{A}}-\mathbf{A}\right)\right\|}{1-\left\|\mathbf{A}^{-1}\left(\tilde{\mathbf{A}}-\mathbf{A}\right)\right\|} .$$

Now we are ready to develop our first *a priori* error estimate.

Lemma 3.6.2 *Suppose that* $\mathbf{A}$ *is a nonsingular* $n \times n$ *matrix and* $\mathbf{b}$ *is a nonzero* n*-vector. Let* $\|\cdot\|$ *be a norm on* n*-vectors, with the same notation for the subordinate norm on* $n \times n$ *matrices. Assume that* $\tilde{\mathbf{b}}$ *is an* n*-vector and* $\tilde{\mathbf{A}}$ *is an* $n \times n$ *matrix satisfying*

$$\left\|\mathbf{A}^{-1}\left(\tilde{\mathbf{A}}-\mathbf{A}\right)\right\| < 1 . \tag{3.27}$$

Then $\tilde{\mathbf{A}}$ *is nonsingular, and the solution* $\tilde{\mathbf{x}}$ *to the perturbed linear system* $\tilde{\mathbf{A}}\tilde{\mathbf{x}} = \tilde{\mathbf{b}}$ *satisfies*

$$\|\mathbf{x}-\tilde{\mathbf{x}}\| \le \frac{\left\|\mathbf{A}^{-1}\left(\tilde{\mathbf{A}}-\mathbf{A}\right)\right\|}{1-\left\|\mathbf{A}^{-1}\left(\tilde{\mathbf{A}}-\mathbf{A}\right)\right\|}\|\mathbf{x}\| + \left[1+\frac{\left\|\mathbf{A}^{-1}\left(\tilde{\mathbf{A}}-\mathbf{A}\right)\right\|}{1-\left\|\mathbf{A}^{-1}\left(\tilde{\mathbf{A}}-\mathbf{A}\right)\right\|}\right]\left\|\mathbf{A}^{-1}\left(\tilde{\mathbf{b}}-\mathbf{b}\right)\right\| . \tag{3.28}$$

Proof First, we will prove that $\tilde{\mathbf{A}}$ is nonsingular. The triangle inequality implies that for any n-vector $\mathbf{y}$ we have

$$\begin{aligned}\left[1-\left\|\mathbf{A}^{-1}\left(\tilde{\mathbf{A}}-\mathbf{A}\right)\right\|\right]\|\mathbf{y}\| &\le \left\|\mathbf{y}+\mathbf{A}^{-1}\left(\tilde{\mathbf{A}}-\mathbf{A}\right)\mathbf{y}\right\| = \left\|\mathbf{A}^{-1}\tilde{\mathbf{A}}\mathbf{y}\right\| \\ &\le \left\|\mathbf{A}^{-1}\right\|\left\|\tilde{\mathbf{A}}\mathbf{y}\right\| .\end{aligned}$$

This inequality and assumption (3.27) imply that whenever $\mathbf{y} \neq \mathbf{0}$ we must have $\tilde{\mathbf{A}}\mathbf{y} \neq \mathbf{0}$. Then Lemma 3.2.9 implies that $\tilde{\mathbf{A}}$ is nonsingular.

Next, we will prove (3.28). Since

$$\tilde{\mathbf{b}}-\mathbf{b} = \tilde{\mathbf{A}}\tilde{\mathbf{x}}-\mathbf{A}\mathbf{x} = \mathbf{A}\left(\tilde{\mathbf{x}}-\mathbf{x}\right)+\left(\tilde{\mathbf{A}}-\mathbf{A}\right)\tilde{\mathbf{x}} ,$$

it follows that

$$\begin{aligned}\mathbf{A}^{-1}\left(\tilde{\mathbf{b}}-\mathbf{b}\right) &= \tilde{\mathbf{x}}-\mathbf{x}+\mathbf{A}^{-1}\left(\tilde{\mathbf{A}}-\mathbf{A}\right)\tilde{\mathbf{x}} \\ &= \tilde{\mathbf{x}}-\mathbf{x}+\mathbf{A}^{-1}\left(\tilde{\mathbf{A}}-\mathbf{A}\right)\tilde{\mathbf{A}}^{-1}\left[\mathbf{A}\mathbf{x}+\tilde{\mathbf{b}}-\mathbf{b}\right] .\end{aligned}$$

This implies that

$$\begin{aligned}\mathbf{x}-\tilde{\mathbf{x}} &= \mathbf{A}^{-1}\left(\tilde{\mathbf{A}}-\mathbf{A}\right)\tilde{\mathbf{A}}^{-1}\left(\mathbf{A}\mathbf{x}+\tilde{\mathbf{b}}-\mathbf{b}\right)-\mathbf{A}^{-1}\left(\tilde{\mathbf{b}}-\mathbf{b}\right) \\ &= \mathbf{A}^{-1}\left(\tilde{\mathbf{A}}-\mathbf{A}\right)\tilde{\mathbf{A}}^{-1}\mathbf{A}\mathbf{x}-\mathbf{A}^{-1}\left[\mathbf{I}-\left(\tilde{\mathbf{A}}-\mathbf{A}\right)\tilde{\mathbf{A}}^{-1}\right]\left(\tilde{\mathbf{b}}-\mathbf{b}\right) \\ &= \mathbf{A}^{-1}\left(\tilde{\mathbf{A}}-\mathbf{A}\right)\left[\mathbf{I}+\mathbf{A}^{-1}\left(\tilde{\mathbf{A}}-\mathbf{A}\right)\right]^{-1}\mathbf{x} \\ &\quad -\left[\mathbf{I}-\mathbf{A}^{-1}\left(\tilde{\mathbf{A}}-\mathbf{A}\right)\left[\mathbf{I}+\mathbf{A}^{-1}\left(\tilde{\mathbf{A}}-\mathbf{A}\right)\right]^{-1}\right]\mathbf{A}^{-1}\left(\tilde{\mathbf{b}}-\mathbf{b}\right) .\end{aligned}$$

We can take norms and apply Lemma 3.6.1 to get

$$\begin{aligned}\|\mathbf{x}-\tilde{\mathbf{x}}\| \leq& \left\|\mathbf{A}^{-1}\left(\tilde{\mathbf{A}}-\mathbf{A}\right)\right\| \left\|\left(\mathbf{I}+\mathbf{A}^{-1}\left(\tilde{\mathbf{A}}-\mathbf{A}\right)\right)^{-1}\right\| \|\mathbf{x}\| \\ &+ \left\|\mathbf{I}-\mathbf{A}^{-1}\left(\tilde{\mathbf{A}}-\mathbf{A}\right)\left\{\mathbf{I}+\mathbf{A}^{-1}\left(\tilde{\mathbf{A}}-\mathbf{A}\right)\right\}^{-1}\right\| \left\|\mathbf{A}^{-1}\left(\tilde{\mathbf{b}}-\mathbf{b}\right)\right\| \\ \leq& \frac{\left\|\mathbf{A}^{-1}\left(\tilde{\mathbf{A}}-\mathbf{A}\right)\right\|}{1-\left\|\mathbf{A}^{-1}\left(\tilde{\mathbf{A}}-\mathbf{A}\right)\right\|}\|\mathbf{x}\| + \left[1+\frac{\left\|\mathbf{A}^{-1}\left(\tilde{\mathbf{A}}-\mathbf{A}\right)\right\|}{1-\left\|\mathbf{A}^{-1}\left(\tilde{\mathbf{A}}-\mathbf{A}\right)\right\|}\right]\left\|\mathbf{A}^{-1}\left(\tilde{\mathbf{b}}-\mathbf{b}\right)\right\| .\end{aligned}$$

Exercise 3.6.1 Suppose that $\mathbf{u}$ and $\mathbf{v}$ are n-vectors with $\mathbf{u}\cdot\mathbf{v} \neq -1$, and let $\mathbf{E} = \mathbf{u}\mathbf{v}^H$. Use Lemma 3.4.1 to compute $\|(\mathbf{I}-\mathbf{E})^{-1}\|_p$ for $p = 1, 2$ and ∞. Then compare the terms on either side of the inequalities in Lemma 3.6.1.

3.6.2 *Condition Number*

In order to develop our second *a priori* error estimate, we will make use of the following important definition.

Definition 3.6.1 If $\mathbf{A}$ is a nonsingular matrix and $\|\cdot\|$ is a subordinate matrix norm, then the **condition number** of $\mathbf{A}$ is defined to be

$$\kappa(\mathbf{A}) \equiv \|\mathbf{A}\| \, \|\mathbf{A}^{-1}\| .$$

Note that Definition 3.5.3 and Lemma 3.5.3 imply that

$$1 = \|\mathbf{I}\| = \left\|\mathbf{A}\mathbf{A}^{-1}\right\| \leq \|\mathbf{A}\| \left\|\mathbf{A}^{-1}\right\| = \kappa(\mathbf{A}) .$$

Thus the condition number is always greater than or equal to one.

In MATLAB, the condition number of a matrix $\mathbf{A}$ can be computed by the cond command. Note that *MATLAB computes the true condition number*, and therefore *must compute the matrix inverse*.

On the other hand, *LAPACK estimates condition numbers* in order to *avoid computing matrix inverses*. In LAPACK, subroutines `_gecon` estimate either the one or infinity condition number of a general matrix. See, for example, dgecon.f. Other LAPACK routines can compute condition numbers for matrices having a special forms: `_trcon` for triangular matrices, `_sycon` for symmetric matrices, `_pocon` for symmetric positive matrices, `_gtcon` for tridiagonal matrices, and `_ptcon` for symmetric positive tridiagonal matrices. In LAPACK, the condition number with respect to the 2-norm could be computed as the ratio of the largest to smallest singular value, obtained from routines `_gesvd`. For more information about condition number estimates, see Sect. 3.8.7.

Exercise 3.6.2 If

$$\mathbf{A} = \begin{bmatrix} \alpha & \beta \\ \gamma & \delta \end{bmatrix}$$

and $\alpha\delta \neq \beta\gamma$, compute $\kappa_1(\mathbf{A})$ and $\kappa_\infty(\mathbf{A})$.

Exercise 3.6.3 Compute the condition number of an $n \times n$ unitary matrix with respect to the 1, 2 and ∞ norms.

3.6.3 *Relative Error*

We are usually more interested in the **relative error** in the solution $\mathbf{x}$, namely $\|\mathbf{x} - \tilde{\mathbf{x}}\| / \|\mathbf{x}\|$, than in the absolute error $\|\mathbf{x} - \tilde{\mathbf{x}}\|$. Using the condition number, we can derive the following estimate for the relative error in the solution of a system of linear equations.

Theorem 3.6.1 (Relative Error Estimate for Linear Systems) *Suppose that* $\mathbf{A}$ *is a nonsingular* $n \times n$ *matrix and* $\mathbf{b}$ *is a nonzero n-vector. Let* $\| \cdot \|$ *be a norm on n-vectors, with the same notation for the corresponding subordinate norm on* $n \times n$ *matrices. Assume that* $\tilde{\mathbf{b}}$ *is an n-vector, and* $\tilde{\mathbf{A}}$ *is an* $n \times n$ *matrix satisfying*

$$\|\mathbf{A}^{-1}\| \left\|\tilde{\mathbf{A}} - \mathbf{A}\right\| < 1 \,.$$

Then $\tilde{\mathbf{A}}$ *is nonsingular, and the solution* $\tilde{\mathbf{x}}$ *to the perturbed linear system* $\tilde{\mathbf{A}}\tilde{\mathbf{x}} = \tilde{\mathbf{b}}$ *satisfies*

$$\frac{\|\mathbf{x} - \tilde{\mathbf{x}}\|}{\|\mathbf{x}\|} \leq \frac{\kappa(\mathbf{A}) \left\|\tilde{\mathbf{A}} - \mathbf{A}\right\| / \|\mathbf{A}\|}{1 - \kappa(\mathbf{A}) \left\|\tilde{\mathbf{A}} - \mathbf{A}\right\| / \|\mathbf{A}\|} + \left[1 + \frac{\kappa(\mathbf{A}) \left\|\tilde{\mathbf{A}} - \mathbf{A}\right\| / \|\mathbf{A}\|}{1 - \kappa(\mathbf{A}) \left\|\tilde{\mathbf{A}} - \mathbf{A}\right\| / \|\mathbf{A}\|}\right] \kappa(\mathbf{A}) \frac{\left\|\tilde{\mathbf{b}} - \mathbf{b}\right\|}{\|\mathbf{b}\|} \,. \tag{3.29}$$

Proof Because

$$\left\|\mathbf{A}^{-1}\left(\tilde{\mathbf{A}} - \mathbf{A}\right)\right\| \leq \|\mathbf{A}^{-1}\| \left\|\tilde{\mathbf{A}} - \mathbf{A}\right\| < 1 \,,$$

Lemma 3.6.2 proves that $\tilde{\mathbf{A}}$ is nonsingular. Since

$$1 - \left\|\mathbf{A}^{-1}\left(\tilde{\mathbf{A}} - \mathbf{A}\right)\right\| \geq 1 - \|\mathbf{A}^{-1}\| \left\|\tilde{\mathbf{A}} - \mathbf{A}\right\| > 0 \,,$$

Lemma 3.6.2 also proves that

$$
\begin{aligned}
\frac{\|\mathbf{x}-\tilde{\mathbf{x}}\|}{\|\mathbf{x}\|} &\leq \frac{\left\|\mathbf{A}^{-1}\left(\tilde{\mathbf{A}}-\mathbf{A}\right)\right\|}{1-\left\|\mathbf{A}^{-1}\left(\tilde{\mathbf{A}}-\mathbf{A}\right)\right\|}+\left[1+\frac{\left\|\mathbf{A}^{-1}\left(\tilde{\mathbf{A}}-\mathbf{A}\right)\right\|}{1-\left\|\mathbf{A}^{-1}\left(\tilde{\mathbf{A}}-\mathbf{A}\right)\right\|}\right]\frac{\left\|\mathbf{A}^{-1}\left(\tilde{\mathbf{b}}-\mathbf{b}\right)\right\|}{\|\mathbf{x}\|} \\
&\leq \frac{\|\mathbf{A}^{-1}\|\left\|\tilde{\mathbf{A}}-\mathbf{A}\right\|}{1-\|\mathbf{A}^{-1}\|\left\|\tilde{\mathbf{A}}-\mathbf{A}\right\|} \\
&+\left[1+\frac{\|\mathbf{A}^{-1}\|\left\|\tilde{\mathbf{A}}-\mathbf{A}\right\|}{1-\|\mathbf{A}^{-1}\|\left\|\tilde{\mathbf{A}}-\mathbf{A}\right\|}\right]\|\mathbf{A}^{-1}\|\frac{\left\|\tilde{\mathbf{b}}-\mathbf{b}\right\|}{\|\mathbf{x}\|}\frac{\|\mathbf{A}\|\,\|\mathbf{x}\|}{\|\mathbf{b}\|} \\
&= \frac{\kappa\left(\mathbf{A}\right)\left\|\tilde{\mathbf{A}}-\mathbf{A}\right\|/\|\mathbf{A}\|}{1-\kappa\left(\mathbf{A}\right)\left\|\tilde{\mathbf{A}}-\mathbf{A}\right\|/\|\mathbf{A}\|} \\
&+\left[1+\frac{\kappa\left(\mathbf{A}\right)\left\|\tilde{\mathbf{A}}-\mathbf{A}\right\|/\|\mathbf{A}\|}{1-\kappa\left(\mathbf{A}\right)\left\|\tilde{\mathbf{A}}-\mathbf{A}\right\|/\|\mathbf{A}\|}\right]\kappa\left(\mathbf{A}\right)\frac{\|\tilde{\mathbf{b}}-\mathbf{b}\|}{\|\mathbf{b}\|} \; .
\end{aligned}
$$

Inequality (3.29) shows that the relative errors in the data can be magnified by the condition number to produce the relative error in the solution. In other words, linear systems involving matrices with large condition numbers can produce large perturbations in their solutions due to small changes in their data.

Example 3.6.1 Suppose that

$$\mathbf{A}=\begin{bmatrix}0 & 1\\ 1 & 1\end{bmatrix} \text{ and } \mathbf{b}=\begin{bmatrix}1\\ 2\end{bmatrix} .$$

Note that

$$\mathbf{A}^{-1}=\begin{bmatrix}-1 & 1\\ 1 & 0\end{bmatrix} ,$$

so

$$\|\mathbf{A}\|_1 = 2 = \|\mathbf{A}\|_\infty \text{ and } \left\|\mathbf{A}^{-1}\right\|_1 = 2 = \left\|\mathbf{A}^{-1}\right\|_\infty .$$

It follows that

$$\kappa_1\left(\mathbf{A}\right) = 4 = \kappa_\infty\left(\mathbf{A}\right) .$$

The solution of $\mathbf{Ax} = \mathbf{b}$ is

$$\mathbf{x} = \begin{bmatrix} 1 \\ 1 \end{bmatrix} .$$

Next, let

$$\tilde{\mathbf{A}} = \begin{bmatrix} \delta & 1 \\ 1 & 1 \end{bmatrix} .$$

Then the solution of $\tilde{\mathbf{A}}\tilde{\mathbf{x}} = \mathbf{b}$ is

$$\tilde{\mathbf{x}} = \begin{bmatrix} 1 \\ 1 - 2\delta \end{bmatrix} \frac{1}{1 - \delta} ,$$

The exact error is

$$\mathbf{x} - \tilde{\mathbf{x}} = \begin{bmatrix} -1 \\ 1 \end{bmatrix} \frac{\delta}{1 - \delta} .$$

Taking norms, we find that the relative error is

$$\frac{\|\mathbf{x} - \tilde{\mathbf{x}}\|_1}{\|\mathbf{x}\|_1} = \frac{2\delta}{1 - \delta} \text{ and } \frac{\|\mathbf{x} - \tilde{\mathbf{x}}\|_\infty}{\|\mathbf{x}\|_\infty} = \frac{\delta}{1 - \delta} .$$

On the other hand, for either choice of vector norm the error bound from Theorem 3.6.1 is

$$\frac{\|\mathbf{x} - \tilde{\mathbf{x}}\|}{\|\mathbf{x}\|} \le \frac{\kappa\left(\mathbf{A}\right) \left\|\tilde{\mathbf{A}} - \mathbf{A}\right\| / \|\mathbf{A}\|}{1 - \kappa\left(\mathbf{A}\right) \left\|\tilde{\mathbf{A}} - \mathbf{A}\right\| / \|\mathbf{A}\|} = \frac{2\delta}{1 - 2\delta} .$$

Example 3.6.2 Readers sometimes assume that matrices with large components in their inverse should cause solutions of linear systems to have entries that are large relative to the right-hand side. To see that this is not necessarily the case, consider the linear system $\mathbf{Ax} = \mathbf{b}$ with

$$\mathbf{A} = \begin{bmatrix} 1 & 1 \\ -\delta & \delta \end{bmatrix} \text{ and } \mathbf{b} = \begin{bmatrix} 1 \\ 0 \end{bmatrix}$$

where δ is assumed to be small. The solution is

$$\mathbf{x} = \begin{bmatrix} 1/2 \\ 1/2 \end{bmatrix} .$$

In this case we have

$$\mathbf{A}^{-1} = \begin{bmatrix} 1/2 & -1/(2\delta) \\ 1/2 & 1/(2\delta) \end{bmatrix} ,$$

so the condition number is

$$\kappa(\mathbf{A}) = 1 + \frac{1}{\delta}$$

with respect to either the 1-norm or the ∞-norm. If δ is small, the condition number is large.

Next, suppose that we solve the nearby linear system $\mathbf{A}\tilde{\mathbf{x}} = \tilde{\mathbf{b}}$ with

$$\tilde{\mathbf{b}} = \begin{bmatrix} 1 \\ \delta \end{bmatrix} .$$

The solution of the linear system is

$$\tilde{\mathbf{x}} = \begin{bmatrix} 0 \\ 1 \end{bmatrix} .$$

For either the 1-norm or the ∞-norm, the relative error in the right-hand side is

$$\frac{\left\|\tilde{\mathbf{b}} - \mathbf{b}\right\|}{\|\mathbf{b}\|} = \delta ,$$

and the relative error in the solution is

$$\frac{\|\tilde{\mathbf{x}} - \mathbf{x}\|}{\|\mathbf{x}\|} = 1 .$$

The large condition number produced a large relative error in the solution, without producing a large solution.

Exercise 3.6.4 For $p = 1$ and ∞, compute the condition number of the matrix $\mathbf{A}$ in Example 3.6.1, as well as $\|\mathbf{A}\|_p$, $\|\mathbf{E}\|_p$, $\|\mathbf{x}\|_p$ and $\|\mathbf{x} - \tilde{\mathbf{x}}\|_p$. Then verify that these terms satisfy the error estimate (3.29).

Exercise 3.6.5 Use MATLAB to compute $\kappa_1(\mathbf{H})$ where $\mathbf{H}$ is the 12×12 **Hilbert matrix** with entries

$$\mathbf{H}_{ij} = \left[\frac{1}{i+j-1}\right] \text{ for } 1 \le i,j \le 12 .$$

3.6.4 A Posteriori *Error Estimate*

The error estimates (3.28) and (3.29) are *a priori* error estimates, meaning that they can be conducted *before* solving the linear system $\mathbf{A}\mathbf{x} = \mathbf{b}$, provided that we can bound the perturbations to $\mathbf{A}$ and $\mathbf{b}$. Our goal in this section is to develop *a posteriori* error estimates, to use *after* we compute an approximate solution to a system of linear equations.

We will develop two kinds of error estimates, which are described by the following definition.

Definition 3.6.2 Suppose that $\mathbf{f}$ is a function, $\mathbf{x}$ is an argument for $\mathbf{f}$, and $\tilde{\mathbf{x}}$ is an approximation to $\mathbf{x}$. Then the forward error **forward error** $\tilde{\mathbf{x}} - \mathbf{x}$ is measured in terms of $\mathbf{f}(\tilde{\mathbf{x}}) - \mathbf{f}(\mathbf{x})$. If $\tilde{\mathbf{f}}$ is an approximation to $\mathbf{f}(\mathbf{x})$, then the **backward error** is the smallest value of $\tilde{\mathbf{x}} - \mathbf{x}$ for which

$$\mathbf{f}(\tilde{\mathbf{x}}) = \tilde{\mathbf{f}} ,$$

and is measured in terms of $\tilde{\mathbf{f}} - \mathbf{f}(\mathbf{x})$.

For specific problems, we will develop appropriate methods for estimating forward and backward errors. However, we can provide the following general discussion. If there is a constant ℓ so that for all $\tilde{\mathbf{x}}$ near $\mathbf{x}$ we have

$$\|\mathbf{f}(\tilde{\mathbf{x}}) - \mathbf{f}(\mathbf{x})\| \geq \ell \|\tilde{\mathbf{x}} - \mathbf{x}\| ,$$

then we can estimate the absolute forward error by

$$\|\tilde{\mathbf{x}} - \mathbf{x}\| \leq \frac{1}{\ell} \|\mathbf{f}(\tilde{\mathbf{x}}) - \mathbf{f}(\mathbf{x})\| .$$

If there is a constant L so that for all $\tilde{\mathbf{x}}$ near $\mathbf{x}$ we have

$$\|\mathbf{f}(\tilde{\mathbf{x}}) - \mathbf{f}(\mathbf{x})\| \leq L \|\tilde{\mathbf{x}} - \mathbf{x}\| ,$$

then the absolute backward error can be estimated by

$$\|\tilde{\mathbf{x}} - \mathbf{x}\| \geq \frac{1}{L} \left\|\tilde{\mathbf{f}} - \mathbf{f}(\mathbf{x})\right\| .$$

In this case, the constant L would be a **Lipschitz continuity constant** for f, and ℓ would be a Lipschitz continuity constant for f^{-1}.

Theorem 3.6.2 (A Posteriori Error Estimation for Linear Systems) *Let $\|\cdot\|$ denote a vector norm, or a corresponding consistent matrix norm. Suppose that* $\mathbf{A}$ *is a nonsingular $n \times n$ matrix,* $\mathbf{b}$ *is a nonzero n-vector, and* $\mathbf{x}$ *solves* $\mathbf{A}\mathbf{x} = \mathbf{b}$*. Given an n-vector $\tilde{\mathbf{x}}$, define the* ***residual*** *by*

$$\mathbf{r} = \mathbf{b} - \mathbf{A}\tilde{\mathbf{x}} .$$

Then $\tilde{\mathbf{x}}$ *satisfies the* forward error estimate

$$\frac{\|\mathbf{x}-\tilde{\mathbf{x}}\|}{\|\mathbf{x}\|} \leq \kappa(\mathbf{A}) \frac{\|\mathbf{r}\|}{\|\mathbf{b}\|} . \tag{3.30}$$

If all of the entries of $\tilde{\mathbf{x}}$ *are nonzero, then there is a diagonal matrix* $\mathbf{E}$ *such that*

$$\mathbf{A}(\mathbf{I}+\mathbf{E})\tilde{\mathbf{x}} = \mathbf{b} .$$

Furthermore, $\mathbf{E}\tilde{\mathbf{x}} = \mathbf{x}-\tilde{\mathbf{x}}$, *and the diagonal entries of* $\mathbf{E}$ *provide the* ***backward error estimate***

$$\max_j \left|\mathbf{E}_{jj}\right| \geq \max_{1\leq i\leq n} \frac{|\mathbf{r}_i|}{\sum_{j=1}^n \left|\mathbf{A}_{ij}\right| \left|\tilde{\mathbf{x}}_j\right|} \tag{3.31}$$

Proof Note that $\mathbf{A}^{-1}\mathbf{r} = \mathbf{A}^{-1}\mathbf{b} - \tilde{\mathbf{x}} = \mathbf{x} - \tilde{\mathbf{x}}$, so

$$\|\mathbf{x}-\tilde{\mathbf{x}}\| = \left\|\mathbf{A}^{-1}\mathbf{r}\right\| \leq \left\|\mathbf{A}^{-1}\right\| \|\mathbf{r}\| .$$

Since $\|\mathbf{b}\| = \|\mathbf{A}\mathbf{x}\| \leq \|\mathbf{A}\| \|\mathbf{x}\|$, we notice that

$$\|\mathbf{x}\| \geq \|\mathbf{b}\|/\|\mathbf{A}\| ,$$

and conclude that

$$\frac{\|\mathbf{x}-\tilde{\mathbf{x}}\|}{\|\mathbf{x}\|} \leq \frac{\left\|\mathbf{A}^{-1}\right\| \|\mathbf{r}\|}{\|\mathbf{b}\|/\|\mathbf{A}\|} = \kappa(\mathbf{A}) \frac{\|\mathbf{r}\|}{\|\mathbf{b}\|} .$$

This proves the forward error estimate (3.30).

It is easy to see that

$$\mathbf{A}\mathbf{E}\tilde{\mathbf{x}} = \mathbf{b} - \mathbf{A}\tilde{\mathbf{x}} = \mathbf{r} ,$$

so

$$\mathbf{E}\tilde{\mathbf{x}} = \mathbf{A}^{-1}\mathbf{r} .$$

If all of the components of $\tilde{\mathbf{x}}$ are nonzero, we can use this equation to define the diagonal entries of $\mathbf{E}$. Next, let us prove the backward error estimate (3.31). For $1 \leq i \leq n$, we have

$$|\mathbf{r}_i| = \left|\sum_j \mathbf{A}_{ij}\mathbf{E}_{jj}\tilde{\mathbf{x}}_j\right| \leq \max_{1\leq j\leq n} \left|\mathbf{E}_{jj}\right| \sum_{j=1}^n \left|\mathbf{A}_{ij}\right| \left|\tilde{\mathbf{x}}_j\right| .$$

Inequality (3.31) now follows easily.

The forward error estimate (3.30), is commonly found in numerical linear algebra texts. See, for example, Stewart [97, Theorem 4.3, p. 196]. This error estimate bounds the largest relative error in the solution that could result from a residual of the given size. This error estimate shows that the relative error in the solution may not be small whenever the condition number is large, or whenever the residual is not small relative to the right-hand side. Note that *this forward error estimate assumes that the residual is computed exactly*. In practice, this error estimate is a bit too optimistic; a better *a posteriori* forward error estimate will be found in inequality (3.49).

The backward error estimate determines a bound on the *smallest* relative change in the approximate solution that would produce the exact solution. Consequently, in this error estimate it is not necessary to worry about rounding errors in the computation of the residual. If all components of the residual are small compared to the corresponding entry of $|\mathbf{A}|\ |\mathbf{x}|$, then the error estimate (3.31) shows that the backward error is small. Here we have used the notation that $|\mathbf{M}|$ denotes the array whose entries are the absolute values of the corresponding entries of $\mathbf{M}$.

The backward- and forward-error estimates (3.31) and (3.30) can be computed by LAPACK routines `_lacn2`, in this situation as called from LAPACK routines `_gerfs`. The LAPACK forward error estimate includes the effect of rounding errors in computing the residual, as described in Theorem 3.8.1.

Example 3.6.3 The linear system in Example 3.6.2 has

$$\mathbf{A} = \begin{bmatrix} 1 & 1 \\ -\delta & \delta \end{bmatrix} \text{ and } \mathbf{b} = \begin{bmatrix} 1 \\ 0 \end{bmatrix} .$$

The exact solution to this linear system is

$$\mathbf{x} = \begin{bmatrix} 1 \\ 1 \end{bmatrix} .$$

If we choose the approximate solution

$$\tilde{\mathbf{x}} = \begin{bmatrix} 0 \\ 1 \end{bmatrix} ,$$

then the corresponding residual is

$$\mathbf{r} = \begin{bmatrix} 1 \\ 0 \end{bmatrix} - \begin{bmatrix} 1 & 1 \\ -\delta & \delta \end{bmatrix} \begin{bmatrix} 0 \\ 1 \end{bmatrix} = \begin{bmatrix} 0 \\ -\delta \end{bmatrix} .$$

If δ is small compared to one, then the residual will be small, even though the error in the approximate solution is not small.

Recall that

$$\mathbf{A}^{-1} = \begin{bmatrix} 1/2 & -1/(2\delta) \\ 1/2 & 1/(2\delta) \end{bmatrix} .$$

It is easy to see that

$$\kappa(\mathbf{A}) = 1/\delta + 1$$

with respect to either the 1-norm or the ∞-norm. This large condition number will allow approximate solutions with small residuals to have large relative errors.

The forward error estimate (3.30) gives us

$$\frac{\|\mathbf{x} - \tilde{\mathbf{x}}\|}{\|\mathbf{x}\|} \leq \kappa(\mathbf{A}) \frac{\|\mathbf{r}\|}{\|\mathbf{b}\|} = 1 + \delta$$

for both the 1-norm and the ∞-norm. For the backward error estimate, we compute

$$|\mathbf{A}| \, |\mathbf{x}| = \begin{bmatrix} 1 & 1 \\ \delta & \delta \end{bmatrix} \begin{bmatrix} 0 \\ 1 \end{bmatrix} = \begin{bmatrix} 1 \\ \delta \end{bmatrix} .$$

Then the backward error estimate (3.31) gives us

$$\max_j \left|\mathbf{E}_{jj}\right| \geq \frac{\delta}{\delta} = 1 .$$

Since we are fortunate enough to know the true solution to this problem, we can see that $\|\mathbf{x} - \tilde{\mathbf{x}}\|_1 = 1 = \|\mathbf{x} - \tilde{\mathbf{x}}\|_\infty$. These true errors are consistent with both error estimates.

Exercise 3.6.6 Let

$$\mathbf{A} = \begin{bmatrix} 1 & -2 \\ \delta & \delta \end{bmatrix} , \; \mathbf{b} = \begin{bmatrix} 6 \\ 12\delta \end{bmatrix} \text{ and } \tilde{\mathbf{x}} = \begin{bmatrix} 8 \\ 4 \end{bmatrix} .$$

1. Compute $\mathbf{A}^{-1}$, $\kappa_1(\mathbf{A})$ and $\kappa_\infty(\mathbf{A})$.
2. Compute $\mathbf{r} = \mathbf{b} - \mathbf{A}\tilde{\mathbf{x}}$ and $|\mathbf{A}| \, |\tilde{\mathbf{x}}|$.
3. Determine the backward error estimate for the relative error in $\tilde{\mathbf{x}}$.
4. Determine the forward error estimate for the relative error in $\tilde{\mathbf{x}}$ with respect to the 1-norm and the ∞-norm.
5. Find the exact solution $\mathbf{x}$ to $\mathbf{A}\mathbf{x} = \mathbf{b}$.
6. Compare the forward and backward error estimates to the true error in the solution.

3.7 Gaussian Factorization

Let us review our progress in applying scientific computing to linear algebra. At this point in the chapter, we have studied fundamental concepts of linear algebra, examined some easily solved linear systems, and developed error estimates. In particular, we have seen that a system of linear equations is well-posed if its matrix is nonsingular and has a condition number that is not large. According to our discussion in Sect. 1.3, we have performed the first two steps in using scientific computing to solve a system of linear equations.

Our next step is to develop numerical methods for solving linear systems. We have already developed some computational methods for simple linear systems in Sect. 3.4. Our goal in this section is to develop an effective and efficient numerical method to solve general systems of linear equations. We will accomplish this goal by finding a way to transform general linear systems into easily solved triangular or trapezoidal linear systems. The latter can be solved by the methods discussed in Sects. 3.4.4 and 3.4.5.

We will present three alternative methods for transforming a system of linear equations to triangular form. These will vary substantially in cost and numerical stability. In this section, we will compare the alternatives for cost; comparative discussions of numerical stability will be presented in Sect. 3.8.

3.7.1 No Pivoting

We will begin by showing that we can convert most $m \times n$ matrices into products of left-trapezoidal and a right-trapezoidal matrices. This process is called **Gaussian factorization**.

3.7.1.1 Algorithm

In the first step of Gaussian factorization, we partition the matrix to separate the first row and column, and then factor:

$$\begin{bmatrix} \alpha_{11} & \mathbf{a}_{12}^\top \\ \mathbf{a}_{21} & \mathbf{A}_{22} \end{bmatrix} = \begin{bmatrix} 1 & \mathbf{0}^\top \\ \boldsymbol{\ell}_{21} & \mathbf{I} \end{bmatrix} \begin{bmatrix} \alpha_{11} & \mathbf{a}_{12}^\top \\ \mathbf{0} & \mathbf{C}_{22} \end{bmatrix} .$$

The form of this factorization has the following implications:

$$\begin{aligned} &\alpha_{11} \neq 0 , \\ &\mathbf{a}_{21} = \boldsymbol{\ell}_{21}\alpha_{11} \Longrightarrow \boldsymbol{\ell}_{21} = \mathbf{a}_{21}/\alpha_{11} \text{ and} \\ &\mathbf{A}_{22} = \boldsymbol{\ell}_{21}\mathbf{a}_{12}^\top + \mathbf{C}_{22} \Longrightarrow \mathbf{C}_{22} = \mathbf{A}_{22} - \boldsymbol{\ell}_{21}\mathbf{a}_{12}^\top . \end{aligned}$$

At this first step, α_{11} is called the **pivot**. It is the only number involved in a division at this step.

Next, we will describe the general step in Gaussian factorization. Let us assume that $k-1$ steps of Gaussian factorization have been successful, where $k \geq 2$. At the k-th step, we modify the partitioning of the previous factorization (by moving the partitioning lines over another row and column), and then update the factorization:

$$\begin{bmatrix} \mathbf{A}_{11} & \mathbf{a}_{12} & \mathbf{A}_{13} \\ \mathbf{a}_{21}{}^\top & \alpha_{22} & \mathbf{a}_{23}{}^\top \\ \mathbf{A}_{31} & \mathbf{a}_{32} & \mathbf{A}_{33} \end{bmatrix} = \begin{bmatrix} \mathbf{L}_{11} & \mathbf{0} & \mathbf{0} \\ \boldsymbol{\ell}_{21}{}^\top & 1 & \mathbf{0} \\ \mathbf{L}_{31} & \mathbf{0} & \mathbf{I} \end{bmatrix} \begin{bmatrix} \mathbf{R}_{11} & \mathbf{r}_{12} & \mathbf{R}_{13} \\ \mathbf{0} & \beta_{22} & \mathbf{b}_{23}{}^\top \\ \mathbf{0} & \mathbf{b}_{32} & \mathbf{B}_{33} \end{bmatrix}$$

$$= \begin{bmatrix} \mathbf{L}_{11} & \mathbf{0} & \mathbf{0} \\ \boldsymbol{\ell}_{21}{}^\top & 1 & \mathbf{0} \\ \mathbf{L}_{31} & \boldsymbol{\ell}_{32} & \mathbf{I} \end{bmatrix} \begin{bmatrix} \mathbf{R}_{11} & \mathbf{r}_{12} & \mathbf{R}_{13} \\ \mathbf{0} & \beta_{22} & \mathbf{b}_{23}{}^\top \\ \mathbf{0} & \mathbf{0} & \mathbf{C}_{33} \end{bmatrix} .$$

Here $\mathbf{A}_{11}$, $\mathbf{L}_{11}$ and $\mathbf{R}_{11}$ are all $(k-1) \times (k-1)$ matrices. The k-th step has the following implications:

$$\begin{aligned} &k \leq \min\{m, n\} , \\ &\beta_{22} \neq \mathbf{0} , \\ &\mathbf{b}_{32} = \boldsymbol{\ell}_{32}\beta_{22} \Longrightarrow \boldsymbol{\ell}_{32} = \mathbf{b}_{32}/\beta_{22} , \text{ and} \\ &\mathbf{B}_{33} = \boldsymbol{\ell}_{32}\mathbf{b}_{23}{}^\top + \mathbf{C}_{33} \Longrightarrow \mathbf{C}_{33} = \mathbf{B}_{33} - \boldsymbol{\ell}_{32}\mathbf{b}_{23}{}^\top . \end{aligned}$$

In this step, β_{22} is the pivot.

For the implementation of Gaussian factorization in a computer program, it is unnecessary to store the entries that are known to be zero or one. As a result, we can represent the first step in Gaussian factorization by the **tableau**

$$\begin{bmatrix} \alpha_{11} & \mathbf{a}_{12}{}^\top \\ \mathbf{a}_{21} & \mathbf{A}_{22} \end{bmatrix} \longrightarrow \begin{bmatrix} \alpha_{11} & \mathbf{a}_{12}{}^\top \\ \boldsymbol{\ell}_{21} & \mathbf{C}_{22} \end{bmatrix} \text{ where } \begin{array}{l} \boldsymbol{\ell}_{21} = \mathbf{a}_{21}/\alpha_{11} \text{ and} \\ \mathbf{C}_{22} = \mathbf{A}_{22} - \boldsymbol{\ell}_{21}\mathbf{a}_{12}{}^\top \end{array} .$$

The k-th tableau step takes the form

$$\begin{bmatrix} \mathbf{L}_{11}\backslash\mathbf{R}_{11} & \mathbf{r}_{12} & \mathbf{R}_{13} \\ \boldsymbol{\ell}_{21}{}^\top & \beta_{22} & \mathbf{b}_{23}{}^\top \\ \mathbf{L}_{31} & \mathbf{b}_{32} & \mathbf{B}_{33} \end{bmatrix} \longrightarrow \begin{bmatrix} \mathbf{L}_{11}\backslash\mathbf{R}_{11} & \mathbf{r}_{12} & \mathbf{R}_{13} \\ \boldsymbol{\ell}_{21}{}^\top & \beta_{22} & \mathbf{b}_{23}{}^\top \\ \mathbf{L}_{31} & \boldsymbol{\ell}_{32} & \mathbf{C}_{33} \end{bmatrix} \text{ where } \begin{array}{l} \boldsymbol{\ell}_{32} = \mathbf{b}_{32}/\beta_{22} \text{ and} \\ \mathbf{C}_{33} = \mathbf{B}_{33} - \boldsymbol{\ell}_{32}\mathbf{b}_{23}{}^\top \end{array} .$$

If Gaussian factorization never finds a zero pivot, then the result is the factorization $\mathbf{A} = \mathbf{L}\mathbf{R}$, where $\mathbf{L}$ is unit left trapezoidal and $\mathbf{R}$ is right trapezoidal.

If $\mathbf{A}$ has entries α_{ij}, it is easy to write Gaussian factorization as the following algorithm:

Algorithm 3.7.1 (Gaussian Factorization)

$$
\begin{array}{l}
\text{for } 1 \le k \le \min\{m,n\} \\
\quad \text{if } \alpha_{kk} = 0 \text{ break} \\
\quad \text{for } k+1 \le i \le m \\
\quad\quad \alpha_{i,k} = \alpha_{i,k}/\alpha_{k,k} \\
\quad \text{for } k+1 \le j \le n \\
\quad\quad \text{for } k+1 \le i \le m \\
\quad\quad\quad \alpha_{i,j} = \alpha_{i,j} - \alpha_{i,k} * \alpha_{k,j}
\end{array}
$$

After we have employed Gaussian factorization to factor $\mathbf{A}$, the linear system $\mathbf{Ax} = \mathbf{b}$ can be solved in two steps. First we forward-solve $\mathbf{Ly} = \mathbf{b}$, and then we back-solve $\mathbf{Rx} = \mathbf{y}$. Recall that the solution of such trapezoidal systems was discussed in Sect. 3.4.5.

Example 3.7.1 Suppose we want to solve

$$
\begin{aligned}
2\xi_1 + 3\xi_2 &= 3 \\
-2\xi_1 + \xi_2 &= -7
\end{aligned}
\;. \tag{3.32}
$$

First, we factor the matrix of coefficients in this linear system:

$$
\begin{bmatrix} 2 & 3 \\ -2 & 1 \end{bmatrix} = \begin{bmatrix} 1 & 0 \\ -1 & 1 \end{bmatrix} \begin{bmatrix} 2 & 3 \\ 0 & 4 \end{bmatrix} .
$$

If we choose, we can multiply $\mathbf{LR}$ to check that $\mathbf{LR} = \mathbf{A}$. Next, we forward-solve $\mathbf{Ly} = \mathbf{b}$:

$$
\begin{bmatrix} 1 & 0 \\ -1 & 1 \end{bmatrix} \begin{bmatrix} \eta_1 \\ \eta_2 \end{bmatrix} = \begin{bmatrix} 3 \\ -7 \end{bmatrix} \Longrightarrow \mathbf{y} = \begin{bmatrix} 3 \\ -4 \end{bmatrix} .
$$

We may multiply $\mathbf{Ly}$ to check that $\mathbf{Ly} = \mathbf{b}$. Finally, we back-solve $\mathbf{Rx} = \mathbf{y}$:

$$
\begin{bmatrix} 2 & 3 \\ 0 & 4 \end{bmatrix} \begin{bmatrix} \xi_1 \\ \xi_2 \end{bmatrix} = \begin{bmatrix} 3 \\ -4 \end{bmatrix} \Longrightarrow \mathbf{x} = \begin{bmatrix} 3 \\ -1 \end{bmatrix} .
$$

We may multiply $\mathbf{Rx}$ to check that $\mathbf{Rx} = \mathbf{y}$. Even better, we should check that $\mathbf{Ax} = \mathbf{b}$.

Example 3.7.2 Suppose that we want to solve a linear system with matrix

$$\mathbf{A} = \begin{bmatrix} \delta & 1 \\ 2 & 3 \end{bmatrix} .$$

If δ is a very small power of two, Gaussian factorization in floating-point arithmetic will factor

$$\begin{bmatrix} \delta & 1 \\ 2 & 3 \end{bmatrix} = \begin{bmatrix} 1 & 0 \\ fl(2/\delta) & 1 \end{bmatrix} \begin{bmatrix} \delta & 1 \\ 0 & fl(3 - 2/\delta) \end{bmatrix} = \begin{bmatrix} 1 & 0 \\ 2/\delta & 1 \end{bmatrix} \begin{bmatrix} \delta & 1 \\ 0 & -2/\delta \end{bmatrix} = \mathbf{L}\mathbf{R} .$$

It is easy to see that

$$\mathbf{L}\mathbf{R} = \begin{bmatrix} \delta & 1 \\ 2 & 0 \end{bmatrix} \Longrightarrow \mathbf{L}\mathbf{R} - \mathbf{A} = \begin{bmatrix} 0 & 0 \\ 0 & -3 \end{bmatrix} .$$

In this case, the existence of a very large sub-diagonal entry in $\mathbf{L}$ produced a large error in the factorization.

3.7.1.2 Computational Cost

Suppose that this algorithm finds r nonzero pivots. If we add up the operations in this Gaussian Elimination Algorithm, we find that there are

- $\sum_{k=1}^{r}(m-k) = r(2m-r-1)/2$ divisions, and
- $\sum_{k=1}^{r}(m-k)(n-k) = r\left[6mn - (r+1)(3m+3n-2r-1)\right]/6$ multiplications and additions.

If $m = r = n$, then Gaussian elimination involves $n(n-1)(2n-1)/6$ multiplications and additions, as well as $n(n-1)/2$ divisions. For large n, the total work is on the order of $n^3/3$ multiplications and additions.

3.7.1.3 Software

In order to see how Gaussian elimination with pivoting might be programmed, readers may view either the Fortran Gaussian factorization program gaussElimNoPiv.f or the C^{++} program gaussElimNoPiv.C. Better yet, readers may also experiment with the JavaScript Gaussian factorization program gaussElimNoPiv.html. This program requires that the matrix be entered by rows, with row entries separately by commas and rows separated by semi-colons (e.g., `1,2;3,4`). The factorization program will stop whenever a zero pivot is encountered, and act as if all unprocessed array entries are zero.

3.7.1.4 Summary

In summary, we can use Gaussian factorization to solve a system of linear equations. If we want to solve $\mathbf{A}\mathbf{x} = \mathbf{b}$, we simply perform the following

Algorithm 3.7.2 (Solve Linear System via Gaussian Factorization)

factor $\mathbf{A} = \mathbf{L}\mathbf{R}$ (see Algorithm 3.7.1)

solve $\mathbf{L}\mathbf{y} = \mathbf{b}$ for $\mathbf{y}$ (see Algorithm 3.4.3)

solve $\mathbf{R}\mathbf{x} = \mathbf{y}$ for $\mathbf{x}$ (see Algorithm 3.4.5)

For a nonsingular $n \times n$ matrix, the first step (Gaussian factorization) costs on the order of $n^3/3$ multiplications and additions. For large n, this is much more work than is performed in the subsequent forward- and back-substitution steps, which, as the reader should recall, each involve on the order of $n^2/2$ multiplications and additions. Thus it is advantageous to save the Gaussian factorization $\mathbf{A} = \mathbf{L}\mathbf{R}$, in case more than one linear system involving the matrix $\mathbf{A}$ will need to be solved.

Exercise 3.7.1 Make a table of the total number of multiply-adds in forward-substitution (either Algorithm 3.4.2 or 3.4.3) for various values of n, ranging in powers of 10 from 10^1 to 10^9. In the same table, enter the number of multiply-adds for Gaussian factorization for each value of n, Assuming that a computer can perform 10^8 multiply-adds in one second, estimate the number of seconds required by forward-substitution and by Gaussian factorization for each entry in your table.

Exercise 3.7.2 Implement two forms of Gaussian factorization in the same programming language, by switching the order of the loops over i and j in Algorithm 3.7.1. Compare the timings for each. (See Sect. 2.6.1 for computer architecture issues to consider in your comparison.)

3.7.2 *Full Pivoting*

Example 3.7.2 demonstrated that a well-conditioned system of linear equations is not necessarily solved accurately by Gaussian factorization. Fortunately, there is a simple fix for such a difficulty in most situations.

Example 3.7.3 Suppose that we want to solve

$$\begin{bmatrix} \delta & 1 \\ 2 & 3 \end{bmatrix} \begin{bmatrix} \xi_1 \\ \xi_2 \end{bmatrix} = \begin{bmatrix} 1 \\ 0 \end{bmatrix} .$$

Here, we assume that δ is so small that $fl(2-3\delta)=2$. The analytical solution of the linear system is

$$\mathbf{x}=\begin{bmatrix}\xi_1\\\xi_2\end{bmatrix}=\begin{bmatrix}-3\\2\end{bmatrix}\frac{1}{2-3\delta}\approx\begin{bmatrix}-3/2\\1\end{bmatrix}.$$

Recall that the matrix in this linear system appeared in Example 3.7.2, and produced significant floating point errors in Gaussian factorization without pivoting.

Suppose that we reorder the equations and unknowns to get the linear system

$$\begin{bmatrix}3&2\\1&\delta\end{bmatrix}\begin{bmatrix}\xi_2\\\xi_1\end{bmatrix}=\begin{bmatrix}0\\1\end{bmatrix}.$$

Next, let us apply the same Gaussian factorization algorithm in floating-point arithmetic:

$$\begin{bmatrix}3&2\\1&\delta\end{bmatrix}\approx\begin{bmatrix}1&\\1/3&1\end{bmatrix}\begin{bmatrix}3&2\\&-2/3\end{bmatrix},$$

$$\begin{bmatrix}1&\\1/3&1\end{bmatrix}\begin{bmatrix}\eta_1\\\eta_2\end{bmatrix}=\begin{bmatrix}0\\1\end{bmatrix}\Rightarrow\mathbf{y}=\begin{bmatrix}0\\1\end{bmatrix}\text{ and}$$

$$\begin{bmatrix}3&2\\&-2/3\end{bmatrix}\begin{bmatrix}\xi_2\\\xi_1\end{bmatrix}=\begin{bmatrix}0\\1\end{bmatrix}\Rightarrow\mathbf{x}=\begin{bmatrix}-3/2\\1\end{bmatrix}.$$

In this case, reordering the equations and unknowns, and then applying Gaussian factorization produces a computational result that is close to the correct answer.

3.7.2.1 Algorithm

The basic idea in **Gaussian factorization with (full) pivoting** is to reorder both equations and unknowns to improve the numerical stability of the algorithm. At the k'th step of Gaussian factorization with pivoting, we will search in rows k through m and columns k through n to find the first entry i,j with largest absolute value, then interchange the k-th row with the i-th, and k-th column with the j-th (if necessary) to bring this largest entry to the k,k position. After the interchange, we perform the factorization.

Here is a mathematical description of the process. Suppose that the matrix $\mathbf{A}$ has entries α_{ij}. At the first step, we find the **pivot** entry α_{ij}, which is the first entry (searching by columns) with largest absolute value. We interchange the first and i-th rows, and interchange the first and j-th columns, then partition and factor:

$$\mathbf{I}_{1i}\mathbf{A}\mathbf{I}_{1j}=\begin{bmatrix}\alpha&\mathbf{a}^\top\\\mathbf{b}&\mathbf{B}\end{bmatrix}=\begin{bmatrix}1&\mathbf{0}^\top\\\boldsymbol{\ell}&\mathbf{I}\end{bmatrix}\begin{bmatrix}\alpha&\mathbf{a}^\top\\\mathbf{0}&\mathbf{C}\end{bmatrix}.$$

Here $\mathbf{I}_{1i}$ and $\mathbf{I}_{1j}$ are interchange matrices, described by Definition 3.4.5. Also, $\boldsymbol{\ell} = \mathbf{b}/\alpha$ and $\mathbf{C} = \mathbf{B} - \boldsymbol{\ell}\mathbf{a}^\top$. By the choice of pivot, if $\alpha = 0$ then we must have $\mathbf{A} = \mathbf{0}$.

Inductively, suppose that after $k-1$ steps we have found

- an $m \times m$ matrix $\mathbf{Q}_{k-1}$, which is a product of $k-1$ interchange matrices,
- an $n \times n$ matrix $\mathbf{P}_{k-1}$, which is a product of $k-1$ interchange matrices,
- a $(k-1) \times (k-1)$ unit left-triangular matrix $\mathbf{L}_{11}$ and a $(m-k+1) \times (k-1)$ matrix $\mathbf{L}_{21}$,
- a $(k-1) \times (k-1)$ right-triangular matrix $\mathbf{R}_{11}$ and a $(k-1) \times (n-k+1)$ matrix $\mathbf{R}_{12}$ and
- a $(m-k+1) \times (n-k+1)$ matrix $\mathbf{B}$

such that

$$\mathbf{Q}_{k-1}{}^\top \mathbf{A}\mathbf{P}_{k-1} = \begin{bmatrix} \mathbf{L}_{11} & \mathbf{0} \\ \mathbf{L}_{21} & \mathbf{I} \end{bmatrix} \begin{bmatrix} \mathbf{R}_{11} & \mathbf{R}_{12} \\ \mathbf{0} & \mathbf{B} \end{bmatrix} .$$

At the k'th step, we find the first entry i, j of $\mathbf{B}$ with largest absolute value, and apply row and column interchanges to move it to the first diagonal entry:

$$\mathbf{Q}_k{}^\top \mathbf{A}\mathbf{P}_k \equiv \begin{bmatrix} \mathbf{I} & \mathbf{0} \\ \mathbf{0} & \mathbf{I}_{1i} \end{bmatrix} \mathbf{Q}_{k-1}{}^\top \mathbf{A}\mathbf{P}_{k-1} \begin{bmatrix} \mathbf{I} & \mathbf{0} \\ \mathbf{0} & \mathbf{I}_{1j} \end{bmatrix} = \begin{bmatrix} \mathbf{I} & \mathbf{0} \\ \mathbf{0} & \mathbf{I}_{1i} \end{bmatrix} \begin{bmatrix} \mathbf{L}_{11} & \mathbf{0} \\ \mathbf{L}_{21} & \mathbf{I} \end{bmatrix} \begin{bmatrix} \mathbf{R}_{11} & \mathbf{R}_{12} \\ \mathbf{0} & \mathbf{B} \end{bmatrix} \begin{bmatrix} \mathbf{I} & \mathbf{0} \\ \mathbf{0} & \mathbf{I}_{1j} \end{bmatrix}$$

$$= \begin{bmatrix} \mathbf{L}_{11} & \mathbf{0} \\ \mathbf{I}_{1i}\mathbf{L}_{21} & \mathbf{I}_{1i} \end{bmatrix} \begin{bmatrix} \mathbf{R}_{11} & \mathbf{R}_{12}\mathbf{I}_{1j} \\ \mathbf{0} & \mathbf{B}\mathbf{I}_{1j} \end{bmatrix} = \begin{bmatrix} \mathbf{L}_{11} & \mathbf{0} \\ \mathbf{I}_{1i}\mathbf{L}_{21} & \mathbf{I} \end{bmatrix} \begin{bmatrix} \mathbf{R}_{11} & \mathbf{R}_{12}\mathbf{I}_{1j} \\ \mathbf{0} & \mathbf{I}_{1i}\mathbf{B}\mathbf{I}_{1j} \end{bmatrix} .$$

Next, we partition

$$\mathbf{I}_{1i}\mathbf{L}_{21} = \begin{bmatrix} \boldsymbol{\ell}_{21}{}^\top \\ \mathbf{L}_{31} \end{bmatrix} , \ \mathbf{R}_{12}\mathbf{I}_{1j} = \begin{bmatrix} \mathbf{r}_{12} & \mathbf{R}_{13} \end{bmatrix} \text{ and } \mathbf{I}_{1i}\mathbf{B}\mathbf{I}_{1j} = \begin{bmatrix} \beta_{22} & \mathbf{b}_{23}{}^\top \\ \mathbf{b}_{32} & \mathbf{B}_{33} \end{bmatrix}$$

and factor

$$\mathbf{Q}_k{}^\top \mathbf{A}\mathbf{P}_k = \begin{bmatrix} \mathbf{L}_{11} & \mathbf{0} & \mathbf{0} \\ \boldsymbol{\ell}_{21}{}^\top & 1 & \mathbf{0} \\ \mathbf{L}_{31} & \mathbf{0} & \mathbf{I} \end{bmatrix} \begin{bmatrix} \mathbf{R}_{11} & \mathbf{r}_{12} & \mathbf{R}_{13} \\ \mathbf{0} & \beta_{22} & \mathbf{b}_{23}{}^\top \\ \mathbf{0} & \mathbf{b}_{32} & \mathbf{B}_{33} \end{bmatrix} = \begin{bmatrix} \mathbf{L}_{11} & \mathbf{0} & \mathbf{0} \\ \boldsymbol{\ell}_{21}{}^\top & 1 & \mathbf{0} \\ \mathbf{L}_{31} & \boldsymbol{\ell}_{32} & \mathbf{I} \end{bmatrix} \begin{bmatrix} \mathbf{R}_{11} & \mathbf{r}_{12} & \mathbf{R}_{13} \\ \mathbf{0} & \beta_{22} & \mathbf{b}_{23}{}^\top \\ \mathbf{0} & \mathbf{0} & C_{33} \end{bmatrix} ,$$

where

$$\boldsymbol{\ell}_{32} = \mathbf{b}_{32}/\beta_{22} \text{ and } C_{33} = \mathbf{B}_{33} - \boldsymbol{\ell}_{32}\mathbf{b}_{23}{}^\top .$$

These computations can be organized into the following

Algorithm 3.7.3 (Gaussian Factorization with Full Pivoting)

for $1 \leq k \leq \min\{m, n\}$

$I = k; \quad J = k; \quad \varrho = |\alpha_{kk}|$

for $k \leq j \leq n$

for $k \leq i \leq m$

if $|\alpha_{ij}| > \varrho$ then

$I = i; J = j; \varrho = |\alpha_{ij}|$

if $\varrho = 0$ break

if $I > k$ then

for $1 \leq j \leq n$

swap α_{kj} with α_{Ij}

if $J > k$ then

for $1 \leq i \leq m$

swap α_{ik} with α_{iJ}

for $k + 1 \leq i \leq m$

$\alpha_{ik} = \alpha_{ik}/\alpha_{kk}$

for $k + 1 \leq j \leq n$

for $k + 1 \leq i \leq m$

$\alpha_{ij} = \alpha_{ij} - \alpha_{ik} * \alpha_{kj}$

Example 3.7.4 Suppose we want to apply Gaussian factorization with full pivoting to

$$\mathbf{A} = \begin{bmatrix} 0 & 4 & -3 \\ 2 & 0 & 4 \\ 3 & -5 & 3 \end{bmatrix} .$$

The entry of largest magnitude is in the 3,2 position, so we interchange rows 1 and 3, and columns 1 and 2 before performing the first step of the factorization:

$$\mathbf{I}_{13}\mathbf{A}\mathbf{I}_{12} = \begin{bmatrix} -5 & 3 & 3 \\ 0 & 2 & 4 \\ 4 & 0 & -3 \end{bmatrix} = \begin{bmatrix} 1 & 0 & 0 \\ 0 & 1 & 0 \\ -4/5 & 0 & 1 \end{bmatrix} \begin{bmatrix} -5 & 3 & 3 \\ 0 & 2 & 4 \\ 0 & 12/5 & -3/5 \end{bmatrix} .$$

Now we look in rows and columns 2 and 3 of the right-hand factor to find the first entry with largest magnitude. This is in the 2,3 entry. We interchange columns 2 and 3 to get

$$\mathbf{I}_{13}\mathbf{A}\mathbf{I}_{12}\mathbf{I}_{23} = \begin{bmatrix} 1 & 0 & 0 \\ 0 & 1 & 0 \\ -4/5 & 0 & 1 \end{bmatrix} \begin{bmatrix} -5 & 3 & 3 \\ 0 & 4 & 2 \\ 0 & -3/5 & 12/5 \end{bmatrix} = \begin{bmatrix} 1 & 0 & 0 \\ 0 & 1 & 0 \\ -4/5 & -3/20 & 1 \end{bmatrix} \begin{bmatrix} -5 & 3 & 3 \\ 0 & 4 & 2 \\ 0 & 0 & 27/10 \end{bmatrix} .$$

At this point, we can see that our final factorization is

$$\mathbf{I}_{13}\mathbf{A}\mathbf{I}_{12}\mathbf{I}_{23} = \begin{bmatrix} 1 & 0 & 0 \\ 0 & 1 & 0 \\ -4/5 & -3/20 & 1 \end{bmatrix} \begin{bmatrix} -5 & 3 & 3 \\ 0 & 4 & 2 \\ 0 & 0 & 27/10 \end{bmatrix} .$$

3.7.2.2 LR Theorem

There are three termination situations in Gaussian factorization with row and column pivoting. If the algorithm terminates because there are no more rows to process, then $m = r \leq n$, and at the end of the factorization we have

$$\mathbf{Q}^\top \mathbf{A}\mathbf{P} = \mathbf{L}_1 \begin{bmatrix} \mathbf{R}_1 & \mathbf{R}_2 \end{bmatrix} ,$$

where $\mathbf{L}_1$ is $r \times r$ unit left triangular and $\mathbf{R}_1$ is $r \times r$ right triangular with nonzero diagonal entries. If the algorithm terminates because there are no more columns to process, then $n = r \leq m$, and at the end of the factorization we have

$$\mathbf{Q}^\top \mathbf{A}\mathbf{P} = \begin{bmatrix} \mathbf{L}_1 & \mathbf{0} \\ \mathbf{L}_2 & \mathbf{I} \end{bmatrix} \begin{bmatrix} \mathbf{R}_1 \\ \mathbf{0} \end{bmatrix} = \begin{bmatrix} \mathbf{L}_1 \\ \mathbf{L}_2 \end{bmatrix} \mathbf{R}_1 ,$$

where $\mathbf{L}_1$ is $r \times r$ unit left triangular and $\mathbf{R}_1$ is $r \times r$ right triangular with nonzero diagonal entries. If the algorithm terminates because there is no nonzero pivot, then $r < m, n$, and at the end of the factorization we have

$$\mathbf{Q}^\top \mathbf{A}\mathbf{P} = \begin{bmatrix} \mathbf{L}_1 & \mathbf{0} \\ \mathbf{L}_2 & \mathbf{I} \end{bmatrix} \begin{bmatrix} \mathbf{R}_1 & \mathbf{R}_2 \\ \mathbf{0} & \mathbf{0} \end{bmatrix} = \begin{bmatrix} \mathbf{L}_1 \\ \mathbf{L}_2 \end{bmatrix} \begin{bmatrix} \mathbf{R}_1 & \mathbf{R}_2 \end{bmatrix} ,$$

where $\mathbf{L}_1$ is $r \times r$ unit left triangular and $\mathbf{R}_1$ is $r \times r$ right triangular with nonzero diagonal entries.

Thus Gaussian factorization with full pivoting proves the following important theorem.

Theorem 3.7.1 (LR) *Suppose that* $\mathbf{A}$ *is a nonzero* $m \times n$ *matrix. Then there is an integer* r *satisfying* $1 \leq r \leq \min\{m, n\}$, *an* $m \times m$ *permutation matrix* $\mathbf{Q}$, *an* $n \times n$

permutation matrix **P**, *an* $m \times r$ *unit left trapezoidal matrix* **L**, *and an* $r \times n$ *right trapezoidal matrix* **R** *with nonzero diagonal entries so that* $\mathbf{Q}^\top \mathbf{A}\mathbf{P} = \mathbf{L}\mathbf{R}$.

3.7.2.3 Computational Cost

If r nonzero pivots are found, then Gaussian factorization with full pivoting involves the following computational work:

- $\sum_{k=1}^{r}(m-k) = r(2m-r-1)/2$ divisions,
- $\sum_{k=1}^{r}(m-k)(n-k) = r\left[6mn-(r+1)(3m+3n-2r-1)\right]/6$ multiplications and additions,
- $\sum_{k=1}^{r}(m+1-k)(n+1-k) = r\left[6mn-(r-1)(3m+3n-2r+1)\right]/6$ comparisons, and
- at most $\sum_{k=1}^{r}(m+n) = (m+n)r$ swaps.

If $m = r = n$, then this algorithm involves on the order of $n^3/3$ multiplications, additions and comparisons. Each comparison is comparable in cost to an addition. For large n, the total work in Gaussian factorization with (full) pivoting is roughly 50% more than the work in Gaussian factorization with no pivoting.

3.7.2.4 Summary

The following algorithm summarizes the steps in solving a system of linear equations by Gaussian factorization with full pivoting.

Algorithm 3.7.4 (Solve Linear System via Gaussian Factorization with Full Pivoting)

factor $\mathbf{Q}^\top \mathbf{A}\mathbf{P} = \mathbf{L}\mathbf{R}$ (see Algorithm 3.7.3)

permute $\mathbf{c} = \mathbf{Q}^\top \mathbf{b}$ (see Sect. 3.4.7)

solve $\mathbf{L}\mathbf{y} = \mathbf{c}$ for $\mathbf{y}$ (see Algorithm 3.4.3)

solve $\mathbf{R}\mathbf{w} = \mathbf{y}$ for $\mathbf{w}$ (see Algorithm 3.4.5)

permute $\mathbf{x} = \mathbf{P}\mathbf{w}$ (see Sect. 3.4.7)

The dominant cost is Gaussian factorization, which costs on the order of $n^3/3$ multiplications, additions and comparisons. The two permutations cost at most n swaps each, and the two triangular systems cost on the order of $n^2/2$ multiplications and additions each.

The MATLAB command

```
[L,U,P,Q] = lu(A)
```

will perform Gaussian factorization with (full) pivoting for *sparse* matrices A. The resulting factors L, U, P and Q will all be sparse, and will correspond to the factorization $\mathbf{PAQ} = \mathbf{LU}$. For more information, please study the MATLAB command lu. After this factorization has been computed, a linear system $\mathbf{Ax} = \mathbf{b}$ can be solved by the MATLAB commands

```
optl.LT = true; optl.TRANS = false;
optu.UT = true; optu.TRANS = false;
x = Q * linsolve( U, linsolve( L, P * b, optl ), optu )
```

Note that full matrices can be converted to sparse matrices in MATLAB by the sparse command.

In LAPACK, Gaussian factorization with full pivoting is computed by subroutines `_getc2`, and routines `_gesc2` solve linear systems using this factorization. See, for example, dgetc2.f and dgesc2.f. Interested readers may discover how to call these LAPACK routines by viewing either the Fortran program gaussElimFullPivF.f, the Fortran 90 program gaussElimFullPivF90.f90, the C program gaussElimFullPivC.c or the C++ program gaussElimFullPivCpp.C. In addition, the C++ program demonstrates how to use LAPACK++ to solve a linear system by Gaussian factorization with full pivoting.

Readers may also experiment with the following JavaScript Gaussian factorization with (full) pivoting program gaussElimFullPiv.html.

Exercise 3.7.3 Examine loops 10 and 20 of LAPACK routine dgetc2.f. Is the access of the array data being performed at stride one? Do these loops find the *first* largest pivot? If pivoting is necessary, is it better for efficient data access to exchange a row or a column?

Exercise 3.7.4 Modify LAPACK routine `dgetc2` to factor rectangular matrices.

Exercise 3.7.5 Show that full pivoting applied to the matrix

$$\mathbf{A} = \begin{bmatrix} 1 & & & & 1 \\ -1 & 1 & & & 1 \\ -1 & -1 & 1 & & 1 \\ -1 & -1 & -1 & 1 & 1 \\ -1 & -1 & -1 & -1 & 1 \end{bmatrix}$$

produces $\mathbf{Q}^\top \mathbf{AP} = \mathbf{LR}$, where the entries of $\mathbf{L}$ have magnitude at most one, and the entries of $\mathbf{R}$ have magnitude at most two. It is interesting to Gaussian factorization with full pivoting for this matrix with partial pivoting (see Example 3.7.5).

Exercise 3.7.6 The 2×2 **Hadamard matrix** is

$$\mathbf{H}_2 = \begin{bmatrix} 1 & 1 \\ 1 & -1 \end{bmatrix} .$$

1. Compute the Gaussian factorization with full pivoting of $\mathbf{H}_2$ and show that the right triangular factor has an entry of magnitude two.
2. The 4×4 Hadamard matrix is

$$\mathbf{H}_4 = \begin{bmatrix} \mathbf{H}_2 & \mathbf{H}_2 \\ \mathbf{H}_2 & -\mathbf{H}_2 \end{bmatrix} .$$

   Compute the Gaussian factorization with full pivoting of $\mathbf{H}_4$ and show that the right triangular factor has an entry of magnitude four.
3. If n is a power of two, then the $2n \times 2n$ Hadamard matrix is

$$\mathbf{H}_{2n} = \begin{bmatrix} \mathbf{H}_n & \mathbf{H}_n \\ \mathbf{H}_n & -\mathbf{H}_n \end{bmatrix} .$$

   Read Cryer [25] to see why Gaussian factorization with full pivoting of $\mathbf{H}_n$ produces a right-triangular factor with an entry of magnitude at least n.

Exercise 3.7.7 Show that the integer r in the LR Theorem 3.7.1 is the rank of $\mathbf{A}$.

Exercise 3.7.8 Use the LR Theorem 3.7.1 to prove the Fundamental Theorem of Linear Algebra 3.2.3

Exercise 3.7.9 LAPACK subroutine `dgetc2` calls `dswap` to interchange rows and columns. Consider avoiding these interchanges by using **indirect addressing** to refer to matrix entries via permutations. If $\mathbf{Q} = [\mathbf{e}_{\omega(i)}]$ and $\mathbf{P} = [\mathbf{e}_{\pi(j)}]$ are permutation matrices and $\tilde{\mathbf{A}} = \mathbf{Q}^\top \mathbf{A} \mathbf{P} = [\tilde{\alpha}_{i,j}]$ is a re-ordering of the rows and columns of $\mathbf{A} = [\alpha_{i,j}]$, then the entries of $\tilde{\mathbf{A}}$ are

$$\tilde{\alpha}_{i,j} \equiv \mathbf{e}_i{}^\top \tilde{\mathbf{A}} \mathbf{e}_j = \mathbf{e}_{\omega(i)}{}^\top \mathbf{A} \mathbf{e}_{\pi(j)} \equiv \alpha_{\omega(i),\pi(j)} .$$

Reprogram `dgetc2` to store the row and column permutations (rather than the row and column interchanges), and use indirect addressing rather than data swaps. Then compare the execution speed of the modified full pivoting algorithm to `dgetc2`.

3.7.3 Partial Pivoting

In Sect. 3.7.1, we saw that Gaussian factorization without pivoting sometimes produces unnecessarily inaccurate solutions to linear systems. Afterward, in Sect. 3.7.2, we found that Gaussian factorization with (full) pivoting is more numerically stable than Gaussian factorization without pivoting, but costs roughly 50% more floating point operations. In this section, we will develop a less expensive alternative, that costs about the same as Gaussian factorization without pivoting, but produces computational results that are generally as good as Gaussian factorization with full pivoting.

3.7.3.1 Algorithm

Let us provide a mathematical description of the process. Suppose that the matrix **A** has entries α_{ij}. At the first step, we find the **pivot** entry α_{i1}, which is the first entry in columns 1 with largest absolute value. We interchange the first and i-th rows, then factor:

$$\mathbf{I}_{1i}\mathbf{A} = \begin{bmatrix} \alpha & \mathbf{a}^\top \\ \mathbf{b} & \mathbf{B} \end{bmatrix} = \begin{bmatrix} 1 & \mathbf{0}^\top \\ \boldsymbol{\ell} & \mathbf{I} \end{bmatrix} \begin{bmatrix} \alpha & \mathbf{a}^\top \\ \mathbf{0} & \mathbf{C} \end{bmatrix} .$$

Here $\mathbf{I}_{1i}$ is an interchange matrix, described by Definition 3.4.5. Also, $\boldsymbol{\ell} = \mathbf{b}/\alpha$ and $\mathbf{C} = \mathbf{B} - \boldsymbol{\ell}\mathbf{a}^\top$. By the choice of pivot, if $\alpha = 0$ then we must have $\mathbf{b} = \mathbf{0}$; $\boldsymbol{\ell}$ would be arbitrary.

Inductively, suppose that after $k-1$ steps we have found

- an $m \times m$ matrix $\mathbf{Q}_{k-1}$, which is a product of $k-1$ interchange matrices,
- a $(k-1) \times (k-1)$ unit left-triangular matrix $\mathbf{L}_{11}$ and a $(m-k+1) \times (k-1)$ matrix $\mathbf{L}_{21}$,
- a $(k-1) \times (k-1)$ right-triangular matrix $\mathbf{R}_{11}$ and a $(k-1) \times (n-k+1)$ matrix $\mathbf{R}_{12}$ and
- a $(m-k+1) \times (n-k+1)$ matrix $\mathbf{B}$

such that

$$\mathbf{Q}_{k-1}{}^\top\mathbf{A} = \begin{bmatrix} \mathbf{L}_{11} & \mathbf{0} \\ \mathbf{L}_{21} & \mathbf{I} \end{bmatrix} \begin{bmatrix} \mathbf{R}_{11} & \mathbf{R}_{12} \\ \mathbf{0} & \mathbf{B} \end{bmatrix} .$$

At the k'th step, we find the first entry $i, 1$ of the first column of **B** with largest absolute value, and apply row interchanges to move it to the first diagonal entry:

$$\begin{aligned} \mathbf{Q}_k{}^\top\mathbf{A} &\equiv \begin{bmatrix} \mathbf{I} & \mathbf{0} \\ \mathbf{0} & \mathbf{I}_{1i} \end{bmatrix} \mathbf{Q}_{k-1}{}^\top\mathbf{A} = \begin{bmatrix} \mathbf{I} & \mathbf{0} \\ \mathbf{0} & \mathbf{I}_{1i} \end{bmatrix} \begin{bmatrix} \mathbf{L}_{11} & \mathbf{0} \\ \mathbf{L}_{21} & \mathbf{I} \end{bmatrix} \begin{bmatrix} \mathbf{R}_{11} & \mathbf{R}_{12} \\ \mathbf{0} & \mathbf{B} \end{bmatrix} \\ &= \begin{bmatrix} \mathbf{L}_{11} & \mathbf{0} \\ \mathbf{I}_{1i}\mathbf{L}_{21} & \mathbf{I}_{1i} \end{bmatrix} \begin{bmatrix} \mathbf{R}_{11} & \mathbf{R}_{12} \\ \mathbf{0} & \mathbf{B} \end{bmatrix} = \begin{bmatrix} \mathbf{L}_{11} & \mathbf{0} \\ \mathbf{I}_{1i}\mathbf{L}_{21} & \mathbf{I} \end{bmatrix} \begin{bmatrix} \mathbf{R}_{11} & \mathbf{R}_{12} \\ \mathbf{0} & \mathbf{I}_{1i}\mathbf{B} \end{bmatrix} . \end{aligned}$$

Next, we partition

$$\mathbf{I}_{1i}\mathbf{L}_{21} = \begin{bmatrix} \boldsymbol{\ell}_{21}{}^\top \\ \mathbf{L}_{31} \end{bmatrix} , \ \mathbf{R}_{12} = \begin{bmatrix} \mathbf{r}_{12} & \mathbf{R}_{13} \end{bmatrix} \text{ and } \mathbf{I}_{1i}\mathbf{B} = \begin{bmatrix} \beta_{22} & \mathbf{b}_{23}{}^\top \\ \mathbf{b}_{32} & \mathbf{B}_{33} \end{bmatrix}$$

and factor

$$\mathbf{Q}_k{}^\top\mathbf{A} = \begin{bmatrix} \mathbf{L}_{11} & \mathbf{0} & \mathbf{0} \\ \boldsymbol{\ell}_{21}{}^\top & 1 & \mathbf{0} \\ \mathbf{L}_{31} & \mathbf{0} & \mathbf{I} \end{bmatrix} \begin{bmatrix} \mathbf{R}_{11} & \mathbf{r}_{12} & \mathbf{R}_{13} \\ \mathbf{0} & \beta_{22} & \mathbf{b}_{23}{}^\top \\ \mathbf{0} & \mathbf{b}_{32} & \mathbf{B}_{33} \end{bmatrix} = \begin{bmatrix} \mathbf{L}_{11} & \mathbf{0} & \mathbf{0} \\ \boldsymbol{\ell}_{21}{}^\top & 1 & \mathbf{0} \\ \mathbf{L}_{31} & \boldsymbol{\ell}_{32} & \mathbf{I} \end{bmatrix} \begin{bmatrix} \mathbf{R}_{11} & \mathbf{r}_{12} & \mathbf{R}_{13} \\ \mathbf{0} & \beta_{22} & \mathbf{b}_{23}{}^\top \\ \mathbf{0} & \mathbf{0} & C_{33} \end{bmatrix} ,$$

where

$$\ell_{32} = \mathbf{b}_{32}/\beta_{22} \text{ and } C_{33} = \mathbf{B}_{33} - \ell_{32}\mathbf{b}_{23}{}^\top .$$

These computations can be described by the following

Algorithm 3.7.5 (Gaussian Factorization with Partial Pivoting)

for $1 \le k \le \min\{m, n\}$
  $I = k; \quad \varrho = |A_{kk}|$
  for $k \le i \le m$
    if $|A_{ik}| > \varrho$ then
      $I = i; \quad \varrho = |A_{ik}|$
  if $\varrho = 0$ break
  if $I > k$ then
    for $1 \le j \le n$
      swap A_{kj} with A_{Ij}
  for $k + 1 \le i \le m$
    $A_{ik} = A_{ik}/A_{kk}$
  for $k + 1 \le j \le n$
    for $k + 1 \le i \le m$
      $A_{ij} = A_{ij} - A_{ik} * A_{kj}$

Note that the only part of this algorithm that does not operate at stride 1 is the row interchange.

This process continues until there are no more rows or columns to process, or some pivot is zero. In practice, rounding errors prevent the pivots from becoming zero, and the algorithm typically terminates when there are no more rows or columns to process.

Example 3.7.5 Suppose we use Gaussian factorization with partial pivoting to factor the following matrix [111, p. 212]:

$$\mathbf{A} = \begin{bmatrix} 1 & & & & -1 \\ -1 & 1 & & & -1 \\ -1 & -1 & 1 & & -1 \\ -1 & -1 & -1 & 1 & -1 \\ -1 & -1 & -1 & -1 & -1 \end{bmatrix} .$$

It is easy to perform the factorization to see that no pivoting will be used, and

$$\mathbf{A} = \begin{bmatrix} 1 & & & & \\ -1 & 1 & & & \\ -1 & 1 & 1 & & \\ -1 & 1 & 1 & 1 & \\ -1 & 1 & 1 & 1 & 1 \end{bmatrix} \begin{bmatrix} 1 & 0 & 0 & 0 & -1 \\ & 1 & 1 & 0 & -2 \\ & & 1 & 1 & -4 \\ & & & 1 & -8 \\ & & & & -16 \end{bmatrix} .$$

In this case, we find exponential growth in the right factor. Such behavior is uncommon, in general.

3.7.3.2 Computational Cost

Gaussian factorization with partial pivoting involves the following computational work.

- $\sum_{k=1}^{r}(m-k) = r(2m-r-1)/2$ divisions,
- $\sum_{k=1}^{r}(m-k)(n-k) = r\left[6mn-(r+1)(3m+3n-2r-1)\right]/6$ multiplications and additions,
- $\sum_{k=1}^{r}(m+1-k) = r\left[2m+1-r\right]/2$ comparisons, and
- at most $\sum_{k=1}^{r} n = nr$ swaps.

If $m = r = n$, then this algorithm involves on the order of $n^3/3$ multiplications and additions. The number of multiplications and additions is the same as in Gaussian factorization with full pivoting, but the number of comparisons involves a lower power of n.

3.7.3.3 Row Echelon Form Theorem

Gaussian **factorization with row pivoting** can be summarized either by Algorithm 3.7.5, or by a mathematical theorem. Given a nonsingular $m \times n$ matrix $\mathbf{A}$, if $r = \min\{m, n\}$, the partial pivoting algorithm determines a $m \times m$ permutation matrix $\mathbf{Q}$, an $m \times r$ unit left trapezoidal matrix $\mathbf{L}$ and a $r \times n$ right trapezoidal matrix $\mathbf{R}$ with nonzero diagonal entries, so that

$$\mathbf{Q}^\top \mathbf{A} = \mathbf{L}\mathbf{R} .$$

For singular matrices, partial pivoting with exact arithmetic can be used to compute something reminiscent of a right trapezoidal matrix.

Definition 3.7.1 Let $\mathbf{R}$ be an $r \times n$ matrix, and for each row index i let j_i be the column index of the first nonzero entry in row i. Then $\mathbf{R}$ is a **row echelon form** if and only if for each row index i we have $j_i < j_{i+1}$.
The partial pivoting algorithm proves the following theorem.

Theorem 3.7.2 (Row Echelon Form) *For every nonzero $m \times n$ matrix **A** there is an integer r satisfying $1 \le r \le \min\{m, n\}$, an $m \times m$ permutation matrix **Q**, an $m \times r$ unit left trapezoidal matrix **L** and an $r \times n$ matrix **R** in row echelon form with no zero rows, so that*

$$\mathbf{Q}^\top \mathbf{A} = \mathbf{L}\mathbf{R} .$$

Proof See Strang [99, p. 73].

3.7.3.4 Block Algorithm

LAPACK routines `_getf2` perform the partial pivoting factorization for small matrices, while routines `_getrf` call the former to compute the factorization of larger matrices that have been partitioned into blocks. The block form `_getrf` of Gaussian factorization is preferable because it is able to use Level 3 BLAS subroutines, thereby gaining execution speed [2, pp. 37–39].

Let us describe the block computations in LAPACK routines `_getrf`. In the beginning, we choose a block size k and partition

$$\mathbf{A} = \begin{bmatrix} \mathbf{A}_{11} & \mathbf{A}_{12} \\ \mathbf{A}_{21} & \mathbf{A}_{22} \end{bmatrix} ,$$

where $\mathbf{A}_{11}$ is $k \times k$. Next, we call `_getf2` to find an $m \times m$ permutation matrix $\mathbf{Q}_1$, a $k \times k$ unit left triangular matrix $\mathbf{L}_{11}$, a $(m-k) \times k$ matrix $\mathbf{L}_{21}$ and a $k \times k$ right triangular matrix $\mathbf{R}_{11}$ so that

$$\mathbf{Q}_1{}^\top \begin{bmatrix} \mathbf{A}_{11} \\ \mathbf{A}_{21} \end{bmatrix} \equiv \begin{bmatrix} \mathbf{B}_{11} \\ \mathbf{B}_{21} \end{bmatrix} = \begin{bmatrix} \mathbf{L}_{11} \\ \mathbf{L}_{21} \end{bmatrix} \mathbf{R}_{11} .$$

Afterward, we permute the remainder of the original matrix:

$$\mathbf{Q}_1{}^\top \begin{bmatrix} \mathbf{A}_{12} \\ \mathbf{A}_{22} \end{bmatrix} \equiv \begin{bmatrix} \mathbf{B}_{12} \\ \mathbf{B}_{22} \end{bmatrix} .$$

Then we solve

$$\mathbf{L}_{11}\mathbf{R}_{12} = \mathbf{B}_{12}$$

for $\mathbf{R}_{12}$, and use the result to compute

$$\mathbf{C}_{22} = \mathbf{B}_{22} - \mathbf{L}_{21}\mathbf{R}_{12} .$$

At this point, we have factored

$$Q_1{}^\top \begin{bmatrix} \mathbf{A}_{11} & \mathbf{A}_{12} \\ \mathbf{A}_{21} & \mathbf{A}_{22} \end{bmatrix} \equiv \begin{bmatrix} \mathbf{B}_{11} & \mathbf{B}_{12} \\ \mathbf{B}_{21} & \mathbf{B}_{22} \end{bmatrix} = \begin{bmatrix} L_{11} & \mathbf{0} \\ L_{21} & \mathbf{I} \end{bmatrix} \begin{bmatrix} \mathbf{R}_{11} & \mathbf{R}_{12} \\ \mathbf{0} & \mathbf{C}_{22} \end{bmatrix} .$$

After s block steps, we are given the current factorization

$$Q_{s-1}{}^\top \mathbf{A} \equiv \begin{bmatrix} \mathbf{B}_{11} & \mathbf{B}_{12} \\ \mathbf{B}_{21} & \mathbf{B}_{22} \end{bmatrix} = \begin{bmatrix} L_{11} & \mathbf{0} \\ L_{21} & \mathbf{I} \end{bmatrix} \begin{bmatrix} \mathbf{R}_{11} & \mathbf{R}12 \\ \mathbf{0} & \mathbf{C}_{22} \end{bmatrix} .$$

Here $\mathbf{C}_{22}$ is $(m - [k-1]s) \times (n - [k-1]s)$. We partition

$$\mathbf{C}_{22} = \begin{bmatrix} \mathbf{D}_{22} & \mathbf{D}_{23} \\ \mathbf{D}_{32} & \mathbf{D}_{33} \end{bmatrix}$$

where $\mathbf{D}_{22}$ is at most $k \times k$. Next, we call `_getf2` to find a permutation matrix P_s at most $(m - [k-1]s) \times (m - [k-1]s)$ in size, a unit left triangular matrix L_{22} at most $k \times k$ in size, a $(m - ks) \times k$ matrix L_{32} (if it has any rows), and a right triangular matrix $\mathbf{R}_{22}$ at most $k \times k$ in size, so that

$$P_s{}^\top \begin{bmatrix} \mathbf{D}_{22} \\ \mathbf{D}_{32} \end{bmatrix} \equiv \begin{bmatrix} \mathbf{E}_{22} \\ \mathbf{E}_{32} \end{bmatrix} = \begin{bmatrix} L_{22} \\ L_{32} \end{bmatrix} \mathbf{R}_{22} .$$

Noting that

$$\begin{bmatrix} \mathbf{I} & \mathbf{0} \\ \mathbf{0} & P_s{}^\top \end{bmatrix} Q_{s-1}{}^\top \mathbf{A} = \begin{bmatrix} L_{11} & \mathbf{0} \\ P_s{}^\top L_{21} & P_s{}^\top \end{bmatrix} \begin{bmatrix} \mathbf{R}_{11} & \mathbf{R}_{12} \\ \mathbf{0} & \mathbf{C}_{22} \end{bmatrix} = \begin{bmatrix} L_{11} & \mathbf{0} \\ P_s{}^\top L_{21} & \mathbf{I} \end{bmatrix} \begin{bmatrix} \mathbf{R}_{11} & \mathbf{R}_{12} \\ \mathbf{0} & P_s{}^\top \mathbf{C}_{22} \end{bmatrix} ,$$

we permute the previous columns, and partition

$$P_s{}^\top L_{21} = \begin{bmatrix} M_{21} \\ M_{31} \end{bmatrix} \text{ and } \mathbf{R}_{12} = \begin{bmatrix} \mathbf{S}_{12} & \mathbf{S}_{13} \end{bmatrix} .$$

We also permute the remaining columns, if they exist:

$$P_s{}^\top \begin{bmatrix} \mathbf{D}_{23} \\ \mathbf{D}_{33} \end{bmatrix} = \begin{bmatrix} \mathbf{E}_{23} \\ \mathbf{E}_{33} \end{bmatrix} .$$

Next, we solve

$$L_{22} \mathbf{R}_{23} = \mathbf{E}_{23}$$

for $\mathbf{R}_{23}$, and then we compute

$$\mathbf{F}_{33} = \mathbf{E}_{33} - \mathbf{L}_{32}\mathbf{R}_{23} .$$

As a result, we have the new factorization

$$\begin{bmatrix} \mathbf{I} & \mathbf{0} \\ \mathbf{0} & \mathbf{P}_s{}^\top \end{bmatrix} \mathbf{Q}_{s-1}{}^\top \mathbf{A} = \begin{bmatrix} \mathbf{L}_{11} & \mathbf{0} & \mathbf{0} \\ \mathbf{M}_{21} & \mathbf{I} & \mathbf{0} \\ \mathbf{M}_{31} & \mathbf{0} & \mathbf{I} \end{bmatrix} \begin{bmatrix} \mathbf{R}_{11} & \mathbf{S}_{12} & \mathbf{S}_{13} \\ \mathbf{0} & \mathbf{E}_{22} & \mathbf{E}_{23} \\ \mathbf{0} & \mathbf{E}_{32} & \mathbf{E}_{33} \end{bmatrix} = \begin{bmatrix} \mathbf{L}_{11} & \mathbf{0} & \mathbf{0} \\ \mathbf{M}_{21} & \mathbf{L}_{22} & \mathbf{0} \\ \mathbf{M}_{31} & \mathbf{L}_{32} & \mathbf{I} \end{bmatrix} \begin{bmatrix} \mathbf{R}_{11} & \mathbf{S}_{12} & \mathbf{S}_{13} \\ \mathbf{0} & \mathbf{R}_{22} & \mathbf{R}_{23} \\ \mathbf{0} & \mathbf{0} & \mathbf{F}_{33} \end{bmatrix} .$$

The block form of Gaussian factorization with partial pivoting leads to the following

Algorithm 3.7.6 (Block Gaussian Factorization with Partial Pivoting)

$k = 1$

while $k \leq \min\{m, n\}$

  $s = \min\{\min\{m, n\} - k + 1, b\}$

  factor $\mathbf{P}_k{}^\top \mathbf{A}[k : m; k : k + s - 1] = \mathbf{L}[k : m; k : k + s - 1]\mathbf{R}[k : k + s - 1; k : k + s - 1]$

  if $k > 1$

    permute $\mathbf{P}_k{}^\top \mathbf{A}[k : m; 1 : k - 1]$

  if $k + s \leq n$

    permute $\mathbf{P}_k{}^\top \mathbf{A}[k : m; k + s : n]$

    solve $\mathbf{L}[k : k + 1 - 1; k : k + s - 1]\mathbf{R}[k : k + s - 1; k + s : n] = \mathbf{A}[k : k + s - 1; k + s : n]$

      for $\mathbf{R}[k : k + s - 1; k + s : n]$

    if $k + 1 \leq m$

      $\mathbf{A}[k + s : m; k + s : n] = \mathbf{A}[k + s : m; k + s : n]$

        $- \mathbf{L}[k : m; k : k + s - 1]\mathbf{R}[k : k + s - 1; k + s : n]$

  $k = k + b$

3.7.3.5 Summary

The following algorithm summarizes how to use Gaussian factorization with partial pivoting to solve a system of linear equations.

Table 3.1 Work in Gaussian factorization

	Solve $\mathbf{Ly} = \mathbf{b}$	Solve $\mathbf{Rx} = \mathbf{y}$	Factor $\mathbf{A} = \mathbf{LU}$	Factor $\mathbf{Q}^\top\mathbf{A} = \mathbf{LU}$	Factor $\mathbf{Q}^\top\mathbf{AP} = \mathbf{LU}$
Multiplications	$n^2/2$	$n^2/2$	$n^3/3$	$n^3/3$	$n^3/3$
Additions or subtractions	$n^2/2$	$n^2/2$	$n^3/3$	$n^3/3$	$n^3/3$
Divisions	0	n	$n^2/2$	$n^2/2$	$n^2/2$
Comparisons			0	$n^2/2$	$n^3/3$
Swaps			0	n^2	$2n^2$
Total work	n^2	n^2	$2n^3/3$	$2n^3/3$	n^3
Pivot growth			∞	2^{n-1}	$\sqrt{n \prod_{k=2}^{n} k^{1/k-1}}$

Algorithm 3.7.7 (Solve Linear System via Gaussian Factorization with Partial Pivoting)

factor $\mathbf{Q}^\top\mathbf{A} = \mathbf{LR}$ (see Algorithm 3.7.5) or 3.7.6

permute $\mathbf{c} = \mathbf{Q}^\top\mathbf{b}$ (see Sect. 3.4.7)

solve $\mathbf{Ly} = \mathbf{c}$ for $\mathbf{y}$ (see Algorithm 3.4.3)

solve $\mathbf{Rw} = \mathbf{y}$ for $\mathbf{w}$ (see Algorithm 3.4.5)

The dominant cost is Gaussian factorization, which costs on the order of $n^3/3$ multiplications, additions and swaps. The permutation of the right-hand side costs at most n swaps, and the triangular system solves cost on the order of $n^2/2$ multiplications and additions each.

Table 3.1 summarizes the computational costs in our various algorithms for solving linear systems. The operation counts for the various algorithms were determined in Sects. 3.4.4, 3.7.1, 3.7.2 and 3.7.3. The pivot growth estimates will be developed in Corollary 3.8.1.

In MATLAB, the reduced row echelon form can be computed by

```
R = rref(A)
```

and Gaussian factorization with partial pivoting $\mathbf{PA} = \mathbf{LU}$ can be computed by the command

```
[L,U,P] = lu(A)
```

For more information, view the MATLAB documentation on rref and lu. We remind the reader that in practice, rounding errors prevent the pivots from becoming zero, and the computation of the reduced row echelon form typically terminates when there are no more rows or columns to process. In MATLAB, the triangular factors `L` and `U` are returned as separate *rectangular arrays*, and the permutation is returned as a *full square array*. After the factorization has been computed, MATLAB can solve a linear system $\mathbf{Ax} = \mathbf{b}$ by performing the following commands:

```
optl.LT = true; optl.TRANS = false;
optu.UT = true; optu.TRANS = false;
x = linsolve( U, linsolve( L, P * b, optl ), optu )
```

If the Gaussian factorization will be used only once to solve a linear system, then the simple MATLAB command

```
x = A \ b
```

will factor $\mathbf{A}$ using partial pivoting, and then solve $\mathbf{Ax} = \mathbf{b}$. Afterward, the factorization is forgotten.

In LAPACK, Gaussian factorization with partial pivoting is performed by subroutines `_getrf` and `_getf2`. For example, interested readers can view dgetf2.f and dgetrf.f. Subsequent linear systems can be solved by routines `_getrs`; see, for example, dgetrs.f. To see how to call these LAPACK routines, readers may examine the Fortran program gaussElimPartialPivF.f, the Fortran 90 program gaussElimPartialPivF90.f90, the C program gaussElimPartialPivC.c and the C++ program gaussElimPartialPivCpp.C. In addition, the C++ program demonstrates how to use LAPACK++ to solve a linear system by Gaussian factorization with partial pivoting.

Readers may also examine the JavaScript Gaussian factorization with partial pivoting program gaussElimPartPiv.html.

Exercise 3.7.10 For the linear system in Example 3.7.2, apply Gaussian factorization with and without pivoting for various small values of ε and compare the results.

Exercise 3.7.11 Apply Gaussian factorization with no pivoting, full pivoting and partial pivoting to the 16×16 **Hadamard matrix**, defined by

$$\mathbf{H}_2 = \begin{bmatrix} 1 & 1 \\ 1 & -1 \end{bmatrix}$$

and

$$\mathbf{H}_{2n} = \begin{bmatrix} \mathbf{H}_n & \mathbf{H}_n \\ \mathbf{H}_n & -\mathbf{H}_n \end{bmatrix} .$$

Then compute the error in each factorization. For example, for Gaussian factorization with no pivoting, compute $\mathbf{H}_{16} - \mathbf{LR}$.

Exercise 3.7.12 Apply Gaussian factorization with no pivoting, full pivoting and partial pivoting to the 10×10 **Hilbert matrix**, defined by

$$\mathbf{H}_{ij} = \frac{1}{i+j-1} \ , \ 1 \le i,j \le 10 .$$

Then compute the error in each factorization.

Exercise 3.7.13 Apply Gaussian factorization with no pivoting, full pivoting and partial pivoting to the 10×10 **Vandermonde matrix**, defined by

$$\mathbf{V}_{ij} = \mathbf{v}_i^{10-j} ,$$

where

$$\mathbf{v}_i = 1 + i/10 \; , \; 1 \le i \le 10 .$$

Then compute the error in each factorization.

Exercise 3.7.14 Explain why it is not possible to develop a block form of Gaussian factorization with full pivoting.

3.8 Rounding Errors

Let us review our progress in applying the five steps of scientific computing to the solution of systems of linear equations. In Sects. 3.3 and 3.4 we described systems of linear equations mathematically in terms of matrices and vectors. We determined circumstances under which linear systems have a unique solution in Sect. 3.2.8, and we analyzed the sensitivity of the solution of linear systems to perturbations in the data in Sects. 3.6.1 and 3.6.3. We constructed numerical methods for linear systems in Sect. 3.7, and provided references to computer implementations in both LAPACK and MATLAB. The final step is to analyze the algorithms themselves.

This analysis will examine the accumulation of rounding errors in both Gaussian factorization, and in the solution of triangular systems of linear equations. We will begin in Sect. 3.8.1 by studying how rounding errors accumulate in general. This will lead to particular discussions of rounding error accumulation for inner products in Sect. 3.8.2, and for matrix products in Sect. 3.8.3. In Sect. 3.8.4, we will modify our *a posteriori* forward error estimate to include rounding errors due to the evaluation of the residual. Rounding errors in Gaussian factorization will be examined in Sect. 3.8.5, and rounding errors in solving triangular systems will be analyzed in Sect. 3.8.6.

Both the *a posteriori* error estimates and the *a priori* relative error estimates depend on knowledge of the condition number of the matrix in the linear system. In order to avoid computing the inverse of the matrix just to find its norm for a condition number, we will examine ways to approximate the condition number in Sect. 3.8.7.

3.8.1 Error Accumulation

In this section, we will develop several bounds for accumulated errors. These bounds will be used to study rounding errors in computing inner products (see Sect. 3.8.2), matrix products (see Sect. 3.8.3), *a posteriori* forward error estimates for linear systems (see Sect. 3.8.4), gaussian factorizations (see Sect. 3.8.5) and solutions of triangular systems (see Sect. 3.8.6).

We begin our discussion of rounding error accumulation with the following fundamental lemma, which is similar to the discussion in Higham [56, p. 63]. In our version of these inequalities, we will take more care to separate first-order and higher-order terms in ε, whenever possible.

Lemma 3.8.1 *Let $\varepsilon \in (0,1)$ and let n be a positive integer. Suppose that the sequence $\{\varepsilon_i\}_{i=1}^{2n}$ of scalars is such that for all i*

$$|\varepsilon_i| \leq \varepsilon \; .$$

If $(n-1)\varepsilon < 2$ then

$$\left| \prod_{i=1}^{n} (1+\varepsilon_i) - 1 \right| \leq \frac{n\varepsilon}{1-(n-1)\varepsilon/2} \; . \tag{3.33}$$

If $(n+1)\varepsilon < 2$ then

$$\left| \prod_{i=1}^{n} \frac{1}{1+\varepsilon_i} - 1 \right| \leq \frac{n\varepsilon}{1-(n+1)\varepsilon/2} \; . \tag{3.34}$$

And, if $n\varepsilon < 1$ then

$$\left| \prod_{i=1}^{n} \frac{1+\varepsilon_{2i-1}}{1+\varepsilon_{2i}} - 1 \right| \leq \frac{2n\varepsilon}{1-n\varepsilon} \; . \tag{3.35}$$

Proof We will prove (3.33) by induction. The claim is obvious for $n = 1$. Suppose that it is true for $n-1 \geq 1$. Then

$$\left| \prod_{i=1}^{n} (1+\varepsilon_i) - 1 \right| = \left| \left\{ \prod_{i=1}^{n-1} (1+\varepsilon_i) - 1 \right\} (1+\varepsilon_n) + \varepsilon_n \right|$$
$$\leq \frac{(n-1)\varepsilon}{1-(n-2)\varepsilon/2}(1+\varepsilon) + \varepsilon = n\varepsilon \frac{1+\varepsilon/2}{1-(n-2)\varepsilon/2} \leq n\varepsilon \frac{1}{1-(n-1)\varepsilon/2} \, ,$$

where the final inequality is true because

$$1-(n-2)\varepsilon/2-(1+\varepsilon/2)[1-(n-1)\varepsilon/2]$$
$$=1-(n-2)\varepsilon/2-1+(n-1)\varepsilon/2-\varepsilon/2+(n-1)\varepsilon^2/4=(n-1)\varepsilon^2/4\geq 0\,.$$

Next, we will prove (3.34). Note that this inequality is obvious for $n = 1$. Inductively, we will assume that the claim is true for $n-1\geq 1$. Then

$$\left|\prod_{i=1}^{n}\frac{1}{1-\varepsilon_i}-1\right|=\left|\left\{\prod_{i=1}^{n-1}\frac{1}{1-\varepsilon_i}-1\right\}\frac{1}{1-\varepsilon_n}+\frac{\varepsilon_n}{1-\varepsilon_n}\right|$$
$$\leq\frac{(n-1)\varepsilon}{1-n\varepsilon/2}\frac{1}{1-\varepsilon}+\frac{\varepsilon}{1-\varepsilon}=\frac{\varepsilon}{1-\varepsilon}\frac{n-1+1-n\varepsilon/2}{1-n\varepsilon/2}=n\frac{\varepsilon}{1-\varepsilon}\frac{1-\varepsilon/2}{1-n\varepsilon/2}$$
$$\leq\frac{n\varepsilon}{1-(n+1)\varepsilon/2}\,,$$

where the final inequality is true because

$$(1-\varepsilon)(1-n\varepsilon/2)-(1-\varepsilon/2)[1-(n+1)\varepsilon/2]$$
$$=1-\varepsilon-n\varepsilon/2+n\varepsilon^2/2-1+\varepsilon/2+(n+1)\varepsilon/2-(n+1)\varepsilon^2/4=(n-1)\varepsilon^2/4\geq 0\,.$$

Finally, we will prove (3.35). For $n = 1$, we have

$$\left|\frac{1+\varepsilon_1}{1+\varepsilon_2}-1\right|=\left|\frac{\varepsilon_1-\varepsilon_2}{1+\varepsilon_2}\right|\leq\frac{2\varepsilon}{1-\varepsilon}\,.$$

Inductively, we will assume that the claim is true for $n-1\geq 1$. Then

$$\left|\prod_{i=1}^{n}\frac{1+\varepsilon_{2i-1}}{1+\varepsilon_{2i}}-1\right|=\left|\left\{\prod_{i=1}^{n-1}\frac{1+\varepsilon_{2i-1}}{1+\varepsilon_{2i}}-1\right\}\frac{1+\varepsilon_{2n-1}}{1+\varepsilon_{2n}}+\frac{1+\varepsilon_{2n-1}}{1-\varepsilon_{2n}}-1\right|$$
$$\leq\frac{2(n-1)\varepsilon}{1-(n-1)\varepsilon}\frac{1+\varepsilon}{1-\varepsilon}+\frac{2\varepsilon}{1-\varepsilon}=\frac{2n\varepsilon}{1-n\varepsilon+(n-1)\varepsilon^2}\leq\frac{2n\varepsilon}{1-n\varepsilon}\,.$$

The previous lemma estimates the accumulation of errors in products, quotients and ratios. Our goal in the next four lemmas is examine operations with error bounds of the form found in Lemma 3.8.1. We will develop simple bounds that separate first-order and higher-order terms in ε), by allowing the coefficients in the numerator and denominator to be distinct. This is different from the approach in Higham [56, p. 67], who required the same coefficients in both numerator and denominator.

Lemma 3.8.2 *Let $\varepsilon \in (0,1)$, and suppose that m_1, m_2, n_1 and n_2 are nonnegative numbers. If $m_1 \leq n_1$, $m_2 \leq n_2$ and $n_2\varepsilon < 1$ then*

$$\frac{m_1\varepsilon}{1-m_2\varepsilon} \leq \frac{n_1\varepsilon}{1-n_2\varepsilon} . \tag{3.36}$$

Alternatively, if $\max\{m_2,n_2\}\varepsilon < 1$ we have

$$\frac{m_1\varepsilon}{1-m_2\varepsilon} + \frac{n_1\varepsilon}{1-n_2\varepsilon} \leq \frac{(m_1+n_1)\varepsilon}{1-\max\{m_2,n_2\}\varepsilon} . \tag{3.37}$$

Proof To prove (3.36), we note that

$$\frac{m_1\varepsilon}{1-m_2\varepsilon} \leq \frac{n_1\varepsilon}{1-m_2\varepsilon} \leq \frac{n_1\varepsilon}{1-n_2\varepsilon} .$$

Inequality (3.37) is also easy to prove:

$$\frac{m_1\varepsilon}{1-m_2\varepsilon} + \frac{n_1\varepsilon}{1-n_2\varepsilon} \leq \frac{m_1\varepsilon}{1-\max\{m_2,n_2\}\varepsilon} + \frac{n_1\varepsilon}{1-\max\{m_2,n_2\}\varepsilon} = \frac{(m_1+n_1)\varepsilon}{1-\max\{m_2,n_2\}\varepsilon} .$$

The next lemma provides an example of a difference between our approach to error bounds and that in Higham [56]. In particular, Higham would replace inequality (3.38) with the following:

$$\frac{m\varepsilon}{1-m\varepsilon}\left\{1+\frac{n\varepsilon}{1-n\varepsilon}\right\} \leq \frac{(m+n)\varepsilon}{1-(m+n)\varepsilon} .$$

Lemma 3.8.3 *Suppose that $\varepsilon > 0$, and that m_1, m_2, n_1 and n_2 are nonnegative numbers. If $\max\{m_2+n_1,n_2\}\varepsilon < 1$ then*

$$\frac{m_1\varepsilon}{1-m_2\varepsilon}\left\{1+\frac{n_1\varepsilon}{1-n_2\varepsilon}\right\} \leq \frac{m_1\varepsilon}{1-\max\{m_2+n_1,n_2\}\varepsilon} . \tag{3.38}$$

Alternatively, if $(n_1+\max\{m_2,n_2\})\varepsilon < 1$ then

$$\frac{(m_1\varepsilon)/(1-m_2\varepsilon)}{1-(n_1\varepsilon)/(1-n_2\varepsilon)} \leq \frac{m_1\varepsilon}{1-(n_1+\max\{m_2,n_2\})\varepsilon} . \tag{3.39}$$

Proof Inequality (3.38) is true because

$$\begin{aligned}
&(1-m_2\varepsilon)(1-n_2\varepsilon) - [1+(n_1-n_2)\varepsilon][1-\max\{m_2+n_1,n_2\}\varepsilon] \\
&= 1-m_2\varepsilon-n_2\varepsilon+m_2n_2\varepsilon^2 \\
&-1-(n_1-n_2)\varepsilon+\max\{m_2+n_1,n_2\}\varepsilon+(n_1-n_2)\max\{m_2+n_1,n_2\}\varepsilon^2 \\
&= \left\{\begin{array}{ll} n_1(m_2+n_1-n_2)\varepsilon^2, & m_2+n_1 \geq n_2 \\ (n_2-m_2-n_1)\varepsilon(1-n_2\varepsilon), & m_2+n_1 < n_2 \end{array}\right. \geq 0 .
\end{aligned}$$

Let us prove (3.39):

$$\frac{(m_1\varepsilon)/(1-m_2\varepsilon)}{1-(n_1\varepsilon)/(1-n_2\varepsilon)} \le \frac{m_1\varepsilon/\{1-\max\{m_2,n_2\}\varepsilon\}}{1-n_1\varepsilon/\{1-\max\{m_2,n_2\}\varepsilon\}} = \frac{m_1\varepsilon}{1-(n_1+\max\{m_2,n_2\})\varepsilon} .$$

Lemma 3.8.4 *Suppose that* $\varepsilon > 0$, *and that* m_1, m_2, n_1 *and* n_2 *are nonnegative numbers. If* $m_1+n_1 > 0$ *and*

$$\max\left\{m_2, n_2, \frac{m_1m_2+m_1n_1+n_1n_2}{m_1+n_1}\right\}\varepsilon < 1 ,$$

then

$$\left(1+\frac{m_1\varepsilon}{1-m_2\varepsilon}\right)\left(1+\frac{n_1\varepsilon}{1-n_2\varepsilon}\right)-1 \le \frac{(m_1+n_1)\varepsilon}{1-\max\left\{m_2,n_2,\frac{m_1m_2+m_1n_1+n_1n_2}{m_1+n_1}\right\}\varepsilon} . \tag{3.40}$$

Alternatively, if $k \ge 1$ *is an integer and* $[m_2+m_1(k-1)/2]\varepsilon < 1$, *then*

$$\left(1+\frac{m_1\varepsilon}{1-m_2\varepsilon}\right)^k - 1 \le \frac{km_1\varepsilon}{1-[m_2+m_1(k-1)/2]\varepsilon} . \tag{3.41}$$

Proof Let us prove (3.40). Without loss of generality, we assume that $m_2 \le n_2$. Since this claim is trivial when $m_1 = 0$, we assume that $m_1 > 0$. Note that

$$\frac{m_1m_2+m_1n_1+n_1n_2}{m_1+n_1} \ge n_2$$

is equivalent to

$$0 \le m_1m_2+m_1n_1+n_1n_2-m_1n_2-n_1n_2 = m_1(m_2+n_1-n_2) .$$

In the case when $m_2 \le n_2 \le m_2+n_1$ we have

$$\begin{aligned}
0 &\le m_1n_1(n_1+m_2-n_2)(n_2+m_1-m_2)\varepsilon^2 \\
&= (m_1+n_1)^2(1-m_2\varepsilon)(1-n_2\varepsilon) \\
&- [m_1+n_1+(m_1n_1-m_1n_2-m_2n_1)\varepsilon][m_1+n_1-(m_1m_2+m_1n_1+n_1n_2)\varepsilon] .
\end{aligned}$$

This is equivalent to

$$\frac{m_1+n_1+(m_1n_1-m_1n_2-m_2n_1)\varepsilon}{(1-m_2\varepsilon)(1-n_2\varepsilon)} \le \frac{(m_1+n_1)^2}{m_1+n_1-(m_1m_2+m_1n_1+n_1n_2)\varepsilon} ,$$

which is in turn equivalent to (3.40). Otherwise, we have $n_2 > m_2 + n_1$. In this case, we have

$$
\begin{aligned}
0 &< m_1(n_2 - m_2 - n_1)\varepsilon \\
&= (m_1 + n_1)(1 - m_2\varepsilon) - [m_1 + n_1 + (m_1 n_1 - m_1 n_2 - m_2 n_1)\varepsilon] \,.
\end{aligned}
$$

This is equivalent to

$$
\frac{m_1 + n_1 + (m_1 n_1 - m_1 n_2 - m_2 n_1)\varepsilon}{(1 - m_2\varepsilon)(1 - n_2\varepsilon)} < \frac{m_1 + n_1}{1 - n_2\varepsilon} \,,
$$

which in turn implies (3.40).

We will prove (3.41) by induction. This inequality is obvious if $k = 1$. Inductively, assume that the inequality is true for $k - 1 \geq 1$. Then

$$
\begin{aligned}
0 &\leq \frac{1}{4} m_1 (k-1)\varepsilon^2 \\
&= \{1 - [m_2 + m_1(k-2)/2]\varepsilon\}(1 - m_2\varepsilon) \\
&\quad - [1 + (m_1/2 - m_2)\varepsilon]\{1 - [m_2 + m_1(k-1)/2]\,\varepsilon\} \,.
\end{aligned}
$$

This is equivalent to

$$
\begin{aligned}
\frac{k m_1 \varepsilon}{1 - [m_2 + m_1(k-1)/2]\varepsilon} &> \frac{k m_1 \varepsilon [1 + (m_1/2 - m_2)\varepsilon]}{\{1 - [m_2 + m_1(k-2)/2]\varepsilon\}(1 - m_2\varepsilon)} \\
&= \left\{1 + \frac{(k-1) m_1 \varepsilon}{1 - [m_2 + m_1(k-2)/2]\varepsilon}\right\}\left\{1 + \frac{m_1\varepsilon}{1 - m_2\varepsilon}\right\} - 1 \\
&\geq \left(1 + \frac{m_1\varepsilon}{1 - m_2\varepsilon}\right)^{k-1}\left(1 + \frac{m_1\varepsilon}{1 - m_2\varepsilon}\right) - 1 = \left(1 + \frac{m_1\varepsilon}{1 - m_2\varepsilon}\right)^{k} - 1 \,.
\end{aligned}
$$

Inequality (3.40) is "tight", in that the numerator and the third option in the denominator are consistent with Taylor expansions around $\varepsilon = 0$. The following lemma produces an inequality that is not tight, but does have a simpler form than (3.40).

Lemma 3.8.5 *Suppose that* $\varepsilon > 0$, *and that* m *and* n *are nonnegative numbers. If* m *and* n *are positive scalars and* $(m + n)\varepsilon < 1$, *then*

$$
\left(1 + \frac{m\varepsilon}{1 - m\varepsilon}\right)\left(1 + \frac{n\varepsilon}{1 - n\varepsilon}\right) - 1 \leq \frac{(m + m)\varepsilon}{1 - (m + n)\varepsilon} \,. \tag{3.42}
$$

Proof Since

$$
0 \leq mn\varepsilon \,,
$$

we see that

$$m + n - mn\varepsilon - (m+n)^2\varepsilon + mn(m+n)\varepsilon \le m + n - (m+n)^2\varepsilon + mn(m+n)\varepsilon \; .$$

This can be factored as

$$[m + n - mn\varepsilon][1 - (m+n)\varepsilon] \le (1 - m\varepsilon)(1 - n\varepsilon)(m+n) \; .$$

Because $(m+n)\varepsilon < 1$, we can divide to get

$$\frac{(m+n)\varepsilon}{1-(m+n1)\varepsilon} \ge \frac{(m+n)\varepsilon - mn\varepsilon^2}{(1-m\varepsilon)(1-n\varepsilon)} = \frac{1-(1-m\varepsilon)(1-n\varepsilon)}{(1-m\varepsilon)(1-n\varepsilon)}$$
$$= \left[1 + \frac{m\varepsilon}{1-m\varepsilon}\right]\left[1 + \frac{n\varepsilon}{1-n\varepsilon}\right] - 1 \; .$$

Inequality (3.42) leads to simpler expressions when the product of a large number of terms of the form $1 + m_1\varepsilon/(1 - m_2\varepsilon)$ must be bounded.

Lemma 3.8.6 *Let $\varepsilon \in (0,1)$ and let m and n be positive integers. Suppose that the sequences $\{\varepsilon_i\}_{i=1}^m$ and $\{\delta_j\}_{j=1}^n$ of scalars are such that*

$$|\varepsilon_i| \le \varepsilon \text{ for all } 1 \le i \le m \text{ and } |\delta_j| \le \varepsilon \text{ for all } 1 \le j \le n \; .$$

Then

$$\left| \frac{\prod_{i=1}^m (1+\varepsilon_i)}{\prod_{j=1}^n (1+\delta_j)} - 1 \right| \le \frac{(m+n)\varepsilon}{1 - \max\{m,n\}\varepsilon} \; . \tag{3.43}$$

Proof If $m = n$, then the claim is the same as inequality (3.35). Suppose that $m > n$. Then

$$\left| \frac{\prod_{i=1}^m (1+\varepsilon_i)}{\prod_{j=1}^n (1+\delta_j)} - 1 \right| = \left| \left\{ \prod_{i=1}^n \frac{1+\varepsilon_i}{1+\delta_j} - 1 \right\} \prod_{j=1}^{m-n} (1+\varepsilon_{n+j}) + \left\{ \prod_{j=1}^{m-n} (1+\varepsilon_{n+j}) - 1 \right\} \right|$$

then inequalities (3.35) and (3.33) imply

$$\le \frac{2n\varepsilon}{1-n\varepsilon}\left\{1 + \frac{(m-n)\varepsilon}{1-(m-n-1)\varepsilon/2}\right\} + \frac{(m-n)\varepsilon}{1-(m-n-1)\varepsilon/2}$$

then inequality (3.38) gives us

$$\le \frac{2n\varepsilon}{1-\max\{m,(m-n-1)/2\}\varepsilon} + \frac{(m-n)\varepsilon}{1-(m-n-1)\varepsilon/2}$$

then inequality (3.37) and the fact that $(m-n-1)/2 < m$ yield

$$\leq \frac{(m+n)\varepsilon}{1-m\varepsilon} \ .$$

On the other hand, if $m < n$ then

$$\left| \frac{\prod_{i=1}^{m}(1+\varepsilon_i)}{\prod_{j=1}^{n}(1+\delta_j)} - 1 \right| = \left| \left\{ \prod_{i=1}^{m} \frac{1+\varepsilon_i}{1+\delta_j} - 1 \right\} \prod_{j=1}^{n-m} \frac{1}{1+\delta_{m+j}} + \left\{ \prod_{j=1}^{n-m} \frac{1}{1+\delta_{m+j}} - 1 \right\} \right|$$

then inequalities (3.35) and (3.34) imply

$$\leq \frac{2m\varepsilon}{1-m\varepsilon} \left\{ 1 + \frac{(n-m)\varepsilon}{1-(n-m+1)\varepsilon/2} \right\} + \frac{(n-m)\varepsilon}{1-(n-m+1)\varepsilon/2}$$

then inequality (3.38) gives us

$$\leq \frac{2m\varepsilon}{1-\max\{n, (n-m+1)/2\}\varepsilon} + \frac{(n-m)\varepsilon}{1-(n-m+1)\varepsilon/2}$$

then inequality (3.37) and the fact that $(n-m+1)/2 < n$ yield

$$\leq \frac{(m+n)\varepsilon}{1-n\varepsilon} \ .$$

The following lemma can also be found in Higham [56, p. 73].

Lemma 3.8.7 *Let $\|\cdot\|$ be a consistent matrix norm on square matrices and let k be a positive integer. Suppose that $\{\delta_j\}_{j=1}^{k}$ is a sequence of nonnegative real numbers. Assume that $\{\mathbf{A}_j\}_{j=1}^{k}$ and $\|\triangle\mathbf{A}_j\}_{j=1}^{k}$ are two sequences of square matrices of the same size satisfying*

$$\|\triangle\mathbf{A}_j\| \leq \delta_j \|\mathbf{A}_j\| \text{ for all } 1 \leq j \leq k \ .$$

Then

$$\left\| \prod_{j=1}^{k}(\mathbf{A}_j + \triangle\mathbf{A}_j) - \prod_{j=1}^{k}\mathbf{A}_j \right\| \leq \left\{ \prod_{j=1}^{k}(1+\delta_j) - 1 \right\} \prod_{j=1}^{k} \|\mathbf{A}_j\| \ . \tag{3.44}$$

Proof The claim is obviously true for $k = 1$. Assume inductively that the claim is true for $k - 1 \geq 1$. Then

$$\left\| \prod_{j=1}^{k} (\mathbf{A}_j + \triangle\mathbf{A}_j) - \prod_{j=1}^{k} \mathbf{A}_j \right\| = \left\| \left\{ \prod_{j=1}^{k-1} (\mathbf{A}_j + \triangle\mathbf{A}_j) \right\} (\mathbf{A}_k + \triangle\mathbf{A}_k) - \left\{ \prod_{j=1}^{k-1} \mathbf{A}_j \right\} \mathbf{A}_k \right\|$$

$$= \left\| \left\{ \prod_{j=1}^{k-1} (\mathbf{A}_j + \triangle\mathbf{A}_j) - \prod_{j=1}^{k-1} \mathbf{A}_j \right\} (\mathbf{A}_k + \triangle\mathbf{A}_k) + \left\{ \prod_{j=1}^{k-1} \mathbf{A}_j \right\} \triangle\mathbf{A}_k \right\|$$

$$\leq \left\{ \prod_{j=1}^{k-1} (1 + \delta_j) - 1 \right\} \left\{ \prod_{j=1}^{k-1} \|\mathbf{A}_j\| \right\} \|\mathbf{A}_k + \triangle\mathbf{A}_k\| + \left\{ \prod_{j=1}^{k} \|\mathbf{A}_j\| \right\} \delta_k$$

$$\leq \left\{ \prod_{j=1}^{k-1} (1 + \delta_j) - 1 \right\} \left\{ \prod_{j=1}^{k} \|\mathbf{A}_j\| \right\} (1 + \delta_k) + \left\{ \prod_{j=1}^{k} \|\mathbf{A}_j\| \right\} \delta_k$$

$$\leq \left\{ \prod_{j=1}^{k} (1 + \delta_j) - 1 - \delta_k + \delta_k \right\} \prod_{j=1}^{k} \|\mathbf{A}_j\| = \left\{ \prod_{j=1}^{k} (1 + \delta_j) - 1 \right\} \prod_{j=1}^{k} \|\mathbf{A}_j\|$$

3.8.2 Inner Products

Next, we will find bounds for rounding errors in computing inner products, similar to the results stated in Wilkinson [111, p. 114], or proved in Higham [56, p. 62–3].

Lemma 3.8.8 *Assume that the relative errors in floating point addition or multiplication satisfy*

$$fl(a + b) = (a + b)(1 + \varepsilon_+) \text{ and}$$
$$fl(a \times b) = (a \times b)(1 + \varepsilon_\times) ,$$

and that there is an upper bound ε *on all such relative errors:*

$$\max\{|\varepsilon_\times|, |\varepsilon_+|\} \leq \varepsilon .$$

Suppose that $\mathbf{x}$ *and* $\mathbf{y}$ *are m-vectors. Then the absolute error in the floating point computation of* $\mathbf{x} \cdot \mathbf{y}$ *satisfies the* ***forward error estimate***

$$|fl(\mathbf{x} \cdot \mathbf{y}) - \mathbf{x} \cdot \mathbf{y}| \leq \frac{m\varepsilon}{1 - (m-1)\varepsilon/2} \|\mathbf{x}\|_2 \, \|\mathbf{y}\|_2 \tag{3.45}$$

and the ***backward error estimates***

$$fl(\mathbf{x}\cdot\mathbf{y}) = \mathbf{x}\cdot(\mathbf{y}+\triangle\mathbf{y}) \text{ where } \|\triangle\mathbf{y}\|_2 \le \frac{m\varepsilon}{1-(m-1)\varepsilon/2}\|\mathbf{y}\|_2 \text{ or} \tag{3.46}$$

$$fl(\mathbf{x}\cdot\mathbf{y}) = \mathbf{x}\cdot(\mathbf{y}+\triangle\mathbf{y}) \text{ where } \|\triangle\mathbf{y}\|_\infty \le \frac{m\varepsilon}{1-(m-1)\varepsilon/2}\|\mathbf{y}\|_\infty \,. \tag{3.47}$$

Proof The inner product can be computed in floating point arithmetic by the linear recurrence

$$\tilde{\sigma}_0 = 0\,,\ \tilde{\sigma}_i = \ \mathrm{fl}\,(\tilde{\sigma}_{i-1} + \ \mathrm{fl}\,(\overline{\mathbf{x}_i}\mathbf{y}_i)) = [\tilde{\sigma}_{i-1} + \overline{\mathbf{x}_i}\mathbf{y}_i(1+\varepsilon_{\times i})]\,(1+\varepsilon_{+i}) \text{ for } 1 \le i \le m\,.$$

We can solve this recurrence to get

$$\mathrm{fl}(\mathbf{x}\cdot\mathbf{y}) \equiv \tilde{\sigma}_m = \sum_{i=1}^m \left[(1+\varepsilon_{\times i})\prod_{j=i+1}^m (1+\varepsilon_{+j})\right]\overline{\mathbf{x}_i}\mathbf{y}_i\,.$$

Inequality (3.33) implies that

$$|\mathrm{fl}(\mathbf{x}\cdot\mathbf{y}) - \mathbf{x}\cdot\mathbf{y}| \le \max_{1\le i\le m}\left|(1+\varepsilon_{\times i})\prod_{j=i+1}^m(1+\varepsilon_{+j}) - 1\right|\sum_{i=1}^m |\mathbf{x}_i|\ |\mathbf{y}_i|$$

$$\le \frac{m\varepsilon}{1-(m-1)\varepsilon/2}\sum_{i=1}^m |\mathbf{x}_i|\ |\mathbf{y}_i| \le \frac{m\varepsilon}{1-(m-1)\varepsilon/2}\|\mathbf{x}\|_2\|\mathbf{y}\|_2\,.$$

This proves the forward error estimate (3.45).

Alternatively, we note that

$$\mathrm{fl}(\mathbf{x}\cdot\mathbf{y}) = \mathbf{x}\cdot(\mathbf{y}+\triangle\mathbf{y})\,,$$

where the vector $\triangle\mathbf{y}$ can be defined component-wise by

$$\triangle\mathbf{y}_i = \left[(1+\varepsilon_{\times i})\prod_{j=i+1}^m(1+\varepsilon_{+j}) - 1\right]\mathbf{y}_i\,,$$

Then it is easy to see from the definition of $\triangle\mathbf{y}$ that

$$\|\triangle\mathbf{y}\|_2^2 \le \sum_{i=1}^m\left[\frac{(m-i+1)\varepsilon}{1-(m-i)\varepsilon/2}\right]^2\mathbf{y}_i^2 \le \left[\frac{m\varepsilon}{1-(m-1)\varepsilon/2}\right]^2\sum_{i=1}^m\mathbf{y}_i^2$$

and

$$\|\triangle \mathbf{y}\|_\infty \leq \frac{m\varepsilon}{1-(m-1)\varepsilon/2}\|\mathbf{y}\|_\infty \; .$$

This proves the backward error estimates
Note that in older texts, such as Wilkinson [110], it was common to bound terms that are greater than first order in ε by some positive constant times ε. This approach lead to claims with messy constants, and assumptions that were often hard to find.

Exercise 3.8.1 For $n = 2^{10}, \ldots, 2^{20}$, let $\tilde{\mathbf{x}}$ and $\tilde{\mathbf{y}}$ be n-vectors with randomly chosen entries in single precision. Also let $\mathbf{x}$ and $\mathbf{y}$ be n-vectors with the same values stored in double precision. Compute

$$\tilde{\delta} = \tilde{\mathbf{x}} \cdot \tilde{\mathbf{y}}$$

in single precision, and

$$\delta = \mathbf{x} \cdot \mathbf{y}$$

in double precision. Then plot $\log(|\tilde{\delta} - \delta|/\varepsilon)$ versus $\log n$. What is the slope of this graph? Discuss how this slope compares to the results in Lemma 3.8.8.

3.8.3 *Matrix Products*

Suppose that we are given an $m \times n$ matrix $\mathbf{A}$ with columns $\mathbf{a}_j$. Let $\mathbf{x}$ be an n-vector with components ξ_j. Let $\mathbf{y}$ be an m-vector, and let α and β be scalars. Consider computing $\mathbf{z} = \mathbf{A}\mathbf{x}\alpha + \mathbf{y}\beta$ by the following algorithm:

Algorithm 3.8.1 (LAPACK _gemv)

$\mathbf{z} = \mathbf{y}$

if $\beta \neq 1$

  if $\beta = 0$ then $\mathbf{z} = \mathbf{0}$

  else $\mathbf{z} = \mathbf{z} * \beta$

for $1 \leq j \leq n$

  $\mathbf{z} = \mathbf{z} + \mathbf{a}_j(\xi_j \alpha)$

The following lemma bounds the rounding errors in this algorithm, and is similar to a result in Wilkinson [111, p. 115] or Higham [56, p. 70].

Lemma 3.8.9 *Assume that the relative errors in floating point addition or multiplication satisfy*

$$fl(a+b) = (a+b)(1+\varepsilon_+) \text{ and}$$
$$fl(a \times b) = (a \times b)(1+\varepsilon_\times) \,,$$

and that there is an upper bound ε *on all such relative errors:*

$$\max\{|\varepsilon_\times|, |\varepsilon_+|\} \leq \varepsilon \,.$$

Suppose that $\mathbf{A}$ *is an* $m \times n$ *matrix,* $\mathbf{x}$ *is an* n*-vector,* $\mathbf{y}$ *is an* m*-vector, and* α *and* β *are scalars. Assume that* $n\varepsilon < 2$*. Then the rounding errors*

$$\triangle\mathbf{z} = fl(\mathbf{A}\mathbf{x}\alpha + \mathbf{y}\beta) - (\mathbf{A}\mathbf{x}\alpha + \mathbf{y}\beta)$$

in Algorithm 3.8.1 for computing $\mathbf{z} = \mathbf{A}\mathbf{x}\alpha + \mathbf{y}\beta$ *satisfies*

$$\|\triangle\mathbf{z}\|_1 \leq \frac{(n+1)\varepsilon}{1-n\varepsilon/2}\{\|\mathbf{y}\|_1|\beta| + \|\mathbf{A}\|_1\|\mathbf{x}\|_1|\alpha|\} \,, \tag{3.48a}$$

$$\|\triangle\mathbf{z}\|_\infty \leq \frac{(n+1)\varepsilon}{1-n\varepsilon/2}\{\|\mathbf{y}\|_\infty|\beta| + \|\mathbf{A}\|_\infty\|\mathbf{x}\|_\infty|\alpha|\} \text{ and} \tag{3.48b}$$

$$\|\triangle\mathbf{z}\|_2 \leq \frac{(n+1)\varepsilon}{1-n\varepsilon/2}\{\|\mathbf{y}\|_2|\beta| + \|\mathbf{A}\|_F\|\mathbf{x}\|_2|\alpha|\} \,. \tag{3.48c}$$

Proof Let $\mathbf{a}_j$ denote the jth column of $\mathbf{A}$ and ξ_j denote the jth component of $\mathbf{x}$. We compute $\mathbf{z} = \mathbf{Z}\mathbf{x}\alpha + \mathbf{y}\beta$ by the algorithm

$$\mathbf{z}^{(0)} = \mathbf{y}\beta \,,\; \mathbf{z}^{(j)} = \mathbf{z}^{(j-1)} + \mathbf{a}_j(\xi_j\alpha) \text{ for } 1 \leq j \leq n \,.$$

However, floating-point computation produces

$$\mathrm{fl}\left(\mathbf{z}^{(0)}\right) = (\mathbf{I} + \mathbf{E}_{\times 0})\mathbf{y}\beta \text{ and}$$
$$\mathrm{fl}\left(\mathbf{z}^{(j)}\right) = (\mathbf{I} + \mathbf{E}_{+j})\left\{\mathbf{z}^{(j-1)} + (\mathbf{I} + \mathbf{E}_{\times j})\mathbf{a}_j\xi_j\alpha(1+\varepsilon_{\times j})\right\} \,.$$

Here $\mathbf{E}_{\times j}$ and $\mathbf{E}_{+j}$ are diagonal matrices with diagonal entries bounded above by ε. We can solve this recurrence to get

$$\mathrm{fl}(\mathbf{z}) = \mathrm{fl}(\mathbf{z}^{(n)})$$
$$= \left\{\prod_{j=1}^{n}(\mathbf{I} + \mathbf{E}_{+j})\right\}(\mathbf{I} + \mathbf{E}_{\times 0})\mathbf{y}\beta + \sum_{j=1}^{n}\prod_{k=j+1}^{n}(\mathbf{I} + \mathbf{E}_{+k})(\mathbf{I} + \mathbf{E}_{\times j})(1+\varepsilon_{\times j})\mathbf{a}_j\xi_j\alpha \,.$$

In particular, we can bound

$$\begin{aligned}\|\triangle\mathbf{z}\|_1 &= \|\mathrm{f\,l}(\mathbf{z})-\mathbf{z}\|_1 \\ &= \left\| \left[\left\{\prod_{j=1}^{n}(\mathbf{I}+\mathbf{E}_{+j})\right\}(\mathbf{I}+\mathbf{E}_{\times 0})-\mathbf{I}\right]\mathbf{y}\beta \right. \\ &\left. +\sum_{j=1}^{n}\left[\left\{\prod_{k=j+1}^{n}(\mathbf{I}+\mathbf{E}_{+k})\right\}(\mathbf{I}+\mathbf{E}_{\times j})(1+\varepsilon_{\times j})-\mathbf{I}\right]\mathbf{a}_j\xi_j\alpha\right\|_1\end{aligned}$$

then inequality (3.44) gives us

$$\leq \left[(1+\varepsilon)^{n+1}-1\right]\|\mathbf{y}\|_1\,|\beta| + \sum_{j=1}^{n}\left[(1+\varepsilon)^{n-j+2}-1\right]\|\mathbf{a}_j\|_1\,\left|\xi_j\right|\ |\alpha|$$

then inequality (3.41) implies that

$$\begin{aligned}&\leq \frac{(n+1)\varepsilon}{1-n\varepsilon/2}\|\mathbf{y}\|_1\,|\beta| + \frac{(n+1)\varepsilon}{1-n\varepsilon/2}\left\{\max_{1\leq j\leq j}\left\|\mathbf{a}_j\right\|_1\right\}\sum_{j=1}^{n}\left|\xi_j\right|\,|\alpha| \\ &= \frac{(n+1)\varepsilon}{1-n\varepsilon/2}\left\{\|\mathbf{y}\|_1|\beta| + \|\mathbf{A}\|_1\|\mathbf{x}\|_1|\alpha|\right\}\ .\end{aligned}$$

This proves (3.48a).

The proof of (3.48b) is similar. First, we note that

$$\begin{aligned}\|\triangle\mathbf{z}\|_\infty &\leq \left\|\left\{\prod_{j=0}^{n}(\mathbf{I}+\mathbf{E}_{+j})(\mathbf{I}+\mathbf{E}_{\times 0})-\mathbf{I}\right\}\mathbf{y}\beta\right\|_\infty \\ &+\left\|\sum_{j=1}^{n}\left\{\prod_{k=j+1}^{n}(\mathbf{I}+\mathbf{E}_{+k})(\mathbf{I}+\mathbf{E}_{\times j})(1+\varepsilon_{\times j})-\mathbf{I}\right\}\mathbf{a}_j\xi_j\alpha\right\|_\infty\end{aligned}$$

then inequalities (3.44) and (3.41) imply that

$$\begin{aligned}&\leq \frac{(n+1)\varepsilon}{1-n\varepsilon/2}\|\mathbf{y}\|_\infty\,|\beta| + \max_{1\leq i\leq m}\sum_{j=1}^{n}\frac{(n+2-j)\varepsilon}{1-(n+1-j)\varepsilon/2}\left|\mathbf{e}_i\cdot\mathbf{a}_j\xi_j\alpha\right| \\ &\leq \frac{(n+1)\varepsilon}{1-n\varepsilon/2}\|\mathbf{y}\|_\infty\,|\beta| + \frac{(n+1)\varepsilon}{1-n\varepsilon/2}\max_{1\leq i\leq m}\sum_{j=1}^{n}\left|\mathbf{e}_i\cdot\mathbf{a}_j\right|\ \|\mathbf{x}\|_\infty\,|\alpha| \\ &= \frac{(n+1)\varepsilon}{1-n\varepsilon/2}\left\{\|\mathbf{y}\|_\infty|\beta| + \|\mathbf{A}\|_\infty\|\mathbf{x}\|_\infty|\alpha|\right\}\ .\end{aligned}$$

Finally,

$$\|\triangle \mathbf{z}\|_2 = \|\mathrm{f}\,1(\mathbf{z}) - \mathbf{z}\|_2 \le \frac{(n+1)\varepsilon}{1-n\varepsilon/2}\|\mathbf{y}\|_2\,|\beta| + \frac{(n+1)\varepsilon}{1-n\varepsilon/2}\sum_{j=1}^{n}\|\mathbf{a}_j\|_2\,|\xi_j|\;\;|\alpha|$$

then the Cauchy inequality (3.15) gives us

$$\le \frac{(n+1)\varepsilon}{1-n\varepsilon/2}\|\mathbf{y}\|_2\,|\beta| + \frac{(n+1)\varepsilon}{1-n\varepsilon/2}\left(\sum_{j=1}^{n}\|\mathbf{a}_j\|_2^2\right)^{1/2}\left(\sum_{j=1}^{n}|\xi_j|^2\right)^{1/2}|\alpha|$$
$$= \frac{(n+1)\varepsilon}{1-n\varepsilon/2}\left\{\|\mathbf{y}\|_2|\beta| + \|\mathbf{A}\|_F\|\mathbf{x}\|_2|\alpha|\right\} .$$

3.8.4 *Forward Error Estimates*

In Theorem 3.6.2, we developed forward and backward error *a posteriori* estimates for the errors in an approximate solution to a system of linear equations. In this section, we will amend the forward error estimate to account for rounding errors in computing the residual.

Theorem 3.8.1 (*a Posteriori* Forward Error Estimation with Rounding Errors) *Suppose that* $\mathbf{A}$ *is a nonsingular* $n \times n$ *matrix,* $\mathbf{b}$ *is a nonzero* n*-vector, and* $\mathbf{x}$ *solves* $\mathbf{A}\mathbf{x} = \mathbf{b}$*. Given an* n*-vector* $\tilde{\mathbf{x}}$ *with entries* ξ_j*, define the* ***residual*** *by*

$$\mathbf{r} = \mathbf{b} - \mathbf{A}\tilde{\mathbf{x}} .$$

Then the forward error in $\mathbf{x}$ *satisfies*

$$\frac{\|\mathbf{x}-\tilde{\mathbf{x}}\|_\infty}{\|\tilde{\mathbf{x}}\|_\infty} \le \frac{\max_{1\le i\le n}\left\{|\mathbf{A}^{-1}|\left(|\mathbf{r}| + \frac{n_z\varepsilon}{1-(n_z-1)\varepsilon/2}\left\{|\mathbf{A}|\,|\tilde{\mathbf{x}}| + |\mathbf{b}|\right\}\right)\right\}_i}{\max_{1\le j\le n}\left|\tilde{\xi}_j\right|} . \tag{3.49}$$

Here $|\mathbf{A}|$ *represents the matrix of absolute values of the corresponding entries of* $\mathbf{A}$*,* n_z *is the maximum number of nonzeros in any row of* $\mathbf{A}$*, and* ε *is machine precision.*

Proof The discussion in the proof of Lemma 3.8.9 shows that

$$\tilde{\mathbf{r}} \equiv \mathrm{f}\,1(\mathbf{b}-\mathbf{A}\tilde{\mathbf{x}}) = (\mathbf{I}+\mathbf{E}_\mathbf{b})\mathbf{b} - (\mathbf{A}+\mathbf{E}_\mathbf{A})\tilde{\mathbf{x}} = \mathbf{A}(\mathbf{x}-\tilde{\mathbf{x}}) + \mathbf{E}_\mathbf{b}\mathbf{b} + \mathbf{E}_\mathbf{A}\tilde{\mathbf{x}} ,$$

where the diagonal matrix $\mathbf{E}_\mathbf{b}$ has the form

$$\mathbf{E}_\mathbf{b} = \prod_{j=1}^{n}(\mathbf{I}+\mathbf{E}_{+j}) - \mathbf{I} ,$$

and

$$\mathbf{E_A}\tilde{\mathbf{x}} = \sum_{j=1}^{n} \left[\prod_{k=j+1}^{n} (\mathbf{I} + \mathbf{E}_{+k})(\mathbf{I} + \mathbf{E}_{\times j}) - \mathbf{I} \right] \mathbf{a}_j \tilde{\xi}_j \;.$$

We can solve for the error in $\mathbf{x}$ to get

$$\mathbf{x} - \tilde{\mathbf{x}} = \mathbf{A}^{-1} \left[\tilde{\mathbf{r}} - \mathbf{E_b b} - \mathbf{E_A}\tilde{\mathbf{x}} \right] \;.$$

Using inequalities (3.44) and (3.33), we see that

$$\frac{\|\mathbf{x} - \tilde{\mathbf{x}}\|_\infty}{\|\tilde{\mathbf{x}}\|_\infty} = \frac{\max_{1 \le i \le n} \left\{ \left| \left(\mathbf{A}^{-1} \left[\tilde{\mathbf{r}} - \mathbf{E_b b} - \mathbf{E_A}\tilde{\mathbf{x}} \right] \right)_i \right| \right\}}{\max_{1 \le j \le n} \left| \tilde{\xi}_j \right|}$$

$$\le \frac{\max_{1 \le i \le n} \left\{ \left(|\mathbf{A}^{-1}| \left[|\tilde{\mathbf{r}}| + \frac{n_z \varepsilon}{1-(n_z-1)\varepsilon/2} \left(|\mathbf{b}| + |\mathbf{A}||\tilde{\mathbf{x}}| \right) \right] \right)_i \right\}}{\max_{1 \le j \le n} \left| \tilde{\xi}_j \right|} \;.$$

The forward error estimate (3.49) is computed in LAPACK routines `_lacn2`, which are called from routines `_gerfs`. In these LAPACK routines, the denominator $1 - (n_z - 1)\varepsilon/2$ is replaced by 1.

3.8.5 *Factorization Errors*

It is useful to analyze the errors produced in the numerical solution of linear systems by Gaussian factorization. Such analysis can explain the relative benefits of competing numerical schemes, such as partial and complete pivoting. In practice, these *a priori* error estimates are seldom used to measure errors in the solution of a linear system. Rather *a posteriori* estimates of the accuracy in a particular numerical solution are made via Theorem 3.6.2 or 3.8.1.

A priori error estimates are typically provided by Wilkinson's **backward error analysis** [109–111]. This analysis shows that floating point operations during the computation of a Gaussian factorization $\mathbf{Q}^\top \mathbf{A} \mathbf{P} = \mathbf{L}\mathbf{R}$ can be interpreted as producing a matrix of errors $\mathbf{E}$ so that the computed Gaussian factors $\tilde{\mathbf{L}}$ and $\tilde{\mathbf{R}}$ satisfy $\mathbf{Q}^\top (\mathbf{A} + \mathbf{E}) \mathbf{P} = \tilde{\mathbf{L}}\tilde{\mathbf{R}}$. We will reproduce Wilkinson's backward analysis in the following lemma, theorem and corollary.

Lemma 3.8.10 *Suppose that $\tilde{\mathbf{A}}$ is an $n \times n$ matrix with entries $\tilde{\alpha}_{i,j}$, and that $\tilde{\mathbf{A}} = \tilde{\mathbf{L}}\tilde{\mathbf{R}}$ where $\tilde{\mathbf{L}}$ is unit left triangular and $\tilde{\mathbf{R}}$ is right triangular. Assume that the sub-diagonal entries $\tilde{\lambda}_{ij}$ of $\tilde{\mathbf{L}}$ satisfy*

$$\left| \tilde{\lambda}_{ij} \right| \le 1$$

for all $1 \leq j < i \leq n$*. Then for all* $1 \leq i \leq j \leq n$*, the entries* $\tilde{\varrho}_{ij}$ *of* $\tilde{\mathbf{R}}$ *satisfy*

$$\left|\tilde{\varrho}_{ij}\right| \leq 2^{i-1} \max_{1 \leq i,j \leq n} \left|\tilde{\alpha}_{i,j}\right| . \tag{3.50}$$

Next, suppose that at the k-th step of the factorization we partition

$$\tilde{\mathbf{A}} = \begin{bmatrix} \tilde{\mathbf{L}}_{11}^{(k)} & \mathbf{0} \\ \tilde{\mathbf{L}}_{21}^{(k)} & \mathbf{I} \end{bmatrix} \begin{bmatrix} \tilde{\mathbf{R}}_{11}^{(k)} & \tilde{\mathbf{R}}_{12}^{(k)} \\ \mathbf{0} & \tilde{\mathbf{A}}^{(k)} \end{bmatrix} ,$$

where $\tilde{\mathbf{L}}_{11}^{(k)}$ *and* $\tilde{\mathbf{R}}_{11}^{(k)}$ *are* $(k-1) \times (k-1)$*, and suppose that the absolute values of the entries in the* ***reduced matrix*** $\tilde{\mathbf{A}}^{(k)}$ *are always no larger than the first diagonal entry of this matrix. Then the entries of* $\tilde{\mathbf{R}}$ *satisfy*

$$\left|\tilde{\varrho}_{ij}\right| \leq \sqrt{i \prod_{k=2}^{i} k^{1/k-1}} \max_{1 \leq i,j \leq n} \left|\tilde{\alpha}_{i,j}\right| . \tag{3.51}$$

Here the empty product (i.e. $\prod_{k=2}^{1}$*) is assumed to be one.*

Proof For notational simplicity, we will use the same subscripts for the entries of the reduced matrices $\tilde{\mathbf{A}}^{(k)}$ as the original matrix $\tilde{\mathbf{A}}$.

We will prove the first claim (3.50) by showing that for all $1 \leq k \leq n$ the entries of the reduced matrix $\tilde{\mathbf{A}}^{(k)}$ satisfy

$$\left|\tilde{\alpha}_{ij}^{(k)}\right| \leq 2^{k-1} \max_{1 \leq i,j \leq n} \left|\tilde{\alpha}_{i,j}\right| . \tag{3.52}$$

Since $\tilde{\mathbf{A}}^{(1)} = \tilde{\mathbf{A}}$, the claim (3.52) is obvious for $k = 1$. Assume inductively that (3.52) is true for $k-1$. At the k-th step, for $k \leq i,j \leq n$ the entries of the reduced matrix satisfy the equation

$$\tilde{\alpha}_{i,j}^{(k)} = \tilde{\alpha}_{i,j}^{(k-1)} - \tilde{\lambda}_{i,k}\tilde{\alpha}_{k,j}^{(k-1)} .$$

Since the entries of $\tilde{\mathbf{L}}$ are assumed to have magnitude at most one,

$$\left|\tilde{\alpha}_{i,j}^{(k)}\right| \leq \left|\tilde{\alpha}_{i,j}^{(k-1)}\right| + \left|\tilde{\lambda}_{i,k-1}\right| \left|\tilde{\alpha}_{k-1,j}^{(k-1)}\right| \leq 2 \max_{k-1 \leq r,s \leq n} \left|\tilde{\alpha}_{r,s}^{(k-1)}\right| \leq 2^{k-1} \max_{k-1 \leq r,s \leq n} \left|\tilde{\alpha}_{r,s}\right| .$$

Since for $1 \leq i \leq j \leq n$ the entries of $\tilde{\mathbf{R}}$ are copied directly from the corresponding location in the reduced matrix, we have

$$\tilde{\varrho}_{i,j} = \alpha_{i,j}^{(i)} .$$

This equation together with (3.52) implies (3.50).

Next, we will prove (3.51). First, we note that the partial and final Gaussian factorization produces

$$\tilde{\mathbf{A}} = \begin{bmatrix} \tilde{\mathbf{L}}_{11}^{(k)} & \mathbf{0} \\ \tilde{\mathbf{L}}_{21}^{(k)} & \mathbf{I} \end{bmatrix} \begin{bmatrix} \tilde{\mathbf{R}}_{11}^{(k)} & \tilde{\mathbf{R}}_{12}^{(k)} \\ \mathbf{0} & \tilde{\mathbf{A}}^{(k)} \end{bmatrix} = \begin{bmatrix} \tilde{\mathbf{L}}_{11}^{(k)} & \mathbf{0} \\ \tilde{\mathbf{L}}_{21}^{(k)} & \tilde{\mathbf{L}}_{22}^{(k)} \end{bmatrix} \begin{bmatrix} \tilde{\mathbf{R}}_{11}^{(k)} & \tilde{\mathbf{R}}_{12}^{(k)} \\ \mathbf{0} & \tilde{\mathbf{R}}_{22}^{(k)} \end{bmatrix} ,$$

from which it follows that the reduced matrix will eventually be factored

$$\tilde{\mathbf{A}}^{(k)} = \tilde{\mathbf{L}}_{22}^{(k)} \tilde{\mathbf{R}}_{22}^{(k)} .$$

Since $\tilde{\mathbf{L}}_{22}^{(k)}$ is unit left triangular and $\tilde{\mathbf{R}}_{22}^{(k)}$ is right triangular, Lemma 3.2.17 and the Laplace expansion (3.7) show that for $1 \le k \le n$ we have

$$\left| \det \tilde{\mathbf{A}}^{(k)} \right| = \prod_{j=k}^{n} |\tilde{\varrho}_{jj}| .$$

Since we assumed that the entries of $\tilde{\mathbf{A}}^{(k)}$ are bounded in magnitude by the first diagonal entry, the columns of $\tilde{\mathbf{A}}^{(k)}$ for $k \le j \le n$ satisfy

$$\left\| \tilde{\mathbf{A}}^{(k)} \mathbf{e}_j \right\|_2^2 = \sum_{i=k}^{n} \left| \tilde{\alpha}_{ij}^{(k)} \right|^2 \le (n-k+1) \left| \tilde{\varrho}_{k,k}^2 \right| .$$

Then Hadamard's inequality (3.22) implies that

$$\left[\prod_{j=k}^{n} \tilde{\varrho}_{jj} \right]^2 = \left[\det \tilde{\mathbf{A}}^{(k)} \right]^2 \le \prod_{j=k}^{n} \left\| \tilde{\mathbf{A}} \mathbf{e}_j \right\|_2^2 \le \left[(n-k+1) \, |\tilde{\varrho}_{k,k}|^2 \right]^{n-k+1} . \tag{3.53}$$

If we take the logarithm of this inequality, we obtain

$$2 \sum_{j=k}^{n} \log |\tilde{\varrho}_{jj}| \le (n-k+1) \left[\log(n-k+1) + 2 \log |\tilde{\varrho}_{k,k}| \right] .$$

Canceling the first term in the sum from both sides and dividing by two produces

$$\sum_{j=k+1}^{n} \log |\tilde{\varrho}_{jj}| \le \frac{n-k+1}{2} \log(n-k+1) + (n-k) \log |\tilde{\varrho}_{k,k}| . \tag{3.54}$$

We also have

$$\sum_{j=1}^{n} \log |\tilde{\varrho}_{jj}| = \log \left| \det \tilde{\mathbf{A}} \right| . \tag{3.55}$$

Next, for $2 \leq k < n$ we divide (3.54) by $(n-k)(n-k+1)$, we divide (3.55) by $n-1$, and add all of these together:

$$\sum_{k=2}^{n-1} \left[\frac{\log(n-k+1)}{2(n-k)} + \frac{\log|\tilde{\varrho}_{k,k}|}{n-k+1} \right] + \frac{\log\left|\det \tilde{\mathbf{A}}\right|}{n-1}$$

$$\geq \sum_{k=2}^{n-1} \frac{1}{(n-k)(n-k+1)} \sum_{j=k+1}^{n} \log|\tilde{\varrho}_{j,j}| + \frac{1}{n-1} \sum_{j=1}^{n} \log|\tilde{\varrho}_{j,j}|$$

$$= \sum_{j=3}^{n} \log|\tilde{\varrho}_{j,j}| \sum_{k=2}^{j-1} \frac{1}{(n-k)(n-k+1)} + \frac{1}{n-1} \sum_{j=1}^{n} \log|\tilde{\varrho}_{j,j}|$$

$$= \sum_{j=3}^{n} \log|\tilde{\varrho}_{j,j}| \sum_{k=2}^{j-1} \left[\frac{1}{n-k} - \frac{1}{n-k+1} \right] + \frac{1}{n-1} \sum_{j=1}^{n} \log|\tilde{\varrho}_{j,j}|$$

$$= \sum_{j=3}^{n} \log|\tilde{\varrho}_{j,j}| \left[\frac{1}{n-j+1} - \frac{1}{n-1} \right] + \frac{1}{n-1} \sum_{j=1}^{n} \log|\tilde{\varrho}_{j,j}|$$

$$= \sum_{j=3}^{n} \frac{\log|\tilde{\varrho}_{j,j}|}{n-j+1} + \frac{\log|\tilde{\varrho}_{1,1}| + \log|\tilde{\varrho}_{2,2}|}{n-1} = \sum_{j=2}^{n} \frac{\log|\tilde{\varrho}_{j,j}|}{n-j+1} + \frac{\log|\tilde{\varrho}_{1,1}|}{n-1} \,.$$

Canceling the sum from 2 to $n-1$ on both sides of the inequality leads to

$$\log|\tilde{\varrho}_{n,n}| + \frac{\log|\tilde{\varrho}_{1,1}|}{n-1} \leq \sum_{k=2}^{n-1} \frac{\log(n-k+1)}{2(n-k)} + \frac{\log\left|\det \tilde{\mathbf{A}}\right|}{n-1} \,,$$

and the logarithm of the Hadamard inequality (3.53) for $k=1$ implies that

$$\log\left|\det \tilde{\mathbf{A}}\right| \leq \frac{n \log n + 2n \log|\tilde{\varrho}_{1,1}|}{2(n-1)} \,.$$

Together, these two inequalities imply that

$$\log|\tilde{\varrho}_{n,n}| - \log|\tilde{\varrho}_{1,1}| \leq \frac{n}{2(n-1)} \log n + \sum_{k=2} n-1 \frac{\log(n-k+1)}{2(n-k)}$$

$$= \frac{1}{2} \sum_{k=1} n-1 \frac{\log(n-k+1)}{n-k} + \frac{1}{2} \log n \,.$$

This is equivalent to (3.51).

We are now ready to analyze the rounding errors in Gaussian factorization.

Theorem 3.8.2 (Wilkinson's Backward Error Analysis) *Suppose that* $\mathbf{A}$ *is a nonsingular* $n \times n$ *matrix. Also suppose that on a floating point machine we compute an approximate Gaussian factorization* $\tilde{\mathbf{L}}\tilde{\mathbf{R}}$ *of* $\mathbf{A}$*, and that the entries of* $\tilde{\mathbf{L}}$ *all have absolute value at most one. Assume that during the factorization process, the relative errors in any multiplication, division or subtraction satisfy*

$$fl(a \times b) = (a \times b)\left(1 + \delta^{[\times]}\right) ,$$
$$fl(a \div b) = (a \div b)\left(1 + \delta^{[\div]}\right) , \text{and}$$
$$fl(a - b) = (a - b)/\left(1 + \delta^{[-]}\right) ,$$

and that there is an upper bound δ *on all such relative errors involved in the factorization:*

$$\max\left\{\left|\delta^{[\times]}\right|, \left|\delta^{[\div]}\right|, \left|\delta^{[-]}\right|\right\} \leq \delta .$$

Let $\tilde{\mathbf{A}}^{(k)}$ *be the approximate reduced matrix computed at step* k *during the factorization, with entries* $\tilde{\alpha}_{i,j}^{(k)}$*. Then the product of the computed factors is equal to a perturbation of the original matrix:*

$$\tilde{\mathbf{L}}\tilde{\mathbf{R}} = \mathbf{A} + \mathbf{E} ,$$

where the perturbation matrix $\mathbf{E}$ *satisfies*

$$\|\mathbf{E}\|_\infty \leq n^2 \delta \max_{1 \leq i,j,k \leq n} \left|\tilde{\alpha}_{i,j}^{(k)}\right| .$$

Proof For $1 \leq k \leq n$, in the k-th step of Gaussian factorization we compute the k-th row of $\tilde{\mathbf{R}}$, the k-th column of $\tilde{\mathbf{L}}$ and the entries of the reduced matrix $\tilde{\mathbf{A}}^{(k)}$ by

$$\tilde{\varrho}_{kj} = \tilde{\alpha}_{kj}^{(k-1)} \text{ for } k \leq j \leq n ,$$
$$\tilde{\lambda}_{ik} = \left(1 + \delta_{ik}^{[\div]}\right)\tilde{\alpha}_{ik}^{(k-1)}/\tilde{\varrho}_{kk} \text{ for } k \leq i \leq n , \text{ and}$$
$$\tilde{\alpha}_{ij}^{(k)} = \frac{\tilde{\alpha}_{ij}^{(k-1)} - \left(1 + \delta_{ik}^{[\times]}\right)\tilde{\lambda}_{ik}\tilde{\varrho}_{kj}}{1 + \delta_{ij}^{[-]}} \text{for } k \leq i,j \leq n .$$

As a result,

$$\tilde{\alpha}_{ij}^{(k)} + \delta_{ij}^{[-]}\tilde{\alpha}_{ij}^{(k)} = \tilde{\alpha}_{ij}^{(k-1)} - \left(1 + \delta_{ik}^{[\times]}\right)\tilde{\lambda}_{ik}\tilde{\varrho}_{kj} ,$$

which implies that for $k \le i,j \le n$

$$\varepsilon_{ij}^{(k)} \equiv \tilde{\alpha}_{ij}^{(k)} - \left[\tilde{\alpha}_{ij}^{(k-1)} - \tilde{\lambda}_{ik}\tilde{\varrho}_{kj}\right] = -\delta_{ij}^{(\times)}\tilde{\lambda}_{ik}\tilde{\varrho}_{kj} - \delta_{ij}^{[-]}\tilde{\alpha}_{ij}^{(k)} . \tag{3.56}$$

For $k < i \le n$ we also have

$$\varepsilon_{ik}^{(k)} \equiv -\tilde{\alpha}_{ik}^{(k-1)} + \tilde{\lambda}_{ik}\tilde{\varrho}_{kk} = \delta_{ik}^{[\div]}\tilde{\alpha}_{ij}^{(k-1)} .$$

Since we assume that $\left|\tilde{\lambda}_{ik}\right| \le 1$, the former equation implies that for all $k \le i,j \le n$ we have

$$\left|\varepsilon_{ij}^{(k)}\right| \le \delta\left[\left|\tilde{\varrho}_{kj}\right| + \left|\tilde{\alpha}_{ij}^{(k)}\right|\right] = \delta\left[\left|\tilde{\alpha}_{kj}^{(k-1)}\right| + \left|\tilde{\alpha}_{ij}^{(k)}\right|\right] \le 2\delta \max_{1\le r,s,t\le n} \left|\tilde{\alpha}_{r,s}^{(t)}\right| .$$

We also have

$$\left|\varepsilon_{ik}^{(k)}\right| \le \delta\left|\tilde{\alpha}_{ij}^{(k-1)}\right| \le \delta \max_{1\le r,s,t\le n} \left|\tilde{\alpha}_{r,s}^{(t)}\right| .$$

Equation (3.56) can be written as the linear recurrence

$$\tilde{\alpha}_{ij}^{(k)} = \tilde{\alpha}_{ij}^{(k-1)} - \tilde{\lambda}_{ik}\tilde{\varrho}_{kj} + \varepsilon_{ij}^{(k)} ,$$

which for $1 \le p < \min\{i,j\}$ has the solution

$$\tilde{\alpha}_{ij}^{(p)} = \alpha_{ij} - \sum_{k=1}^{p}\tilde{\lambda}_{ik}\tilde{\varrho}_{kj} + \sum_{k=1}^{p}\varepsilon_{ij}^{(p)} .$$

Below the diagonal (i.e. for $i > j$) we have

$$\tilde{\alpha}_{ij}^{(j-1)} = \tilde{\lambda}_{ij}\tilde{\varrho}_{jj} - \varepsilon_{ij}^{(j)} = \alpha_{ij} - \sum_{k=1}^{j-1}\tilde{\lambda}_{ik}\tilde{\varrho}_{kj} + \sum_{k=1}^{j-1}\varepsilon_{ij}^{(k)} ,$$

so

$$\alpha_{ij} = \sum_{k=1}^{j}\tilde{\lambda}_{ik}\tilde{\varrho}_{kj} - \sum_{k=1}^{j}\varepsilon_{ij}^{(k)} ,$$

which suggests that for $i > j$ we define

$$\varepsilon_{ij} = \sum_{k=1}^{j}\varepsilon_{ij}^{(k)} .$$

On or above the diagonal (i.e. for $i \leq j$) we have

$$\tilde{\alpha}_{ij}^{(i-1)} = \tilde{\varrho}_{ij} = \tilde{\lambda}_{ii}\tilde{\varrho}_{ij} = \alpha_{ij} - \sum_{k=1}^{i-1} \tilde{\lambda}_{ik}\tilde{\varrho}_{kj} + \sum_{k=1}^{i-1} \varepsilon_{ij}^{(k)} ,$$

so

$$\alpha_{ij} = \sum_{k=1}^{i} \tilde{\lambda}_{ik}\tilde{\varrho}_{kj} - \sum_{k=1}^{i-1} \varepsilon_{ij}^{(k)} ,$$

which suggests that for $i \leq j$ we define

$$\varepsilon_{ij} = \sum_{k=1}^{i-1} \varepsilon_{ij}^{(k)} .$$

If we define $\mathbf{E}$ to have entries ε_{ij}, then we have shown that $\mathbf{A} + \mathbf{E} = \tilde{\mathbf{L}}\tilde{\mathbf{R}}$. All that remains is to bound the entries of $\mathbf{E}$.

For $i > j$ we have

$$\left|\varepsilon_{ij}\right| = \left|\sum_{k=1}^{j} \varepsilon_{ij}^{(k)}\right| \leq \sum_{k=1}^{j-1} \left|\varepsilon_{ij}^{(k)}\right| + \left|\varepsilon_{ij}^{(j)}\right| \leq 2(j-1)\delta \max_{1\leq r,s,t\leq n} \left|\tilde{\alpha}_{r,s}^{(t)}\right| + \delta \max_{1\leq r,s,t\leq n} \left|\tilde{\alpha}_{r,s}^{(t)}\right|$$
$$\leq (2j-1)\delta \max_{1\leq r,s,t\leq n} \left|\tilde{\alpha}_{r,s}^{(t)}\right| .$$

For $i \leq j$ we have

$$\left|\varepsilon_{ij}\right| = \left|\sum_{k=1}^{i-1} \varepsilon_{ij}^{(k)}\right| \leq \sum_{k=1}^{i-1} \left|\varepsilon_{ij}^{(k)}\right| \leq 2(i-1)\delta \max_{1\leq r,s,t\leq n} \left|\tilde{\alpha}_{r,s}^{(t)}\right| .$$

It is easy to see that the bounds on the last row of $\mathbf{E}$ dominate the bounds on other rows, so

$$\|\mathbf{E}\|_\infty = \max_{1\leq i\leq n} \sum_{j=1}^{n} \left|\varepsilon_{i,j}\right| = \sum_{j=1}^{n} \left|\varepsilon_{n,j}\right| = \sum_{j=1}^{n-1} \left|\varepsilon_{n,j}\right| + \left|\varepsilon_{n,n}\right|$$
$$\leq \left[\sum_{j=1}^{n-1} (2j-1)\right] \delta \max_{1\leq r,s,t\leq n} \left|\tilde{\alpha}_{r,s}^{(t)}\right| + 2(n-1)\delta \max_{1\leq r,s,t\leq n} \left|\tilde{\alpha}_{r,s}^{(t)}\right|$$
$$= \left(n^2 - 1\right) \delta \max_{1\leq r,s,t\leq n} \left|\tilde{\alpha}_{r,s}^{(t)}\right| < n^2\delta \max_{1\leq r,s,t\leq n} \left|\tilde{\alpha}_{r,s}^{(t)}\right| .$$

This theorem shows that the accumulated rounding errors in Gaussian factorization can grow in two ways. The errors are bounded by the number of entries in the matrix, and by the size of the entries of the reduced matrices $\tilde{\mathbf{A}}^{(k)}$ produced during the factorization process. The following corollary shows how various pivoting strategies control the size of entries of $\tilde{\mathbf{A}}^{(k)}$.

Corollary 3.8.1 *Suppose that* $\mathbf{A}$ *is a nonsingular* $n \times n$ *matrix, and that we apply Gaussian factorization with either full or partial pivoting to* $\mathbf{A}$*, using floating-point arithmetic satisfying the hypotheses of Theorem 3.8.2. If* $2^{n-1} n^2 \delta < 1$*, then Gaussian elimination with* partial pivoting *finds a permutation matrix* $\mathbf{Q}$ *and approximate triangular factors* $\tilde{\mathbf{L}}$, $\tilde{\mathbf{R}}$ *so that the error*

$$\mathbf{E} \equiv \mathbf{Q}\tilde{\mathbf{L}}\tilde{\mathbf{R}} - \mathbf{A}$$

satisfies

$$\|\mathbf{E}\|_\infty \leq \frac{2^{n-1} n^2 \delta}{1 - 2^{n-1} n^2 \delta} \|\mathbf{A}\|_\infty \,. \tag{3.57}$$

Alternatively, if

$$g_n \equiv \sqrt{n \prod_{k=2}^{n} k^{1/k-1}}$$

and $g_n n^2 \delta < 1$*, then Gaussian elimination with* full pivoting *finds permutation matrices* $\mathbf{P}$, $\mathbf{Q}$ *and approximate triangular factors* $\tilde{\mathbf{L}}$, $\tilde{\mathbf{R}}$ *so that the error*

$$\mathbf{E} \equiv \mathbf{Q}\tilde{\mathbf{L}}\tilde{\mathbf{R}}\mathbf{P}^\top - \mathbf{A}$$

satisfies

$$\|\mathbf{E}\|_\infty \leq \frac{g_n n^2 \delta}{1 - g_n n^2 \delta} \|\mathbf{A}\|_\infty \,. \tag{3.58}$$

Proof We will prove the claim for partial pivoting; the proof of the claim for full pivoting is similar. First, we note that Theorem 3.8.2 implies that

$$\|\mathbf{E}\|_\infty \leq n^2 \delta \max_{1 \leq i,j,k \leq n} \left| \tilde{\alpha}_{i,j}^{(k)} \right| \,,$$

where $\tilde{\alpha}_{i,j}^{(k)}$ are the entries of the computed reduced matrix $\tilde{\mathbf{A}}^{(k)}$ during the factorization process. Next, the proof of Lemma 3.8.10 shows that for all $1 \leq k \leq n$ and all

$k \le i,j \le n$ we have

$$\left|\tilde{\alpha}_{i,j}^{(k)}\right| \le 2^{k-1} \max_{1\le i,j\le n} \left|\tilde{\alpha}_{i,j}\right| ,$$

where $\tilde{\alpha}_{i,j}$ is the corresponding entry of $\tilde{\mathbf{L}}\tilde{\mathbf{R}} = \mathbf{A} + \mathbf{E}$. Combining these two results, we obtain

$$\begin{aligned}\|\mathbf{E}\|_\infty &\le 2^{n-1}n^2\delta \max_{1\le i,j\le n} \left|\tilde{\alpha}_{i,j}\right| \le 2^{n-1}n^2\delta \, \|\mathbf{A}+\mathbf{E}\|_\infty \\ &\le 2^{n-1}n^2\delta \left[\|\mathbf{A}\|_\infty + \|\mathbf{E}\|_\infty\right] .\end{aligned}$$

We can solve this inequality to obtain the claimed result.

Some simple calculations will show that the bound (3.57) for rounding errors with partial pivoting is substantially larger than the bound (3.58) for rounding errors with full pivoting. Typically, both bounds are much larger than what occurs in practice. For most linear systems, partial pivoting produces accurate results. If, however, we notice substantial growth in the reduced matrices during Gaussian factorization, then we should consider using full pivoting.

Exercise 3.8.2 In [110], Wilkinson shows that

$$g_n \equiv \sqrt{n \prod_{k=2}^{n} k^{1/k-1}} \le 1.8 n^{(\log n)/4} .$$

Compute g_n for various large values of n and compare to Wilkinson's bound.

Exercise 3.8.3 For $n = 10^k$ with $k = 1, 2, 3$ define $\mathbf{A}$ to be an $n \times n$ matrix with entries chosen to be independent and identically distributed random numbers chosen from a uniform distribution on the unit interval. Perform Gaussian factorization with both full and partial pivoting, and compute the observed growth in the triangular factors. Compare the observed growth to the theoretical bounds.

3.8.6 Triangular System Errors

The Algorithms 3.7.4 and 3.7.7 show that the solution of systems of linear equations involves both matrix factorization, and the solution of triangular systems of linear equations. We analyzed the rounding errors due to Gaussian factorization in Corollary 3.8.1. Our goal in this section is to examine the rounding errors in solving triangular systems of linear equations.

We will begin by bounding the rounding errors in solving right-triangular systems.

Lemma 3.8.11 *Suppose that* $\mathbf{R}$ *is a nonsingular* $n \times n$ *right-triangular matrix with entries* $\varrho_{i,j}$, $\mathbf{y}$ *is an n-vector, and the n-vector* $\mathbf{x}$ *solves* $\mathbf{Rx} = \mathbf{y}$. *Suppose that the back substitution Algorithm 3.4.5 and floating point arithmetic are used to compute an approximate solution vector* $\tilde{\mathbf{x}}$. *Assume that all floating point errors are of the form*

$$
\begin{aligned}
\mathit{float}(a \times b) &= (a \times b)\left(1 + \varepsilon^{[\times]}\right) , \\
\mathit{float}(a + b) &= (a + b)\left(1 + \varepsilon^{[+]}\right) , \\
\mathit{float}(a - b) &= (a - b)/\left(1 + \varepsilon^{[-]}\right) \text{ and} \\
\mathit{float}(a / b) &= (a / b)/\left(1 + \varepsilon^{[\div]}\right) ,
\end{aligned}
$$

where

$$
\max\left\{\left|\varepsilon^{[\times]}\right| , \left|\varepsilon^{[+]}\right| , \left|\varepsilon^{[-]}\right| , \left|\varepsilon^{[\div]}\right|\right\} \le \varepsilon .
$$

Then $\tilde{\mathbf{x}}$ *solves the approximate linear system* $(\mathbf{R} + \triangle\mathbf{R})\,\tilde{\mathbf{x}} = \mathbf{y}$, *where the entries* $\triangle\rho_{ij}$ *of* $\triangle\mathbf{R}$ *satisfy*

$$
\left|\triangle\rho_{ij}\right| \le \begin{cases} |\varrho_{ii}|\frac{(n-i+1)\varepsilon}{1-(n-i+1)\varepsilon} , & j = i \\ |\varrho_{ij}|\frac{(j-i)\varepsilon}{1-(j-i)\varepsilon} , & j > i \end{cases} .
$$

Consequently,

$$
\|\triangle\mathbf{R}\|_\infty \le \frac{n\varepsilon}{1 - \max\{1 , n-1\}\varepsilon}\|\mathbf{R}\|_\infty .
$$

Also, if $\mathbf{r}_j$ *is the jth column of* $\mathbf{R}$ *and* $\triangle\mathbf{r}_j$ *is the jth column of* $\triangle\mathbf{R}$, *then*

$$
\left\|\triangle\mathbf{r}_j\right\|_2 \le \frac{\max\{j-1 , n+1-j\}\varepsilon}{1 - \max\{1 , j-2 , n-j\}\varepsilon}\|\mathbf{r}_j\|_2 .
$$

Proof The back substitution Algorithm 3.4.5 computes the ith entry of $\mathbf{x}$ to be

$$
\tilde{\xi}_i = \frac{\eta_i \prod_{j=i+1}^n \left(1 + \varepsilon_{ij}^{[-]}\right) - \sum_{j=i+1}^n \varrho_{ij}\tilde{\xi}_j \left(1 + \varepsilon_{ij}^{[\times]}\right) \prod_{\ell=j}^n \left(1 + \varepsilon_{i\ell}^{[-]}\right)}{\varrho_{ii}\left(1 + \varepsilon_i^{[\div]}\right)} ,
$$

We can rewrite this equation in the form

$$\varrho_{ii}\left(1+\varepsilon_i^{[\div]}\right)\tilde{\xi}_i = \eta_i \prod_{j=i+1}^{n}\left(1+\varepsilon_{ij}^{[-]}\right) - \sum_{j=i+1}^{n} \varrho_{ij}\tilde{\xi}_j\left(1+\varepsilon_{ij}^{[\times]}\right)\prod_{\ell=j}^{n}\left(1+\varepsilon_{i\ell}^{[-]}\right) ,$$

and then solve for η_i to get

$$\eta_i = \frac{\varrho_{ii}\left(1+\varepsilon_i^{[\div]}\right)}{\prod_{j=i+1}^{n}\left(1+\varepsilon_{ij}^{[-]}\right)}\tilde{\xi}_i + \sum_{j=i+1}^{n}\left\{\frac{\varrho_{ij}\left(1+\varepsilon_{ij}^{[\times]}\right)}{\prod_{\ell=i+1}^{j-1}\left(1+\varepsilon_{i\ell}^{[-]}\right)}\right\}\tilde{\xi}_j .$$

This equation shows that $\tilde{\mathbf{R}} = \mathbf{R} + \triangle\mathbf{R}$, where the entries of $\triangle\mathbf{R}$ satisfy

$$\left|\triangle\rho_{ij}\right| = \begin{cases} |\varrho_{ii}|\left|\frac{1+\varepsilon_i^{[\div]}}{\prod_{j=i+1}^{n}\left(1+\varepsilon_{ij}^{[-]}\right)} - 1\right| , j = i \\ |\varrho_{ij}|\left|\frac{1+\varepsilon_{ij}^{[\times]}}{\prod_{\ell=i+1}^{j-1}\left(1+\varepsilon_{i\ell}^{[+]}\right)} - 1\right| , j > i \end{cases}$$

then inequality (3.43) gives us

$$\leq \begin{cases} |\varrho_{ii}|\frac{(n-i+1)\varepsilon}{1-\max\{1\ ,\ n-i\}\varepsilon} , j = i \\ |\varrho_{ij}|\frac{(j-i)\varepsilon}{1-\max\{1\ ,j-i-1\})\varepsilon} , j > i \end{cases} .$$

Since the infinity matrix norm is the maximum row sum, we get

$$\begin{aligned} \|\triangle\mathbf{R}\|_\infty &= \max_{1\leq i\leq n}\sum_{j=i}^{n}\left|\triangle\rho_{ij}\right| \\ &\leq \max\{1 \leq i \leq n\}\left\{|\varrho_{ii}|\frac{(n-i+1)\varepsilon}{1-\max\{1\ ,\ n-i\}\varepsilon} + \sum_{j=i+1}^{n}|\varrho_{ij}|\frac{(j-i)\varepsilon}{1-\max\{1\ ,\ j-i-1\}\varepsilon}\right\} \\ &\leq \max\{1 \leq i \leq n\}\left\{\left[\sum_{j=i}^{n}|\varrho_{ij}|\right]\frac{(n-i+1)\varepsilon}{1-\max\{1\ ,\ n-i\}\varepsilon}\right\} \leq \frac{n\varepsilon}{1-\max\{1\ ,\ n-1\}\varepsilon}\|\mathbf{R}\|_\infty . \end{aligned}$$

Similarly,

$$\begin{aligned} \left\|\triangle\mathbf{r}_j\right\|_2^2 &= \sum_{i=1}^{j}\left|\triangle\rho_{ij}\right|^2 \\ &\leq \sum_{i=1}^{j-1}\left|\varrho_{ij}\right|^2\left[\frac{(j-i)\varepsilon}{1-\max\{1\ ,\ j-i-1\}\varepsilon}\right]^2 + \left|\varrho_{jj}\right|^2\left[\frac{(n-j+1)\varepsilon}{1-\max\{1\ ,\ n-j\}\varepsilon}\right]^2 \end{aligned}$$

$$\leq \max\left\{\frac{(j-1)\varepsilon}{1-\max\{1\,,\,j-2\}\varepsilon}\,,\,\frac{(n+1-j)\varepsilon}{1-\max\{1\,,\,n-j\}\varepsilon}\right\}^2 \sum_{i=1}^{j} |\varrho_{ij}|^2$$

$$= \max\left\{\frac{(j-1)\varepsilon}{1-\max\{1\,,\,j-2\}\varepsilon}\,,\,\frac{(n+1-j)\varepsilon}{1-\max\{1,n-j\}\varepsilon}\right\}^2 \|\mathbf{r}_j\|_2^2\;.$$

Together with inequality (3.36), this inequality implies the claimed result.

Note that if **R** is obtained by Gaussian factorization with either full or partial pivoting, then the maximum entry of **R** can be bounded by Lemma 3.8.10. Also note that Lemma 3.8.11 shows that the rounding errors due to back substitution grow proportionally to n, while Corollary 3.8.1 shows that the errors in Gaussian factorization grow proportionally to n^2 times a growth factor that depends on the pivoting strategy. We conclude that the rounding errors in back substitution are negligible in comparison to the rounding errors in Gaussian factorization.

The following describes the errors in forward substitution.

Lemma 3.8.12 *Suppose that* $\mathbf{L}$ *is a unit* $n \times n$ *left-triangular matrix with entries* $\lambda_{i,j}$, $\mathbf{b}$ *is an* n*-vector and* $\mathbf{L}\mathbf{y} = \mathbf{b}$. *Suppose that forward substitution and floating point arithmetic are used to compute an approximate solution vector* $\tilde{\mathbf{y}}$. *Assume that all floating point errors are of the form*

$$\mathit{float}(a \times b) = (a \times b)\,(1 + \varepsilon_\times)\;,$$
$$\mathit{float}(a + b) = (a + b)\,(1 + \varepsilon_+)\;\; \textit{and}$$
$$\mathit{float}(a - b) = (a - b)/\,(1 + \varepsilon_-)\;,$$

where

$$\max\{|\varepsilon_\times|\;,\;|\varepsilon_+|\;,\;|\varepsilon_-|\} \leq \varepsilon\;.$$

Then $\tilde{\mathbf{y}}$ *solves the approximate linear system* $(\mathbf{L} + \triangle\mathbf{L})\,\tilde{\mathbf{y}} = \mathbf{b}$ *where the entries* λ_{ij} *of* $\mathbf{L}$ *and the entries* $\delta\lambda_{ij}$ *of* $\triangle\mathbf{L}$ *satisfy*

$$|\delta\lambda_{ij}| \leq \begin{cases} \frac{(i-1)\varepsilon}{1-(i-1)\varepsilon}\,, & j = i \\ |\lambda_{ij}|\frac{(i-j+1)\varepsilon}{1-\max\{1\,,\,i-j\}\varepsilon}\,, & j < i \end{cases}\;.$$

Consequently,

$$\|\triangle\mathbf{L}\|_\infty \leq \frac{(n-1)\varepsilon}{1-\max\{1\,,\,n/2\,,\,n-2\}\varepsilon}\|\mathbf{L}\|_\infty\;.$$

Also, if $\boldsymbol{\ell}_j$ is the jth column of $\mathbf{L}$ *and* $\triangle\boldsymbol{\ell}_j$ *is the jth column of* $\triangle\mathbf{L}$*, then*

$$\|\triangle\boldsymbol{\ell}_j\|_2 \leq \frac{\max\{j-1\,,\,n-j\}\varepsilon}{1-\max\{1\,,\,j/2\,,\,n-j-1\}\varepsilon}\|\boldsymbol{\ell}_j\|_2\,.$$

Proof The forward substitution Algorithm 3.4.3 computes the ith entry of $\mathbf{y}$ to be

$$\tilde{\eta}_i = \beta_i \prod_{j=1}^{i-1}\left(1+\varepsilon_{ij}^{[-]}\right) - \sum_{j=1}^{i-1}\lambda_{ij}\tilde{\eta}_j\left(1+\varepsilon_{ij}^{[\times]}\right)\prod_{\ell=1}^{j}\left(1+\varepsilon_{i\ell}^{[-]}\right)\,.$$

We can rewrite this equation in the form

$$\beta_i = \frac{1}{\prod_{j=1}^{i-1}\left(1+\varepsilon_{ij}^{[-]}\right)}\tilde{\eta}_i + \sum_{j=1}^{i-1}\left\{\frac{\lambda_{ij}\left(1+\varepsilon_{ij}^{[\times]}\right)}{\prod_{\ell=j+1}^{i-1}\left(1+\varepsilon_{i\ell}^{[-]}\right)}\right\}\tilde{\eta}_j\,.$$

This equation shows that $\tilde{\mathbf{L}} = \mathbf{L} + \triangle\mathbf{L}$, where the entries of $\triangle\mathbf{L}$ satisfy

$$|\triangle\lambda_{ij}| = \begin{cases} \left|\frac{1}{\prod_{j=1}^{i-1}\left(1+\varepsilon_{ij}^{[-]}\right)} - 1\right|, & j = i \\ |\lambda_{ij}|\left|\frac{1+\varepsilon_{ij}^{[\times]}}{\prod_{\ell=j+1}^{i-1}\left(1+\varepsilon_{i\ell}^{[+]}\right)} - 1\right|, & j < i \end{cases}$$

then inequalities (3.34) and (3.43) give us

$$\leq \begin{cases} \frac{(i-1)\varepsilon}{1-i\varepsilon/2}, & j = i \\ |\lambda_{ij}|\frac{(i-j)\varepsilon}{1-\max\{1\,,\,i-j-1\})\varepsilon}, & j < i \end{cases}\,.$$

Since the infinity matrix norm is the maximum row sum, we get

$$\begin{aligned}\|\triangle\mathbf{L}\|_\infty &= \max_{1\leq i\leq n}\sum_{j=1}^{i}|\triangle\lambda_{ij}| \\ &\leq \max\{1\leq i\leq n\}\left\{\frac{(i-1)\varepsilon}{1-i\varepsilon/2} + \sum_{j=1}^{i-1}|\lambda_{ij}|\,\frac{(i-j)\varepsilon}{1-\max\{1\,,\,i-j-1\}\varepsilon}\right\} \\ &\leq \max\{1\leq i\leq n\}\left\{\left[1+\sum_{j=1}^{i-1}|\lambda_{ij}|\right]\frac{(i-1)\varepsilon}{1-\max\{1\,,\,i/2\,,\,i-2\}\varepsilon}\right\} \\ &\leq \frac{(n-1)\varepsilon}{1-\max\{1\,,\,n/2\,,\,n-2\}\varepsilon}\|\mathbf{L}\|_\infty\,.\end{aligned}$$

Similarly,

$$\|\triangle \boldsymbol{\ell}_j\|_2^2 = \sum_{i=j}^{n} |\triangle\lambda_{ij}|^2$$
$$\leq \left[\frac{(j-1)\varepsilon}{1-j\varepsilon/2}\right]^2 + \sum_{i=j+1}^{n} |\lambda_{ij}|^2 \left[\frac{(i-j)\varepsilon}{1-\max\{1\,,\,i-j-1\}\varepsilon}\right]^2$$
$$\leq \max\left\{\frac{(j-1)\varepsilon}{1-j\varepsilon/2}\,,\,\frac{(n-j)\varepsilon}{1-\max\{1\,,\,n-j-1\}\varepsilon}\right\}^2 \left\{1+\sum_{i=j+1}^{n}|\lambda_{ij}|^2\right\}$$
$$= \max\left\{\frac{(j-1)\varepsilon}{1-j\varepsilon/2}\,,\,\frac{(n-j)\varepsilon}{1-\max\{1,n-j-1\}\varepsilon}\right\}^2 \|\boldsymbol{\ell}_j\|_2^2\,.$$

Together with inequality (3.36), this inequality implies the claimed result. Note that with either full or partial pivoting, the entries of $\tilde{\mathbf{L}}$ have magnitude at most one.

Generally speaking, the errors due to either forward or backward substitution are insignificant compared to the errors in Gaussian factorization.

3.8.7 Condition Number Estimates

In order to estimate the errors in solving a system of linear equations, both the *a priori* error estimate (3.29) and the *a posteriori* error estimate (3.30) employ the condition number of the matrix. In Definition 3.6.1, the condition number was defined as $\kappa(\mathbf{A}) = \|\mathbf{A}\|\,\|\mathbf{A}^{-1}\|$. Although Lemma 3.5.4 shows that both the one-norm and infinity-norm of a matrix are easily computed, the inverse of a matrix is not easy to determine. We would like to develop some technique for estimating the norm of the inverse, hopefully without incurring the full cost of computing the inverse beforehand.

According to Definition 3.5.3, for any subordinate matrix norm we have

$$\|\mathbf{A}^{-1}\| = \max_{\mathbf{z}\neq\mathbf{0}} \frac{\|\mathbf{A}^{-1}\mathbf{z}\|}{\|\mathbf{z}\|} = \max_{\|\mathbf{z}\|\leq 1} \frac{\|\mathbf{A}^{-1}\mathbf{z}\|}{\|\mathbf{z}\|}\,.$$

Thus, given any nonzero vector $\mathbf{z}$ with norm at most one, we can solve $\mathbf{Az} = \mathbf{z}$ for $\mathbf{z}$ and find that

$$\kappa(\mathbf{A}) \geq \frac{\|\mathbf{A}\|\,\|\mathbf{z}\|}{\|\mathbf{z}\|}\,.$$

The difficulty lies in choosing a good value for $\mathbf{z}$ so that the right-hand side of this inequality is close to the condition number.

Originally, LINPACK [34, p. 1.12] employed a technique developed by Cline et al. [23] to estimate the condition number. For a factorization $\mathbf{A} = \mathbf{LR}$, this technique chose the entries of $\mathbf{z}$ to have magnitude one, and while solving $\mathbf{R}^\top \mathbf{w} = \mathbf{z}$ chose the values of the entries of $\mathbf{z}$ to make $\|\mathbf{w}\|$ as large as possible. This approach was implemented in LINPACK routines `_geco`. Interested readers may examine loop 100 in dgeco.f to see an implementation of this method.

LAPACK currently uses a technique due to Hager [50], and later improved by Higham [55], to estimate the one-norm of the inverse by means of a convex minimization problem. Here is a summary of the basic ideas. As suggested by the formula (3.18) for the one-norm of a matrix, it is sufficient to search for the largest column norm; in other words, the minimization problem for the matrix norm need only look at matrix products with axis vectors. In particular, the norm of a complex matrix can be estimated through products with real vectors. We will show that the functional

$$\phi(\mathbf{x}) = \|\mathbf{A}\mathbf{x}\|$$

is convex, so this functional lies above its tangent space at a point. If at some point there are no axis vectors that produce larger objective value than the tangent space value, then the maximum has been found. Finally, if we choose an axis vector to produce the largest possible value in the tangent space, the norm of the objective applied to that axis vector cannot be lower than the value at the point of tangency. This leads to an algorithm that typically converges in 5 or fewer iterations.

The following definitions and lemmas contain the essential facts that support this algorithm.

Definition 3.8.1 The set $\mathscr{S}$ of n-vectors is **convex** if and only if for all $\mathbf{s}_1, \mathbf{s}_2 \in \mathscr{S}$ and for all $0 < \alpha < 1$ we have $\mathbf{s}_1\alpha + \mathbf{s}_2(1-\alpha) \in \mathscr{S}$. The n-vector $\mathbf{v}$ is an **extreme point** of the convex set $\mathscr{S}$ if and only if there are no two distinct points $\mathbf{s}_1, \mathbf{s}_2 \in \mathscr{S}$ and no scalar $0 < \alpha < 1$ so that $\mathbf{v} = \mathbf{s}_1\alpha + \mathbf{s}_2(1-\alpha)$.
In other words, a set is convex if and only if all points on the line segment between any two points in the set are also in the set, and an extreme point of a set is not interior to any line segment in the set.

Definition 3.8.2 The functional ϕ mapping n-vectors to real scalars is **convex** if and only if for all n-vectors $\mathbf{z}_1, \mathbf{z}_2$ and all scalars $0 < \alpha < 1$ we have

$$\phi\left(\mathbf{z}_1\alpha + \mathbf{z}_2[1-\alpha]\right) \leq \phi\left(\mathbf{z}_1\right)\alpha + \phi\left(\mathbf{z}_2\right)[1-\alpha] \; . \tag{3.59}$$

In other words, the value of a convex functional at a point on a line segment lies below the line between the functional values at the ends of the line segment.

We can use these definitions to obtain the following results. First, we characterize the points where convex functionals achieve their maximum values.

Lemma 3.8.13 *A convex functional defined on a closed, bounded and convex set attains its maximum at an extreme point of the set.*

Proof See Luenberger [78, p. 119].
Next, we will characterize the extreme points of the set of vectors in our maximization problem for the matrix norm.

Lemma 3.8.14 *The unit ball* $\mathscr{B}_1 = \{n\text{-vectors } \mathbf{z} : \|\mathbf{z}\|_1 \leq 1\}$ *is convex, and the extreme points of* $\mathscr{B}_1$ *are of the form* $\mathbf{e}_j\zeta$ *where* $1 \leq j \leq n$ *and* ζ *is a scalar of magnitude one.*

Proof Let us show that $\mathscr{B}_1$ is convex. If $\mathbf{z}_1, \mathbf{z}_2 \in \mathscr{B}_1$, then $\|\mathbf{z}_1\|_1 \leq 1$ and $\|\mathbf{z}_1\|_1 \leq 1$, so for all $0 < \alpha < 1$ we have

$$\begin{aligned}&\|\mathbf{z}_1\alpha + \mathbf{z}_2(1-\alpha)\|_1 \leq \|\mathbf{z}_1\alpha\|_1 + \|\mathbf{z}_2(1-\alpha)\|_1 = \|\mathbf{z}_1\|_1\alpha + \|\mathbf{z}_2\|_1(1-\alpha)\\ &\leq \alpha + (1-\alpha) = 1\;.\end{aligned}$$

This shows that $\mathbf{z}_1\alpha + \mathbf{z}_2(1-\alpha) \in \mathscr{B}_1$, so $\mathscr{B}_1$ is convex.

Next, we will make an important observation. If $\mathbf{z} \in \mathscr{B}_1$, then the sum of the components of $\mathbf{z}$ is at most one, so all of the components of $\mathbf{z}$ have magnitude at most one. If one of the components of $\mathbf{z}$ has magnitude one, then all of the other components must be zero. In other words, if $\mathbf{z} \in \mathscr{B}_1$ and $\|\mathbf{z}\|_\infty = 1$, then $\mathbf{z} = \mathbf{e}_j\zeta$ for some index $1 \leq j \leq n$ and some unit scalar ζ.

Let us show that the extreme points of the unit ball are axis vectors times unit scalars. Let $\mathbf{e}_j$ be an axis vector and $|\zeta| = 1$. Suppose that there exist $\mathbf{z}_1, \mathbf{z}_2 \in \mathscr{B}_1$ and $0 < \alpha < 1$ so that $\mathbf{e}_j\zeta = \mathbf{z}_1\alpha + \mathbf{z}_2(1-\alpha)$. Taking magnitudes of both sides of the j-th equation in this system, we get

$$1 = \left|{\mathbf{e}_j}^\top\mathbf{z}_1\alpha + {\mathbf{e}_j}^\top\mathbf{z}_2(1-\alpha)\right| \leq \left|{\mathbf{e}_j}^\top\mathbf{z}_1\right|\alpha + \left|{\mathbf{e}_j}^\top\mathbf{z}_2\right|(1-\alpha) \leq \alpha + (1-\alpha) = 1\;,$$

so we must have $\left|{\mathbf{e}_j}^\top\mathbf{z}_1\right| = 1 = \left|{\mathbf{e}_j}^\top\mathbf{z}_2\right|$. Since $\|\mathbf{z}_1\|_1 \leq 1$ and $\|\mathbf{z}_2\|_1 \leq 1$, the other components of $\mathbf{z}_1$ and $\mathbf{z}_2$ must be zero. Returning to the j-th components, we have

$$\zeta = {\mathbf{e}_j}^\top\mathbf{z}_1\alpha + {\mathbf{e}_j}^\top\mathbf{z}_2(1-\alpha)\;.$$

This says that the unit complex number ζ is a convex combination of the unit complex numbers ${\mathbf{e}_j}^\top\mathbf{s}_1$ and ${\mathbf{e}_j}^\top\mathbf{s}_2$, so all three must be equal. This implies that $\mathbf{e}_j\zeta$ is an extreme point.

Next, we will prove some results regarding the maximization problem for the matrix one-norm.

Lemma 3.8.15 *If* $\mathbf{C}$ *is an* $n \times n$ *matrix, then* $\phi_\mathbf{C}(\mathbf{z}) = \|\mathbf{C}\mathbf{z}\|_1$ *is a convex functional on n-vectors. Furthermore, for all n-vectors* $\mathbf{z}$ *and all scalars* ζ *of magnitude one, we have* $\phi_\mathbf{C}(\mathbf{z}\zeta) = \phi_\mathbf{C}(\mathbf{z})$. *The functional* $\phi_\mathbf{C}$ *achieves its maximum at an axis vector, and the maximum value of* $\phi_\mathbf{C}$ *over n-vectors* $\mathbf{z}$ *with* $\|\mathbf{z}\|_1 \leq 1$ *is equal to the maximum value of* $\phi_\mathbf{C}$ *over real n-vectors* $\mathbf{x}$ *with* $\|\mathbf{x}\|_1 \leq 1$. *Given any real n-vector*

x *and given any scalar* σ *of magnitude one, for* $1 \leq i \leq n$ *define the components of the n-vector* $\mathbf{s}(\mathbf{x})$ *by*

$$\mathbf{e}_i{}^\top \mathbf{s}(\mathbf{x}) = \begin{cases} \frac{\mathbf{e}_i{}^\top \mathbf{C}\mathbf{x}}{|\mathbf{e}_i{}^\top \mathbf{C}\mathbf{x}|}, & \mathbf{e}_i{}^\top \mathbf{C}\mathbf{x} \neq 0 \\ \sigma, & \mathbf{e}_i{}^\top \mathbf{C}\mathbf{x} = 0 \end{cases}, \tag{3.60}$$

and define the real n-vector

$$\mathbf{g}(\mathbf{x}) = \Re e \left\{ \mathbf{C}^H \mathbf{s}(\mathbf{x}) \right\} .$$

Then for all real n-vectors **x** *we have*

$$\phi_{\mathbf{C}}(\mathbf{x}) = \mathbf{g}(\mathbf{x})^\top \mathbf{x} . \tag{3.61}$$

Also, for all real n-vectors **x** *and* **y** *we have*

$$\phi_{\mathbf{C}}(\mathbf{y}) \geq \mathbf{g}(\mathbf{x})^\top \mathbf{y} . \tag{3.62}$$

In particular, for all $1 \leq j \leq n$ *and all real n-vectors* **x** *we have*

$$\phi_{\mathbf{C}}\left(\mathbf{e}_j\right) - \phi_{\mathbf{C}}\left(\mathbf{x}\right) \geq \left|\mathbf{e}_j{}^\top \mathbf{g}(\mathbf{x})\right| - \mathbf{g}(\mathbf{x})^\top \mathbf{x} . \tag{3.63}$$

Finally, if $\|\mathbf{x}\| = 1$ *and* $\|\mathbf{g}(\mathbf{x})\|_\infty = \left|\mathbf{e}_j{}^\top \mathbf{g}(\mathbf{x})\right|$, *then*

$$\phi_{\mathbf{C}}\left(\mathbf{e}_j\right) \geq \phi_{\mathbf{C}}(\mathbf{x}) . \tag{3.64}$$

Proof We will begin by showing that $\phi_{\mathbf{C}}$ is convex. If $\mathbf{z}_1$ and $\mathbf{z}_2$ are n-vectors and $0 < \alpha < 1$, then

$$\begin{aligned} \phi_{\mathbf{C}}\left(\mathbf{z}_1\alpha + \mathbf{z}_2[1-\alpha]\right) &= \|\mathbf{C}\mathbf{z}_1\alpha + \mathbf{C}\mathbf{z}_2(1-\alpha)\|_1 \\ &\leq \|\mathbf{C}\mathbf{z}_1\|_1 \alpha + \|\mathbf{C}\mathbf{z}_2\|_1 (1-\alpha) = \phi_{\mathbf{C}}\left(\mathbf{z}_1\right)\alpha + \phi_{\mathbf{C}}\left(\mathbf{z}_2\right)[1-\alpha] . \end{aligned}$$

Next, we will prove the second claim. If **z** is an n-vector and $|\zeta| = 1$, then

$$\phi_{\mathbf{C}}(\mathbf{z}\zeta) = \|\mathbf{C}\mathbf{z}\zeta\|_1 = \|\mathbf{C}\mathbf{z}\|_1 \; |\zeta| = \|\mathbf{C}\mathbf{z}\|_1 = \phi_{\mathbf{C}}(\mathbf{z}) .$$

Let us prove the third claim. Since $\phi_{\mathbf{C}}$ is convex, Lemma 3.8.13 shows that $\phi_{\mathbf{C}}$ achieves its maximum at an extreme point of the unit ball $\mathscr{B}_1$, and Lemma 3.8.14 shows that the extreme points of the unit ball are the axis vectors times scalars of magnitude one. The previous sentence is true for complex **C** whether or not the unit ball consists of real or complex vectors. The second claim in this lemma shows that if $\phi_{\mathbf{C}}$ achieves its maximum at $\mathbf{e}_j\zeta$ in the unit ball, then it achieves the same maximum at $\mathbf{e}_j$.

Now we will prove the fourth claim, namely inequality (3.61). By the definition of $\mathbf{g}(\mathbf{x})$, we have

$$\mathbf{g}(\mathbf{x})^\top \mathbf{x} = \mathbf{x}^\top \Re e\left\{\mathbf{C}^H \mathbf{s}(\mathbf{C}\mathbf{x})\right\} = \Re e\left\{\mathbf{s}(\mathbf{C}\mathbf{x})^H \mathbf{C}\mathbf{x}\right\} = \sum_{i=1}^n \Re e\left\{\overline{\mathbf{e}_i{}^\top \mathbf{s}(\mathbf{C}\mathbf{x})}\mathbf{e}_i{}^\top \mathbf{C}\mathbf{x}\right\}$$

$$= \sum_{i=1}^n \begin{cases} \Re e\left\{\dfrac{\overline{\mathbf{e}_i{}^\top \mathbf{C}\mathbf{x}}}{|\mathbf{e}_i{}^\top \mathbf{C}\mathbf{x}|}\mathbf{e}_i{}^\top \mathbf{C}\mathbf{x}\right\}, & \mathbf{e}_i{}^\top \mathbf{C}\mathbf{x} \neq 0 \\ 0, & \mathbf{e}_i{}^\top \mathbf{C}\mathbf{x} = 0 \end{cases} = \sum_{i=1}^n \left|\mathbf{e}_i{}^\top \mathbf{C}\mathbf{x}\right| = \|\mathbf{C}\mathbf{x}\|_1 .$$

Our next task is to prove (3.62). If $\mathbf{x}$ and $\mathbf{y}$ are real n-vectors, then the definitions of $\mathbf{g}(\mathbf{x})$ and $\mathbf{s}(\mathbf{x})$ imply that

$$\begin{aligned}\phi_{\mathbf{C}}(\mathbf{y}) - \mathbf{y}^\top \mathbf{g}(\mathbf{x}) &= \phi_{\mathbf{C}}(\mathbf{y}) - \mathbf{y}^\top \Re e\left\{\mathbf{C}^H \mathbf{s}(\mathbf{C}\mathbf{x})\right\} \\ &= \sum_{i=1}^n \left[\left|\mathbf{e}_i{}^\top \mathbf{C}\mathbf{y}\right| - \Re e\left\{\overline{\mathbf{e}_i{}^\top \mathbf{s}(\mathbf{C}\mathbf{x})}\mathbf{e}_i{}^\top \mathbf{C}\mathbf{y}\right\}\right] . \end{aligned} \quad (3.65)$$

Each term in this sum can be written in the form $\|\mathbf{w}\|_2 - \mathbf{t}^\top \mathbf{w}$, where the components of the real 2-vector $\mathbf{w}$ are the real and imaginary parts of the scalar $\mathbf{e}_i{}^\top \mathbf{C}\mathbf{y}$, and the components of the real 2-vector $\mathbf{t}$ are the real and imaginary parts of the scalar $\overline{\mathbf{e}_i{}^\top \mathbf{s}(\mathbf{C}\mathbf{x})}$. Then the Cauchy inequality (3.15) implies that

$$\mathbf{t}^\top \mathbf{w} \leq \|\mathbf{t}\|_2 \|\mathbf{w}\|_2 = \|\mathbf{w}\|_2 .$$

It follows that each term in the final sum of (3.65) is nonnegative, proving the claim (3.62).

Next, let us prove inequality (3.63). First, we take $\mathbf{y} = \mathbf{e}_j$ in inequality (3.62) to get

$$\phi_{\mathbf{C}}(\mathbf{e}_j) \geq \mathbf{e}_j{}^\top \mathbf{g}(\mathbf{x}) .$$

Next, we take $\mathbf{y} = -\mathbf{e}_j$ to get

$$\phi_{\mathbf{C}}(-\mathbf{e}_j) \geq -\mathbf{e}_j{}^\top \mathbf{g}(\mathbf{x}) .$$

Since $\phi_{\mathbf{C}}(-\mathbf{e}_j) = \phi_{\mathbf{C}}(\mathbf{e}_j)$, we can combine these two inequalities to prove (3.63).

All that remains is to prove (3.64). Equation (3.61) and the Hölder inequality (3.16) imply that

$$\begin{aligned}\phi_{\mathbf{C}}(\mathbf{x}) &= \mathbf{g}(\mathbf{x})^\top \mathbf{x} \leq \|\mathbf{g}(\mathbf{x})\|_\infty \|\mathbf{x}\|_1 = \|\mathbf{g}(\mathbf{x})\|_\infty = \left|\mathbf{e}_j{}^\top \mathbf{g}(\mathbf{x})\right| = \left|\Re e\left\{\mathbf{e}_j{}^\top \mathbf{C}^H \mathbf{s}(\mathbf{x})\right\}\right| \\ &= \left|\Re e\left\{\mathbf{s}(\mathbf{x})^H \mathbf{C}\mathbf{e}_j\right\}\right| \leq \left|\mathbf{s}(\mathbf{x})^H \mathbf{C}\mathbf{e}_j\right| \leq \|\mathbf{s}(\mathbf{x})\|_\infty \left\|\mathbf{C}\mathbf{e}_j\right\|_1 = \left\|\mathbf{C}\mathbf{e}_j\right\|_1 = \phi_{\mathbf{C}}(\mathbf{e}_j) .\end{aligned}$$

The vector $\mathbf{g}$ was derived by Higham [55] as the gradient of ϕ_C. Inequality (3.64). shows that if we design an iteration to choose the extreme point associated with the entry of $\mathbf{g}(\mathbf{x})$ of maximum magnitude, then the values of the objective cannot decrease.

Our estimate for the norm of the inverse of an $n \times n$ matrix $\mathbf{A}$ is provided by the following

Algorithm 3.8.2 (Inverse One-Norm Estimation)

choose a real n-vector x with $\|\mathbf{x}\|_1 = 1$

until convergence

solve $\mathbf{A}\mathbf{z} = \mathbf{x}$ for $\mathbf{z}$

for $1 \leq i \leq n$

$$\mathbf{e}_i{}^\top \mathbf{s} = \begin{cases} \frac{\mathbf{e}_i{}^\top \mathbf{z}}{|\mathbf{e}_i{}^\top \mathbf{z}|}, & \mathbf{e}_i{}^\top \mathbf{z} \neq 0 \\ \sigma, & \mathbf{e}_i{}^\top \mathbf{z} = 0 \end{cases}$$

solve $\mathbf{A}^H \mathbf{w} = \mathbf{s}$ for $\mathbf{w}$

$\mathbf{g} = \Re e\{\mathbf{w}\}$

find j so that $|\mathbf{e}_j{}^\top \mathbf{g}| = \|\mathbf{g}\|_\infty$

if $\|\mathbf{g}\|_\infty \leq \mathbf{g}^\top \mathbf{x}$ then return $\|\mathbf{z}\|_1$

$\mathbf{x} = \mathbf{e}_j$

Normally, this iteration is begun with $\mathbf{x} = \mathbf{e}/n$, where $\mathbf{e}$ is the vector of ones. Typically, this iteration produces an acceptable approximation in five iterations. For $n > 1$, Higham [55] suggests that at the end of the algorithm that the final objective value be compared to the value at the vector with whose entries are $(-1)^{j+1}[1 + (j-1)/(n-1)]$. This reduces the chance that large entries in $\mathbf{A}^{-1}$ might not be noticed.

LAPACK implements this matrix norm estimation in routines `_lacn2`. For Gaussian factorization with partial pivoting, the norm of the matrix inverse is estimated by routines `_gecon`, which call `_lacn2` in an iteration. Readers may view, for example, dlacn2.f or dgecon.f.

3.8.8 *Case Study*

3.8.8.1 Cylindrical Shells

We would like to conclude our discussion of rounding errors in solving systems of linear equations with a case study involving the matrix s1rmq4m1.mtx, which is available from MatrixMarket, This matrix was generated by a finite element analysis with cylindrical shells. It is 5489×5489, so it has over 30 million entries, most of which are zero. We would like to estimate the rounding errors in solving systems of linear equations involving this matrix, and determine which pivoting strategies are appropriate.

In Corollary 3.8.1, we developed an *a priori* estimate that bounded the rounding errors in Gaussian factorization by the product of machine precision times the number of entries in the matrix, times a growth factor that depends on the pivoting strategy. In Theorem 3.6.1, we developed another *a priori* estimate showing that the relative error in the solution of a linear system is essentially bounded by the relative error in the matrix times the condition number of the matrix. The MatrixMarket web page estimates the condition number of this matrix to be 3.21×10^6, but does not specify which norm is used in this estimate.

3.8.8.2 Single Precision

Suppose that we use single-precision floating point arithmetic to solve a linear system involving this matrix. Even if the factorization errors were on the order of machine precision, which is about 1.19×10^{-7}, our relative error estimate (3.29) would give an *a priori* relative error estimate of about 0.38 for the solution. In other words, *for this matrix we have no reason to expect any accurate digits* in the solution to a linear system computed by single-precision arithmetic.

However, error bounds must deal with the worst-case scenarios, so it is possible that a particular single-precision computation may be more accurate than predicted by this *a priori* bound. Since the MatrixMarket does not provide right-hand sides for linear systems, we will create one with a known solution. We can fill a vector $\mathbf{x}$ with random numbers, and compute $\mathbf{b} = \mathbf{A}\mathbf{x}$. Afterward, we can use Gaussian factorization of $\mathbf{A}$ to compute an approximate solution $\tilde{\mathbf{x}}$ to the linear system $\mathbf{A}\mathbf{x} = \mathbf{b}$. Then we can compute $\mathbf{r} = \mathbf{b} - \mathbf{A}\tilde{\mathbf{x}}$ and use the *a posteriori* error estimate (3.30) to estimate the error. Our program GaussianCaseStudy.C computed

$$\|\mathbf{r}\|_\infty = 5.87 \times 10^8 \ , \ \|\mathbf{b}\|_\infty = 5.25 \times 10^{14} \text{ and } \kappa_\infty(\mathbf{A}) = 2.22 \times 10^6 \ ,$$

so the *a posteriori* error estimate is

$$\frac{\|\mathbf{x} - \tilde{\mathbf{x}}\|_1}{\|\mathbf{x}\|_1} \le \kappa_1(\mathbf{A}) \frac{\|\mathbf{r}\|_1}{\|\mathbf{b}\|_1} = 2.49 \ .$$

This error estimate also indicates that the computed solution may have no significant digits.

Since we created a problem with known solution, we can compute the relative error directly. In this case, we have

$$\frac{\|\mathbf{x}-\tilde{\mathbf{x}}\|_1}{\|\mathbf{x}\|_1} = 1.05 \times 10^{-3} \; .$$

Although this true relative error is much smaller than the *a posteriori* error estimate, we have no reason to hope that the actual relative errors will be so much smaller than the *a posteriori* error bounds for all linear systems involving this matrix. Since we do not know the "exact" solution in general, we will usually have to rely on the *a posteriori* error estimate to judge the accuracy of our computed solution.

3.8.8.3 Double Precision

Next, suppose that we solve the same linear system in double-precision arithmetic. In this case, machine precision is about 2.22×10^{-16}, and there are about 3.01×10^7 matrix entries. If the growth factor in Gaussian factorization is g_n, the *a priori* error estimates in Corollary 3.8.1 indicate that the rounding errors in Gaussian factorization will essentially correspond to a relative error of

$$\frac{\|\tilde{\mathbf{A}}-\mathbf{A}\|_\infty}{\|\mathbf{A}\|_\infty} \leq g_n n^2 \varepsilon = (3.01 \times 10^7)(2.22 \times 10^{-16}) g_n = (6.69 \times 10^{-9}) g_n \; .$$

If we insert this result in the *a priori* relative error estimate (3.29), then we essentially have the error estimate

$$\frac{\|\mathbf{x}-\tilde{\mathbf{x}}\|_\infty}{\|\mathbf{x}\|_\infty} \leq \kappa_\infty(\mathbf{A}) \frac{\|\tilde{\mathbf{A}}-\mathbf{A}\|_\infty}{\|\mathbf{A}\|_\infty} = (2.29{\times}10^6)(6.69{\times}10^{-9}) g_n = (1.53{\times}10^{-2}) g_n \; .$$

Even if we use double precision and there is little growth in the matrix factor $\mathbf{R}$ (i.e. $g_n \approx 1$), we have no reason to expect more than two digits accuracy in advance of actual computation.

But, *a priori* error estimates must deal with the worst possible cases, and may be pessimistic. We can compute $\mathbf{r} = \mathbf{b} - \mathbf{A}\tilde{\mathbf{x}}$ and use the *a posteriori* error estimate (3.30) to estimate the error. Our program computed

$$\|\mathbf{r}\|_\infty = 1.27 \text{ and } \|\mathbf{b}\|_\infty = 5.79 \times 10^{14} \; ,$$

so the *a posteriori* double precision error estimate is

$$\frac{\|\mathbf{x}-\tilde{\mathbf{x}}\|_\infty}{\|\mathbf{x}\|_\infty} \leq \kappa_\infty(\mathbf{A}) \frac{\|\mathbf{r}\|_\infty}{\|\mathbf{b}\|_\infty} = 4.87 \times 10^{-9} \; .$$

This error estimate indicates that about nine digits in the computed solution $\tilde{\mathbf{x}}$ can be trusted in the solution of this particular linear system. Since we know the exact solution, we can also compute the relative error directly to get

$$\frac{\|\mathbf{x} - \tilde{\mathbf{x}}\|_\infty}{\|\mathbf{x}\|_\infty} = 1.36 \times 10^{-12} .$$

In other words, the *a priori* error estimate predicted at least two accurate digits, the *a posteriori* error estimate guaranteed at least nine accurate digits, and direct computation of the error found 12 accurate digits.

3.8.8.4 Timing

Let us make some remarks about computational times for this case study. Gaussian factorization with partial pivoting in single precision took 8.73 s to factor the matrix, solve three linear systems and perform three matrix-vector multiplications. The same work in double precision took 16.29 s. However, *Gaussian factorization with full pivoting in single precision took* 1073 s. The time required for full pivoting seems excessive, since our work estimates in Sects. 3.7.2.3 and 3.7.3.2 indicate that full pivoting involves roughly 50% more floating point operations than partial pivoting.

In our timings, we used threaded ATLAS LAPACK and BLAS libraries on a dual-core laptop to perform the computations. As we remarked in Sect. 2.11, the ATLAS libraries use Posix threads to make maximum use of the CPUs, and use block algorithms to reduce cache misses. Recall that in Sect. 3.7.3.4, we showed that Gaussian factorization with partial pivoting can be implemented as a block algorithm. However, Gaussian factorization with full pivoting cannot be implemented as a block algorithm. This helps to explain why *full pivoting took more than two orders of magnitude more computational time than partial pivoting*.

Exercise 3.8.4 Let $\mathbf{A}$ be the 10×10 Hilbert matrix

$$\mathbf{A}_{ij} = \frac{1}{1+i+j} , \quad 0 \le i,j < 10 .$$

This matrix arises in least squares approximation by monomials.

1. Use LAPACK routine dgecon to estimate the condition number of this matrix.
2. If $\mathbf{b} = \mathbf{e}$ is the vector of ones, use LAPACK routine dgesv to solve $\mathbf{A}\mathbf{x} = \mathbf{b}$.
3. Use the *a posteriori* error estimate (3.30) to estimate the error in the `dgesv` value for $\mathbf{x}$.
4. Compare the *a posteriori* error estimate to the *a priori* error estimate provided by Wilkinson's Backward Error Estimate Corollary 3.8.1, the condition number estimate from `dgecon`, and the *a priori* error estimate in Theorem 3.6.1.

3.9 Improvements

In this section, we will develop two techniques designed to improve the solution of systems of linear equations via Gaussian factorization. Scaling will modify the linear system in the hope of reducing the condition number. The second technique, iterative refinement, will reduce the rounding errors due to Gaussian factorization. We will illustrate the ideas in this section with another case study.

3.9.1 Scaling

Example 3.7.3 illustrated that pivoting can sometimes be used to avoid numerical problems in Gaussian factorization. Theorem 3.8.2 showed that partial pivoting and full pivoting can control the size of the rounding errors in Gaussian factorization, relative to the scale of the entries in the original matrix. However, the next example shows that poor scaling can prevent pivoting from overcoming numerical obstacles.

Example 3.9.1 Consider the linear system

$$\begin{aligned} \varepsilon \xi_1 + \xi_2 &= 1 \, , \\ \xi_1 + \xi_2 &= 2 \, . \end{aligned}$$

If we scale the first equation by $1/\varepsilon$, we get the linear system

$$\begin{aligned} \xi_1 + (1/\varepsilon)\, \xi_2 &= 1/\varepsilon \, , \\ \xi_1 + \xi_2 &= 2 \, . \end{aligned}$$

If ε is sufficiently small, Gaussian factorization with partial pivoting using floating point operations will produce the approximate factorization

$$\mathbf{A} \equiv \begin{bmatrix} 1 & 1/\varepsilon \\ 1 & 1 \end{bmatrix} \approx \begin{bmatrix} 1 & 0 \\ 1 & 1 \end{bmatrix} \begin{bmatrix} 1 & 1/\varepsilon \\ 0 & -1/\varepsilon \end{bmatrix} \equiv \tilde{\mathbf{L}}\tilde{\mathbf{R}} \, .$$

In other words, no pivoting will be used, and

$$\mathbf{E} \equiv \tilde{\mathbf{L}}\tilde{\mathbf{R}} - \mathbf{A} = \begin{bmatrix} 0 & 0 \\ 0 & -1 \end{bmatrix} \, .$$

This error is small relative to the size of $\mathbf{A}$. Forward-solving using floating-point arithmetic will produce

$$\tilde{\mathbf{y}} = \begin{bmatrix} 1/\varepsilon \\ -1/\varepsilon \end{bmatrix} \, ,$$

and back-solving will produce

$$\tilde{\mathbf{x}} = \begin{bmatrix} 0 \\ 1 \end{bmatrix} .$$

However, the exact solution is

$$\mathbf{x} = \begin{bmatrix} 1 \\ 1 - 2\varepsilon \end{bmatrix} \frac{1}{1 - \varepsilon} \approx \begin{bmatrix} 1 \\ 1 \end{bmatrix} .$$

Note that this approximate solution is *not* close to the true solution. If we examine the situation closely, we will find that we have no reason to expect that it should be close. The hidden difficulty is that the row scaling of this example has changed the condition number of the matrix. In this example, we have

$$\|\mathbf{A}\|_\infty = 1 + 1/\varepsilon = \|\mathbf{A}\|_1 \text{ and } \left\|\mathbf{A}^{-1}\right\|_\infty = (1+\varepsilon)/(1-\varepsilon) = \left\|\mathbf{A}^{-1}\right\|_1 ,$$

so the condition number is

$$\kappa_\infty(\mathbf{A}) = \frac{1}{\varepsilon} \frac{(1+\varepsilon)^2}{1-\varepsilon} = \kappa_1(\mathbf{A}) .$$

This is large for small values of ε.

Example 3.9.1 shows that row scaling can turn a well-conditioned linear system into a poorly conditioned system. The more practical question is to ask whether row and/or column scaling can turn a poorly conditioned linear system into a reasonably well-conditioned system. The answer is not clear. We refer the reader to van der Sluis [105] for more discussion. Instead, we will limit our discussion to some illustrative examples.

The next example shows that some scaling choices are significantly better than others, although the best choice may not be obvious.

Example 3.9.2 Consider the following matrix suggested by Forsythe and Moler [41, p. 45]:

$$\mathbf{A} = \begin{bmatrix} 1 & 1 & 2/\varepsilon \\ 2 & -1 & 1/\varepsilon \\ 1 & 2 & 0 \end{bmatrix} .$$

This has inverse

$$\mathbf{A}^{-1} = \begin{bmatrix} -2/9 & 4/9 & 1/3 \\ 1/9 & -2/9 & 1/3 \\ 5\varepsilon/9 & -\varepsilon/9 & -\varepsilon/3 \end{bmatrix} .$$

For $0 < \varepsilon < 3/4$, it is easy to see that $\|\mathbf{A}\|_1 = 3/\varepsilon$ and $\|\mathbf{A}^{-1}\|_1 = 2/3 + \varepsilon/3$, so the condition number is $\kappa_1(\mathbf{A}) = 2/\varepsilon + 1$.

If we scale the rows so that the maximum entry has absolute value one, we get

$$\mathbf{S}^{-1}\mathbf{A} = \begin{bmatrix} \varepsilon/2 & \varepsilon/2 & 1 \\ 2\varepsilon & -\varepsilon & 1 \\ 1/2 & 1 & 0 \end{bmatrix} .$$

This has inverse

$$\mathbf{A}^{-1}\mathbf{S} = \begin{bmatrix} -4/(9\varepsilon) & 4/(9\varepsilon) & 2/3 \\ 2/(9\varepsilon) & -2/(9\varepsilon) & 2/3 \\ 10/9 & -1/9 & -2/3 \end{bmatrix} .$$

For small values of ε, scaling improved the norm of the matrix to be $\|\mathbf{S}^{-1}\mathbf{A}\|_1 = 2$, but now the norm of the inverse after scaling is $\|\mathbf{A}^{-1}\mathbf{S}\|_1 = 2/(3\varepsilon) + 10/9$. This means that the condition number of the **scaled matrix** is $\kappa_1\left(\mathbf{S}^{-1}\mathbf{A}\right) = 4/(3\varepsilon) + 20/9$, which is on the order of the condition number of the original matrix.

Subsequent scaling of the columns to achieve entries with maximum magnitude of one produces

$$\mathbf{S}^{-1}\mathbf{A}\boldsymbol{\Gamma}_{\mathbf{S}}^{-1} = \begin{bmatrix} \varepsilon & \varepsilon/2 & 1 \\ 4\varepsilon & -\varepsilon & 1 \\ 1 & 1 & 0 \end{bmatrix} ,$$

which has inverse

$$\boldsymbol{\Gamma}_{\mathbf{S}}\mathbf{A}^{-1}\mathbf{S} = \begin{bmatrix} -2/(9\varepsilon) & 2/(9\varepsilon) & 1/3 \\ 2/(9\varepsilon) & -2/(9\varepsilon) & 2/3 \\ 10/9 & -1/9 & -2/3 \end{bmatrix} .$$

We now have $\|\mathbf{S}^{-1}\mathbf{A}\boldsymbol{\Gamma}_{\mathbf{S}}^{-1}\|_1 = 2$ and $\|\boldsymbol{\Gamma}_{\mathbf{S}}\mathbf{A}^{-1}\mathbf{S}\|_1 = 4/(9\varepsilon) + 10/9$, so the condition number is $\kappa\left(\mathbf{S}^{-1}\mathbf{A}\boldsymbol{\Gamma}_{\mathbf{S}}^{-1}\right) = 8/(9\varepsilon) + 20/9$. In this case, column scaling after row scaling makes no substantial improvement in the condition number.

On the other hand, if we first scale the columns to have maximum entry of one, we get

$$\mathbf{A}\boldsymbol{\Gamma}^{-1} = \begin{bmatrix} 1/2 & 1/2 & 1 \\ 1 & -1/2 & 1/2 \\ 1/2 & 1 & 0 \end{bmatrix} .$$

This has inverse

$$\boldsymbol{\Gamma}\mathbf{A}^{-1} = \begin{bmatrix} -4/9 & 8/9 & 2/3 \\ 2/9 & 4/9 & 2/3 \\ 10/9 & -2/9 & -2/3 \end{bmatrix} .$$

In this case, we have $\|\mathbf{A}\boldsymbol{\Gamma}^{-1}\|_1 = 2$ and $\|\boldsymbol{\Gamma}\mathbf{A}^{-1}\|_1 = 2$, so $\kappa_1\left(\mathbf{A}\boldsymbol{\Gamma}^{-1}\right) = 4$. This is substantially smaller than the condition number achieved by initially performing row scaling.

Unfortunately, it is not easy to determine which choice of row or column scaling will produce the smallest condition number, without actually performing the scaling and matrix factorizations of both alternatives.

In LAPACK routine `_geequ`, row and column scalings for matrices are determined roughly as follows. If desired we first compute the maximum magnitude in each row:

$$\varrho_i = \max_{1 \le j \le n} \left|\alpha_{i,j}\right| \text{ for } 1 \le i \le m .$$

These give us the diagonal entries of the matrix $\mathbf{S}$ for row scaling. Then we compute the maximum magnitude of the entries of the columns of $\mathbf{S}^{-1}\mathbf{A}$:

$$\gamma_j = \max_{1 \le i \le m} \frac{\left|\alpha_{i,j}\right|}{\varrho_i} \text{ for } 1 \le j \le n .$$

These give us the diagonal entries of the matrix $\boldsymbol{\Gamma}_{\mathbf{S}}$ for column scaling. With row and/or column scaling, we solve the linear system $\mathbf{A}\mathbf{x} = \mathbf{b}$ by performing the following

Algorithm 3.9.1 (LAPACK Scaling)

equilibrate $\tilde{\mathbf{A}} = \mathbf{S}^{-1}\mathbf{A}\boldsymbol{\Gamma}_{\mathbf{S}}^{-1}$

factor $\mathbf{Q}^{-1}\tilde{\mathbf{A}} = \mathbf{L}\mathbf{R}$

scale $\tilde{\mathbf{b}} = \mathbf{S}^{-1}\mathbf{b}$

forward-solve $\mathbf{L}\mathbf{y} = \tilde{\mathbf{b}}$

back-solve $\mathbf{R}\tilde{\mathbf{x}} = \mathbf{y}$ and

scale $\mathbf{x} = \boldsymbol{\Gamma}_{\mathbf{S}}^{-1}\tilde{\mathbf{x}}$.

Readers may view, for example, dgeequ.f to see an implementation of this scaling.

3.9.2 *Iterative Improvement*

We have developed several techniques (such as pivoting and scaling) for attempting to control rounding errors in the solution of linear systems by Gaussian factorization. However, in large linear systems it is still possible that the accumulation of rounding errors in either the factorization process itself, or the subsequent triangular system solves, could lead to unnecessarily large errors in the computed solution. It would be nice to find an inexpensive method to overcome some of those errors. One such method is called **iterative improvement**, and may also be called **iterative refinement** or **residual correction**.

Corollary 3.8.1 provided an error estimate for rounding errors in gaussian factorization, while Lemmas 3.8.11 and 3.8.12 estimated the errors in solving the triangular systems. Together, these results show that the true solution $\mathbf{x}$ to

$$\mathbf{A}\mathbf{x} = \mathbf{b}$$

is approximated by the solution $\tilde{\mathbf{x}}$ of a perturbed linear system

$$\tilde{\mathbf{A}}\tilde{\mathbf{x}} = \mathbf{b} .$$

As a result, we expect the residual

$$\mathbf{r} = \mathbf{b} - \mathbf{A}\tilde{\mathbf{x}}$$

to be small but nonzero. Suppose that we compute a correction $\mathbf{d}$ to $\tilde{\mathbf{x}}$ by attempting to solve

$$\mathbf{A}\mathbf{d} = \mathbf{r} .$$

Then we should have

$$\mathbf{A}\left(\tilde{\mathbf{x}} + \mathbf{d}\right) = \mathbf{A}\tilde{\mathbf{x}} + \mathbf{r} = \mathbf{b} .$$

However, we would have to use our Gaussian factorization to compute $\mathbf{d}$, and its computation would be subject to rounding errors in computing the residual, and in solving triangular systems for $\mathbf{d}$. As a result, we obtain an approximation $\tilde{\mathbf{d}}$ that solves a perturbed equation

$$\tilde{\mathbf{A}}\tilde{\mathbf{d}} = \tilde{\mathbf{r}} .$$

This process will give us the following

Algorithm 3.9.2 (Iterative Improvement)

$$
\begin{aligned}
&\text{given } \tilde{\mathbf{x}}^{(0)} \\
&\text{for } 0 \le k \text{ until convergence} \\
&\quad \tilde{\mathbf{r}}^{(k)} = \mathbf{b} - \mathbf{A}\tilde{\mathbf{x}}^{(k)} + \mathbf{s}^{(k)} \\
&\quad \text{solve } \tilde{\mathbf{A}}\tilde{\mathbf{d}}^{(k)} = \tilde{\mathbf{r}}^{(k)} \text{ for } \tilde{\mathbf{d}}^{(k)} \\
&\quad \tilde{\mathbf{x}}^{(k+1)} = \tilde{\mathbf{x}}^{(k)} + \tilde{\mathbf{d}}^{(k)} .
\end{aligned}
$$

For each step k, we hope that $\tilde{\mathbf{x}}^{(k+1)}$ is closer to the true solution $\mathbf{x}$ than $\tilde{\mathbf{x}}^{(k)}$.

Lemma 3.8.1 showed that the relative errors in Gaussian factorization of an $n \times n$ matrix are bounded by a factor of n^2, while the rounding errors associated with computing a residual were shown in Lemma 3.8.9 to be bounded by a factor of n. The next lemma will show that, under reasonable assumptions, iterative improvement can reduce the error in the numerical solution to the order of the rounding error in computing the residual. For sufficiently large matrices, this may make a substantial improvement in the available accuracy of the numerical solution of linear systems.

Lemma 3.9.1 *Let $\|\cdot\|$ represent some norm on n-vectors, or the corresponding subordinate matrix norm. Suppose that $\mathbf{A}$ and $\tilde{\mathbf{A}}$ are $n \times n$ matrices, and $\tilde{\mathbf{A}}$ is nonsingular. Given an n-vector $\mathbf{b}$, let $\mathbf{x}$ solve $\mathbf{A}\mathbf{x} = \mathbf{b}$, and let $\tilde{\mathbf{x}}^{(0)}$ solve $\tilde{\mathbf{A}}\tilde{\mathbf{x}}^{(0)} = \mathbf{b}$. Define the sequence of vectors $\left\{\tilde{\mathbf{x}}^{(k)}\right\}_{k=0}^{\infty}$ by the iterative improvement Algorithm 3.9.2. Then the errors satisfy the absolute error estimate*

$$
\begin{aligned}
\left\|\tilde{\mathbf{x}}^{(k)} - \mathbf{x}\right\| \le{}& \left\|\tilde{\mathbf{A}}^{-1}\left(\tilde{\mathbf{A}} - \mathbf{A}\right)\right\|^{k} \left\|\tilde{\mathbf{x}}^{(0)} - \mathbf{x}\right\| \\
&+ \sum_{j=0}^{k-1} \left\|\tilde{\mathbf{A}}^{-1}\left(\tilde{\mathbf{A}} - \mathbf{A}\right)\right\|^{k-j-1} \left\|\tilde{\mathbf{A}}^{-1}\left(\tilde{\mathbf{r}}^{(j)} - \mathbf{r}^{(j)}\right)\right\| .
\end{aligned} \tag{3.66}
$$

Proof From the statement 3.9.2 of the iterative improvement algorithm, we see that

$$
\begin{aligned}
&\tilde{\mathbf{A}}\tilde{\mathbf{d}}^{(k)} = \tilde{\mathbf{r}}^{(k)} = \mathbf{b} - \mathbf{A}\tilde{\mathbf{x}}^{(k)} + \left(\tilde{\mathbf{r}}^{(k)} - \mathbf{r}^{(k)}\right) = \mathbf{A}\left(\mathbf{x} - \tilde{\mathbf{x}}^{(k)}\right) + \left(\tilde{\mathbf{r}}^{(k)} - \mathbf{r}^{(k)}\right) \text{ and} \\
&\tilde{\mathbf{A}}\left(\tilde{\mathbf{x}}^{(k+1)} - \mathbf{x}\right) = \tilde{\mathbf{A}}\left(\tilde{\mathbf{x}}^{(k)} + \tilde{\mathbf{d}}^{(k)} - \mathbf{x}\right) = \left(\tilde{\mathbf{A}} - \mathbf{A}\right)\left(\tilde{\mathbf{x}}^{(k)} - \mathbf{x}\right) + \left(\tilde{\mathbf{r}}^{(k)} - \mathbf{r}^{(k)}\right) .
\end{aligned}
$$

The last equation implies that

$$
\left\|\tilde{\mathbf{x}}^{(k+1)} - \mathbf{x}\right\| \le \left\|\tilde{\mathbf{A}}^{-1}\left(\tilde{\mathbf{A}} - \mathbf{A}\right)\right\| \left\|\tilde{\mathbf{x}}^{(k)} - \mathbf{x}\right\| + \left\|\tilde{\mathbf{A}}^{-1}\left(\tilde{\mathbf{r}}^{(k)} - \mathbf{r}^{(k)}\right)\right\| .
$$

This is a linear recurrence inequality, with solution given by (3.66).

Corollary 3.9.1 *Let $\|\cdot\|$ represent some norm on n-vectors, or the corresponding subordinate matrix norm. Suppose that* $\mathbf{A}$ *and* $\tilde{\mathbf{A}}$ *are* $n \times n$ *matrices, and* $\tilde{\mathbf{A}}$ *is nonsingular. Given an n-vector* $\mathbf{b}$*, let* $\mathbf{x}$ *solve* $\mathbf{A}\mathbf{x} = \mathbf{b}$*, and let* $\tilde{\mathbf{x}}^{(0)}$ *solve* $\tilde{\mathbf{A}}\tilde{\mathbf{x}}^{(0)} = \mathbf{b}$*. Define the sequence of vectors* $\left\{\tilde{\mathbf{x}}^{(k)}\right\}_{k=0}^{\infty}$ *by the iterative improvement Algorithm 3.9.2. If*

$$\varrho \equiv \left\|\tilde{\mathbf{A}}^{-1}\left(\tilde{\mathbf{A}} - \mathbf{A}\right)\right\| < 1$$

and for $j \geq 0$ the errors in computing the residual satisfy

$$\left\|\tilde{\mathbf{r}}^{(j)} - \mathbf{r}^{(j)}\right\| \leq \sigma \|\mathbf{b}\| \,,$$

then the absolute errors in iterative improvement satisfy

$$\left\|\tilde{\mathbf{x}}^{(k)} - \mathbf{x}\right\| \leq \varrho^k \left\|\tilde{\mathbf{x}}^{(0)} - \mathbf{x}\right\| + \frac{\left\|\tilde{\mathbf{A}}^{-1}\right\|}{1-\varrho} \sigma \|\mathbf{b}\| \,, \tag{3.67}$$

and the relative errors satisfy

$$\frac{\left\|\tilde{\mathbf{x}}^{(k)} - \mathbf{x}\right\|}{\|\mathbf{x}\|} \leq \varrho^k \|\mathbf{A}\| \frac{\left\|\tilde{\mathbf{x}}^{(0)} - \mathbf{x}\right\|}{\|\mathbf{b}\|} + \|\mathbf{A}\| \left\|\tilde{\mathbf{A}}^{-1}\right\| \frac{\sigma}{1-\varrho} \,. \tag{3.68}$$

Proof We can use the definitions of ϱ and σ to simplify the bound (3.66):

$$\left\|\tilde{\mathbf{x}}^{(k)} - \mathbf{x}\right\| \leq \varrho^k \left\|\tilde{\mathbf{x}}^{(0)} - \mathbf{x}\right\| + \sum_{j=0}^{k-1} \varrho^{k-j-1} \left\|\tilde{\mathbf{A}}^{-1}\right\| \sigma \|\mathbf{b}\|$$

$$= \varrho^k \left\|\tilde{\mathbf{x}}^{(0)} - \mathbf{x}\right\| + \frac{1-\varrho^k}{1-\varrho} \left\|\tilde{\mathbf{A}}^{-1}\right\| \sigma \|\mathbf{b}\| \leq \varrho^k \left\|\tilde{\mathbf{x}}^{(0)} - \mathbf{x}\right\| + \frac{1}{1-\varrho} \left\|\tilde{\mathbf{A}}^{-1}\right\| \sigma \|\mathbf{b}\| \,.$$

This proves (3.67). Noting that $\|\mathbf{b}\| = \|\mathbf{A}\mathbf{x}\| \leq \|\mathbf{A}\| \, \|\mathbf{x}\|$, we see that $\|\mathbf{x}\| \geq \|\mathbf{b}\| / \|\mathbf{A}\|$. Then we can divide the absolute error estimate (3.67) by $\|\mathbf{x}\|$ to obtain

$$\frac{\left\|\tilde{\mathbf{x}}^{(k)} - \mathbf{x}\right\|}{\|\mathbf{x}\|} \leq \varrho^k \|\mathbf{A}\| \frac{\left\|\tilde{\mathbf{x}}^{(0)} - \mathbf{x}\right\|}{\|\mathbf{b}\|} + \|\mathbf{A}\| \left\|\tilde{\mathbf{A}}^{-1}\right\| \frac{\sigma}{1-\varrho} \,.$$

This proves (3.68).

For sufficiently large k or sufficiently small ϱ, the first term on the right in inequality (3.68) will be small compared to the second term. Thus this lemma says that iterative improvement will reduce the error roughly to the size associated with computing a residual, down from the original size associated with performing Gaussian factorization. Typically the error is reduced quickly to this size, usually in a single iteration. For more details about iterative improvement, see Skeel [95].

The LAPACK routines `_gerfs` implement iterative improvement for Gaussian factorization with partial pivoting. Readers may view, for example, dgerfs.f. These routines are called from the expert linear system solver routines `_gesvx`; see, for example, dgesvx.f. In addition, the routines `_gesvx` will perform scaling, and estimate the backward errors (3.31) and forward errors (3.49) in the solution of the linear system.

3.9.3 *Case Study*

In this section, we would like to see if scaling or iterative refinement can improve the numerical solution of linear systems. We will consider the Vandermonde matrix

$$\mathbf{V} = \begin{bmatrix} 1 & 0.1 & (0.1)^2 & (0.1)^3 \\ 1 & 0.1\overline{3} & (0.1\overline{3})^2 & (0.1\overline{3})^3 \\ 1 & 0.1\overline{6} & (0.1\overline{6})^2 & (0.1\overline{6})^3 \\ 1 & 0.2 & (0.2)^2 & (0.2)^3 \end{bmatrix} .$$

This matrix arises in determining a cubic polynomial

$$p(x) = a_0 + a_1 x + a_2 x^2 + a_3 x^3$$

that interpolates some given data at $x = 0.1, 0.1\overline{3}, 0.1\overline{6}$ and 0.2. Later in Sect. 1.2 of Chap. 1 in Volume III, we will develop ways to compute a polynomial interpolant without solving a linear system.

The ∞-norm condition number of $\mathbf{V}$ is approximately 4.00×10^5. Since machine precision for single precision arithmetic is approximately 1.19×10^{-7}, Theorem 3.6.1 suggests that we should expect a relative error on the order of $(4.00 \times 10^5)(1.19 \times 10^{-7}) = 4.76 \times 10^{-2}$ in the solution of linear systems involving this matrix.

The C^{++} program ImprovementCaseStudy.C uses Gaussian factorization with row pivoting to solve a linear system involving this Vandermonde matrix. The *a posteriori* backward error estimate (3.31) for the solution of this system is 3.49×10^{-8}, and the *a posteriori* forward error estimate (3.49) is 7.14×10^{-3}. The actual relative error in the solution is 2.56×10^{-4}. One step of iterative improvement makes no change in the solution or its residual.

The problem, as noted by Stewart [97, p. 203f], is that the residual in these computations has been determined by floating point arithmetic using the same precision as the solution of the linear system. Suppose that the residual is computed in *double precision* and then rounded to single precision, before solving a single precision linear system for the correction and adding that to the previous solution. If we compute the backward and forward error estimates after this improvement, we find no change. However, the true error is now 0.

Next, suppose that we use row and/or column scaling to improve the condition number of $\mathbf{V}$. The resulting condition number is improved to 2.16×10^2. The *a posteriori* backward error estimate for the scaled system is 3.49×10^{-8}, and the *a posteriori* forward error estimate is 6.50×10^{-3}. Although scaling improved the condition number of the matrix, it made little change in the *a posteriori* error estimates for the solution. However, we would like to note that the *a posteriori* error estimate (3.30) produces a value of 1.45×10^{-5}. The true relative error in this case is 2.59×10^{-4}. The error estimate (3.30) is too small because it does not account for rounding errors that occur in computing the residual; as a result, this estimate should not be trusted *in general*.

3.10 Inverses

Introductory linear algebra books often write the solution of a linear system $\mathbf{A}\mathbf{x} = \mathbf{b}$ in the form $\mathbf{x} = \mathbf{A}^{-1}\mathbf{b}$. As a result, readers often enter a scientific computing course with the habit of computing $\mathbf{A}^{-1}$ and then multiplying it times $\mathbf{b}$. Unfortunately, in solving linear systems the use of *an inverse almost always involves unnecessary computational expense*. We make an exception for very small linear systems, such as the 2×2 linear systems described in Sect. 3.4.2. We will recommend some uses for matrix inverses, such as in quasi-Newton methods for solving systems of nonlinear equations (see Sect. 3.7 of Chap. 3 in Volume II), or some implementations of the simplex method for solving linear programming problems. However, even in these situations there may be better ways to avoid working with an inverse matrix.

Nevertheless, let us examine how we could compute the inverse of a nonsingular matrix. First, we would use Gaussian factorization with partial pivoting to produce

$$\mathbf{A} = \mathbf{Q}\mathbf{L}\mathbf{R} \,,$$

where $\mathbf{Q}$ is a permutation matrix, $\mathbf{L}$ is unit left triangular and $\mathbf{R}$ is right triangular. Then

$$\mathbf{A}^{-1} = \mathbf{R}^{-1}\mathbf{L}^{-1}\mathbf{Q}^{-1} \,.$$

This means that the inverse matrix can be computed by the following

Algorithm 3.10.1 (Matrix Inverse)

factor $\mathbf{Q}^\top\mathbf{A} = \mathbf{L}\mathbf{R}$ (see Algorithm 3.7.5)

solve $\mathbf{L}\mathbf{Y} = \mathbf{I}$ for $\mathbf{Y}$ (see Algorithm 3.4.3)

solve $\mathbf{R}\mathbf{X} = \mathbf{Y}$ for $\mathbf{X}$ (see Algorithm 3.4.5)

permute $\mathbf{A}^{-1} = \mathbf{X}\mathbf{Q}^\top$ (see Sect. 3.4.7)

Table 3.2 Work in solving linear system via Gaussian factorization

	Factor $\mathbf{Q}^\top\mathbf{A}=\mathbf{L}\mathbf{R}$	Permute $\mathbf{Q}^\top\mathbf{B}$	Solve $\mathbf{L}\mathbf{Y}=\mathbf{B}$	Solve $\mathbf{R}\mathbf{X}=\mathbf{Y}$	Total
Multiplications	$n^3/3$	0	$sn^2/2$	$sn^2/2$	sn^2
Additions or subtractions	$n^3/3$	0	$sn^2/2$	$sn^2/2$	sn^2
Divisions	$n^2/2$	0	0	sn	sn
Comparisons	$n^2/2$	0	0	0	0
Swaps	n^2	sn	0	0	sn
Total	$2n^3/3$				$2sn^2$

Table 3.3 Work in solving linear system via matrix inverse

	Factor $\mathbf{Q}^\top\mathbf{A}=\mathbf{L}\mathbf{R}$	Solve $\mathbf{L}\mathbf{Y}=\mathbf{I}$	Solve $\mathbf{R}\mathbf{X}=\mathbf{Y}$	Permute $\mathbf{X}\mathbf{Q}^\top$	Total $\mathbf{A}^{-1}$	Multiply $\mathbf{A}^{-1}\mathbf{B}$
Multiplications	$n^3/3$	$n^3/6$	$n^3/2$	0	n^3	sn^2
Additions or subtractions	$n^3/3$	$n^3/6$	$n^3/2$	0	n^3	sn^2
Divisions	$n^2/2$	0	n^2	0	$3n^2/2$	sn
Comparisons	$n^2/2$	0	0	0	$n^2/2$	0
Swaps	n^2	0	0	0	n^2	0
Total					$2n^3$	$2sn^2$

Tables 3.2 and 3.3 summarize the work involved in solving s linear systems by either Gaussian factorization or inverse multiplication. These tables show that the work in computing a matrix inverse is essentially three times the work involved in Gaussian factorization. For each subsequent linear system, the total work involved in using the Gaussian factors to solve the linear system is essentially the same as the work involved in multiplying by a matrix inverse. In other words, the extra cost of computing the inverse generally cannot be recovered in subsequent linear system solves.

LAPACK provides routines `_trtri` to compute the inverse of a triangular matrix, and routines `_getri` to compute general matrix inverses. Interested readers can view, for example, dtrtri.f to see an algorithm for computing the inverse of a triangular matrix, or dgetri.f to see an implementation of Algorithm 3.10.1. In MATLAB, the command `inv(A)` will compute the inverse of the matrix `A`.

Exercise 3.10.1 Determine the number of floating point operations in performing Gaussian factorization with partial pivoting on a 2×2 matrix. Also determine the number of floating point operations in using such a factorization to solve a system of two linear equations. Compare this work with that involved in solving a system of two linear equations as in Sect. 3.4.2.

Exercise 3.10.2 It is well-known [99, p. 231] that for $n > 1$ the entries of the inverse of an $n \times n$ matrix can be written explicitly in terms of determinants:

$$\left[\mathbf{A}^{-1}\right]_{ij} = \frac{(-1)^{i+j} \det \mathbf{A}_{ji}}{\det \mathbf{A}} .$$

Here $\mathbf{A}_{ij}$ is the $(n-1) \times (n-1)$ matrix formed by deleting row i and column j of $\mathbf{A}$. We are interested in using this equation and expansion by minors (3.8) to compute the inverse of a 3×3 matrix.

1. Determine the number of floating point operations needed to compute $\mathbf{A}^{-1} \det \mathbf{A}$, which has entries

$$\mu_{ij} = (-1)^{i+j} \det \mathbf{A}_{ji} \; .$$

2. Determine the number of additional floating point operations needed to determine $\det \mathbf{A}$, and $\mathbf{A}^{-1}$.
3. Compare the total work in computing $\mathbf{A}^{-1}$ in this way with the cost of performing Gaussian factorization with partial pivoting on a 3×3 matrix.
4. Is it less expensive to solve systems of three linear equations via Gaussian factorization with partial pivoting, or via inverses?

3.11 Determinants

In elementary linear algebra, readers are often taught that the determinant is useful in testing nonsingularity of a square matrix. Such a test is acceptable for $n = 2$ or 3, since such small linear systems can be solved efficiently via inverses computed by means of determinants. However, there is no practical value in computing determinants for large matrices, and some substantial risk of overflow or underflow. For example, suppose that we compute the determinant of a 40×40 single precision diagonal matrix with all diagonal entries equal to 0.1. The determinant will be the product of the diagonal entries. On most machines the product will underflow and the value will be reported as zero. Clearly, such a matrix is nonsingular, and using its determinant to test nonsingularity is misleading.

If it really is necessary to compute the determinant of an $n \times n$ matrix with $n > 3$, the most efficient way is via Gaussian factorization. Suppose that we have factored $\mathbf{Q}^\top \mathbf{A} = \mathbf{L}\mathbf{R}$. Then Lemma 3.2.17 implies that $\det \mathbf{A} = \det \mathbf{L} \cdot \det \mathbf{R} / \det \mathbf{Q}$. Lemmas 3.2.15 and 3.2.16 imply that $\det \mathbf{Q} = (-1)^{N(\omega)}$, where $N(\omega)$ is the number of transpositions used to represent the permutation ω associated with the permutation matrix $\mathbf{Q}$. The Laplace expansion (3.7) implies that $\det \mathbf{L} = 1$, and that

$$\det \mathbf{R} = \prod_{j=1}^{n} \varrho_{jj}$$

is the product of the diagonal entries of $\mathbf{R}$. In other words,

$$\det \mathbf{A} = \pm \prod_{j=1}^{n} \varrho_{jj} \; ,$$

with the plus sign being chosen if and only if Gaussian factorization involved an even number of row interchanges. The work involved in computing the determinant in this way is n multiplications, plus the much larger work to perform Gaussian factorization.

The MATLAB command `det(A)` will compute the determinant of the matrix `A`. LAPACK does not provide a routine to compute a determinant. However, its precursor, LINPACK, provided routines `_gedi` to compute determinants. Interested readers may view, for example, **dgedi.f.**

Exercise 3.11.1 Elementary linear algebra books typically suggest computing determinants by means of **expansion by minors**, namely Eq. (3.8). If W_n is the number of multiplications involved in computing the determinant of $\mathbf{A}$ by this formula, show that $W_2 = 2$, and $W_n = n + nW_{n-1}$. Then show that the solution of this linear recurrence is

$$W_n = \sum_{j=1}^{n-1} \frac{n!}{j!} .$$

How does this work compare with the work required to compute the determinant via Gaussian factorization?

3.12 Object-Oriented Programming

While the LAPACK Fortran routines produce highly reliable and efficient numerical results, their programming interfaces are complicated. The LAPACK project designers chose Fortran 77 for the programming language because of its innate support for arrays, but Fortran 77 has no data structures that could bind array data together with array dimensions. Another problem is somewhat historical: subroutine names in LAPACK have generally been limited to at most 6 letters (due to limitations in early versions of Fortran), and are somewhat cryptic.

We briefly discussed the advantages and disadvantages of various programming languages in Sect. 2.4.1. In this section, we will discuss the use of the **object-oriented programming** to achieve certain programming objectives related to linear algebra. Interested readers may view the Wikipedia.org web page on object-oriented programming to learn more about this subject.

3.12.1 Class Design

Object-oriented programming is usually achieved by means of a programming feature called a **class**. A class (whether in C^{++}, Python, Java, or JavaScript) is a special kind of data structure. A class has both member data and procedures; this is the principal requirement for object-oriented programming. Class members in C++ also have **protection levels**. **Private** members are not accessible outside this class, **protected** members are accessible only in this and derived classes, and **public** members are accessible anywhere. Protection levels can eliminate side effects, caused by data changes occurring in other classes and functions. However, use of private and protected protection generally requires the programmer to write access routines that provide copies of the data.

C++ classes have **constructors** and **destructors**. The former are called when a class instance is created (for example, by dynamic memory allocation through `operator new`). Destructors are called whenever a class instance goes out of scope, or when `operator delete` is called. Constructors and destructors are much less available in Java or JavaScript. In C++, member procedures and arguments can be declared `const` to prevent changes to class data within the procedure. We will use C++ to produce our object-oriented programs.

For object-oriented implementations of linear algebra software, it is common to design classes for vectors and matrices. However, object-oriented programmers can have very different opinions about how to design these classes, and none of the publicly available software projects has achieved dominance.

The LAPACK++ and CPPLAPACK projects were designed to call LAPACK directly from C++. Unfortunately, LAPACK++ and CPPLAPACK avoided generic programming, (i.e. C++ templates) by using a large number of class names for different data types. For example, LAPACK++ has classes `LaGenMatFloat`, `LaGenMatDouble`, `LaGenMatFComplex` and `LaGenMatDComplex`. CPPLAPACK has classes `dgematrix` and `zgematrix`, each of which is templated by the numbers of rows and columns. Unlike LAPACK++, CPPLAPACK has implemented a matrix algebra: it defines addition, subtraction and multiplication of pairs of matrices and vectors.

On the other hand, the Template Numerical Toolkit (TNT) does use generic programming. This package has a templated class `Array2D<T>`, but the `LU` method assumes that the template type is real.

These various object-oriented linear algebra packages illustrate a common problem with object-oriented programming, namely that *beauty is in the eye of the programmer*. Often, publicly available software packages are organized in ways that do not please some of the users. We will provide our own C++ linear algebra software, principally to serve as a wrapper for calls to LAPACK routines. The reader is free to choose to use our software, to adopt other publicly available objected-oriented linear algebra software, or to write their own.

We will employ the following C++ style conventions for object-oriented programming.

- Words in the name of a C++ class will all begin with capital letters (e.g., `SymmetricMatrix`).
- Words in class procedure names will begin with capital letters, except for the first word (e.g., `Matrix::copyFrom`).
- Words in class data member names are all lowercase separated by underscores.

3.12.2 Vector Class

It is generally best to find the core elements of the program structure and develop these as base classes. More complicated ideas should be developed from simpler classes through the C++ feature of class inheritance. Derived classes inherit the data and functions of parent classes, and then may add data and functions of their own.

Suppose that we want to develop a C++ Vector class.[1] First, we should ask ourselves what the Vector class data should be. Clearly, we need to store the size of the Vector, as well as an array of Vector entries. Next, we need to think about the operations we should be able to perform with a Vector. We need to construct and destruct Vectors. We want to refer to individual entries. We will probably want to add Vectors and multiply Vectors by scalars. More generally, we need to perform the BLAS routines, which were described in Sect. 2.10.1.

Our implementation of the Vector class is available in the files Vector.H and Vector.C. The implementation of this class requires two template parameters, corresponding to the scalar data type for Vector entries, and to the scalar data type for the absolute value of Vector entries. In order to wrap around the LAPACK routines, which are programmed in Fortran, we also provide files LaFloat.C, LaDouble.C, LaComplex.C and LaDoubleComplex.C. For example, Vector<double,double>::dot is programmed in LaDouble.C to call LAPACK routine ddot, and Vector<double,complex<double> >::dotc is programmed in LaDoubleComplex.C to call LAPACK routine zdotc.

As in CPPLAPACK, we also define some C++ operators to enable simple algebraic operations on Vectors. In particular, we can add and subtract Vectors, and multiply or divide by scalars. This allows us to write C++ code with a high level program syntax that is similar to mathematics or MATLAB.

3.12.3 Matrix Class

After developing a Vector class, it is natural to develop a class to represent matrices. The data should include the number of rows and columns, as well as an array of matrix entries. Our implementation of the Matrix class is available in the files Matrix.H and Matrix.C. This implementation contains functions to

[1]The C++ language include files now contain a vector class from the **Standard Template Library**. This vector class has no connection to LAPACK; rather, the emphasis in this class is on the use of **iterators** for random access to members of an array, and to handle resize operations. Our needs for iterators are very simple, and resize operations should be unnecessary. To avoid confusion, we use a capital "V" for our Vector class, and a lower-case "v" for the standard template library vector class.

add and subtract matrices, or to multiply and divide them by scalars. There are also LAPACK routines to multiply matrices times vectors or matrices, to compute various matrix norms, to equilibrate matrices, and to solve linear systems. Note that the `Matrix::solve` routine uses either Gaussian factorization for square matrices, or techniques from least-squares methods (described in Chap. 6) for general rectangular matrices.

We would also like to remark that there is no `Matrix::transpose` function in our implementation. Instead, routines such as `solve`, `gemv` and `gemm` have arguments that direct the use of `Matrix` transposes. This class design helps to prevent readers from using mathematical notation to produce inefficient computer algorithms.

3.12.4 *GaussianFactorization Class*

If we want to solve a system of linear equations, we could call the `Matrix::solve` routines. Each call to one of these routines would involve a separate call to an LAPACK Gaussian factorization. Since Gaussian factorization is the dominant cost in solving a linear system, it would be advantageous to compute the factorization once and use it to solve multiple linear systems as they arise. Our solution is to develop a new C^{++} `GaussianFactorization` class to hold the factorization. Its data consist of a `Matrix` to hold the left and right trapezoidal factors, two arrays of `int`s to hold the row and column ordering for pivoting, and a flag to indicate which kind of pivoting is being performed. The class constructor calls LAPACK to compute the factorization. Afterward, users can call routines to solve linear systems, to determine condition numbers, or to perform iterative improvement and estimate the error in the computed solution. For more details, the reader should consult files GaussianFactorization.H and GaussianFactorization.C.

Exercise 3.12.1 Compare the implementations of `Vector` classes in LAPACK++ LaVectorDouble and Template Numerical Toolkit (TNT). Which of these implementations provide a parent class for `Vector`? How do the implementations differ in the class data? What member functions are common to all implementations?

3.13 Special Matrices

In C^{++}, **class inheritance** allows a derived class to access the data members and functions of its base classes. In Fig. 3.5, we show the `Matrix` class inheritance tree for our C^{++} linear algebra classes.

Matrices with special structure can take advantage of specialized algorithms to save work in linear algebra computations. LAPACK has a number of such algorithms, which we will use to design particular C^{++} classes in the sections

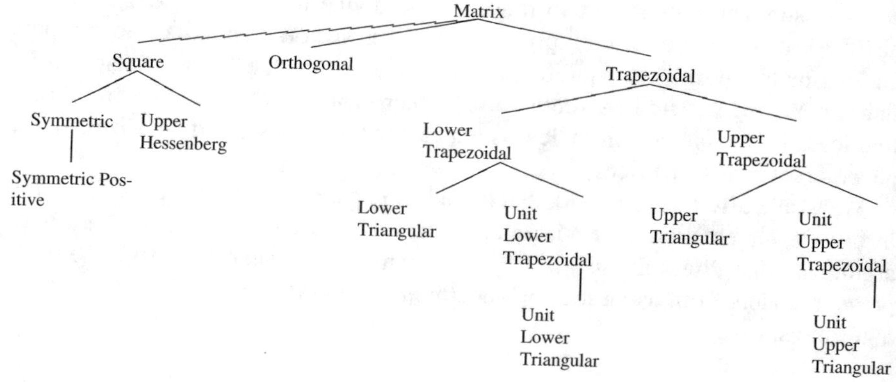

Fig. 3.5 Matrix class derivation tree

below. For example, in Sects. 3.13.2 and 3.13.3 we will examine two matrix factorizations that take advantage of matrix symmetry. MATLAB recognizes some of these special structures in its linsolve command.

In C^{++}, **virtual functions** allows a derived class to provide its own definition of functions originally defined in the base class. For example, a `SymmetricMatrix` class derived from `Matrix` can define its own `solve` procedure, distinct from and presumably more efficient than the `Matrix` class function `solve`. The C^{++} compiler will establish a table to be examined at run time; the internal structure of a `Matrix` reference or pointer will allow the program to choose from the table either the `Matrix::solve` procedure or the `SymmetricMatrix::solve` procedure. This dynamic selection of functions is called **run-time binding** or late binding. Because of run-time binding, C^{++} code can work with generic data structures (e.g., a `Matrix`), and still execute the most efficient algorithms (e.g., for a `SymmetricMatrix`) at run time.

Recall that we discussed how to solve linear systems involving orthogonal matrices in Sect. 3.4.8, triangular linear systems in Sect. 3.4.4, and trapezoidal linear systems in Sect. 3.4.5. Readers who are interested in the corresponding C^{++} class designs can examine the files OrthogonalMatrix.H, OrthogonalMatrix.C, TrapezoidalMatrix.H and TrapezoidalMatrix.C. In the remainder of this section, we will primarily study algorithms for symmetric matrices and banded matrices.

3.13.1 Square

So far, all of the example linear systems we have seen in this chapter have been **square**, meaning that they have the same number of rows as columns. There are a couple of computations that make sense for a square matrix, but not for a general matrix. For example, nonsingular square matrices have meaningful condition numbers, described in Sects. 3.6.2 and 3.8.7. In addition, square matrices have eigenvalues and eigenvectors, which will be discussed in Chap. 1 of Volume II. As a result, it is natural that we form a C++ class, called `SquareMatrix`, derived from class `Matrix`. Our `SquareMatrix` class is implemented in the files SquareMatrix.H and SquareMatrix.C.

3.13.2 *Symmetric*

Recall from Definition 3.2.12 that a matrix $\mathbf{A}$ is **symmetric** if and only if $\mathbf{A}^\top = \mathbf{A}$, and a complex matrix $\mathbf{A}$ is **Hermitian** if and only if $\mathbf{A}^H = \mathbf{A}$. This means that $n \times n$ symmetric or Hermitian matrices can have at most $n(n+1)/2$ independent entries, as opposed to n^2 for a square matrix. Thus it is useful to develop a factorization of symmetric or Hermitian matrices that can take advantage of the symmetry to reduce both the data storage and the arithmetic operations.

3.13.2.1 Factorization

Given an $n \times n$ Hermitian matrix $\mathbf{A}$, let

$$\delta = \max_{1 \le i \le n} |\mathbf{A}_{ii}| \text{ and } \omega = \max_{1 \le i,j \le n} \left|\mathbf{A}_{ij}\right| .$$

These definitions imply that $\delta \le \omega$. Given some number $\tau \in (0,1)$, suppose that $\delta \ge \omega\tau$ and $|\mathbf{A}_{ii}| = \delta$. Then we can interchange rows and columns, partition, and factor to get

$$\mathbf{I}_{1i}\mathbf{A}\mathbf{I}_{1i} = \begin{bmatrix} \alpha & \mathbf{a}^H \\ \mathbf{a} & \mathbf{B} \end{bmatrix} = \begin{bmatrix} 1 & \\ \boldsymbol{\ell} & \mathbf{I} \end{bmatrix} \begin{bmatrix} \alpha & \\ & \mathbf{C} \end{bmatrix} \begin{bmatrix} 1 & \boldsymbol{\ell}^H \\ & \mathbf{I} \end{bmatrix} ,$$

where

$$\boldsymbol{\ell} = \mathbf{a}\frac{1}{\alpha} , \; \mathbf{C} = \mathbf{B} - \boldsymbol{\ell}\alpha\boldsymbol{\ell}^H = \mathbf{B} - \boldsymbol{\ell}\mathbf{a}^H ,$$

and $|\alpha| = \delta$. Note that since $\mathbf{A}$ is Hermitian, α is real.

If $\delta < \omega\tau$, then we should consider a different form of pivoting and factorization. Specifically, if $\left|\mathbf{A}_{ij}\right| = \omega$, we can interchange rows and columns 1 and i, interchange rows and columns 2 and j, then partition and factor to get

$$\mathbf{I}_{2j}\mathbf{I}_{1i}\mathbf{A}\mathbf{I}_{1i}\mathbf{I}_{2j} = \begin{bmatrix} \alpha & \overline{\beta} & \mathbf{a}^H \\ \beta & \gamma & \mathbf{b}^H \\ \mathbf{a} & \mathbf{b} & \mathbf{B} \end{bmatrix} = \begin{bmatrix} 1 & & \\ 0 & 1 & \\ \boldsymbol{\ell} & \mathbf{m} & \mathbf{I} \end{bmatrix} \begin{bmatrix} \alpha & \overline{\beta} & \\ \beta & \gamma & \\ & & \mathbf{C} \end{bmatrix} \begin{bmatrix} 1 & 0 & \boldsymbol{\ell}^H \\ & 1 & \mathbf{m}^H \\ & & \mathbf{I} \end{bmatrix} .$$

Here

$$|\beta| = \omega > \delta/\tau > \delta \geq \max\{|\alpha|\ ,\ |\gamma|\}\ .$$

Note that ℓ^H and m^H satisfy

$$\begin{bmatrix} \alpha & \overline{\beta} \\ \beta & \gamma \end{bmatrix} \begin{bmatrix} \ell^H \\ m^H \end{bmatrix} = \begin{bmatrix} \mathbf{a}^H \\ \mathbf{b}^H \end{bmatrix} \Longrightarrow \begin{bmatrix} \ell^H \\ m^H \end{bmatrix} = \begin{bmatrix} -\gamma & \overline{\beta} \\ \beta & -\alpha \end{bmatrix} \begin{bmatrix} \mathbf{a}^H \\ \mathbf{b}^H \end{bmatrix} \frac{1}{|\beta|^2 - \alpha\gamma}\ ,$$

and

$$\mathbf{C} = \mathbf{B} - \begin{bmatrix} \ell & m \end{bmatrix} \begin{bmatrix} \alpha & \overline{\beta} \\ \beta & \gamma \end{bmatrix} \begin{bmatrix} \ell^H \\ m^H \end{bmatrix} = \mathbf{B} - \begin{bmatrix} \ell & m \end{bmatrix} \begin{bmatrix} \mathbf{a}^H \\ \mathbf{b}^H \end{bmatrix}\ .$$

By design, we can perform at least one of the two forms of pivoting for any nonzero matrix $\mathbf{A}$.

In general, at some stage of the factorization process we have factored

$$\mathbf{P}_{k-1}{}^\top \mathbf{A} \mathbf{P}_{k-1} = \begin{bmatrix} \mathbf{L}_{11} & \\ \mathbf{L}_{21} & \mathbf{I} \end{bmatrix} \begin{bmatrix} \mathbf{D}_1 & \\ & \mathbf{C} \end{bmatrix} \begin{bmatrix} \mathbf{L}_{11}{}^H & \mathbf{L}_{21}{}^H \\ & \mathbf{I} \end{bmatrix}$$

We examine the entries of $\mathbf{C}$ to find a permutation matrix $\mathbf{Q}_k$ involving either one or two interchanges. We interchange

$$\begin{aligned} &\begin{bmatrix} \mathbf{I} & \\ & \mathbf{Q}_k{}^\top \end{bmatrix} \mathbf{P}_{k-1}{}^\top \mathbf{A} \mathbf{P}_{k-1} \begin{bmatrix} \mathbf{I} & \\ & \mathbf{Q}_k \end{bmatrix} = \begin{bmatrix} \mathbf{I} & \\ & \mathbf{Q}_k{}^\top \end{bmatrix} \begin{bmatrix} \mathbf{L}_{11} & \\ \mathbf{L}_{21} & \mathbf{I} \end{bmatrix} \begin{bmatrix} \mathbf{D}_1 & \\ & \mathbf{C} \end{bmatrix} \begin{bmatrix} \mathbf{L}_{11}{}^H & \mathbf{L}_{21}{}^H \\ & \mathbf{I} \end{bmatrix} \begin{bmatrix} \mathbf{I} & \\ & \mathbf{Q}_k \end{bmatrix} \\ &= \begin{bmatrix} \mathbf{L}_{11} & \\ \mathbf{Q}_k{}^\top \mathbf{L}_{21} & \mathbf{Q}_k{}^\top \end{bmatrix} \begin{bmatrix} \mathbf{D}_1 & \\ & \mathbf{C} \end{bmatrix} \begin{bmatrix} \mathbf{L}_{11}{}^H & \mathbf{L}_{21}{}^H \mathbf{Q}_k \\ & \mathbf{Q}_k \end{bmatrix} \\ &= \begin{bmatrix} \mathbf{L}_{11} & \\ \mathbf{Q}_k{}^\top \mathbf{L}_{21} & \mathbf{I} \end{bmatrix} \begin{bmatrix} \mathbf{D}_1 & \\ & \mathbf{Q}_k{}^\top \mathbf{C} \mathbf{Q}_k \end{bmatrix} \begin{bmatrix} \mathbf{L}_{11}{}^H & \mathbf{L}_{21}{}^H \mathbf{Q}_k \\ & \mathbf{I} \end{bmatrix} \end{aligned}$$

then we repartition

$$= \begin{bmatrix} \mathbf{L}_{11} & & \\ \ell_{21} & \mathbf{I} & \\ \mathbf{L}_{31} & \mathbf{0} & \mathbf{I} \end{bmatrix} \begin{bmatrix} \mathbf{D}_1 & & \\ & \mathbf{D}_2 & \mathbf{M}_{32}{}^H \\ & \mathbf{M}_{32} & \mathbf{F} \end{bmatrix} \begin{bmatrix} \mathbf{L}_{11}{}^H & \ell_{21}{}^H & \mathbf{L}_{31}{}^H \\ & \mathbf{I} & \mathbf{0} \\ & & \mathbf{I} \end{bmatrix}$$

and finally we factor

$$= \begin{bmatrix} \mathbf{L}_{11} & & \\ \ell_{21} & \mathbf{I} & \\ \mathbf{L}_{31} & \mathbf{L}_{32} & \mathbf{I} \end{bmatrix} \begin{bmatrix} \mathbf{D}_1 & & \\ & \mathbf{D}_2 & \mathbf{0} \\ & \mathbf{0} & \mathbf{G} \end{bmatrix} \begin{bmatrix} \mathbf{L}_{11}{}^H & \ell_{21}{}^H & \mathbf{L}_{31}{}^H \\ & \mathbf{I} & \mathbf{L}_{32}{}^H \\ & & \mathbf{I} \end{bmatrix}$$

where

$$\mathbf{L}_{32} = \mathbf{M}_{32}\mathbf{D}_2^{-1} \text{ and } \mathbf{G} = \mathbf{F} - \mathbf{L}_{32}\mathbf{M}_{32}^H .$$

The process can be continued, much like Gaussian factorization, until we either run out of rows and columns to process, or at some point the remaining matrix to be processed is all zero. This gives us the following theorem.

Theorem 3.13.1 (Symmetric Indefinite Factorization) *If* $\mathbf{A}$ *is a real nonzero* $n \times n$ *matrix, then there are a permutation matrix* $\mathbf{P}$*, a unit left-trapezoidal matrix* $\mathbf{M}$ *and a block-diagonal matrix* $\mathbf{D}$ *with either* 1×1 *or* 2×2 *diagonal blocks so that*

$$\mathbf{P}^\top \mathbf{A}\mathbf{P} = \mathbf{M}\mathbf{D}\mathbf{M}^H .$$

3.13.2.2 Pivoting Tolerance

Next, we would like to choose a value for the pivoting tolerance τ. Our experience in Wilkinson's Backward Error Analysis Theorem 3.8.2 suggests that we want to pivot so that at each step of the symmetric indefinite factorization the entries in C grow as little in magnitude as possible, relative to the entries in $\mathbf{A}$.

There are two cases. If we choose a 1×1 pivot, then $\omega\tau \leq \delta$,

$$|\boldsymbol{\ell}_i| = \left|\frac{\mathbf{a}_i}{\alpha}\right| \leq \frac{\omega}{\delta} \leq \frac{1}{\tau} , \text{ and } |\mathbf{C}_{ij}| \leq |\mathbf{B}_{ij}| + |\boldsymbol{\ell}_i| \, |\mathbf{a}_j| \leq \omega + \frac{\omega}{\tau} .$$

If we choose a 2×2 pivot, then $\omega\tau > \delta$,

$$|\boldsymbol{\ell}_j| = \left|\frac{\overline{\beta}\mathbf{b}_j - \gamma\mathbf{a}_j}{|\beta|^2 - \alpha\gamma}\right| \leq \frac{\omega^2 + \delta\omega}{\omega^2 - \delta^2} = \frac{\omega}{\omega - \delta} ,$$

$$|\mathbf{m}_j| = \left|\frac{\beta\mathbf{a}_j - \alpha\mathbf{b}_j}{|\beta|^2 - \alpha\gamma}\right| \leq \frac{\omega^2 + \delta\omega}{\omega^2 - \delta^2} = \frac{\omega}{\omega - \delta} ,$$

and

$$|\mathbf{C}_{ij}| = \left|\mathbf{B}_{ij} - \boldsymbol{\ell}_i\overline{\mathbf{a}_j} - \mathbf{m}_i\overline{\mathbf{b}_j}\right| \leq \omega + 2\frac{\omega^2}{\omega - \delta} < \omega\left[1 + \frac{2}{1 - \tau}\right] .$$

A 2×2 pivot corresponds to two 1×1 pivots, and thus the square of a 1×1 growth factor. Thus, we should choose $\tau \in (0, 1)$ to minimize the maximum of the two growth factors $1 + 1/\tau$ and $\sqrt{1 + 2/(1 - \tau)}$. This maximum occurs at the point where $(1 + 1/\tau)^2 = 1 + 2/(1 - \tau)$, and suggests that we take

$$\tau = \frac{1 + \sqrt{17}}{8} \approx 0.64 .$$

This pivoting strategy is comparable to Gaussian factorization with full pivoting, in that all remaining entries of the matrix are searched to select the pivot. For more details, see Bunch and Parlett [20].

Example 3.13.1 Suppose that we want to compute the symmetric indefinite factorization of

$$\mathbf{A} = \begin{bmatrix} 1 & 2 & -1 \\ 2 & 1 & 1 \\ -1 & 1 & 3 \end{bmatrix} .$$

Here $\delta = 3 = \omega$, we begin the factorization with a 1×1 pivot. We interchange rows 1 and 3 and factor

$$\mathbf{I}_{13}\mathbf{A}\mathbf{I}_{13} = \begin{bmatrix} 3 & 1 & -1 \\ 1 & 1 & 2 \\ -1 & 2 & 1 \end{bmatrix} = \begin{bmatrix} 1 & & \\ 1/3 & 1 & 0 \\ -1/3 & 0 & 1 \end{bmatrix} \begin{bmatrix} 3 & & \\ & 2/3 & 7/3 \\ & 7/3 & 2/3 \end{bmatrix} \begin{bmatrix} 1 & 1/3 & -1/3 \\ & 1 & 0 \\ & 0 & 1 \end{bmatrix} .$$

This completes the first step of the symmetric indefinite factorization.

Next, we examine the matrix

$$\mathbf{C} = \begin{bmatrix} 2/3 & 7/3 \\ 7/3 & 2/3 \end{bmatrix}$$

to find that $\delta = 2/3$ and $\omega = 7/3$. Since $\delta < \tau\omega$, we choose a 2×2 pivot. This means that no further factorization operations are needed.

Example 3.13.2 Suppose that we want to compute the symmetric indefinite factorization of

$$\mathbf{A} = \begin{bmatrix} 3 & -1 & 5 \\ -1 & 2 & -10 \\ 5 & -10 & 1 \end{bmatrix} .$$

We compute $\delta = 3$ and $\omega = 10$. Since $\delta < \tau\omega$, we begin with a 2×2 pivot. Since the maximum entry is in the 3, 2 position, we interchange rows and columns 1 and 3, and rows and columns 2 and 2 to get

$$\mathbf{I}_{22}\mathbf{I}_{13}\mathbf{A}\mathbf{I}_{13}\mathbf{I}_{22} = \begin{bmatrix} 1 & -10 & 5 \\ -10 & 2 & -1 \\ 5 & -1 & 3 \end{bmatrix} = \begin{bmatrix} 1 & & \\ 0 & 1 & \\ 0 & -1/2 & 1 \end{bmatrix} \begin{bmatrix} 1 & -10 & \\ -10 & 2 & \\ & & 5/2 \end{bmatrix} \begin{bmatrix} 1 & 0 & 0 \\ & 1 & -1/2 \\ & & 1 \end{bmatrix} .$$

This is a completed symmetric indefinite factorization.

3.13.2.3 Reduced Pivoting

Bunch and Kaufman [19] developed a less-expensive pivoting strategy, which is more analogous to partial pivoting in Gaussian factorization. This symmetric indefinite matrix factorization can be computed by the MATLAB command ldl.

In LAPACK, this pivoting strategy has been implemented in routines `_sytr2`. For example, readers may view dsytf2.f. LAPACK provides subroutines `_sytrf` and `_sysv` to factor and solve with symmetric matrices, and routines `_hetrf` and `_hesv` to factor and solve with Hermitian matrices. Routines `_sytrf` implement block factorizations by calling routines `_sytf2` to factor the blocks.

We have encapsulated these LAPACK operations into two C^{++} classes. For more details, the reader should consult files SymmetricMatrix.H and SymmetricMatrix.C. In order to solve multiple linear systems involving the same symmetric matrix, readers should examine MDMtFactorization.H and MDMtFactorization.C.

3.13.2.4 Linear Systems

We can use a symmetric indefinite factorization to solve a symmetric linear system $\mathbf{A}\mathbf{x} = \mathbf{b}$ by the following

Algorithm 3.13.1 (Symmetric Indefinite Linear System Solve)

$$\text{factor } \mathbf{P}\mathbf{A}\mathbf{P}^\top = \mathbf{M}\mathbf{D}\mathbf{M}^H$$

$$\text{permute } \mathbf{c} = \mathbf{P}^\top\mathbf{b}$$

$$\text{solve } \mathbf{M}\mathbf{z} = \mathbf{c} \text{ for } \mathbf{z}$$

$$\text{compute } \mathbf{w} = \mathbf{D}^{-1}\mathbf{z}$$

$$\text{solve } \mathbf{M}^H\mathbf{y} = \mathbf{w} \text{ for } \mathbf{y}$$

$$\text{permute } \mathbf{x} = \mathbf{P}\mathbf{y}$$

LAPACK provides routines `_sytrs` and `_hetrs` to solve linear systems, given a symmetric indefinite factorization of a matrix. For example, readers may view dsytrs.f.

Exercise 3.13.1 Count the number of arithmetic operations in performing the $\mathbf{M}\mathbf{D}\mathbf{M}^H$ factorization of a real $n \times n$ matrix with full pivoting. Compare the operation counts to the work in performing Gaussian factorization with full pivoting.

3.13.3 Symmetric Positive

As we saw in Sect. 3.13.2, symmetric matrices allow for reduced storage of their data, and more efficient methods of factorization for solving linear systems. In order to guarantee bounded growth of the pivots during symmetric indefinite factorization, it was necessary to use symmetric pivoting. In this section, we will study a special case of symmetric matrices whose factorizations do not require pivoting. Such linear systems will arise in Chap. 6 on least squares problems, and in Chap. 4 of Volume III on boundary value problems.

3.13.3.1 Positive Factorization

Suppose that we are able to factor a nonsingular Hermitian matrix $\mathbf{A}$ by Gaussian factorization *without pivoting*. In other words, assume that we can factor

$$\mathbf{A} = \mathbf{L}\mathbf{R}$$

where $\mathbf{L}$ is unit left-triangular and $\mathbf{R}$ is right-triangular. Since $\mathbf{A} = \mathbf{A}^H$, we also have

$$\mathbf{L}\mathbf{R} = \mathbf{A} = \mathbf{A}^H = \mathbf{R}^H\mathbf{L}^H .$$

This suggests that we solve

$$\mathbf{L}\mathbf{D} = \mathbf{R}^H$$

for the matrix $\mathbf{D}$. Since $\mathbf{L}$ and $\mathbf{R}^H$ are both left-triangular, so is $\mathbf{D}$. Since

$$\mathbf{L}\mathbf{R} = \mathbf{R}^H\mathbf{L}^H = \mathbf{L}\mathbf{D}\mathbf{L}^H ,$$

we must have

$$\mathbf{L}(\mathbf{R} - \mathbf{D}\mathbf{L}^H) = 0 ,$$

and the nonsingularity of $\mathbf{L}$ implies that

$$\mathbf{R} = \mathbf{D}\mathbf{L}^H .$$

This equation shows that $\mathbf{D} = \mathbf{R}\mathbf{L}^{-H}$ is also right-triangular. Since $\mathbf{D}$ is both left- and right-triangular, $\mathbf{D}$ is diagonal. In other words, if we can compute the Gaussian factorization of $\mathbf{A} = \mathbf{A}^H$ without pivoting, then $\mathbf{A} = \mathbf{L}\mathbf{D}\mathbf{L}^H$, where $\mathbf{L}$ is unit left-triangular and $\mathbf{D}$ is diagonal.

Next, let us define a special kind of symmetric matrix.

Definition 3.13.1 A real $n \times n$ symmetric matrix is **positive** if and only if for all nonzero real n-vectors $\mathbf{x}$ we have $\mathbf{x}^\top \mathbf{A}\mathbf{x} > 0$. Similarly, a complex $n \times n$ Hermitian matrix is **positive** if and only if for all nonzero complex n-vectors $\mathbf{x}$ we have $\mathbf{x}^H \mathbf{A}\mathbf{x} > 0$.
This definition suggest the following simple lemma.

Lemma 3.13.1 *Suppose that* $\mathbf{L}$ *is a nonsingular matrix,* $\mathbf{D}$ *is a Hermitian matrix, and* $\mathbf{A} = \mathbf{L}\mathbf{D}\mathbf{L}^H$*. Then* $\mathbf{A}$ *is positive if and only if* $\mathbf{D}$ *is positive.*

Proof First, we note that

$$\mathbf{A}^H = \left(\mathbf{L}\mathbf{D}\mathbf{L}^H\right)^H = \mathbf{L}\mathbf{D}^H\mathbf{L}^H = \mathbf{L}\mathbf{D}\mathbf{L}^H = \mathbf{A} \; ,$$

so $\mathbf{A}$ is Hermitian, and it is possible to discuss whether $\mathbf{A}$ is positive.

Since $\mathbf{L}$ is nonsingular, for any nonzero vector $\mathbf{y}$ we can solve $\mathbf{L}^H\mathbf{x} = \mathbf{y}$ for the nonzero vector $\mathbf{x}$. If $\mathbf{A}$ is positive, it follows that for any nonzero vector $\mathbf{y}$ we have

$$\mathbf{y}^H\mathbf{D}\mathbf{y} = \left(\mathbf{L}^H\mathbf{x}\right)^H\mathbf{D}\left(\mathbf{L}^H\mathbf{x}\right) = \mathbf{x}^H\left(\mathbf{L}\mathbf{D}\mathbf{L}^H\right)\mathbf{x} = \mathbf{x}^H\mathbf{A}\mathbf{x} > 0 \; .$$

This shows that whenever $\mathbf{A}$ is positive, so is $\mathbf{D}$. Conversely, if $\mathbf{D}$ is positive and $\mathbf{x}$ is nonzero, then

$$0 < \left(\mathbf{L}^H\mathbf{x}\right)^H\mathbf{D}\left(\mathbf{L}^H\mathbf{x}\right) = \mathbf{x}^H\left(\mathbf{L}\mathbf{D}\mathbf{L}^H\right)\mathbf{x} = \mathbf{x}^H\mathbf{A}\mathbf{x} \; .$$

3.13.3.2 Cholesky Factorization

In particular, if $\mathbf{A}$ is symmetric positive, and $\mathbf{A} = \mathbf{L}\mathbf{D}\mathbf{L}^H$ where $\mathbf{D}$ is diagonal, then the diagonal entries of $\mathbf{D}$ are all positive. In such a case, we can let $\sqrt{\mathbf{D}}$ be the diagonal matrix whose diagonal entries are the square roots of the diagonal entries of $\mathbf{D}$, and define

$$\tilde{\mathbf{L}} = \mathbf{L}\sqrt{\mathbf{D}} \; .$$

Then

$$\mathbf{A} = \mathbf{L}\mathbf{D}\mathbf{L}^H = (\mathbf{L}\sqrt{\mathbf{D}})(\mathbf{L}\sqrt{\mathbf{D}})^H = \tilde{\mathbf{L}}\tilde{\mathbf{L}}^H \; .$$

This is called the **Cholesky factorization** of a positive Hermitian matrix.

We would like to determine the Cholesky factorization directly. The following lemma will prepare the way.

Lemma 3.13.2 *Suppose that the Hermitian matrix* $\mathbf{A}$ *is partitioned*

$$\mathbf{A} = \begin{bmatrix} \alpha & \mathbf{a}^H \\ \mathbf{a} & \mathbf{B} \end{bmatrix} .$$

Then $\mathbf{A}$ *is positive if and only both if* $\alpha > 0$ *and* $\mathbf{B} - \mathbf{a}\mathbf{a}^H/\alpha$ *is positive.*

Proof First, suppose that $\mathbf{A}$ is positive. Then $\mathbf{x}^H\mathbf{A}\mathbf{x} > 0$ for all nonzero vectors $\mathbf{x}$, and in particular we have that $0 < \mathbf{e}_1{}^H\mathbf{A}\mathbf{e}_1 = \alpha$. Since $\mathbf{A}$ is Hermitian, so is $\mathbf{B}$. Given any $\mathbf{x} \neq 0$, let $\xi = -\mathbf{a}^H\mathbf{x}/\alpha$. Then

$$0 < [\xi \; \mathbf{x}^H] \begin{bmatrix} \alpha & \mathbf{a}^H \\ \mathbf{a} & \mathbf{B} \end{bmatrix} \begin{bmatrix} \xi \\ \mathbf{x} \end{bmatrix} = \alpha|\xi|^2 + \overline{\xi}\mathbf{a}^H\mathbf{x} + \xi\mathbf{x}^H\mathbf{a} + \mathbf{x}^H\mathbf{B}\mathbf{x}$$

$$= \alpha \left| \xi + \frac{\mathbf{a}^H\mathbf{x}}{\alpha} \right|^2 + \mathbf{x}^H \left(\mathbf{B} - \mathbf{a}\frac{1}{\alpha}\mathbf{a}^H \right) \mathbf{x} = \mathbf{x}^H \left(\mathbf{B} - \mathbf{a}\frac{1}{\alpha}\mathbf{a}^H \right) \mathbf{x} .$$

This proves that $\mathbf{B} - \mathbf{a}\mathbf{a}^H/\alpha$ is positive.

It remains to prove the reverse direction of the lemma. Suppose $\alpha > 0$ and $\mathbf{B} - \mathbf{a}\mathbf{a}^H/\alpha$ is positive. Then for any scalar ξ and vector $\mathbf{x}$, we have that

$$[\xi \; \mathbf{x}^H] \begin{bmatrix} \alpha & \mathbf{a}^H \\ \mathbf{a} & \mathbf{B} \end{bmatrix} \begin{bmatrix} \xi \\ \mathbf{x} \end{bmatrix} = \alpha \left| \xi + \frac{\mathbf{a}^H\mathbf{x}}{\alpha} \right|^2 + \mathbf{x}^H \left(\mathbf{B} - \mathbf{a}\frac{1}{\alpha}\mathbf{a}^H \right) \mathbf{x} \geq 0 .$$

If $\mathbf{x} \neq 0$, the positivity of $\mathbf{B} - \mathbf{a}\mathbf{a}^H/\alpha$ shows that

$$[\xi \; \mathbf{x}^H]\, \mathbf{A} \begin{bmatrix} \xi \\ \mathbf{x} \end{bmatrix} > 0 .$$

On the other hand, if $\mathbf{x} = 0$ but $\xi \neq 0$, then

$$[\xi \; \mathbf{0}]\, \mathbf{A} \begin{bmatrix} \xi \\ 0 \end{bmatrix} = \alpha|\xi|^2 > 0 .$$

Either way, we conclude that $\mathbf{A}$ is positive.

3.13.3.3 Cholesky Algorithm

This lemma suggests the following factorization process. We begin by partitioning the first column of $\mathbf{A}$:

$$\mathbf{A}\mathbf{e}_1 = \begin{bmatrix} \alpha \\ \mathbf{a} \end{bmatrix}$$

If $\alpha \leq 0$, then $\mathbf{A}$ is not positive and we stop. Otherwise we factor

$$\mathbf{A}\mathbf{e}_1 = \begin{bmatrix} \lambda \\ \boldsymbol{\ell} \end{bmatrix} \lambda \; ,$$

where

$$\lambda = \sqrt{\alpha} \text{ and } \boldsymbol{\ell} = \mathbf{a}/\lambda \; .$$

Inductively, suppose that after k steps we have partitioned and factored the first $(k-1)$ columns of $\mathbf{A}$ as

$$\begin{bmatrix} \mathbf{A}_{11} \\ \mathbf{A}_{21} \end{bmatrix} = \begin{bmatrix} \mathbf{L}_{11} \\ \mathbf{L}_{21} \end{bmatrix} \mathbf{L}_{11}{}^H \; .$$

Here $\mathbf{L}_{11}$ is $(k-1)\times(k-1)$ and $\mathbf{L}_{21}$ is $(n-k+1)\times(k-1)$. Then we can repartition and factor the first k columns of $\mathbf{A}$ in the form

$$\begin{bmatrix} \mathbf{A}_{11} & \overline{\mathbf{a}_{12}} \\ \mathbf{a}_{12}{}^\top & \alpha_{22} \\ \mathbf{A}_{31} & \mathbf{a}_{32} \end{bmatrix} = \begin{bmatrix} \mathbf{L}_{11} & \\ \boldsymbol{\ell}_{21}{}^\top & \lambda_{22} \\ \mathbf{L}_{31} & \boldsymbol{\ell}_{32} \end{bmatrix} \begin{bmatrix} \mathbf{L}_{11}{}^H & \overline{\boldsymbol{\ell}_{21}} \\ & \lambda_{22} \end{bmatrix} ,$$

where

$$\lambda_{22} = \sqrt{\alpha_{22} - \boldsymbol{\ell}_{21}{}^\top \overline{\boldsymbol{\ell}_{21}}} = \sqrt{\alpha_{22} - \boldsymbol{\ell}_{21} \cdot \boldsymbol{\ell}_{21}} \text{ and } \boldsymbol{\ell}_{32} = \left\{\mathbf{a}_{32} - \mathbf{L}_{31}\overline{\boldsymbol{\ell}_{21}}\right\} / \lambda_{22} \; .$$

We can summarize this factorization with the following

Algorithm 3.13.2 (Cholesky Factorization)

for $1 \leq k \leq n$

$$\alpha_{kk} = \alpha_{kk} - \sum_{j=1}^{k-1} \left|\alpha_{kj}\right|^2$$

if $\alpha_{kk} \leq 0$ stop

$$\alpha_{kk} = \sqrt{\alpha_{kk}}$$

for $k+1 \leq i \leq n$

$$\alpha_{ik} = \alpha_{ik} - \sum_{j=1}^{k-1} \alpha_{ij}\overline{\alpha_{kj}}$$

for $k+1 \leq i \leq n$

$$\alpha_{ik} = \alpha_{ik}/\alpha_{kk}$$

This algorithm only accesses entries on and below the diagonal. Also note that the computation of α_{ik} has been organized as a matrix-vector multiply, in order to allow level 2 BLAS routines to be used.

The process stops whenever we run out of rows and columns, thereby obtaining $\mathbf{A} = \mathbf{L}\mathbf{L}^H$. The factorization also stops whenever one of the pivots satisfies $\alpha \leq 0$, which implies that $\mathbf{A}$ is not positive. The computational work in Cholesky factorization of an $n \times n$ matrix is

- n square roots,
- $\sum_{k=1}^{n-1}(n-k) = n(n-1)/2$ divisions, and
- $\sum_{k=1}^{n-1}(n-k)(k-1) + \sum_{k=2}^{n}(k-1) = n(n-1)(n-2)/6$ multiplications and additions.

The total work is on the order of $n^3/6$ arithmetic operations, or about half the work involved in Gaussian factorization.

The Cholesky factorization process also proves the following theorem.

Theorem 3.13.2 (Cholesky Factorization) *Suppose that* $\mathbf{A}$ *is a Hermitian matrix. Then* $\mathbf{A}$ *is positive if and only if there exists a nonsingular left-triangular matrix* $\mathbf{L}$ *so that*

$$\mathbf{A} = \mathbf{L}\mathbf{L}^H .$$

Example 3.13.3 Suppose $\mathbf{A}$ is the 3×3 **Hilbert matrix**

$$\mathbf{A} = \begin{bmatrix} 1 & 1/2 & 1/3 \\ 1/2 & 1/3 & 1/4 \\ 1/3 & 1/4 & 1/5 \end{bmatrix} .$$

Then the steps in the Cholesky factorization of $\mathbf{A}$ are

$$\mathbf{A} = \begin{bmatrix} 1 & & \\ 1/2 & 1 & \\ 1/3 & & 1 \end{bmatrix} \begin{bmatrix} 1 & 1/2 & 1/3 \\ & 1/12 & 1/12 \\ & 1/12 & 4/45 \end{bmatrix} = \begin{bmatrix} 1 & & \\ 1/2 & \sqrt{1/12} & \\ 1/3 & \sqrt{1/12} & 1 \end{bmatrix} \begin{bmatrix} 1 & 1/2 & 1/3 \\ & \sqrt{1/12} & \sqrt{1/12} \\ & & 1/180 \end{bmatrix}$$

$$= \begin{bmatrix} 1 & & \\ 1/2 & \sqrt{1/12} & \\ 1/3 & \sqrt{1/12} & \sqrt{1/180} \end{bmatrix} \begin{bmatrix} 1 & 1/2 & 1/3 \\ & \sqrt{1/12} & \sqrt{1/12} \\ & & \sqrt{1/180} \end{bmatrix} .$$

3.13.3.4 Block Algorithm

The Cholesky factorization can be organized in block form. The general step takes the form

$$\begin{bmatrix} \mathbf{A}_{11} & {\mathbf{A}_{21}}^H \\ \mathbf{A}_{21} & \mathbf{A}_{22} \\ \mathbf{A}_{31} & \mathbf{A}_{32} \end{bmatrix} = \begin{bmatrix} \mathbf{L}_{11} & \\ \mathbf{L}_{21} & \mathbf{L}_{22} \\ \mathbf{L}_{31} & \mathbf{L}_{32} \end{bmatrix} \begin{bmatrix} {\mathbf{L}_{11}}^H & {\mathbf{L}_{21}}^H \\ & {\mathbf{L}_{22}}^H \end{bmatrix} ,$$

where

$$\mathbf{A}_{22} = \mathbf{L}_{21}{\mathbf{L}_{21}}^H + \mathbf{L}_{22}{\mathbf{L}_{22}}^H \text{ and } \mathbf{A}_{32} = \mathbf{L}_{31}{\mathbf{L}_{21}}^H + \mathbf{L}_{32}{\mathbf{L}_{22}}^H .$$

This leads to the following

Algorithm 3.13.3 (Block Cholesky)

$k = 1$

while $k \leq n$

$s = \min\{b, n - k + 1\}$

$A[k : k + s - 1; k : k + s - 1] = A[k : k + s - 1; k : k + s - 1]$

$- \mathbf{L}[k : k + s - 1; 1 : k - 1]\mathbf{L}[k : k + s - 1; 1 : k - 1]^H$

factor $A[k : k + s - 1; k : k + s - 1]$

$= \mathbf{L}[k : k + s - 1; k : k + s - 1]\mathbf{L}[k : k + s - 1; k : k + s - 1]^H$

$A[k + s : n; k : k + s - 1] = A[k + s : n; k : k + s - 1]$

$- \mathbf{L}[k + s : n; 1 : k - 1]\mathbf{L}[k : k + s - 1; 1 : k - 1]^H$

solve $\mathbf{L}[k + s : n; k : k + s - 1]\mathbf{L}[k : k + s - 1; k : k + s - 1]^H = A[k + s : n; k : k + s - 1]$

for $\mathbf{L}[k + s : n; k : k + s - 1]$

$k = k + b$

In LAPACK, the block Cholesky algorithm is performed by routines `_potrf`. See, for example, dpotrf.f.

3.13.3.5 LDL' Factorization

There is an alternative to the Cholesky factorization, called the $\mathbf{LDL}^\top$ factorization. Its general step, involving partitioning, factorization and repartitioning, looks like

$$\begin{bmatrix} \mathbf{A}_{11} & \overline{\mathbf{a}_{12}} \\ \mathbf{a}_{12}{}^\top & \alpha_{22} \\ \mathbf{A}_{31} & \mathbf{a}_{32} \end{bmatrix} = \begin{bmatrix} \mathbf{L}_{11} & \\ \boldsymbol{\ell}_{21}{}^\top & 1 \\ \mathbf{L}_{31} & \boldsymbol{\ell}_{32} \end{bmatrix} \begin{bmatrix} \mathbf{D}_{11} & \\ & \delta_{22} \end{bmatrix} \begin{bmatrix} \mathbf{L}_{11}{}^H & \overline{\boldsymbol{\ell}_{21}} \\ & 1 \end{bmatrix}$$

where

$$\alpha_{22} = \boldsymbol{\ell}_{21}{}^\top \mathbf{D}_{11} \overline{\boldsymbol{\ell}_{21}} + \delta_{22} \text{ and } \mathbf{a}_{32} = \mathbf{L}_{31} \mathbf{D}_{11} \overline{\boldsymbol{\ell}_{21}} + \boldsymbol{\ell}_{32} \delta_{22} \,.$$

The algorithm stops whenever $\delta_{22} \leq 0$. This process also proves the following

Theorem 3.13.3 ($\mathbf{LDL}^\top$ Factorization) *Suppose that* $\mathbf{A}$ *is a Hermitian matrix. Then* $\mathbf{A}$ *is positive if and only if there exists a unit left-triangular matrix* $\mathbf{L}$ *and a positive diagonal matrix* $\mathbf{D}$ *so that*

$$\mathbf{A} = \mathbf{L}\mathbf{D}\mathbf{L}^H \,.$$

We can summarize this factorization with the following

Algorithm 3.13.4 ($\mathbf{LDL}^\top$ Factorization)

$$\begin{aligned}
&\text{for } 1 \leq k \leq n \\
&\quad \text{for } 1 \leq i < k \\
&\qquad \tau_i = \alpha_{ii} \overline{\alpha_{ki}} \\
&\quad \alpha_{kk} = \alpha_{kk} - \sum_{i=1}^{k-1} \alpha_{ik} \tau_i \\
&\quad \text{if } \alpha_{kk} \leq 0 \text{ stop} \\
&\quad \text{for } k+1 \leq i \leq n \\
&\qquad \alpha_{ik} = \alpha_{ik} - \sum_{j=k+1}^{n} \alpha_{ik} \tau_j \\
&\quad \text{for } k+1 \leq i \leq n \\
&\qquad \alpha_{ik} = \alpha_{ik} / \alpha_{kk}
\end{aligned}$$

The computational work in the $\mathbf{LDL}^\top$ factorization of an $n \times n$ matrix is

- $\sum_{k=1}^{n-1}(n-k) = n(n-1)/2$ divisions, and
- $\sum_{k=1}^{n-1}(n-k)(n-k+1)/2 = n(n^2-1)/6$ multiplications and additions.

The total work is on the order of $n^3/6$ arithmetic operations, or about half the work involved in Gaussian factorization. Unlike the Cholesky factorization, the $\mathbf{LDL}^\top$ involves no square roots. However, it does require work space for the values τ_i.

3.13.3.6 Pivoting Unnecessary

Let us explain why pivoting is not necessary in the Cholesky factorization. Our discussion will use the **spectral radius**

$$\varrho(\mathbf{A}) \equiv \max_{\|\mathbf{x}\|_2=1} \left|\mathbf{x}^H\mathbf{A}\mathbf{x}\right| ,$$

which will be studied in Sect. 1.3 of Chap. 1 in Volume II.

Let us examine the terms in the Cholesky factorization step:

$$\begin{bmatrix} \alpha & \mathbf{a}^H \\ \mathbf{A} & \mathbf{B} \end{bmatrix} = \begin{bmatrix} \lambda & \\ \ell & \mathbf{I} \end{bmatrix} \begin{bmatrix} \lambda & \ell^H \\ & \mathbf{C} \end{bmatrix} \Longrightarrow \lambda = \sqrt{\alpha} \,,\; \ell = \mathbf{a}/\lambda \text{ and } \mathbf{C} = \mathbf{B} - \ell\ell^H \,.$$

First, note that

$$\lambda^2 = \alpha = \mathbf{e}_1{}^\top \mathbf{A}\mathbf{e}_1 \leq \varrho(\mathbf{A}) \Longrightarrow \lambda \leq \sqrt{\varrho(\mathbf{A})} \,.$$

Next, note that the Cauchy inequality (3.15) and the proof of Lemma 3.13.2 imply that

$$\begin{aligned} \|\ell\|_2^2 &= \frac{\|\mathbf{a}\|_2^2}{\alpha} = \max_{\|\mathbf{x}\|_2=1} \frac{\left|\mathbf{x}^H\mathbf{a}\right|^2}{\alpha} = \max_{\|\mathbf{x}\|_2=1} \mathbf{x}^H\mathbf{a}\frac{1}{\alpha}\mathbf{a}^H\mathbf{x} \leq \max_{\|\mathbf{x}\|_2=1} \mathbf{x}^H \left\{\mathbf{a}\frac{1}{\alpha}\mathbf{a}^H + \left[\mathbf{B} - \mathbf{a}\frac{1}{\alpha}\mathbf{a}^H\right]\right\} \mathbf{x} \\ &= \max_{\|\mathbf{x}\|_2=1} \mathbf{x}^H\mathbf{B}\mathbf{x} = \max_{\|\mathbf{x}\|_2=1} \begin{bmatrix} 0 & \mathbf{x}^H \end{bmatrix} \mathbf{A} \begin{bmatrix} 0 \\ \mathbf{x} \end{bmatrix} \leq \varrho(\mathbf{A}) \,, \end{aligned}$$

so $\|\ell\|_2 \leq \sqrt{\varrho(\mathbf{A})}$. Finally,

$$\begin{aligned} \varrho(\mathbf{C}) &= \max_{\|\mathbf{x}\|_2=1} \mathbf{x}^H\mathbf{C}\mathbf{x} = \max_{\|\mathbf{x}\|_2=1} \mathbf{x}^H \left[\mathbf{B} - \mathbf{a}\frac{1}{\alpha}\mathbf{a}^H\right] \mathbf{x} = \max_{\|\mathbf{x}\|_2=1} \left[\mathbf{x}^H\mathbf{B}\mathbf{x} - \left|\mathbf{x}^H\ell\right|^2\right] \\ &\leq \max_{\|\mathbf{x}\|_2=1} \mathbf{x}^H\mathbf{B}\mathbf{x} \leq \varrho(\mathbf{A}) \,. \end{aligned}$$

These inequalities imply that Cholesky factorization does not lead to growth in the factorization terms, and thus pivoting is unnecessary.

Pivoting is also unnecessary in the $\mathbf{L}\mathbf{D}\mathbf{L}^\top$ factorization. In this case, we factor

$$\begin{bmatrix} \alpha & \mathbf{a}^H \\ \mathbf{A} & \mathbf{B} \end{bmatrix} = \begin{bmatrix} 1 & \\ \ell & \mathbf{I} \end{bmatrix} \begin{bmatrix} \alpha & \\ & \mathbf{C} \end{bmatrix} \begin{bmatrix} 1 & \ell^H \\ & \mathbf{I} \end{bmatrix} \Longrightarrow \ell = \mathbf{a}/\alpha \text{ and } \mathbf{C} = \mathbf{B} - \ell\alpha\ell^H .$$

Here, the matrix $\mathbf{C}$ is the same as in the Cholesky factorization. On the other hand,

$$\|\ell\|_2^2 = \frac{\max_{\|\mathbf{x}\|_2=1} \left|\mathbf{x}^H\mathbf{a}\right|_2}{\alpha^2} = \frac{1}{\alpha} \max_{\|\mathbf{x}\|_2=1} \mathbf{x}^H\mathbf{a}\frac{1}{\alpha}\mathbf{a}^H\mathbf{x} \le \frac{1}{\alpha} \max_{\|\mathbf{x}\|_2=1} \mathbf{x}^H\mathbf{B}\mathbf{x} \le \frac{1}{\alpha}\varrho(\mathbf{A}) .$$

Eventually, we will see that α is bounded below by the smallest eigenvalue of $\mathbf{A}$ and $\varrho(\mathbf{A})$ is the largest eigenvalue of $\mathbf{A}$. In other words, we can bound $\|\ell\|_2$ by the square root of the ratio of the largest to smallest eigenvalues of $\mathbf{A}$.

3.13.3.7 Software

MATLAB programmers can use the command chol to perform a Cholesky factorization. Also, the MATLAB command ldl will compute the $\mathbf{L}\mathbf{D}\mathbf{L}^\top$ factorization of a symmetric matrix $\mathbf{A}$. For a symmetric positive matrix, this MATLAB command will perform unnecessary comparisons for pivoting.

The LAPACK Cholesky factorization of a symmetric positive matrix is computed by routines `_potrf`; see, for example, dpotrf.f. Once the Cholesky factorization of a symmetric positive matrix has been computed, linear systems can be solved by routines `_potrs`; see, for example, dpotrs.f. Readers may also examine the following interactive JavaScript **Cholesky factorization.**

In order to provide a user-friendly interface in C++ to the LAPACK Fortran routines, we have developed both a `SymmetricPositiveMatrix` class and a `CholeskyFactorization` class. The former are implemented in the files SymmetricMatrix.H and SymmetricMatrix.C. The latter class is implemented in the files CholeskyFactorization.H and CholeskyFactorization.C. Note that for symmetric positive tridiagonal matrices, this class calls LAPACK routines `_pttrf` to perform an $\mathbf{L}\mathbf{D}\mathbf{L}^\top$ factorization, and for symmetric positive banded matrices, this class calls LAPACK routines `_pbtrf` to perform a Cholesky factorization.

Exercise 3.13.2 ([99, p. 328]) Consider the matrix

$$\mathbf{A} = \begin{bmatrix} 1 & \beta \\ \beta & 9 \end{bmatrix} .$$

1. For which numbers β is $\mathbf{A}$ positive?
2. Find the Cholesky factorization of $\mathbf{A}$ when β is in the range for positivity.

3. Find the minimum value of $\phi(\mathbf{x}) = \frac{1}{2}\mathbf{x}^\top\mathbf{A}\mathbf{x} - \mathbf{e}_2{}^\top\mathbf{x}$ when β is in the range for positivity.
4. What is the minimum value of ϕ if $\beta = 3$?

Exercise 3.13.3 Show that the matrix **A** is Hermitian and positive, then $\mathbf{A}^{-1}$ is also Hermitian and positive.

Exercise 3.13.4 ([97, p. 143]) Show that for a symmetric positive matrix, the entry of maximum absolute value always lies on the diagonal. (Hint: if the i,j entry is maximal, examine $\mathbf{x}^\top\mathbf{A}\mathbf{x}$ where $\mathbf{x} = \mathbf{e}_i \pm \mathbf{e}_j$.)

Exercise 3.13.5 ([99, p. 328]) Let the vector **x** have entries ξ_j. Suppose that

$$\phi_1(\mathbf{x}) = \xi_1^2 + \xi_2^2 + \xi_3^2 - 2\xi_1\xi_2 - 2\xi_1\xi_3 + 2\xi_2\xi_3$$

and

$$\phi_2(\mathbf{x}) = \xi_1^2 + 2\xi_2^2 + 11\xi_3^2 - 2\xi_1\xi_2 - 2\xi_1\xi_3 - 4\xi_2\xi_3 \; .$$

1. Write each of these quadratic forms in terms of matrices and vectors.
2. Show that the matrix $\mathbf{A}_1$ in the quadratic form ϕ_1 is not positive, and find the set of all **x** where ϕ_1 reaches its minimum.
3. Use the Cholesky factorization of the matrix $\mathbf{A}_2$ in the quadratic form ϕ_2 to determine whether $\phi_2(\mathbf{x})$ has a minimum.

Exercise 3.13.6 The $n \times n$ Hilbert matrix **H** has entries

$$\eta_{ij} = \int_0^1 x^{i+j-2} \, dx$$

for $1 \le i,j \le n$. Use this integral formula to show that **H** is symmetric positive.

Exercise 3.13.7 Develop a matrix representation of the block form of the Cholesky factorization. This should be consistent with the code in LAPACK routine dpotrf.f.

3.13.4 Diagonally Dominant

Here is another important class of matrices that do not require pivoting in their factorization.

Definition 3.13.2 An $n \times n$ matrix **A** with entries α_{ij} is **diagonally dominant** if and only if either for all $1 \le i \le n$ we have $|\alpha_{ii}| \ge \sum_{j \ne i} |\alpha_{ij}|$, or for all $1 \le j \le n$ we have $|\alpha_{jj}| \ge \sum_{i \ne j} |\alpha_{ij}|$. If strict inequality occurs for all row indices (or all column indices), then the matrix is said to be **strictly diagonally dominant**.

Diagonally dominant linear systems often arise in the numerical solution of ordinary and partial differential equations; see, for example, Sect. 4.4 of Chap. 4 in Volume III.

The following lemma shows that Gaussian factorization preserves diagonal dominance, and controls the growth of array entries during Gaussian factorization.

Lemma 3.13.3 *Suppose that* **A** *is diagonally dominant with entries* α_{ij}*, and that* **A** *has been partitioned and factored as follows:*

$$\mathbf{A} = \begin{bmatrix} \alpha & \mathbf{a}^H \\ \mathbf{b} & \mathbf{B} \end{bmatrix} = \begin{bmatrix} 1 & \\ \boldsymbol{\ell} & \mathbf{I} \end{bmatrix} \begin{bmatrix} \alpha & \mathbf{a}^H \\ & \mathbf{C} \end{bmatrix} .$$

Then **C** *is diagonally dominant, and the entries* γ_{ij} *of* **C** *satisfy*

$$\max_{1 \le i,j < n} \left|\gamma_{ij}\right| \le 2 \max_{1 \le i,j \le n} \left|\alpha_{ij}\right| .$$

Proof First, we suppose that **A** is row diagonally dominant. Then

$$|\alpha_{11}| \ge \sum_{j=2}^{n} \left|\alpha_{1j}\right| \Longrightarrow 1 \ge \sum_{j=2}^{n} \left|\frac{\alpha_{1j}}{\alpha_{11}}\right| .$$

It follows that for all $2 \le i \le n$,

$$\begin{aligned}
\sum_{\substack{j=2 \\ j \ne i}}^{n} \left|\gamma_{i-1,j-1}\right| &= \sum_{\substack{j=2 \\ j \ne i}}^{n} \left|\alpha_{ij} - \frac{\alpha_{i1}}{\alpha_{11}}\alpha_{1j}\right| \le \sum_{\substack{j=2 \\ j \ne i}}^{n} \left\{ \left|\alpha_{ij}\right| + \left|\frac{\alpha_{i1}}{\alpha_{11}}\alpha_{1j}\right| \right\} \\
&= \left[\sum_{\substack{j=1 \\ j \ne i}}^{n} \left|\alpha_{ij}\right| - |\alpha_{i1}| \right] + \left[\sum_{\substack{j=1 \\ j \ne i}}^{n} \left|\frac{\alpha_{i1}}{\alpha_{11}}\alpha_{1j}\right| - \left|\frac{\alpha_{i1}}{\alpha_{11}}\alpha_{1i}\right| \right] \\
&\le [|\alpha_{ii}| - |\alpha_{i1}|] + \left[|\alpha_{i1}| - \left|\frac{\alpha_{i1}}{\alpha_{11}}\alpha_{1i}\right| \right] = |\alpha_{ii}| - \left|\frac{\alpha_{i1}}{\alpha_{11}}\alpha_{1i}\right| \\
&\le \left|\alpha_{ii} - \frac{\alpha_{i1}}{\alpha_{11}}\alpha_{1i}\right| = |\gamma_{i-1,i-1}| .
\end{aligned}$$

This shows that **C** is also row diagonally dominant. In addition, for all $2 \le i,j \le n$,

$$\begin{aligned}
\left|\gamma_{i-1,j-1}\right| &= \left|\alpha_{ij} - \frac{\alpha_{i1}}{\alpha_{11}}\alpha_{1j}\right| \le \sum_{\substack{k=2 \\ k \ne i}}^{n} \left|\alpha_{ik} - \frac{\alpha_{i1}}{\alpha_{11}}\alpha_{1k}\right| \le \left|\alpha_{ii} - \frac{\alpha_{i1}}{\alpha_{11}}\alpha_{1i}\right| \\
&\le |\alpha_{ii}| + \left|\alpha_{i1}\frac{\alpha_{1i}}{\alpha_{11}}\right| \le |\alpha_{ii}| + |\alpha_{i1}| \le 2 \max_{1 \le i,j \le n} \left|\alpha_{ij}\right| .
\end{aligned}$$

This bounds the growth of entries during Gaussian factorization.

The proof when **A** is column diagonally dominant is similar. Note that

$$|\alpha_{11}| \geq \sum_{i=2}^{n} |\alpha_{i1}| \Longrightarrow 1 \geq \sum_{i=2}^{n} \left|\frac{\alpha_{i1}}{\alpha_{11}}\right| .$$

It follows that for all $2 \leq j \leq n$,

$$\begin{aligned}
\sum_{\substack{i=2 \\ i\neq j}}^{n} |\gamma_{i-1,j-1}| &= \sum_{\substack{i=2 \\ i\neq j}}^{n} \left|\alpha_{ij} - \frac{\alpha_{i1}}{\alpha_{11}}\alpha_{1j}\right| \leq \sum_{\substack{i=2 \\ i\neq j}}^{n} \left\{|\alpha_{ij}| + \left|\frac{\alpha_{i1}}{\alpha_{11}}\alpha_{1j}\right|\right\} \\
&= \left[\sum_{\substack{i=1 \\ i\neq j}}^{n} |\alpha_{ij}| - |\alpha_{1j}|\right] + \left[\sum_{\substack{i=1 \\ i\neq j}}^{n} \left|\frac{\alpha_{i1}}{\alpha_{11}}\alpha_{1j}\right| - \left|\frac{\alpha_{j1}}{\alpha_{11}}\alpha_{1i}\right|\right] \\
&\leq \left[|\alpha_{jj}| - |\alpha_{1j}|\right] + \left[|\alpha_{1j}| - \left|\frac{\alpha_{j1}}{\alpha_{11}}\alpha_{1j}\right|\right] = |\alpha_{jj}| - \left|\frac{\alpha_{j1}}{\alpha_{11}}\alpha_{1j}\right| \\
&\leq \left|\alpha_{jj} - \frac{\alpha_{j1}}{\alpha_{11}}\alpha_{1j}\right| = |\gamma_{j-1,j-1}| .
\end{aligned}$$

This shows that **C** is also column diagonally dominant. In addition, for all $2 \leq i,j \leq n$,

$$\begin{aligned}
|\gamma_{i-1,j-1}| &= \left|\alpha_{ij} - \frac{\alpha_{i1}}{\alpha_{11}}\alpha_{1j}\right| \leq \sum_{\substack{k=2 \\ k\neq j}}^{n} \left|\alpha_{kj} - \frac{\alpha_{k1}}{\alpha_{11}}\alpha_{1j}\right| \leq \left|\alpha_{jj} - \frac{\alpha_{j1}}{\alpha_{11}}\alpha_{1i}\right| \\
&\leq |\alpha_{jj}| + \left|\frac{\alpha_{j1}}{\alpha_{11}}\alpha_{1j}\right| \leq |\alpha_{jj}| + |\alpha_{1j}| \leq 2 \max_{1\leq i,j\leq n} |\alpha_{ij}| .
\end{aligned}$$

Because array entries can grow by at most a factor of two during Gaussian factorization of a diagonally dominant matrix, pivoting is unnecessary. However, it is still possible that the last pivot is zero. On the other hand, if strict inequality occurs all of the diagonal dominance inequalities of Definition 3.13.2, then the matrix is nonsingular. This fact follows easily from the Gerschgorin Circle Theorem 1.2.2 of Chap. 1 in Volume II.

Here is another important subclass of diagonally dominant matrices.

Definition 3.13.3 A square matrix **A** is **irreducibly diagonally dominant matrix** if and only if

- **A** is diagonally dominant,
- strict inequality holds in the diagonal dominance inequalities for at least one row (or column), and

- it is not possible to permute rows and columns of the matrix to obtain a partitioned form

$$\mathbf{Q}^\top \mathbf{A}\mathbf{P} = \begin{bmatrix} \mathbf{A}_{11} & 0 \\ \mathbf{A}_{21} & \mathbf{A}_{22} \end{bmatrix} .$$

The Gerschgorin Circle Theorem 1.2.2 of Chap. 1 in Volume II can also be used to show that an irreducibly diagonally dominant matrix is nonsingular. For a proof, see Axelsson [4, p. 133].

Neither LAPACK nor MATLAB provides any special routines for solving linear systems with diagonally dominant matrices. In fact, neither provides any way to avoid pivoting in Gaussian factorization. However, partial pivoting adds little expense to Gaussian factorization, especially when no rows are actually interchanged. Gaussian factorization routines that employ partial pivoting are generally efficient for solving diagonally dominant linear systems.

Exercise 3.13.8 ([97, p. 143]) Suppose that the matrix $\mathbf{A}$ is symmetric, strictly diagonally dominant, and has positive diagonal entries. Show that Gaussian factorization applied to $\mathbf{A}$ will find only positive pivots. The prove that the Gaussian factorization $\mathbf{A} = \mathbf{L}\mathbf{R}$ can be modified to a factorization of the form $\mathbf{A} = \mathbf{L}\mathbf{D}\mathbf{L}^\top$, and conclude that $\mathbf{A}$ is positive.

3.13.5 Tridiagonal

LAPACK provides special routines to deal with certain matrices that can be represented by significantly reduced computer storage in simple patterns. For example, symmetric matrices need only store the entries on and below (or above) the diagonal. These special algorithms can take advantage of the reduced storage to significantly decrease the computational work in solving linear systems.

Here is a definition of another important class of matrices that can be represented with reduced computer storage.

Definition 3.13.4 A matrix $\mathbf{T}$ is tridiagonal if and only if its entries τ_{ij} are zero for $|i-j| > 1$.

Tridiagonal matrices commonly arise in certain methods for the numerical solution of ordinary differential equations. See, for example, Sect. 4.4 of Chap. 4 in Volume III.

For a strictly diagonally dominant or irreducibly diagonally dominant tridiagonal matrix $\mathbf{T}$ with entries τ_{ij}, Gaussian factorization leads to a very simple algorithm:

for $1 \leq k \leq n$

$$\tau_{k+1,k} = \tau_{k+1,k} / \tau_{kk}$$

$$\tau_{k+1,k+1} = \tau_{k+1,k+1} - \tau_{k+1,k} * \tau_{k,k+1} .$$

However, this algorithm assumes that $\mathbf{T}$ is stored as a full matrix.

Instead, let us store the entries of $\mathbf{T}$ in three vectors $\boldsymbol{\ell}$, $\mathbf{d}$ and $\mathbf{r}$ with entries

$$\lambda_j = \tau_{j+1,j} \;,\; 1 \leq j < n$$
$$\delta_j = \tau_{j,j} \;,\; 1 \leq j \leq n \text{ and}$$
$$\varrho_i = \tau_{i,i+1} \;,\; 1 \leq i < n \;.$$

Then Gaussian factorization without pivoting can be implemented as

Algorithm 3.13.5 (Tridiagonal Factorization)

$$\begin{aligned}
&\text{for } 1 \leq k < n \\
&\quad \lambda_k = \lambda_k/\delta_k \\
&\quad \delta_{k+1} = \delta_{k+1} - \lambda_k \varrho_k \;.
\end{aligned}$$

This algorithm gives us a unit left triangular matrix $\mathbf{L}$ and a right triangular matrix $\mathbf{R}$, both with only two diagonals of nonzero entries. Afterward, the linear system $\mathbf{Tx} = \mathbf{b}$ can be solved in the stages $\mathbf{Ly} = \mathbf{b}$ and $\mathbf{Rx} = \mathbf{y}$ by the following:

$$\begin{aligned}
&\eta_1 = \beta_1 \\
&\text{for } 2 \leq k \leq n \;,\; \eta_k = \beta_k - \lambda_{k-1}\eta_{k-1} \\
&\xi_n = \eta_n/\delta_n \\
&\text{for } n > k \geq 1 \;,\; \xi_k = (\eta_k - \varrho_k \xi_{k+1})/\delta_k \;.
\end{aligned}$$

Typically, the vector $\mathbf{y}$ overwrites $\mathbf{b}$ in computer storage, and then the vector $\mathbf{x}$ overwrites $\mathbf{y}$, giving us the

Algorithm 3.13.6 (Tridiagonal Solution)

$$\begin{aligned}
&\text{for} 2 \leq k \leq n \;,\; \beta_k = \beta_k - \lambda_{k-1}\beta_{k-1} \\
&\beta_n = \beta_n/\delta_n \\
&\text{for} n > k \geq 1 \;,\; \beta_k = (\beta_k - \varrho_k \beta_{k+1})/\delta_k \;.
\end{aligned}$$

It is easy to see that the factorization Algorithm 3.13.5 requires $n - 1$ multiplications and additions, and $n - 1$ divides. However, each subsequent solution of a linear system by Algorithm 3.13.6 requires $2(n - 1)$ multiplications and additions, and n divides. Thus roughly twice as much work is involved in forward- and back-solution of a tridiagonal linear system, as in its matrix factorization.

LAPACK provides routines `_gttrf` and `_gttrs` to factor and solve tridiagonal linear systems with partial pivoting. These routines require an additional vector of length $n - 2$ to hold the nonzero entries of $\mathbf{R}$ that can arise from row interchanges. They also require a vector of integers to represent the row interchanges. Interested readers may view, for example, dgttrf.f and dgttrs.f.

LAPACK provides routines `_pttrf` and `_pttrs` to factor and solve linear systems involving symmetric positive tridiagonal matrices. These routines perform $\mathbf{LDL}^\top$ factorizations of symmetric positive tridiagonal matrices, so that the work is not dominated by square root computations. Interested readers may view, for example, dpttrf.f and dpttrs.f.

We have provided a user-friendly interface to the LAPACK subroutines through a `TridiagonalMatrix` class. Because of the form of the reduced storage, `TridiagonalMatrix` is not derived from any `Matrix` class. we have also provided classes `SymmetricTridiagonalMatrix` and `SymmetricPositiveTridiagonalMatrix`. These have been implemented in BandMatrix.H and BandMatrix.C.

Exercise 3.13.9 Develop the algorithm for Gaussian factorization with partial pivoting of a tridiagonal matrix, and the algorithm for subsequent solution of a linear system.

Exercise 3.13.10 Develop the $\mathbf{LDL}^\top$ factorization for a symmetric positive tridiagonal matrix, and the algorithm for subsequent solution of a linear system.

3.13.6 Band

Here is a definition of another important class of matrices that allow reduced computer storage.

Definition 3.13.5 An $n \times n$ matrix $\mathbf{B}$ is a **band matrix** with k **subdiagonals** and m **super-diagonals** if and only if its entries β_{ij} satisfy

$$\beta_{ij} \neq 0 \text{ implies } -k \leq j - i \leq m \, .$$

Band matrices often arise in the numerical solution of ordinary and partial differential equations.

The definition of a band matrix suggests a reduced storage scheme to represent the matrix. In Fortran, we can declare

```
real A(n,-k:m)
```

and identify `B(i,j) = A(i,j-i)`. Since the array `A` stores $n(m + k + 1)$ data entries and the original matrix `A` stores n^2 data entries, this band storage is more efficient whenever the total number of bands, namely $m + k + 1$, is less than n.

LAPACK provides routines `_gbtrf` and `_pbtrf` to compute factorizations of general band matrices, and symmetric positive band matrices, respectively. The routines `_pbtrf` perform Cholesky factorizations. LAPACK also provides routines `_gbtrs` and `pbtrs` to solve linear systems involving factored band matrices, and factored symmetric positive band matrices. Interested readers may view dgbtrf.f, dgbtrs.f, dpbtrf.f and dpbtrs.f. for more details.

In C++, we have provided three classes named `BandMatrix`, `SymmetricBandMatrix` and `SymmetricPositiveBandMatrix`. These classes provide wrappers around calls to LAPACK Fortran routines. For more details, see BandMatrix.H and BandMatrix.C.

Exercise 3.13.11 Suppose that we want to solve

$$-\frac{\mathrm{d}^2 u}{\mathrm{d}x^2} = 8 \; , \; u(0) = 0 = u(1) \; .$$

We can approximate this ordinary differential equation by the system of difference equations

$$-\left[\frac{u_{i+1} - u_i}{\triangle x} - \frac{u_i - u_{i-1}}{\triangle x}\right] = 8 \; \triangle x \; , \; 0 < i < n$$

$$u_0 = 0 = u_n \; .$$

Here $\triangle x = 1/n$.

1. Rewrite this system of difference equations in the form $\mathbf{A}u = \mathbf{b}$, where $\mathbf{A}$ is symmetric and tridiagonal.
2. Choose the best class in `BandMatrix.H` for solving this problem, then write and run a computer program using this C++ class to solve this problem.
3. Plot the numerical solution and analytical solution as a function of $\mathbf{x}$ or $i\triangle x$ for $n = 9, 17, 33$. Also plot minus the natural logarithm of the error in the solution at the midpoint versus the natural logarithm of n.

Chapter 4
Scientific Visualization

pixel, n.: A mischievous, magical spirit associated with screen displays. The computer industry has frequently borrowed from mythology: Witness the sprites in computer graphics, the demons in artificial intelligence, and the trolls in the marketing department.

Jeff Meyer

There are lies, damned lies and color graphics.

Phillip Colella

Abstract This chapter surveys some basic concepts and methods for scientific visualization. After describing the basic notions of pixels and color, the coordinate systems for 2D and 3D graphics are examined, including orthographic and perspective transformations. Interactive 3D rotations by quaternions and trackballs are discussed. Lighting models are introduced. Techniques for viewing functions of one, two and three variables are presented, along with related graphics software. Object-oriented programming of graphics, and graphical user interfaces are also presented.

4.1 Overview

Most of us understand pictures far more quickly than words or equations. In scientific computing, graphical displays of numerical results are much more effective than tables of numbers in demonstrating the effectiveness of calculations. These facts indicate that computer graphics are an important part of scientific computing.

In this book, we would have preferred to treat graphics as a preliminary topic, along with the other topics in Chap. 2. The difficulties with such a plan are that multidimensional computer graphics use substantial linear algebra, and numerical linear algebra is hard to visualize.

Additional Material: The details of the computer programs referred in the text are available in the Springer website (http://extras.springer.com/2018/978-3-319-69105-3) for authorized users.

© Springer International Publishing AG, part of Springer Nature 2017
J.A. Trangenstein, *Scientific Computing*, Texts in Computational Science and Engineering 18, https://doi.org/10.1007/978-3-319-69105-3_4

Our goals in this chapter are to describe some fundamental aspects of computer graphics, and to discuss the use of computer graphics to visualize functions of one to three variables. We will demonstrate some general techniques for plotting functions, and suggest a variety of software packages for realizing these plots. In some cases, we will show how to make these plots *interactive*, so that the user can select points of interest, or rotate the graph for better understanding. Readers should be expected to produce useful graphs for scientific computing assignments in subsequent chapters.

For general ideas regarding computer graphics, readers may read the books by Foley et al. [40], Hearn et al. [53], and Shirley et al. [93]. Readers should also consult the online Open GL Programming Guide.

For graphical display of scientific computations, many readers will be familiar with the MATLAB 1D plotting commands plot, axis, title and loglog, the 2D plotting commands pcolor, contour, mesh and surf and the 3D plotting commands plot3, slice, isosurface, surfl and view. As an alternative, Scilab is a free and open source alternative to MATLAB that contains a number of routines for graphics. Linux programmers may consider using gnuplot or grace. `Gnuplot` users should examine the Demo scripts for gnuplot version 4.2. Those who wish to insert graphics into their scientific computing programs (rather than write files to be read by separate graphics programs) should consider (GTK), GIMP Toolkit OpenGL, MathGL and the and the Visualization Toolkit (VTK).

4.2 Pixels and Color

Computer graphics are displayed on a screen via tiny dots called **pixels**. Pixels are arranged in a rectangular array, with the number of pixels in each direction depending on the display screen and the computer hardware. For example, the screen on my Dell laptop is 1920 by 1080 pixels, with a resolution of 96 dots per inch. The pixel dimensions in this laptop are the same as in 1080p high-definition television displays. Standard definition televisions have screens that are only 640 by 480 pixels.

Linux users can find the number of pixels on their screen by typing

```
xdpyinfo | grep dimensions
```

There are dimensions for each screen, but there is probably only one screen. In the windows operation system, users may be able to right click with the mouse, pull down to screen resolution and read the current values. On a Mac, click on the apple icon on the top right, pull down to System Preferences, and click on Display.

The fact that computer graphics are displayed via an array of pixels leads to interesting issues in drawing lines, filling polygons, clipping, and so on. For example, the precise mathematical description of a line will almost surely involve infinitesimal points that fall between the pixels. Consequently, in order to display a line on a computer screen we will have to choose which nearby pixels to excite. This problem might become noticeable on low resolution screens, or when graphs are enlarged without being redrawn. For more details on this **scan conversion** problem, see Fig. 4.1, or consult Foley et al. [40, p. 72ff].

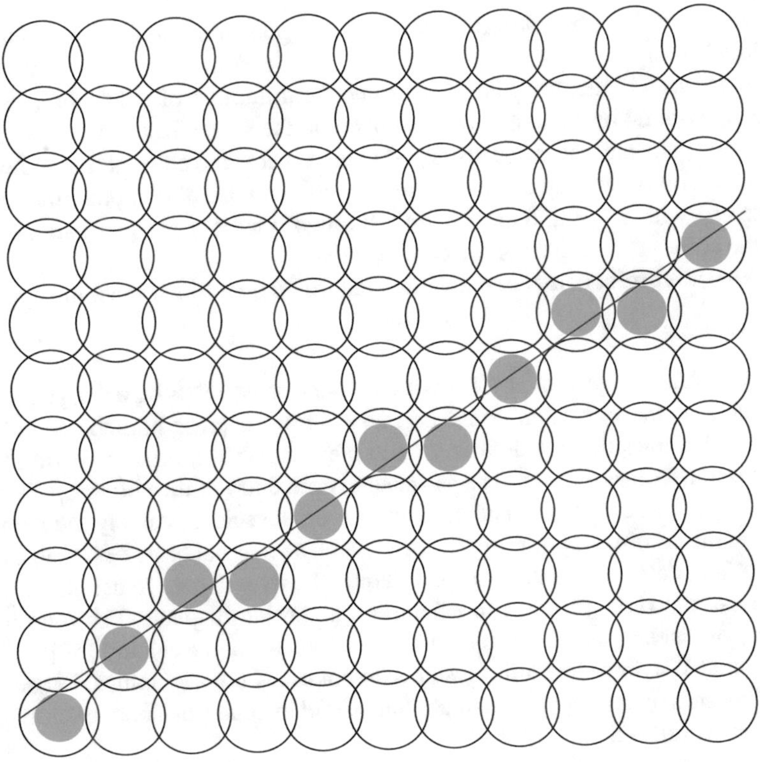

Fig. 4.1 Scan conversion

Colors can be represented by a triple (r, g, b), where r stands for *red*, g for *green*, and b for *blue*. There are several ways to represent the colors in computer memory, depending on how much memory is available to store the colors. These color representations are called **visuals**. In the X Window system, there are six possible visuals listed in file `X.h`:

StaticGray: statically allocated black and white display,
GrayScale: dynamically variable black and white display,
StaticColor: statically allocated colormap,
PseudoColor: dynamically variable colormap,
TrueColor: statically allocated colors, and
DirectColor: dynamically variable colors.

This file is normally located in the Linux directory `/usr/include/X11`. Depending on hardware and software, a programmer may have several of these visual options available. To find out which visuals are available under the Linux operating system, type `xdpyinfo` and look for the `visual` information under each screen. On my laptop, there are a number of TrueColor and DirectColor visuals. The TrueColor visuals generally use 8 bits in the color specification, meaning that there are $2^8 = 256$ possible values for each component in the (r, g, b) triple for an individual color.

In **colormap** mode, a visual stores some fixed number of colors in a colormap (typically with 64 or 256 entries). Each pixel on the screen has an associated index in the colormap, and the color displayed at the pixel is the color at that index in the colormap. Thus colormap display make very efficient use of computer memory; for a colormap with $2^8 = 256$ entries, only 8 bits of memory per pixel are needed to store the color information for the entire screen.

To see the current colors in the default colormap on a Linux machine, type

```
xcmap
```

In some operating systems the correct command may be `xshowcmap`. Note that each entry in the colormap could be chosen from a large number of colors. If `xdpyinfo` shows that the default `visual` has `depth` n, then there are likely to be 2^n possible colors that could be assigned to each entry of the colormap. However, only a few of these colors can be used at once on the screen, namely the number of entries in the colormap. Some applications, such as internet browsers, grab as much of the default colormap as they can. Consequently, we will tend to use `TrueColor` and `DirectColor` displays in scientific computing displays. For more details about color display, see Brennan et al. [15] or Johnson and Reichard [59].

Note that Windows operating systems do not use X color graphics. Readers who use a Windows operating system might find useful information about colormaps and visuals in Kruglinski et al. [70].

We can think of the (r, g, b) color representation in terms of a Cartesian coordinate system. If the maximum values of r, g and b are normalized to lie between 0 and 1, then the corners of the unit cube correspond to certain colors as described in Table 4.1 or Fig. 4.2.

Sometimes color is described via (h, s, v), or **hue**, **saturation** and **value**. This color representation uses a **hexacone** coordinate system [40, p. 590ff]. The following algorithm to convert an (rgb) triple to (hsv) can also be found in Foley et al. [40, p. 592].

Table 4.1 Red-green-blue color components

Black	$(0, 0, 0)$
White	$(1, 1, 1)$
Red	$(1, 0, 0)$
Green	$(0, 1, 0)$
Blue	$(0, 0, 1)$
Yellow	$(1, 1, 0)$
Cyan	$(0, 1, 1)$
Magenta	$(1, 0, 1)$

Fig. 4.2 RGB color cube

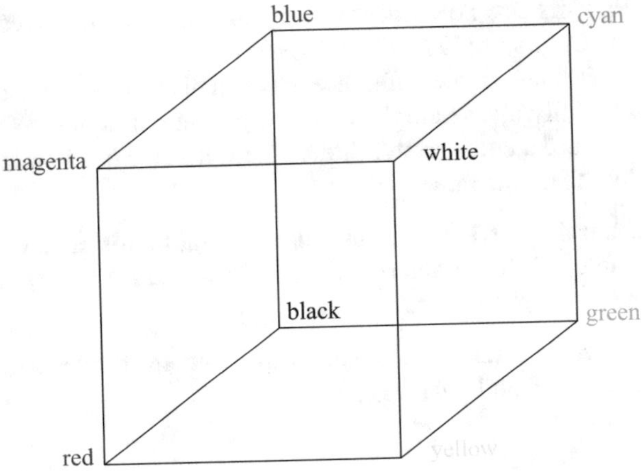

Algorithm 4.2.1 (Compute HSV from RGB)

value $= \max\{r, g, b\}$

$\mu = \min\{r, g, b\}$

$\delta =$ value $- \mu$

if value > 0

  saturation $=$ (value $- \mu$)/value

else saturation $= 0$

if saturation $= 0$

  hue $=$ undefined

else

  if $r =$ value

    hue $= (g - b)/\delta$

  else if $g =$ value

    hue $= 2 + (b - r)/\delta$

  else

    hue $= 4 + (r - g)/\delta$

hue $=$ hue $\times 60$

if hue < 0

  hue $=$ hue $+ 360°$

Readers can experiment with colors via (r, g, b) values or (h, s, v) values online at the Colorspire RGB Color Wheel.

Because of the difference between display mechanisms, it is best to choose colors carefully. For example, on color printers, it is best to use colors with $value = 1$, because a color printer dithers the full-value color with white to obtain colors with less than full value.

Exercise 4.2.1 Determine the hue, saturation and value for the eight colors in Table 4.1. Then determine the (h, s, v) values for arbitrary points along the edges of the (r, g, b) cube.

Exercise 4.2.2 Write an algorithm to convert hue, saturation and value (h, s, v) to red, green and blue (r, g, b).

4.3 Coordinate Systems

In this section, we will describe various coordinate systems used in graphical display. These are fairly simple to describe in two dimensions, and a bit more difficult in three dimensions.

4.3.1 2D

Displaying two-dimensional graphics on a computer screen is relatively straightforward. Given some user-specified viewing region $(\mathbf{b}_0, \mathbf{t}_0) \times (\mathbf{b}_1, \mathbf{t}_1)$, the linear transformation

$$\mathbf{s} = \begin{bmatrix} w & \\ & h \end{bmatrix} \begin{bmatrix} \mathbf{t}_0 - \mathbf{b}_0 & \\ & \mathbf{b}_1 - \mathbf{t}_1 \end{bmatrix}^{-1} \left(\mathbf{u} - \begin{bmatrix} \mathbf{b}_0 \\ \mathbf{t}_1 \end{bmatrix} \right)$$

that maps a user coordinate vector $\mathbf{u}$ to a screen coordinate vector $\mathbf{s}$ in the rectangle of screen coordinates $(0, w) \times (0, h)$. Then the process of drawing a line between points in the user's viewing region can be easily transformed into drawing a line between points on the screen (Fig. 4.3).

4.3.2 3D

Three-dimensional graphics are more difficult, because it is necessary to project points in the user's three-dimensional coordinate system onto a two-dimensional computer screen. Moreover, for viewing purposes the user often wants to rotate or translate the image interactively before the screen projection is performed.

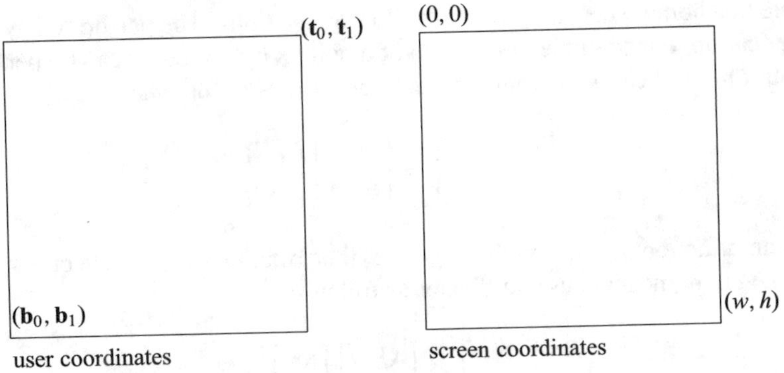

Fig. 4.3 User and screen coordinates

Three-dimensional coordinates in the user's view of the world are called **user coordinates**, typically specified as a box $(\mathbf{b}_0, \mathbf{t}_0) \times (\mathbf{b}_1, \mathbf{t}_1) \times (\mathbf{b}_2, \mathbf{t}_2)$. For interactive rotation of the displayed image, it is useful to transform these coordinates via a translation and scaling to a cube centered around the origin. These are called **object coordinates**, and we will call the mapping from user coordinates to object coordinates the **window transformation**.

Object coordinates are manipulated by a **model view matrix** to produce **eye coordinates**, which typically lie in the cube $(-1, 1) \times (-1, 1) \times (-1, 1)$. Typically, the model view matrix involves an initial orientation, possibly to arrange for the third axis to point toward the top of the screen rather than toward the viewer, followed by an interactive rotation, relative to some camera view.

Eye coordinates are modified by a **projection matrix** to produce **clip coordinates**, within the camera's optical field. Finally, a **viewport transformation** maps clip coordinates to the computer **screen coordinates**. Readers may read more about these coordinate systems by viewing the OpenGL Transformation webpage.

4.3.3 *Homogeneous Coordinates*

In three-dimensional computer graphics software, object coordinates, eye coordinates and clip coordinates are typically stored as four-vectors, called **homogeneous coordinates**. For any nonzero scalar w and three-vector $\mathbf{x}$, the four-vector of homogeneous coordinates $(\mathbf{x}, w)$ corresponds to spatial coordinates $\mathbf{x}/w$. Transforming from four-dimensional homogeneous coordinates to three-dimensional spatial coordinates in this way is called **homogenizing**.

Note that homogeneous coordinates allow translations to be performed by matrix multiplication. For example, translation of a point $\mathbf{x}$ by a vector $\mathbf{t}$ can be performed by matrix multiplication in homogeneous coordinates as follows:

$$\begin{bmatrix} \mathbf{x}w + \mathbf{t}w \\ w \end{bmatrix} = \begin{bmatrix} \mathbf{I} & \mathbf{t} \\ \mathbf{0}^\top & 1 \end{bmatrix} \begin{bmatrix} \mathbf{x}w \\ w \end{bmatrix} .$$

If $\mathbf{Q}$ is an orthogonal matrix with $\det \mathbf{Q} = 1$, then rotation by $\mathbf{Q}$ of a vector $\mathbf{x}$ can be performed in homogeneous coordinates as follows:

$$\begin{bmatrix} \mathbf{Q}\mathbf{x}w \\ w \end{bmatrix} = \begin{bmatrix} \mathbf{Q} & \mathbf{0} \\ \mathbf{0}^\top & 1 \end{bmatrix} \begin{bmatrix} \mathbf{x}w \\ w \end{bmatrix} .$$

If $\mathbf{S}$ is a diagonal matrix with positive diagonal entries, then inhomogeneous scaling by $\mathbf{S}$ of a vector $\mathbf{x}$ can be performed in homogeneous coordinates as follows:

$$\begin{bmatrix} \mathbf{S}\mathbf{x}w \\ w \end{bmatrix} = \begin{bmatrix} \mathbf{S} & \mathbf{0} \\ \mathbf{0}^\top & 1 \end{bmatrix} \begin{bmatrix} \mathbf{x}w \\ w \end{bmatrix} .$$

If s is a nonzero scalar, then homogeneous scaling of the vector $\mathbf{x}$ by the scalar s can be performed in homogeneous coordinates as follows:

$$\begin{bmatrix} \mathbf{x}w \\ w/s \end{bmatrix} = \begin{bmatrix} \mathbf{I} & \mathbf{0} \\ \mathbf{0}^\top & 1/s \end{bmatrix} \begin{bmatrix} \mathbf{x}w \\ w \end{bmatrix} .$$

4.3.4 Window Transformation

If user coordinates $\mathbf{u}$ are contained in a box with lowest coordinates $\boldsymbol{\ell}$ and highest coordinates $\mathbf{h}$, then the user coordinates are related to the object coordinates $\boldsymbol{\omega}$ by translation and scaling. We can represent this transformation via matrix multiplication and vector addition on three-vectors:

$$\boldsymbol{\omega} = \mathbf{S}(\mathbf{u} - \mathbf{c}) .$$

In other words, we translate by the center $\mathbf{c} = (\mathbf{h} + \boldsymbol{\ell})/2$ of the user box, and then scale by means of the matrix

$$\mathbf{S} = \begin{bmatrix} 2/(\mathbf{h}_0 - \boldsymbol{\ell}_0) & 0 & 0 \\ 0 & 2/(\mathbf{h}_1 - \boldsymbol{\ell}_1) & 0 \\ 0 & 0 & 2/(\mathbf{h}_2 - \boldsymbol{\ell}_2) \end{bmatrix} .$$

In homogeneous coordinates, these same operations can be written

$$\begin{bmatrix} \omega \\ 1 \end{bmatrix} = \begin{bmatrix} \mathbf{S} & \mathbf{St} \\ \mathbf{0}^\top & 1 \end{bmatrix} \begin{bmatrix} \mathbf{u} \\ 1 \end{bmatrix} \equiv \mathbf{W} \begin{bmatrix} \mathbf{u} \\ 1 \end{bmatrix} .$$

In other words, the window transformation matrix is

$$\mathbf{W} = \begin{bmatrix} 2/(\mathbf{h}_0 - \boldsymbol{\ell}_0) & 0 & 0 & -(\mathbf{h}_0 + \boldsymbol{\ell}_0)/(\mathbf{h}_0 - \boldsymbol{\ell}_0) \\ 0 & 2/(\mathbf{h}_1 - \boldsymbol{\ell}_1) & 0 & -(\mathbf{h}_1 + \boldsymbol{\ell}_1)/(\mathbf{h}_1 - \boldsymbol{\ell}_1) \\ 0 & 0 & 2/(\mathbf{h}_2 - \boldsymbol{\ell}_2) & -(\mathbf{h}_2 + \boldsymbol{\ell}_2)/(\mathbf{h}_2 - \boldsymbol{\ell}_2) \\ 0 & 0 & 0 & 1 \end{bmatrix} .$$

Readers may view an implementation of a window transformation either by viewing line 754 of VolGraphTool.js, or line 63 of VolGraphTool.H.

4.3.5 *Model View Transformation*

Normally, software for viewing three-dimensional images orients the x-axis toward the right of the screen, the y-axis toward the top of the screen, and the z-axis toward the viewer. This is a right-handed system of orthogonal axes, and is convenient for the initial view of surfaces $\{(x, y, f(x, y))\}$.

However, for viewing points (x, y, z) in three-dimensional space, it is convenient to orient the z-axis toward the top of the screen. In such cases, the y-axis is often pointed to the right, and the x-axis is pointed toward the viewer. In homogeneous coordinates, this corresponds to the **model view matrix**

$$\mathbf{M}_y = \begin{bmatrix} 0 & 1 & 0 & 0 \\ 0 & 0 & 1 & 0 \\ 1 & 0 & 0 & 0 \\ 0 & 0 & 0 & 1 \end{bmatrix} . \tag{4.1}$$

Alternatively, we could point the z-axis toward the top of the screen, the x-axis to the right, and the y axis away from the viewer:

$$\mathbf{M}_x = \begin{bmatrix} 1 & 0 & 0 & 0 \\ 0 & 0 & 1 & 0 \\ 0 & -1 & 0 & 0 \\ 0 & 0 & 0 & 1 \end{bmatrix} . \tag{4.2}$$

The product of the window transformation and the model view transformation can be represented by any composition of translation, rotation or scaling. When these transformations are performed by matrices acting on homogeneous coordinates, the product of matrices will always have the partitioned form

$$\mathbf{MW} = \begin{bmatrix} \mathbf{A}_{11} & \mathbf{a}_{12} \\ \mathbf{0}^\top & \alpha_{22} \end{bmatrix}$$

where $\det \mathbf{A}_{11} > 0$ and $\alpha_{22} > 0$.

Readers may view an implementation of a model view transformation either by viewing line 236 of VolGraphTool.js, or lines 71, 193, 196, 917 and 930 of GTKGLWindow.C.

4.3.6 *Projection Transformations*

There are two possible ways to transform from eye coordinates to clip coordinates. The first is commonly used in scientific computing, and the second is commonly used in drawing scenes, such as in computer games.

4.3.6.1 Orthographic Transformation

The simplest projection transformation is called an **orthographic projection**. It maps a given eye coordinate box $(\ell, r) \times (b, t) \times (-n, -f)$ to the clip coordinate unit cube. Here the scalars have the mnemonic values ℓ = "left", r = "right", b = "bottom", t = "top", n = "near" and f = "far". These correspond to a right-handed coordinate system in which the z-axis is oriented toward the viewer.

Orthographic projection can be accomplished in homogeneous coordinates by the **projection matrix**

$$\mathbf{P} = \begin{bmatrix} 2/(r-\ell) & 0 & 0 & -(r+\ell)/(r-\ell) \\ 0 & 2/(t-b) & 0 & -(t+b)/(t-b) \\ 0 & 0 & 2/(n-f) & (n+f)/(n-f) \\ 0 & 0 & 0 & 1 \end{bmatrix} .$$

Here, we assume that $\ell < r$, $b < t$ and $f < n$, so that $\det P > 0$.

Suppose that some window transformation has been used to transform user coordinates to object coordinates in the cube $(-1,1)\times(-1,1)\times(-1,1)$. Subsequent rotation of the unit cube could transform the diagonal $[1,1,1]$ to the vector $[\sqrt{3},0,0]$. To make sure that points in the original object coordinate unit cube are not clipped, we should take the eye coordinate box to be $(-\sqrt{3},\sqrt{3})\times(-\sqrt{3},\sqrt{3})\times(-\sqrt{3},\sqrt{3})$.

Readers may view an implementation of an orthographic projection transformation either by viewing line 172 of VolGraphTool.js, or lines 71, 193, 196, 917 and 930 of GTKGLWindow.C. For JavaScript programs, the transformation to screen coordinates is handled in the `.html` file; see, for example, lines 33 and 34 of testVolGraphTool3.html.

4.3.6.2 Perspective Transformation

Human eyes perceive far objects to be smaller than near objects, and objects arbitrarily far away are perceived to be vanishingly small. This is because the field of vision takes in larger cross-sections at farther distances. **Perspective transformations** are designed to map the eye coordinate box $(\ell, r)\times(b,t)\times(-n,-f)$ to a **frustum**. Perspective transformation can be accomplished in homogeneous coordinates by the **projection matrix**

$$\mathbf{P} = \begin{bmatrix} 2n/(r-\ell) & 0 & (r+\ell)/(r-\ell) & 0 \\ 0 & 2n/(t-b) & (t+b)/(t-b) & 0 \\ 0 & 0 & (n+f)/(n-f) & 2nf/(n-f) \\ 0 & 0 & -1 & 0 \end{bmatrix} .$$

It is easy to see that eye coordinate points with third component equal to $-n$ are mapped to clip coordinate points with third component equal to -1, and eye coordinate points with third component equal to $-f$ are mapped to clip coordinate points with third component equal to 1. For perspective transformations, we assume that $\ell < r$, $b < t$, $f < n$ and $nf > 0$, in order to guarantee that $\det P > 0$. For more detailed information regarding orthographic and perspective transformations, interested readers may view a web page describing the OpenGL Projection Matrix.

4.3.7 *Transforming Normals*

Drawing of three-dimensional surfaces is typically accomplished by drawing a collection of triangles. Exceptions to this rule may include spheres, cylinders, disks, and any surface described by non-uniform rational b-splines. When triangles are drawn, it is important to know the normal to the triangle, both for lighting models and to assist the visualization software in drawing surfaces differently when viewed from below than from above.

If a triangle has vertices $\mathbf{v}_0$, $\mathbf{v}_1$ and $\mathbf{v}_2$, then its normal lies in the direction of the vector $\mathbf{n} = (\mathbf{v}_1 - \mathbf{v}_1) \times (\mathbf{v}_2 - \mathbf{v}_0)$. The ordering of the three vertices determines the sign of the direction of the normal, according to the right-hand rule. A point $\mathbf{x}$ is in the plane of the triangle if and only if

$$0 = \mathbf{n}^\top(\mathbf{x} - \mathbf{v}_0) = \begin{bmatrix} \mathbf{n} \\ -\mathbf{n}^\top \mathbf{v}_0 \end{bmatrix}^\top \begin{bmatrix} \mathbf{x} \\ 1 \end{bmatrix} = \begin{bmatrix} \mathbf{n} \\ -\mathbf{n}^\top \mathbf{v}_0 \end{bmatrix}^\top \begin{bmatrix} \mathbf{x}w \\ w \end{bmatrix}$$

for any nonzero scalar w representing a fourth component for homogeneous coordinates.

If the vertices $\mathbf{v}_j$ of the triangle are specified in homogeneous object coordinates as $[\mathbf{v}_j w_j, w_j]$, then the vertices in clip coordinates are $\mathbf{PM}[\mathbf{v}_j w_j, w_j]$. A point $\mathbf{PM}[\mathbf{x}w, w]$ in clip coordinates is in the plane of the triangle mapped to clip coordinates if and only if

$$0 = \mathbf{n}^\top(\mathbf{x} - \mathbf{v}_0) = \begin{bmatrix} \mathbf{n} \\ -\mathbf{n}^\top \mathbf{v}_0 \end{bmatrix}^\top \begin{bmatrix} \mathbf{x}w \\ w \end{bmatrix} = \left\{ \mathbf{M}^{-\top} \mathbf{P}^{-\top} \begin{bmatrix} \mathbf{n} \\ -\mathbf{n}^\top \mathbf{v}_0 \end{bmatrix} \right\}^\top \left\{ \mathbf{PM} \begin{bmatrix} \mathbf{x}w \\ w \end{bmatrix} \right\} .$$

Thus the normal to the triangle in clip coordinates lies in the direction of the first three homogeneous coordinates of $\mathbf{P}^{-\top}\mathbf{M}^{-\top}[\mathbf{n}, -\mathbf{n}^\top \mathbf{v}_0]$.

Exercise 4.3.1 Suppose that a user selects a point $\mathbf{c}$ in clip coordinates by clicking on the computer screen. Recall that in clip coordinates, the third axis is oriented toward the viewer. As a result, the user has actually selected a line $\{\mathbf{c} + \mathbf{e}_2\alpha\}$ in clip coordinates. Determine the corresponding line in object coordinates.

Exercise 4.3.2 Suppose that a graphics programmer has chosen some line $\{\boldsymbol{\omega} + \mathbf{t}\alpha\}$ in object coordinates, and a graphics user selects some point $\mathbf{c}$ in clip coordinates by clicking on the computer screen. Describe how to determine the minimal distance in object coordinates between the graphics programmer's line and the line selected by the user.

Exercise 4.3.3 Suppose that a graphics programmer has selected a line $\{\boldsymbol{\omega} + \mathbf{t}\alpha\}$ in object coordinates. Let a user move from some old point $\mathbf{c}_o$ to some new point $\mathbf{c}_n$ in clip coordinates, by attempting to move the mouse along the drawing of this line on the computer screen in clip coordinates. Describe how to determine the best value for the corresponding change in positions along the line in object coordinates.

4.4 Interactive Rotations

Let $\mathbf{a}$ be a unit three-vector representing an axis of rotation, and let α be some angle of rotation. We would like to represent the rotation of an arbitrary three-vector $\mathbf{v}$ about the axis $\mathbf{a}$ clockwise by the angle α, using the right-hand rule. We can decompose $\mathbf{v}$ into components parallel and perpendicular to $\mathbf{a}$ as follows:

$$\mathbf{v} = \mathbf{a}\mathbf{a} \cdot \mathbf{v} + [\mathbf{v} - \mathbf{a}\mathbf{a} \cdot \mathbf{v}] \equiv \mathbf{v}_\parallel + \mathbf{v}_\perp .$$

Then the rotation of $\mathbf{v}$ by an angle α with respect to the axis $\mathbf{a}$ produces the vector

$$\begin{aligned} \mathbf{Q}\mathbf{v} &= \mathbf{v}_\parallel + \mathbf{v}_\perp \cos\alpha + \mathbf{a} \times \mathbf{v}_\perp \sin\alpha \\ &= \mathbf{a}\mathbf{a} \cdot \mathbf{v} + [\mathbf{v} - \mathbf{a}\mathbf{a} \cdot \mathbf{v}] \cos\alpha + \mathbf{a} \times \mathbf{v} \sin\alpha \\ &= \mathbf{a}\mathbf{a} \cdot \mathbf{v}[1 - \cos\alpha] + \mathbf{v}\cos\alpha + \mathbf{a} \times \mathbf{v} \sin\alpha . \end{aligned} \tag{4.3}$$

4.4.1 Quaternions

Quaternions provide a useful way to represent such rotations. If $\mathbf{w}$ is a three-vector and ω is a scalar, the corresponding set of quaternions

$$\mathbf{q} = \begin{bmatrix} \mathbf{w} \\ \omega \end{bmatrix}$$

can be combined with addition

$$\mathbf{q} + \tilde{\mathbf{q}} \equiv \begin{bmatrix} \mathbf{w} + \tilde{\mathbf{w}} \\ \omega + \tilde{\omega} \end{bmatrix}$$

and a multiplication operator

$$\mathbf{q} \times \tilde{\mathbf{q}} \equiv \begin{bmatrix} \mathbf{w} \times \tilde{\mathbf{w}} + \mathbf{w}\tilde{\omega} + \tilde{\mathbf{w}}\omega \\ \omega\tilde{\omega} - \mathbf{w} \cdot \tilde{\mathbf{w}} \end{bmatrix}$$

to produce a **Clifford algebra**. The additive identity is $[\mathbf{0}, 0]$, and the multiplicative identity is $\mathbf{u} = [\mathbf{0}, 1]$ because for any quaternion $\mathbf{q}$ we have

$$\mathbf{u} \times \mathbf{q} = \begin{bmatrix} \mathbf{0} \times \mathbf{w} + \mathbf{0}\omega + \mathbf{w}1 \\ 1\omega - \mathbf{0} \cdot \mathbf{w} \end{bmatrix} = \begin{bmatrix} \mathbf{w} \\ \omega \end{bmatrix} = \mathbf{q} \,,$$

as well as $\mathbf{q} \times \mathbf{u} = \mathbf{q}$. The additive inverse of $\mathbf{q}$ is

$$-\mathbf{q} = \begin{bmatrix} -\mathbf{w} \\ -\omega \end{bmatrix} ,$$

and if $\mathbf{q}$ is nonzero its multiplicative inverse is

$$\mathbf{q}^{-1} \equiv \begin{bmatrix} -\mathbf{w} \\ \omega \end{bmatrix} \frac{1}{\mathbf{w} \cdot \mathbf{w} + \omega^2} .$$

Note that addition is commutative, since in general we have $\mathbf{q}+\tilde{\mathbf{q}} = \tilde{\mathbf{q}}+\mathbf{q}$. However, multiplication is not commutative because $\mathbf{q}\times\tilde{\mathbf{q}} \neq \tilde{\mathbf{q}}\times\mathbf{q}$ in general. Although vector cross products are not associative, quaternion multiplication is associative; we have left proof of this claim to the exercises.

A **unit quaternion** $\mathbf{q} = [\mathbf{w}, \omega]^\top$ satisfies $\mathbf{w} \cdot \mathbf{w} + \omega^2 = 1$. Note that if $\mathbf{q}$ is a unit quaternion, then its multiplicative inverse is

$$\mathbf{q}^{-1} = \begin{bmatrix} -\mathbf{w} \\ \omega \end{bmatrix} .$$

Thus the multiplicative inverse of a unit quaternion is easy to compute.

Our ultimate objective is to relate quaternion multiplication to rotation. Suppose that **a** is a unit 3-vector and

$$\beta = \alpha/2 \;,\; \mathbf{q} = \begin{bmatrix} \mathbf{a}\sin\beta \\ \cos\beta \end{bmatrix} \text{ and } \mathbf{p} = \begin{bmatrix} \mathbf{v} \\ 1 \end{bmatrix} .$$

Then **q** is a unit quaternion. We can compute

$$\begin{aligned}
\mathbf{q}\times\mathbf{p}\times\mathbf{q}^{-1} &= \begin{bmatrix} \mathbf{a}\sin\beta \\ \cos\beta \end{bmatrix} \times \begin{bmatrix} -\mathbf{v}\times\mathbf{a}\sin\beta + \mathbf{v}\cos\beta - \mathbf{a}\sin\beta \\ \cos\beta + \mathbf{v}\cdot\mathbf{a}\sin\beta \end{bmatrix} \\
&= \begin{bmatrix} -\mathbf{v}\mathbf{a}\cdot\mathbf{a}\sin^2\beta + \mathbf{a}\mathbf{a}\cdot\mathbf{v}\sin^2\beta + \mathbf{a}\times\mathbf{v}2\sin\beta\cos\beta + \mathbf{a}\mathbf{a}\cdot\mathbf{v}\sin^2\beta + \mathbf{v}\cos^2\beta \\ \cos^2\beta + \sin^2\beta \end{bmatrix} \\
&= \begin{bmatrix} \mathbf{a}\mathbf{a}\cdot\mathbf{v}2\sin^2\beta + \mathbf{v}(\cos^2\beta - \sin^2\beta) + \mathbf{a}\times\mathbf{v}2\sin\beta\cos\beta \\ 1 \end{bmatrix} \\
&= \begin{bmatrix} \mathbf{v}_\parallel + \mathbf{v}_\perp\cos\alpha + \mathbf{a}\times\mathbf{v}_\perp\sin\alpha \\ 1 \end{bmatrix} = \begin{bmatrix} \mathbf{Q}\mathbf{v} \\ 1 \end{bmatrix} .
\end{aligned}$$

Here **Q** is the **rotation matrix** in Eq. (4.3). Thus this triple product of quaternions produces the rotation of the vector **v** in the leading three components of the resulting quaternion.

Suppose that we have a quaternion $\mathbf{q} = [\mathbf{a}\sin\beta, \cos\beta] = [\mathbf{w}, \omega]$. Let us find a matrix representation for the corresponding rotation matrix **Q**, in terms of the components of **q**. Since

$$\begin{bmatrix} \mathbf{Q}\mathbf{v} \\ 1 \end{bmatrix} = \mathbf{q}\times\mathbf{p}\times\mathbf{q}^{-1} = \begin{bmatrix} \mathbf{w} \\ \omega \end{bmatrix} \times \begin{bmatrix} \mathbf{v} \\ 1 \end{bmatrix} \times \begin{bmatrix} -\mathbf{w} \\ \omega \end{bmatrix} = \begin{bmatrix} \mathbf{w} \\ \omega \end{bmatrix} \times \begin{bmatrix} -\mathbf{v}\times\mathbf{w} + \mathbf{v}\omega - \mathbf{w} \\ \omega + \mathbf{v}\cdot\mathbf{w} \end{bmatrix}$$

We have

$$\begin{aligned}
\mathbf{Q}\mathbf{v} &= \mathbf{w}\times(-\mathbf{v}\times\mathbf{w} + \mathbf{v}\omega - \mathbf{w}) + \mathbf{w}(\omega + \mathbf{v}\cdot\mathbf{w}) + (-\mathbf{v}\times\mathbf{w} + \mathbf{v}\omega - \mathbf{w})\omega \\
&= \mathbf{v}(\omega^2 - \mathbf{w}\cdot\mathbf{w}) + 2(\mathbf{w}\mathbf{w}\cdot\mathbf{v} + \mathbf{w}\times\mathbf{v}\omega) \\
&= \left\{ \mathbf{I}(\omega^2 - \mathbf{w}_0^2 - \mathbf{w}_1^2 - \mathbf{w}_2^2) + \mathbf{w}2\mathbf{w}^\top + \begin{bmatrix} 0 & -\mathbf{w}_2 & \mathbf{w}_1 \\ \mathbf{w}_2 & 0 & -\mathbf{w}_0 \\ -\mathbf{w}_1 & \mathbf{w}_0 & 0 \end{bmatrix} 2\omega \right\} \mathbf{v} \\
&= \begin{bmatrix} \omega^2 + \mathbf{w}_0^2 - \mathbf{w}_1^2 - \mathbf{w}_2^2 & 2\mathbf{w}_0\mathbf{w}_1 - 2\mathbf{w}_2\omega & 2\mathbf{w}_0\mathbf{w}_2 + 2\mathbf{w}_1\omega \\ 2\mathbf{w}_1\mathbf{w}_0 + 2\mathbf{w}_2\omega & \omega^2 + \mathbf{w}_1^2 - \mathbf{w}_2^2 - \mathbf{w}_0^2 & 2\mathbf{w}_1\mathbf{w}_2 - 2\mathbf{w}_0\omega \\ 2\mathbf{w}_2\mathbf{w}_0 - 2\mathbf{w}_1\omega & 2\mathbf{w}_2\mathbf{w}_1 + 2\mathbf{w}_0\omega & \omega^2 + \mathbf{w}_2^2 - \mathbf{w}_0^2 - \mathbf{w}_1^2 \end{bmatrix} \mathbf{v} .
\end{aligned} \tag{4.4}$$

In particular, we note that the rotation matrix $\mathbf{M}_y$ in Eq. (4.1) corresponds to the unit quaternion

$$\mathbf{q}_y = \begin{bmatrix} -1/2 \\ -1/2 \\ -1/2 \\ 1/2 \end{bmatrix} ,$$

and the rotation matrix $\mathbf{M}_x$ in Eq. (4.2) corresponds to the unit quaternion

$$\mathbf{q}_x = \begin{bmatrix} -1/\sqrt{2} \\ 0 \\ 0 \\ 1/\sqrt{2} \end{bmatrix} .$$

For more information about quaternions, interested readers may view the Wikipedia web page on Quaternions and spatial rotation.

Exercise 4.4.1 Show that quaternion multiplication is associative. In other words, show that $\mathbf{q}_1 \times (\mathbf{q}_2 \times \mathbf{q}_3) = (\mathbf{q}_1 \times \mathbf{q}_2) \times \mathbf{q}_3$.

Exercise 4.4.2 Determine the quaternion that corresponds to the rotation that rotates the z axis from pointing toward the user to pointing right, and leaves the y axis pointing up.

4.4.2 Trackballs

In order for users to rotate three-dimensional computer images interactively, it is helpful for the user to act as if the mouse were moving the viewing object inside a transparent ball. This corresponds to implementing a virtual **trackball**.

The first step in implementing a virtual trackball is to map screen coordinates to a sphere of some predetermined radius r. Given screen coordinates $\mathbf{s} = [\mathbf{s}_x, \mathbf{s}_y]$, we use the screen width w and height h to map $\mathbf{s}$ a unit square by computing the two-vector

$$\mathbf{t} = \begin{bmatrix} 2\mathbf{s}_x/w - 1 \\ 2\mathbf{s}_y/h - 1 \end{bmatrix} .$$

(If the screen coordinates place the origin at the top of the screen, then the second coordinate of $\mathbf{t}$ should be reversed in sign.) Given some radius r (generally taken to be close to one), we map $\mathbf{t}$ to a unit three-vector $\mathbf{v}$ by computing

$$\sigma^2 = \min\left\{r^2, \|\mathbf{t}\|_2^2\right\} , \; \varrho = \sqrt{r^2 - \sigma^2} \text{ and } \mathbf{v} = \begin{bmatrix} \mathbf{t} \\ \varrho \end{bmatrix} \frac{1}{\sqrt{\|\mathbf{t}\|_2^2 + \varrho^2}} .$$

Next, we want to use an old screen location to determine an approximate angle of rotation that leads to some new screen location. The two screen locations can be associated with three-vectors on a sphere of radius r, as in the previous paragraph. Let us denote these two vectors as $\mathbf{v}_o$ and $\mathbf{v}_n = \mathbf{Q}\mathbf{v}_o$. The axis of rotation from $\mathbf{v}_o$ to $\mathbf{v}_n$, oriented by the right-hand rule, is given by the unit vector

$$\mathbf{a} = \frac{\mathbf{v}_o \times \mathbf{v}}{\|\mathbf{v}_o \times \mathbf{v}_n\|_2} .$$

Let

$$\mathbf{v}_\parallel = \mathbf{a}\mathbf{a} \cdot \mathbf{v}_0 \text{ and } \mathbf{v}_\perp = \mathbf{v}_0 - \mathbf{v}_\parallel .$$

be the components of $\mathbf{v}$ parallel and perpendicular to $\mathbf{a}$. Formula (4.3) for the rotation of a three-vector can be written in the current circumstances as

$$\mathbf{v}_n = \mathbf{Q}\mathbf{v}_o = \mathbf{v}_\parallel + \mathbf{v}_\perp \cos\alpha + \mathbf{a} \times \mathbf{v}_\perp \sin\alpha .$$

Since $\mathbf{a} \times \mathbf{v}_\perp$ is orthogonal to both $\mathbf{v}_\parallel$ and $\mathbf{v}_\perp$, we see that

$$\sin\alpha = \frac{\mathbf{v}_n \cdot \mathbf{a} \times \mathbf{v}_\perp}{\|\mathbf{a} \times \mathbf{v}_\perp\|_2^2} .$$

This formula allows us to determine the rotation angle with its appropriate sign. The quaternion associated with this rotation is

$$\mathbf{q} = \begin{bmatrix} \mathbf{a} \sin\alpha/2 \\ \cos\alpha/2 \end{bmatrix} .$$

From this quaternion $\mathbf{q}$, we can use Eq. (4.4) to determine the rotation matrix associated with this rotation. The model view transformation for the rotation is then

$$\mathbf{M} = \begin{bmatrix} \mathbf{Q} & \mathbf{0} \\ \mathbf{0}^\top & 1 \end{bmatrix} .$$

We have implemented a virtual trackball in both C^{++} and JavaScript. See either function `trackBall` in the C^{++} Quaternion class, or function `rotateImage` in the JavaScript VolGraphTool class.

Although MATLAB is often favored by readers because it contains graphics, it does not have any built-in way for users to interactively rotate a three-dimensional image. Some MATLAB users have developed programs for trackballs. See, for example, the Tobias Maier MATLAB trackball program.

Exercise 4.4.3 Download the Tobias Maier trackball module for MATLAB. Then use the trackball to interactively rotate the `peaks` surface built into MATLAB.

4.5 Lighting Models

In nature, points on surfaces have colors and those surface colors are illuminated by colored light sources. The illumination at a point on a surface is proportional to the intensity of the light source, the material's reflectivity, and the angle between the direction of the light and the normal to the surface at that point.

Let **a** be the three-vector of red, green and blue components of the **ambient light**, which is typically reflected off the collection of other objects in some scene. Let **d** be the three-vector of red, green and blue components of the **directional light**, which is provided by some light source and acts in the same direction $\boldsymbol{\ell}$ at all points in the scene. For example, sunlight on the earth is for all practical purposes directional light. Let **A** and **D** be the diagonal matrices whose diagonal components are given by **a** and **d**, respectively. Then at surface point with normal **n** and color three-vector **c** the illuminated color is

$$\mathbf{i} = (\mathbf{A} + \mathbf{D}\max\{-\mathbf{n} \cdot \boldsymbol{\ell}, 0\})\mathbf{c} \; .$$

More complicated illumination models may be used in computer graphics for games and movies, but this model will suffice for our purposes in this course.

Readers who are interested in a more detailed discussion of this topic may read Chap. 16 of Foley et al. [40].

4.6 Viewing Functions

Computer graphics are useful in understanding many topics in scientific computing. These topics include iterations to find zeros of functions, approximation of functions, integration, differentiation, and the solution of differential equations. Of course, our techniques for visualizing functions will vary with the number of function arguments.

4.6.1 *Functions of One Variable*

Scalar-valued functions of a single variable are the easiest to visualize. Typically, we are given a function $f(x)$ and some interval $[a, b]$ on which f is defined. We can visualize f on this interval by attempting to draw the curve $\{(x, f(x)) : x \in [a, b]\}$. Because this curve involves infinitely many points, we can only draw it approximately on the computer screen. Typically, we will choose some mesh $a = x_0 < x_1 < \ldots < x_n = b$ and approximate the curve by drawing the collection of line segments between the points in the set $\{[x_i, f(x_i)] : 0 \leq i < n\}$. Note that if the interval $[a, b]$ corresponds to a screen width of N pixels, then there is no advantage in choosing the number of mesh points n to be more than N.

All computer graphics software contains commands to draw straight lines. Often, there is a function that begins drawing at some point on the screen, and a function that draws a line from the previous point to some new point. Color is usually associated with individual functions, or with individual components of vector-valued functions, rather than with the values of individual functions.

As an example, the reader may experiment with the following JavaScript program for **Newton's method.** This program draws a given function in blue, and the tangent at a user-selected point in red. The tangent line can be updated by clicking on the button labeled "Perform another Newton iteration". Our goal in presenting this program is to illustrate visualization of a function of a single variable. Details regarding how the program operates will be presented later.

Example 4.6.1 The function

$$f(x) = 3x^2 + \tan(x)$$

has poles at $(n + 1/2)\pi$ for arbitrary integers n. In order to control the range of the axes for plotting this function, we can type inside `gnuplot` the command

```
plot [-3:3] [-30:30] 3*x**2 + tan(x)
```

This command will plot values of the function for $-3 \le x \le 3$ and $-30 \le f(x) \le 30$. In MATLAB, we could type

```
x=-3:6/100:3; f=3*x.^2+tan(x); plot(x,f)
```

These commands generate the graphs shown in Fig. 4.4.

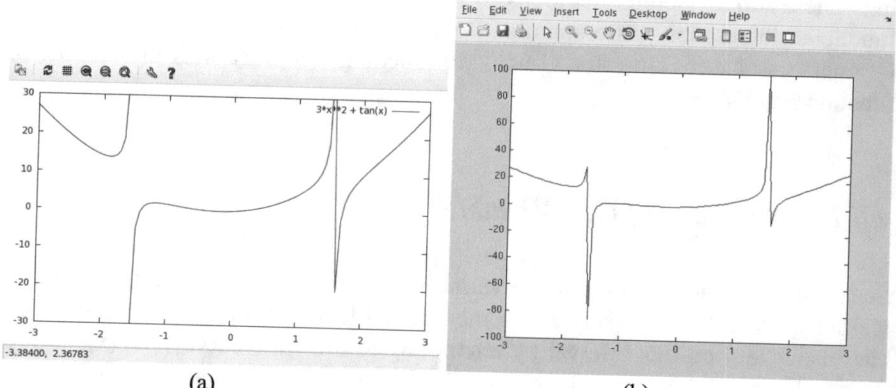

(a) (b)

Fig. 4.4 Graphs of $3x^2 + \tan(x)$; (**a**) gnuplot; (**b**) MATLAB

Some readers may enjoy Helping you with gnuplot. Other Linux users may prefer to read the Grace Tutorials. MATLAB users should study Plotting Functions.

Exercise 4.6.1 Plot the function $1/x - \tan(x)$ on $(.2, 1.4)$.

Exercise 4.6.2 Use appropriate graphics to decide if the function

$$f(x) = \frac{x^3 + 4x^2 + 3x + 5}{2x^3 - 9x^2 + 18x - 2}$$

has a zero in $(0, .2)$.

4.6.2 *Functions of Two Variables*

Scalar-valued functions of two variables can be viewed in either two or three dimensions. Typically, we are given a function $f(x, y)$ and some rectangle $[a^{(0)}, b^{(0)}] \times [a^{(1)}, b^{(1)}]$ on which f is defined. We can choose meshes $a^{(d)} = x_0^{(d)} < x_1^{(d)} < \ldots < x_{n_d}^{(d)} = b^{(d)}$ for $d = 0, 1$. These coordinate meshes determine mesh rectangles $[x_i^{(0)}, x_{i+1}^{(0)}] \times [x_j^{(1)}, x_{j+1}^{(1)}]$. The values of the function at the mesh corners or centers could be used to generate a graph of the function.

4.6.2.1 Color Fill

For two-dimensional graphics, we can fill the mesh rectangles with color depending on the function values at the corners of the mesh rectangles. It is common to choose the colors from a **hot-cold colormap**, in which the colors range from blue at the low end to red at the high end. As an example, readers may execute a JavaScript **color fill program.**

The mental interpretation of a color fill image depends strongly on the colormap used to associate functions values with colors. In regions where the color varies slightly, oscillations in the function will not be visible. On the other hand, if colors vary rapidly in some interval of function values, viewers may conclude that the function has a sharp gradient. Honest presentations will display the colormap together with the color fill plot, so that the viewer can better appraise the image.

Example 4.6.2 To plot the function

$$f(\mathbf{x}) = \sin(\mathbf{x}_1 * \mathbf{x}_2/20) \ ,$$

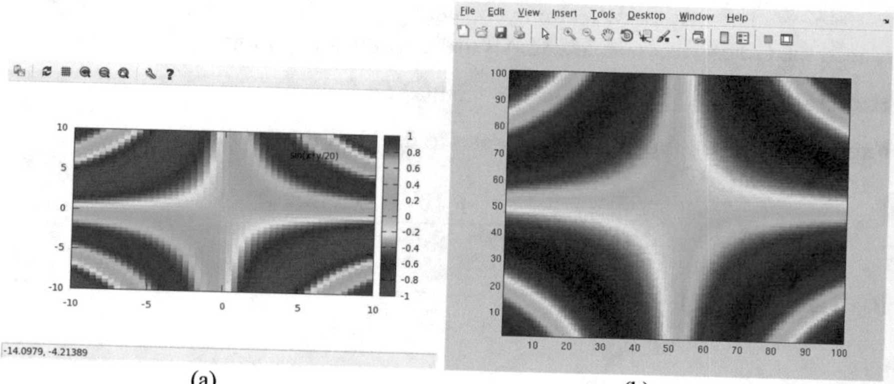

(a) (b)

Fig. 4.5 Graphs of $\sin(\mathbf{x}_1\mathbf{x}_2/20)$; (**a**) gnuplot; (**b**) MATLAB

we can use the gnuplot commands

```
set view map
set samples 50, 50
set isosamples 50, 50
unset surface
set palette defined ( 0 1 0 1, 0.2 1 0 0, 0.4 1 1 0,
  0.6 0 1 0,\ 0.8 0 1 1, 1 0 0 1 )
set pm3d implicit at b
splot sin(x*y/20)
```

In MATLAB, we could type

```
x=-10:.2:10; y=-10:.2:10; z=sin(x'*y/20); pcolor(z)
shading flat
```

These commands generate the graphs shown in Fig. 4.5.

4.6.2.2 Contouring

Suppose that we are given a function $f(\mathbf{x})$ of a 2-vector $\mathbf{x}$ defined on a rectangular mesh. Given some **contour** value γ, we can loop over all mesh rectangles $(\mathbf{x}_{0,i}, \mathbf{x}_{0,i+1}) \times (\mathbf{x}_{1,j}, \mathbf{x}_{1,j+1})$ and plot the line segments within this rectangle connecting points on the boundary of the rectangle where $f(\mathbf{x}) = \gamma$. Usually, these points on the boundary are determined by linear interpolation of the function values at the corners. We can draw contours representing the iso-values of f on the mesh rectangles. Suppose that $\mathbf{x}$ and $\mathbf{x} + \triangle\mathbf{x}$ are two neighboring vertices of a rectangle with function values on opposite sides of γ. Then the model function

$$\phi(\tau) = [f(\mathbf{x} + \triangle\mathbf{x}) - f(\mathbf{x})]\,\tau + f(\mathbf{x})$$

has value $\phi(\tau) = \gamma$ at

$$\tau = \frac{f(\mathbf{x}) - \gamma}{[f(\mathbf{x}) - \gamma] - [f(\mathbf{x} + \triangle\mathbf{x}) - \gamma]} .$$

Since $f(\mathbf{x}) - \gamma$ and $f(\mathbf{x} + \triangle\mathbf{x}) - \gamma$ are assumed to have opposite signs, no cancellation occurs in evaluating the denominator, and we must have $\tau \in [0, 1]$. One end of the contour line segment would be positioned at $\mathbf{x} + \triangle\mathbf{x}\tau$. Each mesh rectangle must have an even number of sides that contain an endpoint for a contour line segment. As an example, readers may execute a JavaScript **contour program.**

Example 4.6.3 To plot the function

$$f(\mathbf{x}) = \sin(\mathbf{x}_0 * \mathbf{x}_1/20) ,$$

we can use the `gnuplot` commands

```
set view map
set samples 50, 50
set isosamples 50, 50
unset surface
unset clabel
set palette rgbformulae 33,13,10
set contour base
set cntrparam levels auto 30
splot sin(x*y/20) palette
```

In MATLAB, we could type

```
x=-10:.2:10; y=-10:.2:10; z=sin(x'*y/20); contour(z,30)
```

These commands generate the graphs shown in Fig. 4.6.

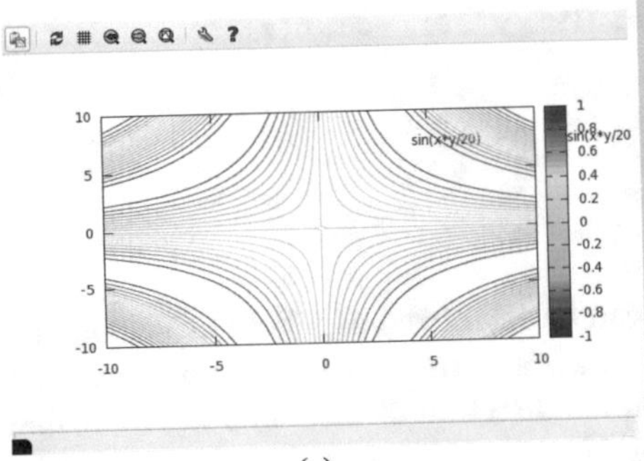

(a)

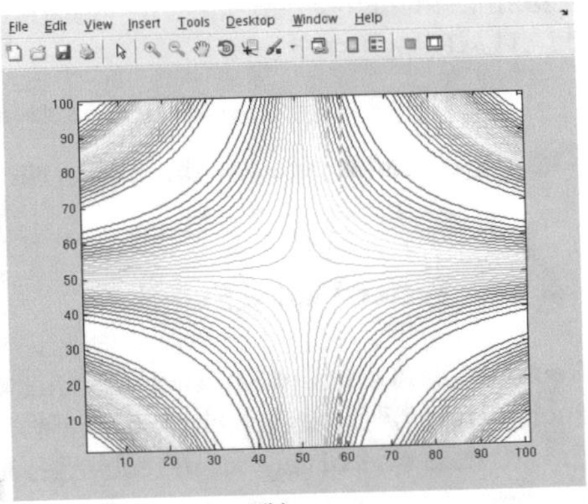

(b)

Fig. 4.6 Graphs of $\sin(\mathbf{x}_0\mathbf{x}_1/20)$; (**a**) gnuplot; (**b**) MATLAB

4.6.2.3 3D Surface

Suppose that we are given a function f of a 2-vector $\mathbf{x}$ defined on a rectangular mesh. We can loop over all mesh rectangles $(\mathbf{x}_{0,i}, \mathbf{x}_{0,i+1}) \times (\mathbf{x}_{1,j}, \mathbf{x}_{1,j+1})$ and plot the surface

$$\left\{(\mathbf{x}_0, \mathbf{x}_1, f(\mathbf{x}_0, \mathbf{x}_1)) : (\mathbf{x}_0, \mathbf{x}_1) \in [\mathbf{x}_{0,i}, \mathbf{x}_{0,i+1}] \times [\mathbf{x}_{1,j}, \mathbf{x}_{1,j+1}]\right\} .$$

Such a surface might be approximated by various kinds of polynomial interpolations. The simplest interpolations approximate f by a linear interpolation on a triangle. For example, we could use the center of the rectangle to create four triangles, each having the center and two neighboring vertices of the rectangle as their vertices. The color of each triangle should be determined by the values of f at the vertices.

It is very useful to be able to rotate the surface interactively, and possibly to use a **lighting model**, in order to comprehend the shape of the surface. As an example, readers may execute a JavaScript **2D surface program.**

Example 4.6.4 To plot the function

$$f(\mathbf{x}) = \sin(\mathbf{x}_0 * \mathbf{x}_1/20) ,$$

we can use the `gnuplot` commands

```
set samples 50, 50
set isosamples 50, 50
unset surface
set pm3d implicit at s
set pm3d scansbackward
set palette rgbformulae 33,13,10
splot sin(x*y/20) palette
```

The resulting `gnuplot` image can be rotated interactively (not within this text, but within the window generated by `gnuplot`) by means of the left mouse button. In MATLAB, we could type

```
x=-10:.2:10; y=-10:.2:10; z=sin(x'*y/20); surf(z)
```

These commands generate the graphs shown in Fig. 4.7.

4.6.2.4 Vector-Valued Functions

Vector-valued functions of two variables can be visualized as a field of vectors. At each vertex of a mesh rectangle, we can draw an arrow of length no more than the shortest side of any mesh rectangle, in the direction of the value of f. For fine meshes, it becomes hard to see the arrow heads, so these vector plots cannot be refined to the same degree as color fill or contour plots.

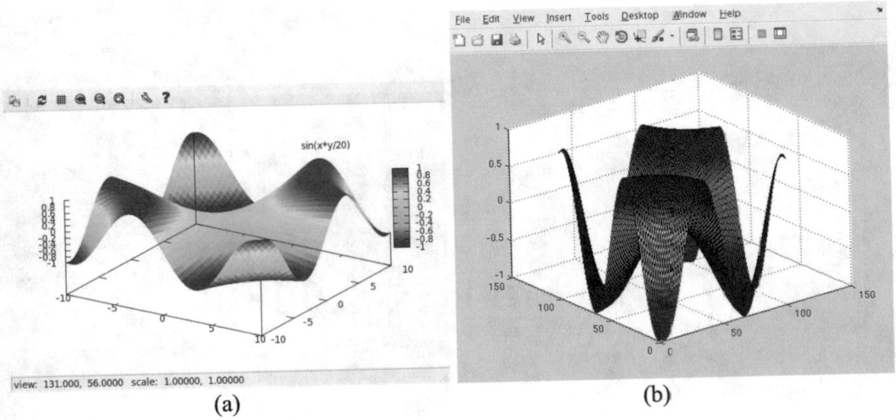

Fig. 4.7 Graphs of $\sin(\mathbf{x}_0\mathbf{x}_1/20)$; (**a**) gnuplot; (**b**) MATLAB

Example 4.6.5 To plot the function

$$\mathbf{f}(\mathbf{x}) = \begin{bmatrix} \mathbf{x}_1 \\ 2\mathbf{x}_2 \end{bmatrix} ,$$

we can use the `gnuplot` commands

```
set xrange [-5:5]
set yrange [-5:5]
set samples 11
set isosamples 11
dx(x)  = coef*x
dy(y)  = coef*2*y
plot "++" using 1:2:(dx($1)):(dy($2)) with vectors
```

In MATLAB, we could type

```
a=-5:.5:5; b=-5:.5:5; [x,y]=meshgrid(a,b); u=x; v=2*y;
quiver(x,y,u,v)
```

These commands generate the graphs shown in Fig. 4.8.

Exercise 4.6.3 Plot the function

$$f(\mathbf{x}) = \frac{1}{2}(\mathbf{x}_1^2 - \mathbf{x}_2)^2 + \frac{1}{2}(1 - \mathbf{x}_1)^2$$

on $[-2, 2] \times [-2, 2]$.

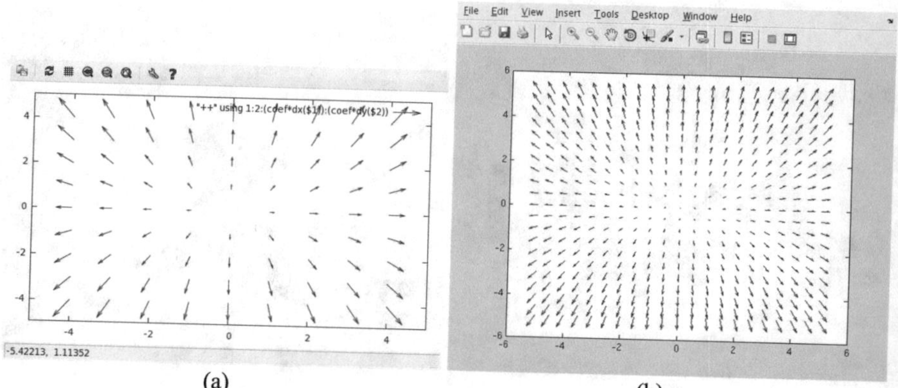

(a) (b)

Fig. 4.8 Graphs of $[\mathbf{x}_1, 2\mathbf{x}_2]$; (**a**) gnuplot; (**b**) MATLAB

Exercise 4.6.4 Plot the vector-valued function

$$\mathbf{f}(\mathbf{x}) = \begin{bmatrix} \mathbf{x}_2 \\ -\mathbf{x}_1 \end{bmatrix}$$

on $[-10, 10] \times [-10, 10]$.

Exercise 4.6.5 Vector fields on fine meshes are hard to visualize, because the vectors are small and the arrow heads are even smaller. Instead, it may be useful to plot **pathlines**. Given a vector-valued function $\mathbf{f}(\mathbf{x})$ and a starting point $\mathbf{s}_0$, a pathline $\mathbf{s}(t)$ could be defined by

$$\mathbf{s}'(t) = \mathbf{f}(\mathbf{s}(t)) \text{ and } \mathbf{s}(0) = \mathbf{s}_0 .$$

Use the MATLAB command streamline to plot **streamlines** for

$$\mathbf{f}(\mathbf{x}) = \begin{bmatrix} \mathbf{x}_2 \\ -\mathbf{x}_1 \end{bmatrix} .$$

4.6.3 *Functions of Three Variables*

4.6.3.1 Cut Planes

Suppose that we are given a function f of a 3-vector $\mathbf{x}$ defined on a rectangular mesh. One way to visualize f on the box is to draw color fill plots of f on planes that intersect the box. For example, we could draw a color fill plot in the plane consisting of all points within the box with some prescribed value for $\mathbf{x}_1$. User comprehension of f could be aided by allowing interactive selection of the locations of the planes. As an example, readers may execute the following JavaScript **3D cut plane program.**

Example 4.6.6 To plot the function

$$f(\mathbf{x}) = \mathbf{x} \cdot \mathbf{x}$$

we can use the MATLAB commands

```
[x,y,z] = meshgrid(-2:.2:2,-2:.25:2,-2:.16:2);
v = x.^2+y.^2+z.^2;
xslice = [-1.2,.8,2]; yslice = 2; zslice = [-2,0];
colormap hsv;
slice(x,y,z,v,xslice,yslice,zslice)
```

These commands generate the graph shown in Fig. 4.9.

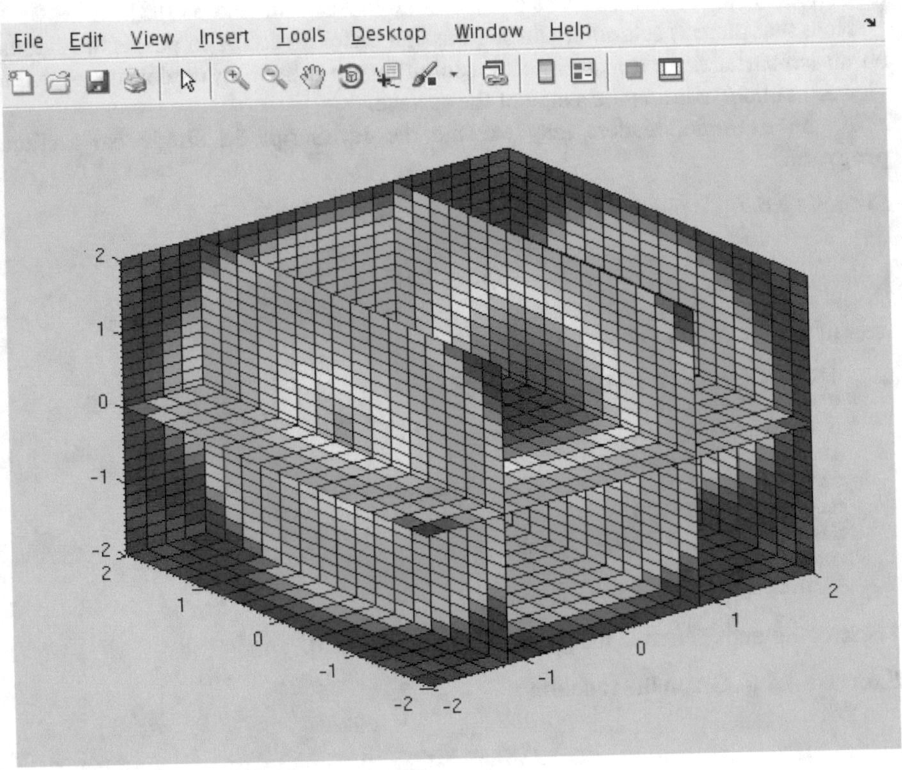

Fig. 4.9 Slices of $\mathbf{x} \cdot \mathbf{x}$

4.6.3.2 Iso-Surfaces

Another way to visualize scalar-valued functions of three variables is to draw iso-surfaces, on which the given function f takes a constant value. Within each mesh box, portions of the iso-surface are constructed to intersect the edges of the box at the appropriate points. The portion of the iso-surface within a mesh box is represented as a collection of triangles with vertices on the mesh box edges. Since there are eight vertices in a mesh box, each of which could have function value either above or below the iso-value, there are potentially $2^8 = 256$ cases to consider in drawing the iso-surface within each mesh box. If we use rotation and/or reflection to identify equivalent arrangements of the iso-surface within a mesh cube, there are only 15 distinct cases to consider. Usually, these 15 distinct cases are expanded into a table of the 256 possible cases, for easy look up. The algorithm is known as **marching cubes**, and was developed by Lorensen and Cline [77]. Some mistakes in this paper have been corrected in later work, and other refinements have been identified. A survey of the results can be found in Newman and Yi [83].

Note that there is seldom a reason to assign different colors to points or regions on an iso-surface. As a result, it is essential to use a lighting model to assist the viewer in understanding the shape of the surface.

As an example, readers may execute the JavaScript **3d single iso-surface program.**

Example 4.6.7 To plot the function

$$f(\mathbf{x}) = \mathbf{x} \cdot \mathbf{x}$$

we can use the MATLAB commands

```
[x,y,z] = meshgrid(-2:.2:2,-2:.25:2,-2:.16:2);
v = x.^2+y.^2+z.^2;
p = patch(isosurface(x,y,z,v,1));
isonormals(x,y,z,v,p)
set(p,'FaceColor','red','EdgeColor','none');
daspect([1,1,1])
view(3); axis tight
camlight
lighting gouraud
```

These commands generate the graph shown in Fig. 4.10.

Exercise 4.6.6 Graph the function

$$f(\mathbf{x}) = \mathbf{x}_1 e^{-\mathbf{x} \cdot \mathbf{x}}$$

on $[-2, 2] \times [-2, 2] \times [-2, 2]$.

Exercise 4.6.7 Suppose that have a scalar-valued function whose values are given at the vertices of a collection of tetrahedra. Show that there are $2^4 = 16$ possible

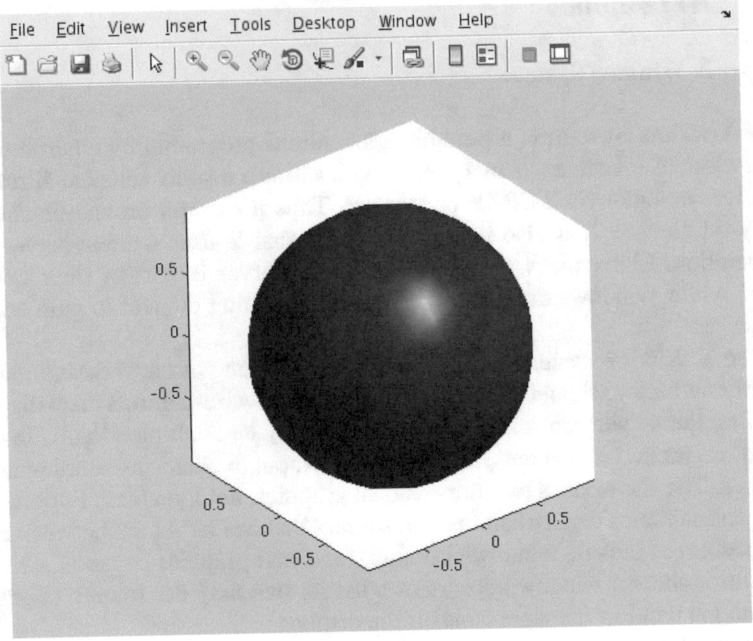

Fig. 4.10 Isosurface of $\mathbf{x} \cdot \mathbf{x}$

cases for describing how an iso-surface of the function could intersect an individual tetrahedron. Show that there after rotation, there are three distinct cases: no intersection with the isosurface, one vertex on the opposite side of the isosurface from the others, and two vertices on either side of the surface. Describe how to develop a marching tetrahedra algorithm.

4.7 Graphics Software

In Sect. 4.6, we described various ways to view functions outside of a programming language, using either `gnuplot` or MATLAB. Our goal in this section is to describe graphics software that can be called from inside a computer program.

For a list of high-level software packages that provide scientific visualization primarily in a programming environment, see Wikipedia : Scientific Visualization. These software packages differ in ease of use and operation. We will survey several important software packages, and distinguish them by their ability to project three-dimensional images onto a computer screen.

4.7.1 2D Graphics

4.7.1.1 X Window System

The **X Window System** is a portable applications programming environment for a wide class of machines, namely those with **bitmap** display screens. **X** runs over computer networks via **TCP-IP** or **DecNet**. Thus it can run on one machine and display on another. It is also important to note that **X** uses a client-server model for operation. Linux users will find that the **X** server is part of their operating system, while Windows users should consider installing cygwin to gain access to an **X** server.

In the X Window system, a **display** is a workstation with keyboard, mouse and (possibly multiple) screens. The X Window system server controls each display. In particular, the server controls access to the display by multiple clients, interprets network messages from clients, and passes user input to clients by sending network messages. The server does two-dimensional graphics, not the client. For efficiency, the server maintains data structures as resources that can be shared between clients. These resources include windows, cursors, fonts and graphics contexts. Clients do not control where a window appears or what its size may be. Rather, clients give hints and the window manager controls the display.

Clients respond to **events**, such as `KeyPress`, `MotionNotify`, `Expose`, `ResizeRequest` and `DestroyNotify`; for a full list of events, see X.h. User applications must search the event queue provided by the server and take appropriate action.

X applications communicate with the server via **Xlib** calls in C. There are Xlib calls to connect to the server, create windows, draw graphics, and respond to events. These functions can be found in Xlib.h after the `_XFUNCPROTOBEGIN` line. Usually, the definition of the function is also available in the online manual pages. Xlib calls are translated to protocol requests via TCP/IP. Typically, Xlib procedures are very primitive and many applications will use higher-level libraries that build on Xlib, such as Xt, XMotif and OpenGL.

The X protocol specifies what makes up each packet of information moving between the server and Xlib. Xlib saves its requests in a buffer. Consequently, its operation may be asynchronous with the application. This saves network communication time by grouping requests. The Xlib buffer is cleared when the application is waiting for an event, or the Xlib call involves a reply, or the client application requests a flush of the queue.

The server maintains a number of abstractions referenced by an identification number. These abstractions include windows, fonts, **pixmaps**, **colormaps** and **graphics contexts**. The X protocol sends the identification number, not the entire data structures, in its network messages. Note that access to a data structure is unprotected once its resource identification number becomes known.

A **window** is a rectangular area on the screen. Each window has a parent, and is contained within the bounds of the parent. A window does not normally respond to input from outside its bounds. The **root window** is created by the server and fills the screen.

Unfortunately for conventional mathematics, the X Window coordinate system is oriented from the top left corner. This means that x coordinates increase from left to right, but y coordinates increase from top to bottom. Each window has a position relative to its parent's coordinate system. The window **geometry** consists of its width, height and position. To get information about an individual window in X, type

```
xwininfo
```

in any window, and click a mouse button in any other window for which you would like the information.

For more information about the X Window System, see the X.org Foundation webpage, or read the programming manual by Nye [84].

4.7.1.2 XToolkit

The X Window System enables a programmer to interface with the X server, and thereby enable displays on local or remote screens. However, the Xlib library does not provide functions for implementing widgets such as buttons or menus. **XToolkit** supports the creation and use of widgets, but does not provide any widgets of its own. As a result, XToolkit is useful for 2D graphics, but not directly useful for graphical user interfaces. Generally, scientific computing programs implementing 2D graphics in XToolkit run very fast.

For more information about XToolkit, see the book by Asente and Swick [3].

4.7.1.3 XMotif

The Xlib functions are primitive and difficult to use, and the XToolkit widgets are very simple in appearance. On the other hand, **XMotif** provides a somewhat more friendly user interface with greatly enhanced features. The enhancements are primarily for window decorations and graphical user interfaces; there are essentially no improvements X Motif that make scientific visualization easier.

Motif provides mechanisms for communication between a **graphical user interface** and an applications program. These mechanisms appear as objects on a screen, such as buttons or menus. Motif uses the X Window system for client-server communication. It is also important to note that Motif is written in ANSI C. As a result, it may be tricky to request Motif to call a user application written in C^{++}.

Event processing with Motif is asynchronous. Typically, the client will enter an infinite applications loop. Inside that loop, the client must be careful about making assumptions regarding the order of events relative to requests. For example, it is possible to generate a client request, then read an event generated before the request.

For more information about the open source Motif implementation, see the Open Motif webpage, or read the programming manual by Brennan et al. [15].

4.7.1.4 GTK

The **GNU Image Manipulation Project (GIMP) Toolkit**, often abbreviated as **GTK**, is more recent than Motif. It runs on Linux, Windows and Mac operating systems, and interfaces with Xlib directly. GTK provides the functionality of XMotif, but allows the user or operating system to change the look and feel of the widgets.

For more information about GTK, see The GTK Project webpage, or read the programming manual by Logan [75].

4.7.1.5 Qt

Qt is similar to GTK, in that it offers widgets that interface directly with Xlib. However, Qt offers programmer interfaces via C^{++}, while GTK uses C. For more information about Qt, see the Basic Qt Programming Tutorial website, or read the programming manual by Blanchette and Summerfield [13].

4.7.1.6 HTML5 and JavaScript

The canvas element in **HTML5** provides commands to draw lines and fill polygons. JavaScript provides an interpreted programming language to interface to these HTML5 commands. Using these commands, programmers can display simple scientific computing results on web pages. For more information, either read Chaps. 35 and 36 of the HTML5 programming manual by Freeman [42], or read Chap. 21 of the JavaScript programming manual by Flanagan [39].

4.7.2 3D Graphics

4.7.2.1 OpenGL

Silicon Graphics Incorporated developed **OpenGL** for easy 3D programming. It provides utilities for coordinate transformations, screen projection, lighting and textures. This software is publicly available at the Mesa 3D Graphics Library webpage. For a programmer guide, read Munshi et al. [80].

4.7.2.2 MathGL

Alexey Balakin has developed C^{++} classes for 1D, 2D and 3D graphics. His software package is called **MathGL**, and is publicly available at MathGL - library for scientific data visualization. This software works on a variety of operating systems and compilers. It has functions to parse character strings describing mathematical functions, and to perform a wide array of scientific operations and transforms. Documentation is provided online. The example programs all write to `.png` files, rather than by opening screen windows. However, it is possible to draw MathGL plots in windows by providing the graphics routines with a viewer, such as the Linux `xv` command. For example, readers can view the example MathGL programs testMGLContour.C, testMGLColor.C, testMGLSurface.C and testMGLIsoSurface.C.

4.7.2.3 WebGL

Recently, software for 3D visualization on web pages has become available via **WebGL**. This software uses the HTML5 canvas element, and provides most of the OpenGL functionality. Some features, such as user-defined clip planes, are not yet available. For more information, see either the Kronos Group WebGL website, the Mozilla Developer Network WebGL online documentation with tutorials, or the programmer guide by Parisi [86]. For lighting and shading, WebGL programs depend on the **OpenGL Shading Language**, which has acronym **GLSL**. More information about GLSL is available at the OpenGL Shading Language website, or in The OpenGL Shading Language manuscript.

4.8 Graphics Classes

At this point, we have discussed several graphics packages. Next, we would like to compare them for suitability in scientific computing visualization. To get some idea of the differences between the packages for scientific visualization, readers might examine programs to plot a simple function in VTK, MathGL and MATLAB. The **(Visualization Tookit (VTK)** and MathGL programs write to files, but the writing can be piped to a viewer command in the operating system. The MATLAB example is slightly less complicated than the MathGL example, and the VTK example is far more complicated. Both MATLAB and MathGL have routines to visualize functions of multiple variables. MathGL and VTK allow programmers to plot data directly from within a compiled language program, while MATLAB would require such a programmer to write a data file to be read by MATLAB for visualization.

In the remainder of this section, we will describe some C^{++} and JavaScript classes for scientific visualization. The C^{++} classes were originally written for XToolkit and OpenGL, then modified for XMotif and GTK. The JavaScript classes were made possible by the recent development of WebGL. These classes lack many features found in MathGL and MATLAB, such as elaborate coordinate systems and parsing of functions. However, they do provide the ability to interact with the image, both for orientation and slicing of 3D plots, and for selecting points within an image in order to modify the plot. Because we will use these classes in example programs throughout the remaining chapters, the following brief description may be helpful the reader.

4.8.1 2D Graphics

4.8.1.1 C++

We have developed three basic C++ classes that interface with available software libraries for one- and two-dimensional graphics, namely Palette, VirtualColormap and VirtualWindow. The `VirtualColormap` is specialized through class inheritance into a GTKColormap, and the `VirtualWindow` is specialized into a GTKWindow. In other programming environments, the `VirtualWindow` has been specialized into an `XWindow`, or a `DistributedWindow` for plotting data on a distributed memory machine.

A `Palette` consists of a character string for the name, an integer number of colors, an array of character strings for the color names, and an array of locations of the colors in the XColormap. A `Palette` can be used to create a `GTKColormap`. It is important that the `Palette` work with color names, to enforce uniformity of color choices across hardware vendors. Under Linux, a list of color names and their (r, g, b) values can typically be found in the file rgb.txt.

A `VirtualColormap` contains a `Palette` pointer and some desired number of colors. The `VirtualColormap` constructor forms a colormap by interpolating the (r, g, b) values of the colors in a `Palette`. Further specialization must be done in the class `GTKColormap` to allow the colormap to interact with the graphics library.

A `VirtualWindow` is a pure virtual class. It has no data, just pure virtual functions corresponding to functions in a `GTKWindow`. The latter class translates unit coordinates into screen coordinates, color names into graphics library colors, and applies library-specific drawing commands. For example, there are member functions to draw lines, color rectangles, put strings in the window, set the foreground and background colors, clear the window and get the window coordinates of the mouse.

Users interact with these C++ classes through a VirtualGraphTool. This class is specialized through C++ inheritance into either an XYGraphTool or an AreaGraphTool. An `XYGraphtool` owns a `VirtualWindow` pointer, user coordinates for the graphics window corners and a pointer to a `VirtualColormap`. This class has member functions to get the user coordinates of the mouse position, draw axes, write strings, set foreground and background colors, and draw lines and symbols. An

`AreaGraphTool` has the same data, but has additional functions to color polygons and draw vectors.

For examples using these C^{++} graphics classes in scientific computing, see newton.C, testAreaGraphToolColor.C and testAreaGraphToolContour.C.

4.8.1.2 JavaScript

We have used similar ideas to develop JavaScript graphics classes for 1D and 2D plotting. The Palette and Colormap implement the same kind of functionality that was previously described in the C^{++} classes. There is no need for a virtual window class in JavaScript, because JavaScript uses only HTML5 for its display. There is an XYGraphTool class, and an AreaGraphTool class. These have functionality very similar to the corresponding C^{++} classes. Example programs using these JavaScript graphics classes were provided in Sects. 4.6.1, 4.6.2.1 and 4.6.2.2.

Unlike the GTK graphic library, there is no need to enter an **event loop** in JavaScript or HTML5. Each HTML5 web page already handles an event loop. Instead, the JavaScript graph tool classes enable handlers to respond to mouse and keyboard events, through user functions.

4.8.2 3D Graphics

4.8.2.1 C^{++}

We have also written several different C^{++} classes for 3D graphics. Color is handled by a GDKColorArray. This uses a `Palette` to create an array of colors, as appropriate for the graphics library. The interface with the library drawing functions is handled by a VirtualGLWindow. This class is specialized via C^{++} class inheritance into a GTKGLWindow. A `VirtualGLWindow` maps coordinates from the cube $(-1, 1) \times (-1, 1) \times (-1, 1)$ to coordinates that are appropriate for the graphics libraries. This window class contains functions to draw lines, color triangles and polygons, draw vectors, and handle mouse events.

The VolGraphTool class provides a user-friendly interface to the 3D window classes. The `VolGraphTool` functions transform user coordinates to window coordinates, draw lines, fill triangles and polygons, and provide specific actions for mouse events. In a `VolGraphTool`, a trackball works with the left mouse button to rotate the image. The middle mouse button is used to select and move clip planes. These can be used to restrict drawing to boxes aligned within the drawing region, so that the user can examine the interior of the drawing region. The right mouse button draws **crosshairs**, which can be used to determine the location of a point within the drawn object.

Some additional C^{++} classes enable visualization of scalar-valued functions of three variables. The AxisClipPlane class holds information about the location and orientation of clip planes. The TypedPlotObj class provides functions to draw a data set in clip planes or iso-surfaces. Finally, SurfaceDataSet uses a user-provided function to perform a marching cubes algorithm for drawing iso-surfaces. All of these C++ classes must encapsulate information to allow the C-language functions in the graphics libraries to call C++ class member functions within graphics event loops.

For examples using these C++ graphics classes in scientific computing, see testVolGraphTool2.C, testVolGraphTool3.C, testVolGraphTool4.C and testVolGraphTool5.C.

4.8.2.2 JavaScript

The VolGraphTool class handles WebGL drawing in an HTML5 canvas. This class has a function to draw a surface for a scalar-valued function of two variables. For a scalar-valued function of three variables, this class also has functions to color clip planes or draw iso-surfaces. Users can use the mouse to interact with the image, either by rotation, clipping or crosshairs. Example programs using this JavaScript `VolGraphTool` class were provided in Sects. 4.6.2.3, 4.6.3.1 and 4.6.3.2.

Exercise 4.8.1 If you have GTK graphics installed on your machine, make a directory called `graphics` and put a copy of the graphics tarfile in your graphics directory. Extract the files, and use the `GNUmakefile` to build the various libraries on your machine. Then build the test executables in the graphics directory. Run each to verify that it works properly on your machine.

4.9 Graphical User Interfaces

In Sect. 4.8, we described various C++ classes for graphical output displays. In this section, we will describe some C++ classes for graphical user input. Such user input devices clearly list for the user the necessary input parameters, and allow for these parameters to be adjusted before computations are performed.

Some programming languages, such as HTML5, make graphical user input techniques available as part of the language. We will make use of this feature in our JavaScript programs that are called from HTML5. On the other hand, graphical display environments such as GTK are more primitive. Their event loop might be expected to handle both a parameter change and the execution of the user program. In this environment, the graphical user input is more difficult, and the user program is run as a program thread. Debugging the user program is more difficult.

We will use our graphical user interface to read input data, warn us about bad input, run and interrupt the program during execution, and rerun the program after changing input. All of these will take place within a single executable, without writing plot files and without running a separate graphics program.

There are three kinds of C^{++} classes that we will use for our graphical user interface. The simplest is VirtualInput, from which we will derive `InputString`, `InputIFStream`, `InputOFStream` and InputParameter. The data for a `VirtualInput` consists of a character string to use as a label in the graphical user interface. A `VirtualInput` knows how to read formatted or unformatted input values from an `ifstream`, and how to write unformatted values to an `ofstream`. It can return a character string representing the value of the `VirtualInput` to write in a graphical user interface.

The `InputParameter` class is intended to handle gui input of `int`, `float`, `double` and so on. Every `InputParameter` owns a pointer to its stored value, which is presumably an input parameter in the user's program. Each `InputParameter` also knows lower and upper bounds on allowed input values, to allow the gui to check input before storing it.

The second kind of C^{++} class in our graphical user interface is the VirtualThread. This class allows us to form separate processes, and link them for communication. For example, the graphical user interface can form a `VirtualThread` to run the user's routine to integrate an ordinary differential equation and plot the results. Since the gui initiated the run `Thread`, it can pause the program execution to interact with the graph at intermediate times. A `VirtualThread` can perform such functions as `run`, `suspend`, `resume` and `stop`.

We derive several class from `VirtualThread`. A Thread is a `Virtual Thread` with a pointer to a function with no arguments and that returns a void. Other classes derived from `VirtualThread`, namely `ClassThread` and `ClassThread1` store pointers to member functions in some class, possibly with an argument. When the `VirtualThread` is `run`, the function being pointed to is executed by the `pthread` library.

The third kind of class is a GUI. Its data consists of various widgets, a `ClassThread` to run the user's computational routine, various function pointers for callbacks, and a list of `InputParameters`. We derive various classes from `GUI` for different graphical toolkits. For example, the `GTKGUI` constructor initializes GTK and opens a graphical user interface from which we can enter input or run a program; see GTKGUI.C.

Rather than concentrate on the very complicated contents of the `GUI` member functions, let us examine how to use the `GUI` in an applications program. Consider the file guiivp.C. The main program now looks very different.

1. It calls `processCommandLine` to read the name of an optional input file from the command line.
2. Next, it calls `makeMainList`. This is specific to individual scientific computing problems.
3. If there is an input file on the command line, the main program calls `readMainInput` to read the `InputParameters`, and `checkMainInput` to make sure that the input values are acceptable.
4. Next, the main program calls the `GUI` constructor, and creates various pull-down menus. Notice that the main program passes pointers to various procedures to run

the application program, check the input, cleanup temporary arrays and shutdown the program.

5. Finally, the main program calls `GUI::eventLoop`. When the `eventLoop` reads input data, `checkMainInput` is called. When the `eventLoop` runs the user application, `runMain` is called from a `Thread`. When the `eventLoop` completes a calculation, `cleanup` is called. When the `eventLoop` terminates, `shutdown` is called.

In summary, a user must provide the following problem-specific routines to the graphical user interface:

- `makeMainList` to define the `InputParameters`,
- `checkMainInput` to check for incompatibility between values of different `InputParameters`,
- `runMain` to solve the scientific computing problem,
- `cleanup` to delete temporary pointers created during `runMain` but kept alive (for example, so that the user can view graphics at the end of the run), and
- `userShutdown` to delete pointers created before calling the `GUI` constructor.

It is strongly suggested that readers not run a graphical user interface while debugging their computational routine. Any floating point exceptions will occur in a pthread called from the `GUI`. When the debugger traces back through the execution, it may not trace back to that individual process. Instead, put the line

```
skip_gui 1
```

in the command line input file, so that the main program never calls the `GUI` constructor.

The graphical user interfaces for `guiivp.C` are shown in Fig. 4.11.

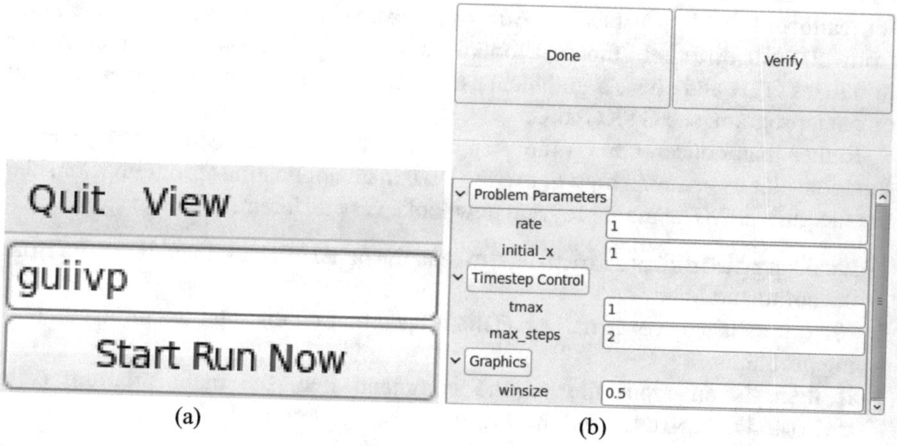

Fig. 4.11 Graphical user interface. (**a**) Main gui. (**b**) Input parameter gui

Exercise 4.9.1 If you have GTK graphics installed on your machine, make a directory called `gui` and put a copy of the gui tarfile in your gui directory. Extract the files, and use the `GNUmakefile` to build the various libraries on your machine. Then build the test executable `guiivp` in the gui directory. Run it to verify that it works properly on your machine.

Chapter 5
Nonlinear Equations

... [E]quation solving is impossible in general, however necessary it may be in particular cases of practical interest. Therefore, ask not whether SOLVE can fail: rather ask, 'When will it succeed?'

William Kahan *[61]*

Abstract This chapter examines the problems, of solving a nonlinear equation in a single unknown, and of minimizing a nonlinear function of a single variable. First, the existence, uniqueness and perturbation theory for a nonlinear equation is analyzed. The bisection algorithm is developed and analyzed next. Newton's method is also developed and analyzed; this analysis includes its convergence rate, maximum attainable accuracy in the presence of rounding errors, termination criteria and the use of approximate derivatives. A similar analysis of the secant method follows. Afterward, the chapter presents a variety of techniques for assuring global convergence for a nonlinear equation. Then the focus turns to nonlinear minimization. First the well-posedness of the minimization problem is analyzed. Then a variety of deterministic methods for nonlinear minimization are developed and analyzed. The chapter ends with a discussion of stochastic optimization.

5.1 Overview

At this point in the book, we have studied floating point arithmetic, numerical solution of systems of linear equations, and computer graphics. These previous chapters have provided the tools needed to develop our next topic, namely iterative methods for solving nonlinear equations. If motivation for this new topic is needed, recall that in Chap. 1, we considered some simple initial value problems for nonlinear ordinary differential equations. We saw that Euler's method could be used

Additional Material: The details of the computer programs referred in the text are available in the Springer website (http://extras.springer.com/2018/978-3-319-69105-3) for authorized users.

© Springer International Publishing AG, part of Springer Nature 2017
J.A. Trangenstein, *Scientific Computing*, Texts in Computational Science and Engineering 18, https://doi.org/10.1007/978-3-319-69105-3_5

to solve to solve such problems, but for stiff problems the timestep with this method might need to be very small. We mentioned that we could use the backward Euler method to solve stiff problem; this method would require that we solve a nonlinear equation at each timestep.

In this chapter, we will consider two basic problems. The first problem is the **root-finding** problem: given a nonlinear scalar-valued function f, find (if possible) a zero z so that $f(z) = 0$. The second problem is **optimization**: given a nonlinear scalar-valued function f, find (if possible) an argument z so that $f(z)$ is as large (or as small) as possible.

In studying the root-finding and optimization problems, we will follow the steps in solving a scientific computing problem, as described in Sect. 1.3. We will begin by studying the circumstances under which the root-finding or optimization problem is well-posed, meaning that a solution z exists, is unique, and depends continuously on the data f. The analytical issues here are more difficult than they were for systems of linear equations. Afterward, we will develop numerical algorithms for solving these problems, and implement these algorithms in computer programs. Necessarily, the numerical algorithms are also more complicated than algorithms for linear algebra. An important step will be to analyze the root-finding and optimization algorithms, to understand their rates of convergence and sensitivity to rounding errors.

Our first goal in this chapter is to use some basic tools of calculus to determine circumstances under which the root-finding problem has a unique solution, and to describe how a root is changed by perturbations in the function. Afterward, we will examine how various local models for the function lead to distinct algorithms for the root-finding problem. If a root-finding problem has a solution, then the basic algorithms will typically converge whenever the algorithm is started sufficiently close to a solution. To avoid local convergence limitations, we will develop techniques that guarantee convergence from starting points that are not necessarily close to a root.

Our experiences with the root-finding problem will help us to develop algorithms for the optimization problem. We will see that it is very difficult to find the global optimum of a function with multiple local minima; in this case, stochastic optimization methods will be helpful.

For more information about the material in this chapter, we recommend the numerical analysis books by Dahlquist and Björck [26], Greenbaum [48], Henrici [54], Kincaid and Cheney [66] and Ralston and Rabinowitz [90]. Readers may also be interested in books devoted to nonlinear equations and optimization by Dennis and Schnabel [31], Gill et al. [44], Kelley [63] and Ortega and Rheinboldt [85].

For nonlinear equation software, we recommend the GNU Scientific Library (GSL) One dimensional Root-Finding, which implements most of the basic locally convergent methods (bisection, Newton's method, the secant method, regular falsi and Steffensen's method) as well as Brent's method. We also recommend netlib for Larkin [71] routines zero1 and zero2, Brent routine zeroin, Dekker's routine dfzero and TOMS 748. Readers might also consider various Harwell routines, such as nb01 and nb02. In MATLAB, we recommend function fzero, which uses Brent's method. In Scilab we recommend the command fsolve.

For nonlinear minimization software, we recommend the GSL One dimensional Minimization, which provides routines to perform golden search and Brent's method. Other routines can be found at netlib, such as the Fortran routine fmin or the C++ routine fminbr. For stochastic optimization, we recommend GSL Simulated Annealing. In MATLAB, we recommend function fzero which uses golden search and parabolic interpolation, as well as the GlobalSearch class, the function simulannealbnd and ga.

5.2 Well-Posedness

Let us recall the scientific computing steps, and apply them to the root-finding problem. Once we have been given a function, our first step is to examine the well-posedness of finding a zero or a minimum of that function. Recall that a problem is well-posed if a solution exists, is unique and depends continuously on the data. We will examine these three issues in the subsections below.

5.2.1 *Existence*

Suppose that we have a model in the following form:

$$\text{Given a real-valued function } f \text{ defined on } [a,b]$$
$$\text{Find } z \in [a,b] \text{ so that } f(z) = 0\;.$$

There could be alternative formulations of the root-finding problem. For example, if we want to solve $g(z) = c$, we can rewrite the problem in terms of $f(x) \equiv g(x) - c$ and find a zero of f. Similarly, if we want to find a minimum or maximum of a continuously differentiable function g, we can look for a zero of $f = g'$.

Unlike linear equations, nonlinear equations may have no solutions, one solution or multiple solutions. For example, $f(x) = \tan^{-1}(x)$ has exactly one zero $z = 0$, while $f(x) = \sec(x)$ has no real zeros. The nonlinear function $f(x) = \sin(x)$ has infinitely many zeros. For functions such as this with multiple zeros, it may be desirable to design an algorithm to select a particular solution, often within some specified interval.

Let us begin by discussing circumstances under which we can guarantee that a nonlinear equation $f(z) = 0$ has a solution. We will begin with the following important definition.

Definition 5.2.1 Let $[a,b]$ be a closed interval of real numbers, and let f be defined on $[a,b]$. Then f is **continuous** at $x \in [a,b]$ if and only if for all $\varepsilon > 0$ there exists $\delta > 0$ so that for all $\xi \in (x-\delta, x+\delta) \cap [a,b]$ we have $|f(\xi) - f(x)| < \varepsilon$. Also, f is continuous in $[a,b]$ if and only if f is continuous at all points in $[a,b]$.

The following theorem expresses the intuitive notion of continuity for most people, and can be found in most advanced calculus texts.

Theorem 5.2.1 (Intermediate Value) *If f is continuous on $[a, b]$, then for all numbers y between $f(a)$ and $f(b)$ there exists $x \in [a, b]$ so that $f(x) = y$.*

Proof This theorem is commonly proved by contradiction. See, for example, Buck [18, p. 75].

One immediate consequence of the intermediate value theorem is the following existence theorem for nonlinear equations.

Theorem 5.2.2 *If f is continuous on the closed interval $[a, b]$ and $f(a)f(b) \leq 0$, then there exists $z \in [a, b]$ so that $f(z) = 0$.*

Proof If $f(a) = 0$ then we take $z = a$; ditto for $f(b) = 0$. If both $f(a)$ and $f(b)$ are nonzero, then zero is between $f(a)$ and $f(b)$, so the intermediate value theorem implies that there exists $z \in [a, b]$ so that $f(z) = 0$.

Theorem 5.2.2 provides the most common circumstance under which we can guarantee that a nonlinear equation has a solution.

We will also make use of the following two important theorems, both of which are proved in essentially all calculus books.

Theorem 5.2.3 (Integral Mean Value) *If f is continuous on $[a, b]$, then there exists $c \in (a, b)$ so that*

$$\int_a^b f(x) \, dx = f(c)(b - a) .$$

Theorem 5.2.4 (Fundamental Theorem of Calculus) *Let f be continuous on the interval $[a, b]$, and define*

$$F(x) = \int_a^x f(t) \, dt$$

for $x \in [a, b]$. Then F is differentiable on (a, b), and for all $x \in (a, b)$ we have

$$F'(x) = f(x) .$$

Furthermore,

$$\int_a^b F'(x) \, dx = F(b) - F(a) .$$

Exercise 5.2.1 Suppose that f is continuous on $[a, b]$, and the mean value

$$\bar{f} = \frac{1}{b - a} \int_a^b f(x) \, dx$$

has sign opposite that of $f(a)$. Show that f has a zero in $[a, b]$.

5.2.2 Uniqueness

Next, let us consider whether a zero of a nonlinear function is unique. Here is a useful definition that will help our discussion.

Definition 5.2.2 Let the real-valued function f be defined on the closed interval $[a, b]$. Then f is **increasing** on $[a, b]$ if and only if for all $x_1, x_2 \in [a, b]$ with $x_1 \leq x_2$ we have $f(x_1) \leq f(x_2)$. Similarly, f is **decreasing** on $[a, b]$ if and only if for all $x_1, x_2 \in [a, b]$ with $x_1 \leq x_2$ we have $f(x_1) \geq f(x_2)$. The function f is **monotonic** on $[a, b]$ if and only if it is either increasing on $[a, b]$, or decreasing on $[a, b]$. Also, f is **strictly increasing** on $[a, b]$ if and only if for all $x_1, x_2 \in [a, b]$ with $x_1 < x_2$ we have $f(x_1) < f(x_2)$. Similarly, f is **strictly decreasing** on $[a, b]$ if and only if for all $x_1, x_2 \in [a, b]$ with $x_1 < x_2$ we have $f(x_1) < f(x_2)$. The function f is **strictly monotonic** on $[a, b]$ if and only if it is either strictly increasing on $[a, b]$ or strictly decreasing on $[a, b]$.

Definition 5.2.2 leads to the following uniqueness result.

Theorem 5.2.5 *If f is strictly monotonic on $[a, b]$, then for any number y between $f(a)$ and $f(b)$ there is at most one point $x \in [a, b]$ so that $f(x) = y$.*

Proof Suppose that f is strictly monotonic on $[a, b]$, y is between $f(a)$ and $f(b)$, and $f(x_1) = y = f(x_2)$. Suppose that $x_1 \neq x_2$. Without loss of generality, we may assume that $x_1 < x_2$; otherwise, we may interchange x_1 and x_2. Also without loss of generality, we may assume that f is strictly increasing; otherwise we can work with $-f$. Since f is strictly increasing and $x_1 < x_2$, we must have $f(x_1) < f(x_2)$, which contradicts the assumption that $f(x_1) = y = f(x_2)$. We conclude that $x_1 = x_2$.

5.2.3 Perturbation Theory

Finally, let us examine how the zero z of a nonlinear equation $f(z) = 0$ is affected by perturbations to the function f. We begin with the following definition.

Definition 5.2.3 Suppose that the function f is defined on the closed interval $[a, b]$. Then f is **differentiable** at $x \in [a, b]$ if and only if there is a number $f'(x)$, called the **derivative** of f at x, such that for all $\varepsilon > 0$ there exists $\delta > 0$ so that for all $\tilde{x} \in (x - \delta, x + \delta) \cap [a, b]$ we have

$$\left| f(\tilde{x}) - f(x) - f'(x)(\tilde{x} - x) \right| \leq \varepsilon |\tilde{x} - x| \,.$$

The function f is **continuously differentiable** on $[a, b]$ if and only if it is differentiable at every $x \in [a, b]$ and the derivative function f' is continuous on $[a, b]$.

Note that this definition implies that the tangent line

$$\tilde{f}(\tilde{x}) = f(x) - f'(x)(\tilde{x} - x)$$

is a very good approximation to $f(\tilde{x})$ for $\tilde{x}$ near x. The error $|f(\tilde{x}) - \tilde{f}(\tilde{x})|$ goes to zero faster than the difference in arguments $|\tilde{x} - x|$ as $\tilde{x}$ approaches x.

Definition 5.2.3 of the derivative has the following useful consequence.

Theorem 5.2.6 *Suppose that f is defined on a closed interval $[a, b]$ and that there exists $z \in (a, b)$ so that $f(z) = 0$. Let f be continuously differentiable at z, and assume that $f'(z) \neq 0$. Given $\tau \in (0, 1)$, suppose that $\delta > 0$ is such that for all $x \in (z - \delta, z + \delta) \cap (a, b)$ we have*

$$\left|f(x) - f(z) - f'(z)(x - z)\right| < \tau \left|f'(z)\right| |x - z| \; .$$

Then for all $\tilde{z} \in (z - \delta, z + \delta) \cap (a, b)$ we have

$$|\tilde{z} - z| < \frac{1}{1 - \tau} \frac{|f(\tilde{z})|}{|f'(z)|} \; .$$

Proof Note that Definition 5.2.3 of a differentiable function implies that δ exists. It follows that

$$\tau \left|f'(z)\right| |\tilde{z} - z| > \left|f(\tilde{z}) - f(z) - f'(z)(\tilde{z} - z)\right| = \left|f(\tilde{z}) - f'(z)(\tilde{z} - z)\right|$$
$$> \left|f'(z)(\tilde{z} - z)\right| - |f(\tilde{z})| \; ,$$

which in turn implies that

$$|f(\tilde{z})| > (1 - \tau) \left|f'(z)\right| \; |\tilde{z} - z| \; .$$

This is equivalent to the conclusion of the theorem.

Theorem 5.2.6 says that if the perturbation in the zero is small enough, then the error in the solution of the nonlinear equation is essentially bounded by the size of the function at the perturbed point divided by the derivative of the function at the true zero.

We can modify this perturbation analysis slightly, so that it more closely corresponds to perturbations related to rounding errors in a computation.

Corollary 5.2.1 *Suppose that f is defined on a closed interval $[a, b]$ and that there exists $z \in [a, b]$ so that $f(z) = 0$. Let f be continuously differentiable at z, and assume that $f'(z) \neq 0$. Given $\tau \in (0, 1)$, let $\delta > 0$ be such that for all $x \in (z - \delta, z + \delta) \cap [a, b]$ there exists $\mu > 0$ so that for all $x \in (z - \delta, z + \delta) \cap [a, b]$*

$$\left|f'(x)\right| \geq \mu \; , \; \textit{and}$$

$$\textit{for all } x \in (z - \delta, z + \delta) \cap [a, b] \; , \; \left|f(x) - f(z) - f'(z)(x - z)\right| < \tau\mu|x - z| \; .$$

Let the function $\tilde{f}$ be such that there exists $\varepsilon > 0$ so that for all $x \in (z-\delta, z+\delta) \cap [a,b]$ we have $|\tilde{f}(x) - f(x)| \leq \varepsilon$. Then for all $\tilde{z} \in (z-\delta, z+\delta) \cap [a,b]$

$$|\tilde{z} - z| < \frac{|\tilde{f}(\tilde{z})| + \varepsilon}{\mu(1-\tau)} .$$

Proof Since

$$\tau\mu|\tilde{z} - z| > \left| f(z) - f(\tilde{z}) - f'(\tilde{z})(z - \tilde{z}) \right| \geq \mu|\tilde{z} - z| - |f(\tilde{z})| ,$$

it follows that

$$\mu(1-\tau)|\tilde{z} - z| < |f(\tilde{z})| \leq |\tilde{f}(\tilde{z})| + \varepsilon .$$

This is equivalent to the claim.

Note that as δ tends to zero, we can choose μ arbitrarily close to $|f'(z)|$. Also note that if $\tilde{z}$ is a zero of $\tilde{f}$, then we essentially have $|\tilde{z} - z| < \varepsilon/|f'(z)|$. If our objective function f is subject to rounding errors of size ε in computer algorithms, then this bound expresses the smallest error we can reasonably anticipate in our final numerical evaluation of the zero. Thus the term $\varepsilon/[\mu(1-\tau)]$ represents the **maximum attainable accuracy** of the solution to a perturbed problem.

This corollary provides both **termination criteria** for methods used to solve nonlinear equations, and an estimate of the maximum attainable accuracy. Suppose that in a neighborhood of the zero, rounding errors in evaluating the function f are bounded above by ε, and the derivative of f is bounded below by μ. We can stop an algorithm for finding a zero of f when the computed function value $\tilde{f}$ is small enough that the function value plus rounding error divided by μ is less than the desired tolerance in the computed solution z. Furthermore, even if we find an approximate root $\tilde{z}$ for which the computed function value is $\tilde{f}(\tilde{z}) = 0$, the error in the computed solution is still not guaranteed to be smaller than the rounding error in evaluating f divided by μ. Thus ε/μ is the maximum attainable accuracy in computing z.

Example 5.2.1 In Example 2.3.1.5, we considered the function $f(x) = x^2 - a$ in connection with an iteration to compute $z = \sqrt{a}$. If rounding errors cause us to find the zero of

$$\tilde{f}(x) = x^2 - a - \varepsilon ,$$

then the true zero of the perturbed function will be

$$\tilde{z} = \sqrt{a + \varepsilon} \approx \sqrt{a} + \frac{\varepsilon}{2\sqrt{a}} .$$

Whenever $a > 0$, the perturbation in the zero is proportional to ε. The proportionality constant in the perturbation is $1/(2\sqrt{a})$, which is the derivative of f at $z = a$.

Example 5.2.2 Consider the function $f(x) = x^3$. The only zero of f is $z = 0$. Since $f'(z) = 0$ as well, we should expect that the zero is sensitive to perturbations in f. In fact, the zero of $\tilde{f}(x) = f(x) - \varepsilon$ is $\tilde{z} = \varepsilon^{1/3}$. So, for example, a perturbation of $\varepsilon = 10^{-15}$ in the function f leads to a perturbation of 10^{-5} in the zero z.

Exercise 5.2.2 Let f be twice continuously differentiable. Suppose that $f(z) = 0$ and $f'(z) = 0$, but $f''(z) \neq 0$. Show that $g(x) = f(x)/f'(x)$ satisfies $g(z) = 0$ but $g'(z) \neq 0$. What are the pros and cons of finding a zero of g instead of f?

5.3 Bisection

The existence Theorem 5.2.2 provides motivation for a simple but useful algorithm. If we have two points a and b where the function f has values of opposite sign, then a zero of f must lie in the interval between them. We can compute the midpoint of the interval, evaluate f and continue the process with whichever of the two sub-intervals still brackets the zero.

5.3.1 Development

We can formalize these steps in the following algorithm. Suppose that we are given points $a < b$ with $f(a)f(b) < 0$, and we know the sign of the function value $f_a = f(a)$. Then we perform

Algorithm 5.3.1 (Bisection)

while not converged

  $z = (a + b)/2$

  $f_z = f(z)$

  if $f_z = 0$ break

  if $\text{sign}(f_z) \neq \text{sign}(f_a)$ then

    $b = z$

  else

    $a = z$

    $\text{sign}(f_a) = \text{sign}(f_z)$

return z

At the end of the iteration, z is the best approximation to the zero.

5.3.2 *Analysis*

It is easy to show that bisection converges.

Lemma 5.3.1 *Suppose that f is a continuous function on $[a, b]$ and $f(a)f(b) < 0$. Then the bisection Algorithm 5.3.1 converges to a zero of f.*

Proof At the nth step of bisection, we find an interval $[a_n, b_n]$ in which $f(a_n)f(b_n) \leq 0$. Consequently,

$$a_0 \leq a_1 \leq \ldots a_n \leq b_n \leq \ldots \leq b_1 \leq b_0 ,$$

and $f(a_i)f(b_j) \leq 0$ for all $i, j \geq 0$. Since the sequence $\{a_n\}$ is increasing and is bounded above by b_0, it has a limit $a_\infty = \lim_{n\to\infty} a_n \leq b_0$. Similarly, there exists a limit $b_\infty = \lim_{n\to\infty} b_n \geq a_0$, and $b_\infty \geq a_\infty$. Since

$$b_n - a_n = \frac{1}{2}(b_{n-1} - a_{n-1}) = \ldots = 2^{-n}(b_0 - a_0) ,$$

we have that

$$b_\infty - a_\infty = \lim_{n\to\infty} b_n - \lim_{n\to\infty} a_n = \lim_{n\to\infty} \{2^{-n}(b_0 - a_0)\} = 0 .$$

Since $f(a_\infty)f(b_\infty) \leq 0$ and $b_\infty = a_\infty$, we conclude that $f(a_\infty)^2 \leq 0$. Thus $b_\infty = z = a_\infty$ is a zero of f.

This proof shows that, with each iteration, bisection reduces the length of the interval in which the solution can lie by a factor of 2.

In summary, the convergence of bisection is slow, but guaranteed. We gain one bit of accuracy in the solution with each iteration, or one digit of accuracy with each $\log_2 10 \approx 3.3$ iterations. The slow performance is due to the fact that bisection does not use all of the available information: it uses only the sign of the function evaluations, not the magnitude.

5.3.3 *Termination Criteria*

In order to allow the bisection iteration to terminate as soon as it finds an acceptable approximation to the solution, we should provide at least two additional parameters. First, we should always provide a bound on the maximum number of steps in the iteration. This guarantees that the iteration must stop, even if there is a programming error in the computation of the bisection method itself. If the allowed number of steps is too small, then the iteration could stop before the desired accuracy is reached. If the allowed number of steps is too large, the iteration is likely to be terminated by the test in the next paragraph.

Secondly, we should provide a **convergence tolerance** on the solution z. For example, we could provide an **absolute error tolerance** δ_a, meaning that we stop when

$$|b_n - a_n| \leq 2\delta_a \ .$$

When this test is satisfied, we know that a zero of the computed function lies within an interval of width δ_a of the midpoint $z_n = (b_n + a_n)/2$. Alternatively, we could provide a **relative error tolerance** δ_r, and stop when

$$|b_n - a_n| \leq 2\delta_r|z_n| \ .$$

If z_n is nonzero, then Lemma 5.3.1 implies that the number of iterations n required to guarantee the relative error condition satisfies

$$2^{-n}(b-a) \leq 2|x_n|\delta_r \ .$$

Thus the number of bisection iterations required for the relative error condition is approximately

$$n \geq 1 + \log_2\left(\frac{b-a}{|z|\delta_r}\right)$$

If the initial upper and lower bounds a and b, as well as the root z are all machine numbers with the same exponent, and if the mantissa of these numbers contains m bits, then bisection will need at most $m+1$ iterations to find all of the bits in the zero. At this point, the machine numbers for a_n and b_n are either identical or consecutive. The exception to this statement occurs when the root is $z = 0$. In this case *bisection will proceed until underflow occurs*, after reducing all the bits in both the mantissa and exponent to the lowest possible values.

As a third alternative, we may provide a convergence tolerance δ_f on the absolute value of f computed during bisection. If we stop when we find that $|f(z)| \leq \delta_f$, then we know from Corollary 5.2.1 that a zero of the computed solution lies within an interval of approximate width $2\delta_f/\mu$, where μ is a lower bound on the absolute value of the derivative near the solution z. However, since bisection does not require or compute the derivative of f, it is hard to know how to choose the function tolerance for this test. Even worse, since bisection only requires the sign of f for its operation, even the function value may not be available for this convergence test.

It is possible to specify convergence tolerances that are too small for the available precision. If the convergence tolerance on the solution is smaller than machine precision times $\max\{|a|\ ,\ |b|\}$, then the convergence test based on the width $b - a$ of the interval will not be satisfied unless b is equal to a. On the other hand, if the convergence tolerance on the function value f is smaller than the size of the rounding errors that occur in the evaluation of f, then a convergence test based on the absolute value of f may never be satisfied.

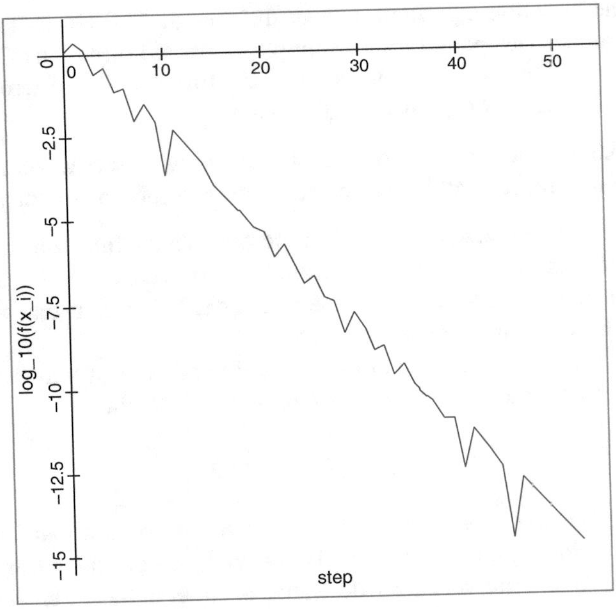

Fig. 5.1 Error in bisection algorithm: $log_{10}(f)$ versus iterations

Fortunately, an appropriate **maximum attainable accuracy** test will always terminate the algorithm. For bisection, we test maximum attainable accuracy by checking whether the midpoint z satisfies $a < z < b$. If this test is not satisfied, then a and b are consecutive machine numbers, and there is no point in continuing the bisection iteration.

Interested readers may experiment with a JavaScript **bisection program.** Alternatively, readers may view a Fortran bisection program, or a C^{++} bisection program that plots the function during the iteration. This C^{++} program also demonstrates how to use the GSL bisection routine gsl_root_fsolver_bisection All of these programs iterate to find a zero of $f(x) = x^2 - 4\sin(x)$. The zeros are $z = 0$ and $z \approx 1.9337573$. A plot of the error in bisection is shown in Fig. 5.1.

Exercise 5.3.1 Suppose that we compute an initial value of $h = b - a$, and inside the bisection iteration we update $h \leftarrow h * 0.5$ followed by taking the midpoint to be $x = a + h$ (instead of $x = (a + b) * .5$). This form of the computation was suggested by Kincaid and Cheney [66, p. 59]. The computation $x = a + h$ guarantees that $a \leq x \leq b$, while $x = (a + b) * 0.5$ could lie outside the interval $[a, b]$. Examine the floating point operations (see Sect. 2.3.1) in computing $x = (a + b) * 0.5$ to see how large $h = b - a$ may be when the computed value of x satisfies $x < a$ or $x > b$. How does this size of h compare to the size of h when the computed value of $x = a + h/2$ satisfies $x <= a$ or $x >= b$?

Exercise 5.3.2 Suppose we would like to find a zero of $f(x) = \frac{1}{x}$. We note that $f(-1) = -1$ and $f(2) = 1/2$, so we begin bisection with $a = -1$ and $b = 1/2$.

What will the bisection algorithm do near the pole of f? What is wrong with this problem, from the viewpoint of the assumptions behind bisection? Can we fix the algorithm to catch this problem? (Make sure that you can still find zeros of functions that get large before they rapidly cross the axis.)

Exercise 5.3.3 Kahan [61] made several interesting observations regarding the difficulty in providing a hand-held calculator button to solve a nonlinear equation.

1. Consider $f(x) = x+2.*(x-5.)$. (It is important that the function be programmed in this form.) Using the bisection algorithm to find a zero of f in $(1., 20./3.)$, show that $f(x)$ is never zero for any machine number x. This means that we cannot use $|f(x)| = 0.$ as a termination requirement.
2. Let $c = f(z)$ where z is the ultimate machine number at which the bisection method stops after attempting to find a zero of $f(x)$. Define

$$g(x) = 1 - 2\exp(-|f(x)|/c^2) \ .$$

Show that $g(x)$ has two zeros at $10/3 \pm (c^2/3)\ln 2$. Also show that for all machine numbers the floating-point value of g is always 1. According to our perturbation theory, since $g(x)$ has a very large derivative at either of these zeros, this problem should have very well-conditioned zeros. Why is the perturbation theory useless in this example? Can you suggest a change of variables that will make this problem better for computation?
3. Consider the two functions $h_1(x) = 1./f(x)$ and

$$h_2(x) = \frac{1/c}{(f(x)/c) + c^2/(f(x)/c)} \ .$$

Show that these two functions have identical values, after rounding, for all machine numbers. Also show that, analytically speaking, $h_1(x)$ has a pole at the zero of $f(x)$ while $h_2(x)$ has a zero at the zero of $f(x)$. This means that it is impossible for the machine to distinguish between finding roots and finding poles. Since $h_2(x)$ has a very large derivative at its zero, our perturbation theory again suggests that the zero of h_2 is well-conditioned. Is there a numerical difficulty in computing the zero of h_2 by bisection?

Exercise 5.3.4 Use the bisection algorithm to prove the intermediate value theorem.

5.4 Newton's Method

Bisection converges slowly, because it uses only the sign of the function value to compute a new guess for the zero of the function. As a result, it cannot make a simplified model of the given function, nor can it approximate the zero of the true solution by computing a zero of some model function.

Newton's method works by approximating a given function f by its tangent line at an approximate zero. Consequently, this method approximates the true zero of f by the zero of the tangent line to f at some approximate zero z_k. Hopefully, the distance from the zero of this tangent line to the true zero z of f is less than the distance from z_k. Of course, the tangent line may not have a zero if $f'(z_k) = 0$. Also, Newton's method requires both the function value $f(z_k)$ and the derivative $f'(z_k)$ to construct the tangent line. The need for the derivative introduces extra cost per iteration, when compared to bisection. The derivative evaluation also requires programming care, to make sure that the program for the derivative f' is consistent with the program for the function f.

5.4.1 *Algorithm Development*

Mathematically speaking, Newton's method approximates the true function $f(x)$ by the tangent line

$$\tilde{f}(x) = f(z_k) + f'(z_k)(x - z_k) .$$

We can solve $\tilde{f}(\tilde{z}) = 0$ for $\tilde{z}$ to obtain the following

Algorithm 5.4.1 (Newton's Method)

while not converged

$$z_{k+1} = z_k - f(z_k)/f'(z_k)$$

Exercise 5.4.1 Describe Newton's method to find a zero of $f(x) = x^2 - a$ for some given number a. Describe how to use the exponent in the hexadecimal representation of a to provide an initial guess for zero of f.

Exercise 5.4.2 Describe Newton's method to find a zero of $f(x) = x^{-2} - a$ for some given number a. Describe how to use the exponent in the hexadecimal representation of a to provide an initial guess for zero of f.

Exercise 5.4.3 Describe Newton's method to find a zero of $f(x) = 1/x - a$ for some given number a. Describe how to use the exponent in the hexadecimal representation of a to provide an initial guess for zero of f.

Exercise 5.4.4 Describe Newton's method to find a zero of $f(x) = \tan^{-1}(x)$. Show that $f(x) \uparrow \pi/2$ as $x \uparrow \infty$, $f(x) \downarrow -\pi/2$ as $x \downarrow -\infty$, and that $f'(x) > 0$ for all x. Explain why these facts imply that f has a unique zero. Show that Newton's method converges for any initial guess $\tilde{z}$ satisfying $|\tilde{z}| < 1.391745\ldots$, and diverges otherwise.

5.4.2 Convergence Behavior

The next lemma will analyze the numerical performance of Newton's method.

Lemma 5.4.1 *Suppose that f is continuously differentiable on some interval $(z-\alpha, z+\alpha)$, where $f(z) = 0 \neq f'(z)$. Also let f' be* **Lipschitz continuous** *on $(z-\alpha, z+\alpha)$, meaning that there is a constant $\gamma > 0$ so that for all x and $\tilde{x}$ in this interval we have*

$$|f(x) - f(\tilde{x})| \leq \gamma |x - \tilde{x}| \ .$$

Suppose that $z_k \in (z-\alpha, z+\alpha)$ is an approximate zero of f. If $f'(z_k) \neq 0$, and

$$|z_k - z| < 2\frac{|f'(z_k)|}{\gamma} \ ,$$

then the Newton iterate $z_{k+1} = z_k - f(z_k)/f'(z_k)$ exists, $z_{k+1} \in (z-\alpha, z+\alpha)$, and its error satisfies

$$|z_{k+1} - z| \leq \frac{\gamma}{2\,|f'(z_k)|}\,|z_k - z|^2 < |z_k - z| \ . \tag{5.1}$$

Alternatively, if

$$|z_k - z| < \frac{2}{3}\frac{|f'(z)|}{\gamma} \ ,$$

then the Newton iterate $z_{k+1} = z_k - f(z_k)/f'(z_k)$ exists, $z_{k+1} \in (z-\alpha, z+\alpha)$, its error satisfies

$$|z_{k+1} - z| \leq \frac{3}{2}\frac{\gamma}{|f'(z)|}\,|z_k - z|^2 < |z_k - z| \ . \tag{5.2}$$

If, in addition, f'' is continuous at z, then

$$\lim_{z_k \to z} \frac{z_{k+1} - z}{[z_k - z]^2} = \frac{f''(z)}{2f'(z)} \ .$$

Proof Algorithm 5.4.1 for Newton's method implies that

$$z_{k+1} - z = z_k - z - \frac{f(z_k)}{f'(z_k)} = \frac{f(z) - f(z_k) - f'(z_k)(z - z_k)}{f'(z_k)} \ .$$

The fundamental theorem of calculus 5.2.4 implies that

$$\left|f(z)-f(z_k)-f'(z_k)(z-z_k)\right| = \left|\int_{z_k}^{z} f'(\zeta)-f'(z_k)\,\mathrm{d}\zeta\right|$$

then Lipschitz continuity of f' implies that

$$\le \left|\int_{z_k}^{z} \gamma\,|\zeta - z_k|\,\mathrm{d}\zeta\right| = \frac{\gamma}{2}\,|z_k - z|^2 \,. \tag{5.3}$$

Thus

$$|z_{k+1}-z| \le \frac{\gamma\,|\,|z_k - z|^2}{2\,|f'(z_k)|}\,|z_k - z|^2 \,.$$

If

$$|z_k - z| < 2\frac{|f'(z_k)|}{\gamma}\,,$$

then it is easy to see that (5.1) is satisfied. This proves the first claim.

Lipschitz continuity also implies that

$$\left|f'(z_k)\right| = \left|f'(z) + [f'(z_k)-f'(z)]\right| \ge \left|f'(z)\right| - \gamma\,|z_k - z| \,.$$

If

$$|z_k - z| < \frac{2}{3}\frac{|f'(z)|}{\gamma}\,,$$

then (5.4.2) shows that

$$\left|f'(z_k)\right| \ge \left|f'(z)\right| - \gamma\,|z_k - z| > \frac{1}{3}\left|f'(z)\right| > 0\,.$$

Then this inequality and (5.3) give us

$$|z_k - z| < \frac{\frac{\gamma}{2}\,|z_k - z|^2}{\frac{1}{3}\,|f'(z)|} < |z_k - z| \,.$$

This proves the second claim (5.2).

Let us prove the third claim. If f'' is continuous at z, then

$$\lim_{z_k \to z} \frac{z_{k+1}-z}{(z_k - z)^2} = \lim_{z_k\to z} \frac{f(z)-f(z_k)-f'(z_k)(z-z_k)}{f'(z_k)(z-z_k)^2}$$

$$= \lim_{z_k\to z} \frac{\int_{z_k}^{z} f'(\zeta)-f'(z_k)\,\mathrm{d}\zeta}{f'(z_k)(z-z_k)^2} = \lim_{z_k\to z} \frac{\int_{z_k}^{z}\int_{z_k}^{\zeta} f''(\eta)\,\mathrm{d}\eta\,\mathrm{d}\zeta}{f'(z_k)(z-z_k)^2} = \frac{1}{2}\frac{f''(z)}{f'(z)}\,.$$

Table 5.1 Newton iterates for zero of $f(x) = x^2 - 2$

n	z_n
0	1.
1	1.5
2	1.416...
3	1.414215...
4	1.414213562374...
5	1.4142135623730951

This lemma shows that under normal circumstances, as the Newton iterate z_k approaches the true solution z, we double the number of correct digits (or bits) in the solution with each iteration.

Example 5.4.1 Suppose that we want to find the zero of

$$f(x) = x^2 - 2 .$$

Newton's method gives us the algorithm

$$z_{k+1} = z_k - f(z_k)/f'(z_k) = (z_k + 2/z_k)/2 .$$

If we choose $z_0 = 1.$, then the sequence of Newton approximations to the zero of f are given in Table 5.1. The results in this table show that the number of correct digits in the Newton approximation to $z = \sqrt{2} \approx 1.4142135623730950488\dots$ essentially doubles with each iteration, until limiting accuracy of the machine is reached.

Exercise 5.4.5 Suppose that f is twice continuously differentiable, and assume that $f(z) = 0, f'(z) = 0$ but $f''(z) \neq 0$. Show how to bound the errors $e_k = z_k - z$ in Newton's method for f.

Exercise 5.4.6 Suppose that f is three times continuously differentiable, and assume that $f(z) = 0, f'(z) \neq 0, f''(z) = 0$ and $f'''(z) \neq 0$. Show how to bound the errors $e_k = z_k - z$ in Newton's method for f.

5.4.3 Convergence Rates

Our analysis of the convergence of Newton's method in Sect. 5.4.2 suggests the following definition.

Definition 5.4.1 Suppose that the sequence $\{e_k\}_{k=0}^{\infty}$ of nonzero scalars converges to zero. Then $\{e_k\}_{k=0}^{\infty}$ has **order of convergence** $p \geq 1$ and **asymptotic error constant** c if and only if

$$\lim_{k\to\infty} \frac{|e_{k+1}|}{|e_k|^p} = c .$$

The sequence is **linearly convergent** if and only if the order of convergence is one and the asymptotic error constant satisfies $c \in (0, 1)$. The sequence is **quadratically convergent** if and only if the order of convergence is two. The sequence is **superlinearly convergent** if and only if

$$\lim_{k\to\infty} \frac{|e_{k+1}|}{|e_k|} = 0 .$$

Finally, a sequence $\{e_k\}_{k=0}^{\infty}$ is **convergent with rate** $r > 1$ if and only if there is an integer $K > 0$ and a scalar $c \geq 0$ such that for all $k > K$

$$|e_{k+1}| \leq c|e_k|^r .$$

Here are some examples of sequences that have different orders of convergence to zero.

Example 5.4.2 Suppose that $a > 0$.

1. The sequence $e_k = a/(k+1)$ is convergent with order one, but is not linearly convergent.
2. For any $\varrho > 1$ the sequence $e_k = a\varrho^{-k}$ is linearly convergent, but not superlinearly convergent.
3. For any $\varrho > 1$ the sequence $e_k = a\varrho^{-k}/k!$ is superlinearly convergent, but does not have an order of convergence greater than one, and is not convergent with a rate greater than one.
4. For any $\varrho > 1$ the sequence $e_k = a\varrho^{-2^k}$ is quadratically convergent.
5. For any $\varrho > 1$ and $r > 1$ the sequence $e_k = a\varrho^{-r^k}$ is convergent with rate r.

In general, the error $e_k = z_k - z$ in Newton's method has order 2 and asymptotic error constant

$$c = \frac{1}{2}\frac{f''(z)}{f'(z)} .$$

On the other hand, the error in bisection does not have an order of convergence, and does not have a convergence rate.

Exercise 5.4.7 If $x_k = \log(k)/k$, is the sequence $\{x_k\}_{k=1}^{\infty}$ convergent? If so, is it linearly convergent?

Exercise 5.4.8 If $\mathbf{x}_k = k^{-k}$, determine whether $\{\mathbf{x}_k\}_{k=0}^{\infty}$ is convergent, linearly convergent, superlinearly convergent and/or quadratically convergent.

Exercise 5.4.9 Suppose that f is twice continuously differentiable, and assume that $f(z) = 0, f'(z) = 0$ but $f''(z) \neq 0$. What is the convergence rate of Newton's method for such a function?

Exercise 5.4.10 Suppose that f is three times continuously differentiable, and assume that $f(z) = 0, f'(z) \neq 0, f''(z) = 0$ and $f'''(z) \neq 0$. What is the convergence rate of Newton's method for such a function?

5.4.4 Convex Functions

For certain kinds of functions, we can show that Newton's method will converge from any initial guesses that are not necessarily close to the zero. Here is a mathematical definition of the functions we have in mind.

Definition 5.4.2 Suppose that f is defined on the interval (a, b). Then f is **convex** on this interval if and only if for all $x, \tilde{x} \in (a, b)$ and all $\alpha \in [0, 1]$ we have

$$f\left(x\alpha + \tilde{x}[1-\alpha]\right) \leq f(x)\alpha + f(\tilde{x})[1-\alpha] \,. \tag{5.4}$$

Geometrically speaking, this definition says that the chord between the two points $(x, f(x))$ and $(\tilde{x}, f(\tilde{x}))$ lies above the graph of the function f.

Using this definition, we can prove the next lemma, which shows that a continuously differentiable function is convex if and only if its tangent lines lie strictly below its graph.

Lemma 5.4.2 *Suppose that f is continuously differentiable on the interval (a, b). Then f is convex on (a, b) if and only if for all $x, \tilde{x} \in (a, b)$ we have*

$$f(\tilde{x}) \geq f(x) + f'(x)(\tilde{x} - x) \,. \tag{5.5}$$

Proof Let us begin by proving the forward direction. Suppose that f is convex. Then for all $\alpha \in (0, 1]$ we have

$$\frac{f\left(\tilde{x} + [x - \tilde{x}]\alpha\right) - f(\tilde{x})}{\alpha} \leq f(x) - f(\tilde{x}) \,.$$

In the limit as $\alpha \downarrow 0$ we obtain a result equivalent to the claim.

Next, we will prove the reverse direction. Suppose that (5.5) is satisfied for $x, \tilde{x} \in (a, b)$. Then for all $\xi_1, \xi_2 \in (a, b)$ and all $\alpha \in [0, 1]$ we can choose $x = \xi_1\alpha + \xi_2(1-\alpha)$. By choosing either $\tilde{x} = \xi_1$ or $\tilde{x} = \xi_2$, we obtain

$$f(\xi_1) \geq f(x) + f'(x)[\xi_1 - x] \text{ and}$$
$$f(\xi_2) \geq f(x) + f'(x)[\xi_2 - x] \,.$$

We can multiply the former inequality by α, the latter by $1 - \alpha$ and add to obtain

$$f(\xi_1)\alpha + f(\xi_2)[1-\alpha] \geq f(x) + f'(x)[\xi_1\alpha + \xi_2(1-\alpha) - x] = f(x) = f(\xi_1\alpha + \xi_2[1-\alpha]) \,.$$

This proves that f is convex.

The previous lemma leads us to the following convergence theorem.

Theorem 5.4.1 *Suppose that f is convex and continuously differentiable on the interval (a, b). Also suppose that f has a zero in (a, b). Then for any initial guess $z_0 \in (a, b)$ with $f(z_0) > 0$, Newton's method converges monotonically to a zero of f.*

Proof Suppose that z_0 is less than the smallest zero z of f in (a, b), and $f(z_0) > 0$. Then Lemma 5.4.2 implies that

$$0 = f(z) \geq f(z_0) + f'(z_0)(z - z_0) ,$$

from which we conclude that

$$f'(z_0) \leq -\frac{f(z_0)}{z - z_0} < 0 .$$

It follows that the Newton iterate $z_1 = z_0 - f(z_0)/f'(z_0)$ satisfies

$$z_0 < z_0 + \frac{f(z_0)}{-f'(z_0)} = z_1 \leq z_0 + \frac{f(z_0)}{f(z_0)/(z - z_0)} = z_0 + (z - z_0) = z .$$

The same argument shows that in general the sequence of Newton iterates is strictly increasing and bounded above by the smallest zero of f in (a, b). Because this sequence is increasing and bounded above, it must converge to some point $r \leq z$. At that point r, the Newton iteration implies that

$$r = r - \frac{f(r)}{f'(r)} .$$

Since f is continuously differentiable on $[z_0, z]$, it is bounded above and below on this interval, and cannot be infinite at r. We conclude that $f(r) = 0$. Since r is less than or equal to the smallest zero z of f, we must have $r = z$.

The proof when z_0 is greater than the largest zero of f in (a, b) and $f(z_0) > 0$ is similar.

Note that if f is convex on the entire real line and we begin Newton's iteration at a point where $f(z_0) < 0$, then the next Newton will lie at a point where f is positive. On the other hand, if the convexity of f is restricted to some interval (a, b), such a Newton step may take us beyond this interval of convexity.

Exercise 5.4.11 Suppose that f is continuously differentiable on $[a, b]$. Also assume that f' is nonzero and that f'' has constant sign on $[a, b]$. Let $f(a)f(b) < 0$, and suppose that the Newton steps begun at either a or b take us to points within (a, b). Show that Newton's method converges from any initial guess in $[a, b]$.

5.4.5 *Kantorovich Theorem*

Kantorovich [62] showed that under certain circumstances, it is possible to relate the sequence of Newton iterates for some given function to the Newton iterates for a convex quadratic function. By showing that the Newton iterates for the quadratic function converge at a certain rate, Kantorovich was able to prove that the Newton iterates for the original function converge at least as fast.

We will begin by studying the Kantorovich quadratic function.

Lemma 5.4.3 *Let γ, β and σ be positive numbers satisfying*

$$\gamma\sigma\beta \leq 1/2 \, .$$

Then

$$\phi(\xi) \equiv \frac{\gamma\xi^2}{2} - \frac{\xi}{\beta} + \frac{\sigma}{\beta}$$

is convex and has two positive zeros. The smaller zero is

$$\zeta = \frac{1 - \sqrt{1 - 2\gamma\beta\sigma}}{\gamma\beta} \, .$$

If $\zeta_0 = 0$ and $\{\zeta_k\}$ is the sequence of Newton iterates for ϕ, then $\zeta_k \uparrow \zeta$, and for all $k \geq 0$ we have both $\phi(\zeta_k) > 0$ and $\phi'(\zeta_k) < 0$. Furthermore, for all $k \geq 0$ we have

$$\frac{\phi'(\zeta_{k+1})}{\phi'(\zeta_k} \geq \frac{1}{2} \tag{5.6}$$

and

$$\frac{1}{-\phi'(\zeta_k)} \leq 2^k \beta \, . \tag{5.7}$$

Finally, for all $k \geq 0$ we have

$$\zeta - \zeta_{k+1} \leq 2^{k-1} \beta\gamma (\zeta - \zeta_k)^2 \tag{5.8}$$

and

$$\zeta - \zeta_k \leq \frac{(\gamma\beta\zeta)^{2^k}}{2^k \gamma\beta} = \frac{\left[1 - \sqrt{1 - 2\gamma\beta\sigma}\right]^{2^k}}{2^k \gamma\beta} \, . \tag{5.9}$$

Proof We can easily compute $\phi'(\xi) = \gamma\xi - 1/\beta$. The smaller zero ζ of ϕ is easily found from the quadratic formula, and the assumption $\gamma\sigma\beta \leq 1/2$ shows that this smaller zero is real and positive. Furthermore, $\phi(\zeta_0) = \phi(0) = \sigma/\beta > 0$. Since $\phi''(\xi) = \gamma > 0$, ϕ is convex.

Theorem 5.4.1 proves that the sequence $\{\zeta_k\}$ computed by Newton's method increases monotonically to ζ. Since $\zeta_k < \zeta$ for all k, it follows (say, from factoring

ϕ in terms of its zeros) that $\phi(\zeta_k) > 0$ for all k. The facts that $\phi'(1/[\gamma\beta]) = 0$, that ϕ' is strictly increasing and that $\zeta < 1/(\gamma\beta)$ imply that $\phi'(\zeta_k) < 0$ for all k.

Next, we will prove (5.6). Since $\gamma\beta\sigma \leq 1/2$, we have

$$\gamma\phi(\zeta_k) = \gamma\left[\frac{\gamma}{2}\zeta_k^2 - \frac{\zeta_k}{\beta} + \frac{\sigma}{\beta}\right] \leq \frac{\gamma^2}{2}\zeta_k^2 - \frac{\gamma}{\beta}\zeta_k + \frac{1}{2\beta^2} = \frac{1}{2}\left[\gamma\zeta_k - \frac{1}{\beta}\right]^2 = \frac{1}{2}\phi'(\zeta_k)^2 .$$

It follows that

$$\frac{\phi'(\zeta_{k+1})}{\phi'(\zeta_k)} = \frac{\gamma\zeta_{k+1} - 1/\beta}{\gamma\zeta_k - 1/\beta} = \frac{\gamma\left[\zeta_k - \phi(\zeta_k)/\phi'(\zeta_k)\right] - 1/\beta}{\gamma\zeta_k - 1/\beta} = 1 - \frac{\gamma\phi(\zeta_k)}{\phi'(\zeta_k)^2} \geq \frac{1}{2} .$$

Let us prove (5.7) by induction. Since $\phi'(\zeta_0) = -1/\beta$, this inequality is obviously true for $k = 0$. Assume inductively that it is true for $k \geq 0$. Then inequality (5.6) implies that

$$-\phi'(\zeta_{k+1}) \geq -\frac{1}{2}\phi'(\zeta_k) \geq \frac{1}{2}\frac{1}{2^k\beta} = \frac{1}{2^{k+1}\beta} .$$

Our next goal is to prove (5.8). Note that for all $k \geq 0$ we have

$$\zeta - \zeta_{k+1} = \zeta - \zeta_k + \frac{\phi(\zeta_k)}{\phi'(\zeta_k)} = \frac{\phi(\zeta) - \phi(\zeta_k) - \phi'(\zeta_k)(\zeta - \zeta_k)}{-\phi'(\zeta_k)} = \frac{\int_{\zeta_k}^{\zeta} \phi'(\xi) - \phi'(\zeta_k)\, d\xi}{-\phi'(\zeta_k)}$$

$$= \frac{\gamma\int_{\zeta_k}^{\zeta} \xi - \zeta_k\, d\xi}{-\phi'(\zeta_k)} = \frac{1}{2}\frac{\gamma(\zeta - \zeta_k)^2}{-\phi'(\zeta_k)} \leq 2^{k-1}\gamma\beta(\zeta - \zeta_k)^2 .$$

Finally, we will prove (5.9) by induction. This inequality is obvious for $k = 0$. Assume inductively that it is true for $k \geq 0$. Then (5.8) implies that

$$\zeta - \zeta_{k+1} \leq 2^{k-1}\gamma\beta(\zeta - \zeta_k)^2 \leq 2^{k-1}\gamma\beta\left[\frac{1}{2^k\gamma\beta}(\gamma\beta\zeta)^{2^k}\right]^2 = \frac{1}{2^{k+1}\gamma\beta}(\gamma\beta\zeta)^{2^{k+1}} .$$

Now we are ready to prove

Theorem 5.4.2 (Kantorovich) *Suppose that $\alpha > 0$ and the real-valued function f is continuously differentiable in the interval $(z_0 - \alpha, z_0 + \alpha)$. Further assume that f' is Lipschitz continuous in this interval, meaning that there is a positive number $\gamma > 0$ so that for all $x, \tilde{x} \in (z_0 - \alpha, z_0 + \alpha)$ we have*

$$\left|f'(x) - f'(\tilde{x})\right| \leq \gamma|x - \tilde{x}| . \tag{5.10}$$

Also assume that $f'(z_0) \neq 0$ and that

$$\frac{\gamma\,|f(z_0)|}{|f'(z_0)|^2} \leq \frac{1}{2} . \tag{5.11}$$

Let

$$\beta = \frac{1}{|f'(z_0)|}, \quad \sigma = \frac{|f(z_0)|}{|f'(z_0)|}.$$

Assume that $\gamma\beta\sigma \geq 1/2$, *and that*

$$\alpha \geq \zeta \equiv \frac{1 - \sqrt{1 - 2\gamma\beta\sigma}}{\gamma\beta}.$$

Then the sequence $\{z_k\}$ *of Newton iterates is well-defined and is contained in the interval* $[z_0 - \zeta, z_0 + \zeta]$. *Furthermore,* $\{z_k\} \to z \in [z_0 - \zeta, z_0 + \zeta]$ *and*

$$|z_k - z| \leq \frac{[1 - \sqrt{1 - 2\gamma\beta\sigma}]^{2^k}}{2^k\gamma\beta}. \tag{5.12}$$

Proof Define

$$\phi(\xi) = \frac{\gamma}{2}\xi^2 - \frac{\xi}{\beta} + \frac{\sigma}{\beta}.$$

Let $\{z_k\}$ be the sequence of Newton iterates determined by the initial guess z_0 for a zero of f, and let $\{\zeta_k\}$ be the sequence of Newton iterates determined by the initial guess $\zeta_0 = 0$ for a zero of ϕ. First, we will show by induction that for all $k \geq 0$ we have

$$\left|f'(z_k)\right| \geq -\phi'(\zeta_k) \tag{5.13}$$

and

$$|z_{k+1} - z_k| \leq \zeta_{k+1} - \zeta_k. \tag{5.14}$$

For $k = 0$, we note that

$$|f'(z_0)| = \frac{1}{\beta} = -\phi'(0) = -\phi'(\zeta_0),$$

and

$$|z_1 - z_0| = \frac{|f(z_0)|}{|f'(z_0)|} = \sigma = -\frac{\phi(0)}{\phi'(0)} = \zeta_1 - \zeta_0.$$

Assume inductively that (5.13) and (5.14) are true for $k \geq 0$. Then

$$\begin{aligned} |f'(z_{k+1})| &\geq |f'(z_k)| - |f'(z_{k+1}) - f'(z_k)| \geq |f'(z_k)| - \gamma||z_{k+1} - z_k| \\ &\geq -\phi'(\zeta_k) - \gamma(\zeta_{k+1} - \zeta_k) = -\phi'(\zeta_{k+1}). \end{aligned}$$

This proves (5.13). Next, we note that

$$\begin{aligned}
&|z_{k+1}-z_k|\left|f'(z_k)\right| = |f(z_k)| = \left|f(z_k)-f(z_{k-1})-f'(z_{k-1})\,(z_k-z_{k-1})\right| \\
&= \left|\int_{z_{k-1}}^{z_k} f'(x)-f'(z_{k-1})\,\mathrm{d}x\right| \le \gamma\left|\int_{z_{k-1}}^{z_k} x-z_{k-1}\,\mathrm{d}x\right| = \frac{1}{2}\gamma|z_k-z_{k-1}|^2 \\
&\le \frac{1}{2}\gamma(\zeta_k-\zeta_{k-1})^2 = \gamma\int_{\zeta_{k-1}}^{\zeta_k} \xi-\zeta_{k-1}\,\mathrm{d}\xi = \int_{\zeta_{k-1}}^{\zeta_k} \phi'(\xi)-\phi'(\zeta_{k-1})\,\mathrm{d}\xi \\
&= \phi(\zeta_k)-\phi(\zeta_{k-1})-\phi'(\zeta_{k-1})(\zeta_k-\zeta_{k-1}) = \phi(\zeta_k)\,.
\end{aligned}$$

It follows that

$$|z_{k+1}-z_k| \le \frac{\phi(\zeta_k)}{|f'(z_k)|} \le -\frac{\phi(\zeta_k)}{\phi'(\zeta_k)} = \zeta_{k+1}-\zeta_k\,.$$

Then (5.14) and (5.9) imply that

$$\sum_{k=0}^{\infty}|z_{k+1}-z_k| \le \sum_{k=0}^{\infty}(\zeta_{k+1}-\zeta_k) = \zeta-\zeta_0 = \zeta\,.$$

This shows that the telescoping series $\sum_{k=0}^{\infty} z_{k+1}-z_k$ converges absolutely. As a result, $z_0+\sum_{k=j}^{\infty}(z_{k+1}-z_k)$ converges to some value z, and $|z-z_0| \le \zeta$. Thus z lies within the interval where f' is Lipschitz continuous. A similar argument shows that all of the Newton iterates z_k lie within this interval. It also follows from (5.13) and (5.7) that

$$\left|f'(z_k)\right| \ge -\phi'(\zeta_k) \ge \frac{1}{2^k\beta}\,.$$

Thus the sequence $\{z_k\}$ of Newton iterates is well-defined and converge to a zero of f.

We remark that the Kantorovich theorem can be amended to show that the Newton iteration converges to a unique zero of f within the given interval. For more details, see either Dennis [30] or Kantorovich [62].

In summary, the advantages of Newton's method are that it converges (or diverges) quickly in most circumstances, and that only one starting guess is needed. The disadvantages are that Newton's method is only **locally convergent**, and requires evaluation of the derivative of f. In particular, note that the programmer must be sure that the program for f' is consistent with the program for f.

Exercise 5.4.12 Determine experimentally the interval of points around the origin for which the Kantorovich Theorem 5.4.2 guarantees the convergence of Newton's method for finding a zero of $\tan^{-1}$ using that point as an initial guess. Compare this interval to the interval $|z_0| < 1.391745\ldots$ found in the exercises of Sect. 5.4.1, for which Newton's method applied to $\tan^{-1}$ converges.

5.4.6 Contractive Mappings

We will present yet another situation in which we can prove the convergence of an iteration.

Definition 5.4.3 Suppose that ϕ is a function defined on some closed interval $[a, b]$. Then ϕ is **contractive** if and only if there exists a constant $\mu \in (0, 1)$ so that for all $x, \tilde{x} \in [a, b]$ we have

$$|\phi(x) - \phi(\tilde{x})| \leq \mu |x - \tilde{x}| .$$

In other words, a contractive function is a Lipschitz continuous function with Lipschitz constant less than one.

Contractive functions are often sought in connection with finding a **fixed point** x of a function ϕ, defined by the equation

$$\phi(x) = x .$$

Often, fixed points are found as the limits of the **fixed point iteration**

$$x_{k+1} = \phi(x_k) .$$

The following important theorem explains why.

Theorem 5.4.3 (Contractive Mapping) *Suppose that ϕ is a contractive function on the closed interval $[a, b]$, and that for all $x \in [a, b]$ we have $\phi(x) \in [a, b]$. Then ϕ has a unique fixed point in $[a, b]$.*

Proof Given any point $x_0 \in [a, b]$, define the sequence $\{x_k\}_{k=0}\infty$ by the fixed point iteration $x_{k+1} = \phi(x_k)$. Since ϕ maps $[a, b]$ into the same interval, this sequence is well-defined and is contained in $[a, b]$. Also, for all $k > 0$ we have

$$|x_{k+1} - x_k| = |\phi(x_k) - \phi(x_{k-1})| \leq \mu |x_k - x_{k-1}| .$$

It follows that for all $n > 0$ we have

$$|x_{k+n} - x_k| = \left| \sum_{j=1}^{n} x_{k+j} - x_{k+j-1} \right| \leq \sum_{j=1}^{n} \left| x_{k+j} - x_{k+j-1} \right| \leq \sum_{j=1}^{n} \mu^{j-1} |x_{k+1} - x_k|$$
$$= \frac{1 - \mu^n}{1 - \mu} |x_{k+1} - x_k| \leq \frac{\mu^k}{1 - \mu} |x_1 - x_0| .$$

Thus for all $\varepsilon > 0$, we can choose $k > \log(\varepsilon[1 - \mu]/|x_1 - x_0|)/\log \mu$ and find that for all $n > 0$ we have $|x_{k+n} - x_k| < \varepsilon$. It follows that $\{x_k\}$ converges to some point $x \in [a, b]$. Since ϕ is continuous, the fixed point iteration shows that x is a fixed point of ϕ.

Note that Newton's method is a fixed point iteration with

$$\phi(x) = x - \frac{f(x)}{f'(x)} .$$

If f is twice continuously differentiable, then

$$\phi'(x) = \frac{f(x)f''(x)}{f'(x)^2} .$$

If for some interval $[a, b]$ this derivative is bounded in absolute value by some constant $\mu \in (0, 1)$, and if ϕ maps this interval into itself, then Newton's method will converge from any initial guess in this interval. Unfortunately, these conditions are harder to verify than the assumptions of the Kantorovich theorem.

5.4.7 Rounding Errors

Next, we would like to discuss the effect of rounding errors in Newton's method.

Lemma 5.4.4 *Suppose that floating-point arithmetic leads to approximate Newton iterates satisfying*

$$\tilde{z}_{k+1} = \left[\tilde{z}_k - \frac{f(\tilde{z}_k) + \phi_k \varepsilon}{f'(\tilde{z}_k)}(1 + \varepsilon_\div)\right](1 + \varepsilon_-) .$$

Here the rounding errors due to subtraction, division or evaluating $f'(\tilde{z}_k)$ are bounded above by ε

$$|\varepsilon_\div| , |\varepsilon_-| \le \varepsilon$$

and the absolute errors involved in the evaluation of f are bounded above:

$$|\phi_k| \le \phi .$$

In this inequality, ϕ_k represents the size of the terms involved in evaluating f, and is sometimes called the **scale of** *f. We also suppose that there is a positive constant M so that for all k we have*

$$\left|f(z) - f(\tilde{z}_k) - f'(\tilde{z}_k)(z - \tilde{z}_k)\right| \le \frac{M}{2} |\tilde{z}_k - z|^2 .$$

Let

$$\Phi = |\tilde{z}_k| \, |f'(\tilde{z}_k)| + (2 + \varepsilon)|f(\tilde{z}_k)| + (1 + \varepsilon)^2 \phi .$$

Then if

$$\frac{2\varepsilon\Phi}{|f'(\tilde{z}_k)| + \sqrt{|f'(\tilde{z}_k)|^2 - 2M\varepsilon\Phi}} < |\tilde{z}_k - z| < \frac{1}{M}\left[|f'(\tilde{z}_k)| + \sqrt{|f'(\tilde{z}_k)|^2 - 2M\varepsilon\Phi}\right], \tag{5.15}$$

it follows that

$$|\tilde{z}_{k+1} - z| < |\tilde{z}_k - z| .$$

Note that we have combined the rounding errors in performing the division and in computing the derivative, into a single relative error $\varepsilon_{\div}$. We did not assume that $f(\tilde{z}_k)$ is computed with relative error proportional to machine precision, because we expect nearly equal terms to cancel when $\tilde{z}_k$ is near the zero z of f.

Proof We can compute

$$\begin{aligned}\tilde{z}_{k+1} - z &= \left[\tilde{z}_k - \frac{f(\tilde{z}_k) + \phi_k\varepsilon}{f'(\tilde{z}_k)}(1+\varepsilon_{\div})\right](1+\varepsilon_{-}) - z \\ &= \frac{f(z) - f(\tilde{z}_k) - f'(\tilde{z}_k)(z - \tilde{z}_k)}{f'(\tilde{z}_k)} + \tilde{z}_k\varepsilon_{-} - \frac{f(\tilde{z}_k)}{f'(\tilde{z}_k)}(\varepsilon_{\div} + \varepsilon_{-} + \varepsilon_{\div}\varepsilon_{-}) \\ &\quad - \frac{\phi_k\varepsilon}{f'(\tilde{z}_k)}(1+\varepsilon_{\div})(1+\varepsilon_{-}) ,\end{aligned}$$

and bound

$$|\tilde{z}_{k+1} - z| \leq \frac{1}{2}\frac{M|\tilde{z}_k - z|^2}{|f'(\tilde{z}_k)|} + |\tilde{z}_k|\,\varepsilon + \frac{|f(\tilde{z}_k)|}{|f'(\tilde{z}_k)|}\varepsilon(2+\varepsilon) + \frac{\phi}{|f'(\tilde{z}_k)|}\varepsilon(1+\varepsilon)^2 .$$

If we require that the right-hand side is less than $|\tilde{z}_k - z|$, we get a quadratic inequality that is equivalent to the conclusion of the lemma.

Lemma 5.4.4 essentially says that as the Newton iterates approach the true solution, the errors will decrease until they are of the size of the left-hand side in inequality (5.15). If we ignore small terms, this inequality essentially says that

$$|\tilde{z}_k - z| > \varepsilon\left\{|\tilde{z}_k| + \frac{\phi}{|f'(z)|}\right\} .$$

Here ϕ represents the size of the terms that are canceling in the evaluation of f near its zero. This lemma shows that the errors in Newton's method can decrease until they are very near the maximum attainable accuracy found in Corollary 5.2.1.

The right-hand bound in inequality (5.15) is similar to a Kantorovich bound. If the error in the current Newton iterate violates the right-hand bound in this inequality, then the current iterate is too far from the true zero to guarantee convergence.

Example 5.4.3 The function $f(x) = 4\sin(x) + 1 - x$ has three roots. One of the roots is negative. Near this negative root, the term $1 - x$ will be positive and the term $4\sin(x)$ must be negative. These two terms must cancel at the root, suggesting that we could take $\phi = |1 - x|$ or $|4\sin(x)|$ to be the scale of f in this case. Other combinations of these terms could cancel at other roots. In general, we could take

$$\phi = \max\{|4\sin(x)|, 1, |x|\}$$

to represent the size (or scale) of f.

Once the maximum attainable accuracy is reached, a different description of rounding errors in Newton's method is needed. If we have

$$|\tilde{z}_k - z| \approx \varepsilon \left[|\tilde{z}_k| + \frac{\phi}{|f'(z)|} \right] ,$$

then we also expect that

$$|f(\tilde{z}_k)| \approx \phi\varepsilon .$$

In this case, we have

$$\begin{aligned} |\tilde{z}_{k+1} - z| &\leq \left| \tilde{z}_k - z + \tilde{z}_k\varepsilon_- - \frac{f(\tilde{z}_k)(1+\varepsilon_\div)(1+\varepsilon_-)}{f'(\tilde{z}_k)} \right| \\ &\leq |\tilde{z}_k - z| + |\tilde{z}_k|\,\varepsilon + \frac{|f(\tilde{z}_k)|}{|f'(\tilde{z}_k)|}(1+\varepsilon)^2 \\ &\approx \varepsilon \left\{ |z| + \frac{\phi}{|f'(z)|} \right\} + \varepsilon\,|z| + \varepsilon \frac{\phi}{|f'(z)|} = 2\varepsilon \left\{ |z| + \frac{\phi}{|f'(z)|} \right\} . \end{aligned}$$

In other words, after the maximum attainable accuracy is reached, the next iterate with Newton's method could have error on the order of twice the maximum attainable accuracy, at worst.

We have provided some simple examples of programs implementing Newton's method. Readers may view a Fortran program for Newton's method, or they may view a C++ program for an interactive Newton's method. The latter C++ program also demonstrates how to use the GSL routine gsl_root_fdfsolver_newton. Readers may also execute the JavaScript program for **Newton's method.** All of these programs iterate to find a zero of $f(x) = x^2 - 4\sin(x)$. The zeros are $z = 0$ and $z \approx 1.9337573$. A plot of the error in Newton's method is shown in Fig. 5.2.

Exercise 5.4.13 Consider the Newton iteration

$$z_{k+1} = (z_k + 2/z_k)/2$$

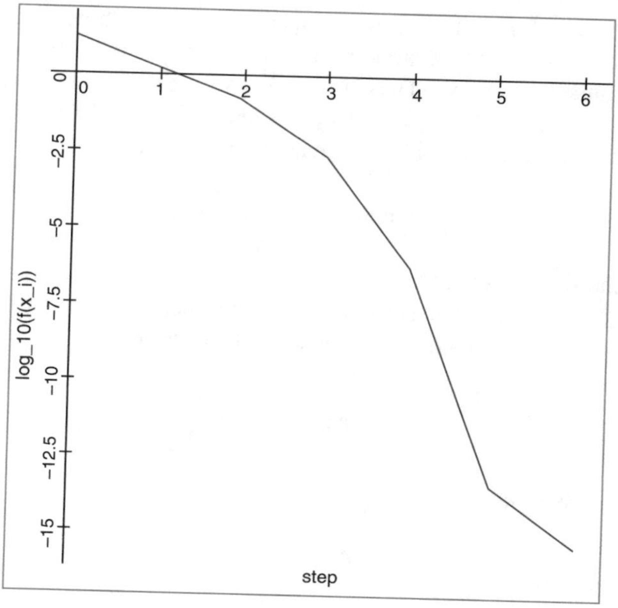

Fig. 5.2 Error in Newton's method: $log_{10}(f)$ versus iterations

for computing a zero of $f(x) = x^2 - 2$. Analyze the effect of rounding errors on this iteration, without using Lemma 5.4.4. What condition must be imposed on the Newton iterate to guarantee smaller error in the next iterate?

5.4.8 *Termination Criteria*

In order to guarantee termination of our computer implementation for Newton's method, we should always provide a bound on the maximum number of steps. Generally, we recommend setting the maximum number of iterations large enough that one of the other termination criteria should be satisfied first, but not so large that the computation could become prohibitively expensive.

Secondly, we should include a test for no change in the solution. If $z_{k+1} = z_k$, then there will be no change in any future iterate, and the iteration should be terminated. This situation could occur if $f(z_k) = 0$, or if the Newton step is so small that $|f(z_k)/f'(z_k)| < \varepsilon |z_k|$.

Thirdly, we should provide a tolerance δ_z on the size of the relative error in the solution. If the true solution has size ζ, then for Newton's method we could require that

$$|z_{k+1} - z_k| \leq \delta_z \zeta \ . \tag{5.16}$$

Here δ_z is a dimensionless tolerance. Because Newton's method normally has a quadratic rate of convergence, the next iterate z_{k+1} should be a very good approximation to the true solution z. Alternatively, we note that

$$|z_{k+1} - z_k| = \frac{|f(z_k)|}{|f'(z_k)|} = \frac{|\int_z^{z_k} f'(\xi)\, d\xi|}{|f'(z_k)|} \geq \frac{\min_{\xi \text{ between } z \text{ and } z_k} |f'(\xi)|}{|f'(z_k)|} |z_k - z| \ .$$

Thus our relative error test implies that

$$|z_k - z| \leq \delta_z \zeta \frac{|f'(z_k)|}{\min_{\xi \text{ between } z \text{ and } z_k} |f'(\xi)|} \ .$$

Once the approximate solution is close enough to the true solution that the derivative of f does not change much, this bound on the change in the solution provides a good estimate for the relative error in the solution. However, note that if $f'(z)$ is small, then the convergence test (5.16) may be difficult to satisfy. This is because rounding errors may limit the smallest value of $|f(z_k)|$ to a number on the order of $\varepsilon\phi$, where ε is machine precision and ϕ is the scale of f. Under such circumstances, if we choose

$$\delta_z < \varepsilon \frac{\phi}{\zeta |f'(z)|}$$

then the convergence test (5.16) may be impossible to satisfy. For a relative error test, we could take $\zeta = \max\{|z_k| \ , \ |z_{k+1}|\}$. For an absolute error test, we could take $\zeta = |z_0|$.

A fourth test stops the iteration when

$$|f(z_k)| < \delta_f \phi \ . \tag{5.17}$$

Here ϕ is the size (or scale) of f and δ_f is a dimensionless tolerance. This convergence criterion implies that

$$\delta_f \phi \geq |f\ (z_k)| = |f(z_k) - f(z)| = \left| \int_z^{z_k} f'(\xi)\, d\xi \right| \geq \min_{\xi \text{ between } z \text{ and } z_k} f'(\xi)|\ |z_k - z| \ ,$$

which in turn implies a bound on the error in the solution. Note that if $f'(z)$ is large, then the convergence test (5.17) may be difficult to satisfy. This is because rounding errors may prevent $|z_k - z|$ from being smaller than $\varepsilon\zeta/2$, where ε is machine precision and ζ is the size of z. Under such circumstances, if we choose

$$\delta_f < \frac{\varepsilon}{2} \frac{\zeta |f'(z)|}{\phi} \ ,$$

then the convergence test (5.17) may be impossible to satisfy.

Because Newton's method converges so rapidly, it is often convenient to allow Newton's method to iterate until its **maximum attainable accuracy** is achieved. Due to quadratic convergence, the step size $|\tilde{z}_{k+1} - \tilde{z}_k|$ is a very good estimate for the error $|z - \tilde{z}_k|$. Again, let ζ be the size of the solution z and ϕ be the size (or scale) of f. If δ is somewhat larger than machine precision (for example, δ could be the square root of machine precision), and we see that

$$|\tilde{z}_{k+1} - \tilde{z}_k| \leq \delta \left[\zeta + \frac{\phi}{|f'(\tilde{z}_k)|} \right] , \tag{5.18a}$$

then we could stop the iteration whenever we observe that

$$|\tilde{z}_{k+1} - \tilde{z}_k| \geq |\tilde{z}_k - \tilde{z}_{k-1}| . \tag{5.18b}$$

Under such circumstances, the maximum attainable accuracy should have been achieved. In other words, our analysis in Lemma 5.4.4 shows that further quadratic convergence has become impossible due to rounding errors, and additional iterates will continue to wander around near the true solution.

Example 5.4.4 Suppose that we want to use Newton's method to find a zero of

$$f(x) = x^2 - 4\sin(x) .$$

In double precision, the zeros of f are $z = 0$ and

$$z \approx 1.9337537628270212 .$$

We will consider three termination tests for this iteration, namely

$$|z_{k+1} - z_k| \leq \delta_z \zeta_k ,$$
$$|f(z_k)| \leq \delta_f \phi(z_k) \text{ and}$$
$$|z_{k+1} - z_k| \geq |z_k - z_{k-1}| \text{ and } |z_k - z_{k-1}| \leq \sqrt{\varepsilon} \max\{|z_k| , |z_{k+1}|\} .$$

In this example, the scale of z is chosen to be

$$\zeta_k = \max\{|z_k|, ||z_{k+1}|\} ,$$

and the scale of f is taken to be

$$\phi_k = \max\{z_k^2, 4|\sin z_k|\} .$$

Table 5.2 shows the computational results for Newton's method with initial guess $z_0 = 2$. The true error appears in the second column. Columns five and six show the values of δ_z and δ_f that would satisfy the convergence tests (5.16) and (5.17), respectively. The final column shows whether the step lengths satisfy the maximum attainable accuracy test (5.18). In this iteration, the first two convergence tests do a good job of of estimating the true error. Since $z_5 = z_4$, this iteration was terminated before the maximum attainable accuracy test could be satisfied.

We note that the computations for Table 5.2 (and the subsequent three tables) were performed by the Fortran program newton.f.

Example 5.4.5 Suppose that we want to use Newton's method to find a zero of

$$f(x) = x^2 - 4\sin(x) ,$$

using initial guess $z_0 = 1$. In this case, Newton's iteration converges to $z = 0$. If we use compute ζ_k and ϕ_k as in Example 5.4.4, then Table 5.3 shows that neither the relative error test (5.16) nor the function value error test (5.17) will ever be satisfied for tolerances less than 0.1. The difficulty is that the true solution $z = 0$ has no scale, and the terms involved in the evaluation of $f(z)$ have no scale. We also remark that the test $z_{12} = z_{11}$ terminates the iteration before the test for maximum attainable accuracy can be satisfied.

Table 5.2 Convergence criteria for $f(x) = x^2 - 4\sin(x)$, $z_0 = 2$

k	$\|z_k - z\|$	ζ_k	ϕ_k	$\|z_{k+1} - z_k\|/\zeta$	$\|f(z_k)\|/\phi$	$\frac{\|z_{k+1}-z_k\|}{\geq \|z_k - z_{k-1}\|}$?
0	6.62×10^{-2}	2.00	4.00	3.20×10^{-2}	9.07×10^{-2}	F
1	2.20×10^{-3}	1.94	3.75	1.13×10^{-3}	3.10×10^{-3}	F
2	2.61×10^{-6}	1.93	3.74	1.35×10^{-6}	3.70×10^{-6}	F
3	3.81×10^{-12}	1.93	3.74	1.97×10^{-12}	5.39×10^{-12}	F
4	0.	1.93	3.74	0.	1.19×10^{-16}	F

Table 5.3 Convergence criteria for $f(x) = x^2 - 4\sin(x)$, $z_0 = 1$, relative error tests

k	$\|z_k - z\|$	ζ_k	ϕ_k	$\|z_{k+1} - z_k\|/\zeta_k$	$\|f(z_k)\|/\phi_k$	$\frac{\|z_{k+1}-z_k\|}{\geq \|z_k - z_{k-1}\|}$?
0	1.00×10^{0}	1.37×10^{1}	3.37×10^{0}	1.07×10^{0}	7.03×10^{-1}	
1	1.37×10^{1}	1.37×10^{1}	1.87×10^{2}	4.78×10^{-1}	1.02×10^{0}	F
2	7.13×10^{0}	7.13×10^{0}	5.09×10^{1}	4.47×10^{-1}	1.06×10^{0}	F
3	3.95×10^{0}	3.95×10^{0}	15.6×10^{1}	6.28×10^{-1}	8.15×10^{-1}	F
4	1.47×10^{0}	1.47×10^{0}	3.98×10^{0}	1.24×10^{0}	1.54×10^{0}	F
5	3.63×10^{-1}	3.63×10^{-1}	1.42×10^{0}	1.18×10^{0}	9.07×10^{-1}	F
6	6.46×10^{-2}	6.46×10^{-2}	2.58×10^{-1}	9.86×10^{-1}	1.02×10^{0}	F
7	9.26×10^{-4}	9.26×10^{-4}	3.70×10^{-3}	1.00×10^{0}	1.00×10^{0}	F
8	2.14×10^{-7}	2.14×10^{-7}	8.55×10^{-7}	1.00×10^{0}	1.00×10^{0}	F
9	1.41×10^{-14}	1.41×10^{-14}	5.62×10^{-14}	1.00×10^{0}	1.00×10^{0}	F
10	4.89×10^{-29}	4.89×10^{-29}	1.96×10^{-28}	1.00×10^{0}	1.00×10^{0}	F
11	0.	0.	0.	*NaN*	*NaN*	F

Table 5.4 Convergence criteria for $f(x) = x^2 - 4\sin(x)$, $z_0 = 1$, absolute error tests

k	$\lvert \tilde{z}_k - z \rvert$	ζ	ϕ	$\frac{\lvert \tilde{z}_{k+1} - \tilde{z}_k \rvert}{\zeta}$	$\frac{\lvert f(\tilde{z}_k) \rvert}{\phi}$	$\lvert \tilde{z}_{k+1} - \tilde{z}_k \rvert \geq \lvert \tilde{z}_k - \tilde{z}_{k-1} \rvert$?
0	1.00×10^0	1.00	2.37	1.47×10^1	1.00×10^0	
1	1.37×10^1	1.00	2.37	6.54×10^0	8.06×10^1	F
2	7.13×10^0	1.00	2.37	3.19×10^0	2.28×10^1	F
3	3.95×10^0	1.00	2.37	2.48×10^0	5.36×10^0	F
4	1.47×10^0	1.00	2.37	1.83×10^0	2.59×10^0	F
5	3.63×10^{-1}	1.00	2.37	4.28×10^{-1}	5.45×10^{-1}	F
6	6.46×10^{-2}	1.00	2.37	6.37×10^{-2}	1.11×10^{-1}	F
7	9.26×10^{-4}	1.00	2.37	9.25×10^{-4}	1.56×10^{-3}	F
8	2.14×10^{-7}	1.00	2.37	2.14×10^{-7}	3.62×10^{-7}	F
9	1.41×10^{-14}	1.00	2.37	1.41×10^{-14}	2.38×10^{-14}	F
10	4.89×10^{-29}	1.00	2.37	4.89×10^{-29}	8.26×10^{-29}	F
11	0.	1.00	2.37	0.	0.	F

If we choose positive values for the scales of z and f, then the convergence tests can be successful. For example, we might choose

$$\zeta = |z_0|$$

for the scale of z. For the scale of f, we might choose

$$\phi = |f(z_0)| \; .$$

The revised results using these scales is shown in Table 5.4.

Example 5.4.6 Suppose that we want to use Newton's method to find a zero of

$$f(x) = 10^{16}x^2 - 4\sin\left(10^8 x\right) \; ,$$

using initial guess $\tilde{z}_0 = 1$. In this case, Newton's iteration converges to $z \approx 1.93 \times 10^{-8}$. If we use $\zeta = |\tilde{z}_0|$ for the scale of the solution, then for $\delta_z > 10^{-8}$ the relative error test (5.16) will terminate Newton's method with no significant digits in the solution. Similarly, if we use $\phi = |f(\tilde{z}_0)|$ for the scale of f, the convergence test (5.17) can terminate the iteration with no significant digits in the solution.

If we really seek some desired number of significant digits in the solution whenever possible, we should use the relative error tests, with the scales of z and f chosen as in Example 5.4.4. The computational results using these scales is shown in Table 5.5.

Table 5.5 Convergence criteria for $f(x) = 10^{16}x^2 - 4\sin(10^8 x)$, $z_0 = 1$

k	$\lvert\tilde{z}_k - z\rvert$	ζ_k	ϕ_k	$\lvert\tilde{z}_{k+1} - \tilde{z}_k\rvert/\zeta_k$	$\lvert f(\tilde{z}_k)\rvert/\phi_k$	$\lvert\tilde{z}_{k+1}-\tilde{z}_k\rvert \ge \lvert\tilde{z}_k - \tilde{z}_{k-1}\rvert$?
0	1.00×10^{0}	1.00×10^{0}	1.00×10^{16}	5.00×10^{-1}	1.00×10^{0}	
⋮	⋮	⋮	⋮	⋮	⋮	⋮
10	9.77×10^{-4}	9.77×10^{-4}	9.54×10^{9}	5.00×10^{-1}	1.00×10^{0}	F
⋮	⋮	⋮	⋮	⋮	⋮	⋮
20	9.43×10^{-7}	9.62×10^{-7}	9.25×10^{3}	4.96×10^{-1}	1.00×10^{0}	F
21	4.65×10^{-7}	4.85×10^{-7}	2.35×10^{3}	4.96×10^{-1}	1.00×10^{0}	F
22	2.25×10^{-7}	2.44×10^{-7}	5.96×10^{2}	5.35×10^{-1}	1.00×10^{0}	F
23	9.41×10^{-8}	1.13×10^{-7}	1.29×10^{2}	5.48×10^{-1}	1.03×10^{0}	F
24	3.20×10^{-8}	5.13×10^{-8}	2.63×10^{1}	6.76×10^{-1}	1.14×10^{0}	F
25	2.74×10^{-9}	1.99×10^{-8}	3.98×10^{0}	1.68×10^{-1}	3.09×10^{-1}	F
26	6.05×10^{-10}	1.99×10^{-8}	3.98×10^{0}	2.94×10^{-2}	8.31×10^{-2}	F
27	1.85×10^{-11}	1.94×10^{-8}	3.75×10^{0}	9.54×10^{-4}	2.61×10^{-3}	F
28	1.85×10^{-14}	1.93×10^{-8}	3.74×10^{0}	9.56×10^{-7}	2.61×10^{-6}	F
29	1.86×10^{-20}	1.93×10^{-8}	3.74×10^{0}	9.61×10^{-13}	2.63×10^{-12}	F
30	0.	1.93×10^{-8}	3.74×10^{0}	1.71×10^{-16}	2.38×10^{-16}	F
31	3.31×10^{-21}	1.93×10^{-8}	3.74×10^{0}	1.71×10^{-16}	4.75×10^{-16}	T

5.4.9 *Approximate Derivatives*

Although Newton's method has significant benefits, such as its quadratic rate of convergence, it comes at the cost of evaluating both the function f and its derivative. For nonlinear equations in a single variable, this additional cost is usually not too difficult to bear. However, for systems of nonlinear equations, the evaluation of all the partial derivatives in the derivative matrix can involve significant cost, in addition to the function evaluation. We will discuss three strategies for avoiding this cost.

5.4.9.1 Fixed Derivative

Suppose that we simply fix the derivative at some value $f'(x)$, which is presumably close to $f'(z)$. The revised Newton's method would be

$$z_{k+1} = z_k - \frac{f(z_k)}{f'(x)} .$$

Let us analyze this method. If f' is Lipschitz continuous with constant γ, then since $f(z) = 0$ we have

$$z_{k+1} - z = z_k - z + \frac{f(z) - f(z_k)}{f'(x)} = \frac{\int_{z_k}^{z} f'(\xi) - f'(x) \, d\xi}{f'(x)}$$

$$= \frac{\int_0^1 f'\left(z_k + \xi[z - z_k]\right) - f'(x) \, d\xi [z - z_k]}{f'(x)} ,$$

so

$$
\begin{aligned}
|z_{k+1} - z| &\leq \frac{|z_k - z|}{|f'(x)|} \int_0^1 \left| f'(z_k + \xi[z - z_k]) - f'(x) \right| \, d\xi \\
&\leq \frac{|z_k - z|}{|f'(x)|} \gamma \int_0^1 |z_k - x + \xi(z - z_k)| \, d\xi \\
&\leq \frac{|z_k - z|}{|f'(x)|} \gamma \left\{ |z_k - x| + \frac{1}{2} |z_k - z| \right\}
\end{aligned}
$$

If $\gamma |z_k - x| / |f'(x)| \geq 1$, then we have no reason to expect this iteration to converge. In other words, the derivative should not be fixed at some point x unless we can be sure that

$$
|z_k - x| < |f'(x)| / \gamma
$$

for all future iterates z_k. Even if this condition were satisfied and the iteration did converge, the convergence rate would be at best linear.

5.4.9.2 Fixed Finite Difference Increment

Suppose that we choose some fixed value for h and apply the following

Algorithm 5.4.2 (Newton's Method with Finite Difference Derivative)

$$
\begin{aligned}
&\text{while not converged} \\
&\quad f_k = f(z_k) \\
&\quad \text{choose } h_k \\
&\quad J_k = (f(z_k + h_k) - f_k)/h \\
&\quad z_{k+1} = z_k - f_k / J_k
\end{aligned}
$$

Then the errors in the iterates satisfy

$$
\begin{aligned}
z_{k+1} - z &= \frac{[f(z) - f(z_k)]h - [f(z_k + h) - f(z_k)][z - z_k]}{f(z_k + h) - f(z_k)} \\
&= \frac{\int_{z_k}^{z} \int_{z_k}^{z_k + h} f'(\zeta) - f'(\eta) \, d\eta \, d\zeta}{\int_{z_k}^{z_k + h} f'(\eta) \, d\eta} \\
&= \frac{\int_0^1 \int_0^1 f'(z_k + \tau[z - z_k]) - f'(z_k + \sigma h) \, d\sigma \, d\tau h[z - z_k]}{f'(z)h + \int_0^1 f'(z_k + \sigma h) - f'(z) \, d\sigma h} \, .
\end{aligned}
$$

We can take absolute values and use Lipschitz continuity to obtain

$$
\begin{aligned}
|z_{k+1} - z| &\leq \gamma\,|z_k - z|\,\frac{\int_0^1 \int_0^1 |\tau[z - z_k] - \sigma h|\,\mathrm{d}\sigma\,\mathrm{d}\tau}{|f'(z)| - \gamma \int_0^1 |z_k - z + \sigma h|\,\mathrm{d}\sigma} \\
&\leq \frac{\gamma}{2}\,|z_k - z|\,\frac{|z_k - z| + h}{|f'(z)| - \gamma(|z_k - z| + h/2)}\,.
\end{aligned} \tag{5.19}
$$

In order for this iteration to have any chance to be contractive, we should require

$$
\frac{\gamma h/2}{|f'(z)| - \gamma h/2} < 1 \Longleftrightarrow h < \frac{|f'(z)|}{\gamma}\,.
$$

In addition, we should choose the initial iterate sufficiently close to z, so that

$$
\frac{\gamma}{2}\,\frac{|z_k - z| + h}{|f'(z)| - \gamma[|z_k - z| + h/2]} < 1 \Longleftrightarrow |z_k - z| < \frac{2}{3}\left[\frac{|f'(z)|}{\gamma} - h\right].
$$

Under such circumstances, the iteration will be contractive, and should converge linearly.

5.4.9.3 Vanishing Finite Difference Increment

Our analysis above indicates that using a fixed increment h to compute a finite difference approximation to the derivative reduces the convergence rate of the approximate Newton iteration to linear. Inequality (5.19) indicates that in order for the convergence rate to be superlinear, the finite difference increment h must converge to zero as the iteration proceeds. Further, in order for the iteration to retain quadratic convergence, we must have $h = O(|z_k - z|)$.

We can accomplish this goal of quadratic convergence by choosing $h = |f(z_k)|/c$ for some constant c. The constant c should be chosen with units of f', so that h has the units of the argument to f. For more information regarding the convergence of Newton methods using approximate derivatives, see either Dennis and Schnabel [31, p. 30] or an online paper by Sidi [94].

Readers may examine C^{++} code for Newton's method via approximate derivatives. This program implements various forms of finite differences in Newton's method to compute a zero of $f(x) = x^2 - 4\sin(x)$. In the first iteration, the slope is fixed as the true derivative at the initial guess. In a second group of iterations, the slope is approximated by finite differences with increments $h = 10^{-n}, n = 1, \ldots, 7$. In the final iteration, the slope is approximated by finite differences with increment $h = f(z_k)/f'(z_0)$.

Some sample results are shown in Fig. 5.3. The graph with fixed slope shows linear convergence. The graph with finite difference increment proportional to the function value shows quadratic convergence. In the graph with fixed increments,

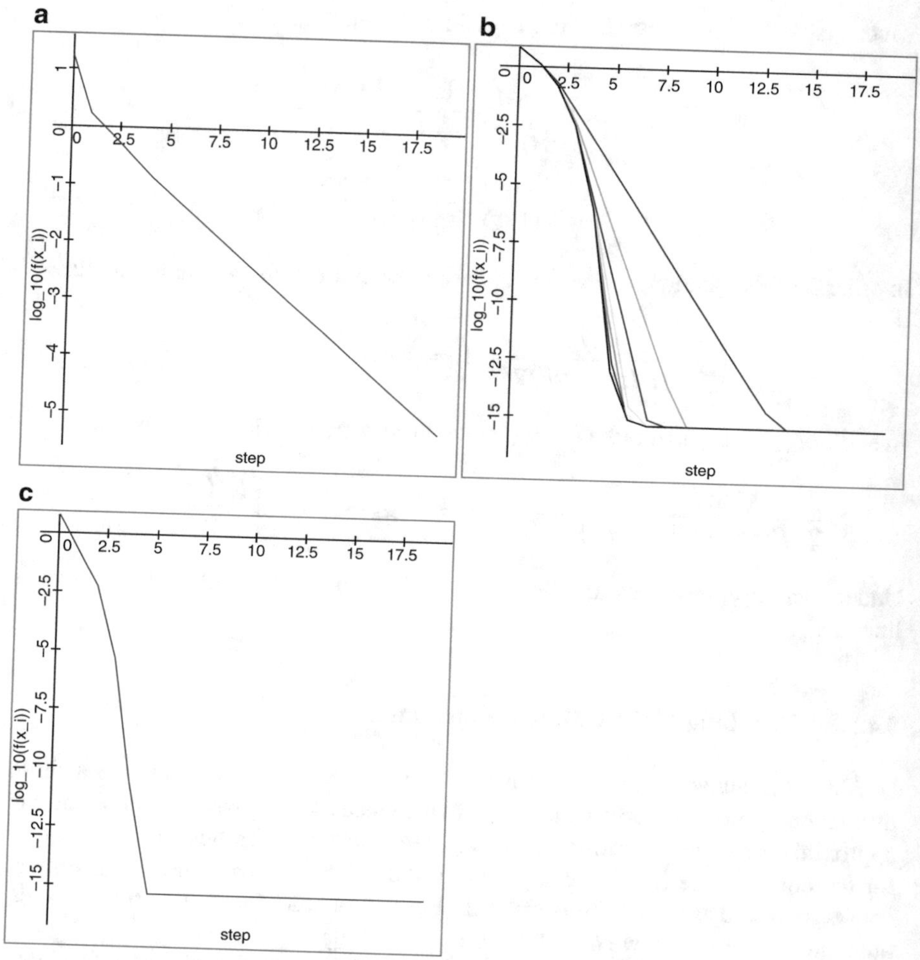

Fig. 5.3 Errors in Newton's method with approximate derivatives: $log_{10}(f)$ versus iterations. (**a**) Fixed slope. (**b**) Fixed increments $h = 10^{-n}, n = 1, \ldots, 7$. (**c**) Vanishing increment

the curves progress from right to left as the increment is decreased. This graph shows linear convergence for large increments, and behavior indistinguishable from quadratic convergence for increments near the square root of machine precision.

Recall from Sect. 2.3.1.2 that a finite difference increment h on the order of the square root of machine precision produces the most accurate derivative approximation in the presence of rounding errors. This claim will be made more precise in Lemma 2.2.3 of Chap. 2 in Volume III.

5.4.9.4 Rounding Errors

The following lemma will analyze the effect of rounding errors in Newton's method when the derivatives are approximated by finite differences using the optimal increment h determined by Lemma 2.2.3 of Chap. 2 in Volume III.

Lemma 5.4.5 *Suppose that f is a continuously differentiable real-valued function, and that f' is Lipschitz continuous with Lipschitz constant γ. Assume that floating-point computations produce Newton iterates using finite difference approximations such that*

$$\tilde{z}_{k+1} = \left[\tilde{z}_k - \frac{f(\tilde{z}_k) + \phi_k}{f(\tilde{z}_k + h) + \phi_h - f(\tilde{z}_k) - \phi_k} h(1 + \varepsilon_\div)\right](1 + \varepsilon_-) ,$$

where the rounding errors are bounded above by ε:

$$|\varepsilon_f| , |\varepsilon_h| , |\varepsilon_\div| , |\varepsilon_-| \le \varepsilon ,$$

and the size of the terms involved in the evaluation of f (also called the **scale of** *f) are bounded above:*

$$|\phi_k| , |\phi_h| \le \phi\varepsilon .$$

In these computations, we assume that the finite difference increment h is given by

$$h = \sqrt{\frac{2\phi\varepsilon}{\gamma}} . \tag{5.20}$$

We also assume that the derivative of f at $\tilde{z}_k$ is not too small:

$$|f'(\tilde{z}_k)| > \frac{3\phi\varepsilon}{h} = \sqrt{\frac{9}{2}\gamma\phi\varepsilon} . \tag{5.21}$$

Define

$$A = \gamma h/2 ,$$
$$B = |f'(\tilde{z}_k)|\, h - 6\phi\varepsilon - \varepsilon(2+\varepsilon)h \min_{\xi \text{ between } z,\, \tilde{z}_k} |f'(\xi)| \text{ and}$$
$$C = |\tilde{z}_k|\,\varepsilon\, |f(\tilde{z}_k + h) - f(\tilde{z}_k)| + \phi\varepsilon h\left[\frac{2}{3} + (1+\varepsilon)^2\right] + |\tilde{z}_k|\, 2\phi\varepsilon^2 .$$

If

$$\frac{2C}{B + \sqrt{B^2 - 4AC}} < |\tilde{z}_k - z| < \frac{B + +\sqrt{B^2 - 4AC}}{2A} , \tag{5.22}$$

then

$$|\tilde{z}_{k+1} - z| < |\tilde{z}_k - z| \; .$$

Proof We can rewrite the error in the computed Newton iteration to get

$$\{\tilde{z}_{k+1} - z\}\,\delta = \{\tilde{z}_k - z + \tilde{z}_k\varepsilon_-\}\,\delta - \{f(\tilde{z}_k) + \phi_k\}\,h(1+\varepsilon_\div)(1+\varepsilon_-) \; , \tag{5.23}$$

where

$$\delta \equiv f(\tilde{z}_k + h) - f(\tilde{z}_k) + \phi_h - \phi_k = \int_{\tilde{z}_k}^{\tilde{z}_k+h} f'(\zeta)\,\mathrm{d}\zeta + \phi_h - \phi_k$$

$$= f'(\tilde{z}_k)\,h + \int_{\tilde{z}_k}^{\tilde{z}_k+h} \left[f'(\zeta) - f'(\tilde{z}_k)\right]\,\mathrm{d}\zeta + \phi_h - \phi_k \; .$$

We can take absolute values of both sides of this expression to see that

$$|\delta| \geq \left|f'(\tilde{z}_k)\right| h - \int_{\tilde{z}_k}^{\tilde{z}_k+h} \left|f'(\zeta) - f'(\tilde{z}_k)\right|\,\mathrm{d}\zeta - 2\phi\varepsilon$$

$$\geq \left|f'(\tilde{z}_k)\right| h - \int_{\tilde{z}_k}^{\tilde{z}_k+h} \gamma\,|\zeta - \tilde{z}_k|\,\mathrm{d}\zeta - 2\phi\varepsilon$$

$$= \left|f'(\tilde{z}_k)\right| h - \gamma h^2/2 - 2\phi\varepsilon$$

$$= \left|f'(\tilde{z}_k)\right| h - 3\phi\varepsilon \; .$$

Note that assumption (5.21) implies that $|\delta| > 0$.

Returning to Eq. (5.23), we now see that

$$\{\tilde{z}_{k+1} - z\}\,\delta = \{\tilde{z}_k - z + \tilde{z}_k\varepsilon_-\}\,\delta - \{f(\tilde{z}_k) + \phi_k\}\,h(1+\varepsilon_\div)(1+\varepsilon_-)$$

$$= \{\tilde{z}_k - z + \tilde{z}_k\varepsilon_-\}\left\{\int_{\tilde{z}_k}^{\tilde{z}_k+h} f'(\zeta)\,\mathrm{d}\zeta + \phi_h - \phi_k\right\}$$

$$-\left\{\int_z^{\tilde{z}_k} f'(\xi)\,\mathrm{d}\xi + \phi_k\right\} h(1+\varepsilon_\div)(1+\varepsilon_-)$$

$$= \int_z^{\tilde{z}_k}\int_{\tilde{z}_k}^{\tilde{z}_{k+1}} f'(\zeta) - f'(\xi)\,\mathrm{d}\zeta\,\mathrm{d}\xi + \tilde{z}_k\varepsilon_-\left[f(\tilde{z}_k + h) - f(\tilde{z}_k)\right]$$

$$+ (\tilde{z}_k - z + \tilde{z}_k\varepsilon_-)(\phi_h - \phi_k) - f(\tilde{z}_k)\,h(\varepsilon_\div + \varepsilon_- + \varepsilon_\div\varepsilon_-) - \phi_k h(1+\varepsilon_\div)(1+\varepsilon_-) \; .$$

We can take the absolute value of this expression to get

$$
\begin{aligned}
&|\tilde{z}_{k+1} - z| \; |\delta| \\
&\leq \left| \int_z^{\tilde{z}_k} \int_{\tilde{z}_k}^{\tilde{z}_k+h} f'(\zeta) - f'(\xi) \, \mathrm{d}\zeta \, \mathrm{d}\xi \right| + |\tilde{z}_k| \, \varepsilon \, |f(\tilde{z}_k + h) - f(\tilde{z}_k)| \\
&+ (|\tilde{z}_k - z| + |\tilde{z}_k| \, \varepsilon) \, 2\phi\varepsilon + |f(\tilde{z}_k)| \, h\varepsilon(2 + \varepsilon) + \phi\varepsilon h(1 + \varepsilon)^2 \, . \qquad (5.24)
\end{aligned}
$$

Next, we would like to bound the double integral in (5.24). We will consider three cases. First, if $\tilde{z}_k \geq z$ then

$$
\begin{aligned}
&\left| \int_z^{\tilde{z}_k} \int_{\tilde{z}_k}^{\tilde{z}_k+h} f'(\zeta) - f'(\xi) \, \mathrm{d}\zeta \, \mathrm{d}\xi \right| \leq \gamma \int_z^{\tilde{z}_k} \int_{\tilde{z}_k}^{\tilde{z}_k+h} \zeta - \xi \, \mathrm{d}\zeta \, \mathrm{d}\xi \\
&= \gamma h \int_z^{\tilde{z}_k} \tilde{z}_k + h/2 - \xi \, \mathrm{d}\xi = \gamma h \int_0^{\tilde{z}_k - z} \tau + h/2 \, \mathrm{d}\tau \\
&= \frac{\gamma h}{2} \left\{ (\tilde{z}_k - z)^2 + h(\tilde{z}_k - z) \right\} = \frac{\gamma h}{2} |\tilde{z}_k - z| \left\{ |\tilde{z}_k - z| + h \right\} \, . \qquad (5.25)
\end{aligned}
$$

Next, if $z \geq \tilde{z}_k + h$ then

$$
\begin{aligned}
&\left| \int_z^{\tilde{z}_k} \int_{\tilde{z}_k}^{\tilde{z}_k+h} f'(\zeta) - f'(\xi) \, \mathrm{d}\zeta \, \mathrm{d}\xi \right| \leq \gamma \int_{\tilde{z}_k}^{z} \int_{\tilde{z}_k}^{\tilde{z}_k+h} |\zeta - \xi| \, \mathrm{d}\zeta \, \mathrm{d}\xi \\
&= \gamma \int_{\tilde{z}_k}^{\tilde{z}_k+h} \int_{\tilde{z}_k}^{\zeta} \zeta - \xi \, \mathrm{d}\xi \, \mathrm{d}\zeta + \gamma \int_{\tilde{z}_k}^{\tilde{z}_k+h} \int_{\zeta}^{z} \xi - \zeta \, \mathrm{d}\xi \, \mathrm{d}\zeta \\
&= \frac{\gamma}{2} \int_{\tilde{z}_k}^{\tilde{z}_k+h} (\zeta - \tilde{z}_k)^2 \, \mathrm{d}\zeta + \frac{\gamma}{2} \int_{\tilde{z}_k}^{\tilde{z}_k+h} (z - \zeta)^2 \, \mathrm{d}\zeta \\
&= \frac{\gamma}{6} h^3 + \frac{\gamma}{6} h \left[(\tilde{z}_k - z)^2 + (\tilde{z}_k - z)(\tilde{z}_k + h - z) + (\tilde{z}_k + h - z)^2 \right] \\
&= \frac{\gamma}{6} h^3 + \frac{\gamma}{6} h \left[3 |\tilde{z}_k - z|^2 + 3h |\tilde{z}_k - z| + h^2 \right] \\
&= \frac{\gamma h}{6} \left[3 |\tilde{z}_k - z|^2 + 3h |\tilde{z}_k - z| + 2h^2 \right] \, . \qquad (5.26)
\end{aligned}
$$

Otherwise, we have $\tilde{z}_k < z < \tilde{z}_k + h$ and

$$
\begin{aligned}
&\left| \int_z^{\tilde{z}_k} \int_{\tilde{z}_k}^{\tilde{z}_k+h} f'(\zeta) - f'(\xi) \, \mathrm{d}\zeta \, \mathrm{d}\xi \right| \\
&\leq \gamma \int_{\tilde{z}_k}^{z} \int_{\tilde{z}_k}^{\xi} \xi - \zeta \, \mathrm{d}\zeta \, \mathrm{d}\xi + \gamma \int_{\tilde{z}_k}^{z} \int_{\xi}^{z} \zeta - \xi \, \mathrm{d}\zeta \, \mathrm{d}\xi + \gamma \int_{\tilde{z}_k}^{z} \int_{z}^{\tilde{z}_k+h} \zeta - \xi \, \mathrm{d}\zeta \, \mathrm{d}\xi
\end{aligned}
$$

$$= \frac{\gamma}{2}\int_{\tilde{z}_k}^{z} (\tilde{z}_k - \xi)^2 \, d\xi + \frac{\gamma}{2}\int_{\tilde{z}_k}^{z} (z-\xi)^2 \, d\xi + \gamma h \int_{\tilde{z}_k}^{z} \frac{\tilde{z}_k + h + z}{2} - \xi \, d\xi$$
$$= \frac{\gamma}{6}(z-\tilde{z}_k)^3 + \frac{\gamma}{6}(z-\tilde{z}_k)^3 + \frac{\gamma h}{2}(z-\tilde{z}_k)\{(\tilde{z}_k + h - z) + (z - \tilde{z}_k)\}$$
$$= \frac{\gamma}{3}(z-\tilde{z}_k)^3 + \frac{\gamma h^2}{2}(z-\tilde{z}_k)$$

and since in this case $|\tilde{z}_k - z| = z - \tilde{z}_k < h$, we get

$$\leq \frac{\gamma h}{3}(z-\tilde{z}_k)^2 + \frac{\gamma h^2}{2}(z-\tilde{z}_k) \; . \tag{5.27}$$

By combining (5.25)–(5.27), we arrive at

$$\left| \int_{z}^{\tilde{z}_k} \int_{\tilde{z}_k}^{\tilde{z}_k+h} f'(\zeta) - f'(\xi) \, d\zeta \, d\xi \right| \leq \frac{\gamma h}{6}\left[3\,|\tilde{z}_k - z|^2 + 3h\,|\tilde{z}_k - z| + 2h^2\right] . \tag{5.28}$$

Next, we combine (5.24) and (5.28) to get

$$|\tilde{z}_{k+1} - z| \leq \frac{1}{|f'(\tilde{z}_k)|\, h - 3\phi\varepsilon} \left\{ \frac{\gamma h}{6}\left[3\,|\tilde{z}_k - z|^2 + 3h\,|\tilde{z}_k - z| + 2h^2\right] \right.$$
$$+ |\tilde{z}_k|\,\varepsilon\,|f(\tilde{z}_k + h) - f(\tilde{z}_k)| + (|\tilde{z}_k - z| + |\tilde{z}_k|\,\varepsilon)\, 2\phi\varepsilon$$
$$\left. + |f(\tilde{z}_k)|\, h\varepsilon(2+\varepsilon) + \phi\varepsilon h(1+\varepsilon)^2 \right\} .$$

The right-hand side of this inequality is less than $|\tilde{z}_k - z|$ if and only if

$$\frac{\gamma h}{6}\left[3\,|\tilde{z}_k - z|^2 + 3h\,|\tilde{z}_k - z| + 2h^2\right] + |\tilde{z}_k|\,\varepsilon\,|f(\tilde{z}_k + h) - f(\tilde{z}_k)| + (|\tilde{z}_k - z| + |\tilde{z}_k|\,\varepsilon)\, 2\phi\varepsilon$$
$$+ |f(\tilde{z}_k)|\, h\varepsilon(2+\varepsilon) + \phi\varepsilon h(1+\varepsilon)^2$$
$$< |\tilde{z}_k - z| \left\{ |f'(\tilde{z}_k)|\, h - -3\phi\varepsilon \right\} ,$$

which is equivalent to

$$0 \geq \frac{\gamma h}{2}|\tilde{z}_k - z|^2 - \left\{ |f'(\tilde{z}_k)|\, h - 5\phi\varepsilon - \frac{\gamma h^2}{2} \right\} |\tilde{z}_k - z| + |f(\tilde{z}_k)|\, h\varepsilon(2+\varepsilon)$$
$$+ \frac{\gamma h^3}{3} + |\tilde{z}_k|\,\varepsilon\,|f(\tilde{z}_k + h) - f(\tilde{z}_k)| + |\tilde{z}_k|\, 2\phi\varepsilon^2 + \phi\varepsilon h(1+\varepsilon)^2$$
$$\geq \frac{\gamma h}{2}|\tilde{z}_k - z|^2 - \left\{ |f'(\tilde{z}_k)|\, h - \varepsilon(2+\varepsilon)h \min_{\xi \text{ between } z\,,\, \tilde{z}_k} |f'(\xi)| - 6\phi\varepsilon \right\} |\tilde{z}_k - z|$$

$$+ |\tilde{z}_k| \, \varepsilon \, |f(\tilde{z}_k + h) - f(\tilde{z}_k)| + |\tilde{z}_k| \, 2\phi\varepsilon^2 + \phi\varepsilon h \left[\frac{2}{3} + (1 + \varepsilon)^2 \right] .$$

$$= A \, |\tilde{z}_k - z|^2 - B \, |\tilde{z}_k - z| + C \, .$$

This quadratic inequality is equivalent to the claim in the lemma.

If we consider only the most significant terms in inequality (5.22), then we can simplify this result to the approximate inequality

$$\frac{5}{3} \frac{\phi\varepsilon}{|f'(\tilde{z}_k)|} + |\tilde{z}_k| \, \varepsilon < |\tilde{z}_k - z| < \frac{2 \, |f'(\tilde{z}_k)|}{\gamma} .$$

The upper bound in this inequality (which is like a Kantorovich bound on the convergence region) is essentially the same as in inequality (5.15) for Newton's method with analytical derivatives. On the other hand, the lower bound in this inequality (which represents the **maximum attainable accuracy**) involves a slightly larger coefficient on the scale of f when compared to inequality (5.15). As a result, we expect that there is little difference between the performance of Newton's method with exact derivatives, and Newton's method with finite difference derivatives using a fixed increment satisfying (5.20).

Exercise 5.4.14 Program Newton's method to find the zeros of $f_1(x) = x^2 - 1$ and $f_2(x) = x^2 - 2x + 1$. Plot $\ln |f(\tilde{z}_k)|$ versus k for each of the two iterations. Explain the differences in the performance of Newton's method for these two problems.

Exercise 5.4.15 Consider $f(x) = \tan^{-1}(x)$.

1. Find the maximum M of $|f''(x)|$.
2. Let $s(x) = -f(x)/f'(x)$, and plot $2|s(x)|M$ and $|f'(x)|$. Use a computable convergence test to determine an interval around $x = 0$ in which Newton's method will converge.
3. Program Newton's method for this problem and show that it converges for $\tilde{z}_0 = 1.3917$ and diverges for $\tilde{z}_0 = 1.3918$. How do these values compare with the interval you computed in the previous part of this problem?
4. Explain why the function values in the Newton iterates alternate in sign for this problem. (In the discussion in Sect. 5.6.1 below, we will argue that eventually the signs will all be the same.)

Exercise 5.4.16 Suppose that $0 = f(z) = f'(z) = \ldots = \frac{d^p f}{dx^p} z \neq \frac{d^{p+1} f}{dx^{p+1}} z$.

1. Show that if Newton's method converges near z, then it has order 1.
2. Let $g(x) = f(x)/f'(x)$. Show that $g(z) = 0 \neq g'(z)$. What should be the order of Newton's method when applied to g? How many derivatives of f does that iteration require that we evaluate?

5.5 Secant Method

A significant disadvantage of Newton's method is the need to program and evaluate the derivative. We have seen that we can avoid programming the derivative by using a finite difference approximation, as described in Sect. 5.4.9. The disadvantage of approximating derivatives by finite differences in this way is that the computation requires an additional function evaluation at a point $z_k + h$ that is not a Newton iterate. A more efficient approach might be to use the two previous iterates z_k and z_{k-1} to construct a difference approximation to f'.

5.5.1 Algorithm Development

The **secant method** estimates the zero of f by finding the zero of the secant line

$$z(x) = f(z_k) + \frac{f(z_k) - f(z_{k-1})}{z_k - z_{k-1}}(x - z_k) .$$

This is different from Newton's method, in which the true function f is approximated by the *tangent line* at some point z_k. The secant method can be summarized by the following

Algorithm 5.5.1 (Secant Method)

$$f_0 = f(z_0)$$

$$f_1 = f(z_1)$$

while not converged

$$j_k = (f_k - f_{k-1})/(z_k - z_{k-1})$$

$$z_{k+1} = z_k - f_k/j_k$$

We have provided some simple examples of programs implementing the secant method. Readers may view an interactive C^{++} secant method. This program iterates to find a zero of $f(x) = x^2 - 4\sin(x)$. The zeros are $z = 0$ and $z \approx 1.9337573$. A plot of the error in the secant method is shown in Fig. 5.4.

This C^{++} program also demonstrates how to use the GSL software to perform a secant iteration. Note that the GSL procedure `gsl_root_fdfsolver_secant` accepts a single starting point for the iteration. The first step of this procedure is a Newton step, which requires the user to provide a function to evaluate f' for this step. This derivative function could use the secant slope instead of the true derivative, as demonstrated in our C^{++} implementation.

Readers may also execute the JavaScript **secant method.**

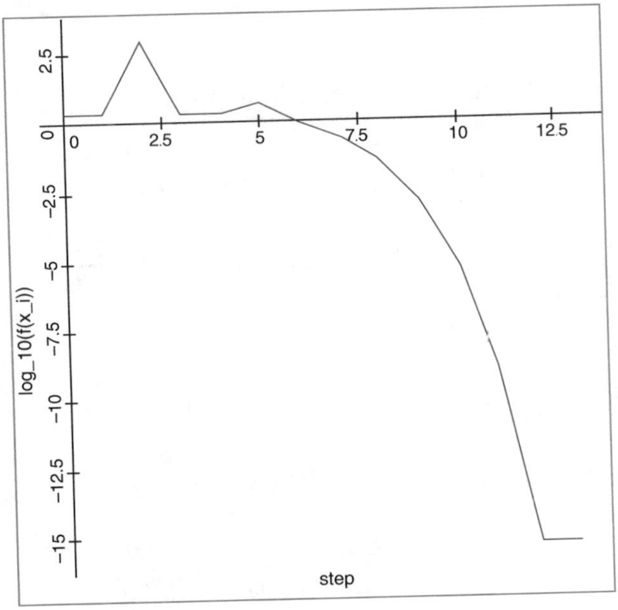

Fig. 5.4 Error in the secant method: $log_{10}(f)$ versus iterations

5.5.2 *Error Analysis*

The following lemma describes the behavior of the iterates in the secant method.

Lemma 5.5.1 *Let f be a real-valued function with $f(z) = 0$. Assume that there is an interval $[z - \alpha, z + \alpha]$ containing the true zero z so that f is twice continuously differentiable on this interval. We also suppose that $f'(z) \neq 0$. Let $z_{k-1}, z_k \in (z - \alpha, z + \alpha)$ be two distinct approximate zeros of f, and let*

$$M = \max_{|\xi - z| \leq \alpha} \left| f''(\xi) \right| .$$

If

$$|z_k - z_{k-1}| + \max\{|z_k - z| \ , \ |z_{k-1} - z|\} < 2\frac{|f'(z_k)|}{M} , \tag{5.29}$$

then the secant iterate

$$z_{k+1} \equiv z_k - f(z_k)\frac{z_k - z_{k-1}}{f(z_k) - f(z_{k-1})}$$

satisfies

$$|z_{k+1} - z| \leq \frac{M|z_k - z|\ |z_{k-1} - z|}{2\,|f'(z_k)| - M|z_k - z_{k-1}|} < \min\{|z_k - z|, |z_{k-1} - z|\}\ , \tag{5.30}$$

so it also lies in the interval $(z - \alpha, z + \alpha)$*. Alternatively, if*

$$|z_k - z| + |z_{k-1} - z| + \max\{|z_k - z|\ ,\ |z_{k-1} - z|\} < 2\frac{|f'(z_k)|}{M}\ , \tag{5.31}$$

then the secant iterate satisfies

$$|z_{k+1} - z| \leq \frac{M|z_k - z|\ |z_{k-1} - z|}{2\,|f'(z)| - M|z_k + z_{k-1} - 2z|} < \min\{|z_k - z|, |z_{k-1} - z|\}\ , \tag{5.32}$$

so z_{k+1} *lies in the interval* $(z - \alpha, z + \alpha)$*. Finally, if* f'' *is continuous at* z*, then*

$$\lim_{z_k, z_{k-1} \to z} \frac{z_{k+1} - z}{(z_k - z)(z_{k-1} - z)} = -\frac{f''(z)}{2f'(z)}\ . \tag{5.33}$$

Proof Note that

$$\begin{aligned} z_{k+1} - z &= z_k - z - \frac{f(z_k)(z_k - z_{k-1})}{f(z_k) - f(z_{k-1})} \\ &= \frac{[f(z_{k-1}) - f(z_k)](z - z_k) - [f(z) - f(z_{k-1})](z_{k-1} - z_k)}{f(z_k) - f(z_{k-1})}\ . \end{aligned}$$

Since f'' is continuous, the mean-value theorem implies that

$$\begin{aligned} &[f(z_{k-1}) - f(z_k)](z - z_k) - [f(z) - f(z_{k-1})](z_{k-1} - z_k) \\ &= \int_{z_k}^{z}\int_{z_k}^{z_{k-1}} f'(\eta) - f'(\zeta)\ \mathrm{d}\eta\ \mathrm{d}\zeta = \int_{z_k}^{z}\int_{z_k}^{z_{k-1}}\int_{\zeta}^{\eta} f''(\xi)\ \mathrm{d}\xi\ \mathrm{d}\eta\ \mathrm{d}\zeta \\ &= f''(\xi_*)\int_{z_k}^{z}\int_{z_k}^{z_{k-1}} \eta - \zeta\ \mathrm{d}\eta\ \mathrm{d}\zeta = f''(\xi_*)\int_{z_k}^{z} \frac{z_{k-1}^2 - z_k^2}{2} - \zeta(z_{k-1} - z_k)\ \mathrm{d}\zeta \\ &= f''(\xi_*)(z_{k-1} - z_k)\left[\frac{z_{k-1} + z_k}{2}(z - z_k) - \frac{z^2 - z_k^2}{2}\right] \qquad (5.34) \\ &= \frac{1}{2}f''(\xi_*)(z_{k-1} - z_k)(z - z_k)(z_{k-1} - z)\ . \qquad (5.35) \end{aligned}$$

Also,

$$
\begin{aligned}
f(z_k) - f(z_{k-1}) &= \int_{z_{k-1}}^{z_k} f'(\eta)\, d\eta = f'(z_k)(z_k - z_{k-1}) + \int_{z_{k-1}}^{z_k} f'(\eta) - f'(z_k)\, d\eta \\
&= f'(z_k)(z_k - z_{k-1}) + \int_{z_{k-1}}^{z_k} \int_{z_k}^{\eta} f''(\xi)\, d\xi\, d\eta \\
&= f'(z_k)(z_k - z_{k-1}) + f''(\xi_{**}) \int_{z_{k-1}}^{z_k} \eta - z_k\, d\eta \\
&= f'(z_k)(z_k - z_{k-1}) + f''(\xi_{**}) \left[\frac{z_k^2 - z_{k-1}^2}{2} - z_k(z_k - z_{k-1}) \right] \\
&= f'(z_k)(z_k - z_{k-1}) - \frac{1}{2} f''(\xi_{**})(z_k - z_{k-1})^2 . \qquad (5.36)
\end{aligned}
$$

Equations (5.34)–(5.36) imply that

$$
\begin{aligned}
|z_{k+1} - z| &= \frac{|[f(z_{k-1}) - f(z_k)](z - z_k) - [f(z) - f(z_{k-1})](z_{k-1} - z_k)|}{|f(z_k) - f(z_{k-1})|} \\
&\le \frac{\frac{1}{2}|f''(\xi_*)|\,|z_k - z_{k-1}|\,|z_k - z|\,|z_{k-1} - z|}{|f'(z_k)|\,|z_k - z_{k-1}| - \frac{1}{2}|f''(\xi_{**})|\,|z_k - z_{k-1}|^2} \\
&\le \frac{M|z_k - z|\,|z_{k-1} - z|}{2\,|f'(z_k)| - M|z_k - z_{k-1}|} . \qquad (5.37)
\end{aligned}
$$

Assumption (5.29) together with inequalities (5.36) and (5.37) give us

$$
\begin{aligned}
|z_{k+1} - z| &= \frac{M|z_k - z|\,|z_{k-1} - z|}{2|f'(z_k)| - M|z_k - z_{k-1}|} < \frac{M|z_k - z|\,|z_{k-1} - z|}{M \max\{|z_k - z|\,,\, |z_{k-1} - z|} \\
&= \min\{|z_k - z|\,,\, |z_{k-1} - z|\} .
\end{aligned}
$$

This proves the first claim (5.30).

Alternatively, we note that

$$
\begin{aligned}
f(z_k) - f(z_{k-1}) &= \int_{z_{k-1}}^{z_k} f'(\eta)\, d\eta = f'(z)(z_k - z_{k-1}) + \int_{z_{k-1}}^{z_k} f'(\eta) - f'(z)\, d\eta \\
&= f'(z)(z_k - z_{k-1}) + \int_{z_{k-1}}^{z_k} \int_{z}^{\eta} f''(\xi)\, d\xi\, d\eta \\
&= f'(z)(z_k - z_{k-1}) + f''(\xi_{**}) \int_{z_{k-1}}^{z_k} \eta - z\, d\eta \\
&= f'(z)(z_k - z_{k-1}) + f''(\xi_{**}) \left[\frac{z_k^2 - z_{k-1}^2}{2} - z(z_k - z_{k-1}) \right] \\
&= f'(z)(z_k z_{k-1}) - \frac{1}{2} f''(\xi_{**})(z_k - z_{k-1})(z_k + z_{k-1} - 2z) . \qquad (5.38)
\end{aligned}
$$

Assumption (5.31) together with inequalities (5.37) and (5.38) give us

$$|z_{k+1} - z| = \frac{M|z_k - z|\,|z_{k-1} - z|}{2|f'(z)| - M|z_k + z_{k-1} - 2z|} < \frac{M|z_k - z|\,|z_{k-1} - z|}{M\max\{|z_k - z|\,,\ |z_{k-1} - z|\}}$$
$$= \min\{|z_k - z|\,,\ |z_{k-1} - z|\}\,.$$

This proves the second claim.

To prove the final claim, we note

$$\lim_{z_k, z_{k-1} \to z} \frac{z_{k+1} - z}{(z_k - z)(z_{k-1} - z)}$$
$$= -\lim_{z_k, z_{k-1} \to z} \frac{\int_{z_k}^{z} \int_{z_{k-1}}^{z_k} \int_{\eta}^{\zeta} f''(\xi)\, \mathrm{d}\xi\, \mathrm{d}\eta\, \mathrm{d}\zeta \,/\, [(z_k - z_{k-1})(z_k - z)(z_{k-1} - z)]}{[f(z_k) - f(z_{k-1})] \,/\, [z_k - z_{k-1}]}$$
$$= -\frac{f''(z)}{2f'(z)}\,.$$

Inequalities (5.29) and (5.31) express the fact that the secant method is only locally convergent. Since Lemma 5.5.1 assumes that the secant method is performed with exact arithmetic, we can allow the iterates z_k and z_{k-1} to be arbitrarily close, provided that they are distinct. Equation (5.33) shows that the error in the next iterate is proportional to the product of the errors in the two previous iterates. In the next section, we will show that this implies a superlinear order of convergence for the secant method.

5.5.3 *Order of Convergence*

Recall that we defined various rates of convergence for a sequence in Definition 5.4.1. In order to discover the order of convergence for the secant method, we will adopt the ansatz

$$|\tilde{z}_{k+1} - z| \approx C|\tilde{z}_k - z|^p\,,$$

for some unknown order p and asymptotic error constant C. Then

$$|\tilde{z}_{k-1} - z| \approx \left(\frac{1}{C}|\tilde{z}_k - z|\right)^{\frac{1}{p}}\,.$$

The error analysis for the secant method says that

$$|\tilde{z}_{k+1} - z| \approx K|\tilde{z}_{k-1} - z|\;|\tilde{z}_k - z|$$

where

$$K = \frac{1}{2}\frac{f''(z)}{f'(z)} \;.$$

We conclude that

$$|\tilde{z}_{k+1} - z| \approx \frac{K}{C^{1/p}} \, |\tilde{z}_k - z|^{1+1/p} \;.$$

Thus $p = 1 + 1/p$; the only positive solution is

$$p = \frac{1+\sqrt{5}}{2} \approx 1.6 \;.$$

Also, the asymptotic error constant C satisfies

$$\frac{K}{C^{1/p}} = C \;.$$

We can solve this equation to get

$$C = K^{p-1} \;.$$

We conclude that the secant method converges at a lower rate than Newton's method, which is quadratically convergent. This lower convergence rate is a consequence of using a finite difference approximation to the derivative, to reduce the cost of the iteration.

5.5.4 *Relative Efficiency*

Although Newton's method has higher order than the secant method, the former requires more work than the latter. We would like to know which of these methods will obtain a given error reduction for less work.

For Newton's method, the error estimate says that

$$|\tilde{z}_{k+1} - z| \approx K|\tilde{z}_k - z|^2 \Longrightarrow K|\tilde{z}_{k+1} - z| \approx (K|\tilde{z}_k - z|)^2 \;.$$

For the secant method, we have $p^2 = p + 1$, $C^p = K$ and

$$|\tilde{z}_{k+1} - z| \approx C|\tilde{z}_k - z|^p \Longrightarrow C^p\,|\tilde{z}_{k+1} - z| \approx (C^p\,|\tilde{z}_k - z|)^p \;.$$

Inductive arguments imply that the Newton iterates satisfy

$$|\tilde{z}_k - z| \approx \frac{1}{K}\,(K|\tilde{z}_0 - z|)^{2^k} \;, \tag{5.39}$$

and the secant iterates satisfy

$$|\tilde{z}_k - z| \approx \frac{1}{C^p}\left(C^p|\tilde{z}_0 - z|\right)^{p^k} = \frac{1}{K}\left(K|\tilde{z}_0 - z|\right)^{p^k} . \tag{5.40}$$

Suppose that Newton's method requires n iterations to achieve the error reduction

$$|\tilde{z}_n - z| \approx \varepsilon|\tilde{z}_0 - z| .$$

Then our asymptotic expression (5.39) leads to

$$2^n \log\left(K|\tilde{z}_0 - z|\right) \approx \log\left(K|\tilde{z}_n - z|\right) \approx \log\varepsilon + \log(K|\tilde{z}_0 - z|) .$$

We conclude that the number n of Newton iterations required to achieve this error reduction satisfies

$$2^n \approx 1 + \frac{\log\varepsilon}{\log(K|\tilde{z}_0 - z|)} \Longrightarrow n \approx \frac{1 + [\log\varepsilon]/[\log(K|\tilde{z}_0 - z|)]}{\log 2} .$$

For the same error reduction, the secant method should require m iterations, where

$$p^m \log\left(K|\tilde{z}_0 - z|\right) \approx \log\left(K|\tilde{z}_m - z|\right) \approx \log\varepsilon + \log\left(K|\tilde{z}_0 - z|\right) .$$

We conclude that

$$m \approx \frac{1 + [\log\varepsilon]/[\log(K|e_0|)]}{\log p} \approx n\frac{\log 2}{\log p} \approx 1.44n .$$

Now, suppose that the cost of evaluating f is ϕ, the cost of evaluating f' is $s\phi$, and that both ϕ and $s\phi$ are much greater than the cost of any other operations in either iteration. Then the cost for Newton's method to reduce the error by a factor of ε is

$$n(\phi + s\phi) = n\phi(1 + s) ,$$

while the cost for the secant method is

$$m\phi \approx n\phi\frac{\log 2}{\log p} .$$

Thus Newton's method is asymptotically more expensive then the secant method whenever

$$s > \frac{\ln 2}{\ln p} - 1 \approx 0.44 .$$

This says that, in general, if the derivative f' costs more than about one-half as much as the function f to evaluate, then it is more efficient to use the secant method.

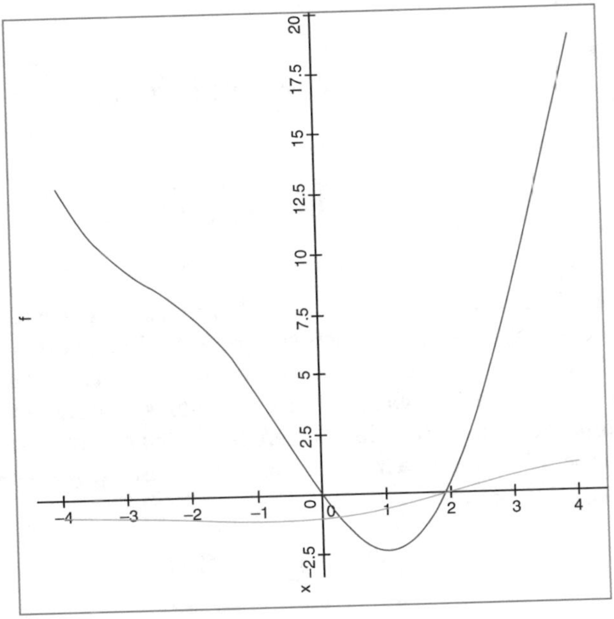

Fig. 5.5 Function for solver efficiency comparison: original function in blue, modified function in green

Example 5.5.1 Suppose that we want to find the zero of

$$f(x) = x^2 - 4\sin(x)$$

that is near $x = 2$. This function is plotted in Fig. 5.5 as the blue curve.

Obviously, $f(0) = 0$, so we would like to avoid this second root. We could do so by modifying the function to remove this root. Let $z = 0$, and that $f'(z) = -4$ is nonzero. Consider

$$g(x) = \frac{f(x)}{f'(z)(x - z)} \; .$$

This function is plotted in Fig. 5.5 as the green curve. The modified function has a single root in the interval $[-4, 4]$. However, the modified function and its derivative are more complicated to compute, so we will work with the original function in our comparison of nonlinear equation solvers.

Next, we note that the rounding error analysis of Newton's method in Lemma 5.4.4 requires a value for the scale of f. Near a zero of f we have

$$\begin{aligned} \mathrm{fl}\,(f(x)) &= \mathrm{fl}\left(x^2(1+\varepsilon_\times) - 4\sin(x)(1+\varepsilon_{\sin})\right) = x^2(1+\varepsilon_\times) - 4\sin(x)(1+\varepsilon_{\sin}) \\ &= f(x) + x^2\varepsilon_\times - 4\sin(x)\varepsilon_{\sin} \; , \end{aligned}$$

so

$$|\mathrm{fl}\,(f(x)) - f(x)| \le 2\max\left\{|x^2|, |4\sin(x)|\right\}\varepsilon\;,$$

where ε is machine precision. Thus we take

$$\phi = 2\max\left\{|x^2|, |4\sin(x)|\right\}$$

to be the scale of f.

Finally, we will perform Newton's method with the initial guess $z_0 = 1.2$, followed by the secant method and bisection with the initial guesses $z_0 = 1.2$ and $z_1 = 2.6$. Each of these methods will compute a zero $z \approx 1.93$. We will monitor the computer time required for each of these algorithms, and plot the accuracy of the solution versus computer time to compare the **efficiency** of the methods. The accuracy of the solution could be measured in either of two ways. The easiest is to compute $\log_{10}|f(z_k)|$, because

$$x - z \approx \frac{f(z_k) - f(z)}{f'(z)} = \frac{f(z_k)}{f'(z)}\;.$$

Once the zero z has been determined by a root-finding process, we could also measure the accuracy of the solution by computing $\log_{10}|z_k - z|$. The results of the comparison are shown in Fig. 5.6.

The results show that for this function and the specified initial guesses, bisection is more efficient for very low accuracy, but for intermediate to high accuracy the secant method is more efficient than Newton's method, which is in turn far more

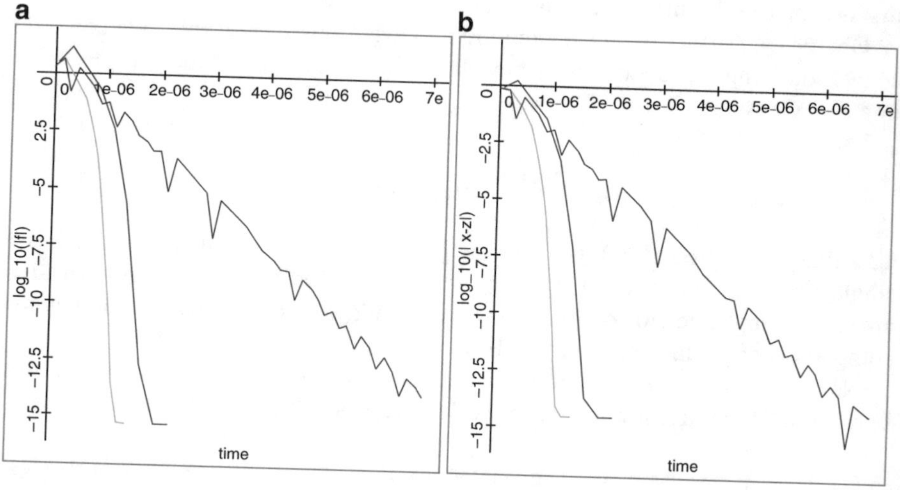

Fig. 5.6 Solver efficiency comparison: Newton iterates in blue, secant iterates in green, bisection iterates in red. (**a**) $log_{10}|f(z_k)|$ vs time. (**b**) $log_{10}|z_k - z|$ vs time

efficient than bisection. Even though the secant method took ten iterations to reach maximum attainable accuracy and Newton's method took only eight iterations, the lower cost per iteration of the secant method allowed it to be more efficient overall.

This graphical form of efficiency comparison for algorithms is an important tool for advertizing and selecting algorithms. Interested readers can view the C++ code that performed these computations by viewing relativeEfficiency.C.

5.5.5 *Rounding Errors*

Let us examine how rounding errors affect the secant method.

Lemma 5.5.2 *Let f be a continuous real-valued function such that $f(z) = 0$, and suppose that f is differentiable at z with $f'(z) \neq 0$. Assume that there exists $\gamma > 0$ and $\alpha > 0$ so that for all $\eta, \zeta \in (z-\alpha, z+\alpha)$ we have*

$$|f'(\eta) - f'(\zeta)| \leq \gamma |\eta - \zeta| .$$

Let $\tilde{z}_k$ and $\tilde{z}_{k-1}$ be distinct points in $(z-\alpha, z+\alpha)$. Suppose that floating-point arithmetic leads to computed secant iterates satisfying

$$\tilde{z}_{k+1} = \left[\tilde{z}_k - \frac{f(\tilde{z}_k) + \phi_k}{f(\tilde{z}_k) + \phi_k - f(\tilde{z}_{k-1}) - \phi_{k-1}} (\tilde{z}_k - \tilde{z}_{k-1})(1 + \varepsilon_{\div}) \right] (1 + \varepsilon_{-}) ,$$

where the rounding errors are bounded above by ε:

$$|\varepsilon_{\div}|, |\varepsilon_{-}| \leq \varepsilon$$

and the size of the errors involved in the evaluation of f are bounded above by $\phi\varepsilon$:

$$|\phi_k|, |\phi_{k-1}| \leq \phi\varepsilon .$$

If

$$|f'(\tilde{z}_k)| \geq 2\sqrt{\gamma\phi\varepsilon} , \tag{5.41}$$

$$\frac{4\phi\varepsilon}{|f'(\tilde{z}_k)| + \sqrt{|f'(\tilde{z}_k)|^2 - 4\gamma\phi\varepsilon}} < |\tilde{z}_k - \tilde{z}_{k-1}| < \frac{|f'(\tilde{z}_k)| + \sqrt{|f'(\tilde{z}_k)|^2 - 4\gamma\phi\varepsilon}}{\gamma} \tag{5.42}$$

and

$$\gamma|\tilde{z}_k - z| \max\{|\tilde{z}_k - z| \, , \, |\tilde{z}_{k-1} - z| \, , \, |\tilde{z}_k - \tilde{z}_{k-1}|\} + 2\phi|\tilde{z}_k - z| \frac{\varepsilon}{|\tilde{z}_k - \tilde{z}_{k-1}|}$$
$$+ |f(\tilde{z}_k)|\varepsilon(2+\varepsilon) + \phi\varepsilon(1+\varepsilon)^2 + |\tilde{z}_k|\varepsilon\left\{|f'(\tilde{z}_k)| + \frac{\gamma}{2}|\tilde{z}_k - \tilde{z}_{k-1}| + 2\phi\frac{\varepsilon}{|\tilde{z}_k - \tilde{z}_{k-1}|}\right\}$$
$$< \min\{|\tilde{z}_k - z| \, , \, |\tilde{z}_{k-1} - z|\}\left\{|f'(\tilde{z}_k)| - \frac{\gamma}{2}|\tilde{z}_k - \tilde{z}_{k-1}| - 2\phi\frac{\varepsilon}{|\tilde{z}_k - \tilde{z}_{k-1}|}\right\} \tag{5.43}$$

then

$$|\tilde{z}_{k+1} - z| \le \min\{|\tilde{z}_k - z| \, , \, |\tilde{z}_{k-1} - z|\} \, .$$

Proof Let

$$\delta = f(\tilde{z}_k) - f(\tilde{z}_{k-1}) + \phi_k - \phi_{k-1} = \int_{\tilde{z}_{k-1}}^{\tilde{z}_k} f'(\zeta)\,\mathrm{d}\zeta + \phi_k - \phi_{k-1}$$
$$= f'(\tilde{z}_k)(\tilde{z}_k - \tilde{z}_{k-1}) + \int_{\tilde{z}_{k-1}}^{\tilde{z}_k} \left[f'(\zeta) - f'(\tilde{z}_k)\right]\,\mathrm{d}\zeta + \phi_k - \phi_{k-1} \, .$$

It follows that

$$|\delta| \le \left|f'(\tilde{z}_k)\right| \, |\tilde{z}_k - \tilde{z}_{k-1}| + \gamma\left|\int_{\tilde{z}_{k-1}}^{\tilde{z}_k} |\zeta - \tilde{z}_k| \,\mathrm{d}\zeta\right| + 2\phi\varepsilon$$
$$\le |f'(\tilde{z}_k)| \, |\tilde{z}_k - \tilde{z}_{k-1}| + \frac{\gamma}{2}|\tilde{z}_k - \tilde{z}_{k-1}|^2 + 2\phi\varepsilon \, .$$

Similarly,

$$|\delta| \ge \left|f'(\tilde{z}_k)\right| \, |\tilde{z}_k - \tilde{z}_{k-1}| - \frac{\gamma}{2}\,|\tilde{z}_k - \tilde{z}_{k-1}|^2 - 2\phi\varepsilon \, .$$

Note that Assumptions 5.41 and 5.42 imply that the right-hand side of this inequality is positive. Next, we note that

$$(\tilde{z}_{k+1} - z)\delta = [\tilde{z}_k - z + \tilde{z}_k\varepsilon_-]\,\delta - [f(\tilde{z}_k) + \phi_k]\,(\tilde{z}_k - \tilde{z}_{k-1})\,(1+\varepsilon_\div)(1+\varepsilon_-)$$
$$= (\tilde{z}_k - z)\left\{\int_{\tilde{z}_{k-1}}^{\tilde{z}_k} f'(\zeta)\,\mathrm{d}\zeta + \phi_k - \phi_{k-1}\right\} - \int_z^{\tilde{z}_k} f'(\eta)\,\mathrm{d}\eta(\tilde{z}_k - \tilde{z}_{k-1})$$
$$+ \tilde{z}_k\varepsilon_-\delta - f(\tilde{z}_k)(\tilde{z}_k - \tilde{z}_{k-1})(\varepsilon_\div + \varepsilon_- + \varepsilon_\div\varepsilon_-) + \phi_k(\tilde{z}_k - \tilde{z}_{k-1})(1+\varepsilon_\div)(1+\varepsilon_-)$$
$$= \int_z^{\tilde{z}_k}\int_{\tilde{z}_{k-1}}^{\tilde{z}_k} \left[f'(\zeta) - f'(\eta)\right]\,\mathrm{d}\zeta\,\mathrm{d}\eta + (\phi_k - \phi_{k-1})(\tilde{z}_k - z) + \tilde{z}_k\varepsilon_-\delta$$
$$- f(\tilde{z}_k)(\tilde{z}_k - \tilde{z}_{k-1})(\varepsilon_\div + \varepsilon_- + \varepsilon_\div\varepsilon_-) + \phi_k(\tilde{z}_k - \tilde{z}_{k-1})(1+\varepsilon_\div)(1+\varepsilon_-) \, .$$

We can take absolute values of both sides to get

$$
\begin{aligned}
&|\tilde{z}_{k+1} - z|\,|\delta| \leq \gamma \left| \int_z^{\tilde{z}_k} \int_{\tilde{z}_{k-1}}^{\tilde{z}_k} |\zeta - \eta| \;\mathrm{d}\zeta\;\mathrm{d}\eta \right| + 2\phi\varepsilon\,|\tilde{z}_k - z| + |\tilde{z}_k|\,\varepsilon\,|\delta| \\
&+ |f(\tilde{z}_k)|\;|\tilde{z}_k - \tilde{z}_{k-1}|\varepsilon(2+\varepsilon) + \phi\varepsilon|\tilde{z}_k - \tilde{z}_{k-1}|(1+\varepsilon)^2 \\
&\leq \gamma|\tilde{z}_k - z|\;|\tilde{z}_k - \tilde{z}_{k-1}|\max\{|\tilde{z}_k - \tilde{z}_{k-1}|\;,\;|\tilde{z}_k - z|\;,\;|\tilde{z}_{k-1} - z|\} + 2\phi\varepsilon|\tilde{z}_k - z| \\
&+ |\tilde{z}_k|\,\varepsilon\left\{|f'(\tilde{z}_k)|\;|\tilde{z}_k - \tilde{z}_{k-1}| + \frac{\gamma}{2}|\tilde{z}_k - \tilde{z}_{k-1}|^2 + 2\phi\varepsilon\right\} \\
&+ |f(\tilde{z}_k)|\;|\tilde{z}_k - \tilde{z}_{k-1}|\varepsilon(2+\varepsilon) + \phi\varepsilon|\tilde{z}_k - \tilde{z}_{k-1}|(1+\varepsilon)^2\;.
\end{aligned}
$$

It is now easy to see that Assumption 5.43 will imply that $|\tilde{z}_{k+1} - z| < \min\{|\tilde{z}_k - z|\;,\;|\tilde{z}_{k-1} - z|\}$.

Generally, we will use the secant method to find zeros of functions with nonzero derivative at the solution. In such cases, the Assumption 5.41 is easily satisfied. The lower bound in Assumption 5.42 represents a **maximum attainable accuracy**; essentially this inequality requires that

$$|\tilde{z}_k - \tilde{z}_{k-1}| \geq \frac{2\phi\varepsilon}{|f'(\tilde{z}_k)|}\;.$$

The right-hand inequality in Assumption 5.42 is like a Kantorovich bound on the convergence region; it essentially requires that

$$|\tilde{z}_k - \tilde{z}_{k-1}| \leq \frac{2|f'(\tilde{z}_k)|}{\gamma}\;.$$

The remaining Assumption 5.43 requires additional elaboration. If we can ignore terms involving machine precision, then this assumption is equivalent to

$$
\begin{aligned}
&|\tilde{z}_k - z|\max\{|\tilde{z}_k - z|\;,\;|\tilde{z}_{k-1} - z|\;,\;|\tilde{z}_k - \tilde{z}_{k-1}|\} \\
&< \min\{|\tilde{z}_k - z|\;,\;|\tilde{z}_{k-1} - z|\}\left\{\frac{|f'(\tilde{z}_k)|}{\gamma} - \frac{1}{2}|\tilde{z}_k - \tilde{z}_{k-1}|\right\}\;.
\end{aligned}
$$

This suggests that a successful start to the secant method needs the increment in the initial steps to satisfy

$$|\tilde{z}_k - \tilde{z}_{k-1}| < 2\frac{|f'(\tilde{z}_k)|}{\gamma}\;.$$

This is the same Kantorovich-like bound we found in the previous paragraph.

Once the secant method begins to make progress, we should expect that $|z|\varepsilon \ll |\tilde{z}_k - z| < |\tilde{z}_{k-1} - z|$, and inequality (5.43) essentially requires that

$$\max\{|\tilde{z}_{k-1} - z| \ , \ |\tilde{z}_k - \tilde{z}_{k-1}|\} < \frac{|f'(\tilde{z}_k)|}{\gamma} - \frac{1}{2}|\tilde{z}_k - \tilde{z}_{k-1}| \approx \frac{|f'(\tilde{z}_k)|}{\gamma} \ .$$

Because of the superlinear convergence of the secant method, we eventually expect to see a step where $|\tilde{z}_k - z|$ is on the order of $|z|\varepsilon$ and $|\tilde{z}_{k-1} - z|$ is significantly larger. For such a step, we should be able to ignore the first three terms on the left of inequality (5.43). Then this assumption essentially requires that

$$|\tilde{z}_k - z| > \varepsilon\left\{|\tilde{z}_k| + \frac{\phi\varepsilon}{|f'(\tilde{z}_k)|}\right\} \ .$$

This provides a second form of the **maximum attainable accuracy** of the secant method, and is the same estimate that we found for Newton's method.

However, once the errors in both $\tilde{z}_k$ and $\tilde{z}_{k-1}$ are both on the order of the maximum attainable accuracy, then a different discussions of the secant method is needed. Suppose that

$$\max\{|\tilde{z}_k - z| \ , \ |\tilde{z}_{k-1} - z|\} \le \varepsilon\left[|z| + \frac{\phi}{\gamma}\right] \text{ and}$$
$$\max\{|f(\tilde{z}_k) + \phi_k| \ , \ |f(\tilde{z}_{k-1}) + \phi_k|\} \le \phi\varepsilon \ .$$

Then

$$\begin{aligned}
&|\tilde{z}_{k+1} - z| \\
&= \left|\tilde{z}_k - z + \tilde{z}_k\varepsilon_- - \frac{f(\tilde{z}_k) + \phi_k}{f(\tilde{z}_k) + \phi_k - f(\tilde{z}_{k-1}) - \phi_{k-1}}(\tilde{z}_k - \tilde{z}_{k-1})(1 + \varepsilon_\div)(1 + \varepsilon_-)\right| \\
&\le \varepsilon\left\{|z| + \frac{\phi}{\gamma}\right\} + \varepsilon\,|z| + \frac{2\phi\varepsilon^2(1+\varepsilon)^2}{|f(\tilde{z}_k) - f(\tilde{z}_{k-1}) + \phi_k - \phi_{k-1}|}\left\{|z| + \frac{\phi}{\gamma}\right\} \ .
\end{aligned}$$

Unfortunately, there is no good lower bound on the term $|f(\tilde{z}_k) - f(\tilde{z}_{k-1}) + \phi_k - \phi_{k-1}|$ in denominator of the right-hand side. As a result, the final term on the right can be arbitrarily large. This means that the secant method will not necessarily stay close to the solution once the two previous iterates are near the maximum attainable accuracy.

Exercise 5.5.1 Some authors suggest computing the secant method iterates in the form

$$\tilde{z}_{k+1} = \frac{\tilde{z}_{k-1}f(\tilde{z}_k) - \tilde{z}_k f(\tilde{z}_{k-1})}{f(\tilde{z}_k) - f(\tilde{z}_{k-1})} \ .$$

Determine how rounding errors affect this computation, and modify the proof of Lemma 5.5.2 to determine conditions on the iterates so that the error in the secant method is reduced. How does the maximum attainable accuracy compare with that we found after Lemma 5.5.2.

5.6 Global Convergence

So far in this chapter, we have studied three different methods for finding a zero z of a nonlinear function f. Of these three methods, only the bisection method is **globally convergent**, meaning that it converges whenever the starting points are useful (meaning that f has opposite sign at those points). The problem is that bisection converges much more slowly than either Newton's method or the secant method. In this section, we will examine techniques for combining these three previous methods, possibly with some new methods, to maintain bracketing. We will also seek to achieve a convergence rate similar to that of either the Newton or secant methods.

5.6.1 *Bracketing*

Suppose that the errors $z_k - z$ in the secant method are small for $k = 0, 1$. Recall that Lemma 5.5.1 proved that the secant method iterates satisfy

$$\lim_{z_k, z_{k-1} \to z} \frac{z_{k+1} - z}{(z_k - z)(z_{k-1} - z)} = -\frac{f''(z)}{2f'(z)} .$$

Thus we expect the ratio inside this limit to have constant sign, once the iterates are close to the solution. Suppose that our initial guesses also bracket the solution; in other words, $(z_0 - z)(z_1 - z) < 0$. Then if $f''(z)/f'(z) < 0$, the sequence of signs of the errors should look like

$$+ - - + - - \ldots \text{ or}$$
$$- + - - + - - \ldots .$$

Otherwise we should have $f''(z)/f'(z) > 0$, and the sequence of signs should look like

$$+ - + + - + + \ldots \text{ or}$$
$$- + + - + + \ldots .$$

Thus, even if we begin with two iterates that bracket the zero, successive iterates of the secant method will not always bracket the zero.

For Newton's method, Lemma 5.4.1 shows that

$$\lim_{z_k \to z} \frac{z_{k+1} - z}{[z_k - z]^2} = \frac{f''(z)}{2f'(z)} .$$

Thus the ratio inside the limit should eventually have constant sign. If $f''(z)/f'(z) > 0$, the sequence of signs of the errors in the iterates should be all positive after the first iterate. If $f''(z)/f'(z) < 0$ then the sequence of signs of should be all negative after the first iterate. Consequently, neither the secant method nor Newton's method will maintain bracketing of the zero.

5.6.2 *Significant Change*

One of the problems shared by global convergence algorithms is that each can fail to make a significant change in the iterates during special steps. The following lemma addresses this problem.

Lemma 5.6.1 *Let a and b be distinct points, and choose a number s > 2. Then*

$$c \in \left(\min\{a, b\} + \frac{|b-a|}{s} , \max\{a, b\} - \frac{|b-a|}{s} \right) \tag{5.44}$$

if and only if

$$\frac{1}{s} < \frac{c - \min\{a, b\}}{|b-a|} < 1 - \frac{1}{s} . \tag{5.45}$$

Furthermore, if $f(a)f(b) < 0$ then the secant iterate

$$c = b - f(b) \frac{b-a}{f(b) - f(a)}$$

satisfies (5.44) *if and only if*

$$\frac{1}{s-1} < -\frac{f(a)}{f(b)} < s - 1 . \tag{5.46}$$

Proof Suppose that $a < b$. Then

$$a + \frac{b-a}{s} < c < b - \frac{b-a}{s}$$

is equivalent to

$$\frac{b-a}{s} < c - a < (b-a)\left(1 - \frac{1}{s}\right) ,$$

and this is easily seen to be equivalent to the claim (5.45). On the other hand, if $b < a$ then

$$b + \frac{a-b}{s} < c < a - \frac{a-b}{s}$$

is equivalent to

$$\frac{a-b}{s} < c - b < (a-b)\left(1 - \frac{1}{s}\right) ,$$

which is also equivalent to the claim (5.45).

Next, let us suppose that c is computed by a secant step in which the previous iterates bracket a zero of f. Consider the case $a < b$. Then

$$a + \frac{b-a}{s} < b - f(b)\frac{b-a}{f(b)-f(a)} < b - \frac{b-a}{s}$$

is equivalent to

$$\frac{1}{s} < 1 - \frac{f(b)}{f(b)-f(a)} < 1 - \frac{1}{s} = \frac{s-1}{s} ,$$

and taking reciprocals shows that this is equivalent to

$$s > 1 - \frac{f(b)}{f(a)} > \frac{s}{s-1} .$$

This statement is equivalent to the claim (5.46). On the other hand, if $b < a$ then

$$b + \frac{a-b}{s} < b - f(b)\frac{b-a}{f(b)-f(a)} < a - \frac{a-b}{s}$$

is equivalent to

$$\frac{1}{s} < \frac{f(b)}{f(b)-f(a)} < 1 - \frac{1}{s} = \frac{s-1}{s} ,$$

and taking reciprocals shows that this is equivalent to

$$s > 1 - \frac{f(a)}{f(b)} > \frac{s}{s-1} .$$

Again, this statement is equivalent to the claim (5.46).

A reasonable choice might be $s = 10$, so that the next iterate lies between 10% and 90% of the distance between a and b. If we attempt a special step and find that it does not produce a significant change in the solution, then we could take a bisection step.

Note that we should never require a significant change in the iterates during a normal secant or Newton step. In such circumstance, as the iterates converge we expect $f(b)$ to be much smaller than $f(a)$ in magnitude, so switching to bisection would destroy the superlinear convergence of the secant or Newton method.

We remark that the Dekker [28] routine dfzero requires that a significant change is produced in iterates that are determined by either bisection or the secant method.

5.6.3 *Regula Falsi*

Suppose that we begin with two points a and b that bracket the zero:

$$a < b \text{ and } f(a)f(b) < 0 \ .$$

In Sect. 5.6.1, We observed that we cannot maintain bracketing with the secant method. If we nevertheless insist on maintaining the bracketing, then we obtain the following

Algorithm 5.6.1 (Regula Falsi)

$$f_a = f(a)$$

$$f_b = f(b)$$

while not converged

$$c = b - \frac{b-a}{f_b - f_a} f_b \ ,$$

$$f_c = f(c)$$

if $\text{sign}(f_c) = \text{sign}(f_a)$ then

$$a = c \text{ and } f_a = f_c$$

else

$$b = c \text{ and } f_b = f_c \ .$$

The trouble with this algorithm is that one of either a or b could be preserved in the iteration indefinitely. Then the method would converge with order one, largely because $[f(b) - f(a)]/[b - a]$ does not approach $f'(z)$.

Let us examine this convergence rate in more detail. Suppose that the retained endpoint is a and the true zero is z. If $a < z < b$ but b is close to z, then since $f(a)$ and $f(b)$ have opposite signs

$$c - z = [b - z] - f(b)\frac{b-a}{f(b) - f(a)} \approx \left[1 - \frac{|f(b)|}{|f(a)|}\frac{z-a}{b-z}\right][b - z] .$$

If

$$|f(a)| > |f(b)|\frac{z-a}{b-z} \approx |f'(z)|\,(z-a) ,$$

then $c-z$ will have the same sign as $b-z$, regula falsi will preserve a in the iteration, and the convergence will have order 1.

5.6.4 *Illinois Algorithm*

"Who controls the past controls the future. Who controls the present controls the past."
George Orwell, "1984"

We have seen that regula falsi can retain an old iterate indefinitely if its function value is too large in magnitude. We can overcome this problem by replacing f_a with αf_a where $0 < \alpha < 1$ is chosen appropriately. If α is made sufficiently small, then eventually regula falsi should replace a in the iteration. A common choice is $\alpha = \frac{1}{2}$, as in the **Illinois algorithm**.

However, it generally pays to give more thought to how we choose α, or in general how would we modify f_a in regula falsi. Note that if $f_a = -f_b$, then the secant method computes the next iterate to be

$$c = b - f_b\frac{b-a}{f_b - f_a} = b - \frac{1}{2}(b-a) = \frac{1}{2}(b+a) .$$

In this case, c is the same result obtained by bisection, which is globally convergent. We could consider this to be a modification of the Illinois algorithm, in which we replace f_a with αf_a, where $\alpha = -f_b/f_a$.

Here is a first attempt at a global convergence strategy. Suppose that we are given a and b so that $f(a)f(b) < 0$, so that a and b bracket a zero of f. We will order these initial points so that $|f(b)| \leq |f(a)|$. The resulting ideas, due to Dowell and Jarratt [35], are contained in the following

Algorithm 5.6.2 (Illinois)

$f_a = f(a)$
$f_b = f(b)$ /* require $f_b f_a < 0$ */
$c = b - f_b * (b-a)/(f_b - f_a)$ /* secant step */
$\zeta = \max\{|a|, |b|\}$
if $|c-b| \leq \delta_z \zeta$ break /* relative error convergence test */
if $|c-b| \leq \sqrt{\varepsilon}\zeta$ and $|c-b| \geq |b-a|$ break /* maximum attainable accuracy */
while not converged
  $f_c = f(c)$ /* also compute $\phi_c =$ scale of f_c */
  if $|f_c| \leq \delta_f \phi_c$ break /* function value convergence test */
  if sign(f_c) $\neq$ sign(f_b) then
    $a = b$ and $f_a = f_b$
    $b = c$ and $f_b = f_c$
    $c = b - f_b * (b-a)/(f_b - f_a)$ /* secant step */
    $\zeta = \max\{|a|, |b|\}$
    if $|c-b| \leq \delta_z \zeta$ break /* relative error convergence test */
    if $|c-b| \leq \sqrt{\varepsilon}\zeta$ and $|c-b| \geq |b-a|$ break /* maximum attainable accuracy */
  else /* special step */
    $f_a = 0.5 * f_a$
    $b = c$ and $f_b = f_c$
    if $1/9 < -f_b/f_a < 9$ then /* significant change */
      $c = b - f_b * (b-a)/(f_b - f_a)$ /* Illinois step */
    else
      $c = (a+b)/2$ /* bisection */
  if $c = a$ or $c = b$ break /* no change */

We have implemented the Illinois algorithm in the interactive C^{++} program illinoisMain.C.

5.6.5 *Muller-Traub Method*

One alternative to the Illinois algorithm is to use the **Muller-Traub** method [58, p. 159] This method begins by finding the quadratic function

$$\tilde{f}(x) = f_c + [f_c - f_b]\frac{x-c}{c-b} + \left[\frac{f_c - f_b}{c-b} - \frac{f_b - f_a}{b-a}\right]\frac{(x-c)(x-b)}{c-a}$$

that interpolates f at $x = a$, b and c. (See Sect. 1.2.2 of Chap. 1 in Volume III for more information about polynomial interpolation.)

Let us use a zero of such a quadratic function to choose a new iterate under the circumstances of a "special step" in the Illinois algorithm In other words, $f(a)f(b) < 0$, c lies between a and b, and $f(b)f(c) > 0$. In such a case, we cannot safely take a secant step using the information at points b and c. However, in this case $\tilde{f}$ must have a unique zero between a and c. A Taylor expansion of $\tilde{f}$ implies that

$$\tilde{f}(x) = f_c + \phi_1(x-c) + \phi_2(x-c)^2 ,$$

where

$$\phi_2 \equiv \frac{1}{2}\tilde{f}''(x) = \left[\frac{f_c - f_b}{c-b} - \frac{f_b - f_a}{b-a}\right]\frac{1}{c-a}$$

and

$$\phi_1 \equiv \tilde{f}'(c) = \frac{f_c - f_b}{c-b} + \left[\frac{f_c - f_b}{c-b} - \frac{f_b - f_a}{b-a}\right]\frac{c-b}{c-a} = \frac{f_c - f_b}{c-b} + \phi_2(c-b) .$$

The zeros of $\tilde{f}$ can be evaluated as

$$d = c - \frac{2f_c}{\phi_1 \pm \sqrt{\phi_1^2 - 4f_c\phi_2}} \text{ or } d = c - \frac{\phi_1 \mp \sqrt{\phi_1^2 - 4f_c\phi_2}}{2\phi_1} .$$

We can choose the sign in these expressions to avoid cancellation, and choose whichever of the two roots lies between a and c. This gives us the following global convergence algorithm.

Algorithm 5.6.3 (Muller-Traub)

$f_a = f(a)$
$f_b = f(b)$ /* require $f_b f)a < 0$ */
$c = b - f_b * (b - a)/(f_b - f_a)$ /* secant step */
$\zeta = \max\{|a| \, , \, |b|\}$
if $|c - b| \le \delta_z \zeta$ break /* relative error convergence test */
if $|c - b| \le \sqrt{\varepsilon}\zeta$ and $|c - b| \ge |b - a|$ break /* maximum attainable accuracy */
special = *false*
while not converged
  $f_c = f(c)$ /* also compute $\phi_c =$ scale of f */
  if $|f_c| \le \delta_f \phi_c$ break /* function value convergence test */
  if $\text{sign}(f_c) \ne \text{sign}(f_b)$ then
    $a = b$ and $f_a = f_b$
    $b = c$ and $f_b = f_c$
    $c = b - f_b * (b - a)/(f_b - f_a)$ /* secant step */

  $\zeta = \max\{|a| \, , \, |b|\}$
  if $|c - b| \le \delta_z \zeta$ break /* relative error convergence test */
  if $|c - b| \le \sqrt{\varepsilon}\zeta$ and $|c - b| \ge |b - a|$ break /* maximum attainable accuracy */
  special = *false*
else /* special step */
  $\sigma = (f_c - f_b)/(c - b)$
  $\phi_2 = (\sigma - (f_b - f_a)/(b - a))/(c - a)$
  $\phi_1 = \sigma + \phi_2(c - b)$
  $\delta = \sqrt{\phi_1^2 - 4 f_c \phi_2}$
  $\tau = (\phi_1 > 0 \; ? \; \phi_1 + \delta : \phi_1 - \delta)$
  $d = c - 2 * f_c/\tau$
  $\varrho = (d - \min\{a \, , \, b\})/|b - a|$
  if $0 < \varrho$ and $\varrho < 1$ then /* first root out of bounds */
    $d = c - 0.5 * \tau/\phi_2$
    $\varrho = (d - \min\{a \, , \, b\})/|b - a|$
  $b = c$ and $f_b = f_c$
  if not special or $.1 < \varrho < .9$ then /* significant change */
    $c = d$ /* Muller-Traub step */
  else
    $c = (a + b)/2$ /* bisection */
  special = *true*
if $c = a$ or $c = b$ break /* no change */

In this algorithm, the flag `special` is designed to require a significant change in the iterates if more than one Muller-Traub step is taken consecutively.

We have implemented the Muller-Traub algorithm in the interactive C^{++} program mullerTraub.C.

5.6.6 Rational Interpolation

Another idea, which can be found in either Householder [58, p. 159] or Dahlquist and Björck [26, p. 233], is to find a rational function

$$\tilde{f}(x) = \frac{x - A}{Bx + C}$$

that goes through the three points (a, f_a), (b, f_b) and (c, f_c). Note that

$$\frac{b - a}{f_b - f_a} = \frac{b - a}{\frac{b-A}{Bb+C} - \frac{a-A}{Ba+C}} = \frac{(Ba + C)(Bb + C)}{AB + C} .$$

It follows that for any fourth point (x, f_x) on this curve, we have

$$\frac{f_x - f_c}{x - c} \frac{x - a}{f_x - f_a} \frac{b - c}{f_b - f_c} \frac{f_b - f_a}{b - a}$$
$$= \frac{AB + C}{(Bx + C)(Bc + C)} \frac{(Bx + C)(Ba + C)}{AB + C} \frac{(Bb + C)(Bc + C)}{AB + C} \frac{AB + C}{(Bb + C)(Ba + C)} = 1.$$

In particular, if d is the zero of this interpolating rational polynomial $\tilde{f}$, then

$$1 = \frac{-f_c}{d - c} \frac{d - a}{-f_a} \frac{b - c}{f_b - f_c} \frac{f_b - f_a}{b - a} .$$

We can use this fact to avoid the determination of the coefficients A, B and C.

The solution of this equation is

$$d = c - f_c \frac{c - a}{f_c - \alpha f_a} .$$

If $\alpha > 0$ and $f_c f_a < 0$, then this equation for d contains a way to choose α in the Illinois Algorithm 5.6.2. One nice side benefit is that the rational function does a reasonable job in representing bad initial guesses or functions with poles. This leads to the following

Algorithm 5.6.4 (Rational Polynomial Interpolation)

```
  f_a = f(a)
  f_b = f(b)                                                      /* require f_bf)a < 0 */
  c = b - f_b * (b - a)/(f_b - f_a)                                /* secant step */
  ζ = max{|a| , |b|}
  if |c - b| ≤ δ_z ζ break                                         /* relative error convergence test */
  if |c - b| ≤ √ε ζ and |c - b| ≥ |b - a| break                    /* maximum attainable accuracy */
  special = false
  while not converged
    f_c = f(c)                                                     /* also compute φ_c = scale of f */
    if |f_c| ≤ δ_f φ_c break                                       /* function value convergence test */
    if sign(f_c) ≠ sign(f_b) then
      a = b and f_a = f_b
      b = c and f_b = f_c
      c = b - f_b * (b - a)/(f_b - f_a)                            /* secant step */
      ζ = max{|a| , |b|}
      if |c - b| ≤ δ_z ζ break                                     /* relative error convergence test */
      if |c - b| ≤ √ε ζ and |c - b| ≥ |b - a| break                /* maximum attainable accuracy */
      special = false


else                                                               /* special step */
  σ = (b - a)/(f_b - f_a)
  α = σ * (f_c - f_b)/(c - b)
  b = c and f_b = f_c
  if f_b - α f_a ≠ 0
    d = b - f_b(b - a)/(f_b - α f_a)                               /* zero of rational interpolant */
    ϱ = (d - min{a , b})/|b - a|
    if (0 < ϱ < 1) and ( not special or .1 < ϱ < .9) then/* rational interpolation acceptable */
      c = d
    else
      c = (a + b)/2                                                /* bisection */
  else                                                             /* rational interpolant undefined */
    c = (a + b)/2                                                  /* bisection */
  special = true
if c = a or c = b break                                            /* no change */
```

We have implemented the rational polynomial interpolation algorithm in the interactive C++ program rationalInterpolation.C. Also note that the Larkin [71] routines zero1 and zero2 use rational interpolation by polynomials of degree *at most eleven* in the context of a global convergence strategy for finding the zero of a function of a single variable.

5.6.7 Inverse Quadratic Interpolation

Yet another global convergence idea, which can be found in Brent [17, Chapter 4] or Epperson [38, pp. 182–185], is to form a quadratic function that interpolates the inverse function of f at the points $(f(a), a)$, $(f(b), b)$ and $(f(c), c)$:

$$\tilde{f}^{-1}(y) = a\frac{y-f(c)}{f(a)-f(c)}\frac{y-f(b)}{f(a)-f(b)} + b\frac{y-f(a)}{f(b)-f(a)}\frac{y-f(c)}{f(b)-f(c)} + c\frac{y-f(a)}{f(c)-f(a)}\frac{y-f(b)}{f(c)-f(b)} .$$

To see how this interpolant was determined, see Sect. 1.2.3 of Chap. 1 in Volume III below.

We would like to consider choosing our new approximate zero of f to be

$$d = \tilde{f}^{-1}(0) = a\frac{f(c)}{f(a)-f(b)}\frac{f(b)}{f(a)-f(c)} + b\frac{f(a)}{f(b)-f(c)}\frac{f(c)}{f(b)-f(a)}$$
$$+ c\frac{f(a)}{f(c)-f(b)}\frac{f(b)}{f(c)-f(a)} .$$

Unlike the Muller-Traub quadratic interpolation, inverse quadratic interpolation avoids the quadratic formula and the need to choose the appropriate solution. On the other hand, inverse quadratic interpolation requires that the function values must be unique, and there is no guarantee that d lies between a and c.

These ideas can be summarized in the following

Algorithm 5.6.5 (Inverse Quadratic Interpolation)

$f_a = f(a)$
$f_b = f(b)$ /* require $f_b f) a < 0$ */
$c = b - f_b * (b-a)/(f_b - f_a)$ /* secant step */
$\zeta = \max\{|a|\,,\,|b|\}$
if $|c-b| \leq \delta_z \zeta$ break /* relative error convergence test */
if $|c-b| \leq \sqrt{\varepsilon}\zeta$ and $|c-b| \geq |b-a|$ break /* maximum attainable accuracy */
special = *false*
while not converged
  $f_c = f(c)$ /* also compute $\phi_c =$ scale of f */
  if $|f_c| \leq \delta_f \phi_c$ break /* function value convergence test */
  if $\text{sign}(f_c) \neq \text{sign}(f_b)$ then
    $a = b$ and $f_a = f_b$
    $b = c$ and $f_b = f_c$
    $c = b - f_b * (b-a)/(f_b - f_a)$ /* secant step */
    $\zeta = \max\{|a|\,,\,|b|\}$
    if $|c-b| \leq \delta_z \zeta$ break /* relative error convergence test */
    if $|c-b| \leq \sqrt{\varepsilon}\zeta$ and $|c-b| \geq |b-a|$ break /* maximum attainable accuracy */
    special = *false*
  else
    if $f_b \neq f_c$ /* special step defined */
      $\varrho_a = f_a/(f_b - f_c)$ /* either sign possible */
      $\varrho_b = f_b/(f_c - f_a)$ /* positive */
      $\varrho_c = f_c/(f_a - f_b)$ /* negative */
      $d = -a * \varrho_c * \varrho_b - b * \varrho_a * \varrho_c - c * \varrho_a * \varrho_b$ /* zero of inverse quadratic interpolant */
      $b = c$ and $f_b = f_c$
      $\varrho = (d - \min\{a\,,\,b\})/|b-a|$
      if $(0 < \varrho < 1)$ and (not special or $.1 < \varrho < .9$) then /* special step acceptable */
        $c = d$
      else
        $c = (a+b)/2$ /* bisection */
    else /* special step undefined */
      $c = (a+b)/2$ /* bisection */
    special = *true*
  if $c = a$ or $c = b$ break /* no change */

We have implemented the inverse quadratic interpolation algorithm in the interactive C++ program inverseQuadraticInterpolation.C. Another implementation of inverse quadratic interpolation is available in the Brent routine zeroin. Brent's method is also implemented in the GSL root bracketing algorithm named `gsl_root_fsolver_brent`. We also remark that the MATLAB function fzero uses a combination of bisection, the secant method and inverse quadratic interpolation to find the zero of a function.

5.6.8 *Summary*

For a comparison of the global convergence strategies above, readers may experiment with the JavaScript **global convergence program.**

For more information about global convergence strategies to find a zero of a function of a single variable, readers should read Brent [17, p. 47ff], Dennis and Schnabel [31, p. 24ff], Epperson [38] or Kincaid and Cheney [66, p. 108ff]. We also recommend the interesting article by Kahan [61], which discusses the ideas behind the design of a program to find the zero of a function on a hand-held calculator.

There are additional publicly available routines for finding zeros of functions of a single variable, either at Netlib, **Harwell** or Netlib TOMS 748. The last of these three is based on ideas due to Le [74].

Exercise 5.6.1 The following functions are relatively flat near their zero. Discuss why a global convergence strategy may be needed to find each zero.

1. $x^7 + 28 * x^4 - 480$ on $(-3, 4)$
2. $x^3 - 2 * x - 5$ on (-3,3)
3. $\cos(x) - x$ on (-4,2)
4. $1 - x - e^{-2x}$ on $(-3, 2)$
5. $(x + 1)e^{x-1} - 1$ on $(-5, 2)$
6. $x^3 - x - 4$ on $(-2, 2.5)$

Exercise 5.6.2 The following functions are very steep near a zero. Describe why this will cause problems for an iterative method.

1. $e^x - x^2 - 2x - 2$ on $(-2, 3)$
2. $2x - e^{-x}$ on $(-2, 8)$
3. $e^{-2x} - 1 + x$ on $(-1, 6)$
4. $1/x - 2^x$ on $(.3, 1)$

Exercise 5.6.3 The following functions have zero derivative at a zero. Discuss why iterative methods may have trouble converging rapidly to the zero.

1. $\sin(x) - x$ on $(-.5, .5)$
2. x^9 on $(-1, 1)$
3. x^{19} on $(-1, 1)$

Exercise 5.6.4 The following functions have multiple zeros. Discuss how you can design an iterative method to converge to a desired zero.

1. $4\sin(x) + 1 - x$ on $(-5, 5)$
2. $x^4 - 4x^3 + 2x^2 - 8$ on $(-2, 4)$
3. $x\cos(x) - \log(x)$ on $(0, 20)$

4. $\sin(x) - \cos(x)$ on $(-10, 10)$
5. $x^4 - 12x^3 + 47x^2 - 60x + 24$ on $(0, 2)$
6. $2^{-x} + e^x + 2\cos(x) - 6$ on $(-4, 3)$

Exercise 5.6.5 The following functions have multiple zeros, some of which are at local maxima. Discuss how you can design an iterative method to converge to a desired zero. How would your strategy differ for zeros at local maxima from the strategy for other zeros?

1. $\sin(x) - \cos(2x)$ on $(-10, 10)$
2. $x^3 - 7x^2 + 11x - 5$ on $(-1, 6)$

Exercise 5.6.6 The following functions have multiple zeros and poles. Discuss how you can design an iterative method to converge to a desired zero.

1. $3x^2 + \tan(x)$ on $(-3, 3)$
2. $1/x - \tan(x)$ on $(.2, 1.4)$

Exercise 5.6.7 The following functions may appear to have a zero but do not. Describe what would happen if you apply an iterative method to find a zero.

1. $e^x + x^2 + x$ on $(-6, 2)$

Exercise 5.6.8 Does the function

$$f(x) = \frac{x^3 + 4x^2 + 3x + 5}{2x^3 - 9x^2 + 18x - 2}$$

have a zero in $(0, .2)$? What will happen if we use an globally convergent iterative method to find a zero of this function?

Exercise 5.6.9 Program the Muller-Traub method, in which z_{k+1} is determined to be a zero of the quadratic interpolant to some function f at z_k, z_{k-1} and z_{k-2}. Choose that zero of the quadratic which is closest to z_k. Apply this iteration to find the zero of

$$f(x) = e^x - 2 .$$

Choose the three initial iterates randomly in the interval $[1/2, 1]$, and order them so that $f(z_0)$ is largest and $f(z_2)$ is smallest. Plot the log of the error in the Muller-Traub iteration versus the number of iterations, and experimentally determine the order of convergence.

Exercise 5.6.10 Perform the previous exercise for the sequence of iterates determined by rational interpolation.

Exercise 5.6.11 Perform the previous exercise for the sequence of iterates determined by inverse quadratic interpolation.

Exercise 5.6.12 Suppose that the sequence $\{z_k\}$ is defined by the Muller-Traub iteration. In other words,

$$0 = f(z_k) + [f(z_k) - f(z_{k-1})] \frac{z_{k+1} - z_k}{z_k - z_{k-1}}$$
$$+ \left\{ \frac{f(z_k) - f(z_{k-1})}{z_k - z_{k-1}} - \frac{f(z_{k-1}) - f(z_{k-2})}{z_{k-1} - z_{k-2}} \right\} \frac{(z_{k+1} - z_k)(z_{k+1} - z_{k-1})}{z_k - z_{k-2}} .$$

Define $e_k = z_k - z$ where $f(z) = 0$. Use Taylor expansions to show that

$$0 \approx f'(z)e_{k+1} + \frac{1}{2}f''(z)e_{k+1}^2$$
$$+ \frac{1}{6}f'''(z) \left\{ e_{k+1}^2(e_k + e_{k-1} + e_{k-2}) - e_{k+1}(e_k e_{k-1} \right.$$
$$\left. + e_{k-1}e_{k-2} + e_{k-2}e_k) + e_k e_{k-1} e_{k-2} \right\} .$$

If $\{e_k\}$ converges with order $p > 1$, show that the dominant terms from the Taylor expansions are

$$0 \approx f'(z)e_{k+1} + \frac{1}{6}f'''(z)e_k e_{k-1} e_{k-2} .$$

Conclude that

$$p^3 = p^2 + p + 1 .$$

Show that the only real solution of this equation is

$$p = \frac{1}{3}\left[19 + 3\sqrt{33}\right]^{1/3} + \frac{4}{3\left[19 + 3\sqrt{33}\right]^{1/3}} + \frac{1}{3} \approx 1.839 .$$

Exercise 5.6.13 Suppose that the sequence $\{z_k\}$ is defined by rational interpolation. In other words,

$$(z_{k+1} - z_{k-2})f(z_k)\frac{f(z_{k-1}) - f(z_{k-2})}{z_{k-1} - z_{k-2}} = (z_{k+1} - z_k)f(z_{k-2})\frac{f(z_k) - f(z_{k-1})}{z_k - z_{k-1}} .$$

Define $e_k = z_k - z$ where $f(z) = 0$. Use Taylor expansions to show that

$$(e_{k+1} - e_{k-2})e_k \left[f'(z) + \frac{1}{2}f''(z)e_k + \frac{1}{6}f'''(z)e_k^2 \right]$$
$$\times \left[f'(z) + \frac{1}{2}f''(z)(e_{k-1} + e_{k-2}) + \frac{1}{6}f'''(z)(e_{k-1}^2 + e_{k-1}e_{k-2}e_{k-2}^2) \right]$$
$$\approx (e_{k+1} - e_k)e_{k-2} \left[f'(z) + \frac{1}{2}f''(z)e_{k-2} + \frac{1}{6}f'''(z)e_{k-2}^2 \right]$$
$$\times \left[f'(z) + \frac{1}{2}f''(z)(e_k + e_{k-1}) + \frac{1}{6}f'''(z)(e_k^2 + e_k e_{k-1} e_{k-1}^2) \right]$$

Collect terms to show that

$$0 \approx (e_k - e_{k-2}) \left\{ f'(z)^2 e_{k+1} + \frac{1}{2}f'(z)f''(z)e_{k+1}(e_k + e_{k-1} + e_{k-2}) \right.$$
$$+ \frac{1}{6}f'(z)f'''(z) \left[e_{k+1}(e_k - e_{k-2})^2 + e_{k+1}e_{k-1}^2 + e_k e_{k-1} e_{k-2} \right]$$
$$\left. + \frac{1}{4}f''(z)^2 \left[e_{k+1}e_k(e_{k-1} + e_{k-2}) - e_k e_{k-1} e_{k-2} \right] \right\}$$

If $\{e_k\}$ converges with order $p > 1$, conclude that the dominant terms from the Taylor expansions show that

$$e_{k+1} \approx \left[\frac{1}{4}\frac{f''(z)^2}{f'(z)^2} - \frac{1}{6}\frac{f'''(z)}{f'(z)} \right] e_k e_{k-1} e_{k-2} \,.$$

Conclude that

$$p^3 = p^2 + p + 1 \,.$$

Show that the only real solution of this equation is

$$p = \frac{1}{3}\left[19 + 3\sqrt{33} \right]^{1/3} + \frac{4}{3\left[19 + 3\sqrt{33} \right]^{1/3}} + \frac{1}{3} \approx 1.839 \,.$$

Exercise 5.6.14 Suppose that the sequence $\{z_k\}$ is defined by inverse quadratic interpolation. In other words,

$$z_{k+1} = z_k - f(z_k)\frac{z_k - z_{k-1}}{f(z_k) - f(z_{k-1})} + f(z_k)f(z_{k-1})\frac{\frac{z_k - z_{k-1}}{f(z_k) - f(z_{k-1})} - \frac{z_{k-1} - z_{k-2}}{f(z_{k-1}) - f(z_{k-2})}}{f(z_k) - f(z_{k-2})} \,.$$

Define $e_k = z_k - z$ where $f(z) = 0$. Use Taylor expansions to show that

$$\frac{z_k - z_{k-1}}{f(z_k - f(z_{k-1})} \approx \frac{1}{f'(z) + \frac{1}{2}f''(z)(e_k + e_{k-1}) + \frac{1}{6}f'''(z)(e_k^2 + e_k e_{k-1} + e_{k-1}^2)} ,$$

that

$$\begin{aligned} e_k - f(z_k)&\frac{z_k - z_{k-1}}{f(z_k) - f(z_{k-1})} \\ &\approx e_k e_{k-1} \frac{\frac{1}{2}f''(z) + \frac{1}{6}f'''(z)(e_k + e_{k-1})}{f'(z) + \frac{1}{2}f''(z)(e_k + e_{k-1}) + \frac{1}{6}f'''(z)(e_k^2 + e_k e_{k-1} + e_{k-1}^2)} \end{aligned}$$

and that

$$\begin{aligned} &\frac{\frac{z_k - z_{k-1}}{f(z_k) - f(z_{k-1})} - \frac{z_{k-1} - z_{k-2}}{f(z_{k-1}) - f(z_{k-2})}}{f(z_k) - f(z_{k-2})} \\ &\approx -\left[\frac{1}{2}f''z + \frac{1}{6}f'''z(e_k + e_{k-1} + e_{k-2})\right]\left[f'(z) + \frac{1}{2}f''(z)(e_k + e_{k-2}) \right. \\ &\left. + \frac{1}{6}f'''(z)(e_k^2 + e_k e_{k-2} + e_{k-2}^2)\right]^{-1} \\ &\times \left[f'(z) + \frac{1}{2}f''(z)(e_{k-1} + e_{k-2}) + \frac{1}{6}f'''(z)(e_{k-1}^2 + e_{k-1}e_{k-2} + e_{k-2}^2)\right]^{-1} \\ &\times \left[f'(z) + \frac{1}{2}f''(z)(e_k + e_{k-1}) + \frac{1}{6}f'''(z)(e_k^2 + e_k e_{k-1} + e_{k-1}^2)\right]^{-1} . \end{aligned}$$

If $\{e_k\}$ converges with order $p > 1$, show that

$$\begin{aligned} &\frac{e_{k+1}}{e_k e_{k-1}} \left[f'(z) + \frac{1}{2}f''(z)(e_k + e_{k-2}) + \frac{1}{6}f'''(z)(e_k^2 + e_k e_{k-2} + e_{k-2}^2)\right] \\ &\times \left[f'(z) + \frac{1}{2}f''(z)(e_{k-1} + e_{k-2}) + \frac{1}{6}f'''(z)(e_{k-1}^2 + e_{k-1}e_{k-2} + e_{k-2}^2)\right] \\ &\times \left[f'(z) + \frac{1}{2}f''(z)(e_k + e_{k-1}) + \frac{1}{6}f'''(z)(e_k^2 + e_k e_{k-1} + e_{k-1}^2)\right] \\ &\approx \left[\frac{1}{2}f'f''(z)^2 - \frac{1}{6}f'(z)^2 f'''(z)\right] e_{k-2} . \end{aligned}$$

Conclude that

$$e_{k+1} \approx \left[\frac{1}{2}f'f''(z)^2 - \frac{1}{6}f'(z)^2 f'''(z)\right] e_k e_{k-1} e_{k-2} .$$

and that

$$p^3 = p^2 + p + 1 \; .$$

Show that the only real solution of this equation is

$$p = \frac{1}{3}\left[19 + 3\sqrt{(33)}\right]^{1/3} + \frac{4}{3\left[19 + 3\sqrt{(33)}\right]^{1/3}} + \frac{1}{3} \approx 1.839 \; .$$

Exercise 5.6.15 Program a globally convergent algorithm to find a zero of a function. You may use bisection to maintain bracketing of the zero, but you must combine it with one or more other methods to try to achieve superlinear convergence. Use it to find zeros of the following functions, and plot $\ln |f(z_k)|$ versus k for each:

1. $f(x) = \sin(x) - \cos(2x), -3.5 < x < 2.5$.
2. $f(x) = e^{-x^2} - 10^{-4}, 0 < x < 5$.
3. $f(x) = x/(1 - 100x) - 1, 0 < x < .009999$.
4. $f(x) = 2x^3 - 9x^2 + 12x - 3.9, 0 < x < 10$.

Exercise 5.6.16 Obtain copies of the Harwell Subroutine Library routines nb01 and nb02 Examine the codes and explain their choices of numerical methods. Test them on the problems in the first exercise above.

Exercise 5.6.17 Obtain a copy of the **NAPACK** routine `root` from netlib. Examine the code and explain its choices of numerical methods. Test it on the problems in the first exercise above.

Exercise 5.6.18 Obtain a copy of the **SLATEC** routine `fzero` from netlib. Examine the code and explain its choices of numerical methods. Test it on the problems in the first exercise above.

Exercise 5.6.19 Obtain a copy of the TOMS routine number 631 named `zero1` from netlib. Examine the code and explain its choices of numerical methods. Test it on the problems in the first exercise above.

5.7 Minimization

Next, we turn to the computation of extrema of functions. We will begin by discussing conditions that guarantee the existence of extrema. For example,, if the given function is continuously differentiable, then it is well-known that local extrema are zeros of the first derivative. Thus, if the derivative is easy to evaluate, we might apply methods to compute a zero of the first derivative. If we are interested only in minima, we will need to distinguish local minima from other critical points in such computations. Furthermore, if the function is not continuously differentiable, then it will be necessary to develop new computational techniques to find its minimizer.

5.7.1 Necessary and Sufficient Conditions

The following result is commonly developed in introductory calculus.

Lemma 5.7.1 (Necessary Condition for Minimum) *Suppose that f if continuously differentiable in an open interval D. If $z \in D$ is a local* ***minimizer*** *of f, then $f'(z) = 0$. If, in addition, f is twice continuously differentiable in D, then $f''(z) \geq 0$.*

Proof Suppose that $f'(z) \neq 0$ and let $\sigma = \text{sign}(f(z))$. Choose $t > 0$ so that $\sigma f'(z - \lambda\sigma) > 0$ for all $\lambda \in (0, t)$, and $z - t\sigma \in D$. Then

$$f(z-\lambda\sigma) - f(z) = \int_z^{z-\lambda\sigma} f'(x)\mathrm{d}x = -\int_0^\lambda f'(z-\alpha\sigma)\sigma\,\mathrm{d}\alpha = -\int_0^\lambda \left|f'(z-\alpha\sigma)\right| \mathrm{d}\alpha < 0\,.$$

Thus there are points arbitrarily close to z where f takes on values smaller than $f(z)$. This contradicts the assumption that $f'(z) \neq 0$.

Next, suppose that $f''(z) < 0$. Then there is a point $\tilde{z}$ near z so that for all ζ between z and $\tilde{z}$ we have $f''(\zeta) < 0$. We can use the fundamental theorem of calculus to show that

$$\begin{aligned}
f(\tilde{z}) &= f(z) + \int_z^{\tilde{z}} f'(\eta)\,\mathrm{d}\eta = f(z) + \int_z^{\tilde{z}} f'(\eta) - f'(z)\,\mathrm{d}\eta \\
&= f(z) + \int_z^{\tilde{z}} f''(\zeta)(\tilde{z}-\zeta)\,\mathrm{d}\zeta \leq f(z) + \max_{\zeta \text{ between } z \text{ and } \tilde{z}} f''(\zeta) \int_z^{\tilde{z}} \tilde{z}-\zeta\,\mathrm{d}\zeta \\
&= f(z) + \frac{1}{2}(\tilde{z}-z)^2 \max_{\zeta \text{ between } z \text{ and } \tilde{z}} f''(\zeta) < f(z)\,.
\end{aligned}$$

This contradicts the assumption that z is a local minimizer of f.

The next lemma provides conditions that will guarantee the existence of a local minimum.

Lemma 5.7.2 (Sufficient Condition for Minimum) *Suppose that f is twice continuously differentiable in an open interval D. Further suppose that at $z \in D$, $f'(z) = 0$ and $0 < f''(z)$. Then there is an open subset $D' \subset D$ containing z such that for all $x \in D'$ with $x \neq z$, $f(x) > f(z)$.*

Proof Since $f''(z) > 0$ and $z \in D$, there is an open subset $D' \subset D$ such that for all $x \in D'$, $f''(x) > 0$. By the mean value theorem, for all $x \in D'$ there exists ξ in the interval $int(z, x) \subset D'$ such that

$$f(x) - f(z) = f'(z)(x-z) + \frac{1}{2}f''(\xi)(x-z)^2\,.$$

Since $f'(z) = 0$, this equations implies that

$$f(x) - f(z) = \frac{1}{2} f''(\xi)(x-z)^2 > f(z) .$$

Note that both of these lemmas assume that f is continuously differentiable, and both discuss only *local* minima. In general, it is difficult to find *global* minima. We will discuss global minimization in Sect. 5.7.7, when we present some stochastic optimization methods.

5.7.2 Perturbation Analysis

In order to perform scientific computing steps to minimization, we would like to determine how perturbations affect the solution of a nonlinear minimization problem. We will begin with the following analogue of Theorem 5.2.6.

Theorem 5.7.1 *Suppose that f is twice continuously differentiable on $[a,b]$. Let $z \in (a,b)$ be a* ***strict local minimizer*** *of f, meaning that $f'(z) = 0$, f'' is continuous at z and $0 < f''(z)$. Given $\tau \in (0,1)$, suppose that there exists $\delta > 0$ so that for all $x \in (z-\delta, z+\delta) \cap (a,b)$ we have*

$$\left| f'(x) - f'(z) - f''(z)(x-z) \right| < \tau f''(z)|x-z| .$$

Then for all $\tilde{z} \in (z-\delta, z+\delta) \cap (a,b)$ we have

$$|\tilde{z} - z| < \frac{|f'(\tilde{z})|}{(1-\tau)f''(z)} . \tag{5.47}$$

Proof Since f'' is continuous at z, for any $\varepsilon > 0$ there exists $\delta > 0$ so that for all $|x-z| < \delta$ we have

$$\left| f''(x) - f''(z) \right| < \varepsilon .$$

Consequently,

$$\left| f'(x) - f'(z) - f''(z)(x-z) \right| = \left| \int_z^x f''(\zeta) - f''(z) \, d\zeta \right| < \varepsilon \left| \int_z^x d\zeta \right| = \varepsilon \, |x-z| .$$

Thus the continuity of f'' at z implies the existence of δ for any given τ.

Next,

$$\tau f''(z) \, |\tilde{z} - z| > \left| f'(\tilde{z}) - f'(z) - f''(z) \, (\tilde{z} - z) \right| \geq f''(z) \, |\tilde{z} - z| - \left| f'(\tilde{z}) \right| .$$

This inequality implies that

$$\left|f'(\tilde{z})\right| > (1-\tau)f''(z)\,|\tilde{z}-z| \;.$$

This inequality is equivalent to the claim.

This theorem implies that the **maximum attainable accuracy** in minimization is $\phi'\varepsilon/f''(z)$, where ε is machine precision and ϕ' is the scale of f' near z.

Our next theorem discusses how perturbations in the objective function f affect the determination of a local minimizer.

Theorem 5.7.2 *Suppose that f is continuously differentiable on $[a,b]$, f' is differentiable on (a,b). Let $z \in (a,b)$ be a* ***strict local minimizer*** *of f. Given $\sigma \in (0,1/2)$, suppose that there exists $\delta > 0$ so that for all $x \in (z-\delta, z+\delta) \cap (a,b)$ we have*

$$\left|f(x)-f(z)--f'(z)-\frac{1}{2}f''(z)(x-z)^2\right| < \sigma f''(z)\,|x-z|^2 \;.$$

Then for all $\tilde{z} \in (z-\delta, z+\delta) \cap (a,b)$ we have

$$|\tilde{z}-z| < \sqrt{\frac{|f(\tilde{z})-f(z)|}{(1/2-\sigma)f''(z)}} \;. \tag{5.48}$$

Proof Since f'' is continuous at z, for any $\varepsilon > 0$ there exists $\delta > 0$ so that for all $|x-z| < \delta$ we have

$$\left|f''(x)-f''(z)\right| < \varepsilon \;.$$

Consequently,

$$\left|f(x)-f(z)-f'(z)(x-z)-\frac{1}{2}f''(z)(x-z)^2\right|$$
$$=\left|\int_z^x f'(\zeta)-f'(z)-\frac{1}{2}f''(z)(x-z)\,\mathrm{d}\zeta\right| = \left|\int_z^x \int_z^\zeta f''(\eta)-f''(z)\,\mathrm{d}\eta\,\mathrm{d}\zeta\right|$$
$$< \varepsilon\left|\int_z^x \int_z^\eta \mathrm{d}\eta\,\mathrm{d}\zeta\right| = \frac{1}{2}\varepsilon\,|\tilde{z}-z|^2 \;.$$

Thus the continuity of f'' at z implies the existence of δ for any given σ.

Next,

$$\sigma f''(z)\,|\tilde{z}-z|^2 > \left|f(\tilde{z})-f(z)-f'(z)\,(\tilde{z}-z)-\frac{1}{2}f''(z)\,(\tilde{z}-z)^2\right|$$
$$\geq \frac{1}{2}f''(z)\,|\tilde{z}-z|^2 - |f(\tilde{z})-f(z)| \;.$$

This inequality implies that

$$|f(\tilde{z}) - f(z)| > (1/2 - \sigma) f''(z) |\tilde{z} - z|^2 .$$

This inequality is equivalent to the claim.
This theorem says that for minimization problems, termination tests based on changes in the objective f are far less sensitive than termination tests based on values of the derivative f'. For problems in which the derivative is unknown or prohibitively expensive to evaluate, we may have to accept that the error in the computed minimizer is on the order of the square root of machine precision.

5.7.3 Newton's Method

If we are willing to compute two derivatives of f, then we can apply Newton's method to find a local minimizer of f. Here we use the local model

$$\tilde{f}(x) = f(z_k) + f'(z_k)(x - z_k) + \frac{1}{2} f''(z_k)(x - z_k)^2 .$$

The extremum of the quadratic model $\tilde{f}(x)$ is

$$z_{k+1} = z_k - f'(z_k)/f''(z_k) .$$

Of course, if $f''(z_k) < 0$, this iteration has dubious value in finding a *minimum* of f. In other words, information about the second derivative of f can also be useful in testing the sufficient condition for a minimizer. Since this algorithm is equivalent to using Newton's method to find a zero of f', this method converges locally to an extremum of f, and the order of convergence is 2.

5.7.4 Line Searches

Suppose that we want to find a local minimizer z of a continuous function f. Let z_k be some approximation to z, and let s_k be proposed increment to improve z_k. Define the **line search** function

$$\phi(\lambda) = f(z_k + s_k \lambda) .$$

We assume that s_k is a **descent step**, meaning that

$$0 > \phi'(0) = f'(z_k) s_k .$$

For example, with Newton's method we have

$$s_k = -f'(z_k)/f''(z_k) .$$

With this choice of step, if $f''(z_k) > 0$ then we have

$$\phi'(0) = -f'(z_k)^2/f''(z_k) < 0 ,$$

and s_k is necessarily a descent step. On the other hand, the **steepest descent method** would choose

$$s_k = -f'(z_k) .$$

The steepest descent step is obviously a descent step. In general, if s_k is nonzero and not a descent step, then $-s_k$ is a descent step.

5.7.4.1 Goldstein-Armijo Principle

Given a descent step, the basic goals of a line search are to select a step length λ_k that obtains a sufficient decrease in the objective, while avoiding small steps. The following lemma shows that these goals are possible. This lemma is due to Wolfe [113, 114], and may also be found Ortega and Rheinboldt [85, p. 257].

Lemma 5.7.3 (Goldstein-Armijo Principle) *Suppose that x is a real number and s is a nonzero real number. Let f be a real-valued function such that $f(x+s\lambda)$ is a continuous function of λ for $\lambda \in [0,\infty)$ and such that for all $\lambda \geq 0$*

$$f(x+s\lambda) \geq \underline{\phi} .$$

Also assume thatf is differentiable at x and $f'(x)s < 0$. Let α and β be scalars satisfying $0 < \alpha < \beta < 1$. Then there exist scalars $0 < \underline{\lambda} < \overline{\lambda}$ such that for all $\lambda \in (\underline{\lambda}, \overline{\lambda})$,

$$-\alpha f'(x)s\lambda \leq f(x) - f(x+s\lambda) \leq -\beta f'(x)s\lambda . \tag{5.49}$$

Proof Note that since $\alpha \in (0,1)$ and $f'(x) \neq 0$, we can choose $\varepsilon < (1-\alpha)|f'(x)|$. Then Definition 5.2.3 of the derivative of f at x implies that there exists $\delta > 0$ so that for all $0 < \lambda|s| \leq \delta$ we have

$$f(x+s\lambda) - f(x) - f'(x)s\lambda \leq \varepsilon|s|\lambda < (1-\alpha)|f'(x)s|\lambda .$$

Since $f'(x)s < 0$, this implies that for all $0 < \lambda < \delta/|s|$ we have

$$f(x+s\lambda) < f(x) + \alpha\lambda f'(x)s .$$

Next, for $\lambda \geq 0$ define

$$\phi(\lambda) \equiv f(x + s\lambda) - f(x) - \alpha f'(x)s\lambda \ .$$

Then for $0 < \lambda < \delta/|s|$ we have $\phi(\lambda) < 0$, and for all

$$\lambda \geq \frac{f(x) - \underline{\phi}}{\alpha|f'(x)s|}$$

we have

$$\phi(\lambda) = f(x+s\lambda) - f(x) - \alpha f'(x)s\lambda \geq f(x+s\lambda) - f(x) + [f(x) - \underline{\phi}] = f(x+s\lambda) - \underline{\phi} > 0.$$

Since $\phi(\lambda)$ is continuous, there exists a smallest λ_1 such that $\phi(\lambda_1) = 0$; in other words,

$$f(x + s\lambda_1) = f(x) + \alpha\lambda_1 f'(x)s \ .$$

Thus any $\lambda \in (0, \lambda_1)$ will satisfy the first conclusion of the lemma with strict inequality. Since $\alpha \in (0, \beta)$, there exists $0 < \lambda_2 < \lambda_1$ so that

$$f(x + s\lambda_2) = f(x) + \beta\lambda f'(x)s \ .$$

Then we can choose $\underline{\lambda} = \lambda_1$ and $\overline{\lambda} = \lambda_2$.

In the Goldstein-Armijo principle (5.49), the parameter α serves to guarantee a decrease in the objective value, and β prevents the step size λ from becoming arbitrarily small. In practice, $\alpha = 10^{-4}$ is commonly used, so that the corresponding test on the decrease in the objective is easy to satisfy. With back-tracking (described in Sect. 5.7.4.2 below), the test involving β is replaced by a lower bound on the step length λ.

An alternative form of the Goldstein-Armijo principle, namely

$$\beta f'(x)s \leq f'(x + s\lambda)s \text{ and}$$
$$f(x + s\lambda) - f(x) \leq \alpha f'(x)s\lambda \ ,$$

is used in Dennis and Schnabel [31, p. 120]. Note that this alternative form requires that we compute the derivative of f during the line search.

The next theorem provides circumstances under which the Goldstein-Armijo principle guarantees convergence of the minimization algorithm. This theorem is due to Wolfe [113, 114], but may also be found in Ortega and Rheinboldt [85, p. 490]. A proof of a similar theorem for the alternative form (5.50) of the Goldstein-Armijo principle may be found in Dennis and Schnabel [31, p. 121].

Theorem 5.7.3 (Convergence with the Goldstein-Armijo Principle) *Assume that f is a real-value function that is continuous on the real line, and assume that there exists a real number $\underline{\phi}$ such that for all real x*

$$f(x) \geq \underline{\phi} \, .$$

Let f be differentiable on the real line, and let f' be Lipschitz continuous with constant γ, meaning that for all x and y

$$|f'(x) - f'(y)| \leq \gamma |x - y| \, .$$

Choose Goldstein-Armijo principle parameters $0 < \alpha < \beta < 1$. Given a real number z_0, define the sequence $\{z_k\}_{k=0}^{\infty}$, the sequence of nonzero numbers $\{s_k\}_{k=0}^{\infty}$ and the sequence of positive numbers $\{\lambda_k\}_{k=0}^{\infty}$ so that for each $k \geq 0$ s_k is a descent step, meaning that

$$f'(z_k)s_k < 0 \, ,$$

λ_k is a step length that satisfies the Goldstein-Armijo principle

$$-\alpha f'(z_k)s_k\lambda_k \leq f(z_k) - f(z_k + s_k\lambda_k) \leq -\beta f'(z_k)s_k\lambda_k \, ,$$

and

$$z_{k+1} = z_k + s_k\lambda_k \, .$$

Then

$$\lim_{k \to \infty} f'(z_k) = 0$$

Proof Note that for all $j > 0$, the lower bound on f and the Goldstein-Armijo principle (5.49) imply that

$$\underline{\phi} - f(z_0) \leq f(z_j) - f(z_0) = \sum_{k=0}^{j-1} \{f(z_{k+1}) - f(z_k)\}$$

$$\leq \alpha \sum_{k=0}^{j-1} f'(z_k)s_k\lambda_k = \alpha \sum_{k=0}^{j-1} f'(z_k)s_k\lambda_k < 0 \, .$$

Since the terms in the final sum are all negative, and this sum is bounded above and below, the final sum must converge. The convergence of this sum implies that $f'(z_k)s_k\lambda_k \to 0$ as $k \to \infty$.

Next, the fundamental theorem of calculus 5.2.4 and Lipschitz continuity of f' imply that

$$f(z_k + s_k\lambda_k) - f(z_k) - f'(z_k)s_k\lambda_k = \int_{z_k}^{z_k+s_k\lambda_k} f'(\xi) - f'(z_k)\,\mathrm{d}\xi$$

$$\leq \int_{z_k}^{z_k+s_k\lambda_k} \gamma|\xi - z_k|\,\mathrm{d}\xi\ \mathrm{sign}(s_k) = \frac{\gamma}{2}\,(s_k\lambda_k)^2\ .$$

Then the Goldstein-Armijo principle implies that

$$\beta f'(z_k)s_k\lambda_k \leq f(z_{k+1}) - f(z_k) \leq f'(z_k)s_k\lambda_k + \frac{\gamma}{2}\,(s_k\lambda_k)^2\ .$$

The last inequality can be rewritten in the form

$$(1-\beta)\left|f'(z_k)s_k\right|\lambda_k \leq \frac{\gamma}{2}\,(s_k\lambda_k)^2\ .$$

Then the Goldstein-Armijo principle and the previous inequality lead to

$$0 \leq \frac{f(z_k) - f(z_{k+1})}{|s_k|\lambda_k} \leq -\beta\frac{f'(z_k)s_k\lambda_k}{|s_k|\lambda_k} = \beta\left|f'(z_k)\right| \leq \frac{\beta}{1-\beta}\frac{\gamma}{2}\,|s_k|\,\lambda_k\ .$$

We can multiply this inequality by $|f'(z_k)|$ to get

$$\left|f'(z_k)\right|^2 \leq \frac{\beta}{1-\beta}\frac{\gamma}{2}\left|f'(z_k)s_k\lambda_k\right|\ .$$

Since the right-hand side of this inequality tends to zero as $k \to \infty$, the theorem is proved.

5.7.4.2 Back-Tracking

Minimization methods such as Newton's are based on local models for the objective f that become very accurate as the minimizer z is approached. With such methods, we expect that as we approach the minimizer the step s_k should be such that $f(z_k+s_k)$ should be close to the minimum value of the objective. Thus, we expect that a line search within such methods would eventually choose $\lambda_k = 1$ for all sufficiently large step numbers k. The following theorem will provide circumstances under which we can replace the step length test in the Goldstein-Armijo principle with a simpler lower bound.

Theorem 5.7.4 (Convergence with Back-Tracking) *Suppose that f is a real-value function defined for all real numbers, and that there exists $\underline{\phi}$ such that for all real x*

$$f(x) \geq \underline{\phi}\ .$$

Also assume that f is continuously differentiable on the real line, and that f' is Lipschitz continuous with constant γ, meaning that for all x and y

$$|f'(x) - f'(y)| \leq \gamma |x - y| .$$

Choose $\alpha \in (0,1)$ and $\underline{\lambda} \in (0, 1/2)$. Given a real number z_0, define the sequence of real numbers $\{z_k\}_{k=0}^{\infty}$, the sequence of nonzero real numbers $\{s_k\}_{k=0}^{\infty}$ and the sequence of positive numbers $\{\lambda_k\}_{k=0}^{\infty}$ so that for each $k \geq 0$ s_k is a descent step:

$$f'(z_k)s_k < 0 ,$$

the step length $\lambda_k > 0$ satisfies

$$f(z_k + s_k\lambda_k) - f(z_k) \leq \alpha f'(z_k)s_k\lambda_k \text{ and} \tag{5.51a}$$

$$\lambda_k \geq \underline{\lambda} , \tag{5.51b}$$

and

$$z_{k+1} = z_k + s_k\lambda_k .$$

Also assume that there is a constant M so that for all k the step s_k satisfies

$$|f'(z_k)| \leq M|s_k| , \tag{5.52}$$

Then

$$\lim_{k\to\infty} f'(z_k) = 0$$

Proof For all $j > 0$, the lower bound on f and the sufficient decrease assumption (5.51a) imply that

$$\underline{\phi} - f(z_0) \leq f(z_j) - f(z_0) = \sum_{k=0}^{j-1}[f(z_{k+1}) - f(z_k)] \leq \alpha \sum_{k=0}^{j-1} f'(z_k)s_k\lambda_k < 0 .$$

Since the terms in the final sum are all negative, and this sum is bounded above and below, the final sum must converge. The convergence of this sum implies that $f'(z_k)s_k\lambda_k \to 0$ as $k \to \infty$.

Since the step lengths are bounded below by $\underline{\lambda}$ and the steps are chosen so that (5.52) is satisfied, we must have that

$$\underline{\lambda}|f'(z_k)|^2 \leq |f'(z_k)|^2\lambda_k \leq M|f'(z_k)s_k\lambda_k| .$$

Since the right-hand side of this inequality tends to zero, we conclude that $f'(z_k) \to 0$.

Note that if s_k is chosen by Newton's method, then assumption (5.52) is easily satisfied by taking M to be some convenient upper bound on the absolute value of f''. If s_k is chosen by steepest descent, then we can take $M = 1$.

Suppose that we have some algorithm for selecting descent steps s_k, and we want to choose lengths λ_k for these steps so that the objective value $f(z_k + s_k\lambda_k)$ is less than the previous objective value $f(z_k)$, and so that the iterates z_k converge to a zero of f'. For many iterations, such as Newton's method, the descent step s_k is designed so that $\lambda_k = 1$ is the right choice as the minimum is approached. Thus, in **backtracking** we begin by choosing $\lambda_k = 1$ and testing for a sufficient decrease. If the Goldstein-Armijo sufficient decrease test

$$f(z_k) - f(z_k + s_k) \le -\alpha f'(z_k)s_k \;,$$

is satisfied, then we take $z_{k+1} = z_k + s_k$ and test convergence before choosing the next descent step.

On the other hand, if

$$f(z_k) - f(z_k + s_k) > -\alpha f'(z_k)s_k \;,$$

then we did not achieve a sufficient decrease with $\lambda_k = 1$. In this case, we can form the quadratic polynomial

$$\tilde{f}(x) = f(z_k) + f'(z_k)(x - z_k) + \left[f(z_k + s_k) - f(z_k) - f'(z_k)s_k\right] \frac{(x - z_k)^2}{s_k^2}$$

that interpolates f at $x = z_k$ and $x = z_k + s_k$, and interpolates f' at $x = z_k$. Note that

$$\frac{s_k^2}{2}\tilde{f}''(x) = f(z_k + s_k) - f(z_k) - f'(z_k)s_k > -f'(z_k)s_k(1 - \alpha) > 0 \;,$$

so $\tilde{f}$ has a local minimizer at

$$\tilde{z} = z_k - s_k \frac{1}{2} \frac{f'(z_k)s_k}{f(z_k + s_k) - f(z_k) - f'(z_k)s_k} \;.$$

Thus our next candidate for the step length is

$$\lambda_k = \frac{1}{2} \frac{-f'(z_k)s_k}{f(z_k + s_k) - f(z_k) - f'(z_k)s_k} \;.$$

In particular, note that this potential new step length satisfies

$$\lambda_k < \frac{1}{2} \frac{-f'(z_k)s_k}{-f'(z_k)s_k(1 - \alpha)} = \frac{1}{2(1 - \alpha)} \;.$$

In order to prevent the step length from becoming too small, our actual choice for the step length is

$$\lambda_k = \min\left\{0.1, \frac{1}{2}\frac{-f'(z_k)s_k}{f(z_k+s_k)-f(z_k)-f'(z_k)s_k}\right\} .$$

In the case that the step length via the minimizer of the quadratic interpolant also fails to generate a sufficient decrease in the objective, then there are several possible strategies. One approach is to continue to halve the step length until a sufficient decrease is achieved. Another approach, described in Dennis and Schnabel [31, p. 126ff], is to choose λ_k so that the next iterate would be located at the unique local minimizer of a cubic polynomial. That cubic is chosen to interpolate f at z_k and the previous two step length attempts from that point, and to interpolate f' at z_k. If $\alpha < 1/4$, such a cubic cannot have complex values for its local extrema. The former strategy leads to the following

Algorithm 5.7.1 (Newton Minimization with Line Search)

$\varepsilon =$ machine precision

$f_0 = f(z_0)$

for $k = 0, 1, \ldots$

 $f_k' = f'(z_k)$

 $f_k'' = f''(z_k)$

 if $f_k'' > 0$ and $|f_k'| < \delta_f f_k'' |z_k|$ break   /* derivative value convergence test */

 if $f_k'' = 0$ break   /* Newton step undefined */

 $s_k = -f_k'/f_k''$

 if $f_k' s_k > 0$ then $s_k = -s_k$ /* descent direction */

 if $\alpha|f_k' s_k| < \varepsilon|f_k|$ break   /* Goldstein-Armijo test impossible to satisfy */

 $z = z_k + s_k$

 $f_z = f(z)$

 $\triangle f = f_z - f_k$

 if $\triangle f \geq \alpha f_k' s_k$ then   /* Goldstein-Armijo not satisfied */

  $\lambda = \max\left\{0.1\ ,\ \frac{1}{2}\frac{-f_k{}' s_k}{\triangle f - f_k{}' s_k}\right\}$ /* minimizer of quadratic interpolant */

  $z = z_k + s_k\lambda$

  $f_z = f(z)$

  $\triangle f = f_z - f_k$

  while $\triangle f \geq \alpha f_k' s_k \lambda$ /* Goldstein-Armijo not satisfied */

   $\lambda = \lambda/2$ /* back-tracking */

   $z = z_k + s_k\lambda$

   $f_z = f(z)$

   $\triangle f = f_z - f_k$

 $z_{k+1} = z$

 $f_{k+1} = f_z$

Readers may experiment with univariate minimization via a JavaScript program for **Newton's method.** This program finds a minimizer of

$$f(x) = (1-x)^{10} - x^2$$

in the interval $(-8, 10)$. Readers may select the initial guess with the computer mouse. Typically, the Newton iteration proceeds reasonably well, unless it reaches a point where $f''(z_k) < 0$. After that, the Newton steps are reversed (so that they will be descent steps), and the Goldstein-Armijo principle is very difficult to satisfy. The reversed Newton steps are very large, and lead to an increase in objective. However, the quadratic interpolant used for the first back-tracking step selects a very small step length. Eventually, the line search moves beyond the region with $f''(z_k) < 0$, and the convergence then progresses rapidly.

5.7.5 *Hermite Cubic Minimization*

In some cases, it may be difficult or inconvenient to compute first and second derivatives of an objective function. If we are still willing and able to compute the first derivative of f, then we have several options for finding a minimizer of f.

Since Lemma 5.7.1 showed that $f'(z) = 0$ is a necessary condition for z to be a minimizer of f, we could view the minimization problem as a root-finding problem for the derivative. Then we could use root-finding methods, such as bisection or the secant method, applied to f'. Unfortunately, these methods would not distinguish between finding local minima or maxima of f, and they would ignore potentially valuable information contained in the values of f at the iterates.

Suppose that we want to find the minimizer of a function f, and we know two points a and b where the derivative f' has opposite signs. In order to avoid bracketing a local maximum of f, we will assume that $a < b$ and $f'(a) < 0 < f'(b)$.

It is reasonable to use the values of f and f' at points a and b to construct a **Hermite cubic interpolant** to f:

$$\begin{aligned}\tilde{f}_{a,b}(x) = &f(a)\left(\frac{b-x}{b-a}\right)^2\frac{2x+b-3a}{b-a} + f'(a)\left(\frac{b-x}{b-a}\right)^2(x-a)\\ &+ f(b)\left(\frac{x-a}{b-a}\right)^2\frac{3b-2x-a}{b-a} - f'(b)\left(\frac{x-a}{b-a}\right)^2(b-x)\,. \end{aligned} \tag{5.53}$$

The derivative of this function is

$$\tilde{f}'_{a,b}(x) = 6\frac{f(b)-f(a)}{b-a}\frac{b-x}{b-a}\frac{x-a}{b-a} + f'(a)\frac{b-x}{b-a}\frac{b+2a-3x}{b-a}$$
$$- f'(b)\frac{x-a}{b-a}\frac{2b+a-3x}{b-a} \; .$$

Note that both $\tilde{f}_{a,b}$ and $\tilde{f}'_{a,b}$ are symmetric in a and b: $\tilde{f}_{a,b} = \tilde{f}_{b,a}$ and $\tilde{f}'_{a,b} = \tilde{f}'_{b,a}$.

If we define

$$\sigma = 6\frac{f(b)-f(a)}{b-a} \; ,$$
$$\sigma_2 = 3\left[f'(b) + f'(a)\right] - \sigma \; ,$$
$$\sigma_1 = 2f'(b) + 4f'(a) - \sigma \; ,$$

then we have

$$\tilde{f}'_{a,b}(a + [b-a]\xi) = \sigma_2\xi^2 - \sigma_1\xi + f'(a) \; ,$$

and the zeros of $\tilde{f}'$ can be evaluated as $\tilde{z} = a + (b-a)\xi$ where

if $\sigma_1 < 0$
$\quad \xi_1 = \frac{2f'(a)}{\sigma_1 - \sqrt{\sigma_1^2 - 4\sigma_2 f'(a)}}$
else
$\quad \xi_1 = \frac{\sigma_1 + \sqrt{\sigma_1^2 - 4\sigma_2 f'(a)}}{2\sigma_2}$
$\xi_2 = f'(a)/(\sigma_2\xi_1)$
if $\xi_1 \in (0,1)$ then $\xi = \xi_1$ else $\xi = \xi_2$

Since we want $\tilde{z} \in (a,b)$, we chose $\xi \in (0,1)$. The formulas were selected to avoid cancellation in computing ξ. These ideas produce the following

Algorithm 5.7.2 (Hermite Minimization)

if $a = b$ return /* bracketing interval empty */
$f_a = f(a)$ and $f'_a = f'(a)$ /* require $f'_a < 0$ */
$f_b = f(b)$ and $f'_b = f'(b)$ /* require $f'_b > 0$ */
while not converged
  if $b - a < \delta_z \max\{|a|\ ,\ |b|\}$ return /* solution convergence test */
  $\sigma = 6\frac{f_b - f_a}{b-a}$
  $\sigma_2 = 3(f'_b - f'_a) - \sigma$
  $\sigma_1 = 2f'_b + 4f'_a) - \sigma$
  $d = \sqrt{\sigma_1^2 - 4\sigma_2 f'_a}$ /* must be nonnegative */
  $\xi_1 = \left(\sigma_1 < 0 ? \frac{2f'_a}{\sigma_1 - d} : \frac{1}{2}\frac{\sigma_1 + d}{\sigma_2}\right)$
  $\xi_2 = f'_a/(\sigma_2 * \xi_1)$
  $z = a + (b - a) * (\xi_1 > 0 \text{ and } \xi_1 < 1 ? \xi_1 : \xi_2)$ /* minimizer of Hermite cubic interpolant */
  if $z \le a$ or $z \ge b$ return
  $f_z = f(z)$ and $f'_z = f'(z)$ /* also compute $\phi' =$ scale of f'_z */
  if $|f'_z| < \delta_f \phi'$ return /* derivative convergence test */
  if $f'_z < 0$
    $a = z$ and $f_a = f_z$ and $f'_a = f'_z$
  else
    $b = z$ and $f_b = f_z$ and $f'_b = f'_z$

If z is the true minimizer, then we define the error in the iterate z_k to be

$$e_k = z_k - z\ .$$

A Taylor expansion produces

$$\begin{aligned}
0 &= \tilde{f}'_{z_{k-1},z_k}(z_{k+1}) \\
&= 6\frac{f(z_k) - f(z_{k-1})}{z_k - z_{k-1}} \frac{z_k - z_{k+1}}{z_k - z_{k-1}} \frac{z_{k+1} - z_{k-1}}{z_k - z_{k-1}} \\
&+ f'(z_{k-1}) \frac{z_k - z_{k+1}}{z_k - z_{k-1}} \frac{z_k + 2z_{k-1} - 3z_{k+1}}{z_k - z_{k-1}} - f'(z_k) \frac{z_{k+1} - z_{k-1}}{z_k - z_{k-1}} \frac{2z_k + z_{k-1} - 3z_{k+1}}{z_k - z_{k-1}} \\
&\approx f'''(z) e_{k+1} + f''''(z) e_{k+1}^2 \\
&+ f'''''(z) \left\{ \frac{1}{4} e_{k+1}^2 (e_k + e_{k-1}) - \frac{1}{12} e_{k+1} \left(e_k^2 + 4 e_k e_{k-1} + e_{k-1}^2\right) \right. \\
&\left. + \frac{1}{12} e_k e_{k-1} \left(e_k + e_{k-1}\right) \right\} .
\end{aligned}$$

If the errors converge to zero with order $p > 1$, then the dominant terms in the Taylor expansion are

$$0 \approx f'''(z)e_{k+1} + \frac{1}{12}f'''''(z)e_k e_{k-1}^2 \cdot$$

If $f'''(z)$ and $f'''''(z)$ have the same sign, then e_{k+1} will have sign opposite that of e_k, and successive iterates will lie on opposite sides of the true minimizer. In such a circumstance, we conclude that $p = 1 + 2/p$, which implies that $p = 2$.

Reader may experiment with the JavaScript program for univariate minimization using **Hermite interpolation.** This program finds the minimizer of

$$f(x) = (1-x)^{10} - x^2$$

in the interval $(-8, 10)$. Here $f'(-8) < 0 < f'(10)$, so the endpoints of this interval bracket a local minimizer of f. Hermite interpolation makes quick progress to find an small interval that brackets the minimizer. However, once the solution interval has width on the order of the square root of machine precision, rounding errors dominate the computation of the secant slope σ, and the iteration makes poor progress in reducing the solution interval.

5.7.6 *Derivative-Free Methods*

We will begin our development of algorithms for minimization by presenting two methods that make minimal assumptions on the function f. The basic idea in those assumptions is contained in the following definition.

Definition 5.7.1 A continuous function f is **unimodal** on the interval $[a, b]$ if and only if there is a unique $z \in (a, b)$ so that for all $x_1 < x_2 \in [a, b]$, if $x_2 < z$ then $f(x_1) > f(x_2)$, else if $x_1 > z$ then $f(x_1) < f(x_2)$.
This definition implies that z is a local minimizer of the unimodal function. Basically, Definition 5.7.1 says that f is strictly decreasing to the left of z, and strictly increasing to the right of z.

Suppose that we have an interval (a, b) in which we believe that a function f has a local minimizer. We will present two algorithms that will find a minimizer of f, provided that it is unimodal on (a, b). Since we may not know if f is unimodal before we search for the minimum, we will need to test this assumption during the algorithms. Our third algorithm will make assumptions about first and second derivatives of f, but will not evaluate those derivatives.

5.7.6.1 Fibonacci Search

Fibonacci search produces the greatest reduction in the interval containing the minimizer for some given number of function evaluations. This algorithm was originally developed by Kiefer [65]. The algorithm examines values of a given function f on a subset of the set $\{a + (b - a)i/\phi_N\}_{i=0}^{\phi_N}$. Here ϕ_N is the Nth **Fibonacci number**, with N chosen to be the smallest index so that

$$\phi_N \leq (b - a)/\tau$$

for some user-specified tolerance τ. It is well-known that the Fibonacci numbers can be generated by the three-term recurrence $\phi_N = \phi_{N-1} + \phi_{N-2}$. This leads to the following algorithm.

Algorithm 5.7.3 (Fibonacci Search)

```
m = ⌈ ((b − a)/τ)⌉                     /* ceiling: greatest number <= (b − a)/τ */
φ_0 = 0
φ_1 = 1
N = 1
while φ_N < m
   N = N + 1
   φ_N = φ_{N−1} + φ_{N−2}                                  /* Fibonacci numbers */
h = (b − a) * φ_{N−1}/φ_N
c = a + h , f_c = f(d)
d = b − h , f_d = f(d)                                  /* Fibonacci search points */
if max{f_c , f_c} ≥ max{f_a , f_b} return                /* function not unimodal */
i = 1
while i < N − 2
   if f_d > f_c
      a = d , f_a = f_d
      d = c , f_d = f_c
      h = (b − a) * φ_{N−i−1}/φ_{N−i}
      c = a + h , f_c = f(c)                                   /* narrow search */
      if f_c > max{f_a , f_b} return                      /* function not unimodal */
   else
      b = c , f_b = f_c
      c = d , f_c = f_d
      h = (b − a) * φ_{N−i−1}/φ_{N−i}
      d = b − h , f_d = f(d)
      if f_d > max{f_a , f_b} return                      /* function not unimodal */
   i = i + 1
z = c
```

The following lemma justifies the computations in this algorithm.

Lemma 5.7.4 *Given an integer N, let $\{\phi_i\}_{i=0}^N$ be the sequence of Fibonacci numbers. For any two real scalars $a^{(0)} < b^{(0)}$, define the sequences $\{a^{(i)}\}_{i=0}^{N-1}$, $\{b^{(i)}\}_{i=0}^{N-1}$, $\{c^{(i)}\}_{i=0}^{N-1}$ and $\{d^{(i)}\}_{i=0}^{N-1}$ by*

for $0 \leq i < N$

$$c^{(i)} = a^{(i)} + \left[b^{(i)} - a^{(i)}\right] \phi_{N-i-1}/\phi_{N-i}$$

$$d^{(i)} = b^{(i)} - \left[b^{(i)} - a^{(i)}\right] \phi_{N-i-1}/\phi_{N-i}$$

either

$$a^{(i+1)} = d^{(i)} \ , \ b^{(i+1)} = b^{(i)} \ , \ d^{(i+1)} = c^{(i)} \ ,$$

$$c^{(i+1)} = a^{(i+1)} + [b^{(i+1)} - a^{(i+1)}]\phi_{N-i-1}/\phi_{N-i}$$

or

$$a^{(i+1)} = a^{(i)} \ , \ b^{(i+1)} = c^{(i)} \ , \ c^{(i+1)} = d^{(i)} \ ,$$

$$d^{(i+1)} = b^{(i+1)} - [b^{(i+1)} - a^{(i+1)}]\phi_{N-i-1}/\phi_{N-i}$$

Then for $1 \leq i < N$ *we have*

$$b^{(i)} - a^{(i)} = \left[b^{(0)} - a^{(0)}\right] \phi_{N-i}/\phi_N \ .$$

As a result, for $0 \leq i < N$ *we also have*

$$b^{(i)} - d^{(i)} = \left[b^{(0)} - a^{(0)}\right] \phi_{N-i-1}/\phi_N = c^{(i)} - a^{(i)} \ .$$

Proof We will prove the claims by induction. For $i = 0$ we have

$$b^{(0)} - d^{(0)} = \left[b^{(0)} - a^{(0)}\right] \phi_{N-1}/\phi_N = c^{(0)} - a^{(0)}$$

and for $i = 1$ we have either

$$b^{(1)} - a^{(1)} = b^{(0)} - d^{(0)} = \left[b^{(0)} - a^{(0)}\right] \phi_{N-1}/\phi_N$$

or

$$b^{(1)} - a^{(1)} = c^{(0)} - a^{(0)} = \left[b^{(0)} - a^{(0)}\right] \phi_{N-1}/\phi_N \ .$$

Inductively, we will assume that

$$b^{(i)} - a^{(i)} = \left[b^{(0)} - a^{(0)}\right] \phi_{N-i}/\phi_N \ .$$

Then this inductive assumption implies that

$$b^{(i)} - d^{(i)} = \left[b^{(i)} - a^{(i)}\right] \phi_{N-i-1}/\phi_{N-i} = \left[b^{(0)} - a^{(0)}\right] \phi_{N-i-1}/\phi_N = c^{(i)} - a^{(i)} .$$

Since $b^{(i+1)} - a^{(i+1)}$ is equal to either $b^{(i)} - d^{(i)}$ or $c^{(i)} - a^{(i)}$, we have verified the inductive hypothesis for $i + 1$.

Readers may experiment with the JavaScript **Fibonacci search program.** Given a specified tolerance, this program searches for a minimizer of the function

$$f(x) = 1 - 2 * \mathrm{e}^{-|x+2*(x-5)|}$$

on the interval $(0, 6)$.

In summary, if ϕ_N is the first Fibonacci number that is at least $(b - a)/\tau$, then the Fibonacci search algorithm will compute an approximate minimizer for a given function f by making $N + 2$ function evaluations. This idea is sometimes used to search a set of discrete data, but is typically less popular than the next method for minimization of a function over a continuum.

5.7.6.2 Golden Section Search

Given a number $\tau \in (0, 1)$, **golden section search** maintains four points $a < d < c < b$ during the search so that

$$\frac{c-a}{b-a} = \frac{b-d}{b-a} = \frac{d-a}{c-a} = \tau .$$

One of the endpoints a or b will be replaced by an interior point during the iteration, in such a way that the new set of four points maintains these proportions. The equations $(c - a)/(b - a) = \tau$ and $(b - d)/(b - a) = \tau$ give us

$$c = a + \tau(b - a) \text{ and } d = b - \tau(b - a) ,$$

and then the equation $(d - a)/(c - a) = \tau$ implies that

$$\tau^2 = 1 - \tau .$$

It follows that

$$\tau = \frac{2}{1 + \sqrt{5}} \approx 0.618 .$$

With each step of the algorithm, the width $b - a$ of the interval will be decreased by a factor of τ.

Suppose that at some stage of the iteration, we have evaluated $f(a)$, $f(b)$, $f(c)$ and $f(d)$. If f is unimodal, then we must have

$$\max\{f(c)\ ,\ f(d)\} < \max\{f(a)\ ,\ f(b)\}\ .$$

We want to replace one of the endpoints a or b with an interior point in such a way that the new four points maintain the same proportions. This leads to the following algorithm

Algorithm 5.7.4 (Golden Section Search)

```
f_a = f(a)
f_b = f(b)                                /* assume a < b */
μ = max{f_a, f_b}
τ = 2/(1 + √5)
c = a + τ(b − a)
f_c = f(c)
if(f_c > μ) stop                          /* f not unimodal */
d = b − τ(b − a)
f_d = f(d)
if(f_d > μ) stop                          /* f not unimodal */
while not converged
  if f_c < f_d
    a = d and f_a = f_d
    d = c and f_d = f_c
    c = a + τ(b − a)
    f_c = f(c)
    if f_c >= max{f_a, f_b} stop /* f not unimodal or d = a */
  else
    b = c and f_b = f_c
    c = d and f_c = f_d
    d = b − τ(b − a)
    f_d = f(d)
    if f_d >= max{f_a, f_b} stop /* f not unimodal or c = b */
```

Readers may experiment with the JavaScript **golden search program.** This program searches for a minimizer of the function

$$f(x) = 1 - 2 * e^{-|x+2*(x-5)|}$$

on the interval $(0, 6)$. Readers may also view a Fortran golden search program. Golden search is also implemented in the GSL minimization algorithm named `gsl_min_fminimizer_goldensection`.

5.7.6.3 Successive Parabolic Interpolation

While golden search is globally convergent for unimodal functions, its convergence is slow because it does not take advantage of the magnitude of the function values. For a convext function f, an alternative is to define a quadratic local model that interpolates f at the points z_{k-2}, z_{k-1} and z_k. (For more information about this type of polynomial interpolation, please see Sect. 1.2.2 of Chap. 1 in Volume III below.) The resulting local model is

$$\tilde{f}(x) = f(z_k) + \frac{f(z_k) - f(z_{k-1})}{z_k - z_{k-1}}(x - z_k) + \left[\frac{f(z_k) - f(z_{k-1})}{z_k - z_{k-1}} - \frac{f(z_{k-1}) - f(z_{k-2})}{z_{k-1} - z_{k-2}}\right] \frac{(x - z_k)(x - z_{k-1})}{z_k - z_{k-2}} .$$

It is also easy to see that the minimizer of $\tilde{f}$ is

$$z_{k+1} = \frac{1}{2}(z_k + z_{k-1}) - \frac{1}{2} \frac{\frac{f(z_k)-f(z_{k-1})}{z_k-z_{k-1}}}{\frac{f(z_k)-f(z_{k-1})}{z_k-z_{k-1}} - \frac{f(z_{k-1})-f(z_{k-2})}{z_{k-1}-z_{k-2}}}(z_k - z_{k-2}) . \tag{5.54}$$

This leads to the following

Algorithm 5.7.5 (Successive Parabolic Minimization)

$f_0 = f(z_0)$
if $z_1 = z_0$ return
$f_1 = f(z_1)$
$\sigma_1 = (f_1 - f_0)/(z_1 - z_0)$
if $z_2 = z_1$ or $z_2 = z_0$ return
for $k = 2, \ldots$
  $\zeta = \max\{|z_k| , |z_{k-1}| , |z_{k-2}|\}$
  if $|z_k - z_{k-1}| \leq \zeta \sqrt[3]{\varepsilon}$ and $|z_k - z_{k-1}| \geq |z_{k-1} - z_{k-2}|$ return /* maximum attainable accuracy */
  if $z_k = z_{k-1}$ or $z_k = z_{k-2}$ return /* no change */
  $f_k = f(z_k)$
  $\sigma_k = (f_k - f_{k-1})/(z_k - z_{k-1})$
  $\tau = (\sigma_k - \sigma_{k-1})/(z_k - z_{k-2})$ /* approximation to $f''/2$ */
  if $\tau > 0$ and $|z_k - z_{k-1}| \leq \delta_z \zeta$ return /* relative error convergence test */
  if $\tau > 0$ and $|f_k - f_{k-1}| \leq \delta_z^2 \tau \zeta$ return /* function value convergence test */
  if $\tau = 0$ return /* minimizer of parabolic undefined */
  $z_{k+1} = (z_k + z_{k-1} - \sigma_k/\tau)/2$

This algorithm has been implemented in the C++ successive parabolic minimization program. This program searches for the minimizer of

$$f(x) = \pi + 20(x - 1)^2 + 8(x - 1)^3 + (x - 1)^4 ,$$

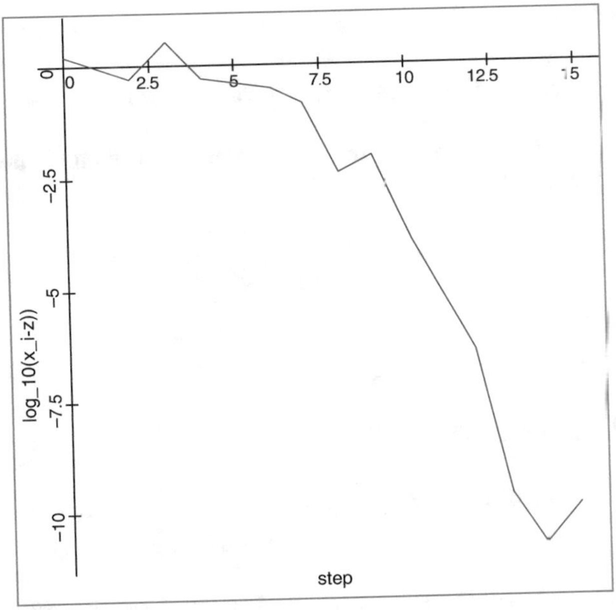

Fig. 5.7 Error in successive parabolic minimization: $log_{10}|z_k - z|$ versus iterations

which occurs at $z = 1$. Because changes in $f(x)$ are proportional to $(x - z)^2$ near the minimizer, Theorem 5.7.2 suggests that this algorithm should not be able to determine the argument at the minimizer with accuracy much greater than the square root of machine precision. This observation is essentially supported by the numerical results in Fig. 5.7.

Let us analyze the convergence behavior of successive parabolic minimization. Define

$$e_k = z_k - z ,$$

where z is the minimizer of f. Thus $f'(z) = 0$ and $f''(z) > 0$. We assume that $f'''(z) \neq 0$. It is easy to show that

$$\frac{f(z_k) - f(z_{k-1})}{z_k - z_{k-1}} \approx \frac{1}{2} f''(z)\left[e_k + e_{k-1}\right] + \frac{1}{6} f'''(z)\left[e_k^2 + e_k e_{k-1} + e_{k-1}^2\right] ,$$

and that

$$\frac{\frac{f(z_k)-f(z_{k-1})}{z_k - z_{k-1}} - \frac{f(z_{k-1})-f(z_{k-2})}{z_{k-1}-z_{k-2}}}{z_k - z_{k-2}} \approx \frac{1}{2} f''(z) + \frac{1}{6} f'''(z)\left[e_k + e_{k-1} + e_{k-2}\right] .$$

Equation (5.54) for the minimizer of the parabolic interpolant implies that

$$\left[z_{k+1} - \frac{z_k + z_{k-1}}{2}\right] \frac{\frac{f(z_k)-f(z_{k-1})}{z_k - z_{k-1}} - \frac{f(z_{k-1})-f(z_{k-2})}{z_{k-1}-z_{k-2}}}{z_k - z_{k-2}} = -\frac{1}{2} \frac{f(z_k) - f(z_{k-1})}{z_k - z_{k-1}} ,$$

which implies that

$$\frac{1}{2}f''(z)e_{k+1} \approx -\frac{1}{12}f'''(z)\left[2e_{k+1}(e_k + e_{k-1} + e_{k-2}) - e_k e_{k-1} - e_{k-1}e_{k-2} - e_{k-2}e_k\right].$$

If the sequence $e_{k}{}_{k=0}^{\infty}$ converges to zero with order $p > 1$, then the dominant terms in this equation imply that

$$\lim_{k\to\infty} \frac{e_{k+1}}{e_{k-1}e_{k-2}} = \frac{f'''(z)}{6f''(z)} .$$

This equation implies that

$$p^3 = p + 1 .$$

The only real solution of this equation is

$$p = \frac{1}{6}\left[108 + 12\sqrt{69}\right]^{1/3} + \frac{2}{\left[108 + 12\sqrt{69}\right]^{1/3}} \approx 1.324 .$$

Rounding errors can pose a significant problem for successive parabolic interpolation. Suppose that

$$|e_k| \ll |e_{k-1}| \ll |e_{k-2}|$$

and

$$\delta \ll e_{k-1}^2 ,$$

where δ is the magnitude of the rounding errors made in evaluating the function f. Then floating point computation of the first divided difference leads to

$$\begin{aligned}
&\frac{f(\tilde{z}_k)(1+\delta_0) - f(\tilde{z}_{k-1})(1+\delta_1)}{\tilde{z}_k - \tilde{z}_{k-1}}(1+\varepsilon_0) \\
&\approx \frac{\left[f(z) + \frac{1}{2}f''(z)e_k^2\right](1+\delta_0) - \left[f(z) + \frac{1}{2}f''(z)e_{k-1}^2\right](1+\delta_1)}{e_k - e_{k-1}}(1+\varepsilon_0) \\
&\approx f(z)\frac{\delta_0 - \delta_1}{e_k - e_{k-1}} + f''(z)\frac{e_k + e_{k-1}}{2} ,
\end{aligned}$$

and the floating point computation of the second divided difference gives us

$$\begin{aligned}
&\frac{\frac{f(\tilde{z}_k)(1+\delta_0) - f(\tilde{z}_{k-1})(1+\delta_1)}{\tilde{z}_k - \tilde{z}_{k-1}}(1+\varepsilon_0) - \frac{f(\tilde{z}_k)(1+\delta_0) - f(\tilde{z}_{k-1})(1+\delta_1)}{\tilde{z}_k - \tilde{z}_{k-1}}(1+\varepsilon_1)}{\tilde{z}_k - \tilde{z}_{k-2}}(1+\varepsilon_2) \\
&\approx \frac{\left\{f(z)\frac{\delta_0 - \delta_1}{e_k - e_{k-1}} + f''(z)\frac{e_k + e_{k-1}}{2}\right\} - \left\{f(z)\frac{\delta_1 - \delta_2}{e_{k-1} - e_{k-2}} + f''(z)\frac{e_{k-1} + e_{k-2}}{2}\right\}}{e_k - e_{k-2}}
\end{aligned}$$

$$\approx f(z)\frac{\frac{\delta_0-\delta_1}{e_k-e_{k-1}}-\frac{\delta_1-\delta_2}{e_{k-1}-e_{k-2}}}{e_k-e_{k-2}}+\frac{1}{2}f''(z)\ldots$$

In this case, the floating point computation of $\tilde{z}_{k+1}$ will yield

$$e_{k+1}=\tilde{z}_{k+1}-z=\frac{\tilde{z}_k+\tilde{z}_{k-1}}{2}(1+\varepsilon_3)-z$$
$$-\frac{1}{2}\frac{\frac{f(\tilde{z}_k)(1+\delta_0)-f(\tilde{z}_{k-1})(1+\delta_1)}{\tilde{z}_k-\tilde{z}_{k-1}}(1+\varepsilon_0)}{\frac{\frac{f(\tilde{z}_k)(1+\delta_0)-f(\tilde{z}_{k-1})(1+\delta_1)}{\tilde{z}_k-\tilde{z}_{k-1}}(1+\varepsilon_0)-\frac{f(\tilde{z}_k)(1+\delta_0)-f(\tilde{z}_{k-1})(1+\delta_1)}{\tilde{z}_k-\tilde{z}_{k-1}}(1+\varepsilon_1)}{\tilde{z}_k-\tilde{z}_{k-2}}(1+\varepsilon_2)}(1+\varepsilon_4)$$
$$\approx\frac{e_k+e_{k-1}}{2}-\frac{\frac{1}{2}f(z)\frac{\delta_0-\delta_1}{e_k-e_{k-1}}+\frac{1}{2}f''(z)\frac{e_k+e_{k-1}}{2}}{\frac{1}{2}f''(z)+f(z)\frac{\frac{\delta_0-\delta_1}{e_k-e_{k-1}}-\frac{\delta_1-\delta_2}{e_{k-1}-e_{k-2}}}{e_k-e_{k-2}}}$$
$$\approx\frac{e_k+e_{k-1}}{2}-\frac{e_k+e_{k-1}}{2}\left\{1+\frac{f(z)}{f''(z)}\frac{\delta_0-\delta_1}{e_k-e_{k-1}}-\frac{2f(z)}{f''(z)}\frac{\frac{\delta_0-\delta_1}{e_k-e_{k-1}}-\frac{\delta_1-\delta_2}{e_{k-1}-e_{k-2}}}{e_k-e_{k-2}}\right\}$$
$$=\frac{f(z)}{f''(z)}\left\{\frac{\delta_0-\delta_1}{e_k-e_{k-1}}-2\frac{\frac{\delta_0-\delta_1}{e_k-e_{k-1}}-\frac{\delta_1-\delta_2}{e_{k-1}-e_{k-2}}}{e_k-e_{k-2}}\right\}\approx-2\frac{f(z)}{f''(z)}\frac{\delta_0-\delta_1}{e_{k-1}e_{k-2}}\,.$$

While the iteration is converging, we would expect this computation to satisfy $|e_{k+1}|<|e_k|$. If $e_k\approx Ce_{k-1}^p$, then the rounding error analysis suggests that we must have

$$4\frac{|f(z)|}{f''(z)}\frac{\delta}{|e_{k-1}|\,|e_{k-2}|}<|e_k|\,,$$

and thus that

$$|e_k|^{1+1/p+1/p^2}>O(\delta)\,.$$

Since $p\approx 1.324$, we have $1+1/p+1/p^2\approx 2.326$. Thus the **maximum attainable accuracy** of piecewise parabolic iteration is on the order of $\delta^{1/2.326}$, where δ is the order of the relative error in evaluating $f(z)$.

5.7.6.4 Hybrid Minimization

There are at least two difficulties with successive parabolic interpolation. First, it does not necessarily maintain the bracketing of the minimizer. Second, it does not necessarily distinguish between searching for a minimum or a maximum. Brent [17, p. 61ff] has suggested a hybridization of successive parabolic interpolation and golden search that maintains bracketing of the minimizer and obtains global convergence to the minimizer. Given approximate minimizers $a < b$, this algorithm takes the following form.

Algorithm 5.7.6 (Brent's Hybrid Minimization)

$\varepsilon =$ machine precision

$r = 2/(3+\sqrt{5})$ /* $= \left[2/(1+\sqrt{5})\right]^2$ */

$c = a + r(b-a)$ /* golden section */

$f_c = f(c)$

$x = d = c$ and $f_x = f_d = f_c$

while not converged

  $m = (a+b)/2$ /* midpoint */

  $\delta = |x|\sqrt{\varepsilon} + \delta_z/3$ /* $\delta_z =$ user tolerance */

  if $|x-m| + (b-a)/2 \le 2\delta$ return x /* absolute error convergence */

  $\triangle x = r(x < m ? b - x : a - x)$ /* default: golden section */

  if $|x-d| \ge \delta$ /* try parabolic interpolation */

    $t_d = (x-d)(f_x - f_c)$

    $t_c = (x-c)(f_x - f_d)$

    $p = (x-c)t_c - (x-d)t_d$

    $q = 2(t_c - t_d)$

    if $q > 0$ then $p = -p$ else $q = -q$ /* adjust p/q so that $q \ge 0$ */

    if $|p| < |\triangle x * q|$ and $p > q(a - x + 2\delta)$ and $p < q(b - x + 2\delta)$

      $\triangle x = p/q$ /* min of parabolic */

    if $|\triangle x| < \delta$

      $\triangle x = (\triangle x > 0 ? \delta : -\delta)$ /* avoid small step */

  $z = x + \triangle x$

  $f_z = f(z)$

  if $f_z \le f_x$ /* z is better guess than x */

    if $z < x$ then $b = x$ else $a = x$

    $c = d$ and $f_c = f_d$

    $d = x$ and $f_d = f_x$

    $x = z$ and $f_x = f_z$

  else

    if $z < x$ then $a = z$ else $b = z$ /* shrink interval around x */

    if $f_z \le f_d$ or $d = x$

      $c = d$ and $f_c = f_d$

      $d = z$ and $f_d = f_z$

    else if $f_z \le f_c$ or $c = x$ or $c = d$

      $c = z$ and $f_c = f_z$

Brent's program is available either as Fortran routine fmin.f or C^{++} routine fminbr. In the GSL, Brent's hybrid minimization routine is available as gsl_min_fminimizer_brent. The MATLAB routine fminbnd also uses Brent's hybrid algorithm.

A similar algorithm, due to Gill and Murray [43], [44, p. 92] is available from the Object-oriented Scientific Computing Library (O_2scl) as class gsl_min_quad_golden. This algorithm is also available in the GSL as gsl_min_fminimizer_quad_golden.

5.7.7 *Stochastic Optimization*

The minimization methods above have all been designed to find a local minimizer, given some specified starting point or points. In order to find a global minimizer in a problem with many local minima, the specified domain needs to be searched fairly carefully. For more information about global optimization, we recommend the book by Gill et al. [44, p. 60], and the two books by Dixon and Szegö [32, 33]. We will discuss two methods for conducting a search for a global minimum.

5.7.7.1 Random Search

Given some interval $[a, b]$, suppose that we randomly select independent points $x \in [a, b]$ from a uniform probability distribution, and evaluate $f(x)$ for each point. Then we can remember the point x_* with smallest objective value, and approximate the global minimum of f by $f(x_*)$.

Here is a brief discussion of the errors in random search. Suppose that $\{x_k\}_{k=1}^N$ is a sequence of randomly chosen points in an interval $[a, b]$, and with probability $1 - \varepsilon$ we want to determine the minimizer z of some function f with error $|z - \min_{1 \le k \le N} x_k| \le \delta$. To do so, we must choose the number N of points to satisfy

$$N \ge -\frac{2\delta}{b-a} \log \varepsilon \; .$$

For a more careful discussion of the errors in random search, see Sect. 3.10.1 of Chap. 3 in Volume II.

Readers may experiment with the JavaScript **random search program.** This program searches for the minimizer of

$$f(x) = \frac{\sin(x) - \cos(2x)}{1 + (x/10)^2}$$

on the interval $(-10, 10)$. The MATLAB GlobalSearch class is similar.

5.7.7.2 Simulated Annealing

One problem with random search is that it continues to search everywhere with equal probability, even after finding that improvements in the objective are only found near the current best guess. It is reasonable to try to concentrate the search near the current best guess, provided that we continue to take some time to look around widely for other local minima.

In the **simulated annealing** algorithm, described by Kirkpatrick et al. [67], we are given an interval (a, b) in which to search for a minimizer of a function f. We select some initial point $x_0 \in (a, b)$. As the step index k increases, we want the expected step length $|x_{k+1} - x_k|$ to approach zero. To connect the step selection with simulated annealing concepts, we can define the "temperature" to be some function T so that $T(k) \to 0$ as $k \to \infty$. For example, we could use $T(k) = 1/k$ or $T(k) = e^{-k}$. Our objective function f corresponds to the "energy," and its argument x corresponds to the "state." We can determine a new state by selecting a random number $r_k \in [0, 1]$ and defining

$$x_{k+1} = \max\left\{a, \min\left\{b, x_k + (b-a)(r_k - 1/2)\right\}\right\} .$$

Next, we need to define the "probability" $P(x_k, x_{k+1}, T(k)))$ of accepting a move from one state x_k to a new state x_{k+1}. This probability function $P(x_k, x_{k+1}, T)$ must be such that

$$\lim_{T\downarrow 0} P(\alpha, \beta, T) = \begin{cases} 0, f(\beta) \geq f(\alpha) \\ c, f(\beta) < f(\alpha) \end{cases} ,$$

where $c > 0$. In practice, it may be difficult to guarantee that the integral of P over its state space is one, as would be required for a true probability function. However, we can easily guarantee that P takes values between zero and one. If Δ_f is some estimate for the difference between the highest and lowest possible values for f, then one possible choice of the function P is the following:

$$P(x_k, x_{k+1}, T(k)) = \begin{cases} 1 & , f(x_{k+1}) < f(x_k) \\ \exp\left(-\frac{f(x_{k+1})-f(x_k)}{T(k)\Delta_f}\right) & , f(x_{k+1}) \geq f(x_k \end{cases} .$$

A proof that the simulated annealing algorithm converges can be found in Granville et al. [47].

Readers may experiment with the JavaScript **simulated annealing program,** which searches for the minimizer of

$$f(x) = \frac{\sin(x) - \cos(2x)}{1 + (x/10)^2}$$

on the interval $(-10, 10)$. Simulated annealing is also implemented in the GSL simulated annealing function `gsl_siman_solve`. MATLAB users should consider the simulated annealing function `simulannealbnd`.

Motivated readers may also be interested in pgapack, which provides a parallel implementation of the genetic algorithm for global optimization. MATLAB users should consider the Genetic Algorithm function `ga`.

Exercise 5.7.1 Consider the function $f(x) = sin(x)$ on the interval $[\pi, 2\pi]$. The minimizer is obviously $z = \pi$. Over what subinterval do we have $f(x) < -1 + \varepsilon$ where ε represents machine precision?

Exercise 5.7.2 Recall from Sect. 5.7.5 that Hermite cubic minimization converges with order 2, and requires the evaluation of both the objective function f and its derivative. On the other hand, Sect. 5.7.6.3 showed that successive parabolic minimization converges with order 1.324, but requires only the evaluation of f with each iteration. Use the ideas in Sect. 5.5.4 to determine how the relative cost of the derivative affects the determination of the relative efficiency of these two minimization methods.

Exercise 5.7.3 Use Brent's hybrid minimization algorithm to find the minimizer of $f(x) = (1-x)^{10} - x^2$ on $[-8, 10]$. Plot $\log_{10} |f'(z_k)|$ versus k during the iteration, to measure the error in the minimizer.

Chapter 6
Least Squares Problems

"It's always good to take an orthogonal view of something. It develops ideas." Unix and Beyond, an Interview with Ken Thompson

Abstract This chapter examines the linear optimization problem of finding the best solution to over-determined, underdetermined or rank-deficient systems of linear equations. The existence theory for these least squares problems is necessarily more complicated than the theory for nonsingular linear systems. Pseudo-inverses, perturbation analysis and *a posteriori* error estimates are developed. Successive reflection factorization is developed and analyzed for over-determined, under-determined and rank-deficient problems. The effect of rounding errors and the use iterative improvement are also discussed. Alternative factorizations by successive orthogonal projection and successive orthogonal rotation are presented next. The singular value decomposition is introduced, but its full development depends on a detailed knowledge of eigenvalues, which are discussed in a later chapter. The chapter ends with a discussion of quadratic programming problems.

6.1 Overview

In data fitting problems, it is common to provide more observations than there are parameters in the model, with the hope that the over-sampling will reduce random errors in the data measurement. This process leads to systems of linear equations with more equations than unknowns. Typically, such linear systems have no mathematical solution, but there should be a unique vector of unknowns that minimizes some useful measure of the error in the linear system.

Additional Material: The details of the computer programs referred in the text are available in the Springer website (http://extras.springer.com/2018/978-3-319-69105-3) for authorized users.

© Springer International Publishing AG, part of Springer Nature 2017
J.A. Trangenstein, *Scientific Computing*, Texts in Computational Science and Engineering 18, https://doi.org/10.1007/978-3-319-69105-3_6

The goals in this chapter are to develop effective numerical methods for solving least-squares problems. Least squares problems mainly come in two forms. The **overdetermined least-squares** problem takes the form

Given an $m \times n$ matrix $\mathbf{A}$ with rank $(\mathbf{A}) = n$ and an m-vector b

Minimize $\|\mathbf{b} - \mathbf{A}\mathbf{x}\|_2^2$ over all n-vectors $\mathbf{x}$.

On the other hand, the **underdetermined least squares** problem has the form

Given an $m \times n$ matrix $\mathbf{A}$ with rank $(\mathbf{A}) = m$ and an m-vector b

Minimize $\|\mathbf{x}\|_2$ over all n-vectors $\mathbf{x}$ such that $\mathbf{A}\mathbf{x} = \mathbf{b}$.

Both of these problems are simplifications of the **general least squares** problem

Given a nonzero $m \times n$ matrix $\mathbf{A}$ and an m-vector b

Find an n-vector $\mathbf{x}$ in order to

Minimize $\|\mathbf{x}\|_2$ over all n-vectors $\mathbf{x}$ that minimize $\|\mathbf{A}\mathbf{x} - \mathbf{b}\|_2^2$.

The underdetermined least squares problem is a simple form of the more general **quadratic programming problem**,

Given a Hermitian $n \times n$ matrix $\mathbf{A}$, an $m \times n$ matrix $\mathbf{B}$ with rank $(\mathbf{B}) = m$,

an n-vector $\mathbf{a}$ and an m-vector $\mathbf{b}$

Minimize $\mathbf{x} \cdot \mathbf{A}\mathbf{x} - \mathbf{a} \cdot \mathbf{x}$ over all n-vectors $\mathbf{x}$ that satisfy $\mathbf{B}\mathbf{x} = \mathbf{b}$.

In order to prove the existence and uniqueness of solutions to these least squares problems, we will use the fundamental theorem of linear algebra 3.2.3. The key concept from that theorem needed for this application is the orthogonality of the fundamental subspaces associated with the range and nullspace of a matrix and its Hermitian.

Orthogonality will be an important concept in the development of numerical methods for solving the least squares problems. Some of our numerical methods for these problems will depend on matrix factorizations, as was our approach in solving systems of linear equations in Chap. 3. In solving systems of linear equations, we used Gaussian factorization, which employed **successive oblique projection**. One classical technique for solving least square problems uses **successive orthogonal projection**, while other approaches use **successive reflection** or **successive plane rotation**. We will also supplement these matrix factorizations with iterative improvement methods.

The topics in this chapter will be used for solving more complicated problems in later chapters. For example, in Chap. 2 of Volume II,we will use some ideas from least squares problems to develop gradient and minimum residual methods for the iterative solution of systems of linear equations. Also, in Chap. 3 of Volume II,we will use least squares problems to develop globally convergent algorithms for solving systems of nonlinear equations. Some ideas in this chapter will be useful in discussing quadratic programming problems in Chap. 4 of Volume II.Other kinds of least squares approximations by polynomials will be examined in Chap. 1 of Volume III.

For additional information about the material in this chapter, we recommend books by Björck [11], Demmel [29], Dongarra et al. [34], Golub and van Loan [46], Higham [56], Lawson and Hanson [73], Stewart [97] and [98], Trefethen and Bau [103], Wilkinson [111] and Wilkinson and Reinsch [112]. Readers who are interested in least squares problems motivated by statistical regression and the analysis of variance might enjoy the books by Draper and Smith [36], and Neter, Wasserman and Kutner [82]

For least squares software, we recommend LAPACK (written in Fortran), CLAPACK (translated from Fortran to C), numpy.linalg (written in Python), GNU Scientific Library (GSL) Linear Algebra and GSL Least-Squares Fitting (written in C). We also recommend MATLAB commands qr, mldivide, svd, rank and quadprog. We suggest that readers minimize their use of the MATLAB command pinv. Scilab provides the lsq command to solve least-squares problems, as well as the commands qr and rankqr to perform QR factorizations, and svd to compute the singular value decomposition.

6.2 Example Problems

Here is a common example of a least squares problem in data fitting, albeit in generic form.

Example 6.2.1 Suppose that we want to find the straight line that provides the best fit to some data. Specifically, we are given m pairs (a_i, b_i) of values for the independent and dependent variables, respectively. Our model is

$$b = \xi_0 + a\xi_1 \; .$$

Because the model is not necessarily exact, we expect that we have some error in each of the observations:

$$b_i = \xi_0 + a_i\xi_1 + \varepsilon_i \text{ for all } 1 \leq i \leq m \; .$$

If each of these errors are independent and identically distributed via the normal distribution, then well-known results from statistics (see Casella and Berger [21, p. 560ff]) show that we should choose the intercept ξ_0 and slope ξ_1 to minimize

$$\sum_{i=1}^{m} \varepsilon_i^2 = \sum_{i=1}^{m} (b_i - \xi_0 - a_i\xi_1)^2 \; .$$

We can use matrices and vectors to represent this problem. Let

$$\mathbf{A} = \begin{bmatrix} 1 & a_1 \\ \vdots & \vdots \\ 1 & a_m \end{bmatrix}, \quad \mathbf{b} = \begin{bmatrix} b_1 \\ \vdots \\ b_m \end{bmatrix}, \quad \mathbf{x} = \begin{bmatrix} \xi_0 \\ \xi_1 \end{bmatrix}.$$

Then we want to find a 2-vector $\mathbf{x}$ to minimize $\|\mathbf{A}\mathbf{x} - \mathbf{b}\|_2^2$, where the vector 2-norm was defined in Eq. (3.14c).

Here is a more concrete example of the previous general problem.

Example 6.2.2 Consider Table 6.1 of heights and weights for a group of individuals: In order to model weight as a linear function of height with errors that are independent and randomly distributed according to a normal distribution, we should find ξ_0 and ξ_1 to minimize the residual sum of squares

$$\varrho^2 = \sum_{i=1}^{8} (w_i - \xi_0 - h_i \xi_1)^2 .$$

Suppose that we collect the data with more accuracy, giving us Table 6.2. We might ask how the extra accuracy in the data changes the model. In particular, we might want to know whether the model can now represent the data more accurately, or determine the unknown coefficients ξ_0 and ξ_1 to greater precision.

Here is yet another example of a least squares problem.

Table 6.1 Lower accuracy table of individual heights and weights

Height (m)	Weight (kg)
1.65	48
1.65	57
1.57	50
1.70	54
1.75	64
1.65	61
1.55	43
1.70	59

Table 6.2 Higher accuracy table of individual heights and weights

Height (m)	Weight (kg)
1.651	47.6
1.653	56.7
1.575	49.9
1.702	54.4
1.753	63.5
1.649	61.2
1.549	43.1
1.701	59.0

Example 6.2.3 Consider Table 6.3 of molecular weights for various molecules combining nitrogen and oxygen:

These data suggest the model

Table 6.3 Nitrogen-oxygen compounds and molecular weights

NO	30.006
N_2O_3	76.012
N_2O	44.013
N_2O_3	108.010
NO_2	46.006
N_2O_4	92.011

$$\begin{aligned} \text{molecular weight} &= \text{number atoms of nitrogen} \times \text{atomic weight of nitrogen} \\ &+ \text{number atoms of oxygen} \times \text{atomic weight of oxygen} . \end{aligned}$$

We can write this as a least squares problem to minimize $\|\mathbf{Ax} - \mathbf{b}\|_2$ with

$$\mathbf{A} = \begin{bmatrix} 1 & 1 \\ 2 & 3 \\ 2 & 1 \\ 2 & 5 \\ 1 & 2 \\ 2 & 4 \end{bmatrix} \text{ and } \mathbf{b} = \begin{bmatrix} 30.006 \\ 76.012 \\ 44.013 \\ 108.010 \\ 46.006 \\ 92.011 \end{bmatrix} .$$

The unknown atomic weights are the entries of $\mathbf{x}$.

6.3 Existence and Uniqueness

The fundamental features of the solution of the least squares problem are described in the following lemma.

Lemma 6.3.1 *Suppose that* $\mathbf{A}$ *is an* $m \times n$ *matrix and* $\mathbf{b}$ *is an* m*-vector. Then there is a unique* n*-vector* $\mathbf{x}$ *such that both* $\|\mathbf{b} - \mathbf{Ax}\|_2$ *is minimized and* $\|\mathbf{x}\|_2$ *is as small as possible. Furthermore, there is a unique* m*-vector* $\mathbf{r}$ *(called the* **residual***) and an* m*-vector* $\mathbf{s}$ *(not necessarily unique) so that*

$$\mathbf{r} + \mathbf{Ax} = \mathbf{b} \; , \; \mathbf{A}^H\mathbf{r} = \mathbf{0} \text{ and } \mathbf{x} - \mathbf{A}^H\mathbf{s} = \mathbf{0} \; .$$

If $\text{rank}\,(\mathbf{A}) = n$ *then the least squares problem is said to be* **overdetermined***, the solution* $\mathbf{x}$ *and residual* $\mathbf{r}$ *vectors satisfy*

$$\begin{bmatrix} \mathbf{I} & \mathbf{A} \\ \mathbf{A}^H & \mathbf{0} \end{bmatrix} \begin{bmatrix} \mathbf{r} \\ \mathbf{x} \end{bmatrix} = \begin{bmatrix} \mathbf{b} \\ \mathbf{0} \end{bmatrix} , \tag{6.1}$$

the matrix in this linear system is nonsingular, and **s** *is not needed. If rank* $(\mathbf{A}) = m$ *then the least squares problem is said to be* **undetermined**, *the solution* **x** *and auxiliary* **s** *vectors satisfy*

$$\begin{bmatrix} \mathbf{I} & \mathbf{A}^H \\ \mathbf{A} & \mathbf{0} \end{bmatrix} \begin{bmatrix} \mathbf{x} \\ -\mathbf{s} \end{bmatrix} = \begin{bmatrix} \mathbf{0} \\ \mathbf{b} \end{bmatrix} , \tag{6.2}$$

the matrix in this linear system is nonsingular, and **r** *is not needed.*

Proof If $\mathbf{A} = \mathbf{0}$, then we take $\mathbf{x} = \mathbf{0}$. For all n-vectors $\boldsymbol{\xi}$, we have $\|\mathbf{A}\boldsymbol{\xi} - \mathbf{b}\|_2 = \|\mathbf{b}\|_2 = \|\mathbf{A}\mathbf{x} - \mathbf{b}\|_2$ and $\|\boldsymbol{\xi}\|_2 \geq 0 = \|\mathbf{x}\|_2$. Thus this choice of **x** solves the general linear least squares problem. In the remainder of the proof, we will assume that **A** is nonzero.

Since $\mathbf{A} \neq \mathbf{0}$, the fundamental theorem of linear algebra 3.2.3 implies that there is an n-vector **x** and an m-vector **r** so that

$$\mathbf{b} = \mathbf{A}\mathbf{x} + \mathbf{r} \text{ and } \mathbf{A}^H\mathbf{r} = \mathbf{0} \; .$$

Furthermore, this theorem implies that both $\mathbf{A}\mathbf{x}$ and **r** are unique. Since $\mathbf{A}\mathbf{x} = \mathbf{b} - \mathbf{r}$ has a solution, Lemma 3.2.3 implies that

$$\mathbf{x} + \mathscr{N}(\mathbf{A}) = \{\mathbf{x} + \mathbf{z} : \mathbf{A}\mathbf{z} = \mathbf{0}\}$$

is the set of all solutions to $\mathbf{A}(\mathbf{x} + \mathbf{z}) = \mathbf{b} - \mathbf{r}$. The fundamental theorem of linear algebra also shows that for our chosen vector **x** there a unique $\mathbf{z} \in \mathscr{N}(\mathbf{A})$ and an m-vector **s** so that

$$\mathbf{x} = \mathbf{A}^H\mathbf{s} - \mathbf{z} \; .$$

We will write

$$\mathbf{x}_* \equiv \mathbf{x} + \mathbf{z} = \mathbf{A}^H\mathbf{s} \; ,$$

and note that $\mathbf{x}_* = \mathbf{x} + \mathbf{z}$ is uniquely determined. These ideas are illustrated in Fig. 6.1.

Next, we want to show that $\mathbf{x}_*$ is the shortest solution to the least squares problem. First, let us show that $\mathbf{x}_*$ minimizes $\|\mathbf{A}\mathbf{x} - \mathbf{b}\|_2^2$. Since $\mathscr{R}(\mathbf{A}) \perp \mathscr{N}(\mathbf{A}^H)$, $\mathbf{A}\mathbf{x}_* - \mathbf{b} \in \mathscr{N}(\mathbf{A}^H)$, and for any n-vector $\boldsymbol{\xi}$ we have $\mathbf{A}(\boldsymbol{\xi} - \mathbf{x}_*) \in \mathscr{R}(\mathbf{A})$, the Pythagorean Theorem 3.2.2 implies that

$$\|\mathbf{A}\boldsymbol{\xi} - \mathbf{b}\|_2^2 = \|(\mathbf{A}\mathbf{x}_* - \mathbf{b}) + \mathbf{A}(\boldsymbol{\xi} - \mathbf{x}_*)\|_2^2 = \|\mathbf{A}\mathbf{x}_* - \mathbf{b}\|_2^2 + \|\mathbf{A}(\boldsymbol{\xi} - \mathbf{x}_*)\|_2^2 \geq \|\mathbf{A}\mathbf{x}_* - \mathbf{b}\|_2^2 \; .$$

Next, let us show that $\mathbf{x}_* = \mathbf{A}^H\mathbf{s}$ is shortest solution to the least squares problem. Recall that an arbitrary solution to the least squares problem has the form $\mathbf{x}_* + \mathbf{z}$

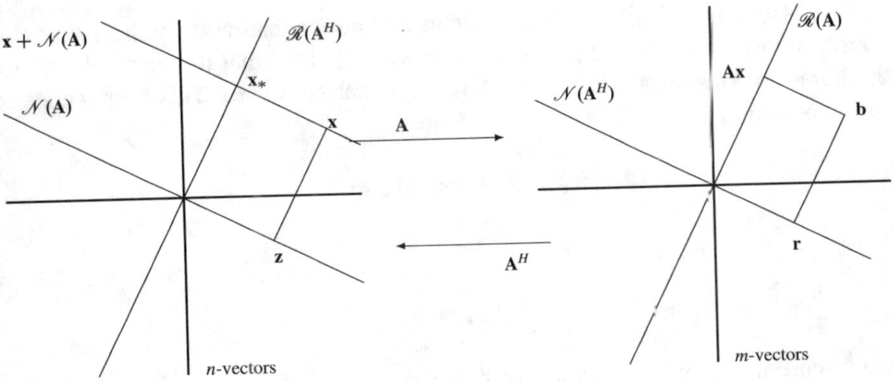

Fig. 6.1 Least squares problem and fundamental subspaces

where $\mathbf{z} \in \mathscr{N}(\mathbf{A})$. Also recall that $\mathbf{x}_* = \mathbf{A}^H\mathbf{s}$ for some m-vector $\mathbf{s}$. Since $\mathscr{R}\left(\mathbf{A}^H\right) \perp \mathscr{N}(\mathbf{A})$, the Pythagorean theorem implies that

$$\|\mathbf{x}_* + \mathbf{z}\|_2^2 = \|\mathbf{A}^H\mathbf{s} + \mathbf{z}\|_2^2 = \|\mathbf{A}^H\mathbf{s}\|_2^2 + \|\mathbf{z}\|_2^2 \geq \|\mathbf{A}^H\mathbf{s}\|_2^2 = \|\mathbf{x}_*\|_2^2 \,.$$

At this point, we have proved the first two claims in the lemma.

In the case when rank $(\mathbf{A}) = n$, we must have $\mathscr{N}(\mathbf{A}) = \{0\}$. In this case, there is only one vector in $\mathbf{x} + \mathscr{N}(\mathbf{A})$. Thus there is no need to find the shortest vector in this set, and thus no need to find the vector $\mathbf{s}$. Under these circumstances, we have

$$\begin{bmatrix} \mathbf{I} & \mathbf{A} \\ \mathbf{A}^H & \mathbf{0} \end{bmatrix} \begin{bmatrix} \mathbf{r} \\ \mathbf{x}_* \end{bmatrix} = \begin{bmatrix} \mathbf{b} \\ \mathbf{0} \end{bmatrix} .$$

This is a Hermitian indefinite system of linear equations. To show that the matrix in this linear system is nonsingular, Lemma 3.2.9 shows that it is sufficient to show that the nullspace of this matrix is zero. If $\mathbf{b} = \mathbf{0}$, then

$$\mathbf{r} = -\mathbf{A}\mathbf{x}_* \text{ and } \mathbf{0} = \mathbf{A}^H\mathbf{r} = -\mathbf{A}^H\mathbf{A}\mathbf{x}_* \,.$$

We can pre-multiply the second equation by ${\mathbf{x}_*}^H$ to get

$$0 = -{\mathbf{x}_*}^H\mathbf{A}^H\mathbf{A}\mathbf{x}_* = \|\mathbf{A}\mathbf{x}_*\|_2^2 \,.$$

This implies that $\mathbf{x}_* \in \mathscr{N}(\mathbf{A})$, so $\mathbf{x}_* = \mathbf{0}$ and consequently $\mathbf{r} = -\mathbf{A}\mathbf{x}_* = \mathbf{0}$. We conclude that the matrix in Eq. (6.1) is nonsingular.

If *rank*$(\mathbf{A}) = m$, then $\mathscr{R}(\mathbf{A})$ is the set of all m-vectors. In this case, there is no need to compute the residual $\mathbf{r}$, because it is zero. Under these circumstances we have

$$\begin{bmatrix} \mathbf{I} & \mathbf{A}^H \\ \mathbf{A} & \mathbf{0} \end{bmatrix} \begin{bmatrix} \mathbf{x}_* \\ -\mathbf{s} \end{bmatrix} = \begin{bmatrix} \mathbf{0} \\ \mathbf{b} \end{bmatrix} .$$

This is also a Hermitian indefinite system of linear equations. To show that the matrix in this linear system is nonsingular, we will show that the only solution to the homogeneous linear system involving this matrix is zero. This homogeneous linear system gives us

$$\mathbf{x}_* = \mathbf{A}^H\mathbf{s} \text{ and } \mathbf{A}\mathbf{x}_* = \mathbf{0} \; .$$

Thus

$$\mathbf{0} = \mathbf{A}\mathbf{x}_* = \mathbf{A}\mathbf{A}^H\mathbf{s} \; .$$

We can pre-multiply this equation by $\mathbf{s}^H$ to get

$$0 = \mathbf{s}^H\mathbf{A}\mathbf{A}^H\mathbf{s} = \|\mathbf{A}^H\mathbf{s}\|_2^2 \; .$$

Thus $\mathbf{s} \in \mathscr{N}\left(\mathbf{A}^H\right)$. However, since $\text{rank}\left(\mathbf{A}^H\right) = \text{rank}\left(\mathbf{A}\right) = m$, the fundamental theorem of linear algebra implies that $\mathscr{N}\left(\mathbf{A}^H\right) = \{\mathbf{0}\}$. Thus $\mathbf{s} = \mathbf{0}$, which implies that $\mathbf{x}_* = \mathbf{A}^H\mathbf{s} = \mathbf{0}$. We conclude that the matrix in Eq. (6.2) is nonsingular.

When $\mathbf{A}$ has full rank, the linear systems (6.1) and (6.2) that determine the solution of the least squares problem are Hermitian of size $m + n$, but not positive-definite. The following corollary shows that when $\mathbf{A}$ has full rank, we can describe the solution of the least squares problem by a an even smaller positive-definite linear system of size $\min\{m, n\}$.

Corollary 6.3.1 *Suppose that* $\mathbf{b}$ *is an* m*-vector, and* $\mathbf{A}$ *is an* $m \times n$ *matrix with* rank $(\mathbf{A}) = \min\{m, n\}$. *If* $m \geq n$*, then the* n*-vector* $\mathbf{x}$ *that minimizes* $\|\mathbf{A}\mathbf{x} - \mathbf{b}\|_2$ *satisfies the* ***normal equations***

$$\mathbf{A}^H\mathbf{A}\mathbf{x} = \mathbf{A}^H\mathbf{b} \; . \tag{6.3}$$

On the other hand, if $m \leq n$*, then the* n*-vector* $\mathbf{x}$ *that minimizes* $\|\mathbf{A}\mathbf{x} - \mathbf{b}\|_2$ *is* $\mathbf{x} = \mathbf{A}^H\mathbf{s}$ *where*

$$\mathbf{A}\mathbf{A}^H\mathbf{s} = \mathbf{b} \; . \tag{6.4}$$

Proof If $\text{rank}\,(\mathbf{A}) = n$, then Eq. (6.1) implies that $\mathbf{r} = \mathbf{b} - \mathbf{A}\mathbf{x}$ and

$$\mathbf{0} = \mathbf{A}^H\mathbf{r} = \mathbf{A}^H\mathbf{b} - \mathbf{A}^H\mathbf{A}\mathbf{x} \; .$$

This equation is equivalent to (6.3). If $\text{rank}\,(\mathbf{A}) = m$, then Eq. (6.2) implies that $\mathbf{x} = \mathbf{A}^H\mathbf{s}$ and

$$\mathbf{b} = \mathbf{A}\mathbf{x} = \mathbf{A}\mathbf{A}^H\mathbf{s} \; .$$

This equation is equivalent to (6.4).

Exercise 6.3.1 The line through the two n-vectors $\mathbf{a}_1$ and $\mathbf{a}_2$ is $\{\mathbf{a} = \mathbf{a}_1(1-\tau) + \mathbf{a}_2\tau : \tau \text{ is a scalar }\}$. Given another n-vector $\mathbf{b}$, find the point on the given line that is closest to $\mathbf{b}$.

Exercise 6.3.2 Find the point $\mathbf{a}$ on the line through $\mathbf{a}_1$ and $\mathbf{a}_2$, and the point $\mathbf{b}$ on the line through $\mathbf{b}_1$ and $\mathbf{b}_2$ so that $\mathbf{a}$ and $\mathbf{b}$ are as close as possible.

Exercise 6.3.3 Given a unit n-vector $\mathbf{u}$ and an n-vector $\mathbf{a}$, the plane through $\mathbf{a}$ with normal $\mathbf{u}$ is $\{\mathbf{x} : \mathbf{u} \cdot (\mathbf{x} - \mathbf{a}) = 0\}$. If $\mathbf{b}$ is also an n-vector, find the point in the plane that is closest to $\mathbf{b}$.

6.4 Pseudo-Inverses

In Sect. 3.2.11 we showed that every *nonsingular* matrix has an inverse. The next theorem generalizes that result, to matrices that are not necessarily nonsingular, or may not even be square.

Theorem 6.4.1 (Pseudo-Inverse) *For every $m \times n$ matrix $\mathbf{A}$ there is a unique $n \times m$ matrix $\mathbf{A}^\dagger$ so that the* **Penrose pseudo-inverse conditions** *[88] are satisfied:*

$$\mathbf{A}^\dagger \mathbf{A} \mathbf{A}^\dagger = \mathbf{A}^\dagger \;, \tag{6.5a}$$

$$\mathbf{A} \mathbf{A}^\dagger \mathbf{A} = \mathbf{A} \;, \tag{6.5b}$$

$$\left(\mathbf{A} \mathbf{A}^\dagger\right)^H = \mathbf{A} \mathbf{A}^\dagger \text{ and} \tag{6.5c}$$

$$\left(\mathbf{A}^\dagger \mathbf{A}\right)^H = \mathbf{A}^\dagger \mathbf{A} \;. \tag{6.5d}$$

Proof First, we will prove that given $\mathbf{A}$ there is at most one matrix $\mathbf{A}^\dagger$ satisfying Eqs. (6.5). Suppose that we also have $\mathbf{BAB} = \mathbf{B}$, $\mathbf{ABA} = \mathbf{A}$, $(\mathbf{AB})^H = \mathbf{AB}$ and $(\mathbf{BA})^H = \mathbf{BA}$. Condition (6.5b) for $\mathbf{A}^\dagger$ and condition (6.5a) for $\mathbf{B}$ imply that

$$\mathbf{A}^\dagger = \mathbf{A}^\dagger \mathbf{A} \mathbf{A}^\dagger = \mathbf{A}^\dagger \mathbf{ABA} \mathbf{A}^\dagger = \mathbf{A}^\dagger \mathbf{ABABABA} \mathbf{A}^\dagger$$

then Eqs. (6.5d) and (6.5c) for both $\mathbf{A}^\dagger$ and $\mathbf{B}$ give us

$$= \left(\mathbf{A}^\dagger \mathbf{A}\right)^H (\mathbf{BA})^H \mathbf{B} (\mathbf{AB})^H \left(\mathbf{A}\mathbf{A}^\dagger\right)^H = \mathbf{A}^H \left(\mathbf{A}^\dagger\right)^H \mathbf{A}^H \mathbf{B}^H \mathbf{B} \mathbf{B}^H \mathbf{A}^H \left(\mathbf{A}^\dagger\right)^H \mathbf{A}^H$$

and condition (6.5a) for $\mathbf{A}^\dagger$ yields

$$= \left(\mathbf{A}\mathbf{A}^\dagger \mathbf{A}\right)^H \mathbf{B}^H \mathbf{B} \mathbf{B}^H \left(\mathbf{A}\mathbf{A}^\dagger \mathbf{A}\right)^H = \mathbf{A}^H \mathbf{B}^H \mathbf{B} \mathbf{B}^H \mathbf{A}^H$$

and finally Eqs. (6.5c) and (6.5d) for $\mathbf{B}$, followed by (6.5b) for $\mathbf{B}$ lead to

$$= (\mathbf{BA})^H \mathbf{B} (\mathbf{AB})^H = \mathbf{BABAB} = \mathbf{BAB} = \mathbf{B} \;.$$

It is more difficult to prove that a pseudo-inverse exists. If $\mathbf{A} = \mathbf{0}$, then we can take $\mathbf{A}^\dagger = \mathbf{0}$. Otherwise, the singular value decomposition Theorem 1.5.1 of Chap. 1 in Volume II implies that there is an $m \times m$ unitary matrix $\mathbf{U}$, an $n \times n$ unitary matrix $\mathbf{V}$, a positive integer r and a positive diagonal matrix Σ such that

$$\mathbf{A} = \mathbf{U} \begin{bmatrix} \Sigma & \mathbf{0} \\ \mathbf{0} & \mathbf{0} \end{bmatrix} \mathbf{V}^H ,$$

and Lemma 6.11.1 shows that the pseudo-inverse of $\mathbf{A}$ is

$$\mathbf{A}^\dagger = \mathbf{V} \begin{bmatrix} \Sigma^{-1} & \mathbf{0} \\ \mathbf{0} & \mathbf{0} \end{bmatrix} \mathbf{U}^H .$$

The Penrose pseudo-inverse conditions (6.5) uniquely characterize the pseudo-inverse, but do not provide a computational algorithm for finding the pseudo-inverse. The proof of Theorem 6.4.1 contains a method for constructing the pseudo-inverse, but this method depends on the singular value decomposition, which will not be computed until Sect. 1.5 of Chap. 1 in Volume II, because the construction depends on an algorithm to find eigenvalues.

The next lemma shows us how to use the pseudo-inverse to solve the least squares problem. We do not recommend that readers solve least squares problems as described in the next lemma, just as we did not recommend the use of a matrix inverse to solve a system of linear equations.

Lemma 6.4.1 *Suppose that* $\mathbf{A}$ *is an* $m \times n$ *matrix and* $\mathbf{b}$ *is an* m*-vector. Then* $\mathbf{x} = \mathbf{A}^\dagger \mathbf{b}$ *solves the least squares problem for* $\mathbf{A}$ *and* $\mathbf{b}$.

Proof Let $\mathbf{r} = \left(\mathbf{I} - \mathbf{A}\mathbf{A}^\dagger\right)\mathbf{b}$ and $\mathbf{s} = \left(\mathbf{A}^\dagger\right)^H \mathbf{A}^\dagger \mathbf{b}$. Then the definitions of $\mathbf{x}$ and $\mathbf{r}$ imply that $\mathbf{r} + \mathbf{A}\mathbf{x} = \mathbf{b}$. Next, the definition of $\mathbf{r}$ and conditions (6.5c) and (6.5b) for a pseudo-inverse give us

$$\mathbf{A}^H \mathbf{r} = \mathbf{A}^H \left[\mathbf{I} - \left(\mathbf{A}\mathbf{A}^\dagger\right)^H\right] \mathbf{b} = \left(\mathbf{A} - \mathbf{A}\mathbf{A}^\dagger \mathbf{A}\right)^H \mathbf{b} = \mathbf{0} .$$

Finally, the definitions of $\mathbf{s}$ and $\mathbf{x}$, together with conditions (6.5d) and (6.5a) imply that

$$\mathbf{A}^H \mathbf{s} = \mathbf{A}^H \left(\mathbf{A}^\dagger\right)^H \mathbf{A}^\dagger \mathbf{b} = \left(\mathbf{A}^\dagger \mathbf{A}\right)^H \mathbf{A}^\dagger \mathbf{b} = \mathbf{A}^\dagger \mathbf{A} \mathbf{A}^\dagger \mathbf{b} = \mathbf{A}^\dagger \mathbf{b} = \mathbf{x} .$$

Lemma 6.3.1 now shows that $\mathbf{x}$ solves the least squares problem.

The next lemma shows how we can use the normal equations to compute pseudo-inverses in certain circumstances.

Lemma 6.4.2 *If* $\mathbf{A}$ *is an* $m \times n$ *matrix with rank* $(\mathbf{A}) = n$*, then* $\mathbf{A}^\dagger = \left(\mathbf{A}^H \mathbf{A}\right)^{-1} \mathbf{A}^H$. *Similarly, if* $\mathbf{A}$ *is an* $m \times n$ *matrix with rank* $(\mathbf{A}) = m$*, then* $\mathbf{A}^\dagger = \mathbf{A}^H \left(\mathbf{A}\mathbf{A}^H\right)^{-1}$.

Proof Lemma 6.4.1 shows that for any m-vector $\mathbf{b}$ the solution of the least squares problem is $\mathbf{x} = \mathbf{A}^\dagger \mathbf{b}$. Also, Eq. (6.3) shows that when rank $(\mathbf{A}) = n$ the solution of the least square problem is $\mathbf{x} = \left(\mathbf{A}^H \mathbf{A}\right)^{-1} \mathbf{A}^H \mathbf{b}$. The first claim follows from these two results, or by checking the Penrose pseudo-inverse conditions (6.5) directly.

To prove the second claim, we note that Eq. (6.4) implies that the solution of the underdetermined least squares problem is

$$\mathbf{x} = \mathbf{A}^H \mathbf{s} = \mathbf{A}^H \left(\mathbf{A}\mathbf{A}^H\right)^{-1} \mathbf{b} .$$

The next example shows that the use of normal equations to solve least squares problems is not numerically stable.

Example 6.4.1 ([72]) Suppose that α is a small positive machine number such that $\mathrm{fl}(1 + \alpha^2) = 1$, and let

$$\mathbf{A} = \begin{bmatrix} 1 & 1 \\ \alpha & 0 \\ 0 & \alpha \end{bmatrix} .$$

Then

$$\mathit{fl}\left(\mathbf{A}^H \mathbf{A}\right) = \mathit{fl}\left(\begin{bmatrix} 1+\alpha^2 & 1 \\ 1 & 1+\alpha^2 \end{bmatrix}\right) = \begin{bmatrix} 1 & 1 \\ 1 & 1 \end{bmatrix}$$

is singular. Thus the normal equations cannot be used in floating point arithmetic to solve least squares problems involving this matrix. Nevertheless, we will show in Examples 6.7.1 and 6.8.2 that properly-designed numerical methods can accurately solve least squares problems involving this matrix.
In Lemma 6.5.2, we will provide further evidence that the normal equations are numerically unstable.

Here is another useful consequence of the pseudo-inverse.

Lemma 6.4.3 *Suppose that* $\mathbf{A}$ *is an* $m \times n$ *matrix and* $\mathbf{A}^\dagger$ *is the pseudo-inverse of* $\mathbf{A}$. *Then the orthogonal projector onto* $\mathscr{R}(\mathbf{A})$ *is* $\mathscr{P}_{\mathscr{R}(\mathbf{A})} = \mathbf{A}\mathbf{A}^\dagger$, *and the orthogonal projector onto* $\mathscr{N}(\mathbf{A})$ *is* $\mathscr{P}_{\mathscr{N}(\mathbf{A})} = \mathbf{I} - \mathbf{A}^\dagger \mathbf{A}$.

Proof Define

$$\mathbf{P} = \mathbf{A}\mathbf{A}^\dagger .$$

Note that requirement (6.5a) for a pseudo-inverse implies that

$$\mathbf{P}^2 = \mathbf{A}\mathbf{A}^\dagger \mathbf{A}\mathbf{A}^\dagger = \mathbf{A}\mathbf{A}^\dagger = \mathbf{P} .$$

Definition 3.2.21 now shows that $\mathbf{P}$ is a projector. Furthermore, requirement (6.5c) for a pseudo-inverse implies that $\mathbf{P}$ is Hermitian, so Definition 3.2.21 shows that $\mathbf{P}$ is an orthogonal projector.

For any n-vector $\mathbf{x}$, the requirement (6.5b) for a pseudo-inverse implies that

$$\mathbf{PAx} = \mathbf{AA}^\dagger\mathbf{Ax} = \mathbf{Ax} \;,$$

so $\mathbf{P}$ preserves vectors in the range of $\mathbf{A}$. On the other hand, if $\mathbf{r} \in \mathscr{N}\left(\mathbf{A}^H\right)$, then requirement (6.5c) for a pseudo-inverse implies that

$$\mathbf{Pr} = \mathbf{AA}^\dagger\mathbf{r} = \left(\mathbf{AA}^\dagger\right)^H\mathbf{r} = \left(\mathbf{A}^\dagger\right)^H\mathbf{A}^H\mathbf{r} = \left(\mathbf{A}^\dagger\right)^H\mathbf{0} = \mathbf{0} \;.$$

Recall that the fundamental theorem of linear algebra 3.2.3 shows that every m-vector $\mathbf{b}$ can be written as a sum of two orthogonal vectors $\mathbf{Ax} \in \mathscr{R}(\mathbf{A})$, and $\mathbf{r} \in \mathscr{N}\left(\mathbf{A}^H\right)$. Finally, recall that Lemma 6.4.1 showed that the least-square solution is $\mathbf{x} = \mathbf{A}^\dagger\mathbf{b}$. Then for all m-vectors $\mathbf{b}$, it follows that

$$\mathscr{P}_{\mathscr{R}(\mathbf{A})}\mathbf{b} = \mathbf{Ax} = \mathbf{AA}^\dagger\mathbf{b} = \mathbf{Pb} \;.$$

The proof that $\mathbf{I} - \mathbf{A}^\dagger\mathbf{A}$ is the orthogonal projector onto $\mathscr{N}(\mathbf{A})$ is similar.

Exercise 6.4.1 If the $m{\times}n$ matrix $\mathbf{Q}$ has orthonormal columns, show that $\mathbf{Q}^\dagger = \mathbf{Q}^H$.

Exercise 6.4.2 Suppose that the $n \times n$ matrix $\mathbf{P}$ is an orthogonal projector (see Definition 3.2.21). Show that for any n-vector $\mathbf{x}$

$$\|\mathbf{x}\|_2^2 = \|\mathbf{Px}\|_2^2 + \|(\mathbf{I}-\mathbf{P})\mathbf{x}\|_2^2 \;.$$

Exercise 6.4.3 Suppose that $\mathbf{P}$ and $\mathbf{Q}$ are orthogonal projectors of the same size. Show that

$$\mathbf{PQ} = \mathbf{0} = \mathbf{QP}$$

if and only if

$$\mathscr{R}(\mathbf{P}) \cap \mathscr{R}(\mathbf{Q}) = \{\mathbf{0}\} \text{ and}$$
$$\mathscr{R}(\mathbf{P}) \perp \mathscr{R}(\mathbf{Q}) \;.$$

If so, show that $\mathbf{P}+\mathbf{Q}$ is also an orthogonal projector, and find the space onto which it projects.

6.5 Perturbation Analysis

In Sect. 6.3, we showed that the least squares problem has a unique solution. Our goal in this section is to determine the sense in which the least squares problem is well-posed, by determining how the solution of the least squares problem depends on its data.

First, we will examine perturbations due to the vector $\mathbf{b}$ in the least squares problem to minimize $\|\mathbf{A}\mathbf{x} - \mathbf{b}\|_2$.

Lemma 6.5.1 *Suppose that* $\mathbf{A}$ *is an* $m \times n$ *matrix, and let* $\mathscr{P}_{\mathscr{R}(\mathbf{A})}$ *be the projector onto* $\mathscr{R}(\mathbf{A})$. *Assume that* $\mathbf{b}$ *and* $\tilde{\mathbf{b}}$ *are m-vectors, and that* $\mathscr{P}_{\mathscr{R}(\mathbf{A})}\mathbf{b} \neq \mathbf{0}$. *Let* $\mathbf{x}$ *solve the least squares problem for* $\mathbf{A}$ *and* $\mathbf{b}$, *and let* $\tilde{\mathbf{x}}$ *solve the least squares problem for* $\mathbf{A}$ *and* $\tilde{\mathbf{b}}$. *Then*

$$\frac{\|\tilde{\mathbf{x}} - \mathbf{x}\|_2}{\|\mathbf{x}\|_2} \leq \|\mathbf{A}\|_2 \left\|\mathbf{A}^\dagger\right\|_2 \frac{\left\|\mathscr{P}_{\mathscr{R}(\mathbf{A})}(\tilde{\mathbf{b}} - \mathbf{b})\right\|_2}{\|\mathscr{P}_{\mathscr{R}(\mathbf{A})}\mathbf{b}\|_2} . \tag{6.6}$$

Proof Requirement (6.5a) for a pseudo-inverse and Lemma 6.4.3 imply that

$$\mathbf{A}^\dagger = \mathbf{A}^\dagger\mathbf{A}\mathbf{A}^\dagger = \mathbf{A}^\dagger \mathscr{P}_{\mathscr{R}(\mathbf{A})} .$$

Recall from Lemma 6.4.1 that $\mathbf{x} = \mathbf{A}^\dagger\mathbf{b}$ solves the least squares problem for $\mathbf{A}$ and $\mathbf{b}$. Similarly, $\tilde{\mathbf{x}} = \mathbf{A}^\dagger\tilde{\mathbf{b}}$ solves the least squares problem for $\mathbf{A}$ and $\tilde{\mathbf{b}}$. As a result,

$$\|\mathbf{x}-\tilde{\mathbf{x}}\|_2 = \left\|\mathbf{A}^\dagger\left(\mathbf{b} - \tilde{\mathbf{b}}\right)\right\|_2 = \left\|\mathbf{A}^\dagger \mathscr{P}_{\mathscr{R}(\mathbf{A})}\left(\mathbf{b} - \tilde{\mathbf{b}}\right)\right\|_2 \leq \left\|\mathbf{A}^\dagger\right\|_2 \left\|\mathscr{P}_{\mathscr{R}(\mathbf{A})}\left(\mathbf{b} - \tilde{\mathbf{b}}\right)\right\|_2 .$$

Since Lemma 6.4.3 showed that $\mathscr{P}_{\mathscr{R}(\mathbf{A})} = \mathbf{A}\mathbf{A}^\dagger$, we have

$$\left\|\mathscr{P}_{\mathscr{R}(\mathbf{A})}\mathbf{b}\right\|_2 \leq \|\mathbf{A}\|_2 \left\|\mathbf{A}^\dagger\mathbf{b}\right\|_2 \Longrightarrow \left\|\mathbf{A}^\dagger\mathbf{b}\right\|_2 \geq \frac{\|\mathscr{P}_{\mathscr{R}(\mathbf{A})}\mathbf{b}\|_2}{\|\mathbf{A}\|_2} .$$

As a result, we can bound the relative error in the least squares solution as follows:

$$\frac{\|\mathbf{x} - \tilde{\mathbf{x}}\|_2}{\|\mathbf{x}\|_2} = \frac{\left\|\mathbf{A}^\dagger\left(\mathbf{b} - \tilde{\mathbf{b}}\right)\right\|_2}{\|\mathbf{A}^\dagger\mathbf{b}\|_2} \leq \frac{\left\|\mathbf{A}^\dagger\right\|_2 \left\|\mathscr{P}_{\mathscr{R}(\mathbf{A})}\left(\mathbf{b} - \tilde{\mathbf{b}}\right)\right\|_2}{\|\mathscr{P}_{\mathscr{R}(\mathbf{A})}\mathbf{b}\|_2/\|\mathbf{A}\|_2} .$$

This is equivalent to the claim.

Lemma 6.5.1 suggests the following definition.

Definition 6.5.1 The **condition number** of a general (rectangular) matrix $\mathbf{A}$ is

$$\kappa_2(\mathbf{A}) \equiv \|\mathbf{A}\|_2 \left\|\mathbf{A}^\dagger\right\|_2 .$$

Lemma 6.5.1 shows that if either $\|\mathscr{P}_{\mathscr{R}(\mathbf{A})}\mathbf{b}\|_2$ is small or $\kappa_2(\mathbf{A})$ is large, then a small perturbation in $\mathbf{b}$ can make a large relative perturbation in the least squares solution $\mathbf{x}$.

Example 6.5.1 Let

$$\mathbf{A} = \begin{bmatrix} 1 & 0 \\ 0 & 1 \\ 0 & 0 \end{bmatrix}, \ \mathbf{b} = \begin{bmatrix} 0.01 \\ 0 \\ 1 \end{bmatrix} \text{ and } \tilde{\mathbf{b}} = \begin{bmatrix} 0.0101 \\ 0 \\ 1 \end{bmatrix},$$

Then

$$\mathbf{A}^\dagger = \begin{bmatrix} 1 & 0 & 0 \\ 0 & 1 & 0 \end{bmatrix} \text{ and } \mathscr{P}_{\mathscr{R}(\mathbf{A})} = \mathbf{A}\mathbf{A}^\dagger = \begin{bmatrix} 1 & 0 & 0 \\ 0 & 1 & 0 \\ 0 & 0 & 0 \end{bmatrix},$$

so the solutions to the least squares problems are

$$\mathbf{x} = \mathscr{P}_{\mathscr{R}(\mathbf{A})}\mathbf{b} = \begin{bmatrix} 0.01 \\ 0 \end{bmatrix} \text{ and } \tilde{\mathbf{x}} = \mathscr{P}_{\mathscr{R}(\mathbf{A})}\tilde{\mathbf{b}} = \begin{bmatrix} 0.0101 \\ 0 \end{bmatrix}.$$

Thus the relative change in the projected right-hand side is

$$\frac{\left\| \mathscr{P}_{\mathscr{R}(\mathbf{A})}\left(\mathbf{b} - \tilde{\mathbf{b}}\right) \right\|}{\|\mathscr{P}_{\mathscr{R}(\mathbf{A})}\mathbf{b}\|} = 10^{-2}.$$

The relative change in the solution is

$$\frac{\|\mathbf{x} - \tilde{\mathbf{x}}\|}{\|\mathbf{x}\|} = 10^{-2},$$

It is also easy to see that

$$\|\mathbf{A}\|_2 = \sup_{\|\mathbf{x}\|_2 = 1} \frac{\|\mathbf{A}\mathbf{x}\|_2}{\|\mathbf{x}\|_2} = 1 \text{ and } \left\|\mathbf{A}^\dagger\right\|_2 = \sup_{\|\mathbf{x}\|_2 = 1} \frac{\left\|\mathbf{A}^\dagger\mathbf{x}\right\|_2}{\|\mathbf{x}\|_2} = 1.$$

These results allow us to verify the claim in Lemma 6.5.1 for the arrays in this example.

Next, we will show that the normal Eqs. (6.3) or (6.4) are not as well conditioned as the least squares problem, and thus should be avoided in most computations.

Lemma 6.5.2 *Suppose that* $\mathbf{A}$ *is an* $m \times n$ *matrix. If* rank $(\mathbf{A}) = n$ *then*

$$\kappa_2\left(\mathbf{A}^H\mathbf{A}\right) = \kappa_2(\mathbf{A})^2.$$

On the other hand, if rank $(\mathbf{A}) = m$ *then*

$$\kappa_2\left(\mathbf{A}\mathbf{A}^H\right) = \kappa_2(\mathbf{A})^2 .$$

Proof Since

$$\|\mathbf{A}\mathbf{x}\|_2^2 = \mathbf{x}^H\mathbf{A}^H\mathbf{A}\mathbf{x} \leq \|\mathbf{A}^H\mathbf{A}\|_2\|\mathbf{x}\|_2^2 ,$$

we find that

$$\|\mathbf{A}\|_2^2 \leq \|\mathbf{A}^H\mathbf{A}\|_2 .$$

Lemma 1.3.3 of Chap. 1 in Volume II will show that if we choose $\mathbf{x}$ to be the eigenvector of $\mathbf{A}^H\mathbf{A}$ corresponding to its largest eigenvalue, then this inequality becomes an equality.

If rank $(\mathbf{A}) = n$ then Lemma 6.4.2 showed that $\mathbf{A}^\dagger = \left(\mathbf{A}^H\mathbf{A}\right)^{-1}\mathbf{A}^H$. It follows that

$$\mathbf{A}^\dagger\left(\mathbf{A}^\dagger\right)^H = (\mathbf{A}^H\mathbf{A})^{-1}\mathbf{A}^H\mathbf{A}\left(\mathbf{A}^H\mathbf{A}\right)^{-1} = \left(\mathbf{A}^H\mathbf{A}\right)^{-1} .$$

We can argue as before to show that

$$\left\|\mathbf{A}^\dagger\right\|_2^2 = \left\|(\mathbf{A}^H\mathbf{A})^{-1}\right\|_2 .$$

Then

$$\kappa_2(\mathbf{A})^2 = \|\mathbf{A}\|_2^2\left\|\mathbf{A}^\dagger\right\|_2^2 = \|\mathbf{A}^H\mathbf{A}\|_2\left\|\left(\mathbf{A}^H\mathbf{A}\right)^{-1}\right\|_2 = \kappa_2\left(\mathbf{A}^H\mathbf{A}\right) .$$

The second claim is the same as the first claim applied to $\mathbf{A}^H$.

It is much more difficult to analyze the effect of perturbations in the matrix $\mathbf{A}$ on the solution of the least squares problem. For a proof of the following perturbation result for the full-rank overdetermined least squares problem, see Stewart [96, p. 39, 44].

Theorem 6.5.1 *Suppose that* $\mathbf{A}$ *and* $\tilde{\mathbf{A}}$ *are* $m \times n$ *matrices, and that rank* $(\mathbf{A}) = n$. *Denote the relative errors in the projections of the matrix perturbation onto the range of* $\mathbf{A}$ *and onto the nullspace of* $\mathbf{A}^H$ *by*

$$\varepsilon_{\mathscr{R}(\mathbf{A})} = \frac{\left\|\mathscr{P}_{\mathscr{R}(\mathbf{A})}\left(\tilde{\mathbf{A}} - \mathbf{A}\right)\right\|_2}{\|\mathbf{A}\|_2} \text{ and } \varepsilon_{\mathscr{N}(\mathbf{A}^H)} = \frac{\left\|\left(\mathbf{I} - \mathscr{P}_{\mathscr{R}(\mathbf{A})}\right)\left(\tilde{\mathbf{A}} - \mathbf{A}\right)\right\|_2}{\|\mathbf{A}\|_2}$$

If

$$\kappa_2(\mathbf{A})\,\varepsilon_{\mathscr{R}(\mathbf{A})} < 1 ,$$

then

$$\frac{\left\|\tilde{\mathbf{A}}^\dagger - \mathbf{A}^\dagger\right\|_2}{\|\mathbf{A}^\dagger\|_2}$$

$$\leq \frac{1}{1-\kappa_2(\mathbf{A})\,\varepsilon_{\mathscr{R}(\mathbf{A})}}\left\{\kappa_2(\mathbf{A})\,\varepsilon_{\mathscr{R}(\mathbf{A})} + \frac{\kappa_2(\mathbf{A})\,\varepsilon_{\mathscr{N}(\mathbf{A}^H)}}{\sqrt{\left(1-\kappa_2(\mathbf{A})\,\varepsilon_{\mathscr{R}(\mathbf{A})}\right)^2 + \kappa_2(\mathbf{A})^2\,\varepsilon^2_{\mathscr{N}(\mathbf{A}^H)}}}\right\}.$$

Next, given an m-vector $\mathbf{b}$*, let* $\mathbf{x} = \mathbf{A}^\dagger\mathbf{b}$ *and* $\tilde{\mathbf{x}} = \tilde{\mathbf{A}}^\dagger\mathbf{b}$*. Also define the residual vector* $\mathbf{r} = \mathbf{b} - \mathbf{A}\mathbf{x}$*. Then*

$$\frac{\|\tilde{\mathbf{x}}-\mathbf{x}\|_2}{\|\mathbf{x}\|_2} \leq \frac{\kappa_2(\mathbf{A})\,\varepsilon_{\mathscr{R}(\mathbf{A})}}{1-\kappa_2(\mathbf{A})\,\varepsilon_{\mathscr{R}(\mathbf{A})}}$$

$$+ \frac{\kappa_2(\mathbf{A})}{1-\kappa_2(\mathbf{A})\,\varepsilon_{\mathscr{R}(\mathbf{A})}}\left\{\frac{\kappa_2(\mathbf{A})^2\,\varepsilon^2_{\mathscr{N}(\mathbf{A}^H)}}{\left(1-\kappa_2(\mathbf{A})\,\varepsilon_{\mathscr{R}(\mathbf{A})}\right)^2 + \kappa_2(\mathbf{A})^2\,\varepsilon^2_{\mathscr{N}(\mathbf{A}^H)}}\right.$$

$$\left.+\frac{\kappa_2(\mathbf{A})\,\varepsilon_{\mathscr{N}(\mathbf{A}^H)}}{\sqrt{\left(1-\kappa_2(\mathbf{A})\,\varepsilon_{\mathscr{R}(\mathbf{A})}\right)^2 + \kappa_2(\mathbf{A})^2\,\varepsilon^2_{\mathscr{N}(\mathbf{A}^H)}}}\frac{\|\mathbf{r}\|_2}{\left\|\mathscr{P}_{\mathscr{R}(\mathbf{A})}\mathbf{b}\right\|_2}\right\}. \tag{6.7}$$

Note that the case when $\operatorname{rank}(\mathbf{A}) = m$ satisfies a similar perturbation inequality, which can be determined by appropriate modifications of the ideas in Stewart [96].

Let us interpret the conclusion of Theorem 6.5.1 under some common circumstances. Suppose that the matrix perturbations are small, so that $\kappa_2(\mathbf{A})\,\varepsilon_{\mathscr{R}(\mathbf{A})} \ll 1$ and $[\kappa_2(\mathbf{A})\,\varepsilon_{\mathscr{N}(\mathbf{A}^H)}]^2 \ll 1$. Then inequality (6.7) can be simplified to

$$\frac{\|\tilde{\mathbf{x}}-\mathbf{x}\|_2}{\|\mathbf{x}\|_2} \leq \kappa_2(\mathbf{A})\,\varepsilon_{\mathscr{R}(\mathbf{A})} + \kappa_2(\mathbf{A})^2\frac{\|\mathbf{r}\|_2}{\left\|\mathscr{P}_{\mathscr{R}(\mathbf{A})}\mathbf{b}\right\|_2}\varepsilon_{\mathscr{N}(\mathbf{A}^H)} + \kappa_2(\mathbf{A})^3\,\varepsilon^2_{\mathscr{N}(\mathbf{A}^H)}\,.$$

Next, if the residual vector is so small that

$$\frac{\|\mathbf{r}\|_2}{\left\|\mathscr{P}_{\mathscr{R}(\mathbf{A})}\mathbf{b}\right\|_2} \ll \kappa_2(\mathbf{A})\,\varepsilon_{\mathscr{N}(\mathbf{A}^H)}\,,$$

then the inequality simplifies to

$$\frac{\|\tilde{\mathbf{x}}-\mathbf{x}\|_2}{\|\mathbf{x}\|_2} \leq \kappa_2(\mathbf{A})\,\varepsilon_{\mathscr{R}(\mathbf{A})} + \kappa_2(\mathbf{A})^3\,\varepsilon_{\mathscr{N}(\mathbf{A}^H)}\,.$$

At this point, we see that for least squares problems with small residuals perturbations and suffering small matrix perturbations, any matrix perturbations that lie mostly in the range of **A** are amplified by the condition number of **A**, while perturbations that lie mostly orthogonal to the range of **A** can be amplified by the cube of the condition number of **A**.

Example 6.5.2 ([97, p. 224]) Suppose that

$$\mathbf{A} = \begin{bmatrix} 1 & 1+\delta \\ 1 & 1 \\ 1 & 1 \end{bmatrix} \text{ and } \tilde{\mathbf{A}} = \begin{bmatrix} 1 & 1+\delta+\varepsilon \\ 1 & 1+\varepsilon \\ 1 & 1 \end{bmatrix} .$$

where $0 < \delta \ll 1$ and $\varepsilon = \delta^{5/3}$. Eventually, we will see that $\tilde{\mathbf{A}} - \mathbf{A}$ includes components in $\mathscr{R}(\mathbf{A})$ and its orthogonal complement of roughly equal norm.

Note that

$$\mathbf{A}^H\mathbf{A} = \begin{bmatrix} 3 & 3+\delta \\ 3+\delta & 3+2\delta+\delta^2 \end{bmatrix} \text{ and}$$

$$\tilde{\mathbf{A}}^H\tilde{\mathbf{A}} = \begin{bmatrix} 3 & 3+\delta+2\varepsilon \\ 3+\delta+2\varepsilon & 3+2\delta+4\varepsilon+\delta^2+2\delta\varepsilon+2\varepsilon^2 \end{bmatrix} .$$

so Eq. (3.20) for the matrix 2-norm implies that

$$\|\mathbf{A}\|_2 = \sqrt{\frac{6+2\delta+\delta^2+\sqrt{4(3+\delta)^2+(2\delta+\delta^2)^2}}{2}}$$
$$\approx \sqrt{6+2\delta} \approx \sqrt{6} + \frac{\delta}{\sqrt{6}} \approx \left\|\tilde{\mathbf{A}}\right\|_2 .$$

Also

$$\mathbf{A}^\dagger = \left(\mathbf{A}^H\mathbf{A}\right)^{-1}\mathbf{A}^H = \begin{bmatrix} -2 & 1+\delta & 1+\delta \\ 2 & -1 & -1 \end{bmatrix} \frac{1}{2\delta} \text{ and}$$

$$\tilde{\mathbf{A}}^\dagger = \begin{bmatrix} -2-\varepsilon/\delta-\varepsilon & 1-\varepsilon/\delta+\delta+\varepsilon & 1+2\varepsilon/\delta+\delta+2\varepsilon+2\varepsilon^2/\delta \\ 2+\varepsilon/\delta & -1+\varepsilon/\delta & -1-2\varepsilon/\delta \end{bmatrix} \frac{1}{2(\delta+\varepsilon+\varepsilon^2/\delta)} ,$$

so

$$\left\|\mathbf{A}^\dagger\right\|_2 = \sqrt{\frac{12+4\delta+2\delta^2+\sqrt{4(6+2\delta)^2+(4\delta+2\delta^2)^2}}{8\delta^2}}$$
$$\approx \frac{\sqrt{3+\delta}}{\delta} \approx \frac{\sqrt{3}}{\delta} + \frac{1}{\sqrt{12}} \approx \left\|\tilde{\mathbf{A}}^\dagger\right\|_2 .$$

Next, note that

$$\mathscr{P}_{\mathscr{R}(\mathbf{A})}\left(\tilde{\mathbf{A}}-\mathbf{A}\right)=\mathbf{A}\mathbf{A}^{\dagger}\left(\tilde{\mathbf{A}}-\mathbf{A}\right)=\begin{bmatrix}0 & \varepsilon \\ 0 & \varepsilon/2 \\ 0 & \varepsilon/2\end{bmatrix} \text{ and}$$

$$\mathscr{P}_{\mathscr{N}(\mathbf{A}^H)}\left(\tilde{\mathbf{A}}-\mathbf{A}\right)=\left(\mathbf{I}-\mathbf{A}\mathbf{A}^{\dagger}\right)\left(\tilde{\mathbf{A}}-\mathbf{A}\right)=\begin{bmatrix}0 & 0 \\ 0 & \varepsilon/2 \\ 0 & -\varepsilon/2\end{bmatrix} ,$$

so

$$\left\|\mathscr{P}_{\mathscr{R}(\mathbf{A})}\left(\tilde{\mathbf{A}}-\mathbf{A}\right)\right\|_2=\varepsilon\sqrt{3/2} \text{ and } \left\|\mathscr{P}_{\mathscr{N}(\mathbf{A}^H)}\left(\tilde{\mathbf{A}}-\mathbf{A}\right)\right\|_2=\varepsilon\sqrt{1/2}\,.$$

Thus

$$\varepsilon_{\mathscr{R}(\mathbf{A})}\equiv\frac{\left\|\mathscr{P}_{\mathscr{R}(\mathbf{A})}\left(\tilde{\mathbf{A}}-\mathbf{A}\right)\right\|_2}{\|\mathbf{A}\|_2}\approx\frac{\varepsilon\sqrt{3/2}}{\sqrt{6}+\delta/\sqrt{6}}\approx\frac{\varepsilon}{2}\left(1-\frac{\delta}{6}\right) \text{ and}$$

$$\varepsilon_{\mathscr{N}(\mathbf{A}^H)}\equiv\frac{\left\|\mathscr{P}_{\mathscr{N}(\mathbf{A}^H)}\left(\tilde{\mathbf{A}}-\mathbf{A}\right)\right\|_2}{\|\mathbf{A}\|_2}\approx\frac{\varepsilon\sqrt{1/2}}{\sqrt{6}+\delta/\sqrt{6}}\approx\frac{\varepsilon}{\sqrt{12}}\left(1-\frac{\delta}{6}\right)\,.$$

In order to apply Theorem 6.5.1 to this problem, we require that

$$1>\kappa_2(\mathbf{A})\,\varepsilon_{\mathscr{R}(\mathbf{A})}\approx\left(\sqrt{6}+\frac{\delta}{\sqrt{6}}\right)\left(\frac{\sqrt{3}}{\delta}+\frac{1}{\sqrt{12}}\right)\left(\frac{\varepsilon}{2}-\frac{\delta\varepsilon}{12}\right)\approx\sqrt{\frac{9}{2}}\frac{\varepsilon}{\delta}\,.$$

This is easily satisfied for small values of δ.

Next, let

$$\mathbf{b}_1=\begin{bmatrix}2+\delta \\ 2 \\ 2\end{bmatrix} \text{ and } \mathbf{b}_2=\begin{bmatrix}2+\delta \\ 3/2 \\ 5/2\end{bmatrix}\,.$$

Then

$$\mathscr{P}_{\mathscr{R}(\mathbf{A})}\mathbf{b}_1=\mathbf{A}\mathbf{A}^{\dagger}\mathbf{b}_1=\begin{bmatrix}2+\delta \\ 2 \\ 2\end{bmatrix}=\mathbf{b}_1 \text{ but } \mathscr{P}_{\mathscr{R}(\mathbf{A})}\mathbf{b}_2=\begin{bmatrix}2+\delta \\ 2 \\ 2\end{bmatrix}\neq\mathbf{b}_2\,.$$

We also have

$$\mathbf{x}_1\equiv\mathbf{A}^{\dagger}\mathbf{b}_1=\begin{bmatrix}1 \\ 1\end{bmatrix} \text{ and } \mathbf{x}_2\equiv\mathbf{A}^{\dagger}\mathbf{b}_2=\begin{bmatrix}1 \\ 1\end{bmatrix}\,,$$

and

$$\mathbf{r}_1 = \mathbf{b}_1 - \mathbf{A}\mathbf{x}_1 = \mathbf{0} \text{ and } \mathbf{r}_2 = \mathbf{b}_2 - \mathbf{A}\mathbf{x}_2 = \begin{bmatrix} 0 \\ -1/2 \\ 1/2 \end{bmatrix}.$$

On the other hand,

$$\tilde{\mathbf{x}}_1 \equiv \tilde{\mathbf{A}}^{\dagger}\mathbf{b}_1 = \begin{bmatrix} 2\delta + 3\varepsilon + 4\varepsilon^2/\delta - \delta\varepsilon \\ 2\delta + \varepsilon \end{bmatrix} \frac{1}{2(\delta + \varepsilon + \varepsilon^2/\delta)} \approx \begin{bmatrix} 1 + \varepsilon/(2\delta) \\ 1 - \varepsilon/(2\delta) \end{bmatrix} \text{ and}$$

$$\tilde{\mathbf{x}}_2 \equiv \tilde{\mathbf{A}}^{\dagger}\mathbf{b}_2 = \begin{bmatrix} 3\varepsilon/\delta + 4\delta + 7\varepsilon + 10\varepsilon^2/\delta - 2\delta\varepsilon \\ -3\varepsilon/\delta + 4\delta + 2\varepsilon \end{bmatrix} \frac{1}{4(\delta + \varepsilon + \varepsilon^2/\delta)}$$

$$\approx \begin{bmatrix} 3\varepsilon/(4\delta^2) + 1 \\ -3\varepsilon/(4\delta^2) + 1 \end{bmatrix}.$$

It follows that

$$\frac{\|\tilde{\mathbf{x}}_1 - \mathbf{x}_1\|_2}{\|\mathbf{x}_1\|_2} = \frac{1}{\sqrt{2}} \left\| \begin{bmatrix} 1 + 2\varepsilon/\delta - \delta \\ -1 - 2\varepsilon/\delta \end{bmatrix} \frac{\varepsilon}{2\delta(1 + \varepsilon/\delta + \varepsilon^2/\delta^2)} \right\|_2 \approx \frac{\varepsilon}{2\delta}\left(1 + \frac{\varepsilon}{\delta}\right) \text{ and}$$

$$\frac{\|\tilde{\mathbf{x}}_2 - \mathbf{x}_2\|_2}{\|\mathbf{x}_2\|_2} = \frac{1}{\sqrt{2}} \left\| \begin{bmatrix} 3/\delta + 3 + 6\varepsilon/\delta - 2\delta \\ -3/\delta - 2 - 4\varepsilon/\delta \end{bmatrix} \frac{\varepsilon}{4\delta(1 + \varepsilon/\delta + \varepsilon^2/\delta^2)} \right\|_2 \approx \frac{3\varepsilon}{4\delta^2}\left(1 - \frac{\varepsilon}{\delta}\right).$$

The perturbations in Theorem 6.5.1 essentially say that

$$\frac{\|\tilde{\mathbf{x}}_1 - \mathbf{x}_1\|_2}{\|\mathbf{x}_1\|_2} \le \kappa_2(\mathbf{A})\,\varepsilon_{\mathscr{R}(\mathbf{A})} + \kappa_2(\mathbf{A}) \left\{ \kappa_2(\mathbf{A})^2 \varepsilon_{\mathscr{N}(\mathbf{A}^H)}^2 + \kappa_2(\mathbf{A})\,\varepsilon_{\mathscr{N}(\mathbf{A}^H)} \frac{\|\mathbf{r}_1\|_2}{\|\mathscr{P}_{\mathscr{R}(\mathbf{A})}\mathbf{b}_1\|_2} \right\}$$

$$\approx \left(\sqrt{6} + \frac{\delta}{\sqrt{6}}\right)\left(\frac{3}{\delta} + \frac{1}{\sqrt{12}}\right)\frac{\varepsilon}{2}\left(1 - \frac{\delta}{6}\right)$$

$$+ \left(\sqrt{6} + \frac{\delta}{\sqrt{6}}\right)^3 \left(\frac{3}{\delta} + \frac{1}{\sqrt{12}}\right)^3 \frac{\varepsilon^2}{12}\left(1 - \frac{\delta}{\sqrt{6}}\right)^2$$

$$\approx \frac{9}{2}\frac{\varepsilon^2}{\delta^3} + \frac{3}{\sqrt{2}}\frac{\varepsilon}{\delta} \text{ and}$$

$$\frac{\|\tilde{\mathbf{x}}_2 - \mathbf{x}_2\|_2}{\|\mathbf{x}_2\|_2} \le \kappa_2(\mathbf{A})\,\varepsilon_{\mathscr{R}(\mathbf{A})} + \kappa_2(\mathbf{A}) \left\{ \kappa_2(\mathbf{A})^2 \varepsilon_{\mathscr{N}(\mathbf{A}^H)}^2 + \kappa_2(\mathbf{A})\,\varepsilon_{\mathscr{N}(\mathbf{A}^H)} \frac{\|\mathbf{r}_2\|_2}{\|\mathscr{P}_{\mathscr{R}(\mathbf{A})}\mathbf{b}_2\|_2} \right\}$$

$$\approx \left(\sqrt{6} + \frac{\delta}{\sqrt{6}}\right)\left(\frac{3}{\delta} + \frac{1}{\sqrt{12}}\right)\frac{\varepsilon}{2}\left(1 - \frac{\delta}{6}\right)$$

$$+ \left(\sqrt{6} + \frac{\delta}{\sqrt{6}}\right)^3 \left(\frac{3}{\delta} + \frac{1}{\sqrt{12}}\right)^3 \frac{\varepsilon^2}{12}\left(1 - \frac{\delta}{\sqrt{6}}\right)^2$$

$$+ \left(\sqrt{6} + \frac{\delta}{\sqrt{6}}\right)^2 \left(\frac{3}{\delta} + \frac{1}{\sqrt{12}}\right)^2 \frac{\varepsilon}{\sqrt{12}} \left(1 - \frac{\delta}{6}\right) \frac{\sqrt{2}/2}{\sqrt{12}(1 + \delta/6)}$$
$$\approx \frac{3\sqrt{2}}{4} \frac{\varepsilon}{\delta^2} + \frac{7\sqrt{2}}{4} \frac{\varepsilon}{\delta} .$$

For the least squares problem with right-hand side $\mathbf{b}_1$, the dominant term in the bound (6.7) has a higher order than that in the true error. For the least squares problem with right-hand side $\mathbf{b}_2$, the order of the dominant term is correct, and is due to the large residual.

Stewart [96, p. 33] notes that for an $m \times n$ matrix $\mathbf{A}$ with rank $(\mathbf{A}) < \min\{m, n\}$, the pseudo-inverse is not necessarily a continuous function of the entries of $\mathbf{A}$. The following example illustrates the problem.

Example 6.5.3 Let

$$\mathbf{A} = \begin{bmatrix} 1 & 0 \\ 0 & 0 \end{bmatrix} \text{ and } \tilde{\mathbf{A}} = \begin{bmatrix} 1 & 0 \\ 0 & \varepsilon \end{bmatrix} .$$

Then

$$\mathbf{A}^\dagger = \begin{bmatrix} 1 & 0 \\ 0 & 0 \end{bmatrix} \text{ and } \tilde{\mathbf{A}}^\dagger = \begin{bmatrix} 1 & 0 \\ 0 & 1/\varepsilon \end{bmatrix} .$$

We conclude that $\|\mathbf{A}^\dagger - \tilde{\mathbf{A}}^\dagger\|_2 = 1/\varepsilon$, which becomes arbitrarily large as $\varepsilon \to 0$,

The following theorem, which is proved by Stewart [96, p. 40], discusses pseudo-inverses of perturbations of rank-deficient matrices.

Theorem 6.5.2 *Suppose that* $\mathbf{A}$ *and* $\tilde{\mathbf{A}}$ *are* $m \times n$ *matrices, and that rank* $(\mathbf{A}) = r < \min\{m, n\}$. *Assume that*

$$\gamma \equiv 1 - \|\mathbf{A}^\dagger\|_2 \left\|\mathbf{A}\mathbf{A}^\dagger \left(\tilde{\mathbf{A}} - \mathbf{A}\right) \mathbf{A}^\dagger \mathbf{A}\right\|_2 > 0 .$$

Define

$$\varepsilon_{11} = \frac{\kappa_2(\mathbf{A})}{\gamma} \frac{\left\|\mathbf{A}\mathbf{A}^\dagger \left(\tilde{\mathbf{A}} - \mathbf{A}\right) \mathbf{A}^\dagger \mathbf{A}\right\|_2}{\|\mathbf{A}\|_2} ,$$

$$\varepsilon_{21} = \frac{\kappa_2(\mathbf{A})}{\gamma} \frac{\left\|\left(\mathbf{I} - \mathbf{A}\mathbf{A}^\dagger\right) \left(\tilde{\mathbf{A}} - \mathbf{A}\right) \mathbf{A}^\dagger \mathbf{A}\right\|_2}{\|\mathbf{A}\|_2} ,$$

$$\varepsilon_{12} = \frac{\kappa_2(\mathbf{A})}{\gamma} \frac{\left\|\mathbf{A}\mathbf{A}^\dagger \left(\tilde{\mathbf{A}} - \mathbf{A}\right) \left(\mathbf{I} - \mathbf{A}^\dagger \mathbf{A}\right)\right\|_2}{\|\mathbf{A}\|_2} \text{ and}$$

$$\varepsilon_{22} = \frac{\kappa_2(\mathbf{A})}{\gamma} \frac{\left\|\left(\mathbf{I} - \mathbf{A}\mathbf{A}^\dagger\right) \left(\tilde{\mathbf{A}} - \mathbf{A}\right) \left(\mathbf{I} - \mathbf{A}^\dagger \mathbf{A}\right)\right\|_2}{\|\mathbf{A}\|_2} .$$

If rank $\left(\tilde{\mathbf{A}}\right) = $ *rank* $(\mathbf{A})$*, then*

$$\frac{\left\|\tilde{\mathbf{A}}^{\dagger} - \mathbf{A}^{\dagger}\right\|_2}{\|\mathbf{A}^{\dagger}\|_2} \leq \varepsilon_{11} + \frac{1}{\gamma}\left\{\frac{\varepsilon_{12}}{\sqrt{1+\varepsilon_{12}^2}} + \frac{\varepsilon_{21}}{\sqrt{1+\varepsilon_{21}^2}} + \frac{\varepsilon_{22}}{\sqrt{1+\varepsilon_{22}^2}}\right\},$$

else

$$\left\|\tilde{\mathbf{A}}^{\dagger}\right\|_2 \geq \frac{1}{\left\|\tilde{\mathbf{A}} - \mathbf{A}\right\|_2}.$$

*Example 6.5.4 (**Läuchli** [72])* Suppose that α is a small positive machine number such that $\mathrm{fl}(1+\alpha^2) = 1$. Let

$$\tilde{\mathbf{A}} = \begin{bmatrix} 1 & 1 \\ \alpha & 0 \\ 0 & \alpha \end{bmatrix}.$$

Then

$$\tilde{\mathbf{A}}^H\tilde{\mathbf{A}} = \begin{bmatrix} 1+\alpha^2 & 1 \\ 1 & 1+\alpha^2 \end{bmatrix}.$$

Also,

$$\tilde{\mathbf{A}}^{\dagger} = (\tilde{\mathbf{A}}^H\tilde{\mathbf{A}})^{-1}\tilde{\mathbf{A}}^H = \begin{bmatrix} 1+\alpha^2 & -1 \\ -1 & 1+\alpha^2 \end{bmatrix} \frac{1}{\alpha^2(2+\alpha^2)} \begin{bmatrix} 1 & \alpha & 0 \\ 1 & 0 & \alpha \end{bmatrix}$$

$$= \begin{bmatrix} 1 & 1/\alpha+\alpha & -1/\alpha \\ 1 & -1/\alpha & 1/\alpha+\alpha \end{bmatrix} \frac{1}{2+\alpha^2}.$$

Suppose that we choose

$$\mathbf{A} = \begin{bmatrix} 1 & 1 \\ \alpha & 0 \\ 0 & 0 \end{bmatrix}$$

so that rank $(\mathbf{A}) = 2 = $ rank $\left(\tilde{\mathbf{A}}\right)$. Then

$$\mathbf{A}^H\mathbf{A} = \begin{bmatrix} 1+\alpha^2 & 1 \\ 1 & 1 \end{bmatrix},$$

so

$$\mathbf{A}^\dagger = \left(\mathbf{A}^H\mathbf{A}\right)^{-1}\mathbf{A}^H = \begin{bmatrix} 1 & -1 \\ -1 & 1+\alpha^2 \end{bmatrix} \frac{1}{\alpha^2} \begin{bmatrix} 1 & \alpha & 0 \\ 1 & 0 & 0 \end{bmatrix} = \begin{bmatrix} 0 & 1/\alpha & 0 \\ 1 & -1/\alpha & 0 \end{bmatrix}$$

and

$$\mathbf{A}^\dagger\left(\mathbf{A}^\dagger\right)^H = \begin{bmatrix} 1 & -1 \\ -1 & 1+\alpha^2 \end{bmatrix} \frac{1}{\alpha^2} .$$

It follows that

$$\|\mathbf{A}\|_2^2 = \frac{2+\alpha^2+\sqrt{4+\alpha^4}}{2} \approx 2+\alpha^2/2 \text{ and}$$
$$\|\mathbf{A}^\dagger\|_2^2 = \frac{2+\alpha^2+\sqrt{4+\alpha^4}}{2\alpha^2} \approx \frac{2}{\alpha^2}+\frac{1}{2} ,$$

which implies that

$$\kappa_2(\mathbf{A}) = \sqrt{\frac{2+\alpha^2+\sqrt{4+\alpha^4}}{2}}\sqrt{\frac{2+\alpha^2+\sqrt{4+\alpha^4}}{2\alpha^2}} \approx 2/\alpha+\alpha/2 .$$

We also have

$$\mathbf{A}\mathbf{A}^\dagger = \begin{bmatrix} 1 & 0 & 0 \\ 0 & 1 & 0 \\ 0 & 0 & 0 \end{bmatrix} \text{ and } \mathbf{A}^\dagger\mathbf{A} = \begin{bmatrix} 1 & 0 \\ 0 & 1 \end{bmatrix} .$$

The matrix perturbation is

$$\tilde{\mathbf{A}}^\dagger - \mathbf{A}^\dagger = \begin{bmatrix} \alpha & -1 & -1 \\ -\alpha(1+\alpha^2) & 1+\alpha^2 & 1+\alpha^2 \end{bmatrix} \frac{1}{\alpha(2+\alpha^2)} ,$$

so

$$\left(\tilde{\mathbf{A}}^\dagger - \mathbf{A}^\dagger\right)\left(\tilde{\mathbf{A}}^\dagger - \mathbf{A}^\dagger\right)^H$$
$$= \begin{bmatrix} 1 & 0 \\ 0 & 1+\alpha^2 \end{bmatrix} \begin{bmatrix} \alpha & -1 & -1 \\ -\alpha & 1 & 1 \end{bmatrix} \begin{bmatrix} \alpha & -\alpha \\ -1 & 1 \\ -1 & 1 \end{bmatrix} \begin{bmatrix} 1 & 0 \\ 0 & 1+\alpha^2 \end{bmatrix} \frac{1}{\alpha^2(2+\alpha^2)^2}$$
$$= \begin{bmatrix} 1 & -1-\alpha^2 \\ -1-\alpha^2 & (1+\alpha^2)^2 \end{bmatrix} \frac{1}{\alpha^2(2+\alpha^2)} ,$$

and

$$\left\|\tilde{\mathbf{A}}^\dagger - \mathbf{A}^\dagger\right\|_2^2 = \frac{1}{2\alpha^2(2+\alpha^2)}\left\{1 + (1+\alpha^2)^2 + \sqrt{4(1+\alpha^2)^2 + \alpha^4(2+\alpha^2)^2}\right\}$$
$$\approx \frac{1}{\alpha^2} + \frac{1}{2}$$

and

$$\frac{\left\|\tilde{\mathbf{A}}^\dagger - \mathbf{A}^\dagger\right\|_2}{\|\mathbf{A}^\dagger\|} \approx \frac{\sqrt{1/\alpha^2 + 1/2}}{\sqrt{2/\alpha^2 + 1/2}} \approx \frac{1}{\sqrt{2}} + \frac{\alpha^2}{8\sqrt{2}} .$$

In order to apply Theorem 6.5.2, we compute

$$\mathbf{A}\mathbf{A}^\dagger\left(\tilde{\mathbf{A}} - \mathbf{A}\right)\mathbf{A}^\dagger\mathbf{A} = \mathbf{0} ,$$
$$\left(\mathbf{I} - \mathbf{A}\mathbf{A}^\dagger\right)\left(\tilde{\mathbf{A}} - \mathbf{A}\right)\mathbf{A}^\dagger\mathbf{A} = \begin{bmatrix} 0 & 0 \\ 0 & 0 \\ 0 & \alpha \end{bmatrix} ,$$
$$\mathbf{A}\mathbf{A}^\dagger\left(\tilde{\mathbf{A}} - \mathbf{A}\right)\left(\mathbf{I} - \mathbf{A}^\dagger\mathbf{A}\right) = \mathbf{0} \text{ and}$$
$$\left(\mathbf{I} - \mathbf{A}\mathbf{A}^\dagger\right)\left(\tilde{\mathbf{A}} - \mathbf{A}\right)\left(\mathbf{I} - \mathbf{A}^\dagger\mathbf{A}\right) = \mathbf{0} .$$

These computations imply that

$$\gamma = 1 ,$$
$$\varepsilon 22 = \varepsilon 12 = \varepsilon_{11} = 0 \text{ and}$$
$$\varepsilon_{21} = \left\|\left(\mathbf{I} - \mathbf{A}\mathbf{A}^\dagger\right)\left(\tilde{\mathbf{A}} - \mathbf{A}\right)\mathbf{A}^\dagger\mathbf{A}\right\|_2 \|\mathbf{A}^\dagger\|_2 = \alpha\sqrt{\frac{2}{\alpha^2} + \frac{1}{2}} \approx \sqrt{2} + \frac{\sqrt{2}}{8}\alpha^2 .$$

Thus Theorem 6.5.2 gives us the bound

$$\frac{\left\|\tilde{\mathbf{A}}^\dagger - \mathbf{A}^\dagger\right\|_2}{\|\mathbf{A}^\dagger\|} \leq \frac{1}{\gamma}\frac{\varepsilon_{21}}{\sqrt{1+\varepsilon_{21}^2}} \approx \sqrt{2/3} .$$

It seems more natural to define

$$\mathbf{A} = \begin{bmatrix} 1 & 1 \\ 0 & 0 \\ 0 & 0 \end{bmatrix}$$

and note that

$$\mathbf{A}^\dagger = \begin{bmatrix} 1/2 & 0 & 0 \\ 1/2 & 0 & 0 \end{bmatrix} .$$

Hence,

$$\|\mathbf{A}^\dagger\|_2^2 = \frac{1}{2} .$$

In this case, Theorem 6.5.1 is not applicable, since the rank of $\mathbf{A}$ is different from the rank of $\tilde{\mathbf{A}}$. Since $\text{rank}\,(\mathbf{A}) \neq \text{rank}\left(\tilde{\mathbf{A}}\right)$, Theorem 6.5.2 merely provides the bound

$$\left\|\tilde{\mathbf{A}}^\dagger\right\|_2 \geq \frac{1}{\left\|\tilde{\mathbf{A}} - \mathbf{A}\right\|_2} = \frac{1}{\alpha} .$$

However, for this problem we actually have

$$\tilde{\mathbf{A}}^\dagger - \mathbf{A}^\dagger = \begin{bmatrix} -\alpha^3 & 2(1+\alpha^2) & -2 \\ -\alpha^3 & -2 & 2(1+\alpha^2) \end{bmatrix} \frac{1}{2\alpha(2+\alpha^2)}$$

so

$$\frac{\left\|\tilde{\mathbf{A}}^\dagger - \mathbf{A}^\dagger\right\|_2}{\|\mathbf{A}^\dagger\|_2} \approx \frac{\sqrt{1/\alpha^2}}{\sqrt{1/2}} = \frac{\sqrt{2}}{\alpha} .$$

This relative error becomes infinite as α approaches zero.

Exercise 6.5.1 Find an upper bound on the condition number of

$$\mathbf{A} = \begin{bmatrix} 1 & 1+\alpha \\ 1 & 1-\alpha \\ 1 & 1 \end{bmatrix}$$

as a function of α.

Exercise 6.5.2 Consider the heights and weights given in Table 6.1.

1. Find the coefficients ξ_0 and ξ_1 that minimize the residual sum of squares

$$\varrho^2 = \sum_{i=1}^{8} (w_i - \xi_0 - h_i \xi_1)^2 .$$

This gives us the best fit of weight as a linear function of height.

2. Compute $\|\mathbf{A}\|_2$, $\|\mathbf{A}^\dagger\|_2$ and the condition number $\kappa_2(\mathbf{A})$.
3. Find the coefficients that minimize the residual sum of squares for Table 6.2.
4. What is the matrix $\tilde{\mathbf{A}}$ and right-hand side $\mathbf{b}$ for this perturbed model?
5. Compute $\|\mathscr{P}_{\mathscr{R}(\mathbf{A})}(b - \tilde{\mathbf{b}})\|_2$.
6. Compute $\varepsilon_{\mathscr{R}(\mathbf{A})}$ and $\varepsilon_{\mathscr{N}(\mathbf{A}^H)}$ as in Theorem 6.5.1. Also verify that $\kappa_2(\mathbf{A})\varepsilon_{\mathscr{R}(\mathbf{A})} < 1$.
7. Use Lemma 6.5.1 to estimate the portion of the change in the least square coefficients that is due to the change in the right-hand side.
8. Use Theorem 6.5.1 to estimate the portion of the change in the least square coefficients that is due to the change in the coefficient matrix.

6.6 *A Posteriori* Error Estimates

Section 6.5 developed *a priori* estimates for perturbations to least squares problems, meaning that those estimates did not require the solution of the least squares problem for their application. In this section, we will develop *a posteriori* estimates, that can be used to estimate errors in approximate solutions to least squares problems. As the reader should expect, it is more difficult to develop *a posteriori* error estimates for least squares problems than for linear equations, as in Sect. 3.6.4. We will limit our discussion of *a posteriori* error estimates to full-rank least squares problems.

For the full-rank overdetermined least squares problem, we begin with the linear system in Eq. (6.1), and then apply the *a posteriori* error estimate (3.30). This error estimate requires an estimate for the condition number of the matrix in Eq. (6.1). In other words, since the true residual $\mathbf{r}$ and true solution $\mathbf{x}$ satisfy

$$\begin{bmatrix} \mathbf{I} & \mathbf{A} \\ \mathbf{A}^H & \mathbf{0} \end{bmatrix} \begin{bmatrix} \mathbf{r} \\ \mathbf{x} \end{bmatrix} = \begin{bmatrix} \mathbf{b} \\ \mathbf{0} \end{bmatrix} ,$$

it is easy to see that

$$\begin{bmatrix} \mathbf{I} & \mathbf{A} \\ \mathbf{A}^H & \mathbf{0} \end{bmatrix} \begin{bmatrix} \tilde{\mathbf{r}} - \mathbf{r} \\ \tilde{\mathbf{x}} - \mathbf{x} \end{bmatrix} = \begin{bmatrix} \tilde{\mathbf{r}} - (\mathbf{b} - \mathbf{A}\tilde{\mathbf{x}}) \\ \mathbf{A}^H \tilde{\mathbf{r}} \end{bmatrix} .$$

The *a posteriori* error estimate (3.30) then implies that

$$\frac{\left\| \begin{bmatrix} \tilde{\mathbf{r}} - \mathbf{r} \\ \tilde{\mathbf{x}} - \mathbf{x} \end{bmatrix} \right\|_2}{\left\| \begin{bmatrix} \mathbf{r} \\ \mathbf{x} \end{bmatrix} \right\|_2} \le \kappa_2\left(\begin{bmatrix} \mathbf{I} & \mathbf{A} \\ \mathbf{A}^H & \mathbf{0} \end{bmatrix} \right) \frac{\left\| \begin{bmatrix} \tilde{\mathbf{r}} - (\mathbf{b} - \mathbf{A}\tilde{\mathbf{x}}) \\ \mathbf{A}^H \tilde{\mathbf{r}} \end{bmatrix} \right\|_2}{\|\mathbf{b}\|_2} .$$

Similarly, for the full-rank underdetermined least squares problem we begin with the linear system in Eq. (6.2) and then apply the *a posteriori* error estimate (3.30). In other words, since the true solution $\mathbf{x}$ and the auxiliary vector $\mathbf{s}$ satisfy

$$\begin{bmatrix} \mathbf{I} & \mathbf{A}^H \\ \mathbf{A} & \mathbf{0} \end{bmatrix} \begin{bmatrix} \mathbf{x} \\ \mathbf{s} \end{bmatrix} = \begin{bmatrix} \mathbf{0} \\ \mathbf{b} \end{bmatrix} ,$$

it is easy to see that

$$\begin{bmatrix} \mathbf{I} & \mathbf{A}^H \\ \mathbf{A} & \mathbf{0} \end{bmatrix} \begin{bmatrix} \tilde{\mathbf{x}} - \mathbf{x} \\ -\tilde{\mathbf{s}} + \mathbf{s} \end{bmatrix} = \begin{bmatrix} \tilde{\mathbf{x}} - \mathbf{A}^H \tilde{\mathbf{s}} \\ \mathbf{A}\tilde{\mathbf{x}} - \mathbf{b} \end{bmatrix} .$$

The *a posteriori* error estimate (3.30) then implies that

$$\frac{\left\| \begin{bmatrix} \tilde{\mathbf{x}} - \mathbf{x} \\ -\tilde{\mathbf{s}} + \mathbf{s} \end{bmatrix} \right\|_2}{\left\| \begin{bmatrix} \mathbf{x} \\ -\mathbf{s} \end{bmatrix} \right\|_2} \leq \kappa_2 \left(\begin{bmatrix} \mathbf{I} & \mathbf{A}^H \\ \mathbf{A} & \mathbf{0} \end{bmatrix} \right) \frac{\left\| \begin{bmatrix} \tilde{\mathbf{x}} - \mathbf{A}^H \tilde{\mathbf{s}} \\ \mathbf{A}\tilde{\mathbf{x}} - \mathbf{b} \end{bmatrix} \right\|_2}{\|\mathbf{b}\|_2} .$$

The difficulty with both error estimates arises in estimating the condition number. Lemma 1.5.5 of Chap. 1 in Volume II will show that $\|\mathbf{A}\|_2 = \sigma_{\max}$ and $\|\mathbf{A}^\dagger\|_2 = \sigma_{\min}$, where $\sigma_{\max}$ and $\sigma_{\min}$ are the maximum and minimum singular values of $\mathbf{A}$. We will introduce singular values later in this chapter, in Sect. 6.11, but a careful discussion will be impossible until Sect. 1.5 of Chap. 1 in Volume II in the chapter on eigenvalues. Thus a detailed discussion of either condition number is inappropriate for this section. For those readers who are familiar with eigenvalues, we will sketch the ideas.

Without loss of generality, we will consider the matrix

$$\mathbf{M} = \begin{bmatrix} \mathbf{I} & \mathbf{A} \\ \mathbf{A}^H & \mathbf{0} \end{bmatrix}$$

We will show that the eigenvectors and corresponding eigenvalues of $\mathbf{M}$ are of two types. Let the columns of the $m \times (m - n)$ matrix $\mathbf{Z}$ be an orthonormal basis for $\mathscr{N}\left(\mathbf{A}^H\right)$, and let the columns of the $n \times n$ matrix $\mathbf{X}$ be an orthonormal set of eigenvectors for $\mathbf{A}^H\mathbf{A}$. In other words,

$$\mathbf{A}^H\mathbf{Z} = \mathbf{0} \text{ and } \mathbf{A}^H\mathbf{A}\mathbf{X} = \mathbf{X}\boldsymbol{\Sigma}^2 ,$$

where $\boldsymbol{\Sigma}^2$ is the matrix of eigenvalues of $\mathbf{A}^H\mathbf{A}$. The proof of Theorem 1.5.1 of Chap. 1 in Volume II shows that the entries of $\boldsymbol{\Sigma}$ are the singular values of $\mathbf{A}$. Next, note that the two solutions of the quadratic equation

$$\lambda^2 - \lambda = \sigma^2$$

are

$$\lambda_+ = \frac{1 + \sqrt{1 + 4\sigma^2}}{2} \text{ or } \lambda_- = -\frac{2\sigma^2}{1 + \sqrt{1 + 4\sigma^2}} ,$$

The coefficients in the quadratic equation imply that

$$\lambda_+ + \lambda_- = 1 \text{ and } \lambda_+\lambda_- = -\sigma^2 .$$

It follows that

$$(\lambda_+ - 1)(\lambda_- - 1) = -\sigma^2 .$$

We will let $\boldsymbol{\Lambda}_+$ and $\boldsymbol{\Lambda}_-$ denote the two diagonal matrices of scalars λ for each of the diagonal entries σ_i^2 of $\boldsymbol{\Sigma}^2$. Then it is easy to see that

$$(\boldsymbol{\Lambda}_+ - \mathbf{I})(\boldsymbol{\Lambda}_- - \mathbf{I}) = -\boldsymbol{\Sigma}^2 .$$

Furthermore, we can express the eigenvectors and eigenvalues of the matrix in Eq. (6.1) as follows:

$$\begin{bmatrix} \mathbf{I} & \mathbf{A} \\ \mathbf{A}^H & \mathbf{0} \end{bmatrix} \begin{bmatrix} \mathbf{Z} & \mathbf{AX} & \mathbf{AX} \\ \mathbf{0} & \mathbf{X}(\boldsymbol{\Lambda}_+ - \mathbf{I}) & \mathbf{X}(\boldsymbol{\Lambda}_- - \mathbf{I}) \end{bmatrix} = \begin{bmatrix} \mathbf{Z} & \mathbf{AX} & \mathbf{AX} \\ \mathbf{0} & \mathbf{X}(\boldsymbol{\Lambda}_+ - \mathbf{I}) & \mathbf{X}(\boldsymbol{\Lambda}_- - \mathbf{I}) \end{bmatrix} \begin{bmatrix} \mathbf{I} & & \\ & \boldsymbol{\Lambda}_+ & \\ & & \boldsymbol{\Lambda}_- \end{bmatrix} .$$

It is straightforward to see that the eigenvectors in this equation are mutually orthogonal, as the spectral Theorem 1.3.1 of Chap. 1 in Volume II guarantees for a symmetric matrix.

Since σ is real, it follows that

$$1 \le \frac{1 + \sqrt{1 + 4\sigma^2}}{2} .$$

Our solution of the quadratic equation $\lambda^2 - \lambda = \sigma^2$ and Lemma 1.3.3 of Chap. 1 in Volume II show that the largest eigenvalue of the matrix $\mathbf{M}$ is

$$\left\| \begin{bmatrix} \mathbf{I} & \mathbf{A} \\ \mathbf{A}^H & \mathbf{0} \end{bmatrix} \right\|_2 = \frac{1}{2} \left[1 + \sqrt{1 + 4\|\mathbf{A}\|_2^2} \right] .$$

Because

$$\frac{2}{1 + \sqrt{1 + 4\sigma^2}} < \frac{1 + \sqrt{1 + 4\sigma^2}}{2\sigma^2}$$

and

$$\frac{\mathrm{d}}{\mathrm{d}\sigma^2}\left\{\frac{1+\sqrt{1+4\sigma^2}}{2\sigma^2}\right\} < 0 \,,$$

it follows that the largest eigenvalue of the inverse matrix satisfies

$$\left\|\begin{bmatrix}\mathbf{I} & \mathbf{A}\\ \mathbf{A}^H & \mathbf{0}\end{bmatrix}^{-1}\right\|_2 = \max\left\{1\,,\ \frac{1}{2}\left\|\mathbf{A}^\dagger\right\|_2\left[\left\|\mathbf{A}^\dagger\right\|_2 + \sqrt{4+\|\mathbf{A}^\dagger\|_2^2}\right]\right\} \,.$$

The discussion in Sect. 6.7.3 will show how the norms of $\mathbf{A}$ and $\mathbf{A}^\dagger$ can be estimated from an orthogonal matrix factorization.

6.7 Householder QR Factorization

Now that we have formulated the least squares problem and discussed its well-posedness, we are ready to develop and analyze numerical methods. We will begin with the numerical method favored by LAPACK and MATLAB. This approach uses successive reflection to transform a given set of matrix columns into a right-triangular or right-trapezoidal array. This method is numerically stable, because reflections preserve the 2-norms of the vectors on which they operate. This method also makes efficient use of array storage, by saving the information about the reflections in the original array below the diagonal. This approach does not compute the residual vector, but it does provide an efficient mechanism for computing the norm of the residual.

6.7.1 *Successive Reflection*

Although reflection is not commonly presented in elementary linear algebra texts, it is easy to describe. We will develop a matrix to represent a particular kind of reflection, and then use that matrix to perform our new matrix factorization.

6.7.1.1 Elementary Reflectors

Let us begin by defining a reflector, which depends on the Definition 3.2.19 of a direct sum.

Definition 6.7.1 Suppose that $\mathscr{U}$ and $\mathscr{W}$ are subspaces of the set of all m-vectors, and that every m-vector $\mathbf{a}$ can be written $\mathbf{a} = \mathbf{u} + \mathbf{w}$ where $\mathbf{u} \in \mathscr{U}$ and $\mathbf{w} \in \mathscr{W}$.

Then an $m \times m$ matrix $\mathbf{R}$ is a **reflector** of $\mathscr{U}$ with respect to $\mathscr{W}$ if and only if

$$\mathbf{R}\mathbf{u} = -\mathbf{u} \text{ for all } \mathbf{u} \in \mathscr{U} \text{ and } \mathbf{R}\mathbf{w} = \mathbf{w} \text{ for all } \mathbf{w} \in \mathscr{W} .$$

An elementary reflector chosen to reflect some given vector onto a coordinate axis is called a **Householder reflector**. We will discuss two choices in generating a Householder reflector. In the first alternative, we will choose the elementary reflector to be *Hermitian*, and in the second alternative we will choose the elementary reflector to map to a *real* multiple of the first axis vector. These two alternatives differ only when we operate on complex vectors.

The next two lemmas provide our description of the Hermitian reflector.

Lemma 6.7.1 *If $\mathscr{U} \oplus \mathscr{W}$ is the set of all m-vectors and $\mathscr{U} \perp \mathscr{W}$, then any Hermitian reflector of $\mathscr{U}$ with respect to $\mathscr{W}$ is a unitary matrix.*

Proof Let $\mathbf{R}$ be an Hermitian reflector of $\mathscr{U}$ with respect to $\mathscr{W}$. For any m-vector $\mathbf{a}$, there exist $\mathbf{u} \in \mathscr{U}$ and $\mathbf{w} \in \mathscr{W}$ so that $\mathbf{a} = \mathbf{u} + \mathbf{w}$. Then

$$\left(\mathbf{R}^H\mathbf{R}\right)\mathbf{a} = \mathbf{R}^H\mathbf{R}(\mathbf{u}+\mathbf{w}) = \mathbf{R}^H(-\mathbf{u}+\mathbf{w}) = \mathbf{R}(-\mathbf{u}+\mathbf{w}) = \mathbf{u}+\mathbf{w} = \mathbf{a} .$$

This implies that $\mathbf{R}^H\mathbf{R} = \mathbf{I}$, so $\mathbf{R}$ is a unitary matrix.

Lemma 6.7.2 *If $\mathbf{u}$ is a nonzero m-vector, let $\mathscr{U}$ be the span of $\mathbf{u}$ (i.e., the set of all scalar multiples of $\mathbf{u}$), and let $\mathscr{W} = \mathscr{U}^\perp$ (i.e., the orthogonal complement of $\mathscr{U}$, which is the set of all vectors orthogonal to $\mathbf{u}$). Then*

$$\mathbf{R} = \mathbf{I} - \mathbf{u}\frac{2}{\mathbf{u}\cdot\mathbf{u}}\mathbf{u}^H \tag{6.8}$$

is an $m \times m$ Hermitian reflector of $\mathscr{U}$ with respect to $\mathscr{W}$, and $\mathbf{R}$ is unitary.

Proof If $\mathbf{w} \perp \mathbf{u}$,then

$$\mathbf{R}\mathbf{w} = \mathbf{w} - \mathbf{u}\frac{2}{\mathbf{u}\cdot\mathbf{u}}\mathbf{u}\cdot\mathbf{w} = \mathbf{w} .$$

Also note that

$$\mathbf{R}\mathbf{u} = \mathbf{u} - \mathbf{u}\frac{2}{\mathbf{u}\cdot\mathbf{u}}\mathbf{u}\cdot\mathbf{u} = \mathbf{u} - \mathbf{u}2 = -\mathbf{u} .$$

Definition 6.7.1 now shows that $\mathbf{R}$ is a reflector of $\mathscr{U}$ with respect to $\mathscr{W}$. Since the form of $\mathbf{R}$ shows that it is Hermitian, Lemma 6.7.1 shows that it is a unitary matrix. Note that the matrix $\mathbf{R}$ defined by (6.8) is independent of the scaling of $\mathbf{u}$. It is commonly called an **elementary reflector**.

Given a partitioned m-vector $[\alpha, \mathbf{a}^H]$, we can generate a particular Hermitian elementary reflector $\mathbf{H} = \mathbf{I} - \mathbf{u}\tau\mathbf{u}^H$ by using the following algorithm:

Algorithm 6.7.1 (Generate Hermitian Reflector)

$$\sigma = \begin{cases} \alpha/|\alpha|, \ \alpha \neq 0 \\ \quad\ 1, \ \alpha = 0 \end{cases}$$

$$\beta = -\sigma\sqrt{|\alpha|^2 + \|\mathbf{a}\|_2^2}$$

if $|\beta| > 0$ then

$$\tau = (\beta - \alpha)/\beta$$

$$\mathbf{u} = \begin{bmatrix} 1 \\ \mathbf{a}/(\alpha - \beta) \end{bmatrix}$$

else

$$\tau = 0$$

$$\mathbf{u} = \begin{bmatrix} 1 \\ \mathbf{0} \end{bmatrix} .$$

This particular elementary reflector has special properties that are described in the following lemma.

Lemma 6.7.3 *Let $m > 1$ be an integer, α be a scalar, and $\mathbf{a}$ be an $(m-1)$-vector. Suppose that we compute σ, β, τ and $\mathbf{u}$ by Algorithm 6.7.1, and we define the $m \times m$ matrix*

$$\mathbf{H} = \mathbf{I} - \mathbf{u}\tau\mathbf{u}^H .$$

Then τ is real, $\mathbf{H}$ is a unitary matrix, and

$$\mathbf{H}\begin{bmatrix} \alpha \\ \mathbf{a} \end{bmatrix} = \begin{bmatrix} \beta \\ \mathbf{0} \end{bmatrix} .$$

Furthermore, if $|\beta| > 0$ then $\mathbf{H}$ is an elementary reflector; otherwise $\mathbf{H} = \mathbf{I}$.

Proof First, we will handle the trivial case when $\alpha = 0$ and $\mathbf{a} = 0$. Then Algorithm 6.7.1 computes $\sigma = 1$, $\beta = 0 = \tau$ and $\mathbf{u} = \mathbf{e}_1$. These values imply that $\mathbf{H} = \mathbf{I} - \mathbf{u}\tau\mathbf{u} = \mathbf{I}$, and that

$$\mathbf{H}\begin{bmatrix} \alpha \\ \mathbf{a} \end{bmatrix} = \begin{bmatrix} \alpha \\ \mathbf{a} \end{bmatrix} = \begin{bmatrix} 0 \\ \mathbf{0} \end{bmatrix} = \begin{bmatrix} \beta \\ \mathbf{0} \end{bmatrix} .$$

Thus the claims in the lemma are all true in this trivial case.

Otherwise, Algorithm 6.7.1 computes $|\beta| > 0$. Note that

$$|\alpha - \beta| = \left| \frac{\alpha}{|\alpha|} \left\{ |\alpha| + \sqrt{|\alpha|^2 + \|\mathbf{a}\|_2^2} \right\} \right| = |\alpha| + \sqrt{|\alpha|^2 + \|\mathbf{a}\|_2^2} = |\alpha| + |\beta| \;,$$

and

$$\tau = \frac{\beta - \alpha}{\beta} = \frac{\sigma \left\{ -\sqrt{|\alpha|^2 + \|\mathbf{a}\|_2^2} - |\alpha| \right\}}{-\sigma \sqrt{|\alpha|^2 + \|\mathbf{a}\|_2^2}} = \frac{|\alpha| + \sqrt{|\alpha|^2 + \|\mathbf{a}\|_2^2}}{\sqrt{|\alpha|^2 + \|\mathbf{a}\|_2^2}} = \frac{|\alpha| + |\beta|}{|\beta|} \;.$$

Thus

$$\frac{2}{\mathbf{u} \cdot \mathbf{u}} = \frac{2}{1 + \frac{\|\mathbf{a}\|_2^2}{|\alpha - \beta|^2}} = \frac{2|\alpha - \beta|^2}{|\alpha - \beta|^2 + \|\mathbf{a}\|_2^2} = 2\frac{(|\alpha| + |\beta|)^2}{(|\alpha| + |\beta|)^2 + \|\mathbf{a}\|_2^2} = 2\frac{(|\alpha| + |\beta|)^2}{2|\beta|\,(|\alpha| + |\beta|)}$$

$$= \frac{|\alpha| + |\beta|}{|\beta|} = \tau \;.$$

Lemma 6.7.2 now shows that $\mathbf{H}$ is an elementary reflector.

Next, note that

$$-\frac{1}{\beta} \begin{bmatrix} 1 & \frac{\mathbf{a}^H}{\overline{\alpha} - \overline{\beta}} \end{bmatrix} \begin{bmatrix} \alpha \\ \mathbf{a} \end{bmatrix} = -\frac{1}{\beta} \left\{ \alpha + \frac{\|\mathbf{a}\|_2^2}{\overline{\alpha} - \overline{\beta}} \right\} = -\frac{1}{\beta} \frac{|\alpha|^2 + |\alpha|\,|\beta| + \|\mathbf{a}\|_2^2}{(|\alpha| + |\beta|)\overline{\alpha}/|\alpha|}$$

$$= -\frac{1}{\beta} \frac{|\beta|(|\alpha| + |\beta|)}{(|\alpha| + |\beta|)\overline{\alpha}/|\alpha|} = -\frac{|\beta|\,|\alpha|}{\beta\overline{\alpha}} = \frac{|\beta|\,|\alpha|}{|\beta|\,|\alpha|} = 1 \;.$$

Thus

$$\mathbf{H} \begin{bmatrix} \alpha \\ \mathbf{a} \end{bmatrix} = \begin{bmatrix} \alpha \\ \mathbf{a} \end{bmatrix} - \begin{bmatrix} 1 \\ \mathbf{a}/(\alpha - \beta) \end{bmatrix} \left\{ \tau \begin{bmatrix} 1 & \frac{\mathbf{a}^H}{\overline{\alpha} - \overline{\beta}} \end{bmatrix} \begin{bmatrix} \alpha \\ \mathbf{a} \end{bmatrix} \right\}$$

$$= \begin{bmatrix} \alpha \\ \mathbf{a} \end{bmatrix} - \begin{bmatrix} 1 \\ \mathbf{a}/(\alpha - \beta) \end{bmatrix} \{\alpha - \beta\} = \begin{bmatrix} \alpha \\ \mathbf{a} \end{bmatrix} - \begin{bmatrix} \alpha - \beta \\ \mathbf{a} \end{bmatrix} = \begin{bmatrix} \beta \\ \mathbf{0} \end{bmatrix} .$$

Given a scalar α and a vector $\mathbf{a}$, we also provide the following alternative method for generating an elementary reflector $\mathbf{H} = \mathbf{I} - \mathbf{u}\tau\mathbf{u}^H$.

Algorithm 6.7.2 (Generate Reflector to Real Axis Vector)

$$
\begin{aligned}
&\sigma = \operatorname{sign}(\Re e(\alpha)) \\
&\beta = -\sigma \sqrt{|\alpha|^2 + \|\mathbf{a}\|_2^2} \;, \\
&\text{if } |\beta| > 0 \\
&\quad \tau = (\beta - \alpha)/\beta \;, \\
&\quad \mathbf{u} = \begin{bmatrix} 1 \\ \mathbf{a}/(\alpha - \beta) \end{bmatrix} \\
&\text{else} \\
&\quad \tau = 0 \\
&\quad \mathbf{u} = \begin{bmatrix} 1 \\ \mathbf{0} \end{bmatrix} \;.
\end{aligned}
$$

The following lemma describes the properties of this reflector.

Lemma 6.7.4 *Let $m > 1$ be an integer, α be a scalar, and $\mathbf{a}$ be an $(m-1)$-vector. Suppose that we compute σ, β, τ and $\mathbf{u}$ by Algorithm 6.7.2. Define*

$$
\mathbf{H} = \mathbf{I} - \mathbf{u}\tau\mathbf{u}^H \;.
$$

Then β is real, $\mathbf{H}$ is a unitary matrix, and

$$
\mathbf{H}^H \begin{bmatrix} \alpha \\ \mathbf{a} \end{bmatrix} = \begin{bmatrix} \beta \\ \mathbf{0} \end{bmatrix} \;.
$$

Furthermore, if $|\beta| > 0$ then $\mathbf{H}$ is an elementary reflector; otherwise $\mathbf{H} = \mathbf{I}$.

Proof First, we will handle the trivial case when $\alpha = 0$ and $\mathbf{a} = \mathbf{0}$. Under these circumstances, we have $\sigma = 1$, $\beta = 0 = \tau$ and $\mathbf{u} = \mathbf{e}_1$. Then $\mathbf{H} = \mathbf{I} - \mathbf{u}\tau\mathbf{u} = \mathbf{I}$, and

$$
\mathbf{H} \begin{bmatrix} \alpha \\ \mathbf{a} \end{bmatrix} = \begin{bmatrix} \alpha \\ \mathbf{a} \end{bmatrix} = \begin{bmatrix} 0 \\ \mathbf{0} \end{bmatrix} = \begin{bmatrix} \beta \\ \mathbf{0} \end{bmatrix} \;.
$$

Thus the claims in the lemma are all true in this trivial case.

In the remainder of the proof, we will assume that $|\beta| > 0$. Then

$$
\begin{aligned}
\mathbf{H}^H\mathbf{H} &= \left(\mathbf{I} - \begin{bmatrix} 1 \\ \mathbf{u} \end{bmatrix} \overline{\tau} \begin{bmatrix} 1 & \mathbf{u}^H \end{bmatrix}\right) \left(\mathbf{I} - \begin{bmatrix} 1 \\ \mathbf{u} \end{bmatrix} \tau \begin{bmatrix} 1 & \mathbf{u}^H \end{bmatrix}\right) \\
&= \mathbf{I} - \begin{bmatrix} 1 \\ \mathbf{u} \end{bmatrix} \left\{\tau + \overline{\tau} - |\tau|^2 \left(1 + \mathbf{u}^H\mathbf{u}\right)\right\} \begin{bmatrix} 1 & \mathbf{u}^H \end{bmatrix} \;,
\end{aligned}
$$

where

$$\tau + \overline{\tau} - |\tau|^2 \left(1 + \mathbf{u}^H \mathbf{u}\right) = 2\frac{\beta - \Re e(\alpha)}{\beta} - \frac{|\beta - \alpha|^2}{\beta^2} \left\{1 + \frac{\|\mathbf{a}\|_2^2}{|\alpha - \beta|^2}\right\}$$

$$= 2 - 2\frac{\Re e(\alpha)}{\beta} - \frac{|\beta - \alpha|^2}{\beta^2} + \frac{\|\mathbf{a}\|_2^2}{\beta^2} = 2 - 2\frac{\Re e(\alpha)}{\beta} - 1 + 2\frac{\Re e(\alpha)\beta}{\beta^2} - \frac{|\alpha|^2}{\beta^2} - \frac{\|\mathbf{a}\|_2^2}{\beta^2}$$

$$= 1 - \frac{|\alpha|^2 + \|\mathbf{a}\|_2^2}{\beta^2} = 0 \,.$$

Thus $\mathbf{H}^H\mathbf{H} = \mathbf{I}$, so $\mathbf{H}$ is unitary.

Next, note that

$$\mathbf{H}^H \begin{bmatrix} \alpha \\ \mathbf{a} \end{bmatrix} = \left(\mathbf{I} - \begin{bmatrix} 1 \\ \mathbf{u} \end{bmatrix} \overline{\tau} \left[1 \ \mathbf{u}^H\right]\right) \begin{bmatrix} \alpha \\ \mathbf{a} \end{bmatrix} = \begin{bmatrix} \alpha \\ \mathbf{a} \end{bmatrix} - \begin{bmatrix} 1 \\ \mathbf{u} \end{bmatrix} \frac{\beta - \overline{\alpha}}{\beta} \left(\alpha + \mathbf{u}^H \mathbf{a}\right)$$

$$= \begin{bmatrix} \alpha \\ \mathbf{a} \end{bmatrix} - \begin{bmatrix} 1 \\ \mathbf{u} \end{bmatrix} \frac{\beta - \overline{\alpha}}{\beta} \left(\alpha + \frac{\|\mathbf{a}\|_2^2}{\overline{\alpha} - \beta}\right) = \begin{bmatrix} \alpha \\ \mathbf{a} \end{bmatrix} - \begin{bmatrix} 1 \\ \mathbf{a}/(\alpha - \beta) \end{bmatrix} \frac{1}{\beta} \left(\beta\alpha - |\alpha|^2 - \|\mathbf{a}\|_2^2\right)$$

$$= \begin{bmatrix} (|\alpha|^2 + \|\mathbf{a}\|_2^2)/\beta \\ \mathbf{a}\left(1 - \frac{\beta\alpha - |\alpha|^2 - \|\mathbf{a}\|_2^2}{\beta\alpha - |\beta|^2}\right) \end{bmatrix} = \begin{bmatrix} \beta \\ \mathbf{0} \end{bmatrix} .$$

The computations contained in Algorithm 6.7.2 closely reproduce the computations in LAPACK routines `_larfg`. See, for example, zlarfg.f. This algorithm can be used to transform a Hermitian matrix to real tridiagonal form; see Sect. 1.3.11 of Chap. 1 in Volume II for more information. On the other hand, Algorithm 6.7.1 corresponds to the computations in texts such as Golub and van Loan [46, p. 236], Higham [56, p. 357] or Stewart [97, p. 233].

Finally, we remark that either Algorithm 6.7.1 or 6.7.2 costs

- $2m - 2$ multiplications,
- m additions or subtractions,
- 3 divisions,
- 1 absolute value or complex modulus, and
- 1 square root.

6.7.1.2 Matrix Multiplication

While solving least squares problems, we will also need to multiply an elementary reflector times a matrix or vector. If $\mathbf{H} = \mathbf{I} - \mathbf{u}\tau\mathbf{u}^H$ is an $m \times m$ elementary reflector and $\mathbf{A}$ is an $m \times n$ matrix, then

$$\mathbf{B} \equiv \mathbf{H}\mathbf{A} = \left(\mathbf{I} - \mathbf{u}\tau\mathbf{u}^H\right)\mathbf{A} = \mathbf{A} - \mathbf{u}\tau\left(\mathbf{A}^H\mathbf{u}\right)^H .$$

This can be computed by the following algorithm.

Algorithm 6.7.3 (Reflector-Matrix Product)

$\mathbf{w} = \mathbf{A}^H\mathbf{u}$ /* LAPACK BLAS routine _gemv */
$\mathbf{B} = \mathbf{A} - \mathbf{u}\tau\mathbf{w}^H$ /* LAPACK BLAS routine _ger (real) or _gerc (complex) */

The call to `_gemv` costs mn multiplications and additions, while the call to `_ger` or _gerc costs $mn + \min\{m, n\}$ multiplications and mn additions. In LAPACK, multiplication by an elementary reflector is performed by routines `_larf`; see, for example dlarf.f.

6.7.1.3 Matrix Factorization

Next, let us show how Lemma 6.7.3 can be used to factor a $m \times n$ matrix $\mathbf{A}$ into a product $\mathbf{Q}\begin{bmatrix}\mathbf{R}\\ \mathbf{0}\end{bmatrix}$ of a unitary matrix $\mathbf{Q}$ and a right-trapezoidal matrix $\mathbf{R}$. In the first step, we partition

$$\mathbf{A} = \begin{bmatrix}\mathbf{a}_1 & \mathbf{A}_2\end{bmatrix}$$

where $\mathbf{a}_1$ is an m-vector, and find an elementary reflector $\mathbf{H}_1$ so that $\mathbf{H}_1{}^H\mathbf{a}_1 = \mathbf{e}_1\beta_1$, where β_1 is real. Then

$$\mathbf{H}_1{}^H\mathbf{A} = \begin{bmatrix}\mathbf{H}_1{}^H\mathbf{a}_1 & \mathbf{H}_1{}^H\mathbf{A}_2\end{bmatrix} = \begin{bmatrix}\mathbf{e}_1\beta_1 & \mathbf{H}_1{}^H\mathbf{A}_2\end{bmatrix}$$

has zeros below the diagonal in the first column.

After k steps, we will have factored

$$\mathbf{H}_k{}^H \dots \mathbf{H}_1{}^H\mathbf{A} = \begin{bmatrix}\mathbf{R}_{11} & \mathbf{r}_{12} & \mathbf{R}_{13}\\ \mathbf{0} & \mathbf{a}_{22} & \mathbf{A}_{23}\end{bmatrix} .$$

Here $\mathbf{R}_{11}$ is a $k \times k$ right triangular matrix with real diagonal entries, $\mathbf{r}_{12}$ is a k-vector and $\mathbf{H}_1, \dots, \mathbf{H}_k$ are unitary matrices which are also elementary reflectors. To perform the next step, we use $\mathbf{a}_{22}$ to find a $(m-k) \times (m-k)$ elementary reflector $\mathbf{H} = \mathbf{I} - \mathbf{u}_{22}\tau_2\mathbf{u}_{22}{}^H$ so that

$$\mathbf{H}^H\mathbf{a}_{22} = \mathbf{e}_1\beta_2 .$$

To simplify the notation without affecting the algorithm, we let

$$\mathbf{H}_{k+1} = \begin{bmatrix}I & \mathbf{0}\\ \mathbf{0} & \mathbf{H}\end{bmatrix} = \begin{bmatrix}I & \mathbf{0}\\ \mathbf{0} & \mathbf{I} - \mathbf{u}_{22}\tau_2\mathbf{u}_{22}{}^H\end{bmatrix} = \begin{bmatrix}\mathbf{I} & \\ & \mathbf{I}\end{bmatrix} - \begin{bmatrix}\mathbf{0}\\ \mathbf{u}_{22}\end{bmatrix}\tau_2\begin{bmatrix}\mathbf{0}^H & \mathbf{u}_{22}{}^H\end{bmatrix} .$$

Then $\mathbf{H}_{k+1}$ is a Householder reflector, and

$$\mathbf{H}_{k+1}{}^H \mathbf{H}_k{}^H \ldots \mathbf{H}_1{}^H \mathbf{A} = \begin{bmatrix} I & \mathbf{0} \\ \mathbf{0} & \mathbf{H} \end{bmatrix} \begin{bmatrix} \mathbf{R}_{11} & \mathbf{r}_{12} & \mathbf{R}_{13} \\ \mathbf{0} & \mathbf{a}_{22} & \mathbf{A}_{23} \end{bmatrix} = \begin{bmatrix} \mathbf{R}_{11} & \mathbf{r}_{12} & \mathbf{R}_{13} \\ \mathbf{0} & \mathbf{H}^H \mathbf{a}_{22} & \mathbf{H}^H \mathbf{A}_{23} \end{bmatrix}$$

$$= \begin{bmatrix} \mathbf{R}_{11} & \mathbf{r}_{12} & \mathbf{R}_{13} \\ \mathbf{0} & \mathbf{e}_1 \beta_2 & \mathbf{H}^H \mathbf{A}_{23} \end{bmatrix}$$

has zeros below the diagonal in column $k+1$. Also

$$\mathbf{H}_{k+1}{}^H \mathbf{H}_{k+1} = \begin{bmatrix} \mathbf{I} & \mathbf{0} \\ \mathbf{0} & \mathbf{H}^H \mathbf{H} \end{bmatrix} = \begin{bmatrix} \mathbf{I} & \mathbf{0} \\ \mathbf{0} & \mathbf{I} \end{bmatrix},$$

so $\mathbf{H}_{k+1}$ is unitary.

If $m \geq n$, then after n steps we will have factored

$$\mathbf{H}_n{}^H \ldots \mathbf{H}_1{}^H \mathbf{A} = \begin{bmatrix} \mathbf{R}_{11} \\ \mathbf{0} \end{bmatrix}$$

where $\mathbf{R}_{11}$ is an $n \times n$ right-triangular matrix with real diagonal entries. On the other hand, if $m < n$ then after m steps we have computed

$$\mathbf{H}_n{}^H \ldots \mathbf{H}_1{}^H \mathbf{A} = \begin{bmatrix} \mathbf{R}_{11} & \mathbf{R}_{12} \end{bmatrix}$$

where $\mathbf{R}_{11}$ is an $m \times m$ right-triangular matrix with real diagonal entries. Since each of the matrices $\mathbf{H}_k$ is unitary, so is the $m \times m$ matrix

$$\mathbf{Q}^H = \mathbf{H}_n{}^H \ldots \mathbf{H}_1{}^H .$$

In other words, $\mathbf{Q}^H \mathbf{A} = \mathbf{R}$, which implies that $\mathbf{A} = \mathbf{Q}\mathbf{R}$.

We can summarize the operations in this factorization by the following algorithm.

Algorithm 6.7.4 (Householder QR Factorization)

for $1 \leq k \leq \min\{m, n\}$
  compute β_k, τ_k and $\mathbf{u}_k$ from $\mathbf{A}_{k:m,k}$ /* Algorithm 6.7.2 i.e., LAPACK _larfg */
  $\mathbf{A}_{kk} = 1$ /* store $\mathbf{u}_k$ in $\mathbf{A}_{ik}$ for $k \leq i \leq m$ */
  $\mathbf{A}_{k:m,k+1:n} = \left(\mathbf{I} - \mathbf{u}_k \tau_k \mathbf{u}_k{}^H\right) \mathbf{A}_{k:m,k+1:n}$ /* Algorithm 6.7.3 i.e., LAPACK _larf */
  $\mathbf{A}_{kk} = \beta_k$

In LAPACK, these computations are performed by routines `_geqrf` and `_geqr2`; see, for example dgeqrf.f which calls dgeqr2.f. For a simplified version of the LAPACK Householder QR factorization, readers may view a simple Householder factorization in Fortran.

In the GSL, the Householder QR factorization is performed by routine gsl_linalg_QR_decomp. MATLAB users can obtain the Householder QR factorization of a matrix by the command

```
[Q,R]=qr(A)
```

This command will return the *product of the elementary reflectors* in the unitary matrix $\mathbf{Q}$, and the right-triangular factor in $\mathbf{R}$. The command

```
[Q,R]=qr(A,0)
```

will return only the first n columns of a $\mathbf{Q}$ when $m \geq n$. Please note that the MATLAB `qr` function *also employs column pivoting*, so it is closer to the complete orthogonal decomposition of Sect. 6.7.2.

Finally, readers may experiment with the JavaScript Householder QR **factorization program.** This program computes the Householder QR factorization of a matrix and displays the errors in the factorizations.

Let us determine the computational work in Algorithm 6.7.4. During the k-th step, we perform the following work:

_larfg : $2(m-k+1)$ multiplications, $m-k+1$ additions, 3 divides, 1 modulus and 1 square root, and
_larf : $(2m-2k+3)(n-k+1)$ multiplications and $2(m-k+1)(n-k+1)$ additions.

Thus the total work is

- $\sum_{k=1}^{\min\{m,n\}}\{2(m-k+1)(n-k+2)+(n-k+1)\}$ multiplications,
- $\sum_{k=1}^{\min\{m,n\}}(m-k+1)\{2(n-k+1)+1\}$ additions,
- $3\min\{m,n\}$ divides,
- $\min\{m,n\}$ moduli and
- $\min\{m,n\}$ square roots.

If $m \geq n$, the total work is essentially $(m-n/3)n^2$ multiplications and additions; if $m \leq n$ the total work is essentially $(n-m/3)m^2$ multiplications and additions.

Next, let us comment on the storage requirements for Algorithm 6.7.4. We can store the vectors $\mathbf{u}_k$ in the array $\mathbf{A}$ below the diagonal, because Algorithm 6.7.1 shows that the first component of $\mathbf{u}_k$ is always one. We can store the entries of $\mathbf{R}$ in the array $\mathbf{A}$ on and above the diagonal. We require additional storage for the values of τ_k, which can be held in one additional n-vector.

6.7.1.4 Block Factorization

In LAPACK, the Householder QR factorization is performed by routine `_geqrf`, which uses the following **block algorithm**. Suppose that we have factored

$$\mathbf{Q}_1{}^H\mathbf{A} = \begin{bmatrix} \mathbf{R}_{11} & \mathbf{R}_{12} & \mathbf{R}_{13} \\ \mathbf{0} & \mathbf{A}_{22} & \mathbf{A}_{23} \end{bmatrix} ,$$

where $\mathbf{R}_{11}$ is right-triangular and $\mathbf{Q}_1{}^H$ is a product of Householder reflectors. In this factorization, $\mathbf{A}_{22}$ is assumed to have b columns. The block algorithm begins by calling LAPACK routine `_geqr2` to find elementary reflectors $\mathbf{H}_1, \ldots, \mathbf{H}_b$ so that

$$\mathbf{H}_b{}^H \ldots \mathbf{H}_1{}^H \mathbf{A}_{22} = \mathbf{R}_{22} \; .$$

Next, it calls LAPACK routine `_larft` to find $\mathbf{U}_{22}$ and $\mathbf{T}_{22}$ so that $\mathbf{T}_{22}$ is right triangular and

$$\mathbf{H}_1 \ldots \mathbf{H}_b = \mathbf{I} - \mathbf{U}_{22}\mathbf{T}_{22}\mathbf{U}_{22}{}^H \; . \tag{6.9}$$

The right-hand side in this expression mimics the form $\mathbf{H} = \mathbf{I} - \mathbf{u}\tau\mathbf{u}^H$ of a Householder reflector; its computation is described in the paragraph that follows. Afterward, the block factorization calls LAPACK routine `_larfb` to compute

$$\mathbf{B}_{23} = \left(\mathbf{I} - \mathbf{U}_{22}\mathbf{T}_{22}\mathbf{U}_{22}{}^H\right)^H \mathbf{A}_{23} \; .$$

At this point, we have factored

$$\begin{bmatrix} \mathbf{I} & \\ & \mathbf{I} - \mathbf{U}_{22}\mathbf{T}_{22}\mathbf{U}_{22}{}^H \end{bmatrix} \begin{bmatrix} \mathbf{R}_{11} & \mathbf{R}_{12} & \mathbf{R}_{13} \\ \mathbf{0} & \mathbf{A}_{22} & \mathbf{A}_{23} \end{bmatrix} = \begin{bmatrix} \mathbf{R}_{11} & \mathbf{R}_{12} & \mathbf{R}_{13} \\ \mathbf{0} & \mathbf{R}_{22} & \mathbf{B}_{23} \end{bmatrix} \; .$$

By partitioning the first b rows and columns of $\mathbf{B}_{23}$, the process can continue in the same way. The process stops when all columns or rows have been processed.

Let us show how Eq. (6.9) is obtained. Since

$$\mathbf{H}_1 = \mathbf{I} - \mathbf{u}_1\tau_1\mathbf{u}_1{}^H$$

is a Householder reflector, we can begin with $\mathbf{U}_1 = [\mathbf{u}_1]$ and $\mathbf{T}_1 = [\tau_1]$. suppose that we have determined

$$\mathbf{H}_1 \cdot \ldots \cdot \mathbf{H}_k = \mathbf{I} - \mathbf{U}_k\mathbf{T}_k\mathbf{U}_k{}^H \; ,$$

where the columns of $\mathbf{U}_k$ are the vectors $\mathbf{u}_1, \ldots \mathbf{u}_k$ in the Householder reflectors $\mathbf{H}_1, \ldots, \mathbf{H}_k$, and $\mathbf{T}_k$ is right-triangular. Then

$$\begin{aligned}
(\mathbf{H}_1 \cdot \ldots \cdot \mathbf{H}_k)\,\mathbf{H}_{k+1} &= \left(\mathbf{I} - \mathbf{U}_k\mathbf{T}_k\mathbf{U}_k{}^H\right)\left(\mathbf{I} - \mathbf{u}_{k+1}\tau_{k+1}\mathbf{u}_{k+1}{}^H\right) \\
&= \mathbf{I} - \mathbf{U}_k\mathbf{T}_k\mathbf{U}_k{}^H - \mathbf{u}_{k+1}\tau_{k+1}\mathbf{u}_{k+1}{}^H + \mathbf{U}_k\mathbf{T}_k\mathbf{U}_k{}^H\mathbf{u}_{k+1}\tau_{k+1}\mathbf{u}_{k+1}{}^H \\
&= \mathbf{I} - \left[\mathbf{U}_k \; \mathbf{u}_{k+1}\right] \begin{bmatrix} \mathbf{T}_k\mathbf{U}_k{}^H - \mathbf{T}_k\mathbf{U}_k{}^H\mathbf{u}_{k+1}\tau_{k+1}\mathbf{u}_{k+1}{}^H \\ \tau_{k+1}\mathbf{u}_{k+1}{}^H \end{bmatrix} \\
&= \mathbf{I} - \left[\mathbf{U}_k \; \mathbf{u}_{k+1}\right] \begin{bmatrix} \mathbf{T}_k & -\mathbf{T}_k\mathbf{U}_k{}^H\mathbf{u}_{k+1}\tau_{k+1} \\ \mathbf{0} & \tau_{k+1} \end{bmatrix} \begin{bmatrix} \mathbf{U}_k{}^H \\ \mathbf{u}_{k+1}{}^H \end{bmatrix} \; .
\end{aligned}$$

This process continues until all of the Householder reflectors have been incorporated.

These ideas can be assembled into the following

Algorithm 6.7.5 (Block Householder QR Factorization)

$i = 1$
while $i \le n$
  $\beta = \min\{b, n - i + 1\}$
  factor $\mathbf{A}_{i:m,i:i+\beta-1} = \mathbf{H}_1 \cdot \ldots \cdot \mathbf{H}_\beta R_{i:m,i:i+\beta-1}$ /* LAPACK routine _geqr2 */
  if $i + \beta \le n$
    $\mathbf{U}_1 = [u_1]$, $\mathbf{T}_1 = [\tau_1]$ /* where $\mathbf{H}_j = \mathbf{I} - \mathbf{u}_j \tau_j \mathbf{u}_j{}^H$ */
    for $1 < j \le \beta$
      $\mathbf{I} - \mathbf{U}_j \mathbf{T}_j \mathbf{U}_j{}^H = (\mathbf{I} - \mathbf{U}_{j-1} \mathbf{T}_{j-1} \mathbf{U}_{j-1}{}^H)\mathbf{H}_j$ /* LAPACK routine _larft */
    $\mathbf{A}_{i:m,i:i+\beta:n} = \mathbf{A}_{i:m,i:i+\beta:n} - \mathbf{U}_\beta \mathbf{T}_\beta{}^H \mathbf{U}_\beta{}^H \mathbf{A}_{i:m,i:i+\beta:n}$ /* LAPACK routine _larfb */
  $i = i + \beta$

The advantage of the block form of the algorithm is that it allows level 3 BLAS routines to perform most of the work in the factorization. For more details, see Bischof and van Loan [7].

Interested readers may view LAPACK routine dgeqrf.f to see an example of how to implement the Householder QR factorization in block form. Readers may also examine an implementation of a C++ `HouseholderQRFactorization` class in files HouseholderQRFactorization.H and HouseholderQRFactorization.C. This class calls data type-specific LAPACK routines to perform the operations; see, for example LaDouble.C for the implementation in real double precision.

Exercise 6.7.1 Let $\mathbf{x}$ and $\mathbf{y}$ be n-vectors, with $\|\mathbf{x}\|_2 = \|\mathbf{y}\|_2$. Show that there is an elementary reflector $\mathbf{R}$ so that $\mathbf{R}\mathbf{x} = \mathbf{y}$.

Exercise 6.7.2 Suppose that we can factor the $m \times n$ matrix $\mathbf{A}$ as

$$\mathbf{A} = \mathbf{Q} \begin{bmatrix} \mathbf{R} \\ \mathbf{0} \end{bmatrix}$$

where $\mathbf{Q}$ is unitary and $\mathbf{R}$ is right-triangular and nonsingular. Partition

$$\mathbf{Q} = \begin{bmatrix} \mathbf{Q}_1 & \mathbf{Q}_2 \end{bmatrix}$$

in such a way that

$$\mathbf{A} = \mathbf{Q}_1 \mathbf{R} .$$

Show that

$$\mathbf{A}^\dagger = \mathbf{R}^{-1} \mathbf{Q}_1{}^H .$$

Exercise 6.7.3 Let $\mathbf{A}$ be the 10×5 **Hilbert matrix**

$$\mathbf{A}_{ij} = \frac{1}{1+i+j} \text{ for } 0 \le i < 10 \text{ and } 0 \le j < 5 .$$

Compute the Householder QR factorization of $\mathbf{A}$.

6.7.2 *Pivoting*

In order to compute the Householder QR factorization, Algorithm 6.7.4 calls Algorithm 6.7.1 to determine a Householder reflector from the entries of the kth column on and below the diagonal. If all of these entries are zero, then Algorithm 6.7.1 returns the identity, rather than a Householder reflector. Successive reflection does not fail under such a circumstance, but it could produce a right-triangular matrix $\mathbf{R}$ with one or more zeros on the diagonal.

One way to avoid *unnecessary zeros* on the diagonal of $\mathbf{R}$ is to reorder the columns so that the next reflector is determined by the largest remaining column. Suppose that through a combination of column pivoting and successive reflection we have factored

$$Q^H \mathbf{A}\mathbf{P} = \begin{bmatrix} \mathbf{R}_{11} & \mathbf{R}_{12} \\ \mathbf{0} & \mathbf{A}_{22} \end{bmatrix} .$$

We can examine the subarray $\mathbf{A}_{22}$ to find the index j of the column of largest norm. Then we could interchange the kth and j columns to obtain the factorization

$$Q^H \mathbf{A}\mathbf{P}\mathbf{I}_{kj} = \begin{bmatrix} \mathbf{R}_{11} & \mathbf{R}_{12}\mathbf{I}_{kj} \\ \mathbf{0} & \mathbf{A}_{22}\mathbf{I}_{kj} \end{bmatrix} = \begin{bmatrix} \mathbf{R}_{11} & \mathbf{r}_{12} & \mathbf{R}_{13} \\ \mathbf{0} & \mathbf{a}_{22} & \mathbf{A}_{23} \end{bmatrix} .$$

Then we could use the largest column, $\mathbf{a}_{22}$, to determine an elementary reflector $\mathbf{H}$. Then next step of successive reflection would lead to

$$\begin{bmatrix} \mathbf{I} & \\ & \mathbf{H}^H \end{bmatrix} Q^H \mathbf{A}\mathbf{P}\mathbf{I}_{kj} = \begin{bmatrix} \mathbf{R}_{11} & \mathbf{r}_{12} & \mathbf{R}_{13} \\ & \mathbf{e}_1 \varrho_{22} & \mathbf{B}_{23} \end{bmatrix} .$$

This process could continue until there are no more rows or columns to process, or until $\mathbf{A}_{22} = \mathbf{0}$.

Householder QR factorization with pivoting can be summarized by the following algorithm.

Algorithm 6.7.6 (Householder QR Factorization With Pivoting)

for $1 \leq k \leq \min\{m, n\}$
  $\ell = k$ and $\mu = 0$
  for $k \leq j \leq n$
    $\nu_j = \|\mathbf{A}_{k:m,j}\|_2$ /* LAPACK BLAS routine _nrm2 */
    if $\nu_j > \mu$ then $\ell = j$ and $\mu = \nu_j$
  if $\ell > k$
    interchange columns $\mathbf{A}_{1:m,k} \leftrightarrow \mathbf{A}_{1:m,\ell}$
  compute β_k, τ_k and $\mathbf{u}_k$ from $\mathbf{A}_{k:m,k}$ /* Algorithm 6.7.1, i.e., LAPACK _larfg */
  $\mathbf{A}_{kk} = 1$
   /* store $\mathbf{u}_k$ in $\mathbf{A}_{ik}$ for $k \leq i \leq m$ */
  $\mathbf{A}_{k:m,k+1:n} = \left(\mathbf{I} - \mathbf{u}_k\tau_k\mathbf{u}_k{}^H\right)\mathbf{A}_{k:m,k+1:n}$ /* Algorithm 6.7.3, i.e., LAPACK _larf */
  $\mathbf{A}_{kk} = \beta_k$

This algorithm proves the following theorem.

Theorem 6.7.1 (Householder QR) *Suppose that* $\mathbf{A}$ *is a nonzero* $m \times n$ *matrix. Then there is an integer* r *satisfying* $1 \leq r \leq \min\{m, n\}$, *an* $n \times n$ *permutation matrix* $\mathbf{P}$, *an* $m \times m$ *unitary matrix* $\mathbf{Q}$ *and an* $r \times n$ *right-trapezoidal matrix* $\mathbf{R}$ *so that*

$$\mathbf{Q}^H\mathbf{A}\mathbf{P} = \begin{bmatrix} \mathbf{R} \\ \mathbf{0} \end{bmatrix} .$$

In LAPACK, successive reflection with pivoting is performed by routines `_geqp3`; see, for example, dgeqp3.f. Note that dynamically-determined pivoting of this sort prevents this routine from using the block program ideas in Eq. (6.9). In the GSL, the Householder QR factorization with column pivoting is performed by routines gsl_linalg_QRPT_decomp and gsl_linalg_QRPT_decomp2. In MATLAB, Householder QR factorization with pivoting can be obtained by

```
[Q,R,E]=qr(A)
```

Here E is the $n \times n$ permutation matrix used in the factorization.

Column pivoting in successive reflection was originally described by Golub [45]. For alternative pivoting techniques, see Golub and van Loan [46, pp. 276–280] or Higham [56, pp. 362–363].

6.7.3 Condition Number Estimates

Suppose that we have used elementary reflectors and column pivoting to factor

$$\mathbf{Q}^H\mathbf{A}\mathbf{P} = \begin{bmatrix} \mathbf{R} \\ \mathbf{0} \end{bmatrix} ,$$

where $\mathbf{Q}$ is an $m \times m$ unitary matrix and $\mathbf{R}$ is an $r \times n$ right-trapezoidal matrix. Then the first r columns of $\mathbf{Q}$ form an orthonormal basis for the range of $\mathbf{A}$. Also,

$$\|\mathbf{A}\|_2 = \sup_{\|\mathbf{x}\|_2=1} \|\mathbf{Q}\mathbf{R}\mathbf{x}\|_2 = \sup_{\|\mathbf{x}\|_2=1} \|\mathbf{R}\mathbf{x}\|_2 = \|\mathbf{R}\|_2 \ .$$

Furthermore, if $\mathbf{b} \in \mathscr{R}(\mathbf{A})$, Lemma 6.3.1 shows that there is an r-vector $\mathbf{y}$ so that

$$\mathbf{b} = \mathbf{Q}\begin{bmatrix}\mathbf{y}\\ \mathbf{0}\end{bmatrix} ,$$

and the minimum-norm solution of $\mathbf{A}\mathbf{x} = \mathbf{b}$ satisfies

$$\mathbf{P}^H\mathbf{x} = \mathbf{P}^H\mathbf{A}^H\mathbf{s} = \begin{bmatrix}\mathbf{R}^H & \mathbf{0}\end{bmatrix}\mathbf{Q}^H\mathbf{s} \equiv \mathbf{R}^H\mathbf{t} ,$$

for some m-vector $\mathbf{s}$ and r-vector $\mathbf{t} = \mathbf{Q}^H\mathbf{s}$. It is easy to check that

$$\mathbf{R}^\dagger = \mathbf{R}^H\left(\mathbf{R}\mathbf{R}^H\right)^{-1}$$

and that

$$\mathbf{A}^\dagger = \mathbf{P}^H\mathbf{R}^\dagger\begin{bmatrix}\mathbf{I} & \mathbf{0}\end{bmatrix}\mathbf{Q}^H .$$

It follows that

$$\left\|\mathbf{A}^\dagger\right\|_2 = \sup_{\|\mathbf{b}\|_2=1}\left\|\mathbf{A}^\dagger\mathbf{b}\right\|_2 = \sup_{\|\mathbf{b}\|_2=1}\left\|\mathbf{P}^H\mathbf{R}^\dagger\begin{bmatrix}\mathbf{I} & \mathbf{0}\end{bmatrix}\mathbf{Q}^H\mathbf{b}\right\|_2 = \sup_{\left\|\begin{bmatrix}\mathbf{y}\\ \mathbf{z}\end{bmatrix}\right\|_2=1}\left\|\mathbf{R}^\dagger\begin{bmatrix}\mathbf{I} & \mathbf{0}\end{bmatrix}\begin{bmatrix}\mathbf{y}\\ \mathbf{z}\end{bmatrix}\right\|_2$$

$$= \sup_{\|\mathbf{y}\|_2=1}\left\|\mathbf{R}^\dagger\mathbf{y}\right\|_2 = \left\|\mathbf{R}^\dagger\right\|_2 .$$

Thus the condition number of $\mathbf{A}$ is equal to the condition number of $\mathbf{R}$.

There are at least three alternative methods for estimating the condition number of $\mathbf{R}$. First, we note that Lemma 1.3.3 of Chap. 1 in Volume II will show that $\|\mathbf{R}\|_2$ is equal to the largest eigenvalue of $\mathbf{R}^H\mathbf{R}$. we may apply the **power method** of Sect. 1.4.6 of Chap. 1 in Volume II to find the largest eigenvalue of $\mathbf{R}^H\mathbf{R}$, and the inverse power method to find the largest eigenvalue of $\left(\mathbf{R}^H\mathbf{R}\right)^{-1}$. Either power method could begin with some random initial vector, and take just a few iterations to estimate the maximum eigenvalue.

An alternative is to estimate either the condition number of $\mathbf{R}$ with respect to either the 1-norm or the ∞-norm as in Sect. 3.8.7. Note that $\mathbf{R}^H\mathbf{R}$ is closely related to a Gaussian factorization (the diagonal entries of the left factor are not one in this case), and LAPACK routine `_gecon` will estimate the condition number of a matrix with known Gaussian factorization, with respect to either the 1-norm or the

∞-norm. Afterward, Lemma 3.5.7 can be used to bound the matrix 2-norm in terms of either the matrix 1-norm or the matrix ∞-norm.

A third alternative is to compute the singular value decomposition of $\mathbf{A}$. Then the condition number of $\mathbf{A}$ with respect to the 2-norm is the ratio of the largest to smallest singular values of $\mathbf{A}$.

Exercise 6.7.4 Let $\mathbf{A}$ be the $m \times n$ **Hilbert matrix**

$$\mathbf{A}_{ij} = \frac{1}{1+i+j} \text{ for } 0 \leq i < m \text{ and } 0 \leq j < n\,.$$

Estimate the condition number of $\mathbf{A}$ for $(m,n) = (10,5)$ and $(m,n) = (20,10)$.

6.7.4 Least Squares Problems

Our goal in this section is to show how to use Householder QR factorizations to solve least squares problems. We will consider three cases: full-rank overdetermined problems, full-rank underdetermined problems, and rank-deficient problems.

6.7.4.1 Full-Rank Overdetermined

We can use successive reflection to solve a full-rank overdetermined least squares problem by means of the following algorithm.

Algorithm 6.7.7 (Overdetermined Least Squares via Householder QR)

$$\text{factor } \mathbf{Q}^H\mathbf{A} = \begin{bmatrix} \mathbf{R} \\ \mathbf{0} \end{bmatrix} \qquad /* \text{ Algorithm 6.7.4 } */$$

$$\text{compute and partition } \mathbf{Q}^H\mathbf{b} = \begin{bmatrix} \mathbf{y} \\ \mathbf{z} \end{bmatrix} \qquad /* \text{ Algorithm 6.7.3 } */$$

$$\text{back-solve } \mathbf{R}\mathbf{x} = \mathbf{y} \text{ for } \mathbf{x} \qquad /* \text{ Algorithm 3.4.5 } */$$

The Householder QR factorization only needs to be done once for any given matrix $\mathbf{A}$, but the multiplication of the right-hand side by the Householder reflectors and the back-solve need to be done for each least-squares problem.

Let us justify the last two steps of Algorithm 6.7.7. The Pythagorean Theorem 3.2.2 shows us that

$$\|\mathbf{b}-\mathbf{A}\mathbf{x}\|_2^2 = \left\|\mathbf{Q}^H(\mathbf{b}-\mathbf{A}\mathbf{x})\right\|_2^2 = \left\|\begin{bmatrix} \mathbf{y} \\ \mathbf{z} \end{bmatrix} - \begin{bmatrix} \mathbf{R} \\ 0 \end{bmatrix}\mathbf{x}\right\|_2^2 = \|\mathbf{y}-\mathbf{R}\mathbf{x}\|_2^2 + \|\mathbf{z}\|_2^2\,.$$

This expression shows that the optimal least squares solution $\mathbf{x}$ is determined by solving $\mathbf{R}\mathbf{x} = \mathbf{y}$.

When we use successive reflection, we do not normally compute the residual $\mathbf{r} = \mathbf{b} - \mathbf{A}\mathbf{x}$. However, we can easily find the norm of the residual, because with this optimal choice of $\mathbf{x}$ we have $\|\mathbf{r}\|_2^2 = \|\mathbf{b} - \mathbf{A}\mathbf{x}\|_2^2 = \|\mathbf{z}\|_2^2$.

The combined algorithm, involving all three steps, can be performed by LAPACK routines `_gels`; see, for example, dgels.f. In the GSL, least squares problems can be solved by routine gsl_linalg_QR_lssolve. In MATLAB, an overdetermined least squares problem involving matrix $\mathbf{A}$ and right-hand side $\mathbf{b}$ can be solved by performing

```
x=A\b
```

Although this is the same command that performs Gaussian factorization for square matrices $\mathbf{A}$, this MATLAB command actually uses Householder factorization to solve the least squares problem when $\mathbf{A}$ is not square.

Finally, readers may experiment with the JavaScript successive reflection **overdetermined least squares program.** This program computes the Householder QR factorization of a matrix and displays the errors in the factorizations.

Example 6.7.1 Consider the least squares problem with

$$\mathbf{A} = \begin{bmatrix} 1 & 1 \\ \alpha & 0 \\ 0 & \alpha \end{bmatrix} \text{ and } \mathbf{b} = \begin{bmatrix} 1 \\ 0 \\ 0 \end{bmatrix} .$$

Here α is a small positive machine number. Note that

$$\mathbf{A}^\dagger = \left(\mathbf{A}^H\mathbf{A}\right)^{-1}\mathbf{A}^H = \begin{bmatrix} 1+\alpha^2 & -1 \\ -1 & 1+\alpha^2 \end{bmatrix} \frac{1}{\alpha^2(2+\alpha^2)} \begin{bmatrix} 1 & \alpha & 0 \\ 1 & 0 & \alpha \end{bmatrix}$$

$$= \begin{bmatrix} \alpha & 1+\alpha^2 & -1 \\ \alpha & -1 & 1+\alpha^2 \end{bmatrix} \frac{1}{\alpha(2+\alpha^2)} .$$

The exact solution of this least squares problem is

$$\mathbf{x} = \mathbf{A}^\dagger\mathbf{b} = \begin{bmatrix} \alpha & 1+\alpha^2 & -1 \\ \alpha & -1 & 1+\alpha^2 \end{bmatrix} \frac{1}{\alpha(2+\alpha^2)} \begin{bmatrix} 1 \\ 0 \\ 0 \end{bmatrix} = \begin{bmatrix} 1 \\ 1 \end{bmatrix} \frac{1}{2+\alpha^2} ,$$

and the residual is

$$\mathbf{r} = \mathbf{b} - \mathbf{A}\mathbf{x} = \begin{bmatrix} 1 \\ 0 \\ 0 \end{bmatrix} - \begin{bmatrix} 1 & 1 \\ \alpha & 0 \\ 0 & \alpha \end{bmatrix} \begin{bmatrix} 1 \\ 1 \end{bmatrix} \frac{1}{2+\alpha^2} = \begin{bmatrix} \alpha & -1 & -1 \end{bmatrix} \frac{\alpha}{2+\alpha^2} .$$

Thus

$$\|\mathbf{r}\|_2 = \frac{\alpha}{\sqrt{2+\alpha^2}} \,.$$

Suppose that $\mathrm{fl}(1+\alpha^2) = 1$, and that we use Householder QR factorization to solve this least squares problem in floating point arithmetic. In the first step of Householder QR factorization, we compute

$$\begin{aligned}
\beta_1 &= -\operatorname{sign}\left(\|\mathbf{A}e_1\|_2, \mathbf{A}_{11}\right) = -\mathrm{fl}\left(\sqrt{1+\alpha^2}\right) = -1 \\
\tau_1 &= (\beta_1 - \mathbf{A}_{11})/\beta_1 = 2 \\
\mathbf{u}_1 &= (\mathbf{A}e_1 - \mathbf{e}_1\beta_1)/(\mathbf{A}_{11} - \beta_1) = \begin{bmatrix} 1 \\ \alpha/2 \\ 0 \end{bmatrix} \\
\alpha_{12} &= \tau_1 \mathbf{u}_1 \cdot (\mathbf{A}\mathbf{e}_2) = 2 \\
(\mathbf{I} - \mathbf{u}_1 \tau_1 \mathbf{u}_1{}^H)\mathbf{A}\mathbf{e}_2 &= \mathbf{A}\mathbf{e}_2 - \mathbf{u}_1\alpha_{12} = \begin{bmatrix} -1 \\ -\alpha \\ \alpha \end{bmatrix} \,.
\end{aligned}$$

In these computations, the only rounding error occurred in computing β_1. In the second step of the factorization, we compute

$$\begin{aligned}
\beta_2 &= -\operatorname{sign}\left(\sqrt{\sum_{i=2}^{3} \mathbf{A}_{i2}^2}, \mathbf{A}_{22}\right) = \alpha\sqrt{2} \,, \\
\tau_2 &= (\beta_2 - \mathbf{A}_{22})/\beta_2 = 1 + 1/\sqrt{2} \\
\mathbf{u}_2 &= \begin{bmatrix} 1 \\ -1/\left(1+\sqrt{2}\right) \end{bmatrix}
\end{aligned}$$

At this point, we have

$$\mathbf{R} = \begin{bmatrix} -1 & -1 \\ & \alpha\sqrt{2} \end{bmatrix} \,.$$

Next, we apply the first reflector to $\mathbf{b}$:

$$\alpha = \tau_1 \mathbf{u}_1 \cdot \mathbf{b} = 2 \,, \; \mathbf{b} - \mathbf{u}_1\alpha = \begin{bmatrix} -1 \\ -\alpha \\ 0 \end{bmatrix} = \begin{bmatrix} -1 \\ \mathbf{b}' \end{bmatrix} \,.$$

Afterward, we apply the second reflector to the last two components of the reflected right-hand side:

$$\alpha = \tau_2 \mathbf{u}_2 \cdot \mathbf{b}' = -\alpha(1 + 1/\sqrt{2}) \ , \ \mathbf{b}' - \mathbf{u}_2\alpha = \begin{bmatrix} \alpha/\sqrt{2} \\ \alpha/\sqrt{2} \end{bmatrix} \ .$$

At this point, we have

$$\mathbf{y} = \begin{bmatrix} -1 \\ \alpha/\sqrt{2} \end{bmatrix} \ , \mathbf{z} = \begin{bmatrix} \alpha/\sqrt{2} \end{bmatrix} \ .$$

This implies that the computed norm of the residual is $\|\mathbf{r}\|_2 = \|\mathbf{z}\|_2 = \alpha/\sqrt{2}$. This value is the correctly rounded value of the true residual norm.

All that remains is to back-solve $\mathbf{R}\tilde{\mathbf{x}} = \mathbf{y}$ for $\tilde{\mathbf{x}}$:

$$\begin{bmatrix} -1 & 1 \\ & \alpha\sqrt{2} \end{bmatrix} \begin{bmatrix} \tilde{\mathbf{x}}_1 \\ \tilde{\mathbf{x}}_2 \end{bmatrix} = \mathbf{R}\tilde{\mathbf{x}} = \mathbf{y} = \begin{bmatrix} -1 \\ \alpha/\sqrt{2} \end{bmatrix} \Longrightarrow \tilde{\mathbf{x}} = \begin{bmatrix} 1/2 \\ 1/2 \end{bmatrix} \ .$$

This is equal to the correctly rounded value of the exact solution to this least squares problem.

6.7.4.2 Full-Rank Underdetermined

Suppose that $\mathbf{A}$ is an $m \times n$ matrix with rank $(\mathbf{A}) = m$. Given an m-vector, we want to find the smallest solution $\mathbf{x}$ of $\mathbf{A}\mathbf{x} = \mathbf{b}$. To compute $\mathbf{x}$ by means of successive reflection, we begin by finding Householder reflectors $\mathbf{H}_1, \ldots, \mathbf{H}_m$ so that

$$\mathbf{A}\mathbf{H}_1 \cdot \ldots \cdot \mathbf{H}_n = \begin{bmatrix} \mathbf{L} & \mathbf{0} \end{bmatrix} \ .$$

This requires a new algorithm similar to Algorithm 6.7.4, and is available in LAPACK routine `_gelqf`. See, for example, dgelqf.f.

Let us formally partition

$$\mathbf{H}_n{}^H \cdot \ldots \cdot \mathbf{H}_1{}^H \equiv \mathbf{Q} = \begin{bmatrix} \mathbf{Q}_1 \\ \mathbf{Q}_2 \end{bmatrix} \ ,$$

where $\mathbf{Q}_1$ is $m \times n$. The LQ factorization of the previous paragraph implies that

$$\mathscr{R}\left(\mathbf{A}^H\right) = \mathscr{R}\left(\mathbf{Q}_1{}^H\right) \ .$$

Since $\mathbf{Q}$ is an $n \times n$ unitary matrix, we have

$$\mathbf{I} = \begin{bmatrix} \mathbf{Q}_1 \\ \mathbf{Q}_2 \end{bmatrix} \begin{bmatrix} \mathbf{Q}_1{}^H & \mathbf{Q}_2{}^H \end{bmatrix} = \begin{bmatrix} \mathbf{Q}_1\mathbf{Q}_1{}^H & \mathbf{Q}_1\mathbf{Q}_2{}^H \\ \mathbf{Q}_2\mathbf{Q}_1{}^H & \mathbf{Q}_2\mathbf{Q}_2{}^H \end{bmatrix} \ .$$

The fundamental theorem of linear algebra 3.2.3, together with the factorization

$$\mathbf{A} = \begin{bmatrix} \mathbf{L} & \mathbf{0} \end{bmatrix} \mathbf{Q} = \begin{bmatrix} \mathbf{L} & \mathbf{0} \end{bmatrix} \begin{bmatrix} \mathbf{Q}_1 \\ \mathbf{Q}_2 \end{bmatrix} = \mathbf{L} \mathbf{Q}_1$$

and the equation

$$\mathbf{Q}_1 \mathbf{Q}_2{}^H = \mathbf{0}$$

imply that

$$\mathscr{N}(\mathbf{A}) = \mathscr{R}\left(\mathbf{Q}_2{}^H\right) .$$

Lemma 6.3.1 shows that there is an m-vector $\mathbf{s}$ so that the minimum norm solution $\mathbf{x}$ satisfies

$$\mathbf{x} = \mathbf{A}^H \mathbf{s} = \mathbf{Q}_1{}^H \mathbf{L}^H \mathbf{s} \equiv \mathbf{Q}_1{}^H \mathbf{y} .$$

Here, we have defined

$$\mathbf{y} \equiv \mathbf{L}^H \mathbf{s} .$$

Thus

$$\mathbf{b} = \mathbf{A}\mathbf{x} = \mathbf{L}\mathbf{Q}_1 \mathbf{x} = \mathbf{L}\mathbf{Q}_1 \mathbf{Q}_1{}^H \mathbf{y} = \mathbf{L}\mathbf{y} .$$

This analysis shows that in order to solve the underdetermined least squares problem, we first forward-solve

$$\mathbf{L}\mathbf{y} = \mathbf{b}$$

for $\mathbf{y}$, then we compute

$$\mathbf{x} = \mathbf{Q}_1{}^H \mathbf{y} = \begin{bmatrix} \mathbf{Q}_1{}^H & \mathbf{Q}_2{}^H \end{bmatrix} \begin{bmatrix} \mathbf{y} \\ \mathbf{0} \end{bmatrix} = \mathbf{H}_1 \cdot \ldots \cdot \mathbf{H}_n \begin{bmatrix} \mathbf{y} \\ \mathbf{0} \end{bmatrix} .$$

In other words, the least squares solution $\mathbf{x}$ is computed by extending the vector $\mathbf{y}$ with zeros and then applying the Householder reflectors in the reverse order of the factorization.

In summary, we have the following algorithm for solving an underdetermined least squares problem.

Algorithm 6.7.8 (Underdetermined Least Squares via Householder LQ)

factor $\mathbf{A}\mathbf{H}_1 \cdot \ldots \cdot \mathbf{H}_n = [\mathbf{L}\ \mathbf{0}]$ /* LAPACK _gelqf */
forward-solve $\mathbf{L}\mathbf{y} = \mathbf{b}$ for $\mathbf{y}$ /* Algorithm 3.4.3 */
compute $\mathbf{x} = \mathbf{H}_1 \cdot \ldots \cdot \mathbf{H}_n \begin{bmatrix} \mathbf{y} \\ \mathbf{0} \end{bmatrix}$ /* multiple calls to Algorithm 6.7.3 */

In LAPACK, this entire algorithm is performed by routines `_gels`. As an example, interested readers may view dgels.f. In MATLAB, an underdetermined least squares problem involving matrix $\mathbf{A}$ and right-hand side $\mathbf{b}$ can be solved by performing

```
x=A\b
```

Example 6.7.2 Consider the undetermined least squares problem with

$$\mathbf{A} = \begin{bmatrix} 1 & \alpha & 0 \\ 1 & 0 & \alpha \end{bmatrix} \text{ and } \mathbf{b} = \begin{bmatrix} 1 \\ 1 \end{bmatrix} .$$

It is easy to check that

$$\mathbf{A}^\dagger = \mathbf{A}^H \left(\mathbf{A}\mathbf{A}^H\right)^{-1} = \begin{bmatrix} 1 & 1 \\ \alpha & 0 \\ 0 & \alpha \end{bmatrix} \begin{bmatrix} 1+\alpha^2 & -1 \\ -1 & 1+\alpha^2 \end{bmatrix} \frac{1}{\alpha^2(2+\alpha^2)}$$

$$= \begin{bmatrix} \alpha & \alpha \\ 1+\alpha^2 & -1 \\ -1 & 1+\alpha^2 \end{bmatrix} \frac{1}{\alpha(2+\alpha^2)} .$$

Then the minimum norm solution of $\mathbf{A}\mathbf{x} = \mathbf{b}$ is

$$\mathbf{x} = \mathbf{A}^\dagger \mathbf{b} = \begin{bmatrix} \alpha & \alpha \\ 1+\alpha^2 & -1 \\ -1 & 1+\alpha^2 \end{bmatrix} \frac{1}{\alpha(2+\alpha^2)} \begin{bmatrix} 1 \\ 1 \end{bmatrix} = \begin{bmatrix} 2 \\ \alpha \\ \alpha \end{bmatrix} \frac{1}{2+\alpha^2} .$$

As in Example 6.7.1, assume that α is a small positive machine number such that $\mathrm{fl}(1+\alpha^2) = 1$. Since the matrix in this example is the transpose of the matrix in Example 6.7.1, the LQ factorization of this matrix is the transpose of the QR factorization in the other example. Thus

$$\mathbf{A} = [\mathbf{L}\ \mathbf{0}]\, \mathbf{H}_1 \mathbf{H}_2$$

where

$$\mathbf{L} = \begin{bmatrix} -1 & -1 \\ & \alpha\sqrt{2} \end{bmatrix}^H = \begin{bmatrix} -1 & \\ -1 & \alpha\sqrt{2} \end{bmatrix} ,$$

$$\mathbf{H}_1 = \mathbf{I} - \mathbf{u}_1\tau_1\mathbf{u}_1{}^H \text{ where } \mathbf{u}_1 = \begin{bmatrix} 1 \\ \alpha/2 \\ 0 \end{bmatrix} \text{ and } \tau_1 = 2 \text{ , and}$$

$$\mathbf{H}_2 = \mathbf{I} - \mathbf{u}_2\tau_2\mathbf{u}_2{}^H \text{ where } \mathbf{u}_2 = \begin{bmatrix} 0 \\ 1 \\ -1/(1+\sqrt{2}) \end{bmatrix} \text{ and } \tau_2 = 1 + 1/\sqrt{2} .$$

To solve the underdetermined least squares problem with this LQ factorization, we begin by forward-solving $\mathbf{L}\mathbf{y} = \mathbf{b}$ for $\mathbf{y}$:

$$\begin{bmatrix} -1 & \\ -1 & \alpha\sqrt{2} \end{bmatrix} \begin{bmatrix} \mathbf{y}_1 \\ \mathbf{y}_2 \end{bmatrix} = \begin{bmatrix} 1 \\ 1 \end{bmatrix} \Longrightarrow \mathbf{y} = \begin{bmatrix} -1 \\ 0 \end{bmatrix} .$$

Next, we apply the Householder reflector $\mathbf{H}_2$ to the 3-vector $\begin{bmatrix} \mathbf{y} \\ 0 \end{bmatrix} = -\mathbf{e}_1$. Since $\mathbf{u}_2 \cdot (-\mathbf{e}_1) = 0$, the result is

$$\mathbf{H}_2(-\mathbf{e}_1) = -\mathbf{e}_1 .$$

Finally, we compute

$$\alpha = \tau_1\mathbf{u}_1 \cdot (-\mathbf{e}_1) = -2 \text{ and}$$

$$\tilde{\mathbf{x}} = -\mathbf{e}_1 - \mathbf{u}_1\alpha = \begin{bmatrix} -1 \\ 0 \\ 0 \end{bmatrix} - \begin{bmatrix} 1 \\ \alpha/2 \\ 0 \end{bmatrix} (-2) = \begin{bmatrix} 1 \\ \alpha \\ 0 \end{bmatrix} .$$

This is *not* the correctly rounded value of the exact solution.

The reader can verify that if we have performed Householder QR factorization of $\mathbf{A}^H$ in exact arithmetic, we would have obtained

$$\beta_1 = -\sqrt{1+\alpha^2} \text{ , } \tau_1 = \frac{1+\sqrt{1+\alpha^2}}{\sqrt{1+\alpha^2}} \text{ , } \mathbf{u}_1 = \begin{bmatrix} 1 \\ \frac{\alpha}{1+\sqrt{1+\alpha^2}} \\ 0 \end{bmatrix} \text{ , } \alpha_{12} = \frac{1+\sqrt{1+\alpha^2}}{\sqrt{1+\alpha^2}}$$

$$\mathbf{H}_1\mathbf{A}\mathbf{e}_2 = \begin{bmatrix} -\frac{1}{\sqrt{1+\alpha^2}} \\ -\frac{\alpha}{\sqrt{1+\alpha^2}} \\ \alpha \end{bmatrix} \text{ , } \beta_2 = \alpha\sqrt{\frac{2+\alpha^2}{1+\alpha^2}} \text{ ,}$$

$$\tau_2 = 1 + \frac{1}{\sqrt{2+\alpha^2}} \text{ and , } \mathbf{u}_2 = \begin{bmatrix} 1 \\ -\frac{\sqrt{1+\alpha^2}}{1+\sqrt{2+\alpha^2}} \end{bmatrix} .$$

Thus we would begin the solution of the underdetermined least squares problem by solving

$$\begin{bmatrix} -\sqrt{1+\alpha^2} & -\frac{1}{\sqrt{1+\alpha^2}} & \alpha\sqrt{\frac{2+\alpha^2}{1+\alpha^2}} \end{bmatrix} \begin{bmatrix} \eta_1 \\ \eta_2 \end{bmatrix} = \mathbf{L}\mathbf{y} = \mathbf{b} = \begin{bmatrix} 1 \\ 1 \end{bmatrix} \Longrightarrow \mathbf{y} = \begin{bmatrix} -\frac{1}{\sqrt{1+\alpha^2}} \\ \frac{\alpha}{\sqrt{(1+\alpha^2)(2+\alpha^2)}} \end{bmatrix} .$$

The floating point computation of the Householder QR undetermined least squares solution differs significantly in the second component of $\mathbf{y}$.

However, the computed underdetermined least squares solution does satisfy

$$\tilde{\mathbf{x}} = \mathbf{A}^H \mathbf{e}_1 ,$$

and has residual

$$\mathbf{b} - \mathbf{A}\tilde{\mathbf{x}} = \begin{bmatrix} 1 \\ 1 \end{bmatrix} - \begin{bmatrix} 1 & \alpha & 0 \\ 1 & 0 & \alpha \end{bmatrix} \begin{bmatrix} 1 \\ \alpha \\ 0 \end{bmatrix} = \begin{bmatrix} -\alpha^2 \\ 0 \end{bmatrix} .$$

The vector

$$\begin{bmatrix} \tilde{\mathbf{x}} - \mathbf{A}^H \mathbf{e}_1 \\ \mathbf{A}\tilde{\mathbf{x}} - \mathbf{b} \end{bmatrix} = \begin{bmatrix} 0 \\ 0 \\ 0 \\ \alpha^2 \\ 0 \end{bmatrix}$$

has norm less than machine precision times $\|\mathbf{b}\|_2$. Furthermore, Example 6.5.4 showed that

$$\|\mathbf{A}\|_2^2 \approx 2 + \alpha^2/2 \text{ and } \left\|\mathbf{A}^\dagger\right\|_2^2 \approx \frac{2}{\alpha^2} + \frac{1}{2} ,$$

so

$$\kappa_2(\mathbf{A}) \approx 2/\alpha + \alpha/2 .$$

Thus it is not surprising that perturbations of magnitude at most α^2 in the matrix factorization could produce errors on the order of α in the computed solution.

6.7.4.3 Rank Deficient

If we suspect that a least squares problem involves a matrix $\mathbf{A}$ that is nearly rank-deficient, then we should avoid the use of the full-rank Householder factorizations described in Algorithm 6.7.4 or in Sect. 6.7.4.2. Instead, we could adopt one of two strategies. We could employ the singular value decomposition as in Sect. 6.11.1, or we could use a modification of successive reflection with pivoting, called the **complete orthogonal decomposition**.

We begin the complete orthogonal decomposition by selecting a desired condition number, and then applying successive reflection with pivoting to find successive integers $r \le \min\{m, n\}$ so that

$$\mathbf{U}^H \mathbf{A}\mathbf{P} = \begin{bmatrix} \mathbf{R}_{11} & \mathbf{R}_{12} \\ \mathbf{0} & \mathbf{E}_{22} \end{bmatrix} . \tag{6.10}$$

Here $\mathbf{P}$ is an $n \times n$ permutation matrix, and $\mathbf{R}_{11}$ is $r \times r$ right-triangular. We stop the process at the last value of r for which the condition number of $\mathbf{R}_{11}$ is estimated not to be smaller than some selected value. The entries of $\mathbf{E}_{22}$ are small, in the sense that adding any additional information from $\mathbf{E}_{22}$ would increase the estimated condition number of $\mathbf{R}_{11}$ beyond the given bound. This approach for estimating condition numbers was briefly described in Sect. 6.7.3, and involves the use of eigenvalues.

Nevertheless, suppose that we are able to determine the integer r and the factorization (6.10). Using elementary reflectors, we can factor

$$\begin{bmatrix} \mathbf{R}_{11} & \mathbf{R}_{12} \end{bmatrix} = \begin{bmatrix} \mathbf{R} & \mathbf{0} \end{bmatrix} \mathbf{V}^H ,$$

where $\mathbf{R}$ is $r \times r$ right-triangular and $\mathbf{V}$ is unitary. This factorization begins by computing an elementary reflector $\mathbf{H}'_r$ using the rth diagonal entry and the last $n - r$ entries of the last row of $\mathbf{R}_{11}$, and then applying $\mathbf{H}'_r$ to $\begin{bmatrix} \mathbf{R}_{11} & \mathbf{R}_{12} \end{bmatrix}$ on the right. This will produce zeros in row r beyond column r. Working backwards for $r \ge k \ge 1$, the general step computes an elementary reflector $\mathbf{H}'_k$ using the kth diagonal entry and the last $n - r$ entries of the kth row of $\mathbf{R}_{11}$, and then applying $\mathbf{H}'_{k+1}$ to $\begin{bmatrix} \mathbf{R}_{11} & \mathbf{R}_{12} \end{bmatrix} \mathbf{H}'_r \cdot \ldots \cdot \mathbf{H}'_k$ on the right. This will produce zeros in row k beyond column r, and leave columns $1 \le j \le r$ with $j \ne k$ unchanged. Because $\mathbf{H}'_{k+1}$ is applied to a zero vector when it operates on row j with $k < j \le r$, the leading $r \times r$ matrix remains right-triangular, and zeros remain beyond column r in rows $k + 1$ through r.

Let us define the nearby matrix

$$\tilde{\mathbf{A}} = \mathbf{U} \begin{bmatrix} \mathbf{R}_{11} & \mathbf{R}_{12} \\ \mathbf{0} & \mathbf{0} \end{bmatrix} \mathbf{P}^H .$$

Then

$$\mathbf{A} - \tilde{\mathbf{A}} = \mathbf{U} \begin{bmatrix} \mathbf{0} & \mathbf{0} \\ \mathbf{0} & \mathbf{E}_{22} \end{bmatrix} \mathbf{P} .$$

Theorem 6.5.1 shows that we will make small perturbations in the solution to the least squares problem that uses $\tilde{\mathbf{A}}$ in place of $\mathbf{A}$, provided that the condition number of $\mathbf{A}$ is not too large.

It is straightforward to see that

$$\tilde{\mathbf{A}}^\dagger = \mathbf{PV}\begin{bmatrix} \mathbf{R}^{-1} & \mathbf{0} \\ \mathbf{0} & \mathbf{0} \end{bmatrix} \mathbf{U}^H .$$

Thus, in order to find the minimum norm solution $\tilde{\mathbf{x}}$ to $\tilde{\mathbf{A}}\tilde{\mathbf{x}} = \mathbf{b}$, we select a bound κ on the condition number and perform the following algorithm:

Algorithm 6.7.9 (Rank Deficient Least Squares via Householder)

factor $\mathbf{H}_r{}^H \cdot \ldots \cdot \mathbf{H}_1{}^H \mathbf{AP} = \begin{bmatrix} \mathbf{R}_{11} & \mathbf{R}_{12} \\ \mathbf{0} & \mathbf{E}_{22} \end{bmatrix}$ /* LAPACK _geqp3; here $\kappa_2(\mathbf{R}_{11}) < \kappa$ */

factor $\begin{bmatrix} \mathbf{R}_{11} & \mathbf{R}_{12} \end{bmatrix} \mathbf{H}_r' \cdot \ldots \cdot \mathbf{H}_1' = \begin{bmatrix} \mathbf{R} & \mathbf{0} \end{bmatrix}$ /* LAPACK _tzrzf */

successively reflect $\begin{bmatrix} \tilde{\mathbf{y}} \\ \tilde{\mathbf{z}} \end{bmatrix} = \mathbf{H}_r{}^H \cdot \ldots \cdot \mathbf{H}_1{}^H \mathbf{b}$ /* LAPACK _ormqr */

solve $\mathbf{R}\tilde{\mathbf{v}} = \tilde{\mathbf{y}}$ /* LAPACK BLAS _trsv */

successively reflect $\tilde{\mathbf{x}}' = \mathbf{H}_r' \cdot \ldots \cdot \mathbf{H}_1' \begin{bmatrix} \tilde{\mathbf{v}} \\ \mathbf{0} \end{bmatrix}$ /* LAPACK _ormrz */

permute $\tilde{\mathbf{x}} = \mathbf{P}\tilde{\mathbf{x}}'$

This entire algorithm is available in LAPACK routine `_gelsy`. Interested readers may view, for example, dgelsy.f. Readers may also examine an implementation of a C++ `CompleteOrthogonalDecomposition` class in files CompleteOrthogonalDecomposition.H and CompleteOrthogonalDecomposition.C. This class calls data type-specific LAPACK routines to perform the operations; see, for example LaDouble.C for the implementation in real double precision.

Unlike LAPACK routines `_gelsy`, the MATLAB operation `A \ b` does not accept a user-specified value for the desired condition number of **A**. Instead, the matrix is determined to be rank-deficient whenever a diagonal entry of **R** is smaller than

$$\text{tol}_{\text{MATLAB} \backslash} = \max\{m, n\} |\mathbf{R}_{11}| \varepsilon ,$$

where ε is machine precision.

Exercise 6.7.5 Consider Table 6.3 containing molecular weights for various oxides of nitrogen. Solve the least squares problem to determine the atomic weights of nitrogen and oxygen. How many significant digits do these values have, based on the perturbation theory for least squares problems?

6.7.5 Rounding Errors

A careful discussion of the rounding errors in successive reflection can be found in Wilkinson [111, p. 152ff], or in Higham [56, p. 357ff]. Our discussion will be similar to that in Higham.

We begin by studying the rounding errors in the algorithm for generating an elementary reflector.

Lemma 6.7.5 *Assume that the relative errors in floating point addition, multiplication, division or square roots satisfy*

$$\begin{aligned} fl(\xi + \eta) &= (\xi + \eta)(1 + \varepsilon_+) \, , \\ fl(\xi \times \eta) &= (\xi \times \eta)(1 + \varepsilon_\times) \, , \\ fl(\xi \div \eta) &= (\xi \div \eta)(1 + \varepsilon_\div) \text{ and} \\ fl(\sqrt{\xi}) &= \sqrt{\xi}(1 + \varepsilon_{\sqrt{}}) \, . \end{aligned}$$

Also suppose that there is an upper bound ε *on all such relative errors:*

$$\max\{|\varepsilon_+|, |\varepsilon_\times|, |\varepsilon_\div|, |\varepsilon_{\sqrt{}}|\} \le \varepsilon \, .$$

Given a scalar α *and an* $(m-1)$*-vector* $\mathbf{a}$ *with* $|\alpha|^2 + \|\mathbf{a}\|_2^2 > 0$*, let* β*,* τ *and* $\mathbf{u}$ *be determined by the Algorithm 6.7.2 for computing the components of an elementary reflector. Partition*

$$\mathbf{u} = \begin{bmatrix} 1 \\ \mathbf{w} \end{bmatrix} \, .$$

If $3(m+1)\varepsilon < 2$*, then*

$$\frac{|fl(\beta) - \beta|}{|\beta|} \le \frac{(m+1)\varepsilon}{1 - (m+1)\varepsilon/2} \, , \tag{6.11a}$$

$$|fl(\tau) - \tau| \le \frac{(m+5)\varepsilon}{1 - 3(m+1)\varepsilon/2} \, . \tag{6.11b}$$

$$\frac{\|fl(\mathbf{w}) - \mathbf{w}\|_2}{\|\mathbf{w}\|_2} \le \frac{(m+3)\varepsilon}{1 - 3(m+1)\varepsilon/2} \text{ and}$$

$$\frac{\|fl(\mathbf{u}) - \mathbf{u}\|_2}{\|\mathbf{u}\|_2} \le \frac{(m+3)\varepsilon}{1 - 3(m+1)\varepsilon/2} \, . \tag{6.11c}$$

Proof We begin by bounding the floating point error in computing β. Note that $\sigma = \text{sign}(\Re e(\alpha))$ is computed without rounding error, and multiplication by σ does not cause a rounding error. Recall that inequality (3.45) showed that

$$\mathrm{fl}\left(\begin{bmatrix}\alpha\\ \mathbf{a}\end{bmatrix}\cdot\begin{bmatrix}\alpha\\ \mathbf{a}\end{bmatrix}\right)=\left\|\begin{bmatrix}\alpha\\ \mathbf{a}\end{bmatrix}\right\|_2^2(1+\delta_.)\text{ where }|\delta_.|\le\frac{m\varepsilon}{1-(m-1)\varepsilon/2}\,.$$

Then

$$\begin{aligned}\delta_\beta &\equiv \frac{\mathrm{fl}(\beta)-\beta}{\beta}=\frac{\beta(1+\varepsilon_{\surd})\sqrt{1+\delta}-\beta}{\beta}\\ &=(1+\varepsilon_{\surd})\sqrt{1+\delta_.}-1=(1+\varepsilon_{\surd})\left[\sqrt{1+\delta_.}-1\right]+\varepsilon_{\surd}\,.\end{aligned}$$

Since $|\sqrt{1+\delta}-1|\le|\delta|$ for all $\delta\ge -1$, inequalities (3.38) and (3.37) imply that

$$\begin{aligned}|\delta_\beta| &\le (1+\varepsilon)|\delta|+\varepsilon\le(1+\varepsilon)\frac{m\varepsilon}{1-(m-1)\varepsilon/2}+\varepsilon\\ &\le\frac{m\varepsilon}{1-(m+1)\varepsilon/2}+\varepsilon\le\frac{(m+1)\varepsilon}{1-(m+1)\varepsilon/2}\,.\end{aligned}$$

This proves (6.11a).

Next, let us bound the error in τ. We have

$$\begin{aligned}\frac{\mathrm{fl}(\tau)-\tau}{\tau} &= \frac{\mathrm{fl}(\beta)-\alpha}{\mathrm{fl}(\beta)}(1+\varepsilon_-)(1+\varepsilon_\div)\frac{\beta}{\beta-\alpha}-1\\ &=\frac{\beta(1+\delta_\beta)-\alpha}{(\beta-\alpha)(1+\delta_\beta)}(1+\varepsilon_-)(1+\varepsilon_\div)-1\\ &=\frac{\beta(1+\delta_\beta)\left\{(1+\varepsilon_-)(1+\varepsilon_\div)-1\right\}+\alpha\left\{(1+\delta_\beta)-(1+\varepsilon_-)(1+\varepsilon_\div)\right\}}{(\beta-\alpha)(1+\delta_\beta)}\\ &=\frac{\beta}{\beta-\alpha}\left\{(1+\varepsilon_-)(1+\varepsilon_\div)-1\right\}+\frac{\alpha}{\beta-\alpha}\,\frac{(1+\delta_\beta)-(1+\varepsilon_-)(1+\varepsilon_\div)}{1+\delta_\beta}\,.\end{aligned}$$

Note that $0\le-\alpha/\beta\le 1$, and $\max\{|\beta|,|\alpha|\}\le|\beta-\alpha|$. Consequently,

$$\begin{aligned}\left|\frac{\mathrm{fl}(\tau)-\tau}{\tau}\right| &\le \frac{|\beta|}{|\beta-\alpha|}\,|(1+\varepsilon_-)(1+\varepsilon_\div)-1|\\ &\quad+\frac{|\alpha|}{|\beta-\alpha|}\,\frac{|\delta_\beta|+|(1+\varepsilon_-)(1+\varepsilon_\div)-1|}{|1+\delta_\beta|}\end{aligned}$$

then inequalities (3.33) and (6.11a) lead to

$$\le\frac{2\varepsilon}{1-\varepsilon/2}+\frac{(m+1)\varepsilon/(1-[m+1]\varepsilon/2)+2\varepsilon/(1-\varepsilon/2)}{1-(m+1)\varepsilon/(1-[m+1]\varepsilon/2)}$$

then inequality (3.37) gives us

$$\le \frac{2\varepsilon}{1-2\varepsilon} + \frac{(m+3)\varepsilon/(1-[m+1]\varepsilon/2)}{1-(m+1)\varepsilon/(1-[m+1]\varepsilon/2)}$$

then inequality (3.39) produces

$$\le \frac{2\varepsilon}{1-2\varepsilon} + \frac{(m+3)\varepsilon}{1-3(m+1)\varepsilon/2}$$

and finally inequality (3.37) results in

$$\le \frac{(m+5)\varepsilon}{1-3(m+1)\varepsilon/2} \; .$$

This proves (6.11b).

Finally, we will bound the error in $\mathbf{w}$. There is no error in the first component of $\mathbf{w}$, so we will only consider the other components, and denote them also by $\mathbf{w}$. We have

$$\begin{aligned} fl(\mathbf{w}) &= (\mathbf{I}+\mathbf{E}_\times)\,\mathbf{a}\frac{1+\varepsilon_-}{\alpha - fl(\beta)} = (\mathbf{I}+\mathbf{E}_\times)\,\mathbf{a}\frac{1+\varepsilon_-}{\alpha - \beta(1+\delta_\beta)} \\ &= (\mathbf{I}+\mathbf{E}_\times)\,\mathbf{w}\frac{(1+\varepsilon_-)(\alpha-\beta)}{\alpha-\beta-\beta\delta_\beta} = (\mathbf{I}+\mathbf{E}_\times)\,\mathbf{w}\frac{1+\varepsilon_-}{1+\delta_\beta/\tau} \; , \end{aligned}$$

where $\mathbf{E}_\times$ is a diagonal matrix with entries bounded in modulus by ε. Note that τ is always positive; in fact, $1 \le \tau \le 2$. Thus

$$\begin{aligned} \frac{\|fl(\mathbf{w})-\mathbf{w}\|_2}{\|\mathbf{w}\|_2} &= \left\| \left\{ (\mathbf{I}+\mathbf{E}_\times)\frac{1+\varepsilon_-}{1+\delta_\beta/\tau} - \mathbf{I} \right\} \mathbf{w} \right\|_2 / \|\mathbf{w}\|_2 \\ &\le \frac{\|(\mathbf{I}+\mathbf{E}_\times)(1+\varepsilon_-) - \mathbf{I}(1+\delta_\beta/\tau)\|_2}{1-|\delta_\beta|/\tau} \le \frac{\|(\mathbf{I}+\mathbf{E}_\times)(1+\varepsilon_-) - \mathbf{I}\|_2 + |\delta_\beta|/\tau}{1-|\delta_\beta|/\tau} \end{aligned}$$

then we use inequality (3.33) and the bound (6.11a) for δ_β to get

$$\le \frac{2\varepsilon/(1-\varepsilon/2) + (m+1)\varepsilon/(1-[m+1]\varepsilon/2)}{1-(m+1)\varepsilon/(1-[m+1]\varepsilon/2)}$$

then we use inequality (3.37) to obtain

$$\le \frac{(m+3)\varepsilon/(1-[m+1]\varepsilon/2)}{1-(m+1)\varepsilon/(1-[m+1]\varepsilon/2)} = \frac{(m+3)\varepsilon}{1-3(m+1)\varepsilon/2} \; .$$

Since

$$\mathbf{u} = \begin{bmatrix} 1 \\ \mathbf{w} \end{bmatrix} ,$$

the previous inequality implies that

$$\frac{\| \mathrm{fl}(\mathbf{u}) - \mathbf{u}\|_2}{\|\mathbf{u}\|_2} = \frac{\| \mathrm{fl}(\mathbf{w}) - \mathbf{w}\|_2}{\sqrt{1 + \|\mathbf{w}\|_2^2}} \leq \frac{(m+3)\varepsilon}{1 - 3(m+1)\varepsilon/2} \frac{\|\mathbf{w}\|_2}{\sqrt{1 + \|\mathbf{w}\|_2^2}} \leq \frac{(m+3)\varepsilon}{1 - 3(m+1)\varepsilon/2} .$$

Next, we will examine the error in applying an elementary reflector to a vector.

Lemma 6.7.6 *Assume that the relative errors in floating point addition, multiplication, division or square roots satisfy*

$$\begin{aligned} fl(\xi + \eta) &= (\xi + \eta)(1 + \varepsilon_+) , \\ fl(\xi \times \eta) &= (\xi \times \eta)(1 + \varepsilon_\times) , \\ fl(\xi \div \eta) &= (\xi \div \eta)(1 + \varepsilon_\div) \text{ and} \\ fl(\sqrt{\xi}) &= \sqrt{\xi}(1 + \varepsilon_{\sqrt{}}) . \end{aligned}$$

Also suppose that there is an upper bound ε *on all such relative errors:*

$$\max\{|\varepsilon_+|, |\varepsilon_\times|, |\varepsilon_\div|, |\varepsilon_{\sqrt{}}|\} \leq \varepsilon .$$

Assume that m *is a positive integer satisfying* $(m+5)\varepsilon < 2$. *Let* $\mathbf{u}$ *be an* m*-vector,* $\tau = 2/\mathbf{u}\cdot\mathbf{u}$ *and* $\mathbf{H} = \mathbf{I} - \mathbf{u}\tau\mathbf{u}^H$. *Then for every* m*-vector* $\mathbf{b}$ *there is an* $m \times m$ *matrix* $\triangle\mathbf{H}$ *so that*

$$fl(\mathbf{Hb}) = (\mathbf{H} + \triangle\mathbf{H})\mathbf{b} \text{ , where } \|\triangle\mathbf{H}\|_2 \leq \frac{(2m+7)\varepsilon}{1 - (m+5)\varepsilon/2} . \tag{6.12}$$

Proof Floating point computation of the reflected vector produces

$$\begin{aligned} \mathrm{fl}(\mathbf{Hb}) - \mathbf{Hb} &= (\mathbf{I} + \mathbf{E}_-)\{\mathbf{b} - \mathrm{fl}\,(\mathbf{u}\tau\mathbf{u}\cdot\mathbf{b})\} - \mathbf{b} + \mathbf{u}\tau\mathbf{u}\cdot\mathbf{b} \\ &= \mathbf{E}_-\mathbf{b} - (\mathbf{I} + \mathbf{E}_-)(\mathbf{I} + \mathbf{E}_\times)\mathbf{u}\tau(1 + \varepsilon_\times)(\mathbf{u} + \triangle\mathbf{u})\cdot\mathbf{b} + \mathbf{u}\tau\mathbf{u}\cdot\mathbf{b} \\ &= \left[\mathbf{E}_- + \{\mathbf{I} - (\mathbf{I} + \mathbf{E}_-)(\mathbf{I} + \mathbf{E}_\times)(1 + \varepsilon_\times)\}\,\mathbf{u}\tau\mathbf{u}^H \right. \\ &\quad \left. + (\mathbf{I} + \mathbf{E}_-)(\mathbf{I} + \mathbf{E}_\times)\mathbf{u}\tau(1 + \varepsilon_\times)\triangle\mathbf{u}^H\right]\mathbf{b} . \\ &\equiv \triangle\mathbf{Hb} . \end{aligned}$$

Thus

$$\|\triangle\mathbf{H}\,\mathbf{b}\|_2 \leq \|\mathbf{E}_{-}\mathbf{b}\|_2 + \|\mathbf{I} - (\mathbf{I}+\mathbf{E}_{-})(\mathbf{I}+\mathbf{E}_{\times})(1+\varepsilon_{\times})\|_2\,\|\mathbf{u}\|_2\tau|\mathbf{u}\cdot\mathbf{b}|$$
$$+ \|(\mathbf{I}+\mathbf{E}_{-})(\mathbf{I}+\mathbf{E}_{\times})\|_2\,(1+\varepsilon)\|\mathbf{u}\|_2\tau|\triangle\mathbf{u}\cdot\mathbf{b}|$$

then the Cauchy inequality (3.15) and the backward error estimate (3.46) for inner products yield

$$\leq \varepsilon\|\mathbf{b}\|_2 + \left[(1+\varepsilon)^3 - 1\right]\|\mathbf{u}\|_2^2\tau\|\mathbf{b}\|_2 + (1+\varepsilon)^3\|\mathbf{u}\|_2^2\tau\frac{m\varepsilon}{1-(m-1)\varepsilon/2}\|\mathbf{b}\|_2\,.$$

Next, we note that $\tau = 2/\|\mathbf{u}\|_2^2$ and use inequality (3.33) to obtain

$$\leq \left\{\varepsilon + 2\frac{3\varepsilon}{1-\varepsilon} + 2\left(1+\frac{3\varepsilon}{1-\varepsilon}\right)\frac{m\varepsilon}{1-(m-1)\varepsilon/2}\right\}\|\mathbf{b}\|_2$$

then inequalities (3.37) and (3.38) give us

$$\leq \left\{\frac{7\varepsilon}{1-\varepsilon} + \frac{2m\varepsilon}{1-(m+5)\varepsilon/2}\right\}\|\mathbf{b}\|_2$$

and finally inequality (3.37) produces

$$\leq \frac{(2m+7)\varepsilon}{1-(m+5)\varepsilon/2}\|\mathbf{b}\|_2\,.$$

Our next lemma will examine the rounding errors in Householder QR factorization.

Lemma 6.7.7 *Assume that the relative errors in floating point addition, multiplication, division or square roots satisfy*

$$fl(\xi+\eta) = (\xi+\eta)(1+\varepsilon_{+})\,,$$
$$fl(\xi\times\eta) = (\xi\times\eta)(1+\varepsilon_{\times})\,,$$
$$fl(\xi\div\eta) = (\xi\div\eta)(1+\varepsilon_{\div}) \text{ and}$$
$$fl(\sqrt{\xi}) = \sqrt{\xi}(1+\varepsilon_{\surd})\,.$$

Also suppose that there is an upper bound ε on all such relative errors:

$$\max\{|\varepsilon_{+}|, |\varepsilon_{\times}|, |\varepsilon_{\div}|, |\varepsilon_{\surd}|\} \leq \varepsilon\,.$$

Given an $m\times n$ matrix $\mathbf{A}$ with rank $(\mathbf{A}) = n$, assume that the elementary reflectors $\tilde{\mathbf{H}}_1,\ldots,\tilde{\mathbf{H}}_n$ and right-triangular matrix $\tilde{\mathbf{R}}$ are generated in floating point arithmetic *by the Householder QR factorization Algorithm 6.7.4. Let* $\mathbf{H}_1, \ldots, \mathbf{H}_n$ *and* $\mathbf{R}$ *be the quantities that would be computed by exact arithmetic, and define*

$$\mathbf{Q} = \mathbf{H}_1 \cdot \ldots \cdot \mathbf{H}_n \; .$$

Assume that $(m + 6 + [n-1][2m+7-n])\varepsilon < 1$. *Then*

$$\begin{bmatrix} \tilde{\mathbf{R}} \\ \mathbf{0} \end{bmatrix} = \begin{bmatrix} \mathbf{R} \\ \mathbf{0} \end{bmatrix} + \mathbf{Q}^H \triangle\mathbf{A} \; ,$$

where the jth column $\triangle\mathbf{a}_j$ *of* $\triangle\mathbf{A}$ *and jth column* $\mathbf{a}_j$ *of* $\mathbf{A}$ *satisfy*

$$\left\| \triangle\mathbf{a}_j \right\|_2 \le \frac{j(2m+8-j)\varepsilon}{1 - j(2m+8-j)\varepsilon} \|\mathbf{a}\|_2 \; . \tag{6.13}$$

Consequently,

$$\|\triangle\mathbf{A}\|_F \le \frac{M_{(6.14)}\varepsilon}{1 - M_{(6.14)}\varepsilon} \|\mathbf{A}\|_F \; . \, , \tag{6.14}$$

where

$$M_{(6.14)} = n(2m+8-n) \; .$$

Proof For each $1 \le k \le n$, and for all $k \le j \le n$ let $\tilde{\mathbf{a}}_j^{(k)}$ be the computed value of the jth column of the array during the Householder QR factorization Algorithm 6.7.4. Lemma 6.7.6 shows that

$$\tilde{\mathbf{a}}_j^{(k)} = (\mathbf{H}_k + \triangle\mathbf{H}_k) \cdot \ldots \cdot (\mathbf{H}_1 + \triangle\mathbf{H}_1)\mathbf{a}_j$$

where for $1 \le \ell \le n$ we have

$$\|\triangle\mathbf{H}_\ell\|_2 \le \frac{(2[m-\ell]+9)\varepsilon}{1-(m-\ell+6)\varepsilon/2} \le \frac{(2[m-\ell]+9)\varepsilon}{1-(2[m-\ell]+9)\varepsilon} \equiv \delta_{m-\ell} \; .$$

Since the jth column of $\tilde{\mathbf{R}}$ is $\tilde{\mathbf{a}}_j^{(j)}$, the definition of $\triangle\mathbf{A}$ gives us

$$\left\| \triangle\mathbf{A}\mathbf{e}_j \right\|_2 = \left\| \mathbf{Q}\left(\tilde{\mathbf{R}} - \mathbf{R}\right)\mathbf{e}_j \right\|_2 = \left\| \mathbf{Q}\left(\tilde{\mathbf{a}}_j^{(j)} - \mathbf{a}_j^{(j)}\right) \right\|_2$$

then since $\mathbf{Q}$ is unitary, we have

$$= \left\| \tilde{\mathbf{a}}_j^{(j)} - \mathbf{a}_j^{(j)} \right\|_2 = \left\| \left\{ (\mathbf{H}_j + \triangle\mathbf{H}_j) \cdot \ldots \cdot (\mathbf{H}_1 + \triangle\mathbf{H}_1) - \mathbf{H}_j \cdot \ldots \cdot \mathbf{H}_1 \right\} \mathbf{A}\mathbf{e}_j \right\|_2$$

then Lemma 3.8.7 implies that

$$\leq \left\{ \prod_{\ell=1}^{j} (1 + \delta_{m-\ell}) - 1 \right\} \left\{ \prod_{\ell=1}^{j} \|\mathbf{H}_\ell\|_2 \right\} \|\mathbf{A}\mathbf{e}_j\|_2$$

$$= \left\{ \prod_{\ell=1}^{j} \left(1 + \frac{(2[m-\ell]+9)\varepsilon}{1-(2[m-\ell]+9)\varepsilon} \right) - 1 \right\} \|\mathbf{A}\mathbf{e}_j\|_2$$

and inequality (3.42) gives us

$$\leq \frac{\sum_{\ell=1}^{j}(2[m-\ell]+9)\varepsilon}{1-\sum_{\ell=1}^{j}(2[m-\ell]+9)\varepsilon} \|\mathbf{A}\mathbf{e}_j\|_2 = \frac{j(2m-j+8)\varepsilon}{1-j(2m-j+8)\varepsilon} \|\mathbf{A}\mathbf{e}_j\|_2$$

This proves (6.13). The Frobenius bound (6.14) follows easily.

By treating the right-hand side $\mathbf{b}$ as the $(n+1)$th column, we get the following corollary.

Corollary 6.7.1 *Suppose that the assumptions of Lemma 6.7.7 are satisfied. Let* $\mathbf{b}$ *be an m-vector. Under floating point arithmetic, successive reflection to compute*

$$\begin{bmatrix} \mathbf{y} \\ \mathbf{z} \end{bmatrix} = \mathbf{Q}^H \mathbf{b}$$

produces

$$\begin{bmatrix} \tilde{\mathbf{y}} \\ \tilde{\mathbf{z}} \end{bmatrix} = \begin{bmatrix} \mathbf{y} \\ \mathbf{z} \end{bmatrix} + \mathbf{Q}^H \Delta\mathbf{c} ,$$

where

$$\|\Delta\mathbf{c}\|_2 \leq \frac{M_{(6.14)}\varepsilon}{1-M_{(6.14)}\varepsilon} \|\mathbf{b}\|_2 , \tag{6.15}$$

and $M_{(6.14)}$ *was defined in Eq.* (6.14).

Finally, we will study the rounding errors in solving an overdetermined least squares problem by successive reflection.

Theorem 6.7.2 *Suppose that the assumptions of Lemma 6.7.7 are satisfied. If* $\mathbf{b}$ *is an m-vector, apply successive reflection as in Sect. 6.7.4.1 to find the n-vector* $\tilde{\mathbf{y}}$ *and* $(m-r)$*-vector* $\tilde{\mathbf{z}}$ *using floating point arithmetic. Let* $\tilde{\mathbf{x}}$ *solve the right-triangular system* $\tilde{\mathbf{R}}\tilde{\mathbf{x}} = \tilde{\mathbf{y}}$*. Let* $\mathbf{y}$*,* $\mathbf{z}$ *and* $\mathbf{x}$ *be the quantities that would be computed by exact arithmetic. Assume that*

$$\frac{M_{(6.14)}\varepsilon}{1-M_{(6.14)}\varepsilon} \|\mathbf{A}\|_F \left\|\mathbf{A}^\dagger\right\|_2 < 1 , \tag{6.16}$$

where $M_{(6.14)}$ was defined in Eq. (6.14). *Then the error in the computed solution* $\tilde{\mathbf{x}}$ *satisfies*

$$\|\tilde{\mathbf{x}}-\mathbf{x}\|_2 \leq \frac{M_{(6.14)}\varepsilon}{1-M_{(6.14)}\varepsilon} \frac{\left\|\mathbf{A}^\dagger\right\|_2 \{\|\mathbf{A}\|_F\|\mathbf{x}\|_2+\|\mathbf{b}\|_2\}}{1-\left\|\mathbf{A}^\dagger\right\|_2\|\mathbf{A}\|_F M_{(6.14)}\varepsilon/\left(1-M_{(6.14)}\varepsilon\right)} \tag{6.17}$$

Also, the error in the computed residual norm is

$$\left|\|\tilde{\mathbf{z}}\|_2-\|\mathbf{z}\|_2\right| \leq \frac{M_{(6.14)}\varepsilon}{1-M_{(6.14)}\varepsilon}\|\mathbf{b}\|_2 . \tag{6.18}$$

Finally, the error in the approximate unitary matrix is

$$\left\|\tilde{\mathrm{Q}}-\mathrm{Q}\right\|_F \leq \frac{\sqrt{n}M_{(6.14)}\varepsilon}{1-M_{(6.14)}\varepsilon} . \tag{6.19}$$

Proof We can partition

$$\mathrm{Q}=\left[\mathrm{Q}_1\ \mathrm{Q}_2\right]$$

where Q_1 is $m\times n$. Note that

$$\mathbf{I}=\mathrm{Q}^H\mathrm{Q}=\begin{bmatrix}\mathrm{Q}_1{}^H\mathrm{Q}_1 & \mathrm{Q}_1{}^H\mathrm{Q}_2\\ \mathrm{Q}_2{}^H\mathrm{Q}_1 & \mathrm{Q}_2{}^H\mathrm{Q}_2\end{bmatrix} .$$

Also note that

$$\mathbf{A}^\dagger=\mathbf{R}^{-1}\mathrm{Q}_1{}^H .$$

Let

$$\triangle\mathbf{A}\equiv\mathrm{Q}\begin{bmatrix}\tilde{\mathbf{R}}-\mathbf{R}\\ \mathbf{0}\end{bmatrix} , \text{ and } \triangle\mathbf{c}=\mathrm{Q}\begin{bmatrix}\tilde{\mathbf{y}}-\mathbf{y}\\ \tilde{\mathbf{z}}-\mathbf{z}\end{bmatrix} .$$

Then

$$\tilde{\mathbf{R}}=\mathbf{R}+\mathrm{Q}_1{}^H\triangle\mathbf{A}$$

and

$$\mathbf{R}^{-1}\tilde{\mathbf{R}}=\mathbf{I}+\mathbf{R}^{-1}\mathrm{Q}_1{}^H\triangle\mathbf{A}=\mathbf{I}+\mathbf{A}^\dagger\triangle\mathbf{A} .$$

Inequality (6.14) and assumption (6.16) imply that

$$\left\|\mathbf{A}^\dagger\triangle\mathbf{A}\right\|_2 \leq \left\|\mathbf{A}^\dagger\right\|_2\|\triangle\mathbf{A}\|_2 \leq \frac{M_{(6.14)}\varepsilon}{1-M_{(6.14)}\varepsilon}\|\mathbf{A}\|_F<1 .$$

Then Lemma 3.6.1 proves that $\mathbf{R}^{-1}\tilde{\mathbf{R}} = \mathbf{I} + \mathbf{A}^{\dagger}\triangle\mathbf{A}$ is invertible, and that

$$\left\|\tilde{\mathbf{R}}^{-1}\mathbf{R}\right\|_2 \leq \frac{1}{1 - \|\mathbf{A}^{\dagger}\|_2 \, \|\triangle\mathbf{A}\|_F} \,.$$

Since $\tilde{\mathbf{R}}\tilde{\mathbf{x}} = \tilde{\mathbf{y}} = \mathbf{y} + \mathrm{Q}_1{}^H\triangle\mathbf{c}$,

$$\tilde{\mathbf{x}} - \mathbf{x} = \tilde{\mathbf{R}}^{-1}\tilde{\mathbf{y}} - \mathbf{x} = \tilde{\mathbf{R}}^{-1}\left(\mathbf{y} + \mathrm{Q}_1{}^H\triangle\mathbf{c}\right) - \mathbf{x}$$

and since $\mathbf{y} = \mathbf{R}\mathbf{x}$ we obtain

$$= \tilde{\mathbf{R}}^{-1}\mathbf{R}\mathbf{x} + \tilde{\mathbf{R}}^{-1}\mathrm{Q}_1{}^H\triangle\mathbf{c} - \mathbf{x}$$

then we use the fact that $\mathbf{R} = \tilde{\mathbf{R}} - \mathrm{Q}_1{}^H\triangle\mathbf{A}$ to get

$$= \tilde{\mathbf{R}}^{-1}\left(\tilde{\mathbf{R}} - \mathrm{Q}_1{}^H\triangle\mathbf{A}\right)\mathbf{x} + \tilde{\mathbf{R}}^{-1}\mathrm{Q}_1{}^H\triangle\mathbf{c} - \mathbf{x}$$

and, finally, since $\mathrm{Q}_1{}^H = \mathbf{R}\mathbf{A}^{\dagger}$, we get

$$= -\left(\mathbf{R}^{-1}\tilde{\mathbf{R}}\right)^{-1}\mathbf{A}^{\dagger}\triangle\mathbf{A}\mathbf{x} + \left(\mathbf{R}^{-1}\tilde{\mathbf{R}}\right)^{-1}\mathbf{A}^{\dagger}\triangle\mathbf{c} \,.$$

We can take norms to get

$$\begin{aligned}\|\tilde{\mathbf{x}} - \mathbf{x}\|_2 &\leq \left\|\left(\mathbf{R}^{-1}\tilde{\mathbf{R}}\right)^{-1}\right\|_2 \|\mathbf{A}^{\dagger}\|_2 \, \|\triangle\mathbf{A}\|_2\|\mathbf{x}\|_2 + \left\|\left(\mathbf{R}^{-1}\tilde{\mathbf{R}}\right)^{-1}\right\|_2 \left\|\mathbf{A}^{\dagger}\triangle\mathbf{c}\right\|_2 \\ &\leq \frac{\|\mathbf{A}^{\dagger}\|_2}{1 - \|\mathbf{A}^{\dagger}\|_2 \, \|\triangle\mathbf{A}\|_F}\|\triangle\mathbf{A}\|_F\|\mathbf{x}\|_2 + \frac{\|\mathbf{A}^{\dagger}\|_2}{1 - \|\mathbf{A}^{\dagger}\|_2 \, \|\triangle\mathbf{A}\|_F}\|\triangle\mathbf{c}\|_2\end{aligned}$$

then we use inequalities (6.14) and (6.15) to get

$$\leq \frac{M_{(6.14)}\varepsilon}{1 - M_{(6.14)}\varepsilon}\frac{\|\mathbf{A}^{\dagger}\|_2}{1 - \|\mathbf{A}^{\dagger}\|_2 \, \|\mathbf{A}\|_F M_{(6.14)}\varepsilon / \left(1 - M_{(6.14)}\varepsilon\right)}\left\{\|\mathbf{A}\|_F\|\mathbf{x}\|_2 + \|\mathbf{b}\|_2\right\}$$

This proves claim (6.17).

Since $\tilde{\mathbf{z}} = \mathbf{z} + \mathrm{Q}_2{}^H\triangle\mathbf{c}$, the error in the norm of the residual is

$$\begin{aligned}\left|\|\tilde{\mathbf{z}}\|_2 - \|\mathbf{z}\|_2\right| &= \left|\left\|\mathbf{z} + \mathrm{Q}_2{}^H\triangle\mathbf{c}\right\|_2 - \|\mathbf{z}\|_2\right| \leq \left\|\mathrm{Q}_2{}^H\triangle\mathbf{c}\right\|_2 \leq \|\triangle\mathbf{c}\|_2 \\ &\leq \frac{M_{(6.14)}\varepsilon}{1 - M_{(6.14)}\varepsilon}\|\mathbf{b}\|_2 \,.\end{aligned}$$

This proves claim (6.18).

Finally, suppose that we compute the approximate unitary matrix

$$\tilde{\mathbf{Q}} = \tilde{\mathbf{H}}_1 \cdot \ldots \cdot \tilde{\mathbf{H}}_n$$

by the following recursion:

$$\tilde{\mathbf{Q}}_{n+1} = \mathbf{I}$$
$$\text{for } n \geq k \geq 1 \;,\; \tilde{\mathbf{Q}}_k = \tilde{\mathbf{H}}_k \tilde{\mathbf{Q}}_{k+1}$$

and take $\tilde{\mathbf{Q}} = \tilde{\mathbf{Q}}_1$. Although the order of the application of the Householder reflectors is reversed, we can follow the proof of Lemma 6.7.7 to show that

$$\left\|\tilde{\mathbf{Q}} - \mathbf{Q}\right\|_F \leq \frac{M_{(6.14)}\varepsilon}{1 - M_{(6.14)}\varepsilon} \|\mathbf{I}\|_F \leq \frac{\sqrt{n} M_{(6.14)}\varepsilon}{1 - M_{(6.14)}\varepsilon} \;.$$

This proves claim (6.19).

Theorem 6.7.2 shows that rounding errors in the solution of an overdetermined least squares problem solved by Householder QR factorization grow at a rate proportional to the number of entries in the matrix $\mathbf{A}$, and proportional to the condition number of the matrix $\mathbf{A}$. This means that these computational results should be reliable unless the matrix is very large or very ill-conditioned.

Exercise 6.7.6 Let $\mathbf{A}$ be the 10×5 **Hilbert matrix**

$$\mathbf{A}_{ij} = \frac{1}{1 + i + j} \text{ for } 0 \leq i < 10 \text{ and } 0 \leq j < 5 \;,$$

and let the 10-vector $\mathbf{b}$ have components

$$\mathbf{b}_i = \frac{1}{6 + i} \text{ for } 0 \leq i < 10 \;.$$

1. Estimate the condition number of $\mathbf{A}$.
2. Use Householder QR factorization to factor $\mathbf{A} \approx \tilde{\mathbf{Q}}\tilde{\mathbf{R}}$, and compute $\left\|\tilde{\mathbf{Q}}\tilde{\mathbf{R}} - \mathbf{A}\right\|_F$. Discuss this error in terms of the results in Lemma 6.7.7.
3. Use this approximate Householder QR factorization to solve the overdetermined least squares problem for the approximate solution $\tilde{\mathbf{x}}$, and to find the norm of the approximate residual. Then compute $\tilde{\mathbf{r}} = \mathbf{b} - \mathbf{A}\tilde{\mathbf{x}}$ and compare $\|\tilde{\mathbf{r}}\|_2$ to the residual norm provided by the Householder QR least squares solution process.
4. Compute $\|\mathbf{A}^H\tilde{\mathbf{r}}\|_2$, and use the *a posteriori* error estimates in Sect. 6.6 to estimate the error in $\tilde{\mathbf{x}}$ and $\tilde{\mathbf{r}}$.

6.7.6 *Iterative Improvement*

In Sect. 3.9.2 we discussed how to improve the solution of a linear system by means of an approximate factorization of its matrix. In this section, we will discuss how to use Householder QR factorizations to apply iterative improvement to least squares problems. Motivated readers could apply the rounding error analysis in Lemma 6.7.2 with the study of the convergence of iterative improvement in Lemma 3.9.1 to achieve a fuller understanding of the numerical behavior of iterative improvement. Alternatively, readers may read a detailed study of iterative improvement by Björck [9], a less detailed description by Stewart [97, p. 245ff] or a short discussion by Higham [56, p. 368f].

6.7.6.1 Full-Rank Overdetermined

We begin with the case in which the $m \times n$ matrix $\mathbf{A}$ has rank n. Recall from Lemma 6.3.1 that the true residual $\mathbf{r}$ and the true solution $\mathbf{x}$ satisfy the linear system

$$\begin{bmatrix} \mathbf{I} & \mathbf{A} \\ \mathbf{A}^H & \mathbf{0} \end{bmatrix} \begin{bmatrix} \mathbf{r} \\ \mathbf{x} \end{bmatrix} = \begin{bmatrix} \mathbf{b} \\ \mathbf{0} \end{bmatrix} .$$

Suppose that we are given an approximate residual $\tilde{\mathbf{r}}$ and an approximate solution $\tilde{\mathbf{x}}$. We can use these to compute the corresponding errors in this equation:

$$\begin{bmatrix} \delta\mathbf{b} \\ \delta\mathbf{c} \end{bmatrix} \equiv \begin{bmatrix} \mathbf{b} \\ \mathbf{0} \end{bmatrix} - \begin{bmatrix} \mathbf{I} & \mathbf{A} \\ \mathbf{A}^H & \mathbf{0} \end{bmatrix} \begin{bmatrix} \tilde{\mathbf{r}} \\ \tilde{\mathbf{x}} \end{bmatrix} .$$

If we can solve

$$\begin{bmatrix} \mathbf{I} & \mathbf{A} \\ \mathbf{A}^H & \mathbf{0} \end{bmatrix} \begin{bmatrix} \delta\mathbf{r} \\ \delta\mathbf{x} \end{bmatrix} = \begin{bmatrix} \delta\mathbf{b} \\ \delta\mathbf{c} \end{bmatrix} ,$$

then we should have

$$\begin{bmatrix} \mathbf{I} & \mathbf{A} \\ \mathbf{A}^H & \mathbf{0} \end{bmatrix} \begin{bmatrix} \tilde{\mathbf{r}} + \delta\mathbf{r} \\ \tilde{\mathbf{x}} + \delta\mathbf{x} \end{bmatrix} = \begin{bmatrix} \mathbf{b} - \delta\mathbf{b} \\ \mathbf{0} - \delta\mathbf{c} \end{bmatrix} + \begin{bmatrix} \delta\mathbf{b} \\ \delta\mathbf{c} \end{bmatrix} = \begin{bmatrix} \mathbf{b} \\ \mathbf{0} \end{bmatrix} .$$

Thus $\tilde{\mathbf{x}} + \delta\mathbf{x}$ will hopefully be a better approximation to the true solution, and $\tilde{\mathbf{r}} + \delta\mathbf{r}$ should be closer to the true residual.

Suppose that we have approximately factored

$$\mathbf{A} \approx \tilde{\mathbf{Q}} \begin{bmatrix} \tilde{\mathbf{R}} \\ \mathbf{0} \end{bmatrix}$$

where $\tilde{\mathbf{Q}}$ is nearly a unitary matrix, and $\tilde{\mathbf{R}}$ is right-triangular and nonsingular. This approximate factorization should allow us to solve for approximate increments $\widetilde{\delta\mathbf{r}}$ and $\widetilde{\delta\mathbf{x}}$. Let us define

$$\begin{bmatrix} \widetilde{\delta\mathbf{y}} \\ \widetilde{\delta\mathbf{z}} \end{bmatrix} \equiv \tilde{\mathbf{Q}}^H \delta\mathbf{b} \text{ and } \begin{bmatrix} \widetilde{\delta\mathbf{v}} \\ \widetilde{\delta\mathbf{w}} \end{bmatrix} \equiv \tilde{\mathbf{Q}}^H \widetilde{\delta\mathbf{r}} .$$

Here $\widetilde{\delta\mathbf{y}}$ and $\widetilde{\delta\mathbf{v}}$ are n-vectors. We require the approximate increments to satisfy

$$\delta\mathbf{b} = \widetilde{\delta\mathbf{r}} + \tilde{\mathbf{Q}} \begin{bmatrix} \tilde{\mathbf{R}} \\ \mathbf{0} \end{bmatrix} \widetilde{\delta\mathbf{x}} ,$$

from which it follows that

$$\begin{bmatrix} \widetilde{\delta\mathbf{y}} \\ \widetilde{\delta\mathbf{z}} \end{bmatrix} = \tilde{\mathbf{Q}}^H \widetilde{\delta\mathbf{r}} + \begin{bmatrix} \tilde{\mathbf{R}}\widetilde{\delta\mathbf{x}} \\ \mathbf{0} \end{bmatrix} = \begin{bmatrix} \widetilde{\delta\mathbf{v}} \\ \widetilde{\delta\mathbf{w}} \end{bmatrix} + \begin{bmatrix} \tilde{\mathbf{R}}\widetilde{\delta\mathbf{x}} \\ \mathbf{0} \end{bmatrix} .$$

We should also have

$$\delta\mathbf{c} = \begin{bmatrix} \mathbf{R}^H & \mathbf{0} \end{bmatrix} \tilde{\mathbf{Q}}^H \widetilde{\delta\mathbf{r}} = \tilde{\mathbf{R}}^H \widetilde{\delta\mathbf{v}} .$$

If we do not already have an approximation to the residual vector, we can compute $\tilde{\mathbf{r}} = \mathbf{b} - \mathbf{A}\tilde{\mathbf{x}}$. Then the observations in the previous paragraph lead to the following algorithm:

Algorithm 6.7.10 (Householder Overdetermined Iterative Improvement)

$\delta\mathbf{b} = \mathbf{b} - \tilde{\mathbf{r}} - \mathbf{A}\tilde{\mathbf{x}}$ /* LAPACK BLAS routine _gemv */

$\delta\mathbf{c} = -\mathbf{A}^H\tilde{\mathbf{r}}$ /* LAPACK BLAS routine _gemv */

$\begin{bmatrix} \widetilde{\delta\mathbf{y}} \\ \widetilde{\delta\mathbf{z}} \end{bmatrix} = \tilde{\mathbf{Q}}^H \delta\mathbf{b}$ /* LAPACK routine _ormqr */

solve $\tilde{\mathbf{R}}^H \widetilde{\delta\mathbf{v}} = \delta\mathbf{c}$ /* LAPACK BLAS routine _trsv */

solve $\tilde{\mathbf{R}} \widetilde{\delta\mathbf{x}} = \widetilde{\delta\mathbf{y}} - \widetilde{\delta\mathbf{v}}$ /* LAPACK BLAS routine _trsv */

$\widetilde{\delta\mathbf{r}} = \tilde{\mathbf{Q}} \begin{bmatrix} \widetilde{\delta\mathbf{v}} \\ \widetilde{\delta\mathbf{z}} \end{bmatrix}$ /* LAPACK routine _ormqr */

$\tilde{\mathbf{x}} = \tilde{\mathbf{x}} + \widetilde{\delta\mathbf{x}}$

$\tilde{\mathbf{r}} = \tilde{\mathbf{r}} + \widetilde{\delta\mathbf{r}}$

Note that it may be necessary to compute $\tilde{\mathbf{r}}$, $\delta\mathbf{b}$ and $\delta\mathbf{c}$ in extended precision arithmetic.

Example 6.7.3 Let

$$\tilde{\mathbf{A}} = \begin{bmatrix} 1 & 1 \\ \alpha & 0 \\ 0 & \alpha \end{bmatrix} \text{ and } \mathbf{b} = \begin{bmatrix} 1 \\ 0 \\ 0 \end{bmatrix} ,$$

where $\text{fl}(1+\alpha^2) = 1$. Recall from Example 6.7.1 that Householder QR factorization of $\mathbf{A}$ produced

$$\tilde{\mathbf{R}} = \begin{bmatrix} -1 & -1 \\ & \alpha\sqrt{2} \end{bmatrix} .$$

Suppose that we take

$$\tilde{\mathbf{x}} = \begin{bmatrix} 1 \\ 0 \end{bmatrix} \text{ and } \tilde{\mathbf{r}} = \mathbf{b} - \mathbf{A}\tilde{\mathbf{x}} = \begin{bmatrix} 0 \\ -\alpha \\ 0 \end{bmatrix} .$$

Then iterative improvement would compute $\delta\mathbf{b} = \mathbf{0}$ and

$$\delta\mathbf{c} = -\mathbf{A}^H\mathbf{r} = \begin{bmatrix} \alpha^2 \\ 0 \end{bmatrix} .$$

Thus $\tilde{\delta\mathbf{y}} = \mathbf{0} = \tilde{\delta\mathbf{z}}$,

$$\mathbf{R}^H\tilde{\delta\mathbf{v}} = \delta\mathbf{c} \Longrightarrow \tilde{\delta\mathbf{v}} = \begin{bmatrix} -\alpha^2 \\ -\alpha/\sqrt{2} \end{bmatrix} , \text{ and}$$

$$\mathbf{R}\tilde{\delta\mathbf{x}} = -\tilde{\delta\mathbf{v}} \Longrightarrow \tilde{\delta\mathbf{x}} = \begin{bmatrix} -1/2 \\ 1/2 \end{bmatrix} .$$

$$\tilde{\mathbf{x}} = \tilde{\mathbf{x}} + \tilde{\delta\mathbf{x}} = \begin{bmatrix} 1/2 \\ 1/2 \end{bmatrix} .$$

In this case, iterative improvement produces the correctly rounded solution.

Readers may also examine an implementation of full-rank iterative improvement for overdetermined least squares problems by the C^{++} `HouseholderQRFactorization` class in files HouseholderQRFactorization.H and HouseholderQRFactorization.C. This class calls type-specific LAPACK routines to perform the operations; for example, the double precision implementation is contained in LaDouble.C.

6.7.6.2 Full-Rank Underdetermined

Next, we will study iterative improvement in the case where the $m \times n$ matrix $\mathbf{A}$ has rank m. Recall from Lemma 6.3.1 that the true solution $\mathbf{x}$ and an auxiliary vector $\mathbf{s}$ satisfy the linear system

$$\begin{bmatrix} \mathbf{I} & \mathbf{A}^H \\ \mathbf{A} & \mathbf{0} \end{bmatrix} \begin{bmatrix} \mathbf{x} \\ -\mathbf{s} \end{bmatrix} = \begin{bmatrix} \mathbf{0} \\ \mathbf{b} \end{bmatrix} .$$

Suppose that we are given an approximate solution $\tilde{\mathbf{x}}$ and an approximate auxiliary vector $\tilde{\mathbf{s}}$. We can use these to compute the corresponding errors in the previous equations:

$$\begin{bmatrix} \delta\mathbf{a} \\ \delta\mathbf{b} \end{bmatrix} = \begin{bmatrix} \mathbf{0} \\ \mathbf{b} \end{bmatrix} - \begin{bmatrix} \mathbf{I} & \mathbf{A}^H \\ \mathbf{A} & \mathbf{0} \end{bmatrix} \begin{bmatrix} \tilde{\mathbf{x}} \\ -\tilde{\mathbf{s}} \end{bmatrix} .$$

If we can solve

$$\begin{bmatrix} \mathbf{I} & \mathbf{A}^H \\ \mathbf{A} & \mathbf{0} \end{bmatrix} \begin{bmatrix} \delta\mathbf{x} \\ -\delta\mathbf{s} \end{bmatrix} = \begin{bmatrix} \delta\mathbf{a} \\ \delta\mathbf{b} \end{bmatrix} ,$$

then we should have

$$\begin{bmatrix} \mathbf{I} & \mathbf{A}^H \\ \mathbf{A} & \mathbf{0} \end{bmatrix} \begin{bmatrix} \tilde{\mathbf{x}} + \delta\mathbf{x} \\ -\tilde{\mathbf{s}} - \delta\mathbf{s} \end{bmatrix} = \begin{bmatrix} -\delta\mathbf{a} \\ \mathbf{b} - \delta\mathbf{b} \end{bmatrix} + \begin{bmatrix} \delta\mathbf{a} \\ \delta\mathbf{b} \end{bmatrix} = \begin{bmatrix} \mathbf{0} \\ \mathbf{b} \end{bmatrix} .$$

Thus $\tilde{\mathbf{x}} + \delta\mathbf{x}$ will hopefully be a better approximation to the true solution.

Suppose that we have approximately factored

$$\mathbf{A} \approx \begin{bmatrix} \tilde{\mathbf{L}} & \mathbf{0} \end{bmatrix} \tilde{\mathbf{Q}}$$

where $\tilde{\mathbf{Q}}$ is nearly a unitary matrix, and $\tilde{\mathbf{L}}$ is left-triangular and nonsingular. We can define

$$\begin{bmatrix} \widetilde{\delta\mathbf{f}} \\ \widetilde{\delta\mathbf{g}} \end{bmatrix} \equiv \tilde{\mathbf{Q}}\delta\mathbf{a} \text{ and } \begin{bmatrix} \widetilde{\delta\mathbf{y}} \\ \widetilde{\delta\mathbf{w}} \end{bmatrix} \equiv \tilde{\mathbf{Q}}\widetilde{\delta\mathbf{x}} .$$

Then

$$\delta\mathbf{a} = \widetilde{\delta\mathbf{x}} - \tilde{\mathbf{Q}}^H \begin{bmatrix} \tilde{\mathbf{L}}^H \\ \mathbf{0} \end{bmatrix} \widetilde{\delta\mathbf{s}} ,$$

from which it follows that

$$\begin{bmatrix} \tilde{\mathbf{L}}^H \widetilde{\delta\mathbf{s}} \\ \mathbf{0} \end{bmatrix} = \tilde{\mathbf{Q}}\delta\mathbf{a} - \tilde{\mathbf{Q}}\widetilde{\delta\mathbf{x}} = \begin{bmatrix} \widetilde{\delta\mathbf{f}} \\ \widetilde{\delta\mathbf{g}} \end{bmatrix} - \begin{bmatrix} \widetilde{\delta\mathbf{y}} \\ \widetilde{\delta\mathbf{w}} \end{bmatrix} .$$

We also have

$$\delta\mathbf{b} = [\mathbf{L}\ \mathbf{0}]\,\tilde{\mathbf{Q}}\widetilde{\delta\mathbf{x}} = [\mathbf{L}\ \mathbf{0}]\begin{bmatrix}\widetilde{\delta\mathbf{y}} \\ \widetilde{\delta\mathbf{w}}\end{bmatrix} = \mathbf{L}\widetilde{\delta\mathbf{y}}\ .$$

If we do not already have an approximation to the auxiliary vector $\mathbf{s}$, one could be obtained by solving $\mathbf{L}^H\mathbf{s} = \mathbf{y}$, where $\mathbf{y}$ was determined as the solution of $\mathbf{L}\mathbf{y} = \mathbf{b}$ in Algorithm 6.7.8. Then the observations in the previous paragraph lead to the following algorithm:

Algorithm 6.7.11 (Householder Underdetermined Iterative Improvement)

$\delta\mathbf{a} = \mathbf{A}^H\tilde{\mathbf{s}} - \tilde{\mathbf{x}}$

$\delta\mathbf{b} = \mathbf{b} - \mathbf{A}\tilde{\mathbf{x}}$ /* LAPACK BLAS routine _gemv */

$\begin{bmatrix}\widetilde{\delta\mathbf{f}} \\ \widetilde{\delta\mathbf{g}}\end{bmatrix} = \tilde{\mathbf{Q}}\delta\mathbf{a}$ /* LAPACK routine _ormqr */

solve $\mathbf{L}\widetilde{\delta\mathbf{v}} = \delta\mathbf{b}$ /* LAPACK BLAS routine _trsv */

$\widetilde{\delta\mathbf{x}} = \tilde{\mathbf{Q}}^H\begin{bmatrix}\widetilde{\delta\mathbf{v}} \\ \widetilde{\delta\mathbf{g}}\end{bmatrix}$ /* LAPACK routine _ormqr */

$\tilde{\mathbf{x}} = \tilde{\mathbf{x}} + \widetilde{\delta\mathbf{x}}$

solve $\mathbf{L}^H\widetilde{\delta\mathbf{s}} = \widetilde{\delta\mathbf{f}} + \widetilde{\delta\mathbf{v}}$ /* LAPACK BLAS routine _trsv */

$\tilde{\mathbf{s}} = \tilde{\mathbf{s}} + \widetilde{\delta\mathbf{s}}$

Note that it may be necessary to compute $\delta\mathbf{a}$ and $\delta\mathbf{b}$ in extended precision arithmetic.

For more information regarding the performance of this iterative improvement algorithm, see Björck and Golub [12]. Readers may examine an implementation of full-rank iterative improvement for underdetermined least squares problems by the C++ `HouseholderQRFactorization` class in files HouseholderQRFactorization.H and HouseholderQRFactorization.C. This class calls data type-specific LAPACK routines to perform the operations; for the implementation in double precision, see LaDouble.C.

6.7.6.3 Rank Deficient

Finally, we will consider iterative improvement for rank-deficient least squares problems. Suppose that $\mathbf{A}$ is an $m \times n$ matrix with rank $(\mathbf{A}) = r$ determined by Householder *QR* factorization *with pivoting*, as described in Sect. 6.7.2. Recall from Lemma 6.3.1 that the true residual $\mathbf{r}$, true solution $\mathbf{x}$ and true auxiliary vector $\mathbf{s}$ satisfy the equations

$$\mathbf{x} - \mathbf{A}^H\mathbf{s} = \mathbf{0}\ ,\ \mathbf{r} + \mathbf{A}\mathbf{x} = \mathbf{b} \text{ and } \mathbf{A}^H\mathbf{r} = \mathbf{0}\ .$$

Suppose that we are given an approximate residual $\tilde{\mathbf{r}}$, an approximate solution $\mathbf{x}$ and an approximate auxiliary vector $\tilde{\mathbf{s}}$. We can use these to compute the following errors in the previous equations:

$$\delta\mathbf{a} = \mathbf{A}^H\tilde{\mathbf{s}} - \tilde{\mathbf{x}} \ , \ \delta\mathbf{b} = \mathbf{b} - \tilde{\mathbf{r}} - \mathbf{A}\tilde{\mathbf{x}} \text{ and } \delta\mathbf{c} = -\mathbf{A}^H\tilde{\mathbf{r}} \ .$$

If we can solve the linear system of equations

$$\delta\mathbf{x} - \mathbf{A}^H\delta\mathbf{s} = \delta\mathbf{a} \ , \ \delta\mathbf{r} + \mathbf{A}\delta\mathbf{x} = \delta\mathbf{b} \text{ and } \mathbf{A}^H\delta\mathbf{r} = \delta\mathbf{c} \ ,$$

then we find that

$$(\tilde{\mathbf{x}} + \delta\mathbf{x}) - \mathbf{A}^H(\tilde{\mathbf{s}} + \delta\mathbf{s}) = \left(\mathbf{H}^H\tilde{\mathbf{s}} - \delta\mathbf{a} + \delta\mathbf{x}\right) - \mathbf{A}^H\tilde{\mathbf{s}} + (\delta\mathbf{a} - \tilde{\mathbf{x}}) = \mathbf{0} \ ,$$
$$(\tilde{\mathbf{r}} + \delta\mathbf{r}) + \mathbf{A}(\tilde{\mathbf{x}} + \delta\mathbf{x}) = (\mathbf{b} - \delta\mathbf{b} - \mathbf{A}\tilde{\mathbf{x}} + \delta\mathbf{r}) + \mathbf{A}\delta\mathbf{x} + \delta\mathbf{b} - \delta\mathbf{r} = \mathbf{b} \text{ and}$$
$$\mathbf{A}^H(\tilde{\mathbf{r}} + \delta\mathbf{r}) = -\delta\mathbf{c} + \mathbf{A}^H\delta\mathbf{r} = \mathbf{0} \ .$$

Thus $\tilde{\mathbf{x}} + \delta\mathbf{x}$ will hopefully be a better approximation to the true solution.

Suppose that we have computed the complete orthogonal decomposition

$$\mathbf{A} \approx \tilde{\mathbf{U}} \begin{bmatrix} \tilde{\mathbf{R}} & \mathbf{0} \\ \mathbf{0} & \mathbf{0} \end{bmatrix} \tilde{\mathbf{V}}^H\mathbf{P}^H$$

where $\tilde{\mathbf{U}}$ and $\tilde{\mathbf{V}}$ are approximately unitary matrices, $\mathbf{P}$ is a permutation matrix, and $\tilde{\mathbf{R}}$ is $r \times r$ right triangular and nonsingular. Define

$$\begin{bmatrix} \delta\mathbf{v} \\ \delta\mathbf{w} \end{bmatrix} \equiv \tilde{\mathbf{U}}^H\delta\mathbf{r} \ .$$

Then we have

$$\begin{bmatrix} \tilde{\delta\mathbf{p}} \\ \tilde{\delta\mathbf{q}} \end{bmatrix} \equiv \tilde{\mathbf{V}}^H\mathbf{P}^H\delta\mathbf{c} = -\begin{bmatrix} \tilde{\mathbf{R}}^H & \mathbf{0} \\ \mathbf{0} & \mathbf{0} \end{bmatrix} \tilde{\mathbf{U}}^H\mathbf{r} \equiv -\begin{bmatrix} \tilde{\mathbf{R}}^H & \mathbf{0} \\ \mathbf{0} & \mathbf{0} \end{bmatrix} \begin{bmatrix} \delta\mathbf{v} \\ \delta\mathbf{w} \end{bmatrix} \ .$$

If we define

$$\begin{bmatrix} \delta\mathbf{u} \\ \delta\mathbf{u}' \end{bmatrix} \equiv \tilde{\mathbf{V}}^H\mathbf{P}^H\delta\mathbf{x} \ ,$$

then we also have

$$\begin{bmatrix} \delta\mathbf{y} \\ \delta\mathbf{z} \end{bmatrix} \equiv \tilde{\mathbf{U}}^H\delta\mathbf{b} = \tilde{\mathbf{U}}^H(\delta\mathbf{r} + \mathbf{A}\delta\mathbf{x}) = \tilde{\mathbf{U}}^H\delta\mathbf{r} + \begin{bmatrix} \tilde{\mathbf{R}} & \mathbf{0} \\ \mathbf{0} & \mathbf{0} \end{bmatrix} \tilde{\mathbf{V}}^H\mathbf{P}^H\delta\mathbf{x} \equiv \begin{bmatrix} \delta\mathbf{v} \\ \delta\mathbf{w} \end{bmatrix} + \begin{bmatrix} \tilde{\mathbf{R}} & \mathbf{0} \\ \mathbf{0} & \mathbf{0} \end{bmatrix} \begin{bmatrix} \delta\mathbf{u} \\ \delta\mathbf{u}' \end{bmatrix} \ .$$

Finally, if we define

$$\begin{bmatrix} \delta\mathbf{t} \\ \delta\mathbf{t}' \end{bmatrix} \equiv \tilde{\mathbf{U}}^H \delta\mathbf{s} ,$$

then we also have

$$\begin{bmatrix} \delta\mathbf{f} \\ \delta\mathbf{g} \end{bmatrix} \equiv \tilde{\mathbf{V}}^H \mathbf{P}^H \delta\mathbf{a} = \tilde{\mathbf{V}}^H \mathbf{P}^H \delta\mathbf{x} - \begin{bmatrix} \tilde{\mathbf{R}}^H & \mathbf{0} \\ \mathbf{0} & \mathbf{0} \end{bmatrix} \tilde{\mathbf{U}}^H \delta\mathbf{s} \equiv \begin{bmatrix} \delta\mathbf{u} \\ \delta\mathbf{u}' \end{bmatrix} - \begin{bmatrix} \tilde{\mathbf{R}}^H & \mathbf{0} \\ \mathbf{0} & \mathbf{0} \end{bmatrix} \begin{bmatrix} \delta\mathbf{t} \\ \delta\mathbf{t}' \end{bmatrix} .$$

If we do not already have an approximation to the auxiliary vector $\tilde{\mathbf{s}}$, one could be computed by the following computations:

$$\text{solve } \tilde{\mathbf{R}}^H \mathbf{t} = \tilde{\mathbf{v}} \text{ for } \mathbf{t} \text{ and compute } \tilde{\mathbf{s}} = \tilde{\mathbf{U}} \begin{bmatrix} \mathbf{t} \\ \mathbf{0} \end{bmatrix} = \mathbf{H}_1 \cdot \ldots \cdot \mathbf{H}_r \begin{bmatrix} \mathbf{t} \\ \mathbf{0} \end{bmatrix} .$$

In these equations, we use the vector $\mathbf{v}$ that was computed in Algorithm 6.7.9. The observations in the previous paragraph lead to the following algorithm:

Algorithm 6.7.12

$$\begin{aligned}
&\delta\mathbf{a} = \mathbf{A}^H \tilde{\mathbf{s}} - \tilde{\mathbf{x}} \\
&\delta\mathbf{b} = \mathbf{b} - \tilde{\mathbf{r}} - \mathbf{A}\tilde{\mathbf{x}} \\
&\delta\mathbf{c} = -\mathbf{A}^H \tilde{\mathbf{r}} \\
&\begin{bmatrix} \widetilde{\delta\mathbf{p}} \\ \widetilde{\delta\mathbf{q}} \end{bmatrix} = \tilde{\mathbf{V}}^H \mathbf{P}^H \delta\mathbf{c} \\
&\text{solve } \tilde{\mathbf{R}}^H \delta\mathbf{v} = -\widetilde{\delta\mathbf{p}} \\
&\begin{bmatrix} \delta\mathbf{y} \\ \delta\mathbf{z} \end{bmatrix} = \tilde{\mathbf{U}}^H \delta\mathbf{b} \\
&\text{solve } \tilde{\mathbf{R}} \delta\mathbf{u} = \delta\mathbf{y} - \delta\mathbf{v} \\
&\begin{bmatrix} \delta\mathbf{f} \\ \delta\mathbf{g} \end{bmatrix} = \tilde{\mathbf{V}}^H \mathbf{P}^H \delta\mathbf{a} \\
&\text{solve } \tilde{\mathbf{R}}^H \delta\mathbf{t} = \delta\mathbf{u} - \delta\mathbf{f} \\
&\delta\mathbf{x} = \mathbf{P}\tilde{\mathbf{V}} \begin{bmatrix} \delta\mathbf{u} \\ \delta\mathbf{g} \end{bmatrix} \\
&\delta\mathbf{r} = \tilde{\mathbf{U}} \begin{bmatrix} \delta\mathbf{v} \\ \delta\mathbf{z} \end{bmatrix} \\
&\delta\mathbf{s} = \tilde{\mathbf{U}} \begin{bmatrix} \delta\mathbf{t} \\ \mathbf{0} \end{bmatrix} \\
&\mathbf{x} = \tilde{\mathbf{x}} + \delta\mathbf{x} \\
&\mathbf{r} = \tilde{\mathbf{r}} + \delta\mathbf{r} \\
&\mathbf{s} = \tilde{\mathbf{s}} + \delta\mathbf{s}
\end{aligned}$$

Exercise 6.7.7 Let $\mathbf{A}$ be the 10×5 **Hilbert matrix**

$$\mathbf{A}_{ij} = \frac{1}{1+i+j} \text{ for } 0 \le i < 10 \text{ and } 0 \le j < 5 ,$$

and let the 10-vector $\mathbf{b}$ have components

$$\mathbf{b}_i = \frac{1}{6+i} \text{ for } 0 \le i < 10 .$$

1. Estimate the condition number of $\mathbf{A}$.
2. Use Householder QR factorization to factor $\mathbf{A} \approx \tilde{\mathbf{Q}}\tilde{\mathbf{R}}$, Lemma 6.7.7 in single precision.
3. Use this approximate Householder QR factorization to solve the overdetermined least squares problem for the approximate solution $\tilde{\mathbf{x}}$ in single precision Then compute $\tilde{\mathbf{r}} = \mathbf{b} - \mathbf{A}\tilde{\mathbf{x}}$ in double precision.
4. Perform one step of iterative improvement in single precision.
5. Use the *a posteriori* error estimates in Sect. 6.6 to estimate the error in $\tilde{\mathbf{x}}$ and $\tilde{\mathbf{r}}$. Did iterative improvement improve the numerical solution of this least squares problem?

6.8 Gram-Schmidt QR Factorization

In Sect. 6.7 we described and analyzed successive reflection for the solution of least squares problems. This approach is the most popular in scientific computing, and is the method of choice in software packages such as LAPACK and MATLAB.

Our goal in this section is to develop successive orthogonal projection as an alternative method for solving least squares problems. In theoretical mathematics, this approach is more popular than successive reflection. There are a couple of reasons for such unbalanced popularity. Orthogonal projection is easier to describe by equations, making it simpler for mathematical exposition. On the other hand, since projection is a length-reducing operation, the potential loss of accuracy and orthogonality due to rounding errors makes orthogonal projection less popular for computation. Furthermore, we will see that the factorization produced by successive orthogonal projection requires more memory than the Householder factorization.

6.8.1 *Successive Orthogonal Projection*

Our successive orthogonal projection method is a modification of the **Gram-Schmidt orthogonalization** process, which is familiar to readers of introductory linear algebra. Almost all linear algebra books discuss this process, and almost all of those books present a numerically unstable algorithm; see, for example, Halmos [51, p. 127] or Shifrin [92, p. 250]. In this section, we will present the stable algorithm, which was discovered to be superior to the classical Gram-Schmidt process by Rice [91]. We will contrast the stable algorithm with the classical (unstable) algorithm in Sect. 6.8.2.

6.8.1.1 Matrix Factorization

The reader should recall that we initially discussed orthogonal projection in Sect. 3.4.8.1. There we described how to solve a linear system involving a matrix with mutually orthogonal columns. This solution process was summarized in the Successive Orthogonal Projection Algorithm 3.4.9. In this section, we will show how to factor an $m \times n$ matrix $\mathbf{A}$ with $\text{rank}(\mathbf{A}) = n$ into a product $\mathbf{A} = \mathbf{Q}\mathbf{R}$, where $\mathbf{Q}$ has orthogonal columns and $\mathbf{R}$ is right-triangular. We will show how to use such a factorization to solve least squares problems later in this section.

Suppose that we are given an $m \times n$ matrix $\mathbf{A}$ with $\text{rank}(\mathbf{A}) = n$. This implies that $m \geq n$. We will describe modified Gram-Schmidt factorization $\mathbf{A} = \mathbf{Q}\mathbf{R}$ in two forms, depending on whether we compute the entries of $\mathbf{R}$ by rows or by columns.

In the row-oriented form of modified Gram-Schmidt factorization, we begin by partitioning the first column of $\mathbf{A}$ and factoring

$$\mathbf{A} = \begin{bmatrix} \mathbf{a}_1 & \mathbf{A}_2 \end{bmatrix} = \begin{bmatrix} \mathbf{q}_1 & \mathbf{A}_2' \end{bmatrix} \begin{bmatrix} \varrho_{11} & {\mathbf{r}_{12}}^\top \\ 0 & \mathbf{I} \end{bmatrix} ,$$

where

$$\varrho_{11} = \|\mathbf{a}_1\|_2 \; , \; \mathbf{q}_1 = \mathbf{a}_1 / \varrho_{11} \; , \; \overline{\mathbf{r}_{12}} = {\mathbf{A}_2}^H \mathbf{q}_1 \text{ and } \mathbf{A}_2' = \mathbf{A}_2 - \mathbf{q}_1 {\mathbf{r}_{12}}^\top \; .$$

Note that these computations imply that ${\mathbf{q}_1}^H \mathbf{A}_2' = \mathbf{0}$.

The general step in the row-oriented modified Gram-Schmidt factorization takes the following form:

$$\begin{aligned} \mathbf{A} &= \begin{bmatrix} \mathbf{Q}_{k-1} & \mathbf{a}_k & \mathbf{A}_{k+1} \end{bmatrix} \begin{bmatrix} \mathbf{R}_{k-1,k-1} & \mathbf{r}_{k-1,k} & \mathbf{R}_{k-1,k+1} \\ & 1 & 0 \\ & & \mathbf{I} \end{bmatrix} \\ &= \begin{bmatrix} \mathbf{Q}_{k-1} & \mathbf{q}_k & \mathbf{A}_{k+1}' \end{bmatrix} \begin{bmatrix} \mathbf{R}_{k-1,k-1} & \mathbf{r}_{k-1,k} & \mathbf{R}_{k-1,k+1} \\ & \varrho_{kk} & {\mathbf{r}_{k,k+1}}^\top \\ & & \mathbf{I} \end{bmatrix} , \end{aligned}$$

where

$$\varrho_{kk} = \|\mathbf{a}_k\|_2 \;,\; \mathbf{q}_k = \mathbf{a}_k/\varrho_{kk} \;,\; \overline{\mathbf{r}_{k,k+1}} = \mathbf{A}_{k+1}{}^H\mathbf{q}_k \text{ and } \mathbf{A}'_{k+1} = \mathbf{A}_{k+1} - \mathbf{q}_k\mathbf{r}_{k,k+1}{}^\top \;.$$

(Note that $\mathbf{a}_k$ is not necessarily equal to any column of the original matrix $\mathbf{A}$; rather, it is a vector determined in the midst of the factorization process.) The process continues in this way, until all columns of the matrix have been orthogonalized with respect to the previous columns.

These computations lead to the following algorithm.

Algorithm 6.8.1 (Row-Oriented Modified Gram-Schmidt Factorization)

for $1 \le k \le n$
  $\varrho_{kk} = \|\mathbf{A}_{1:m,k}\|_2$ /* LAPACK BLAS routine _nrm2 */
  if $\varrho_{kk} = 0$ break /* rank deficient matrix */
  $\mathbf{Q}_{1:m,k} = \mathbf{A}_{1:m,k}/\varrho_{kk}$ /* LAPACK BLAS routine _scal or _dscal */
  $\mathbf{w}_{k+1:n} = \mathbf{A}_{1:m,k+1:n}{}^H\mathbf{Q}_{1:m,k}$ /* LAPACK BLAS routine _gemv */
  $\mathbf{R}_{k,k+1:n} = \overline{\mathbf{w}_{k+1:n}}$ /* no work here in real arithmetic */
  $\mathbf{A}_{1:m,k+1:n} = \mathbf{A}_{1:m,k+1:n} - \mathbf{Q}_{1:m,k}\mathbf{R}_{k,k+1:n}$ /* LAPACK BLAS routine _ger or _geru */

Note that we can store the kth column of $\mathbf{Q}$ in the kth column of $\mathbf{A}$.

In the column-oriented form of modified Gram-Schmidt factorization, we successively project the next column against all preceding columns. The first two steps of this form are the same as in the previous form of the algorithm. The general step in this form of successive orthogonal projection looks like the following:

$$\mathbf{A} = \begin{bmatrix} \mathbf{Q}_{k-1} & \mathbf{a}_k & \mathbf{A}_{k+1} \end{bmatrix} \begin{bmatrix} \mathbf{R}_{k-1} & \mathbf{0} & \mathbf{0} \\ & 1 & 0 \\ & & \mathbf{I} \end{bmatrix} = \begin{bmatrix} \mathbf{Q}_{k-1} & \mathbf{q}_k & \mathbf{A}_{k+1} \end{bmatrix} \begin{bmatrix} \mathbf{R}_{k-1} & \mathbf{r}_k & \mathbf{0} \\ & \varrho_k & \mathbf{0} \\ & & \mathbf{I} \end{bmatrix} ,$$

Here, if $\mathbf{Q}_{k-1} = \begin{bmatrix} \mathbf{q}_1 & \dots & \mathbf{q}_{k-1} \end{bmatrix}$ has orthonormal columns, then we compute

$$\mathbf{a}'_k = \left(\mathbf{I} - \mathbf{q}_{k-1}\mathbf{q}_{k-1}{}^H\right) \dots \left(\mathbf{I} - \mathbf{q}_1\mathbf{q}_1{}^H\right) \mathbf{a}_k$$

by applying successive orthogonal projection. Afterward, we compute

$$\varrho_k = \left\|\mathbf{a}'_k\right\|_2 \text{ and } \mathbf{q}_k = \mathbf{a}'_k/\varrho_k \;.$$

When all columns have been orthogonalized with respect to all previous columns, we have factored $\mathbf{A} = \mathbf{Q}\mathbf{R}$ via the following algorithm:

Algorithm 6.8.2 (Column-Oriented Modified Gram-Schmidt Factorization)

for $1 \leq k \leq n$
  for $1 \leq \ell < k$
    $\mathbf{R}_{\ell k} = \mathbf{Q}_{1:m,\ell} \cdot \mathbf{A}_{1:m,k}$ /* LAPACK BLAS routine _dot or _dotc */
    $\mathbf{A}_{1:m,k} = \mathbf{A}_{1:m,k} - \mathbf{Q}_{1:m,\ell}\mathbf{R}_{\ell k}$ /* LAPACK BLAS routine _axpy */
  $\varrho_k = \|\mathbf{A}_{1:m,k}\|_2$ /* LAPACK BLAS routine _nrm2 */
  $\mathbf{Q}_{1:m,k} = \mathbf{A}_{1:m,k}/\varrho_k$ /* LAPACK BLAS routine _scal or _dscal */
  if $\varrho_k = 0$ break /* rank deficient matrix */

This form of the algorithm is useful in cases when columns of **A** are not available at the same time, as in **reorthogonalization** and some iterative methods. However, this alternative modified Gram-Schmidt factorization does not make use of LAPACK level 2 BLAS routines.

Let us determine the computational work involved in successive orthogonal projection. During the k-th step, we make $k - 1$ calls to _dot (m multiplications and additions per call), $k - 1$ calls to _axpy (m multiplications and additions per call), 1 call to _nrm2 (m multiplications and additions plus 1 square root) and 1 call to *_scal* or *_dscal* (m divides). Thus if **A** is not rank-deficient, the total work is

- $\sum_{k=1}^{n} m(2k-1) = mn^2$ multiplications and additions,
- $\sum_{k=1}^{n} m = mn$ divides and
- n square roots.

The total work is on the order of mn^2 multiplications and additions.

6.8.1.2 Least Squares Problems

Suppose that **A** is an $m \times n$ matrix with $rank\mathbf{A} = n$, and assume that we have factored $\mathbf{A} = \mathbf{QR}$ where $\mathbf{Q}$ has orthonormal columns and **R** is right-triangular with positive diagonal entries. It follows that the columns of $\mathbf{Q}$ form an orthonormal basis for the range of **A**. If **b** is an m-vector, then the least squares residual $\mathbf{r} = \mathbf{b} - \mathbf{Ax}$ must be orthogonal to $\mathscr{R}(\mathbf{A})$. Thus we can compute **r** by successive orthogonal projection of **b** using the columns of $\mathbf{Q}$. We save the coefficients obtained by these projections in a vector **y**, as in the successive orthogonal projection Algorithm 3.4.9. At the end of this algorithm, **b** will contain the residual vector **r**.

Since the residual **r** is orthogonal to $\mathscr{R}(\mathbf{A})$, we have

$$\mathbf{0} = \mathbf{Q}^H\mathbf{r} = \mathbf{y} - \mathbf{Rx} .$$

This implies that we can back-solve $\mathbf{Rx} = \mathbf{y}$ to compute the least-squares solution **x**. In summary, we can use the modified Gram-Schmidt process and successive orthogonal projection to solve an overdetermined least squares problem as follows:

Algorithm 6.8.3 (Modified Gram-Schmidt for Overdetermined Least Squares)

factor $\mathbf{A} = \mathbf{QR}$ /* (see Algorithm 6.8.1 or 6.8.2) */
project $\mathbf{r} = \mathbf{b} - \mathbf{Q}\mathbf{y}$ /* (see Algorithm 3.4.9 */
solve $\mathbf{R}\mathbf{x} = \mathbf{y}$ /* (see Algorithm 3.4.4 or 3.4.5) */

Example 6.8.1 Suppose we are given $m > 1$ data points (α_i, β_i) and we want to find a straight line of the form

$$\beta = \xi_1 + \alpha\xi_2$$

that comes as close as possible to matching the data. This suggests a least squares problem with

$$\mathbf{A} = \begin{bmatrix} 1 & \alpha_1 \\ \vdots & \vdots \\ 1 & \alpha_m \end{bmatrix} , \; \mathbf{b} = \begin{bmatrix} \beta_1 \\ \vdots \\ \beta_m \end{bmatrix} \text{ and } \mathbf{x} = \begin{bmatrix} \xi_1 \\ \xi_2 \end{bmatrix} .$$

To solve this least squares problem, we perform the first step of modified Gram-Schmidt factorization by computing

$$\varrho_{11} = \|\mathbf{A}\mathbf{e}_1\|_2 = \sqrt{m} \,, \; \mathbf{q}_1 \equiv \mathbf{A}\mathbf{e}_1/\varrho_{11} = \mathbf{e}_1/\sqrt{m} \,,$$

$$\varrho_{12} = \mathbf{q}_1 \cdot (\mathbf{A}\mathbf{e}_2) = \frac{1}{\sqrt{m}} \sum_{i=1}^{m} \alpha_i \equiv \overline{\alpha}\sqrt{m} \text{ and}$$

$$\mathbf{B}\mathbf{e}_2 = \mathbf{A}\mathbf{e}_2 - \mathbf{q}_1\varrho_{12} = \begin{bmatrix} \alpha_1 - \overline{\alpha} \\ \vdots \\ \alpha_m - \overline{\alpha} \end{bmatrix} .$$

Here

$$\overline{\alpha} \equiv \frac{1}{m} \sum_{i=1}^{m} \alpha_i$$

is the **sample mean** of the α's.

In the second step of modified Gram-Schmidt factorization, we compute

$$\varrho_{22} = \|\mathbf{q}_2\|_2 = \sqrt{\sum_{i=1}^{m} (\alpha_i - \overline{\alpha})^2} = s\sqrt{m-1} \text{ and}$$

$$\mathbf{q}_2 \equiv \mathbf{B}\mathbf{e}_2/\varrho_{22} = \begin{bmatrix} \alpha_1 - \overline{\alpha} \\ \vdots \\ \alpha_m - \overline{\alpha} \end{bmatrix} \frac{1}{s\sqrt{m-1}} ,$$

where

$$s^2 \equiv \frac{1}{m-1}\sum_{i=1}^{m}(\alpha_i - \overline{\alpha})^2$$

is the **sample variance** of the α's. In this way, we have factored

$$\mathbf{A} = \begin{bmatrix} 1/\sqrt{m} & (\alpha_1 - \overline{\alpha})/\left(s\sqrt{m-1}\right) \\ \vdots & \vdots \\ 1/\sqrt{m} & (\alpha_m - \overline{\alpha})/\left(s\sqrt{m-1}\right) \end{bmatrix} \begin{bmatrix} \sqrt{m} & \overline{\alpha}\sqrt{m} \\ 0 & s\sqrt{m-1} \end{bmatrix} = \mathbf{QR}\ .$$

Next, we apply successive orthogonal projection to $\mathbf{b}$:

$$\eta_1 = \mathbf{q}_1 \cdot \mathbf{b} = \frac{1}{\sqrt{m}}\sum_{i=1}^{m}\beta_i = \overline{\beta}\sqrt{m}\ ,$$

$$\mathbf{c} = \mathbf{b} - \mathbf{q}_1\eta_1 = \begin{bmatrix} \beta_1 - \overline{\beta} \\ \vdots \\ \beta_m - \overline{\beta} \end{bmatrix}\ ,$$

$$\eta_2 = \mathbf{q}_2 \cdot \mathbf{c} = \frac{\sum_{i=1}^{m}(\alpha_i - \overline{\alpha})(\beta_i - \overline{\beta})}{s\sqrt{m-1}} = \frac{\mathrm{cov}(\alpha,\beta)}{s}\sqrt{m-1} \text{ and}$$

$$\mathbf{r} = \mathbf{c} - \mathbf{q}_2\eta_2 = \begin{bmatrix} \beta_1 - \overline{\beta} - (\alpha_1 - \overline{\alpha})\mathrm{cov}(\alpha,\beta)/s^2 \\ \vdots \\ \beta_m - \overline{\beta} - (\alpha_m - \overline{\alpha})\mathrm{cov}(\alpha,\beta)/s^2 \end{bmatrix}\ .$$

Here $\mathbf{r}$ is the residual vector and

$$\mathrm{cov}(\alpha,\beta) \equiv \left[\sum_{i=1}^{m}(\alpha_i - \overline{\alpha})\left(\beta_i - \overline{\beta}\right)\right]/(m-1)$$

is the **sample covariance**. Finally, we back-solve

$$\begin{bmatrix} \sqrt{m} & \overline{\alpha}\sqrt{m} \\ & s\sqrt{m-1} \end{bmatrix} \begin{bmatrix} \xi_1 \\ \xi_2 \end{bmatrix} = \begin{bmatrix} \overline{\beta}\sqrt{m} \\ \sqrt{m-1}\,\mathrm{cov}(\alpha,\beta)/s \end{bmatrix}\ .$$

The solution of this linear system shows that the slope of the line is

$$\xi_2 = \mathrm{cov}(\alpha,\beta)/s^2\ ,$$

and the vertical intercept of the line is

$$\xi_1 = \overline{\beta} - \overline{\alpha}\mathrm{cov}(\alpha, \beta)/s^2 \,.$$

Example 6.8.2 ([72])
Recall the least squares problem due to Läuchli in Example 6.5.4. Suppose that α is a small positive machine number such that $\mathrm{fl}(1 + \alpha^2) = 1$. Let

$$\mathbf{A} = \begin{bmatrix} 1 & 1 \\ \alpha & 0 \\ 0 & \alpha \end{bmatrix} \text{ and } \mathbf{b} = \begin{bmatrix} 1 \\ 0 \\ 0 \end{bmatrix} .$$

Lemma 6.4.2 showed that the pseudo-inverse of **A** is

$$\mathbf{A}^\dagger = \left(\mathbf{A}^H\mathbf{A}\right)^{-1}\mathbf{A}^H = \begin{bmatrix} \alpha & 1+\alpha^2 & 0 \\ \alpha & 0 & 1+\alpha^2 \end{bmatrix} \frac{1}{\alpha(2+\alpha^2)} .$$

Then with exact arithmetic, the solution and residual for this least squares problem are

$$\mathbf{x} = \begin{bmatrix} 1 \\ 1 \end{bmatrix} \frac{1}{2+\alpha^2} \text{ and } \mathbf{r} = \begin{bmatrix} \alpha \\ -1 \\ -1 \end{bmatrix} \frac{\alpha}{2+\alpha^2} .$$

For future reference, we also note that the exact Gram-Schmidt factorization of **A** is

$$\mathbf{A} = \mathbf{QR} \equiv \begin{bmatrix} \frac{1}{\sqrt{1+\alpha^2}} & \frac{\alpha}{\sqrt{(1+\alpha^2)(2+\alpha^2)}} \\ \frac{\alpha}{\sqrt{1+\alpha^2}} & \frac{-1}{\sqrt{(1+\alpha^2)(2+\alpha^2)}} \\ 0 & \sqrt{\frac{1+\alpha^2}{2+\alpha^2}} \end{bmatrix} \begin{bmatrix} \sqrt{1+\alpha^2} & \frac{1}{\sqrt{1+\alpha^2}} \\ & \alpha\sqrt{\frac{2+\alpha^2}{1+\alpha^2}} \end{bmatrix} ,$$

and that the exact value of $\mathbf{y} = \mathbf{Q}^H\mathbf{b}$ is

$$\mathbf{y} = \begin{bmatrix} 1/\sqrt{1+\alpha^2} \\ \alpha/\sqrt{(1+\alpha^2)(2+\alpha^2)} \end{bmatrix} .$$

Modified Gram-Schmidt factorization computes

$$\tilde{\varrho}_{11} = \|\mathbf{A}\mathbf{e}_1\|_2 = \mathrm{fl}(\sqrt{1+\alpha^2}) = 1 \,,\; \tilde{\mathbf{q}}_1 = \mathbf{A}\mathbf{e}_1/\tilde{\varrho}_{11} = \begin{bmatrix} 1 \\ \alpha \\ 0 \end{bmatrix} ,\; \tilde{\varrho}_{12} = \tilde{\mathbf{q}}_1 \cdot \mathbf{A}\mathbf{e}_2 = 1 \,,$$

$$\mathbf{B}\tilde{\mathbf{e}}_2 = \mathbf{a}_2 - \tilde{\mathbf{q}}_1\tilde{\varrho}_{12} = \begin{bmatrix} 0 \\ -\alpha \\ \alpha \end{bmatrix} ,\; \tilde{\varrho}_{22} = \|\mathbf{B}\tilde{\mathbf{e}}_2\|_2 = \alpha\sqrt{2} \text{ and } \tilde{\mathbf{q}}_2 = \mathbf{B}\tilde{\mathbf{e}}_2/\tilde{\varrho}_{22} = \begin{bmatrix} 0 \\ -1/\sqrt{2} \\ 1/\sqrt{2} \end{bmatrix} .$$

Note that the only significant rounding error in this process occurred in the computation of $\tilde{\varrho_{11}}$, and that the columns of

$$\tilde{\mathbf{Q}} = \begin{bmatrix} 1 & 0 \\ \alpha & -1/\sqrt{2} \\ 0 & 1/\sqrt{2} \end{bmatrix}$$

are not exactly orthogonal. In fact,

$$\tilde{\mathbf{Q}}^H\tilde{\mathbf{Q}} = \begin{bmatrix} 1+\alpha^2 & -\alpha/\sqrt{2} \\ -\alpha/\sqrt{2} & 1 \end{bmatrix} .$$

It is interesting to note that

$$\mathbf{A} - \tilde{\mathbf{Q}}\tilde{\mathbf{R}} = \begin{bmatrix} 1 & 1 \\ \alpha & 0 \\ 0 & \alpha \end{bmatrix} - \begin{bmatrix} 1 & 0 \\ \alpha & -1/\sqrt{2} \\ 0 & 1/\sqrt{2} \end{bmatrix} \begin{bmatrix} 1 & 1 \\ & \alpha\sqrt{2} \end{bmatrix} = \mathbf{0} ,$$

but

$$\tilde{\mathbf{Q}} - \mathbf{Q} = \begin{bmatrix} 1 & 0 \\ \alpha & -1/\sqrt{2} \\ 0 & 1/\sqrt{2} \end{bmatrix} - \begin{bmatrix} 1 & \alpha \\ \alpha & -1 \\ 0 & 1+\alpha^2 \end{bmatrix} \begin{bmatrix} \sqrt{1+\alpha^2} & \\ & \sqrt{(1+\alpha^2)(2+\alpha^2)} \end{bmatrix}^{-1}$$
$$\approx \begin{bmatrix} \alpha^2/2 & -\alpha/\sqrt{2} \\ \alpha^3/2 & 3\alpha^2/(4\sqrt{2}) \\ 0 & \alpha^2/(4\sqrt{2}) \end{bmatrix}$$

and

$$\tilde{\mathbf{R}} - \mathbf{R} = \begin{bmatrix} 1 & 1 \\ & \alpha\sqrt{2} \end{bmatrix} - \begin{bmatrix} \sqrt{1+\alpha^2} & 1/\sqrt{1+\alpha^2} \\ & \alpha\sqrt{(2+\alpha^2)/(1+\alpha^2)} \end{bmatrix} \approx \begin{bmatrix} -\alpha^2/2 & \alpha^2/2 \\ & \alpha^3/(2\sqrt{2}) \end{bmatrix} .$$

The successive orthogonal projection process computes

$$\tilde{\eta_1} = \mathbf{q}_1 \cdot \mathbf{b} = 1 \; , \; \tilde{\mathbf{c}} = \mathbf{b} - \mathbf{q}_1\tilde{\eta_1} = \begin{bmatrix} 0 \\ -\alpha \\ 0 \end{bmatrix} ,$$

$$\tilde{\eta_2} = \mathbf{q}_2 \cdot \tilde{\mathbf{c}} = \frac{\alpha}{\sqrt{2}} \text{ and } \tilde{\mathbf{r}} = \tilde{\mathbf{c}} - \mathbf{q}_2\tilde{\eta_2} = \begin{bmatrix} 0 \\ -\alpha/2 \\ -\alpha/2 \end{bmatrix} .$$

These vectors have errors

$$\tilde{\mathbf{y}} - \mathbf{y} = \begin{bmatrix} 1 \\ \alpha/\sqrt{2} \end{bmatrix} - \begin{bmatrix} 1/\sqrt{1+\alpha^2} \\ \alpha/\sqrt{(1+\alpha^2)(2+\alpha^2)} \end{bmatrix} \approx \begin{bmatrix} \alpha^2/2 \\ 3\alpha^2/(4\sqrt{2}) \end{bmatrix}$$

and

$$\tilde{\mathbf{r}} - \mathbf{r} = \begin{bmatrix} 0 \\ -\alpha/2 \\ -\alpha/2 \end{bmatrix} - \begin{bmatrix} \alpha^2/(2+\alpha^2) \\ -\alpha/(2+\alpha^2) \\ -\alpha/(2+\alpha^2) \end{bmatrix} \approx \begin{bmatrix} -\alpha^2/2 \\ -\alpha^3/4 \\ -\alpha^3/4 \end{bmatrix} .$$

Finally, we back-solve

$$\tilde{\mathbf{R}}\tilde{\mathbf{x}} = \begin{bmatrix} 1 & 1 \\ & \alpha\sqrt{2} \end{bmatrix} \begin{bmatrix} \tilde{\xi}_1 \\ \tilde{\xi}_2 \end{bmatrix} = \tilde{\mathbf{y}} = \begin{bmatrix} \tilde{\eta}_1 \\ \tilde{\eta}_2 \end{bmatrix} = \begin{bmatrix} 1 \\ \alpha/\sqrt{2} \end{bmatrix} \Longrightarrow \tilde{\mathbf{x}} = \begin{bmatrix} 1/2 \\ 1/2 \end{bmatrix} .$$

The error in the least squares solution is

$$\tilde{\mathbf{x}} - \mathbf{x} = \begin{bmatrix} 1/2 \\ 1/2 \end{bmatrix} - \begin{bmatrix} 1/(2+\alpha^2) \\ 1/(2+\alpha^2) \end{bmatrix} \approx \begin{bmatrix} \alpha^2/4 \\ \alpha^2/4 \end{bmatrix} .$$

Readers may view a Fortran program to perform modified Gram-Schmidt factorization and successive orthogonal projection. Alternatively, readers may view the C++ files GramSchmidtQRFactorization.H and GramSchmidtQRFactorization.C. These describe a class to perform modified Gram-Schmidt QR factorization, solve over- and under-determined least squares problems, and perform iterative improvement. Most of the details of the class member functions may be found in the files LaDouble.C or LaDoubleComplex.C. These two files implement the `GramSchmidtQRFactorization` class member functions for specific data types, in these cases either double precision or complex double precision. Finally, readers may experiment with either the JavaScript **Gram-Schmidt factorization program,** or the JavaScript **successive orthogonal projection program.** The former computes the modified Gram-Schmidt factorization of a matrix, displays the errors in the factorizations, while the latter uses a Gram-Schmidt factorization to solve an overdetermined least squares problem and display the errors in the solution of this problem.

Exercise 6.8.1 Show that the least squares problem with

$$\mathbf{A} = \begin{bmatrix} 1 & 1 & 1 \\ \alpha & 0 & 0 \\ 0 & \alpha & 0 \\ 0 & 0 & \alpha \end{bmatrix} \text{ and } \mathbf{b} = \begin{bmatrix} 1 \\ 0 \\ 0 \\ 0 \end{bmatrix}$$

has solution and residual vector

$$\mathbf{x} = \begin{bmatrix} 1 \\ 1 \\ 1 \end{bmatrix} \frac{1}{3+\alpha^2} \text{ and } \mathbf{r} = \begin{bmatrix} \alpha \\ -1 \\ -1 \\ -1 \end{bmatrix} \frac{\alpha}{3+\alpha^2} .$$

Then show that the modified Gram-Schmidt process with successive orthogonal projection produces the correctly rounded solution, but not the correctly rounded residual.

Exercise 6.8.2 Use the modified Gram-Schmidt process to solve the least squares problem in Example 6.2.3, which involves molecular weights of oxides of nitrogen. Then compute the factorization errors $\mathbf{Q}^H\mathbf{Q} - \mathbf{I}$ and $\mathbf{A} - \mathbf{Q}\mathbf{R}$. How do these errors compare to the size of rounding error?

Exercise 6.8.3 Develop a block form of the Gram-Schmidt factorization Algorithm 6.8.1.

Exercise 6.8.4 Suppose that $\mathbf{A}$ is an $m \times n$ matrix with rank $(\mathbf{A}) = n$, and we have computed its modified Gram-Schmidt factorization $\mathbf{A} = \mathbf{Q}\mathbf{R}$. Show how to use this factorization to solve the underdetermined least squares problem min $\|\mathbf{x}\|_2$ such that $\mathbf{A}^H\mathbf{x} = \mathbf{b}$.

Exercise 6.8.5 If underflow and overflow are not a concern, then it is possible to implement the modified Gram-Schmidt factorization without computing square roots. If $\mathbf{A}$ is an $m \times n$ matrix with rank $(\mathbf{A}) = n$, develop an algorithm to factor

$$\mathbf{A} = \mathbf{Q}\mathbf{R}$$

where $\mathbf{R}$ is unit right triangular and

$$\mathbf{Q}^H\mathbf{Q} = \boldsymbol{\Sigma}^2$$

is diagonal with positive diagonal entries. Then develop another algorithm that uses this factorization to solve an overdetermined least squares problem.

6.8.2 *Simultaneous Orthogonal Projection*

Suppose again that we are given an $m \times n$ matrix $\mathbf{A}$ with rank $(\mathbf{A}) = n$. In the **classical Gram-Schmidt** process, individual columns of the original matrix are projected with respect to all previous columns in a single step. The general step of this classical Gram-Schmidt process takes the following form:

$$\mathbf{A} = \begin{bmatrix} \mathbf{Q}_1 & \mathbf{a}_2 & \mathbf{A}_3 \end{bmatrix} \begin{bmatrix} \mathbf{R}_{11} & 0 & 0 \\ & 1 & 0 \\ & & \mathbf{I} \end{bmatrix} = \begin{bmatrix} \mathbf{Q}_1 & \mathbf{q}_2 & \mathbf{A}_3 \end{bmatrix} \begin{bmatrix} \mathbf{R}_{11} & \mathbf{r}_{12} & 0 \\ & \varrho_{22} & 0 \\ & & \mathbf{I} \end{bmatrix}$$

where

$$\mathbf{r}_{12} = \mathsf{Q}_1{}^H \mathbf{a}_2 \; , \; \mathbf{a}_2' = \mathbf{a}_2 - \mathsf{Q}_1 \mathbf{r}_{12} \; , \; \varrho_{22} = \|\mathbf{a}_2'\|_2 \text{ and } \mathsf{q}_2 = \mathbf{a}_2'/\varrho_{22} \; .$$

The essential difference between the classical Gram-Schmidt algorithm and the modified Gram-Schmidt algorithm is that the former projects the next column as

$$\mathbf{a}_2' = \left(\mathbf{I} - \mathsf{Q}\mathsf{Q}^H\right) \mathbf{a}_2 \; ,$$

while the modified Gram-Schmidt algorithm computes this vector as

$$\mathbf{a}_2' = \left(\mathbf{I} - \mathsf{q}_{k-1}\mathsf{q}_{k-1}{}^H\right) \cdot \ldots \cdot \left(\mathbf{I} - \mathsf{q}_1\mathsf{q}_1{}^H\right) \mathbf{a}_2 \; .$$

The classical Gram-Schmidt algorithm takes the following form.

Algorithm 6.8.4 (Classical Gram-Schmidt Factorization)

```
for 1 ≤ k ≤ n
   if k > 1
      R_{1:k-1,k} = A_{1:m,1:k-1}^H A_{1:m,k}              /* LAPACK BLAS routine _gemv */
      A_{1:m,k} = A_{1:m,k} - A_{1:m,1:k-1} R_{1:k-1,k}     /* LAPACK BLAS routine _gemv */
   ϱ_{kk} = ||A_{1:m,k}||_2                                 /* LAPACK BLAS routine _nrm2 */
   Q_{1:m,k} = A_{1:m,k}/ϱ_{kk}                   /* LAPACK BLAS routine _scal or _dscal */
```

This algorithm is similar, and substantially inferior, to the column-oriented form of the modified Gram-Schmidt factorization Algorithm 6.8.2.

In order to solve a least-squares problem involving some m-vector $\mathbf{b}$, the classical Gram-Schmidt process would begin by performing the following.

Algorithm 6.8.5 (Simultaneous Orthogonal Projection)

```
y = Q^H b      /* LAPACK BLAS routine _gemv */
r = b - Q y    /* LAPACK BLAS routine _gemv */
```

In other words, the residual vector $\mathbf{r}$ is computed by simultaneous orthogonal projection

$$\mathbf{r} = \left(\mathbf{I} - \mathsf{Q}\mathsf{Q}^H\right) \mathbf{b}$$

instead of successive orthogonal projection

$$\mathbf{r} = \left(\mathbf{I} - \mathsf{q}_n\mathsf{q}_n{}^H\right) \cdot \ldots \cdot \left(\mathbf{I} - \mathsf{q}_1\mathsf{q}_1{}^H\right) \mathbf{b} \; . \tag{6.20}$$

At any rate, after simultaneous orthogonal projection of a vector $\mathbf{b}$, we can determine the solution of the overdetermined least squares problem by back-solving $\mathbf{R}\mathbf{x} = \mathbf{y}$.

Simultaneous orthogonal projection and classical Gram-Schmidt factorization are far more sensitive to rounding errors than successive orthogonal projection and modified Gram-Schmidt factorization. The following example will illustrate this claim.

Example 6.8.3 ([72]) Recall the least squares problem in Examples 6.5.4 and 6.8.2. Suppose that the scalar α is small enough that $\text{fl}(1+\alpha^2) = 1$. Let

$$\mathbf{A} = \begin{bmatrix} 1 & 1 \\ \alpha & 0 \\ 0 & \alpha \end{bmatrix} \text{ and } \mathbf{b} = \begin{bmatrix} 1 \\ 0 \\ 0 \end{bmatrix} .$$

The exact Gram-Schmidt QR factorization of this matrix, as well as the vectors involved in the solution of a least squares problem, were computed in Example 6.8.2.

For this matrix, the classical Gram-Schmidt factorization computes

$$\tilde{\mathbf{R}}_{11} = \|\mathbf{a}_1\|_2 = \sqrt{fl(1+\alpha^2)} = 1 \; , \; \tilde{\mathbf{q}}_1 = \mathbf{a}_1/\tilde{\mathbf{R}}_{11} = \begin{bmatrix} 1 \\ \alpha \\ 0 \end{bmatrix} \; , \; \tilde{\mathbf{R}}_{12} = \tilde{\mathbf{q}}_1 \cdot \mathbf{a}_2 = 1 \; ,$$

$$\tilde{\mathbf{a}}_2' = \mathbf{a}_2 - \mathbf{a}_1\tilde{\mathbf{R}}_{12} = \begin{bmatrix} 0 \\ -\alpha \\ \alpha \end{bmatrix} \; , \; \tilde{\mathbf{R}}_{22} = \|\tilde{\mathbf{a}}_2'\|_2 = \alpha\sqrt{2} \text{ and } \tilde{\mathbf{q}}_2 = \tilde{\mathbf{a}}_2'/\tilde{\mathbf{R}}_{22} = \begin{bmatrix} 0 \\ -1/\sqrt{2} \\ 1/\sqrt{2} \end{bmatrix} .$$

Thus, in this case classical Gram-Schmidt factorization produces precisely the same factorization as the modified Gram-Schmidt process in Example 6.8.2.

However, simultaneous orthogonal projection computes

$$\tilde{\mathbf{y}} = \tilde{\mathbf{Q}}^H\mathbf{b} = \begin{bmatrix} 1 \\ 0 \end{bmatrix} \text{ and } \tilde{\mathbf{r}} = \mathbf{b} - \tilde{\mathbf{Q}}\tilde{\mathbf{y}} = \begin{bmatrix} 0 \\ -\alpha \\ 0 \end{bmatrix} .$$

These vectors have errors

$$\tilde{\mathbf{y}} - \mathbf{y} = \begin{bmatrix} 1 \\ 0 \end{bmatrix} - \begin{bmatrix} 1/\sqrt{1+\alpha^2} \\ \alpha/\sqrt{(1+\alpha^2)(2+\alpha^2)} \end{bmatrix} \approx \begin{bmatrix} \alpha^2/2 \\ -\alpha/\sqrt{2} \end{bmatrix}$$

and

$$\tilde{\mathbf{r}} - \mathbf{r} = \begin{bmatrix} 0 \\ -\alpha \\ 0 \end{bmatrix} - \begin{bmatrix} \alpha^2/(2+\alpha^2) \\ -\alpha/(2+\alpha^2) \\ -\alpha/(2+\alpha^2) \end{bmatrix} \approx \begin{bmatrix} -\alpha^2/2 \\ -\alpha/2 \\ \alpha/2 \end{bmatrix} .$$

These errors are significantly larger than the corresponding errors with successive orthogonal projection.

Finally, we back-solve

$$\tilde{\mathbf{R}}\tilde{\mathbf{x}} = \begin{bmatrix} 1 & 1 \\ & \alpha\sqrt{2} \end{bmatrix} \begin{bmatrix} \tilde{\xi}_1 \\ \tilde{\xi}_2 \end{bmatrix} = \tilde{\mathbf{y}} = \begin{bmatrix} 1 \\ 0 \end{bmatrix} \Longrightarrow \tilde{\mathbf{x}} = \begin{bmatrix} 1 \\ 0 \end{bmatrix} .$$

The error in this vector is

$$\tilde{\mathbf{x}} - \mathbf{x} = \begin{bmatrix} 1 \\ 0 \end{bmatrix} - \begin{bmatrix} 1/(2+\alpha^2) \\ 1/(2+\alpha^2) \end{bmatrix} \approx \begin{bmatrix} 1/2 \\ -1/2 \end{bmatrix} .$$

The error in the classical Gram-Schmidt solution vector is on the order of one, whereas the error in the least squares solution computed by modified Gram-Schmidt in Example 6.8.2 was on the order of α^2.

6.8.3 Pivoting

The modified Gram-Schmidt process can fail when a column of zeros is encountered. This difficulty is largely theoretical, since rounding errors will typically prevent such an event from occurring. Nevertheless, we can avoid this problem by pivoting columns during Gram-Schmidt factorization.

Algorithm 6.8.6 (Modified Gram-Schmidt Factorization with Pivoting)

for $1 \le k \le \min\{m, n\}$
  for $k \le j \le n$
    $\sigma_j = \|\mathbf{A}_{1:m,j}\|_2$ /* LAPACK BLAS routine _nrm2 */
  find J so that $\sigma_J = \max_{k \le j \le n} \sigma_j$ /* LAPACK BLAS routine i_amax */
  if $\sigma_J = 0$ break /* remainder of matrix is zero */
  if $J \neq k$
    interchange $\mathbf{A}_{1:m,k}$ with $\mathbf{A}_{1:m,J}$ /* LAPACK BLAS routine _swap */
    interchange $\mathbf{R}_{1:k-1,k}$ with $\mathbf{R}_{1:k-1,J}$ /* LAPACK BLAS routine _swap */
  $\mathbf{R}_{k,k} = \sigma_J$
  $\mathbf{Q}_{1:m,k} = \mathbf{A}_{1:m,k}/\mathbf{R}_{k,k}$ /* LAPACK BLAS routine _scal or _dscal */
  $\mathbf{w}_{k+1:n} = \mathbf{A}_{1:m,k+1:n}{}^H \mathbf{Q}_{1:m,k}$ /* LAPACK BLAS routine _gemv */
  $\mathbf{R}_{k,k+1:n} = \overline{\mathbf{w}_{k+1:n}}$
  $\mathbf{A}_{1:m,k+1:n} = \mathbf{A}_{1:m,k+1:n} - \mathbf{Q}_{1:m,k}\mathbf{R}_{k,k+1:n}$ /* LAPACK BLAS routine _ger or _geru */

This algorithm terminates in one of three ways. For example, the algorithm could terminate because there are no more rows to process. In this case, $m \le n$, and at the end of the factorization we have $\mathbf{AP} = \mathbf{QR}$ where $\mathbf{P}$ is an $n \times n$ permutation matrix, $\mathbf{Q}$ is an $m \times m$ matrix with orthonormal columns, and $\mathbf{R}$ is an $m \times n$ right trapezoidal matrix. The algorithm could also terminate because there are no more columns to process. In this case, $n \leq m$, and at the end of the factorization we have $\mathbf{AP} = \mathbf{QR}$ where $\mathbf{P}$ is an $n \times n$ permutation matrix, $\mathbf{Q}$ is an $m \times n$ matrix with orthonormal columns, and $\mathbf{R}$ is an $n \times n$ right triangular matrix. If the algorithm terminates because at some step $k = r + 1$ columns k through n are zero, then $r < m, n$ and at the end of the factorization we have

$$\mathbf{AP} = \begin{bmatrix} \mathbf{Q}_1 & \mathbf{0} \end{bmatrix} \begin{bmatrix} \mathbf{R}_{11} & \mathbf{R}_{12} \\ & \mathbf{I} \end{bmatrix} = \mathbf{Q}_1 \begin{bmatrix} \mathbf{R}_{11} & \mathbf{R}_{12} \end{bmatrix} .$$

In these equations, $\mathbf{Q}_1$ is an $m \times r$ matrix with orthonormal columns, and $\begin{bmatrix} \mathbf{R}_{11} & \mathbf{R}_{12} \end{bmatrix}$ is an $r \times n$ right-trapezoidal matrix with positive diagonal entries.

Algorithm 6.8.6 proves the following theorem.

Theorem 6.8.1 (Gram-Schmidt QR) *Suppose that* $\mathbf{A}$ *is a nonzero* $m \times n$ *matrix. Then there is an integer* r *satisfying* $1 \leq r \leq \min\{m, n\}$, *an* $n \times n$ *permutation matrix* $\mathbf{P}$, *an* $m \times r$ *matrix* $\mathbf{Q}$ *with orthonormal columns, and an* $r \times n$ *right trapezoidal matrix* $\mathbf{R}$ *with positive diagonal entries so that* $\mathbf{AP} = \mathbf{QR}$.

6.8.4 Householder Equivalence

It is a curious fact that Gram-Schmidt factorization of some given matrix $\mathbf{A}$ is mathematically equivalent to applying Householder factorization to the larger matrix $\widehat{\mathbf{A}}$ where

$$\widehat{\mathbf{A}} \equiv \begin{bmatrix} \mathbf{0} \\ \mathbf{A} \end{bmatrix} .$$

This connection between the Gram-Schmidt and Householder QR factorizations was published by Björck [10], but attributed to Charles Sheffield.

To prove this equivalence, let us partition

$$\widehat{\mathbf{A}} = \begin{bmatrix} \mathbf{0} & \mathbf{0} \\ \mathbf{a}_1 & \mathbf{A}_2 \end{bmatrix} .$$

When we determine the first elementary reflector for $\widehat{\mathbf{A}}$ using Algorithm 6.7.1, we compute

$$\sigma = 0 \, , \; \beta = \|\mathbf{a}_1\|_2 \, , \; \tau = 1 \, , \; \mathbf{q}_1 = \mathbf{a}_1 / \|\mathbf{a}_1\|_2 \text{ and } \mathbf{u}_1 = \begin{bmatrix} \mathbf{e}_1 \\ \mathbf{q}_1 \end{bmatrix} .$$

By applying this elementary reflector to $\widehat{\mathbf{A}}$, we get

$$\mathbf{H}_1\widehat{\mathbf{A}} = \left\{ \begin{bmatrix} \mathbf{I} & \\ & \mathbf{I} \end{bmatrix} - \begin{bmatrix} \mathbf{e}_1 \\ \mathbf{q}_1 \end{bmatrix} \tau \left[\mathbf{e}_1{}^H \; \mathbf{q}_1{}^H \right] \right\} \begin{bmatrix} \mathbf{0} & \mathbf{0} \\ \mathbf{a}_1 & \mathbf{A}_2 \end{bmatrix} = \begin{bmatrix} -\mathbf{e}_1\mathbf{q}_1 \cdot \mathbf{a}_1 & -\mathbf{e}_1\mathbf{q}_1{}^H\mathbf{A}_2 \\ \mathbf{a}_1 - \mathbf{q}_1\mathbf{q}_1 \cdot \mathbf{a}_1 & \mathbf{A}_2 - \mathbf{q}_1\mathbf{q}_1{}^H\mathbf{A}_2 \end{bmatrix}$$

$$= \begin{bmatrix} -\mathbf{e}_1\|\mathbf{a}_1\|_2 & -\mathbf{e}_1\mathbf{q}_1{}^H\mathbf{A}_2 \\ \mathbf{0} & \mathbf{A}_2 - \mathbf{q}_1\mathbf{q}_1{}^H\mathbf{A}_2 \end{bmatrix} \equiv \begin{bmatrix} -\mathbf{e}_1\varrho_{11} & -\mathbf{e}_1\mathbf{r}_{12}{}^H \\ \mathbf{0} & \mathbf{A}_2' \end{bmatrix} .$$

Since $\mathbf{q}_1{}^H\mathbf{A}_2' = \mathbf{q}_1{}^H(\mathbf{A}_2 - \mathbf{a}_1\mathbf{q}_1{}^H\mathbf{A}_2) = \mathbf{0}^H$, it follows that

$$\widehat{\mathbf{A}} = \mathbf{H}_1 \begin{bmatrix} -\mathbf{e}_1\varrho_{11} & -\mathbf{e}_1\mathbf{r}_{12}{}^H \\ \mathbf{0} & \mathbf{A}_2' \end{bmatrix} = \begin{bmatrix} -\mathbf{e}_1\varrho_{11} & -\mathbf{e}_1\mathbf{r}_{12}{}^H \\ \mathbf{0} & \mathbf{A}_2' \end{bmatrix}$$

$$- \begin{bmatrix} \mathbf{e}_1 \\ \mathbf{q}_1 \end{bmatrix} \left[\mathbf{e}_1{}^H \; \mathbf{q}_1{}^H \right] \begin{bmatrix} -\mathbf{e}_1\varrho_{11} & -\mathbf{e}_1\mathbf{r}_{12}{}^H \\ \mathbf{0} & \mathbf{A}_2' \end{bmatrix}$$

$$= \begin{bmatrix} \mathbf{0} & \mathbf{0} \\ \mathbf{q}_1\varrho_{11} & \mathbf{A}_2' + \mathbf{q}_1\mathbf{r}_{12}{}^H \end{bmatrix} = \begin{bmatrix} \mathbf{0} & \mathbf{0} \\ \mathbf{q}_1 & \mathbf{A}_2' \end{bmatrix} \begin{bmatrix} \varrho_{11} & \mathbf{r}_{12}{}^H \\ & \mathbf{I} \end{bmatrix} .$$

Thus the application of the first elementary reflector to the enlarged matrix is equivalent to orthogonal projection.

In later steps, we assume that we have factored

$$\widehat{\mathbf{A}} = \begin{bmatrix} \mathbf{0} & \mathbf{0} & \mathbf{0} \\ \mathbf{Q}_1 & \mathbf{a}_2 & \mathbf{A}_3 \end{bmatrix} \begin{bmatrix} \mathbf{R}_{11} & \mathbf{r}_{12} & \mathbf{R}_{13} \\ & 1 & \mathbf{0} \\ & & \mathbf{I} \end{bmatrix} .$$

When we determine the next elementary reflector for $\widehat{\mathbf{A}}$ using Algorithm 6.7.1, we compute

$$\sigma = 0 \; , \; \beta = \|\mathbf{a}_2\|_2 \; , \; \tau = 1 \; , \; \mathbf{q}_2 = \mathbf{a}_2/\|\mathbf{a}_2\|_2 \text{ and } \mathbf{u}_2 = \begin{bmatrix} \mathbf{e}_k \\ \mathbf{q}_2 \end{bmatrix} .$$

By applying this elementary reflector to our factorization of $\widehat{\mathbf{A}}$, we get

$$\mathbf{H}_k\widehat{\mathbf{A}} = \left\{ \begin{bmatrix} \mathbf{I} & \\ & \mathbf{I} \end{bmatrix} - \begin{bmatrix} \mathbf{e}_k \\ \mathbf{q}_2 \end{bmatrix} \left[\mathbf{e}_k{}^H \; \mathbf{q}_2{}^H \right] \right\} \begin{bmatrix} \mathbf{0} & \mathbf{0} & \mathbf{0} \\ \mathbf{Q}_1 & \mathbf{a}_2 & \mathbf{A}_3 \end{bmatrix} \begin{bmatrix} \mathbf{R}_{11} & \mathbf{r}_{12} & \mathbf{R}_{13} \\ & 1 & \mathbf{0} \\ & & \mathbf{I} \end{bmatrix}$$

$$= \begin{bmatrix} -\mathbf{e}_k\mathbf{q}_2{}^H\mathbf{Q}_1 & -\mathbf{e}_k\mathbf{q}_2 \cdot \mathbf{a}_2 & -\mathbf{e}_k\mathbf{q}_2{}^H\mathbf{A}_3 \\ \mathbf{Q}_1 - \mathbf{q}_2\mathbf{q}_2{}^H\mathbf{Q}_1 & \mathbf{a}_2 - \mathbf{q}_2\mathbf{q}_2 \cdot \mathbf{a}_2 & \mathbf{A}_3 - \mathbf{q}_2\mathbf{q}_2{}^H\mathbf{A}_3 \end{bmatrix} \begin{bmatrix} \mathbf{R}_{11} & \mathbf{r}_{12} & \mathbf{R}_{13} \\ & 1 & \mathbf{0} \\ & & \mathbf{I} \end{bmatrix}$$

$$
= \begin{bmatrix} \mathbf{0} & -\mathbf{e}_k\|\mathbf{a}_2\|_2 & -\mathbf{e}_k \mathbf{q}_2{}^H\mathbf{A}_3 \\ \mathbf{Q}_1 & \mathbf{0} & \mathbf{A}_3 - \mathbf{q}_2\mathbf{q}_2{}^H\mathbf{A}_3 \end{bmatrix} \begin{bmatrix} \mathbf{R}_{11} & \mathbf{r}_{12} & \mathbf{R}_{13} \\ & 1 & \mathbf{0} \\ & & \mathbf{I} \end{bmatrix}
$$

$$
\equiv \begin{bmatrix} \mathbf{0} & -\mathbf{e}_k\varrho_{22} & -\mathbf{e}_k\mathbf{r}_{23}{}^H \\ \mathbf{Q}_1 & \mathbf{0} & \mathbf{A}_3' \end{bmatrix} \begin{bmatrix} \mathbf{R}_{11} & \mathbf{r}_{12} & \mathbf{R}_{13} \\ & 1 & \mathbf{0} \\ & & \mathbf{I} \end{bmatrix} .
$$

it follows that

$$
\widehat{\mathbf{A}} = \mathbf{H}_k \begin{bmatrix} \mathbf{0} & -\mathbf{e}_k\varrho_{22} & -\mathbf{e}_k\mathbf{r}_{23}{}^H \\ \mathbf{Q}_1 & \mathbf{0} & \mathbf{A}_3' \end{bmatrix} \begin{bmatrix} \mathbf{R}_{11} & \mathbf{r}_{12} & \mathbf{R}_{13} \\ & 1 & \mathbf{0} \\ & & \mathbf{I} \end{bmatrix}
$$

$$
= \left\{ \begin{bmatrix} \mathbf{0} & -\mathbf{e}_k\varrho_{22} & -\mathbf{e}_k\mathbf{r}_{23}{}^H \\ \mathbf{Q}_1 & \mathbf{0} & \mathbf{A}_3' \end{bmatrix} - \begin{bmatrix} \mathbf{e}_k \\ \mathbf{q}_2 \end{bmatrix} \begin{bmatrix} \mathbf{e}_k{}^H & \mathbf{q}_2{}^H \end{bmatrix} \begin{bmatrix} \mathbf{0} & -\mathbf{e}_k\varrho_{22} & -\mathbf{e}_k\mathbf{r}_{23}{}^H \\ \mathbf{Q}_1 & \mathbf{0} & \mathbf{A}_3' \end{bmatrix} \right\} \begin{bmatrix} \mathbf{R}_{11} & \mathbf{r}_{12} & \mathbf{R}_{13} \\ & 1 & \mathbf{0} \\ & & \mathbf{I} \end{bmatrix}
$$

$$
= \begin{bmatrix} \mathbf{0} & \mathbf{0} & \mathbf{0} \\ \mathbf{Q}_1 & \mathbf{q}_2\varrho_{22} & \mathbf{A}_3' + \mathbf{q}_2\mathbf{r}_{23}{}^H \end{bmatrix} \begin{bmatrix} \mathbf{R}_{11} & \mathbf{r}_{12} & \mathbf{R}_{13} \\ & 1 & \mathbf{0} \\ & & \mathbf{I} \end{bmatrix}
$$

$$
= \begin{bmatrix} \mathbf{0} & \mathbf{0} & \mathbf{0} \\ \mathbf{Q}_1 & \mathbf{q}_2 & \mathbf{A}_3' \end{bmatrix} \begin{bmatrix} \mathbf{I} & & \\ & \varrho_{22} & \mathbf{r}_{23}{}^H \\ & & \mathbf{I} \end{bmatrix} \begin{bmatrix} \mathbf{R}_{11} & \mathbf{r}_{12} & \mathbf{R}_{13} \\ & 1 & \mathbf{0} \\ & & \mathbf{I} \end{bmatrix} = \begin{bmatrix} \mathbf{0} & \mathbf{0} & \mathbf{0} \\ \mathbf{Q}_1 & \mathbf{q}_2 & \mathbf{A}_3' \end{bmatrix} \begin{bmatrix} \mathbf{R}_{11} & \mathbf{r}_{12} & \mathbf{R}_{13} \\ & \varrho_{22} & \mathbf{r}_{23}{}^H \\ & & \mathbf{I} \end{bmatrix} .
$$

Thus the general step of successive reflection applied to $\widehat{\mathbf{A}}$ produces the general step of orthogonal projection to $\mathbf{A}$.

6.8.5 Rounding Errors

As we stated during the introduction to this book in Chap. 1, the analysis of numerical schemes is an important part of scientific computing. Most recently in Sect. 6.7.5 of this chapter, we discussed rounding errors in successive reflection. That discussion is available in other texts, and is fairly straightforward to present. In this section, we will analyze the rounding errors in modified Gram-Schmidt factorization, and in successive orthogonal projection. This analysis is more complicated than the analysis of the Householder factorization and successive reflection. An indirect rounding error analysis could be performed by using the discussion of Sect. 6.8.4 to connect the Gram-Schmidt factorization to the Householder QR factorization, and then applying the rounding error analysis for the latter. Instead, we will follow the more direct ideas in Björck [8], but we will use the inequalities in Sect. 3.8.1 to avoid the mysterious coefficients of machine precision that were in common usage when the Björck paper was written.

Readers who want to concentrate on the main results of this section should examine the statements of Theorems 6.8.2–6.8.4. These theorems estimate the errors in various aspects of the Gram-Schmidt factorization. Corollary 6.8.1 uses these ideas to estimate the errors in applying successive orthogonal projection to the right-hand side in a least squares problem. Finally, Theorem 6.8.5 estimates the errors in the residual, solution vector and orthonormality of the Gram-Schmidt factorization matrix $\mathbf{Q}$.

We will begin our rounding error analysis of successive orthogonal projection with the following easy inequalities.

Lemma 6.8.1 *Suppose that $\sigma > 0$. If $|\delta| \leq \sigma^2$ then*

$$\left|\sqrt{\sigma^2 + \delta} - \sigma\right| \leq \frac{|\delta|}{\sigma} . \tag{6.21}$$

On the other hand, if $|\delta| \leq \sigma/2$ then

$$\left|\frac{1}{\sigma + \delta} - \frac{1}{\sigma}\right| \leq \frac{2|\delta|}{\sigma^2} . \tag{6.22}$$

Finally, if $|\delta| \leq \sigma^2(\sqrt{5} - 1)/2$ then

$$\left|\frac{1}{\sqrt{\sigma^2 + \delta}} - \frac{1}{\sigma}\right| \leq \frac{|\delta|}{\sigma^3} . \tag{6.23}$$

Proof We will begin by proving (6.21). Since $|\delta| \leq \sigma^2$, it follows that

$$-2\,|\delta| + \frac{\delta^2}{\sigma^2} \leq \delta \leq 2\,|\delta| + \frac{\delta^2}{\sigma^2} .$$

These two inequalities are equivalent to

$$\left(\sigma - \frac{|\delta|}{\sigma}\right)^2 = \sigma^2 - 2\,|\delta| + \frac{\delta^2}{\sigma^2} \leq \sigma^2 + \delta \leq \sigma^2 + 2\,|\delta| + \frac{\delta^2}{\sigma^2} = \left(\sigma + \frac{|\delta|}{\sigma}\right)^2 .$$

Since the left-hand side of this inequality is nonnegative, we can take square roots to get

$$\sigma - \frac{|\delta|}{\sigma} \leq \sqrt{\sigma^2 + \delta} \leq \sigma + \frac{|\delta|}{\sigma} .$$

These inequalities are equivalent to the first claimed result (6.21).

Next, we will prove (6.22). Since $2|\delta|/\sigma \leq 1$, we see that

$$-2 + \text{sign}(\delta) - 2\frac{\delta}{\sigma} \leq 0 \leq 2 + \text{sign}(\delta) + 2\frac{\delta}{\sigma} .$$

We can multiply by $|\delta|/\sigma$ and add 1 to get

$$\left(\frac{1}{\sigma} - 2\frac{|\delta|}{\sigma^2}\right)(\sigma + \delta) = 1 - 2\frac{|\delta|}{\sigma} + \frac{\delta}{\sigma} - 2\frac{\delta|\delta|}{\sigma^2} \leq 1 \leq 1 + 2\frac{|\delta|}{\sigma} + \frac{\delta}{\sigma} + 2\frac{\delta|\delta|}{\sigma^2}$$
$$= \left(\frac{1}{\sigma} + 2\frac{|\delta|}{\sigma^2}\right)(\sigma + \delta) .$$

This inequality is equivalent to the claimed result (6.22)

Finally, we will prove (6.23). We note that if $|\delta| \leq \sigma^2(\sqrt{5} - 1)/2$, then

$$\frac{|\delta|}{\sigma^2}\left\{1 - 2\,\text{sign}(\delta) + \frac{\delta}{\sigma^2}\right\} \leq 2 - \text{sign}(\delta) \text{ and}$$
$$\frac{|\delta|}{\sigma^2}\left\{-1 - 2\,\text{sign}(\delta) - \frac{\delta}{\sigma^2}\right\} \leq 2 + \text{sign}(\delta) .$$

Here, the upper bound on $|\delta|$ is needed in the second inequality for $\delta < 0$. We can multiply by $|\delta|/\sigma^2$ and add one to each expression to get

$$\left\{\frac{1}{\sigma} - \frac{|\delta|}{\sigma^3}\right\}^2 (\sigma^2 + \delta) \leq 1 \leq \left\{\frac{1}{\sigma} + \frac{|\delta|}{\sigma^3}\right\}^2 (\sigma^2 + \delta) .$$

Finally, we take square roots and divide by $\sqrt{\sigma^2 + \delta}$ to get (6.23).

The previous lemma allows us to estimate the errors in computing a unit vector from some given nonzero vector.

Lemma 6.8.2 *Assume that the relative errors in floating point division or square roots satisfy*

$$fl(\alpha/\beta) = (\alpha/\beta)(1 + \varepsilon_{\div}) \text{ and}$$
$$fl(\sqrt{\alpha}) = \sqrt{\alpha}(1 + \varepsilon_{\surd}) .$$

Also suppose that there is an upper bound ε on all such relative errors:

$$\max\{|\varepsilon_{\div}| , |\varepsilon_{\surd}|\} \leq \varepsilon .$$

Let m be a positive integer such that $(3m+5)\varepsilon < 2$, and let $\mathbf{a}$ be a nonzero m-vector. Define

$$\sigma = \|\mathbf{a}\|_2 \text{ and } \mathbf{q} = \mathbf{a}/\sigma .$$

Then floating point computation of these quantities produces errors satisfying

$$|\,fl(\sigma)-\sigma| \leq \frac{(m+1)\varepsilon}{1-(m+1)\varepsilon/2}\sigma \text{ and} \tag{6.24a}$$

$$\|\,fl(\mathbf{q})-\mathbf{q}\|_2 \leq \frac{(m+2)\varepsilon}{1-(3m+5)\varepsilon/2}\,. \tag{6.24b}$$

Proof Lemma 3.8.8 showed that

$$\text{fl}(\|\mathbf{a}\|_2^2) = \|\mathbf{a}\|_2^2 + \mathbf{a}\cdot\Delta\mathbf{a} \text{ where } \|\Delta\mathbf{a}\|_2 \leq \frac{m\varepsilon}{1-(m-1)\varepsilon/2}\|\mathbf{a}\|_2\,.$$

Consequently,

$$|\,\text{fl}(\sigma)-\sigma| = \left|\sqrt{\sigma^2+\mathbf{a}\cdot\Delta\mathbf{a}}(1+\varepsilon_{\surd})-\sigma\right| \leq \left|\sqrt{\sigma^2+\mathbf{a}\cdot\Delta\mathbf{a}}-\sigma\right|(1+\varepsilon)+\sigma\varepsilon$$

then inequality (6.21), the Cauchy inequality (3.15), inequality (3.46) and inequality (3.37) give us

$$\leq \sigma\left[\frac{m\varepsilon}{1-(m-1)\varepsilon/2}(1+\varepsilon)+\varepsilon\right] \leq \sigma\frac{(m+1)\varepsilon}{1-(m+1)\varepsilon/2}\,.$$

This proves the first claim (6.24a).

Next, we will bound the rounding error in computing $\mathbf{q}$. There is a diagonal matrix $\mathbf{E}_{\div}$, with diagonal entries on the order of the floating point relative error, such that

$$\|\,\text{fl}(\mathbf{q})-\mathbf{q}\|_2 = \left\|(\mathbf{I}+\mathbf{E}_{\div})\mathbf{a}\frac{1}{\text{fl}(\sigma)}-\mathbf{q}\right\|_2 \leq \left\|(\mathbf{I}+\mathbf{E}_{\div})\mathbf{q}\left(\frac{\sigma}{\text{fl}(\sigma)}-1\right)\right\|_2 + \|\mathbf{E}_{\div}\mathbf{q}\|_2$$

$$\leq (1+\varepsilon)\frac{|\sigma-\text{fl}(\sigma)|}{|\sigma+[\,\text{fl}(\sigma)-\sigma]|}+\varepsilon \leq (1+\varepsilon)\frac{|\,\text{fl}(\sigma)-\sigma|}{\sigma-|\,\text{fl}(\sigma)-\sigma|}+\varepsilon$$

then inequalities (6.22) and (6.24a) produce

$$\leq (1+\varepsilon)\frac{\frac{(m+1)\varepsilon}{1-(m+1)\varepsilon/2}}{1-\frac{(m+1)\varepsilon}{1-(m+1)\varepsilon/2}}+\varepsilon = (1+\varepsilon)\frac{(m+1)\varepsilon}{1-3(m+1)\varepsilon/2}+\varepsilon$$

and finally inequalities (3.38) and (3.37) yield

$$\leq \frac{(m+2)\varepsilon}{1-(3m+5)\varepsilon/2}\,.$$

The following simple lemma will help us to prove several subsequent results.

Lemma 6.8.3 *Suppose that* α, β *and* γ *are positive scalars and*

$$\phi(\xi) \equiv \alpha\sqrt{\beta^2-\xi^2}+\gamma\xi \ .$$

Then for all $\xi \in [0,\beta]$ *we have*

$$\phi(\xi) \le \beta\sqrt{\alpha^2+\gamma^2} \ . \tag{6.25}$$

Proof Since

$$\phi'(\xi) = -\frac{\alpha\xi}{\sqrt{\beta^2-\xi^2}}+\gamma \text{ and } \phi''(\xi) = -\frac{\alpha\beta^2}{\sqrt{\beta^2-\xi^2}^3} \ ,$$

we see that the sole critical point of ϕ for $\xi \in [0,\beta]$ is $\xi_* = \gamma\beta/\sqrt{\alpha^2+\gamma^2}$, with

$$\phi(\xi_*) = \beta\sqrt{\alpha^2+\gamma^2} \ .$$

Since $\phi(0) = \alpha\beta \le \phi(\xi_*)$ and $\phi(\beta) = \gamma\beta \le \phi(\xi_*)$, we see that $\phi(\xi) \le \phi(\xi_*)$ for all $\xi \in [0,\beta]$.

The following important lemma will bound errors in orthogonal projection. This result is similar to a result in Björck [8, p. 9].

Lemma 6.8.4 *Assume that the relative errors in floating point addition or subtraction, multiplication and square roots satisfy*

$$\begin{aligned} fl(\alpha \pm \beta) &= (\alpha \pm \beta)(1+\varepsilon_\pm) \ , \\ fl(\alpha \times \beta) &= (\alpha \times \beta)(1+\varepsilon_\times) \text{ and} \\ fl(\sqrt{\alpha}) &= \sqrt{\alpha}(1+\varepsilon_{\sqrt{}}) \ , \end{aligned}$$

and assume that there is an upper bound ε *on all such relative errors:*

$$\max\{|\varepsilon_\pm| \ , \ |\varepsilon_\times| \ , \ |\varepsilon_{\sqrt{}}|\} \le \varepsilon \ .$$

Suppose that $m \ge 2$ *is a positive integer satisfying* $(10m+23)\varepsilon < 2$, *and suppose that* $\mathbf{a}$ *and* $\mathbf{b}$ *are m-vectors. Assume that we compute*

$$\sigma = \|\mathbf{a}\|_2 \ , \ \mathbf{q} = \mathbf{a}/\sigma \ , \varrho = \mathbf{q}\cdot\mathbf{b} \text{ and } \mathbf{p} = \mathbf{b} - \mathbf{q}\varrho$$

in floating point arithmetic, obtaining approximate quantities $\tilde{\sigma} = fl(\sigma)$, $\tilde{\mathbf{q}} = fl(\mathbf{q})$, $\tilde{\varrho} = fl(\varrho)$ *and* $\tilde{\mathbf{p}} = fl(\mathbf{p})$, *respectively. Then*

$$\left\|\tilde{\mathbf{p}} - \left(\mathbf{b} - \tilde{\mathbf{q}}\frac{\tilde{\mathbf{q}}\cdot\mathbf{b}}{\tilde{\mathbf{q}}\cdot\tilde{\mathbf{q}}}\right)\right\|_2 \le \frac{M_{(6.26a)}\varepsilon}{1-N_{(6.26a)}\varepsilon}\|\mathbf{b}\|_2 \text{ and} \tag{6.26a}$$

$$\|\tilde{\mathbf{p}} - (\mathbf{b} - \tilde{\mathbf{q}}\tilde{\mathbf{q}}\cdot\mathbf{b})\|_2 \le \frac{M_{(6.26b)}\varepsilon}{1-N_{(6.26b)}\varepsilon}\|\mathbf{b}\|_2 \ , \tag{6.26b}$$

where

$$M_{(6.26a)} = 3m + 2 + \sqrt{2}\,,\ N_{(6.26a)} = (8m + 19)/2\,,$$

$$M_{(6.26b)} = \sqrt{2} \text{ and } N_{(6.26b)} = \max\{(3m + 2)\sqrt{2} + 2, (8m + 15)/2\}\,.$$

Proof Floating point computation produces $\tilde{\mathbf{p}} \equiv \mathrm{fl}(\mathbf{p}) = (\mathbf{I}+\mathbf{E}_{-})\,[\mathbf{b} - (\mathbf{I}+\mathbf{E}_{\times})\,\tilde{\mathbf{q}}\tilde{\varrho}]$, which implies that

$$\boldsymbol{\delta} \equiv \tilde{\mathbf{p}} - [\mathbf{b} - \tilde{\mathbf{q}}\tilde{\varrho}] = \mathbf{E}_{-}\,(\mathbf{I} + \mathbf{E}_{-})^{-1}\,\tilde{\mathbf{p}} - \mathbf{E}_{\times}\tilde{\mathbf{q}}\tilde{\varrho}\,.$$

We can take norms to get

$$\|\boldsymbol{\delta}\|_2 \le \frac{\varepsilon}{1-\varepsilon}\,\|\tilde{\mathbf{p}}\|_2 + \varepsilon\,\|\tilde{\mathbf{q}}\|_2\,|\tilde{\varrho}|\;. \tag{6.27}$$

Since

$$\mathbf{d} \equiv \tilde{\mathbf{p}} - \left(\mathbf{b} - \tilde{\mathbf{q}}\frac{\tilde{\mathbf{q}}\cdot\mathbf{b}}{\tilde{\mathbf{q}}\cdot\tilde{\mathbf{q}}}\right) = \boldsymbol{\delta} - \tilde{\mathbf{q}}\left[\tilde{\varrho} - \frac{\tilde{\mathbf{q}}\cdot\mathbf{b}}{\tilde{\mathbf{q}}\cdot\tilde{\mathbf{q}}}\right]\,,$$

we can take norms to obtain

$$\|\mathbf{d}\|_2 \le \|\boldsymbol{\delta}\|_2 + \|\tilde{\mathbf{q}}\|_2\left|\tilde{\varrho} - \frac{\tilde{\mathbf{q}}\cdot\mathbf{b}}{\tilde{\mathbf{q}}\cdot\tilde{\mathbf{q}}}\right|\;. \tag{6.28}$$

Next, inequality (6.24b) implies that

$$\|\tilde{\mathbf{q}}\|_2 \le 1 + \|\tilde{\mathbf{q}} - \mathbf{q}\|_2 \le 1 + \frac{(m+2)\varepsilon}{1-(3m+5)\varepsilon/2}\,. \tag{6.29}$$

This inequality and inequality (3.38) produce

$$\begin{aligned}
\|\tilde{\mathbf{q}}\|_2^2 - 1 &\le (1 + \|\tilde{\mathbf{q}} - \mathbf{q}\|_2)^2 - 1 = \|\tilde{\mathbf{q}} - \mathbf{q}\|_2\,(2 + \|\tilde{\mathbf{q}} - \mathbf{q}\|_2) \\
&\le \frac{(m+2)\varepsilon}{1-(3m+5)\varepsilon/2}\left[2 + \frac{(m+2)\varepsilon}{1-(3m+5)\varepsilon/2}\right] \le \frac{2(m+2)\varepsilon}{1-(4m+7)\varepsilon/2}\,,
\end{aligned}$$

and

$$\begin{aligned}
1 - \|\tilde{\mathbf{q}}\|_2^2 &\le 1 - (1 - \|\tilde{\mathbf{q}} - \mathbf{q}\|_2)^2 = \|\tilde{\mathbf{q}} - \mathbf{q}\|_2\,(2 - \|\tilde{\mathbf{q}} - \mathbf{q}\|_2) \le 2\,\|\tilde{\mathbf{q}} - \mathbf{q}\|_2 \\
&\le \frac{2(m+2)\varepsilon}{1-(3m+5)\varepsilon/2}\,.
\end{aligned}$$

Combining these two results gives us

$$\left| \|\tilde{\mathbf{q}}\|_2^2 - 1 \right| \leq \frac{2(m+2)\varepsilon}{1-(4m+7)\varepsilon/2} \tag{6.30}$$

Then the error in the projection multiplier together with the triangle inequality imply that

$$\left| \tilde{\varrho} - \frac{\tilde{\mathbf{q}} \cdot \mathbf{b}}{\tilde{\mathbf{q}} \cdot \tilde{\mathbf{q}}} \right| \leq |\tilde{\varrho} - \tilde{\mathbf{q}} \cdot \mathbf{b}| + |\tilde{\mathbf{q}} \cdot \mathbf{b}| \frac{\left| \|\tilde{\mathbf{q}}\|_2^2 - 1 \right|}{\|\tilde{\mathbf{q}}\|_2^2}$$

then inequalities (3.45), (6.30) and the Cauchy inequality (3.15) give us

$$\leq \frac{m\varepsilon}{1-(m-1)\varepsilon/2} \|\tilde{\mathbf{q}}\|_2 \|\mathbf{b}\|_2 + \frac{2(m+2)\varepsilon}{1-(4m+7)\varepsilon/2} \|\tilde{\mathbf{q}}\|_2 \|\mathbf{b}\|_2$$

then inequalities (3.37), (6.29) and (3.38) yield

$$\leq \frac{(3m+2)\varepsilon}{1-(6m+11)\varepsilon/2} \|\mathbf{b}\|_2 \,. \tag{6.31}$$

Since $\tilde{\mathbf{p}} - \mathbf{d} = \mathbf{b} - \tilde{\mathbf{q}} \frac{\tilde{\mathbf{q}} \cdot \mathbf{b}}{\tilde{\mathbf{q}} \cdot \tilde{\mathbf{q}}} \perp \tilde{\mathbf{q}}$, the Pythagorean Theorem 3.2.2 implies that

$$\|\mathbf{b}\|_2^2 = \|\tilde{\mathbf{p}} - \mathbf{d}\|_2^2 + \frac{|\tilde{\mathbf{q}} \cdot \mathbf{b}|^2}{\|\tilde{\mathbf{q}}\|_2^2} \,.$$

This can be rewritten as

$$\|\tilde{\mathbf{p}} - \mathbf{d}\|_2 = \sqrt{\|\mathbf{b}\|_2^2 - \frac{|\tilde{\mathbf{q}} \cdot \mathbf{b}|^2}{\|\tilde{\mathbf{q}}\|_2^2}} \,,$$

and then the triangle inequality implies that

$$\|\tilde{\mathbf{p}}\|_2 \leq \sqrt{\|\mathbf{b}\|_2^2 - \frac{|\tilde{\mathbf{q}} \cdot \mathbf{b}|^2}{\|\tilde{\mathbf{q}}\|_2^2}} + \|\mathbf{d}\|_2 \,. \tag{6.32}$$

Now we recall inequality (6.28)

$$\|\mathbf{d}\|_2 \leq \|\boldsymbol{\delta}\|_2 + \|\tilde{\mathbf{q}}\|_2 \left| \tilde{\varrho} - \frac{\tilde{\mathbf{q}} \cdot \mathbf{b}}{\tilde{\mathbf{q}} \cdot \tilde{\mathbf{q}}} \right| \,.$$

then we use inequality (6.27) and the triangle inequality to get

$$\leq \frac{\varepsilon}{1-\varepsilon} \|\tilde{\mathbf{p}}\|_2 + \|\tilde{\mathbf{q}}\|_2 \left\{ \varepsilon \left| \frac{\tilde{\mathbf{q}} \cdot \mathbf{b}}{\tilde{\mathbf{q}} \cdot \tilde{\mathbf{q}}} \right| + (1+\varepsilon) \left| \tilde{\varrho} - \frac{\tilde{\mathbf{q}} \cdot \mathbf{b}}{\tilde{\mathbf{q}} \cdot \tilde{\mathbf{q}}} \right| \right\}$$

then inequality (6.32) produces

$$\leq \frac{\varepsilon}{1-\varepsilon}\sqrt{\|\mathbf{b}\|_2^2 - \frac{|\tilde{\mathbf{q}}\cdot\mathbf{b}|^2}{\|\tilde{\mathbf{q}}\|_2^2}} + \frac{\varepsilon}{1-\varepsilon}\|\mathbf{d}\|_2 + \varepsilon\frac{|\tilde{\mathbf{q}}\cdot\mathbf{b}|}{\|\tilde{\mathbf{q}}\|_2} + (1+\varepsilon)\,\|\tilde{\mathbf{q}}\|_2\left|\tilde{\varrho} - \frac{\tilde{\mathbf{q}}\cdot\mathbf{b}}{\tilde{\mathbf{q}}\cdot\tilde{\mathbf{q}}}\right| .$$

We can solve this inequality for $\|\mathbf{d}\|_2$ to get

$$\|\mathbf{d}\|_2 \leq \left(1+\frac{\varepsilon}{1-2\varepsilon}\right)\left\{\frac{\varepsilon}{1-\varepsilon}\sqrt{\|\mathbf{b}\|_2^2 - \frac{|\tilde{\mathbf{q}}\cdot\mathbf{b}|^2}{\|\tilde{\mathbf{q}}\|_2^2}} + \varepsilon\frac{|\tilde{\mathbf{q}}\cdot\mathbf{b}|}{\|\tilde{\mathbf{q}}\|_2} + (1+\varepsilon)\,\|\tilde{\mathbf{q}}\|_2\left|\tilde{\varrho} - \frac{\tilde{\mathbf{q}}\cdot\mathbf{b}}{\tilde{\mathbf{q}}\cdot\tilde{\mathbf{q}}}\right|\right\} . \tag{6.33}$$

This inequality suggests that we take $\alpha = \varepsilon/(1-\varepsilon)$, $\beta = \|\mathbf{b}\|_2$ and $\gamma = \varepsilon$ in Lemma 6.8.3 to get

$$\|\mathbf{d}\|_2 \leq \left(1+\frac{\varepsilon}{1-2\varepsilon}\right)\varepsilon\sqrt{\left(\frac{1}{1-\varepsilon}\right)^2 + 1}\|\mathbf{b}\|_2 + \left(1+\frac{\varepsilon}{1-2\varepsilon}\right)(1+\varepsilon)\,\|\tilde{\mathbf{q}}\|_2\left|\tilde{\varrho} - \frac{\tilde{\mathbf{q}}\cdot\mathbf{b}}{\tilde{\mathbf{q}}\cdot\tilde{\mathbf{q}}}\right|$$

$$= \varepsilon\left(1+\frac{\varepsilon}{1-2\varepsilon}\right)\sqrt{\frac{2-\varepsilon(2-\varepsilon)}{(1-\varepsilon)^2}}\|\mathbf{b}\|_2 + \left(1+\frac{\varepsilon}{1-2\varepsilon}\right)(1+\varepsilon)\,\|\tilde{\mathbf{q}}\|_2\left|\tilde{\varrho} - \frac{\tilde{\mathbf{q}}\cdot\mathbf{b}}{\tilde{\mathbf{q}}\cdot\tilde{\mathbf{q}}}\right|$$

then inequalities (6.21), (6.29) and (6.31) produce

$$\leq \varepsilon\left(1+\frac{\varepsilon}{1-2\varepsilon}\right)\frac{\sqrt{2}}{1-\varepsilon}\left(1+\frac{\varepsilon(2-\varepsilon)}{2}\right)\|\mathbf{b}\|_2$$
$$+\left(1+\frac{\varepsilon}{1-2\varepsilon}\right)(1+\varepsilon)\left(1+\frac{(m+2)\varepsilon}{1-(3m+5)\varepsilon/2}\right)\frac{(3m+2)\varepsilon}{1-(6m+11)\varepsilon/2}\|\mathbf{b}\|_2$$

then inequality (3.38) gives us

$$\leq \frac{\sqrt{2}\varepsilon}{1-3\varepsilon}\|\mathbf{b}\|_2 + \frac{(3m+2)\varepsilon}{1-(8m+19)\varepsilon/2}\|\mathbf{b}\|_2$$

and finally inequality (3.37) yields the claimed result (6.26a).

To prove (6.26b), we begin by recalling inequality (6.27)

$$\|\boldsymbol{\delta}\|_2 \leq \frac{\varepsilon}{1-\varepsilon}\|\tilde{\mathbf{p}}\|_2 + \varepsilon\,\|\tilde{\mathbf{q}}\|_2\,|\tilde{\varrho}|$$

then we use inequality (6.32) and the triangle inequality to get

$$\leq \frac{\varepsilon}{1-\varepsilon}\left\{\sqrt{\|\mathbf{b}\|_2^2 - \frac{|\tilde{\mathbf{q}}\cdot\mathbf{b}|^2}{\|\tilde{\mathbf{q}}\|_2^2}} + \|\mathbf{d}\|_2\right\} + \varepsilon\,\|\tilde{\mathbf{q}}\|_2\left\{\frac{|\tilde{\mathbf{q}}\cdot\mathbf{b}|}{\tilde{\mathbf{q}}\cdot\tilde{\mathbf{q}}} + \left|\tilde{\varrho} - \frac{\tilde{\mathbf{q}}\cdot\mathbf{b}}{\tilde{\mathbf{q}}\cdot\tilde{\mathbf{q}}}\right|\right\}$$

then we employ inequality (6.33) to find that

$$\leq \frac{\varepsilon}{1-\varepsilon}\left\{\sqrt{\|\mathbf{b}\|_2^2-\frac{|\tilde{\mathbf{q}}\cdot\mathbf{b}|^2}{\|\tilde{\mathbf{q}}\|_2^2}}+\frac{\varepsilon}{1-\varepsilon}\left(1+\frac{\varepsilon}{1-2\varepsilon}\right)\sqrt{\|\mathbf{b}\|_2^2-\frac{|\tilde{\mathbf{q}}\cdot\mathbf{b}|^2}{\|\tilde{\mathbf{q}}\|_2^2}}\right.$$
$$\left.+ +\varepsilon\left(1+\frac{\varepsilon}{1-2\varepsilon}\right)\frac{|\tilde{\mathbf{q}}\cdot\mathbf{b}|}{\|\tilde{\mathbf{q}}\|_2}+\left(1+\frac{\varepsilon}{1-2\varepsilon}\right)(1+\varepsilon)\,\|\tilde{\mathbf{q}}\|_2\left|\tilde{\varrho}-\frac{\tilde{\mathbf{q}}\cdot\mathbf{b}}{\tilde{\mathbf{q}}\cdot\tilde{\mathbf{q}}}\right|\right\}$$
$$+\varepsilon\frac{|\tilde{\mathbf{q}}\cdot\mathbf{b}|}{\|\tilde{\mathbf{q}}\|_2}+\varepsilon\,\|\tilde{\mathbf{q}}\|_2\left|\tilde{\varrho}-\frac{\tilde{\mathbf{q}}\cdot\mathbf{b}}{\tilde{\mathbf{q}}\cdot\tilde{\mathbf{q}}}\right|$$

then we use inequality (3.38) to produce

$$\leq \frac{\varepsilon}{1-\varepsilon}\left(1+\frac{\varepsilon}{1-2\varepsilon}\right)\sqrt{\|\mathbf{b}\|_2^2-\frac{|\tilde{\mathbf{q}}\cdot\mathbf{b}|^2}{\|\tilde{\mathbf{q}}\|_2^2}}+\varepsilon\left(1+\frac{\varepsilon}{1-2\varepsilon}\right)\frac{|\tilde{\mathbf{q}}\cdot\mathbf{b}|}{\|\tilde{\mathbf{q}}\|_2}$$
$$+2\varepsilon\left(1+\frac{3\varepsilon/2}{1-2\varepsilon}\right)\|\tilde{\mathbf{q}}\|_2\left|\tilde{\varrho}-\frac{\tilde{\mathbf{q}}\cdot\mathbf{b}}{\tilde{\mathbf{q}}\cdot\tilde{\mathbf{q}}}\right|$$
$$\leq \frac{\varepsilon}{1-2\varepsilon}\sqrt{\|\mathbf{b}\|_2^2-\frac{|\tilde{\mathbf{q}}\cdot\tilde{\mathbf{q}}|^2}{\|\tilde{\mathbf{q}}\|_2^2}}+\frac{\varepsilon}{1-2\varepsilon}\frac{|\tilde{\mathbf{q}}\cdot\mathbf{b}|}{\|\tilde{\mathbf{q}}\|_2}+\frac{2\varepsilon}{1-2\varepsilon}\|\tilde{\mathbf{q}}\|_2\left|\tilde{\varrho}-\frac{\tilde{\mathbf{q}}\cdot\mathbf{b}}{\tilde{\mathbf{q}}\cdot\tilde{\mathbf{q}}}\right|$$

then we use inequality (6.25) with $\alpha=\gamma=\varepsilon/(1-2\varepsilon)$ and $\beta=\|\mathbf{b}\|_2$ to find that

$$\leq \frac{\sqrt{2}\varepsilon}{1-2\varepsilon}\|\mathbf{b}\|_2+\frac{2\varepsilon}{1-2\varepsilon}\|\tilde{\mathbf{q}}\|_2\left|\tilde{\varrho}-\frac{\tilde{\mathbf{q}}\cdot\mathbf{b}}{\tilde{\mathbf{q}}\cdot\tilde{\mathbf{q}}}\right|$$

then we employ inequalities (6.29), (6.31) and (3.38) to get

$$\leq \frac{\sqrt{2}\varepsilon}{1-2\varepsilon}\|\mathbf{b}\|_2+\frac{2\varepsilon}{1-2\varepsilon}\left(1+\frac{(m+2)\varepsilon}{1-(3m+5)\varepsilon/2}\right)\frac{(3m+2)\varepsilon}{1-(6m+11)\varepsilon/2}\|\mathbf{b}\|_2$$
$$=\frac{\sqrt{2}\varepsilon}{1-2\varepsilon}\left\{1+\frac{(3m+2)\sqrt{2}\varepsilon}{1-(6m+11)\varepsilon/2}\left(1+\frac{(m+2)\varepsilon}{1-(3m+5)\varepsilon/2}\right)\right\}\|\mathbf{b}\|_2$$

then we use inequality (3.38) to get

$$\leq \frac{\sqrt{2}\varepsilon}{1-2\varepsilon}\left\{1+\frac{(3m+2)\sqrt{2}\varepsilon}{1-(8m+15)\varepsilon/2}\right\}\|\mathbf{b}\|_2\,.$$

Finally, inequality (3.38) gives us the claimed result (6.26b).

The next three theorems bound the error in the modified Gram-Schmidt factorization. Their proofs are similar to the discussion in Björck [8].

Theorem 6.8.2 (Gram-Schmidt Factorization Rounding Errors I) *Assume that the relative errors in floating point addition or subtraction, multiplication or division, and square roots satisfy*

$$fl(\alpha \pm \beta) = (\alpha \pm \beta)(1 + \varepsilon_{\pm})$$
$$fl(\alpha \times \beta) = (\alpha \times \beta)(1 + \varepsilon_{\times})$$
$$fl(\alpha \div \beta) = (\alpha \div \beta)(1 + \varepsilon_{\div})$$
$$fl(\sqrt{\alpha}) = \sqrt{\alpha}(1 + \varepsilon_{\div}) ,$$

and that there is an upper bound ε on all such relative errors:

$$\max\{|\varepsilon_{\pm}| \, , \, |\varepsilon_{\times}| \, , \, |\varepsilon_{\div}| \, , \, |\varepsilon_{\sqrt{}}\} \le \varepsilon \, .$$

Let $m \ge n$ be positive integers, and define

$$M_{(6.34)} = 1 + [n-1]M_{(6.26b)} \text{ and}$$
$$N_{(6.34)} = \max\{(j-1)(3m+4), \max\{N_{(6.26b)}, (9m+9)/2\} - [j-2][3m+4]/2\} \, .$$

We assume that $N_{(6.34)}\varepsilon < 1$. Suppose that **A** *is a nonzero $m \times n$ matrix with rank* (**A**) $= n$*, and such that*

$$\max\{(j-1)(3m+4) \, , \, \max\{N_{(6.26b)} \, , \, (9m+9)/2\} + [j-2][3m+4]/2\}\varepsilon < 1 \, .$$

Let the $m \times n$ matrix $\tilde{\mathbf{Q}}$ with approximately orthogonal columns, and the $n \times n$ right-triangular matrix $\tilde{\mathbf{R}}$ be computed in floating point arithmetic by the modified Gram-Schmidt factorization Algorithm 6.8.1 or 6.8.2. Assume that all of the columns of $\tilde{\mathbf{Q}}$ are nonzero. Let $\|\cdot\|_F$ denote the Frobenius norm, defined in Eq. (3.21)*. Then the error in the Gram-Schmidt factorization satisfies*

$$\|\tilde{\mathbf{Q}}\tilde{\mathbf{R}} - \mathbf{A}\|_F \le \frac{M_{(6.34)}\varepsilon}{1 - N_{(6.34)}\varepsilon} \|\mathbf{A}\|_F \tag{6.34}$$

Proof In step k of the modified Gram-Schmidt factorization Algorithm 6.8.1, for all $k < j \le n$ we compute

$$\tilde{\mathbf{a}}_j^{(k+1)} \equiv \mathrm{fl}\left(\tilde{\mathbf{a}}_j^{(k)} - \tilde{\mathbf{q}}_k \tilde{\varrho}_{kj}\right) = \tilde{\mathbf{a}}_j^{(k)} - \tilde{\mathbf{q}}_k \tilde{\varrho}_{kj} + \boldsymbol{\delta}_j^{(k)} , \tag{6.35}$$

where inequality (6.26b) implies that

$$\left\|\boldsymbol{\delta}_j^{(k)}\right\|_2 \le \frac{M_{(6.26b)}\varepsilon}{1 - N_{(6.26b)}\varepsilon} \left\|\tilde{\mathbf{a}}_j^{(k)}\right\|_2 .$$

Since

$$\tilde{\mathbf{q}}_k = \text{fl}\left(\tilde{\mathbf{a}}_k^{(k)}\frac{1}{\tilde{\varrho}_{kk}}\right) = (\mathbf{I} + \mathbf{E}_{k\div})\,\tilde{\mathbf{a}}_k^{(k)}\frac{1}{\tilde{\varrho}_{kk}}$$

and $\mathbf{a}_j = \tilde{\mathbf{a}}_j^{(1)}$, we have

$$\mathbf{a}_j - \tilde{\mathbf{q}}_j\tilde{\varrho}_{jj} = \tilde{\mathbf{a}}_j^{(1)} - (\mathbf{I} + \mathbf{E}_{k\div})\,\tilde{\mathbf{a}}_j^{(j)} = \sum_{k=1}^{j-1}\left[\tilde{\mathbf{a}}_j^{(k)} - \tilde{\mathbf{a}}_j^{(k+1)}\right] - \mathbf{E}_{k\div}\tilde{\mathbf{a}}_j^{(j)}$$

$$= \sum_{k=1}^{j-1}\left[\tilde{\mathbf{q}}_k\tilde{\varrho}_{kj} - \boldsymbol{\delta}_j^{(k)}\right] - \mathbf{E}_{k\div}\tilde{\mathbf{a}}_j^{(j)}\,.$$

We can rearrange terms to get

$$\sum_{k=1}^{j}\tilde{\mathbf{q}}_k\tilde{\varrho}_{kj} - \mathbf{a}_j = \sum_{k=1}^{j-1}\boldsymbol{\delta}_j^{(k)} + \mathbf{E}_{k\div}\tilde{\mathbf{a}}_j^{(j)}\,,$$

and then take norms to get

$$\left\|\sum_{k=1}^{j}\tilde{\mathbf{q}}_k\tilde{\varrho}_{kj} - \mathbf{a}_j\right\|_2 \le \frac{M_{(6.26b)}\varepsilon}{1 - N_{(6.26b)}\varepsilon}\sum_{k=1}^{j-1}\left\|\tilde{\mathbf{a}}_j^{(k)}\right\|_2 + \left\|\tilde{\mathbf{a}}_j^{(j)}\right\|_2\varepsilon\,. \tag{6.36}$$

Since $\tilde{\mathbf{a}}_j^{(k+1)}$ is the floating point evaluation of an orthogonal projection of $\tilde{\mathbf{a}}_j^{(k)}$, we should be able to bound $\|\tilde{\mathbf{a}}_j^{(k+1)}\|_2$ by a constant close to one times $\|\tilde{\mathbf{a}}_j^{(k)}\|_2$. Let us determine such a constant. For $1 \le k < j \le n$ we can add and subtract terms to get

$$\left\|\tilde{\mathbf{a}}_j^{(k+1)}\right\|_2 \le \left\|\text{fl}\left(\tilde{\mathbf{a}}_j^{(k)} - \tilde{\mathbf{q}}_k\tilde{\varrho}_{jk}\right) - \left(\tilde{\mathbf{a}}_j^{(k)} - \tilde{\mathbf{q}}_k\frac{\tilde{\mathbf{q}}_k\cdot\tilde{\mathbf{a}}_k^{(k)}}{\tilde{\mathbf{q}}_k\cdot\tilde{\mathbf{q}}_k}\right)\right\|_2 + \left\|\tilde{\mathbf{a}}_j^{(k)} - \tilde{\mathbf{q}}_k\frac{\tilde{\mathbf{q}}_k\cdot\tilde{\mathbf{a}}_k^{(k)}}{\tilde{\mathbf{q}}_k\cdot\tilde{\mathbf{q}}_k}\right\|_2$$

then use inequality (6.33) and the Pythagorean Theorem (3.3) to arrive at

$$\le \left(1 + \frac{\varepsilon}{1-2\varepsilon}\right)\left\{\frac{\varepsilon}{1-\varepsilon}\sqrt{\left\|\tilde{\mathbf{a}}_j^{(k)}\right\|_2^2 - \left(\frac{\left|\tilde{\mathbf{q}}_k\cdot\tilde{\mathbf{a}}_j^{(k)}\right|}{\|\tilde{\mathbf{q}}_k\|_2}\right)^2} + \varepsilon\frac{\left|\tilde{\mathbf{q}}_k\cdot\tilde{\mathbf{a}}_j^{(k)}\right|}{\|\tilde{\mathbf{q}}_k\|_2}\right.$$

$$\left. + (1+\varepsilon)\,\|\tilde{\mathbf{q}}_k\|_2\left|\tilde{\varrho}_{kj} - \frac{\tilde{\mathbf{q}}_k\cdot\tilde{\mathbf{a}}_j^{(k)}}{\tilde{\mathbf{q}}_k\cdot\tilde{\mathbf{q}}_k}\right|\right\} + \sqrt{\left\|\tilde{\mathbf{a}}_j^{(k)}\right\|_2^2 - \left(\frac{\left|\tilde{\mathbf{q}}_k\cdot\tilde{\mathbf{a}}_j^{(k)}\right|}{\|\tilde{\mathbf{q}}_k\|_2}\right)^2}$$

then we use inequalities (3.38) and (3.40) to produce

$$= \left(1 + \frac{\varepsilon}{1-2\varepsilon}\right) \sqrt{\left\|\tilde{\mathbf{a}}_j^{(k)}\right\|_2^2 - \left(\frac{\left|\tilde{\mathbf{q}}_k \cdot \tilde{\mathbf{a}}_j^{(k)}\right|}{\|\tilde{\mathbf{q}}_k\|_2}\right)^2} + \frac{\varepsilon}{1-2\varepsilon} \frac{\left|\tilde{\mathbf{q}}_k \cdot \tilde{\mathbf{a}}_j^{(k)}\right|}{\|\tilde{\mathbf{q}}_k\|_2}$$

$$+ \left(1 + \frac{2\varepsilon}{1-2\varepsilon}\right) \left|\tilde{\varrho}_{kj} - \frac{\tilde{\mathbf{q}}_k \cdot \tilde{\mathbf{a}}_j^{(k)}}{\tilde{\mathbf{q}}_k \cdot \tilde{\mathbf{q}}_k}\right|$$

then inequality (6.25) with $\alpha = 1 + \varepsilon/(1-2\varepsilon)$, $\gamma = \varepsilon/(1-2\varepsilon)$ and $\beta = \|\tilde{\mathbf{a}}_j^{(k))}\|_2$ yields

$$\le \frac{\sqrt{(1-\varepsilon)^2 + \varepsilon^2}}{1-2\varepsilon} \left\|\tilde{\mathbf{a}}_j^{(k))}\right\|_2 + \left(1 + \frac{2\varepsilon}{1-2\varepsilon}\right) \left|\tilde{\varrho}_{kj} - \frac{\tilde{\mathbf{q}}_k \cdot \tilde{\mathbf{a}}_j^{(k)}}{\tilde{\mathbf{q}}_k \cdot \tilde{\mathbf{q}}_k}\right|$$

then we use inequalities (6.21), (6.29) and (6.31) to obtain

$$\le \left(1 + \frac{2\varepsilon(1-\varepsilon)}{1-2\varepsilon}\right) \left\|\tilde{\mathbf{a}}_j^{(k))}\right\|_2$$

$$+ \left(1 + \frac{2\varepsilon}{1-2\varepsilon}\right) \left(1 + \frac{(m+2)\varepsilon}{1-(3m+5)\varepsilon/2}\right) \frac{(3m+2)\varepsilon}{1-(6m+11)\varepsilon/2} \left\|\tilde{\mathbf{a}}_j^{(k))}\right\|_2$$

then we employ inequality (3.38) and (3.37) to arrive at

$$\le \left(1 + \frac{2\varepsilon}{1-2\varepsilon}\right) \left\|\tilde{\mathbf{a}}_j^{(k))}\right\|_2 + \frac{(3m+2)\varepsilon}{1-(9m+9)\varepsilon/2} \left\|\tilde{\mathbf{a}}_j^{(k))}\right\|_2$$

and finally we use inequality (3.37) to get

$$\le \left(1 + \frac{(3m+4)\varepsilon}{1-(9m+9)\varepsilon/2}\right) \left\|\tilde{\mathbf{a}}_j^{(k))}\right\|_2 .$$

This inequality leads to a recurrence, with solution

$$\left\|\tilde{\mathbf{a}}_j^{(k)}\right\|_2 \le \left(1 + \frac{(3m+4)\varepsilon}{1-(9m+9)\varepsilon/2}\right)^{k-1} \|\mathbf{a}_j\|_2 \tag{6.37}$$

for $1 \le k \le j$. Returning to inequality (6.36), we now see that

$$\left\|\sum_{k=1}^{j} \tilde{\mathbf{q}}_k \tilde{\varrho}_{kj} - \mathbf{a}_j\right\|_2 = \left\|\sum_{k=1}^{j-1} \boldsymbol{\delta}_j^{(k)} - \mathbf{E}_{\div}\tilde{\mathbf{a}}_k^{(k)}\right\|_2 \le \frac{M_{(6.26b)}\varepsilon}{1 - N_{(6.26b)}\varepsilon} \sum_{k=1}^{j-1} \left\|\tilde{\mathbf{a}}_j^{(k)}\right\|_2 + \varepsilon \left\|\tilde{\mathbf{a}}_j^{(j)}\right\|_2$$

then inequality (6.37) gives us

$$\leq \frac{M_{(6.26b)}\varepsilon}{1 - N_{(6.26b)}\varepsilon} \sum_{k=1}^{j-1} \left(1 + \frac{(3m+4)\varepsilon}{1-(9m+9)\varepsilon/2}\right)^{k-1} \|\mathbf{a}_j\|_2$$

$$+ \varepsilon \left(1 + \frac{(3m+4)\varepsilon}{1-(9m+9)\varepsilon/2}\right)^{j-1} \|\mathbf{a}_j\|_2$$

$$= \frac{M_{(6.26b)}\varepsilon}{1 - N_{(6.26b)}\varepsilon} \frac{\left(1 + \frac{(3m+4)\varepsilon}{1-(9m+9)\varepsilon/2}\right)^{j-1} - 1}{\frac{(3m+4)\varepsilon}{1-(9m+9)\varepsilon/2}} \|\mathbf{a}_j\|_2$$

$$+ \varepsilon \left(1 + \frac{(3m+4)\varepsilon}{1-(9m+9)\varepsilon/2}\right)^{j-1} \|\mathbf{a}_j\|_2$$

then we use inequality (3.41) to get

$$\leq \frac{M_{(6.26b)}\varepsilon}{1 - N_{(6.26b)}\varepsilon} \frac{\frac{(j-1)(3m+4)\varepsilon}{1-(9m+9+[j-2][3m+4])\varepsilon/2}}{\frac{(3m+4)\varepsilon}{1-(9m+9)\varepsilon/2}} \|\mathbf{a}_j\|_2$$

$$+ \varepsilon \left(1 + \frac{(j-1)(3m+4)\varepsilon}{1-(9m+9+[j-2][3m+4])\varepsilon/2}\right) \|\mathbf{a}_j\|_2$$

$$= \frac{(j-1)M_{(6.26b)}\varepsilon}{1 - N_{(6.26b)}\varepsilon} \left(1 + \frac{(j-2)(3m+4)\varepsilon/2}{1-(9m+9+[j-2][3m+4])\varepsilon/2}\right) \|\mathbf{a}_j\|_2$$

$$+ \varepsilon \left(1 + \frac{(j-1)(3m+4)\varepsilon}{1-(9m+9+[j-2][3m+4])\varepsilon/2}\right) \|\mathbf{a}_j\|_2$$

then inequalities (3.38) and (3.37) produce

$$\leq \frac{(j-1)M_{(6.26b)}\varepsilon}{1 - \max\{N_{(6.26b)} + [j-2][3m+4]/2\,,\ 9[m+1]/2 + [j-2][3m+4]/2\}\varepsilon} \|\mathbf{a}_j\|_2$$

$$+ \frac{\varepsilon}{1 - \max\{(j-1)(3m+4)\,,\ (9m+9+[j-2][3m+4])/2\}\varepsilon} \|\mathbf{a}_j\|_2$$

and finally inequality (3.37) yields

$$\leq \frac{(1 + [j-1]M_{(6.26b)})\varepsilon}{1 - \max\{(j-1)(3m+4)\,,\ \max\{N_{(6.26b)}\,,\ (9m+9)/2\} + [j-2][3m+4]/2\}\varepsilon} \|\mathbf{a}_j\|_2$$

provided that the denominator is positive. As a result, the Frobenius norm of the error in the factorization satisfies

$$\left\|\tilde{Q}\tilde{\mathbf{R}} - \mathbf{A}\right\|_F^2 = \sum_{j=1}^{n} \left\|\sum_{k=1}^{j} \tilde{\mathbf{q}}_k \tilde{\varrho}_{kj} - \mathbf{a}_j\right\|_2^2$$

$$\leq \left\{\frac{M_{(6.34)}\varepsilon}{1 - N_{(6.34)}\varepsilon}\right\}^2 \sum_{j=1}^{n} \|\mathbf{a}_j\|_2^2 .$$

The next theorem bounds the errors in the projections of the columns of the original matrix in a least squares problem. Its proof is similar to the discussion in Björck [8].

Theorem 6.8.3 (Gram-Schmidt Factorization Rounding Errors II) *Suppose that the assumptions of Theorem 6.8.2 are satisfied. Define*

$M_{(6.38)} = (n-1)M_{(6.26a)}$ *and* $N_{(6.38)} = \max\{N_{(6.26a)}, [9m+9]/2\} + (n-2)(3m+4)/2$,

and assume that $N_{(6.26a)}\varepsilon < 1$. *If* $\tilde{\mathbf{q}}_k$ *is the kth column of* $\tilde{\mathbf{Q}}$, *define the product of orthogonal projections*

$$\mathscr{P}_k = \left(\mathbf{I} - \frac{\tilde{\mathbf{q}}_1\tilde{\mathbf{q}}_1^H}{\tilde{\mathbf{q}}_1 \cdot \tilde{\mathbf{q}}_1}\right) \cdot \ldots \cdot \left(\mathbf{I} - \frac{\tilde{\mathbf{q}}_k\tilde{\mathbf{q}}_k^H}{\tilde{\mathbf{q}}_k \cdot \tilde{\mathbf{q}}_k}\right) .$$

Then the successive orthogonal projections of the columns of **A** *satisfy*

$$\left\|\mathscr{P}_n{}^H\mathbf{A}\right\|_F \leq \frac{M_{(6.38)}\varepsilon}{1 - N_{(6.38)}\varepsilon}\|\mathbf{A}\|_F . \tag{6.38}$$

Proof As in the proof of Lemma 6.8.4, we define

$$\mathbf{d}_j^{(k)} \equiv \tilde{\mathbf{a}}_j^{(k+1)} - \left\{\tilde{\mathbf{a}}_j^{(k)} - \tilde{\mathbf{q}}_k \frac{\tilde{\mathbf{q}}_k \cdot \tilde{\mathbf{a}}_j^{(k)}}{\tilde{\mathbf{q}}_k \cdot \tilde{\mathbf{q}}_k}\right\} \tag{6.39}$$

for $1 \leq k < j \leq n$, and we also define

$$\widehat{\mathbf{a}}_j^{(k)} \equiv \left[\mathbf{I} - \tilde{\mathbf{q}}_{k-1}\frac{1}{\tilde{\mathbf{q}}_{k-1} \cdot \tilde{\mathbf{q}}_{k-1}}\tilde{\mathbf{q}}_{k-1}^H\right] \cdot \ldots \cdot \left[\mathbf{I} - \tilde{\mathbf{q}}_1\frac{1}{\tilde{\mathbf{q}}_1 \cdot \tilde{\mathbf{q}}_1}\tilde{\mathbf{q}}_1^H\right]\mathbf{a}_j$$

for $1 \leq k < j \leq n$. Here $\widehat{\mathbf{a}}_j^{(k)}$ is the kth successive orthogonal projection, using the approximate orthogonal vectors $\tilde{\mathbf{q}}_j$, of the jth columns of **A**. The definition of $\mathbf{d}_j^{(k)}$ can be rewritten as the recurrence

$$\tilde{\mathbf{a}}_j^{(k+1)} = \left[\mathbf{I} - \tilde{\mathbf{q}}_k\frac{1}{\tilde{\mathbf{q}}_k \cdot \tilde{\mathbf{q}}_k}\tilde{\mathbf{q}}_k^H\right]\tilde{\mathbf{a}}_j^{(k)} + \mathbf{d}_j^{(k)}$$

for the vectors $\tilde{\mathbf{a}}_j^{(k)}$, with $1 \le k < j$. Since $\tilde{\mathbf{a}}_j^{(1)} = \mathbf{a}_j$, the solution of this recurrence for $2 \le k \le j$ is

$$\begin{aligned}\tilde{\mathbf{a}}_j^{(k)} &= \widehat{\mathbf{a}}_j^{(k)} + \mathbf{d}_j^{(k-1)} \\ &+ \sum_{\ell=1}^{k-2} \left[\mathbf{I} - \tilde{\mathbf{q}}_{k-1} \frac{1}{\tilde{\mathbf{q}}_{k-1} \cdot \tilde{\mathbf{q}}_{k-1}} \tilde{\mathbf{q}}_{k-1}^H \right] \cdot \ldots \cdot \left[\mathbf{I} - \tilde{\mathbf{q}}_{\ell+1} \frac{1}{\tilde{\mathbf{q}}_{\ell+1} \cdot \tilde{\mathbf{q}}_{\ell+1}} \tilde{\mathbf{q}}_{\ell+1}^H \right] \mathbf{d}_j^{(\ell)} .\end{aligned} \tag{6.40}$$

Since $\tilde{\mathbf{a}}_k^{(k)} = \tilde{\mathbf{q}}_k \tilde{\varrho}_{kk}$, we have

$$\left[\mathbf{I} - \tilde{\mathbf{q}}_k \frac{1}{\tilde{\mathbf{q}}_k \cdot \tilde{\mathbf{q}}_k} \tilde{\mathbf{q}}_k^H \right] \widehat{\mathbf{a}}_j^{(k)} = \mathbf{0} .$$

Thus

$$\mathbf{0} = \widehat{\mathbf{a}}_j^{(n)} + \sum_{\ell=1}^{j-1} \left[\mathbf{I} - \tilde{\mathbf{q}}_n \frac{1}{\tilde{\mathbf{q}}_n \cdot \tilde{\mathbf{q}}_n} \tilde{\mathbf{q}}_n^H \right] \cdot \ldots \cdot \left[\mathbf{I} - \tilde{\mathbf{q}}_{\ell+1} \frac{1}{\tilde{\mathbf{q}}_{\ell+1} \cdot \tilde{\mathbf{q}}_{\ell+1}} \tilde{\mathbf{q}}_{\ell+1}^H \right] \mathbf{d}_j^{(\ell)} .$$

We can move the first term on the right to the left and take norms to get

$$\begin{aligned}\left\| \widehat{\mathbf{a}}_j^{(n)} \right\|_2 &= \left\| \left[\mathbf{I} - \tilde{\mathbf{q}}_n \frac{1}{\tilde{\mathbf{q}}_n \cdot \tilde{\mathbf{q}}_n} \tilde{\mathbf{q}}_n^H \right] \cdot \ldots \cdot \left[\mathbf{I} - \tilde{\mathbf{q}}_1 \frac{1}{\tilde{\mathbf{q}}_1 \cdot \tilde{\mathbf{q}}_1} \tilde{\mathbf{q}}_1^H \right] \mathbf{a}_j \right\|_2 \\ &= \left\| \sum_{\ell=1}^{j-1} \left[\mathbf{I} - \tilde{\mathbf{q}}_n \frac{1}{\tilde{\mathbf{q}}_n \cdot \tilde{\mathbf{q}}_n} \tilde{\mathbf{q}}_n^H \right] \cdot \ldots \cdot \left[\mathbf{I} - \tilde{\mathbf{q}}_{\ell+1} \frac{1}{\tilde{\mathbf{q}}_{\ell+1} \cdot \tilde{\mathbf{q}}_{\ell+1}} \tilde{\mathbf{q}}_{\ell+1}^H \right] \mathbf{d}_j^{(\ell)} \right\|_2\end{aligned}$$

then we use the triangle inequality and the fact that the 2-norm of an orthogonal projector is at most one to obtain

$$\le \sum_{\ell=1}^{j-1} \left\| \mathbf{d}_j^{(\ell)} \right\|_2$$

then we recall inequality (6.26a) to find that

$$\le \frac{M_{(6.26a)}\varepsilon}{1 - N_{(6.26a)}\varepsilon} \sum_{\ell=1}^{j-1} \left\| \tilde{\mathbf{a}}_j^{(\ell)} \right\|_2$$

then we use inequality (6.37) to obtain

$$\leq \frac{M_{(6.26a)}\varepsilon}{1-N_{(6.26a)}\varepsilon} \sum_{\ell=1}^{j-1} \left(1+\frac{(3m+4)\varepsilon}{1-(9m+9)\varepsilon/2}\right)^{\ell-1} \|\mathbf{a}_j\|_2$$

$$= \frac{M_{(6.26a)}\varepsilon}{1-N_{(6.26a)}\varepsilon} \frac{\{1+(3m+4)\varepsilon/[1-(9m+9)\varepsilon/2]\}^{j-1}-1}{(3m+4)\varepsilon/[1-(9m+9)\varepsilon/2]} \|\mathbf{a}_j\|_2$$

then inequality (3.41) produces

$$\leq \frac{M_{(6.26a)}\varepsilon}{1-N_{(6.26a)}\varepsilon} \frac{(j-1)(3m+4)\varepsilon/[1-(9m+9+[j-2][3m+4])\varepsilon/2]}{(3m+4)\varepsilon/[1-(9m+9)\varepsilon/2]} \|\mathbf{a}_j\|_2$$

$$= \frac{(j-1)M_{(6.26a)}\varepsilon}{1-N_{(6.26a)}\varepsilon} \left(1+\frac{(j-2)(3m+4)\varepsilon/2}{1-(9m+9+[j-2][3m+4])\varepsilon/2}\right) \|\mathbf{a}_j\|_2$$

and inequality (3.38) yields

$$\leq \frac{(j-1)M_{(6.26a)}\varepsilon}{1-\left(\max\{N_{(6.26a)}\,,\,[9m+9]/2\}+[j-2][3m+4]/2\right)\varepsilon} \|\mathbf{a}_j\|_2$$

This implies that

$$\left\|\left[\mathbf{I}-\tilde{\mathbf{q}}_n\frac{1}{\tilde{\mathbf{q}}_n\cdot\tilde{\mathbf{q}}_n}\tilde{\mathbf{q}}_n^H\right]\cdot\ldots\cdot\left[\mathbf{I}-\tilde{\mathbf{q}}_1\frac{1}{\tilde{\mathbf{q}}_1\cdot\tilde{\mathbf{q}}_1}\tilde{\mathbf{q}}_1^H\right]\mathbf{A}\right\|_F^2 = \sum_{j=1}^n \left\|\widehat{\mathbf{a}}_j^{(n)}\right\|_2^2$$

$$\leq \left[\frac{(n-1)M_{(6.26a)}\varepsilon}{1-\left(\max\{N_{(6.26a)}\,,\,[9m+9]/2\}+[n-2][3m+4]/2\right)\varepsilon}\right]^2 \sum_{j=1}^n \|\mathbf{a}_j\|_2^2\,.$$

Our third theorem concerning floating point errors in Gram-Schmidt factorization will bound the errors in the right-triangular factor $\mathbf{R}$.

Theorem 6.8.4 (Gram-Schmidt Factorization Rounding Errors III) *Suppose that the assumptions of Theorem 6.8.2 are satisfied. Define*

$$M_{(6.41)} = \left[(5m+6)^{2/3}+(2m+2)^{2/3}\max\{n-2\,,\,0\}\right]^{3/2} \text{ and}$$
$$N_{(6.41)} = m+2+\max\left\{N_{(6.26a)}+\max\{n-2\,,\,0\}[3m+4]/2\,,\,(n-1)(3m+4)\right\}\,,$$

and assume that $N_{(6.41)}\varepsilon < 1$. Let $\|\cdot\|_F$ denote the Frobenius norm, defined in Eq. (3.21). If $\tilde{\mathbf{q}}_k$ is the kth column of $\tilde{\mathbf{Q}}$, define the product of orthogonal projections

$$\mathscr{P}_k = \left(\mathbf{I}-\tilde{\mathbf{q}}_1\frac{1}{\|\tilde{\mathbf{q}}_1\|_2^2}\tilde{\mathbf{q}}_1^H\right)\cdot\ldots\cdot\left(\mathbf{I}-\tilde{\mathbf{q}}_k\frac{1}{\|\tilde{\mathbf{q}}_k\|_2^2}\tilde{\mathbf{q}}_k^H\right)\,.$$

Let the vector

$$\widehat{\mathbf{q}}_k = \mathscr{P}_{k-1}\tilde{\mathbf{q}}_k$$

to be the kth column of a matrix $\widehat{\mathbf{Q}}$. *If*

$(m+2+[n-1][3m+4])\varepsilon < 1$ *and* $(10m+25+\max\{n-2\ ,\ 0\}[3m+4])\varepsilon < 2$,

then the entries of $\tilde{\mathbf{R}}$ *satisfy*

$$\left\|\tilde{\mathbf{R}} - \widehat{\mathbf{Q}}^H\mathbf{A}\right\|_F \leq \frac{M_{(6.41)}\varepsilon}{1-N_{(6.41)}\varepsilon}\|\mathbf{A}\|_F \ . \tag{6.41}$$

Proof Since

$$\widehat{\mathbf{q}}_k \cdot \mathbf{a}_j = (\mathscr{P}_{k-1}\tilde{\mathbf{q}}_k)\cdot\mathbf{a}_j = \tilde{\mathbf{q}}_k \cdot \widehat{\mathbf{a}}_j^{(k)} \ ,$$

we see that for $1 \leq k \leq j$ we have

$$\tilde{\varrho}_{kj} - \widehat{\mathbf{q}}_k \cdot \mathbf{a}_j = \left\{\tilde{\varrho}_{kj} - \tilde{\mathbf{q}}_k \cdot \tilde{\mathbf{a}}_j^{(k)}\right\} + \tilde{\mathbf{q}}_k \cdot \left\{\tilde{\mathbf{a}}_j^{(k)} - \widehat{\mathbf{a}}_j^{(k)}\right\} \ .$$

By taking absolute values of both sides in this equation, then applying the triangle inequality and the Cauchy inequality (3.15), we find that

$$\left|\tilde{\varrho}_{kj} - \widehat{\mathbf{q}}_k \cdot \mathbf{a}_j\right| \leq \left|\tilde{\varrho}_{kj} - \tilde{\mathbf{q}}_k \cdot \tilde{\mathbf{a}}_j^{(k)}\right| + \|\tilde{\mathbf{q}}_k\|_2 \left\|\tilde{\mathbf{a}}_j^{(k)} - \widehat{\mathbf{a}}_j^{(k)}\right\|$$

then we use Eq. (6.40) to get

$$= \left|\tilde{\varrho}_{kj} - \tilde{\mathbf{q}}_k \cdot \tilde{\mathbf{a}}_j^{(k)}\right|$$

$$+ \|\tilde{\mathbf{q}}_k\|_2 \begin{cases} \left\|\mathbf{d}_j^{(k-1)} + \sum_{\ell=1}^{k-2}\left[\mathbf{I} - -\frac{\tilde{\mathbf{q}}_{k-1}\tilde{\mathbf{q}}_{k-1}^H}{\tilde{\mathbf{q}}_{k-1}\cdot\tilde{\mathbf{q}}_{k-1}}\right]\cdot\ldots\cdot\left[\mathbf{I} - \frac{\tilde{\mathbf{q}}_{\ell+1}\tilde{\mathbf{q}}_{\ell+1}^H}{\tilde{\mathbf{q}}_{\ell+1}\cdot\tilde{\mathbf{q}}_{\ell+1}}\right]\mathbf{d}_j^{(\ell)}\right\|_2 , & 1 < k \leq j \\ 0, & 1 = k < j \end{cases}$$

and since the 2-norm of an orthogonal projector is one, we can use inequality (6.26a) to obtain

$$\leq \left|\tilde{\varrho}_{kj} - \tilde{\mathbf{q}}_k \cdot \tilde{\mathbf{a}}_j^{(k)}\right| + \frac{M_{(6.26\mathrm{a})}\varepsilon}{1-N_{(6.26\mathrm{a})}\varepsilon}\|\tilde{\mathbf{q}}_k\|_2 \sum_{\ell=1}^{k-1}\left\|\tilde{\mathbf{a}}_j^{(\ell)}\right\|_2 \ . \tag{6.42}$$

If $1 \leq k < j$ then $\tilde{\varrho}_{kj} = \ \mathrm{fl}\left(\tilde{\mathbf{q}}_k \cdot \tilde{\mathbf{a}}_j^{(k)}\right)$, so inequality (3.45) implies that

$$\left|\tilde{\varrho}_{kj} - \tilde{\mathbf{q}}_k \cdot \tilde{\mathbf{a}}_j^{(k)}\right| \leq \frac{m\varepsilon}{1-(m-1)\varepsilon/2}\|\tilde{\mathbf{q}}\|_2 \left\|\tilde{\mathbf{a}}_j^{(k)}\right\|_2 \ .$$

Combining this result with (6.42) gives us

$$\left|\tilde{\varrho}_{kj} - \widehat{\mathbf{q}}_k \cdot \mathbf{a}_j\right| \leq \frac{m\varepsilon}{1-(m-1)\varepsilon/2} \|\tilde{\mathbf{q}}\|_2 \left\|\tilde{\mathbf{a}}_j^{(k)}\right\|_2 + \frac{M_{(6.26a)}\varepsilon}{1-N_{(6.26a)}\varepsilon} \|\tilde{\mathbf{q}}_k\|_2 \sum_{\ell=1}^{k-1} \left\|\tilde{\mathbf{a}}_j^{(\ell)}\right\|_2$$

then inequality (6.29) bounding the norm of $\tilde{\mathbf{q}}_k$ gives us

$$\leq \left(1+\frac{(m+2)\varepsilon}{1-(3m+5)\varepsilon/2}\right)\left\{\frac{m\varepsilon}{1-(m-1)\varepsilon/2}\left\|\tilde{\mathbf{a}}_j^{(k)}\right\|_2 + \frac{M_{(6.26a)}\varepsilon}{1-N_{(6.26a)}\varepsilon}\sum_{\ell=1}^{k-1}\left\|\tilde{\mathbf{a}}_j^{(\ell)}\right\|_2\right\}$$

next we use inequality (6.37) to bound the norms of the projected columns in terms of the original columns:

$$\leq \left(1+\frac{(m+2)\varepsilon}{1-(3m+5)\varepsilon/2}\right)\left\{\frac{m\varepsilon}{1-(m-1)\varepsilon/2}\left(1+\frac{(3m+4)\varepsilon}{1-(9m+9)\varepsilon/2}\right)^{k-1}\|\mathbf{a}_j\|_2\right.$$
$$\left. + \frac{M_{(6.26a)}\varepsilon}{1-N_{(6.26a)}\varepsilon}\sum_{\ell=1}^{k-1}\left(1+\frac{(3m+4)\varepsilon}{1-(9m+9)\varepsilon/2}\right)^{\ell-1}\|\mathbf{a}_j\|_2\right\}$$
$$= \left(1+\frac{(m+2)\varepsilon}{1-(3m+5)\varepsilon/2}\right)\left\{\frac{m\varepsilon}{1-(m-1)\varepsilon/2}\left(1+\frac{(3m+4)\varepsilon}{1-(9m+9)\varepsilon/2}\right)^{k-1}\right.$$
$$\left. + \frac{M_{(6.26a)}\varepsilon}{1-N_{(6.26a)}\varepsilon}\frac{\left(1+\frac{(3m+4)\varepsilon}{1-(9m+9)\varepsilon/2}\right)^{k-1}-1}{\frac{(3m+4)\varepsilon}{1-(9m+9)\varepsilon/2}}\right\}\|\mathbf{a}_j\|_2 \,. \tag{6.43}$$

In the case $1 < k < j$, we can use inequality (3.41) to bound the power of a relative error

$$\left|\tilde{\varrho}_{kj} - \widehat{\mathbf{q}}_k \cdot \mathbf{a}_j\right| \leq \left(1+\frac{(m+2)\varepsilon}{1-(3m+5)\varepsilon/2}\right)$$
$$\left\{\frac{m\varepsilon}{1-(m-1)\varepsilon/2}\left(1+\frac{(k-1)(3m+4)\varepsilon}{1-(9m+9+[k-2][3m+4])\varepsilon/2}\right)\right.$$
$$\left. + \frac{M_{(6.26a)}\varepsilon}{1-N_{(6.26a)}\varepsilon}\frac{\frac{(k-1)(3m+4)\varepsilon}{(1-[9m+9+(k-2)(3m+4)]\varepsilon/2)}}{\frac{(3m+4)\varepsilon}{1-[9m+9]\varepsilon/2}}\right\}\|\mathbf{a}_j\|_2$$
$$= \left(1+\frac{(m+2)\varepsilon}{1-(3m+5)\varepsilon/2}\right)$$
$$\left\{\frac{m\varepsilon}{1-(m-1)\varepsilon/2}\left(1+\frac{(k-1)(3m+4)\varepsilon}{1-(9m+9+[k-2][3m+4])\varepsilon/2}\right)\right.$$
$$\left. + \frac{(k-1)M_{(6.26a)}\varepsilon}{1-N_{(6.26a)}\varepsilon}\left(1+\frac{(k-2)(3m+4)\varepsilon/2}{1-(9m+9+[k-2][3m+4])\varepsilon/2}\right)\right\}\|\mathbf{a}_j\|_2$$

then we use inequalities (3.38), (3.37) and (3.38) to produce

$$\leq \frac{(m+[k-1]M_{(6.26a)})\varepsilon}{1-\left(\max\{N_{(6.26a)}\ ,\ [9m+9]/2\}+[m+2]+[k-2][3m+4]/2\right)\varepsilon}\|\mathbf{a}_j\|_2\ . \tag{6.44}$$

Otherwise, $1=k<j$ and inequality (6.43) produces

$$\left|\tilde{\varrho}_{1j}-\widehat{\mathbf{q}}_1\cdot\mathbf{a}_j\right| \leq \left(1+\frac{(m+2)\varepsilon}{1-(3m+5)\varepsilon/2}\right)\frac{m\varepsilon}{1-(m-1)\varepsilon/2}\left\|\mathbf{a}_j\right\|_2$$

then inequality (3.38) gives us

$$\leq \frac{m\varepsilon}{1-(3m+5)\varepsilon/2}\|\mathbf{a}_j\|_2\ . \tag{6.45}$$

On the other hand, if $k=j$, then $\tilde{\varrho}_{jj} = \operatorname{fl}\left(\|\tilde{\mathbf{a}}_j^{(j)}\|_2\right)$ and $\tilde{\mathbf{q}}_j = \operatorname{fl}\left(\tilde{\mathbf{a}}_j^{(j)}/\tilde{\varrho}_{jj}\right)$, so

$$\left|\tilde{\varrho}_{jj}-\tilde{\mathbf{q}}_j\cdot\tilde{\mathbf{a}}_j^{(j)}\right| = \left|\tilde{\varrho}_{jj}-\left(\tilde{\mathbf{a}}_j^{(j)}\right)^H(\mathbf{I}+\mathbf{E}_{j\div})\,\tilde{\mathbf{a}}_j^{(j)}\frac{1}{\tilde{\varrho}_{jj}}\right|$$

and then the triangle inequality implies that

$$\leq \left|\tilde{\varrho}_{jj}-\frac{\left\|\tilde{\mathbf{a}}_j^{(j)}\right\|_2^2}{\tilde{\varrho}_{jj}}\right| + \varepsilon\frac{\left\|\tilde{\mathbf{a}}_j^{(j)}\right\|_2^2}{\tilde{\varrho}_{jj}}$$

$$= \frac{\left|\tilde{\varrho}_{jj}-\left\|\tilde{\mathbf{a}}_j^{(j)}\right\|_2\right|\,\left|\tilde{\varrho}_{jj}+\left\|\tilde{\mathbf{a}}_j^{(j)}\right\|_2\right|}{\left\|\tilde{\mathbf{a}}_j^{(j)}\right\|_2+\left(\tilde{\varrho}_{jj}-\left\|\tilde{\mathbf{a}}_j^{(j)}\right\|_2\right)} + \varepsilon\frac{\left\|\tilde{\mathbf{a}}_j^{(j)}\right\|_2^2}{\left\|\tilde{\mathbf{a}}_j^{(j)}\right\|_2+\left(\tilde{\varrho}_{jj}-\left\|\tilde{\mathbf{a}}_j^{(j)}\right\|_2\right)}$$

and we use inequality (6.24a) to obtain

$$\leq \frac{\frac{(m+1)\varepsilon}{1-(m+1)\varepsilon/2}\left(1+\frac{(m+1)\varepsilon}{1-(m+1)\varepsilon/2}\right)\left\|\tilde{\mathbf{a}}_j^{(j)}\right\|_2^2}{\left(1-\frac{(m+1)\varepsilon}{1-(m+1)\varepsilon/2}\right)\left\|\tilde{\mathbf{a}}_j^{(j)}\right\|_2} + \varepsilon\frac{\left\|\tilde{\mathbf{a}}_j^{(j)}\right\|_2^2}{\left(1-\frac{(m+1)\varepsilon}{1-(m+1)\varepsilon/2}\right)\left\|\tilde{\mathbf{a}}_j^{(j)}\right\|_2}$$

and we find that inequalities (3.38), (3.39) and (3.37) imply that

$$\leq \frac{(m+2)\varepsilon}{1-5(m+1)\varepsilon/2}\left\|\tilde{\mathbf{a}}_j^{(j)}\right\|_2\ . \tag{6.46}$$

We combine inequalities (6.42) and (6.46) to find that

$$\left|\tilde{\varrho}_{jj} - \widehat{\mathbf{q}}_j \cdot \mathbf{a}_j\right| \leq \frac{(m+2)\varepsilon}{1-5(m+1)\varepsilon/2} \left\|\tilde{\mathbf{a}}_j^{(j)}\right\|_2 + \frac{M_{(6.26a)}\varepsilon}{1-N_{(6.26a)}\varepsilon} \|\tilde{\mathbf{q}}_j\|_2 \sum_{\ell=1}^{j-1} \left\|\tilde{\mathbf{a}}_j^{(\ell)}\right\|_2$$

next, we use inequality (6.29) to bound the norm of $\tilde{\mathbf{q}}_j$:

$$\leq \frac{(m+2)\varepsilon}{1-5(m+1)\varepsilon/2} \left\|\tilde{\mathbf{a}}_j^{(j)}\right\|_2 + \frac{M_{(6.26a)}\varepsilon}{1-N_{(6.26a)}\varepsilon} \left(1 + \frac{(m+2)\varepsilon}{1-(3m+5)\varepsilon/2}\right) \sum_{\ell=1}^{j-1} \left\|\tilde{\mathbf{a}}_j^{(\ell)}\right\|_2$$

then we use inequality (6.37) to get

$$\leq \frac{(m+2)\varepsilon}{1-5(m+1)\varepsilon/2} \left(1 + \frac{(3m+4)\varepsilon}{1-(9m+9)\varepsilon/2}\right)^{j-1} \|\mathbf{a}_j\|_2$$
$$+ \frac{M_{(6.26a)}\varepsilon}{1-N_{(6.26a)}\varepsilon} \left(1 + \frac{(m+2)\varepsilon}{1-(3m+5)\varepsilon/2}\right) \sum_{\ell=1}^{j-1} \left(1 + \frac{(3m+4)\varepsilon}{1-(9m+9)\varepsilon/2}\right)^{\ell-1} \|\mathbf{a}_j\|_2$$
$$= \frac{(m+2)\varepsilon}{1-5(m+1)\varepsilon/2} \left(1 + \frac{(3m+4)\varepsilon}{1-(9m+9)\varepsilon/2}\right)^{j-1} \|\mathbf{a}_j\|_2$$
$$+ \frac{M_{(6.26a)}\varepsilon}{1-N_{(6.26a)}\varepsilon} \left(1 + \frac{(m+2)\varepsilon}{1-(3m+5)\varepsilon/2}\right) \frac{\left(1 + \frac{(3m+4)\varepsilon}{1-(9m+9)\varepsilon/2}\right)^{j-1} - 1}{\frac{(3m+4)\varepsilon}{1-(9m+9)\varepsilon/2}} \|\mathbf{a}_j\|_2 \tag{6.47}$$

If $1 < j$, we use inequality (3.41) to obtain

$$\left|\tilde{\varrho}_{jj} - \widehat{\mathbf{q}}_j \cdot \mathbf{a}_j\right| \leq \frac{(m+2)\varepsilon}{1-5(m+1)\varepsilon/2} \left(1 + \frac{(j-1)(3m+4)\varepsilon}{1-(9m+9+[j-2][3m+4])\varepsilon/2}\right) \|\mathbf{a}_j\|_2$$
$$+ \frac{M_{(6.26a)}\varepsilon}{1-N_{(6.26a)}\varepsilon} \left(1 + \frac{(m+2)\varepsilon}{1-(3m+5)\varepsilon/2}\right) \frac{\frac{(j-1)(3m+4)\varepsilon}{(1-[9m+9+(j-2)(3m+4)]\varepsilon/2)}}{\frac{(3m+4)\varepsilon}{1-[9m+9]\varepsilon/2}} \|\mathbf{a}_j\|_2$$
$$= \frac{(m+2)\varepsilon}{1-5(m+1)\varepsilon/2} \left(1 + \frac{(j-1)(3m+4)\varepsilon}{1-(9m+9+[j-2][3m+4])\varepsilon/2}\right) \|\mathbf{a}_j\|_2$$
$$+ \frac{(j-1)M_{(6.26a)}\varepsilon}{1-N_{(6.26a)}\varepsilon} \left(1 + \frac{(m+2)\varepsilon}{1-(3m+5)\varepsilon/2}\right)$$
$$\times \left(1 + \frac{(j-2)(3m+4)\varepsilon/2}{1-[9m+9+(j-2)(3m+4)]\varepsilon/2)}\right) \|\mathbf{a}_j\|_2$$

then we use inequalities (3.38) and (3.37) to obtain

$$\leq \frac{(m+2)\varepsilon}{1-\max\{(5m+5)/2+(j-1)(3m+4)\,,\,(9m+9)/2+(j-2)(3m+4)/2\}\varepsilon}\left\|\mathbf{a}_j\right\|_2$$
$$+\frac{(j-1)M_{(6.26a)}\varepsilon}{1-\left(N_{(6.26a)}+m+2\right)\varepsilon}\left(1+\frac{(j-2)(3m+4)\varepsilon/2}{1-(9m+9+[j-2][3m+4])\varepsilon/2}\right)\left\|\mathbf{a}_j\right\|_2$$

$$\leq \frac{(m+2)\varepsilon}{1-\max\{(5m+5)/2+(j-1)(3m+4)\,,\,(9m+9)/2+(j-2)(3m+4)/2\}\varepsilon}\left\|\mathbf{a}_j\right\|_2$$
$$+\frac{(j-1)M_{(6.26a)}\varepsilon}{1-(N_{(6.26a)}+m+2+[j-2][3m+4]/2)\varepsilon}\left\|\mathbf{a}_j\right\|_2$$

and finally we use inequality (3.37):

$$\leq \|\mathbf{a}_j\|_2\frac{([m+2]+[j-1]M_{(6.26a)})\varepsilon}{1-\max\{m+2+(j-1)(3m+4)\,,\,N_{(6.26a)}+m+2+(j-2)(3m+4)/2\}\varepsilon}\,. \tag{6.48}$$

Otherwise, we have $1=k=j$, and inequality (6.47) simplifies to

$$|\tilde{\varrho}_{11}-\widehat{\mathbf{q}}_1\cdot\mathbf{a}_1|\leq\frac{(m+2)\varepsilon}{1-5(m+1)\varepsilon/2}\|\mathbf{a}_1\|_2 \tag{6.49}$$

Now we can combine the previous results to bound norms of columns of the factorization error. For $j=1$, inequality (6.49) implies that

$$\left\|\left(\tilde{\mathbf{R}}-\widehat{\mathbf{Q}}^H\mathbf{A}\right)\mathbf{e}_1\right\|_2=|\tilde{\varrho}_{11}-\widehat{\mathbf{q}}_1\cdot\mathbf{a}_1|\leq\frac{(m+2)\varepsilon}{1-5(m+1)\varepsilon/2}\|\mathbf{a}_1\|_2\,.$$

For $j=2$, we can use inequalities (6.45) and (6.48) to see that

$$\left\|\left(\tilde{\mathbf{R}}-\widehat{\mathbf{Q}}^H\mathbf{A}\right)\mathbf{e}_2\right\|_2^2=|\tilde{\varrho}_{12}-\widehat{\mathbf{q}}_1\cdot\mathbf{a}_2|^2+|\tilde{\varrho}_{22}-\widehat{\mathbf{q}}_2\cdot\mathbf{a}_2|^2$$
$$\leq\|\mathbf{a}_2\|_2^2\left\{\frac{m\varepsilon}{1-(3m+5)\varepsilon/2}\right\}^2+\|\mathbf{a}_2\|_2^2\left\{\frac{\left([m+2]+M_{(6.26a)}\right)\varepsilon}{1-(N_{(6.26a)}+m+2)\varepsilon}\right\}^2$$
$$\leq\|\mathbf{a}_2\|_2^2\frac{\left(17m^2+(32+8\sqrt{2})m+18+8\sqrt{2}\right)\varepsilon^2}{\left[1-(m+2+N_{(6.26a)})\varepsilon\right]^2}$$
$$\leq\|\mathbf{a}_2\|_2^2\left\{\frac{(5m+6)\varepsilon}{1-(m+2+N_{(6.26a)})\varepsilon}\right\}^2\,.$$

For $j > 2$, we have

$$\left\|\left(\tilde{\mathbf{R}} - \widehat{\mathbf{Q}}^H \mathbf{A}\right)\mathbf{e}_j\right\|_2^2 = \left|\tilde{\varrho}_{1j} - \widehat{\mathbf{q}}_1 \cdot \mathbf{a}_j\right|^2 + \sum_{k=2}^{j-1} \left|\tilde{\varrho}_{kj} - \widehat{\mathbf{q}}_k \cdot \mathbf{a}_j\right|^2 + \left|\tilde{\varrho}_{jj} - \widehat{\mathbf{q}}_j \cdot \mathbf{a}_j\right|^2$$

then inequalities (6.45), (6.44) and (6.48) imply that

$$\le \|\mathbf{a}_j\|_2^2 \left\{\frac{m\varepsilon}{1-(3m+5)\varepsilon/2}\right\}^2$$
$$+ \|\mathbf{a}_j\|_2^2 \sum_{k=2}^{j-1} \left\{\frac{(m+[k-1]M_{(6.26a)})\varepsilon}{1-\max\{N_{(6.26a)}\,,\,[9m+9]/2\} + m + 2 + [k-2][3m+4]/2\}\varepsilon}\right\}^2$$
$$+ \|\mathbf{a}_j\|_2^2 \left\{\frac{(m+2+[j-1]M_{(6.26a)})\varepsilon}{1-\max\{m+2+[j-1][3m+4],\; N_{(6.26a)}+m+2+[j-2][3m+4]/2\}\varepsilon}\|\mathbf{a}_j\|_2\right\}^2$$

then we pull a common denominator out of the terms in the sum, and expand the numerators to get

$$\le \|\mathbf{a}_j\|_2^2 \frac{m^2\varepsilon^2}{[1-(3m+5)\varepsilon/2]^2}$$
$$+ \frac{\|\mathbf{a}_j\|_2^2}{\left[1-\left(\max\{N_{(6.26a)}\,,\,[9m+9]/2\} + m + 2 + [j-3][3m+4]/2\right)\varepsilon\right]^2}$$
$$\times \sum_{k=2}^{j-1} \left[m^2 + 2mM_{(6.26a)}(k-1) + M_{(6.26a)}^2(k-1)^2\right]$$
$$+ \|\mathbf{a}_j\|_2^2 \frac{\left([m+2]^2 + 2[m+2]M_{(6.26a)}[j-1] + M_{(6.26a)}^2[j-1]^2\right)\varepsilon^2}{\left[1-\max\{m+2+[j-1][3m+4]\,,\,N_{(6.26a)}+m+2+[j-2][3m+4]/2\}\varepsilon\right]^2}$$

then we sum the powers of integers to obtain

$$= \|\mathbf{a}_j\|_2^2 \frac{m^2\varepsilon^2}{[1-(3m+5)\varepsilon/2]^2}$$
$$+ \frac{\|\mathbf{a}_j\|_2^2}{\left[1-\left(\max\{N_{(6.26a)}\,,\,[9m+9]/2\} + m + 2 + [j-3][3m+4]/2\right)\varepsilon\right]^2}$$
$$\times \left[m^2(j-2) + mM_{(6.26a)}(j-2)(j-1) + M_{(6.26a)}^2(j-2)(j-1)(2j-3)/6\right]$$
$$+ \|\mathbf{a}_j\|_2^2 \frac{\left([m+2]^2 + 2[m+2]M_{(6.26a)}[j-1] + M_{(6.26a)}^2[j-1]^2\right)\varepsilon^2}{\left[1-\max\{m+2+[j-1][3m+4]\,,\,N_{(6.26a)}+m+2+[j-2][3m+4]/2\}\varepsilon\right]^2}$$

then we select the smallest denominator to combine all terms

$$\leq \frac{\|\mathbf{a}_j\|_2^2 \varepsilon^2}{\left[1-\left(m+2+\max\{N_{(6.26a)}+[j-2][3m+4]/2\,,\;[j-1][3m+4]\}\right)\varepsilon\right]^2}$$
$$\times \{m^2+m^2[j-2]+mM_{(6.26a)}[j-2][j-1]+M_{(6.26a)}^2[j-2][j-1][2j-3]/6$$
$$+[m+2]^2+2[m+2]M_{(6.26a)}[j-1]+M_{(6.26a)}^2[j-1]^2\}$$
$$= \frac{\|\mathbf{a}_j\|_2^2 \varepsilon^2}{[1-(m+2+[j-1][3m+4])\,\varepsilon]^2}$$
$$\times \left\{\left(18+8\sqrt{2}+[32+8\sqrt{2}]m+17m^2\right)\right.$$
$$+\left(21+38\sqrt{2}/3+[44+16\sqrt{2}]m+59m^2/2\right)(j-2)$$
$$+\left(9+6\sqrt{2}+[20+10\sqrt{2}]m+33m^2/2\right)(j-2)^2$$
$$\left.+\left(2+4\sqrt{2}/3+[4+2\sqrt{2}]m+3m^2\right)(j-2)^3\right\}\;.$$

By rounding up coefficients to the next integer, we see that

$$18+8\sqrt{2}+[32+8\sqrt{2}]m+17m^2 \leq (6+5m)^2 \text{ and}$$
$$2+4\sqrt{2}/3+[4+2\sqrt{2}]m+3m^2 \leq (2+2m)^2\;.$$

Using these two inequalities and a symbolic manipulation program to compare coefficients of powers of $j-2$, we find that

$$\left(18+8\sqrt{2}+[32+8\sqrt{2}]m+17m^2\right)$$
$$+\left(21+38\sqrt{2}/3+[44+16\sqrt{2}]m+59m^2/2\right)(j-2)$$
$$+\left(9+6\sqrt{2}+[20+10\sqrt{2}]m+33m^2/2\right)(j-2)^2$$
$$+\left(2+4\sqrt{2}/3+[4+2\sqrt{2}]m+3m^2\right)(j-2)^3$$
$$\leq \left([5m+6]^{2/3}+[2m+2]^{2/3}[j-2]\right)^3\;.$$

Thus for $j > 2$ we have

$$\left\|\left(\tilde{\mathbf{R}}-\widehat{\mathbf{Q}}^H\mathbf{A}\right)\mathbf{e}_j\right\|_2 \leq \|\mathbf{a}_j\|_2$$
$$\times \frac{\left[(5m+6)^{2/3}+(2m+2)^{2/3}(j-2)\right]^{3/2}\varepsilon}{1-\left(m+2+\max\{N_{(6.26a)}+[j-2][3m+4]/2\,,\;[j-1][3m+4]\}\right)\varepsilon}\;.$$

Finally, we can sum the column norms to get the Frobenius norm. If $n = 1$, we get

$$\left\| \tilde{\mathbf{R}} - \widehat{\mathbf{Q}}^H \mathbf{A} \right\|_F \leq \frac{(m+2)\varepsilon}{1 - 5(m+1)\varepsilon/2} \|\mathbf{A}\|_F \,.$$

If $n = 2$, we have

$$\left\| \tilde{\mathbf{R}} - \widehat{\mathbf{Q}}^H \mathbf{A} \right\|_F^2 \leq \left\{ \frac{(m+2)\varepsilon}{1 - 5(m+1)\varepsilon/2} \|\mathbf{a}_1\|_2 \right\}^2 + \left\{ \frac{(5m+6)\varepsilon}{1 - (m+2+N_{(6.26a)})\varepsilon} \|\mathbf{a}_2\|_2 \right\}^2$$

$$\leq \left\{ \frac{(5m+6)\varepsilon}{1 - (m+2+N_{(6.26a)})\varepsilon} \|\mathbf{A}\|_F \right\}^2 .$$

For $n > 2$, we have

$$\left\| \tilde{\mathbf{R}} - \widehat{\mathbf{Q}}^H \mathbf{A} \right\|_F^2 \leq \left\{ \frac{(m+2)\varepsilon}{1 - 5(m+1)\varepsilon/2} \|\mathbf{a}_1\|_2 \right\}^2 + \left\{ \frac{(5m+6)\varepsilon}{1 - (m+2+N_{(6.26a)})\varepsilon} \|\mathbf{a}_2\|_2 \right\}^2$$

$$+ \sum_{j=3}^{n} \left\{ \frac{\left[(5m+6)^{2/3} + (2m+2)^{2/3}(j-2) \right]^{3/2} \varepsilon}{1 - \left(m+2+\max\{N_{(6.26a)} + [j-2][3m+4]/2 \,,\, [j-1][3m+4]\} \right) \varepsilon} \|\mathbf{a}_j\|_2 \right\}^2$$

then we choose common numerators and denominators to obtain

$$\leq \left\{ \frac{\left[(5m+6)^{2/3} + (2m+2)^{2/3} \max\{n-2 \,,\, 0\} \right]^{3/2} \varepsilon}{1 - \left(m+2+\max\left\{ N_{(6.26a)} + \max\{n-2 \,,\, 0\}[3m+4]/2,\ [n-1][3m+4] \right\} \right) \varepsilon} \right\}^2 \|\mathbf{A}\|_F^2 \,.$$

These three inequalities prove the claim in the theorem.

We can treat a right-hand side $\mathbf{b}$ in a least squares problem as column $n+1$ in the modified Gram-Schmidt factorization to get the following corollary.

Corollary 6.8.1 *Suppose that the assumptions of Theorems 6.8.2–6.8.4 are satisfied. Define*

$$M_{(6.50)} = 1 + nM_{(6.26b)} \,,$$

$$N_{(6.50)} = \max\left\{ n[3m+4] \,,\, \max\{N_{(6.26b)} \,,\, [9m+9]/2\} + [n-1][3m+4]/2 \right\} ,$$

$$M_{(6.51)} = nM_{(6.26a)} \,,$$

$$N_{(6.51)} = \max\{N_{(6.26a)} \,,\, [9m+9]/2\} + [n-1][3m+4]/2 \,,$$

$$M_{(6.52)} = \left\{ (5m+6)^{2/3} + (2m+2)^{2/3}(n-1) \right\}^{3/2} \textit{ and}$$

$$N_{(6.52)} = m+2+\max\{N_{(6.26a)} + [n-1][3m+4]/2 \,,\, n[3m+4]\} \,,$$

and assume that

$$\max\{N_{(6.50)}\ ,\ N_{(6.51)}\ ,\ N_{(6.52)}\}\varepsilon < 1\ .$$

Let $\mathbf{b}$ *be an m-vector, and suppose that the n-vector* $\tilde{\mathbf{y}}$ *and residual vector* $\tilde{\mathbf{r}}$ *are computed in floating point arithmetic by the successive orthogonal projection Algorithm 3.4.9. Then the error in successive orthogonal projection satisfies*

$$\|\tilde{\mathbf{Q}}\tilde{\mathbf{y}} - \mathbf{b}\|_2 \leq \frac{M_{(6.50)}\varepsilon}{1 - N_{(6.50)}\varepsilon}\|\mathbf{b}\|_2\ , \tag{6.50}$$

$$\left\|\tilde{\mathbf{r}} - \mathscr{P}_n{}^H\mathbf{b}\right\|_2 \leq \frac{M_{(6.51)}\varepsilon}{1 - N_{(6.51)}\varepsilon}\|\mathbf{b}\|_2 \text{ and} \tag{6.51}$$

$$\left\|\tilde{\mathbf{y}} - \widehat{\mathbf{Q}}^H\mathbf{b}\right\|_2 \leq \frac{M_{(6.52)}\varepsilon}{1 - N_{(6.52)}\varepsilon}\|\mathbf{b}\|_2\ . \tag{6.52}$$

In order to estimate the errors in the solution of a least squares problem by successive orthogonal projection, we will require three auxiliary lemmas. First, let us examine a succession of orthogonal projections determined from our approximate orthogonal projection vectors.

Lemma 6.8.5 *Let* $\tilde{\mathbf{q}}_k$ *be the kth column of the* $m \times n$ *matrix* $\tilde{\mathbf{Q}}$*. Define* $\mathscr{P}_0 = \mathbf{I} = \boldsymbol{\Pi}_{n+1}$*, and the products of orthogonal projections*

$$\mathscr{P}_k = \left[\mathbf{I} - \frac{\tilde{\mathbf{q}}_1\tilde{\mathbf{q}}_1^H}{\tilde{\mathbf{q}}_1 \cdot \tilde{\mathbf{q}}_1}\right] \cdot \ldots \cdot \left[\mathbf{I} - \frac{\tilde{\mathbf{q}}_k\tilde{\mathbf{q}}_k^H}{\tilde{\mathbf{q}}_k \cdot \tilde{\mathbf{q}}_k}\right] \text{ for } 1 \leq k \leq n\ , \text{ and}$$

$$\boldsymbol{\Pi}_k = \left[\mathbf{I} - \frac{\tilde{\mathbf{q}}_n\tilde{\mathbf{q}}_n^H}{\tilde{\mathbf{q}}_n \cdot \tilde{\mathbf{q}}_n}\right] \cdot \ldots \cdot \left[\mathbf{I} - \frac{\tilde{\mathbf{q}}_k\tilde{\mathbf{q}}_k^H}{\tilde{\mathbf{q}}_k \cdot \tilde{\mathbf{q}}_k}\right] \text{ for } n \geq k \geq 1\ .$$

Let

$$\widehat{\mathbf{q}}_k \equiv \mathscr{P}_k\tilde{\mathbf{q}}_k \text{ and } \widehat{\widehat{\mathbf{q}}}_k \equiv \boldsymbol{\Pi}_k\tilde{\mathbf{q}}_k$$

be the kth columns of the $m \times n$ *matrices* $\widehat{\mathbf{Q}}$ *and* $\widehat{\widehat{\mathbf{Q}}}$*, respectively. Define the diagonal matrix* $\tilde{\boldsymbol{\Sigma}}$ *and the strictly upper triangular matrix* $\mathbf{U}$ *by*

$$\tilde{\mathbf{Q}}^H\tilde{\mathbf{Q}} = \tilde{\boldsymbol{\Sigma}}^2 + \mathbf{U} + \mathbf{U}^H\ , \tag{6.53}$$

and let

$$\mathbf{W} = \tilde{\boldsymbol{\Sigma}}^{-1}\mathbf{U}\tilde{\boldsymbol{\Sigma}}^{-1}\ . \tag{6.54}$$

Then

$$\mathscr{P}_n = \mathbf{I} - \widehat{\mathbf{Q}}\tilde{\boldsymbol{\Sigma}}^{-2}\tilde{\mathbf{Q}}^H \text{ and } \boldsymbol{\Pi}_1 = \mathbf{I} - \widehat{\widehat{\mathbf{Q}}}\tilde{\boldsymbol{\Sigma}}^{-2}\tilde{\mathbf{Q}}^H , \tag{6.55}$$

$$\tilde{\mathbf{Q}}\tilde{\boldsymbol{\Sigma}}^{-1} = \widehat{\mathbf{Q}}\tilde{\boldsymbol{\Sigma}}^{-1}\{\mathbf{I}+\mathbf{W}\} \text{ and } \tilde{\mathbf{Q}}\tilde{\boldsymbol{\Sigma}}^{-1} = \widehat{\widehat{\mathbf{Q}}}\tilde{\boldsymbol{\Sigma}}^{-1}\{\mathbf{I}+\mathbf{W}^H\} \text{ and} \tag{6.56}$$

$$\left\|\widehat{\mathbf{Q}}\tilde{\boldsymbol{\Sigma}}^{-1}\right\|_F \leq \sqrt{n} \text{ and } \left\|\widehat{\widehat{\mathbf{Q}}}\tilde{\boldsymbol{\Sigma}}^{-1}\right\|_F \leq \sqrt{n} . \tag{6.57}$$

Proof We begin by proving inductively that

$$\mathscr{P}_k = \mathbf{I} - \sum_{\ell=1}^{k} \frac{\widehat{\mathbf{q}}_\ell\tilde{\mathbf{q}}_\ell^H}{\tilde{\mathbf{q}}_\ell \cdot \tilde{\mathbf{q}}_\ell} \tag{6.58}$$

This claim is obviously true for $k = 0$. Assume that the claim is true for $k - 1 \geq 0$. Then

$$\mathscr{P}_k = \mathscr{P}_{k-1}\left[\mathbf{I} - \frac{\tilde{\mathbf{q}}_k\tilde{\mathbf{q}}_k^H}{\tilde{\mathbf{q}}_k \cdot \tilde{\mathbf{q}}_k}\right] = \mathscr{P}_{k-1} - \mathscr{P}_{k-1}\tilde{\mathbf{q}}_k\frac{1}{\tilde{\mathbf{q}}_k \cdot \tilde{\mathbf{q}}_k}\tilde{\mathbf{q}}_k^H$$
$$= \mathbf{I} - \sum_{\ell=1}^{k-1} \frac{\widehat{\mathbf{q}}_\ell\tilde{\mathbf{q}}_\ell^H}{\tilde{\mathbf{q}}_\ell \cdot \tilde{\mathbf{q}}_\ell} - \frac{\widehat{\mathbf{q}}_k\tilde{\mathbf{q}}_k^H}{\tilde{\mathbf{q}}_k \cdot \tilde{\mathbf{q}}_k} = \mathbf{I} - \sum_{\ell=1}^{k} \frac{\widehat{\mathbf{q}}_\ell\tilde{\mathbf{q}}_\ell^H}{\tilde{\mathbf{q}}_\ell \cdot \tilde{\mathbf{q}}_\ell} .$$

This implies that

$$\widehat{\mathbf{q}}_k \equiv \mathscr{P}_{k-1}\tilde{\mathbf{q}}_k = \left\{\mathbf{I} - \sum_{\ell=1}^{k-1} \frac{\widehat{\mathbf{q}}_\ell\tilde{\mathbf{q}}_\ell^H}{\tilde{\mathbf{q}}_\ell \cdot \tilde{\mathbf{q}}_\ell}\right\}\tilde{\mathbf{q}}_k ,$$

which can be rewritten in the form

$$\tilde{\mathbf{Q}}\mathbf{e}_k = \tilde{\mathbf{q}}_k = \widehat{\mathbf{q}}_k + \sum_{\ell=1}^{k-1} \widehat{\mathbf{q}}_\ell\frac{\tilde{\mathbf{q}}_\ell^H \cdot \tilde{\mathbf{q}}_k}{\tilde{\mathbf{q}}_\ell \cdot \tilde{\mathbf{q}}_\ell} = \widehat{\mathbf{Q}}\mathbf{e}_k + \widehat{\mathbf{Q}}\tilde{\boldsymbol{\Sigma}}^{-2}\mathbf{U}\mathbf{e}_k ,$$

and implies that

$$\tilde{\mathbf{Q}}\tilde{\boldsymbol{\Sigma}}^{-1} = \widehat{\mathbf{Q}}\tilde{\boldsymbol{\Sigma}}^{-1} + \widehat{\mathbf{Q}}\tilde{\boldsymbol{\Sigma}}^{-2}\mathbf{U}\tilde{\boldsymbol{\Sigma}}^{-1} = \widehat{\mathbf{Q}}\tilde{\boldsymbol{\Sigma}}^{-1}\left(\mathbf{I} + \tilde{\boldsymbol{\Sigma}}^{-1}\mathbf{U}\tilde{\boldsymbol{\Sigma}}^{-1}\right) = \widehat{\mathbf{Q}}\tilde{\boldsymbol{\Sigma}}^{-1}(\mathbf{I}+\mathbf{W}) .$$

In a similar fashion, we can prove that

$$\boldsymbol{\Pi}_k = \mathbf{I} - \sum_{\ell=k}^{n} \frac{\widehat{\widehat{\mathbf{q}}}_\ell\tilde{\mathbf{q}}_\ell^H}{\tilde{\mathbf{q}}_\ell \cdot \tilde{\mathbf{q}}_\ell} , \tag{6.59}$$

and that

$$\tilde{\mathbf{Q}}\mathbf{e}_k = \tilde{\mathbf{q}}_k = \widehat{\widehat{\mathbf{q}}}_k + \sum_{\ell=k+1}^{n} \widehat{\widehat{\mathbf{q}}}_\ell \frac{\tilde{\mathbf{q}}_\ell^H \cdot \tilde{\mathbf{q}}_k}{\tilde{\mathbf{q}}_\ell \cdot \tilde{\mathbf{q}}_\ell} = \widehat{\widehat{\mathbf{Q}}}\mathbf{e}_k + \widehat{\widehat{\mathbf{Q}}}\tilde{\boldsymbol{\Sigma}}^{-2}\mathbf{U}^H\mathbf{e}_k \,.$$

The latter implies that

$$\tilde{\mathbf{Q}}\tilde{\boldsymbol{\Sigma}}^{-1} = \widehat{\widehat{\mathbf{Q}}}\tilde{\boldsymbol{\Sigma}}^{-1} + \widehat{\widehat{\mathbf{Q}}}\tilde{\boldsymbol{\Sigma}}^{-2}\mathbf{U}^H\tilde{\boldsymbol{\Sigma}}^{-1} = \widehat{\widehat{\mathbf{Q}}}\tilde{\boldsymbol{\Sigma}}^{-1}\left(\mathbf{I} + \tilde{\boldsymbol{\Sigma}}^{-1}\mathbf{U}^H\tilde{\boldsymbol{\Sigma}}^{-1}\right) = \widehat{\widehat{\mathbf{Q}}}\tilde{\boldsymbol{\Sigma}}^{-1}\left(\mathbf{I} + \mathbf{W}^H\right) \,.$$

Thus we have proved the claim (6.55).

Next, we note that for $1 \le k \le n$

$$\left(\mathbf{I} - \widehat{\mathbf{Q}}\boldsymbol{\Sigma}^{-2}\tilde{\mathbf{Q}}^H\right)\mathbf{e}_k = \mathbf{e}_k - \sum_{\ell=1}^{n} \widehat{\mathbf{Q}}\mathbf{e}_\ell \frac{1}{\tilde{\mathbf{q}}_\ell \cdot \tilde{\mathbf{q}}_\ell}\mathbf{e}_\ell \cdot \tilde{\mathbf{Q}}^H\mathbf{e}_k = \mathbf{e}_k - \sum_{\ell=1}^{n} \widehat{\mathbf{q}}_\ell \frac{\tilde{\mathbf{q}}_\ell \cdot \mathbf{e}_k}{\tilde{\mathbf{q}}_\ell \cdot \tilde{\mathbf{q}}_\ell}$$
$$= \left\{\mathbf{I} - \sum_{\ell=1}^{n} \widehat{\mathbf{q}}_\ell \frac{\widehat{\mathbf{q}}_\ell \tilde{\mathbf{q}}_\ell^H}{\tilde{\mathbf{q}}_\ell \cdot \tilde{\mathbf{q}}_\ell}\right\}\mathbf{e}_k = \mathscr{P}_n\mathbf{e}_k \,.$$

This equation is equivalent to the kth column of the first half of the claim (6.56). We can prove the second half of this claim in a similar fashion.

Finally, note that because $\mathscr{P}_{k-1}$ is a product of orthogonal projections, we have

$$\frac{\|\widehat{\mathbf{q}}_k\|_2}{\|\tilde{\mathbf{q}}_k\|_2} = \left\| \prod_{\ell=1}^{k-1}\left[\mathbf{I} - \tilde{\mathbf{q}}_\ell \frac{1}{\tilde{\mathbf{q}}_\ell \cdot \tilde{\mathbf{q}}_\ell}\tilde{\mathbf{q}}_\ell^H\right]\tilde{\mathbf{q}}_k \right\|_2 \frac{1}{\|\tilde{\mathbf{q}}_k\|_2} \le \|\tilde{\mathbf{q}}_k\|_2 \frac{1}{\|\tilde{\mathbf{q}}_k\|_2} = 1 \,.$$

This shows that the columns of $\widehat{\mathbf{Q}}\tilde{\boldsymbol{\Sigma}}^{-1/2}$ all have norm at most one. The first half of the claim (6.57) now follows from the definition of the Frobenius norm. The second half of this claim is proved in a similar fashion.

We also need to estimate some terms that arise in bounding the errors in the computed residual and least squares solution.

Lemma 6.8.6 *Assume that the hypotheses of Theorem 6.8.3 are satisfied. Suppose that we have factored* $\mathbf{A} = \mathbf{Q}\mathbf{R}$ *where the* $m \times n$ *matrix* $\mathbf{Q}$ *has orthonormal columns and* $\mathbf{R}$ *is an* $n \times n$ *right-triangular matrix. Define*

$$M_{(6.61)} = 2M_{(6.34)} \,,\; N_{(6.61)} = N_{(6.34)} \text{ and } K_{(6.61)} = M_{(6.34)}/2 \,,$$

$$M_{(6.62)} = \sqrt{(n-1)\left\{1 + n[3m + 2 + \sqrt{2}] + n(2n-1)[3m + 2 + \sqrt{2}]^2/6\right\}} \text{ and}$$

$$N_{(6.62)} = [10m + 23]/2 + [n-2][3m + 4] \,,$$

and assume that

$$N_{(6.62)}\varepsilon < 1 \;,$$
$$\left(N_{(6.61)} + K_{(6.61)}\|\mathbf{A}\|_F \left\|\mathbf{A}^\dagger\right\|_2/2\right)\varepsilon < 1 \text{ and}$$
$$\frac{M_{(6.34)}\varepsilon}{1-N_{(6.34)}\varepsilon}\|\mathbf{A}\|_F \left\|\mathbf{A}^\dagger\right\|_2 < \sqrt{2}-1 \;. \tag{6.60}$$

Define

$$\tilde{\mathbf{Q}}^H\tilde{\mathbf{Q}} = \tilde{\boldsymbol{\Sigma}}^2 + \mathbf{U} + \mathbf{U}^H$$

where $\tilde{\boldsymbol{\Sigma}}$ *is diagonal and* $\mathbf{U}$ *is strictly upper triangular, and*

$$\mathbf{W} \equiv \tilde{\boldsymbol{\Sigma}}^{-1}\mathbf{U}\tilde{\boldsymbol{\Sigma}}^{-1} \;.$$

Then

$$\|\mathbf{R}^{-H}\tilde{\mathbf{A}}^H\tilde{\mathbf{A}}\mathbf{R}^{-1} - \mathbf{I}\|_2 \le \left(1 + \frac{M_{(6.34)}\varepsilon}{1-N_{(6.34)}\varepsilon}\|\mathbf{A}\|_F \left\|\mathbf{A}^\dagger\right\|_2\right)^2 - 1$$
$$\le \frac{M_{(6.61)}\varepsilon}{1-\left\{N_{(6.61)} + K_{(6.61)}\|\mathbf{A}\|_F \|\mathbf{A}^\dagger\|_2/2\right\}\varepsilon}\|\mathbf{A}\|_F \left\|\mathbf{A}^\dagger\right\|_2 \text{ and} \tag{6.61}$$
$$\left\|\mathbf{W}\tilde{\boldsymbol{\Sigma}}\tilde{\mathbf{R}}\right\|_F \le \frac{M_{(6.62)}\varepsilon}{1-N_{(6.62)}\varepsilon}\|\mathbf{A}\|_F \;. \tag{6.62}$$

Proof Since $\mathbf{A} = \mathbf{Q}\mathbf{R}$ and $\mathbf{Q}^H\mathbf{Q} = \mathbf{I}$. it follows that $\mathbf{A}^\dagger = \mathbf{R}^{-1}\mathbf{Q}^H$. Thus

$$\left\|\mathbf{A}^\dagger\right\|_2 = \left\|\mathbf{R}^{-1}\right\|_2 \;.$$

Define

$$\begin{aligned}
\triangle\mathbf{A} &\equiv \tilde{\mathbf{Q}}\tilde{\mathbf{R}} - \mathbf{A}\;;, \\
\triangle\mathbf{D} &\equiv \mathbf{R}^{-H}\triangle\mathbf{A}^H \text{ and} \\
\triangle\mathbf{F} &\equiv \mathbf{R}^{-H}(\mathbf{A}+\triangle\mathbf{A})^H(\mathbf{A}+\triangle\mathbf{A})\mathbf{R}^{-1} - \mathbf{I} \\
&= \mathbf{R}^{-H}\left\{\mathbf{R}^H\mathbf{R} + \mathbf{R}^H\mathbf{Q}^H\triangle\mathbf{A} + \triangle\mathbf{A}^H\mathbf{Q}\mathbf{R} + \triangle\mathbf{A}^H\triangle\mathbf{A}\right\}\mathbf{R}^{-1} - \mathbf{I} \\
&= (\triangle\mathbf{D}\mathbf{Q})^H + \triangle\mathbf{D}\mathbf{Q} + \triangle\mathbf{D}\triangle\mathbf{D}^H \;.
\end{aligned} \tag{6.63}$$

Note that $\triangle\mathbf{F}$ is Hermitian. Since $\|\mathbf{Q}\|_2 = 1$, we have

$$\begin{aligned}
\|\triangle\mathbf{F}\|_2 &= \left\|(\triangle\mathbf{D}\mathbf{Q})^H + \triangle\mathbf{D}\mathbf{Q} + \triangle\mathbf{D}\triangle\mathbf{D}^H\right\|_2 \\
&\le 2\|\triangle\mathbf{D}\|_2 + \|\triangle\mathbf{D}\|_2^2 = (1+\|\mathbf{D}\|_2)^2 - 1 \le \left(1 + \|\triangle\mathbf{A}\|_F \left\|\mathbf{A}^\dagger\right\|_2\right)^2 - 1
\end{aligned}$$

then inequality (6.34) gives us

$$\leq \left(1 + \frac{M_{(6.34)}\varepsilon}{1 - N_{(6.34)}\varepsilon} \|\mathbf{A}\|_F \left\|\mathbf{A}^\dagger\right\|_2\right)^2 - 1$$

and finally inequality (3.41) yields the claimed inequalities (6.61). Assumption (6.60) implies that $\|\triangle\mathbf{F}\|_2 < 1$, so Lemma 3.6.1 implies that $\mathbf{I} + \triangle\mathbf{F}$ is invertible. We conclude from the definition of $\triangle\mathbf{F}$ that $\mathbf{A} + \triangle\mathbf{A} = \tilde{\mathbf{Q}}\tilde{\mathbf{R}}$ has rank n.

Our next goal is to compute a bound on the Frobenius norm of $\mathbf{W}\tilde{\boldsymbol{\Sigma}}\tilde{\mathbf{R}}$. Floating point arithmetic produces

$$\tilde{\mathbf{q}}_j = \mathrm{fl}\left(\tilde{\mathbf{a}}_j^{(j)} \frac{1}{\tilde{\varrho}_{jj}}\right) = \left(\mathbf{I} + \mathbf{E}_{j\div}\right) \tilde{\mathbf{a}}_j^{(j)} \frac{1}{\tilde{\varrho}_{jj}} .$$

Then Eq. (6.35) allows us to write

$$\begin{aligned}\left(\mathbf{I} + \mathbf{E}_{j\div}\right)^{-1} \tilde{\mathbf{q}}_j \tilde{\varrho}_{jj} - \tilde{\mathbf{a}}_j^{(i+1)} &= \tilde{\mathbf{a}}_j^{(j)} - \tilde{\mathbf{a}}_j^{(i+1)} = \sum_{k=i+1}^{j-1} \left[\tilde{\mathbf{a}}_j^{(k+1)} - \tilde{\mathbf{a}}_j^{(k)}\right] \\ &= \sum_{k=i+1}^{j-1} \left[-\tilde{\mathbf{q}}_k \tilde{\varrho}_{kj} + \boldsymbol{\delta}_j^{(k)}\right]\end{aligned}$$

for $0 \leq i < j$. We can rewrite this equation in the form

$$\tilde{\mathbf{a}}_j^{(i+1)} = \left(\mathbf{I} + \mathbf{E}_{j\div}\right)^{-1} \tilde{\mathbf{q}}_j \tilde{\varrho}_{jj} + \sum_{k=i+1}^{j-1} \tilde{\mathbf{q}}_k \tilde{\varrho}_{kj} - \sum_{k=i+1}^{j-1} \boldsymbol{\delta}_j^{(k)} . \tag{6.64}$$

Next, for $1 \leq i < j$ we use Eq. (6.39) to write

$$\tilde{\mathbf{q}}_i \cdot \mathbf{d}_j^{(i)} = \tilde{\mathbf{q}}_i \cdot \tilde{\mathbf{a}}_j^{(i+1)}$$

then we use Eq. (6.64) to get

$$\begin{aligned}&= \tilde{\mathbf{q}}_i \cdot \left(\mathbf{I} + \mathbf{E}_{j\div}\right)^{-1} \tilde{\mathbf{q}}_j \tilde{\varrho}_{jj} + \sum_{k=i+1}^{j-1} \tilde{\mathbf{q}}_i \cdot \tilde{\mathbf{q}}_k \tilde{\varrho}_{kj} - \sum_{k=i+1}^{j-1} \tilde{\mathbf{q}}_i \cdot \boldsymbol{\delta}_j^{(k)} \\ &= \sum_{k=i+1}^{j} \tilde{\mathbf{q}}_i \cdot \tilde{\mathbf{q}}_k \tilde{\varrho}_{kj} + \tilde{\mathbf{q}}_i \cdot \mathbf{E}_{j\div} \left(\mathbf{I} + \mathbf{E}_{j\div}\right)^{-1} \tilde{\mathbf{q}}_j \tilde{\varrho}_{jj} - \sum_{k=i+1}^{j-1} \tilde{\mathbf{q}}_i \cdot \boldsymbol{\delta}_j^{(k)} \\ &= \left(\mathbf{U}\tilde{\mathbf{R}}\right)_{ij} + \tilde{\mathbf{q}}_i \cdot \mathbf{E}_{j\div} \tilde{\mathbf{a}}_j^{(j)} - \sum_{k=i+1}^{j-1} \tilde{\mathbf{q}}_i \cdot \boldsymbol{\delta}_j^{(k)}\end{aligned}$$

Thus we have shown that for $1 \leq i < j$,

$$\left(\mathbf{U}\tilde{\mathbf{R}}\right)_{ij} = \tilde{\mathbf{q}}_i \cdot \mathbf{d}_j^{(i)} + \sum_{k=i+1}^{j-1} \tilde{\mathbf{q}}_i \cdot \boldsymbol{\delta}_j^{(k)} - \tilde{\mathbf{q}}_i \cdot \mathbf{E}_{j\div}\tilde{\mathbf{a}}_j^{(j)} .$$

We can divide by $\|\tilde{\mathbf{q}}_i\|_2$, take the absolute value of both sides, apply the Cauchy inequality (3.15), and then use inequalities (6.26a) and (6.26b) to get

$$\left|\left(\mathbf{W}\tilde{\boldsymbol{\Sigma}}\tilde{\mathbf{R}}\right)_{ij}\right| \leq \frac{M_{(6.26a)}\varepsilon}{1-N_{(6.26a)}\varepsilon}\left\|\tilde{\mathbf{a}}_j^{(i)}\right\|_2 + \sum_{k=i+1}^{j-1}\frac{M_{(6.26b)}\varepsilon}{1-N_{(6.26b)}\varepsilon}\left\|\tilde{\mathbf{a}}_j^{(k)}\right\|_2 + \varepsilon\left\|\tilde{\mathbf{a}}_j^{(j)}\right\|_2$$

then we use inequality (6.37) to get

$$\leq \frac{M_{(6.26a)}\varepsilon}{1-N_{(6.26a)}\varepsilon}\left[1+\frac{(3m+4)\varepsilon}{1-(9m+9)\varepsilon/2}\right]^{i-1}\|\mathbf{a}_j\|_2$$
$$+\frac{M_{(6.26b)}\varepsilon}{1-N_{(6.26b)}\varepsilon}\sum_{k=i+1}^{j-1}\left[1+\frac{(3m+4)\varepsilon}{1-(9m+9)\varepsilon/2}\right]^{k-1}\|\mathbf{a}_j\|_2$$
$$+\varepsilon\left[1+\frac{(3m+4)\varepsilon}{1-(9m+9)\varepsilon/2}\right]^{j-1}\|\mathbf{a}_j\|_2$$

then we sum the geometric series to get

$$= \frac{M_{(6.26a)}\varepsilon}{1-N_{(6.26a)}\varepsilon}\left[1+\frac{(3m+4)\varepsilon}{1-(9m+9)\varepsilon/2}\right]^{i-1}\|\mathbf{a}_j\|_2$$
$$+\frac{M_{(6.26b)}\varepsilon}{1-N_{(6.26b)}\varepsilon}\|\mathbf{a}_j\|_2\left[1+\frac{(3m+4)\varepsilon}{1-(9m+9)\varepsilon/2}\right]^{i}\frac{\left[1+\frac{(3m+4)\varepsilon}{1-(9m+9)\varepsilon/2}\right]^{j-i-1}-1}{\frac{(3m+4)\varepsilon}{1-(9m+9)\varepsilon/2}}\|\mathbf{a}_j\|_2$$
$$+\varepsilon\left[1+\frac{(3m+4)\varepsilon}{1-(9m+9)\varepsilon/2}\right]^{j-1}\|\mathbf{a}_j\|_2 \tag{6.65}$$

We will identify six cases. If $1 < i < j-2$, then inequality (6.65) can be combined with inequality (3.41) to bound powers of relative errors, thereby obtaining

$$\left|\left(\mathbf{W}\tilde{\boldsymbol{\Sigma}}\tilde{\mathbf{R}}\right)_{ij}\right| \leq \frac{M_{(6.26a)}\varepsilon}{1-N_{(6.26a)}\varepsilon}\left\{1+\frac{(i-1)(3m+4)\varepsilon}{1-(9m+9+[i-2][3m+4])\varepsilon/2}\right\}\|\mathbf{a}_j\|_2$$
$$+\frac{M_{(6.26b)}\varepsilon}{1-N_{(6.26b)}\varepsilon}\left\{1+\frac{i(3m+4)\varepsilon}{1-(9m+9+[i-1][3m+4])\varepsilon/2}\right\}$$
$$\times\frac{\frac{(j-i-1)((3m+4)\varepsilon}{1-(9m+9+[j-i-2][3m+4])\varepsilon/2}}{\frac{(3m+4)\varepsilon}{1-(9m+9)\varepsilon/2}}\|\mathbf{a}_j\|_2$$

$$
\begin{aligned}
&+ \varepsilon \left[1 + \frac{(j-1)(3m+4)\varepsilon}{1 - (9m+9+[j-2][3m+4])\varepsilon/2} \right] \|\mathbf{a}_j\|_2 \\
&= \frac{M_{(6.26a)}\varepsilon}{1 - N_{(6.26a)}\varepsilon} \left\{ 1 + \frac{(i-1)(3m+4)\varepsilon}{1 - (9m+9+[i-2][3m+4])\varepsilon/2} \right\} \|\mathbf{a}_j\|_2 \\
&+ \frac{(j-i-1)M_{(6.26b)}\varepsilon}{1 - N_{(6.26b)}\varepsilon} \left\{ 1 + \frac{i(3m+4)\varepsilon}{1 - (9m+9+[i-1][3m+4])\varepsilon/2} \right\} \\
&\times \left\{ 1 + \frac{(j-i-2)(3m+4)\varepsilon}{1 - (9m+9+[j-i-2][3m+4])\varepsilon/2} \right\} \|\mathbf{a}_j\|_2 \\
&+ \varepsilon \left[1 + \frac{(j-1)(3m+4)\varepsilon}{1 - (9m+9+[j-2][3m+4])\varepsilon/2} \right] \|\mathbf{a}_j\|_2
\end{aligned}
$$

then we use inequalities (3.38) and (3.37) to produce

$$
\leq \frac{(M_{(6.26a)} + [j-i-1]M_{(6.26b)} + 1)\varepsilon}{1 - (N_{(6.26b)} + [j-2][3m+4])\varepsilon} \|\mathbf{a}_j\|_2 \,. \tag{6.66}
$$

In our second case we have $1 < i = j-2$. Again, inequality (6.65) can be combined with inequality (3.41) to bound powers of relative errors, thereby obtaining

$$
\begin{aligned}
\left| \left(\mathbf{W}\tilde{\boldsymbol{\Sigma}}\tilde{\mathbf{R}} \right)_{j-2,j} \right| &\leq \frac{M_{(6.26a)}\varepsilon}{1 - N_{(6.26a)}\varepsilon} \left[1 + \frac{(j-3)(3m+4)\varepsilon}{1 - (9m+9+[j-4][3m+4])\varepsilon/2} \right] \|\mathbf{a}_j\|_2 \\
&+ \frac{M_{(6.26b)}\varepsilon}{1 - N_{(6.26b)}\varepsilon} \left[1 + \frac{(j-2)(3m+4)\varepsilon}{1 - (9m+9+[j-3][3m+4])\varepsilon/2} \right] \|\mathbf{a}_j\|_2 \\
&+ \varepsilon \left[1 + \frac{(j-1)(3m+4)\varepsilon}{1 - (9m+9+[j-2][3m+4])\varepsilon/2} \right] \|\mathbf{a}_j\|_2
\end{aligned}
$$

then we use inequalities (3.38) and (3.37) to obtain

$$
\leq \frac{(M_{(6.26a)} + M_{(6.26b)} + 1)\varepsilon}{1 - (N_{(6.26b)} + [j-2][3m+4])\varepsilon} \|\mathbf{a}_j\|_2 \,. \tag{6.67}
$$

In our third case, we have $1 < i = j-1$. Then inequality (6.65) can be combined with inequality (3.41) to bound powers of relative errors, thereby obtaining

$$
\begin{aligned}
\left| \left(\mathbf{W}\tilde{\boldsymbol{\Sigma}}\tilde{\mathbf{R}} \right)_{j-1,j} \right| &\leq \frac{M_{(6.26a)}\varepsilon}{1 - N_{(6.26a)}\varepsilon} \left(1 + \frac{(j-2)(3m+4)\varepsilon}{1 - (9m+9+[j-3][3m+4])\varepsilon/2} \right) \|\mathbf{a}_j\|_2 \\
&+ \varepsilon \left(1 + \frac{(j-1)(3m+4)\varepsilon}{1 - (9m+9+[j-2][3m+4])\varepsilon/2} \right) \|\mathbf{a}_j\|_2
\end{aligned}
$$

then we use inequalities (3.38) and (3.37) to get

$$\leq \frac{(M_{(6.26a)} + 1)\varepsilon}{1 - \max\{N_{(6.26a)} + [j-2][3m+4]\ ,\ [9m+9]/2 + [j-2][3m+4]/2\}\varepsilon} \left\|\mathbf{a}_j\right\|_2 . \tag{6.68}$$

In our fourth case, we have $1 = i < j-2$. Then inequality (6.65) can be combined with inequality (3.41) to bound powers of relative errors, thereby obtaining

$$\begin{aligned}
\left|\left(\mathbf{W}\tilde{\boldsymbol{\Sigma}}\tilde{\mathbf{R}}\right)_{1,j}\right| &\leq \frac{M_{(6.26a)}\varepsilon}{1 - N_{(6.26a)}\varepsilon} \left\|\mathbf{a}_j\right\|_2 \\
&+ \frac{M_{(6.26b)}\varepsilon}{1 - N_{(6.26b)}\varepsilon} \left\{1 + \frac{(3m+4)\varepsilon}{1-(9m+9)\varepsilon/2}\right\} \frac{\frac{(j-2)(3m+4)\varepsilon}{1-(9m+9+[j-3][3m+4])\varepsilon/2}}{\frac{(3m+4)\varepsilon}{1-(9m+9)\varepsilon/2}} \left\|\mathbf{a}_j\right\|_2 \\
&+ \varepsilon \left\{1 + \frac{(j-1)(3m+4)\varepsilon}{1-(9m+9+[j-2][3m+4])\varepsilon/2}\right\} \left\|\mathbf{a}_j\right\|_2 \\
&= \frac{M_{(6.26a)}\varepsilon}{1 - N_{(6.26a)}\varepsilon} \left\|\mathbf{a}_j\right\|_2 \\
&+ \frac{(j-2)M_{(6.26b)})\varepsilon}{1 - N_{(6.26b)}\varepsilon} \left\{1 + \frac{(3m+4)\varepsilon}{1-(9m+9)\varepsilon/2}\right\} \\
&\left\{1 + \frac{(j-3)(3m+4)\varepsilon/2}{1-(9m+9+[j-3][3m+4])\varepsilon/2}\right\} \left\|\mathbf{a}_j\right\|_2 \\
&+ \varepsilon \left\{1 + \frac{(j-1)(3m+4)\varepsilon}{1-(9m+9+[j-2][3m+4])\varepsilon/2}\right\} \left\|\mathbf{a}_j\right\|_2
\end{aligned}$$

then we use inequalities (3.38) and (3.37) to see that

$$\leq \frac{(M_{(6.26a)} + (j-2)M_{(6.26b)} + 1)\varepsilon}{1 - \max\{N_{(6.26b)} + [j-3][3m+4]\ ,\ (9m+9)/2 + (j-2)(3m+4)/2\}\varepsilon} \|\mathbf{a}_j\|_2 . \tag{6.69}$$

In our fifth case, we have $1 = i = j-2$. In this case, inequality (6.65) can be combined with inequality (3.41) to bound powers of relative errors, thereby obtaining

$$\begin{aligned}
\left|\left(\mathbf{W}\tilde{\boldsymbol{\Sigma}}\tilde{\mathbf{R}}\right)_{1,3}\right| &\leq \frac{M_{(6.26a)}\varepsilon}{1 - N_{(6.26a)}\varepsilon} \|\mathbf{a}_3\|_2 \\
&+ \frac{M_{(6.26b)}\varepsilon}{1 - N_{(6.26b)}\varepsilon} \left\{1 + \frac{(3m+4)\varepsilon}{1-(9m+9)\varepsilon/2}\right\} \|\mathbf{a}_3\|_2 + \varepsilon \left[1 + \frac{(6m+8)\varepsilon}{1-(15m+17)\varepsilon/2}\right] \|\mathbf{a}_3\|_2
\end{aligned}$$

then we use inequalities (3.38) and (3.37) to find that

$$\leq \frac{(M_{(6.26a)} + M_{(6.26b)} + 1)\varepsilon}{1 - \max\{N_{(6.26b)} + 3m + 4\,,\ (15m + 17)/2\}\varepsilon} \|\mathbf{a}_3\|_2 \,. \tag{6.70}$$

In our sixth and final case, we have $1 = i = j - 1$. Here inequality (6.65) can be combined with inequality (3.41) to bound powers of relative errors, thereby obtaining

$$\left|\left(\mathbf{W}\tilde{\boldsymbol{\Sigma}}\tilde{\mathbf{R}}\right)_{1,2}\right| \leq \frac{M_{(6.26a)}\varepsilon}{1 - N_{(6.26a)}\varepsilon} \|\mathbf{a}_2\|_2 + \varepsilon\left[1 + \frac{(3m+4)\varepsilon}{1 - (9m+9)\varepsilon/2}\right] \|\mathbf{a}_2\|_2$$

then we use inequalities (3.38) and (3.37) to arrive at

$$\leq \frac{(M_{(6.26a)} + 1)\varepsilon}{1 - \max\{N_{(6.26a)}\,,\ (9m + 9)/2\}\varepsilon} \|\mathbf{a}_2\|_2 \,. \tag{6.71}$$

By examining all six cases, we see that in general for $1 \leq i < j$ we have

$$\left|\left(\mathbf{W}\tilde{\boldsymbol{\Sigma}}\tilde{\mathbf{R}}\right)_{ij}\right|$$

$$\leq \frac{(M_{(6.26a)} + (j-i-1)M_{(6.26b)} + 1)\varepsilon}{1 - \max\{N_{(6.26a)} + (j-2)(3m+4)\,,\ (9m+9)/2 + (j-2)(3m+4)/2\}\varepsilon} \|\mathbf{a}_j\|_2 \,. \tag{6.72}$$

Next, we sum entries in column j to get

$$\left\|\mathbf{W}\tilde{\boldsymbol{\Sigma}}\tilde{\mathbf{R}}\mathbf{e}_j\right\|_2^2$$

$$\leq \sum_{i=1}^{j-1}\left\{\frac{(M_{(6.26a)} + (j-i-1)M_{(6.26b)} + 1)\varepsilon}{1 - \max\{N_{(6.26a)} + (j-2)(3m+4)\,,\ (9m+9)/2 + (j-2)(3m+4)/2\}\varepsilon}\right\}^2 \|\mathbf{a}_j\|_2^2$$

$$= \left\{\frac{\varepsilon}{1 - \max\{N_{(6.26a)} + (j-2)(3m+4)\,,\ (9m+9)/2 + (j-2)(3m+4)/2\}\varepsilon}\right\}^2 \|\mathbf{a}_j\|_2^2$$

$$\times (j-1)\left\{\left(M_{(6.26a)} + 1\right)^2 + j\left(M_{(6.26a)} + 1\right)M_{(6.26b)} + j(2j-1)M_{(6.26b)}^2/6\right\} \,.$$

Finally, we sum the norms of the columns to get

$$\left\|\mathbf{W}\tilde{\boldsymbol{\Sigma}}\tilde{\mathbf{R}}\right\|_F^2$$

$$\leq \sum_{j=1}^{n}\left\{\frac{\varepsilon}{1 - \max\{N_{(6.26a)} + (j-2)(3m+4)\,,\ (9m+9)/2 + (j-2)(3m+4)/2\}\varepsilon}\right\}^2 \|\mathbf{a}_j\|_2^2$$

$$\times (j-1)\left\{\left(M_{(6.26a)}+1\right)^2 + j\left(M_{(6.26a)}+1\right)M_{(6.26b)} + j(2j-1)M_{(6.26b)}^2/6\right\}$$

$$\leq \left\{\frac{\varepsilon\sqrt{(n-1)\left\{\left(M_{(6.26a)}+1\right)^2+n\left(M_{(6.26a)}+1\right)M_{(6.26b)}+n(2n-1)M_{(6.26b)}^2/6\right\}}}{1-\max\{N_{(6.26a)}+(n-2)(3m+4),\ (9m+9)/2+(n-2)(3m+4)/2\}\varepsilon}\right\}^2 \|\mathbf{A}\|_F^2 .$$

This proves claim (6.62).

Our third lemma bounds the inverse of the approximate right-triangular factor in Gram-Schmidt factorization.

Lemma 6.8.7 *Assume that the hypotheses of Lemma 6.8.6 are satisfied. Define* **W** *by Eq.* (6.54), *and*

$$\beta \equiv \frac{M_{(6.61)}\varepsilon}{1-\left\{N_{(6.61)}+K_{(6.61)}\|\mathbf{A}\|_F\,\|\mathbf{A}^\dagger\|_2\right\}\varepsilon}$$

$$+\frac{2M_{(6.62)}\varepsilon}{1-N_{(6.62)}\varepsilon}\sqrt{\left\{1+\frac{M_{(6.34)}\varepsilon}{1-N_{(6.34)}\varepsilon}\|\mathbf{A}\|_F\,\|\mathbf{A}^\dagger\|_2\right\}^2+\left\{\frac{M_{(6.62)}\varepsilon}{1-N_{(6.62)}\varepsilon}\|\mathbf{A}\|_F\,\|\mathbf{A}^\dagger\|_2\right\}^2}$$

$$+2\left\{\frac{M_{(6.62)}\varepsilon}{1-N_{(6.62)}\varepsilon}\right\}^2, . \tag{6.73}$$

Suppose that

$$\beta\varepsilon\|\mathbf{A}\|_F\,\left\|\mathbf{A}^\dagger\right\|_2 < 1 .$$

Then

$$\left\|\left(\tilde{\boldsymbol{\Sigma}}\tilde{\mathbf{R}}\right)^{-1}\right\|_2^2 \leq \frac{1}{1-\beta\varepsilon\|\mathbf{A}\|_F\,\|\mathbf{A}^\dagger\|_2}\left\|\mathbf{A}^\dagger\right\|_2^2 \text{ and} \tag{6.74}$$

$$\left\|\left\{\left(\tilde{\boldsymbol{\Sigma}}\tilde{\mathbf{R}}\right)^H\left[\mathbf{I}+\mathbf{W}^H\right]\left(\tilde{\boldsymbol{\Sigma}}\tilde{\mathbf{R}}\right)\right\}^{-1}\right\|_2 \leq \frac{1}{1-\beta\varepsilon\|\mathbf{A}\|_F\,\|\mathbf{A}^\dagger\|_2}\left\|\mathbf{A}^\dagger\right\|_2^2 . \tag{6.75}$$

Proof First, we would like to bound the norm of $\tilde{\boldsymbol{\Sigma}}\tilde{\mathbf{R}}$. We begin by recalling from Eqs. (6.63) and (6.54) that

$$\mathbf{R}^H(\mathbf{I}+\triangle\mathbf{F})\,\mathbf{R} = (\mathbf{A}+\triangle\mathbf{A})^H(\mathbf{A}+\triangle\mathbf{A})$$

$$= \left(\tilde{\mathbf{Q}}\tilde{\mathbf{R}}\right)^H\left(\tilde{\mathbf{Q}}\tilde{\mathbf{R}}\right) = \tilde{\mathbf{R}}^H\left(\tilde{\boldsymbol{\Sigma}}^2+\mathbf{U}+\mathbf{U}^H\right)\tilde{\mathbf{R}}$$

$$= \left(\tilde{\boldsymbol{\Sigma}}\tilde{\mathbf{R}}\right)^H\left(\mathbf{I}+\mathbf{W}+\mathbf{W}^H\right)\left(\tilde{\boldsymbol{\Sigma}}\tilde{\mathbf{R}}\right) .$$

Next, we define

$$\triangle\mathbf{G} \equiv \triangle\mathbf{F} - (\boldsymbol{\Sigma}\mathbf{R})^{-H}\left(\tilde{\boldsymbol{\Sigma}}\tilde{\mathbf{R}}\right)^H \mathbf{W}\left(\tilde{\boldsymbol{\Sigma}}\tilde{\mathbf{R}}\right)(\boldsymbol{\Sigma}\mathbf{R})^{-1} - (\boldsymbol{\Sigma}\mathbf{R})^{-H}\left(\tilde{\boldsymbol{\Sigma}}\tilde{\mathbf{R}}\right)^H \mathbf{W}^H\left(\tilde{\boldsymbol{\Sigma}}\tilde{\mathbf{R}}\right)(\boldsymbol{\Sigma}\mathbf{R})^{-1} .$$

so that

$$\left(\tilde{\boldsymbol{\Sigma}}\tilde{\mathbf{R}}\right)^H\left(\tilde{\boldsymbol{\Sigma}}\tilde{\mathbf{R}}\right) = \mathbf{R}^H(\mathbf{I} + \triangle\mathbf{G})\mathbf{R} . \tag{6.76}$$

Note that $\triangle\mathbf{G}$ is Hermitian, that

$$\left\|\left(\tilde{\boldsymbol{\Sigma}}\tilde{\mathbf{R}}\right)\mathbf{R}^{-1}\right\|_2^2 = \left\|\mathbf{R}^{-H}\left(\tilde{\boldsymbol{\Sigma}}\tilde{\mathbf{R}}\right)^H\left(\tilde{\boldsymbol{\Sigma}}\tilde{\mathbf{R}}\right)\mathbf{R}^{-1}\right\|_2 \leq 1 + \|\triangle\mathbf{G}\|_2 , \tag{6.77}$$

and that

$$\begin{aligned}\|\triangle\mathbf{G}\|_2 &\leq \|\triangle\mathbf{F}\|_2 + 2\left\|\left(\tilde{\boldsymbol{\Sigma}}\tilde{\mathbf{R}}\right)\mathbf{R}^{-1}\right\|_2\left\|\mathbf{W}\left(\tilde{\boldsymbol{\Sigma}}\tilde{\mathbf{R}}\right)\mathbf{R}^{-1}\right\|_2 \\ &\leq \|\triangle\mathbf{F}\|_2 + 2\sqrt{1 + \|\triangle\mathbf{G}\|_2}\left\|\mathbf{W}\left(\tilde{\boldsymbol{\Sigma}}\tilde{\mathbf{R}}\right)\mathbf{R}^{-1}\right\|_2 .\end{aligned} \tag{6.78}$$

Observe that if a, b and x are nonnegative, the inequality $x \leq a + 2b\sqrt{1+x}$ implies that

$$x \leq a + 2b\sqrt{1 + a + b^2} + 2b^2 .$$

This observation implies that

$$\|\triangle\mathbf{G}\|_2 \leq \|\triangle\mathbf{F}\|_2 + 2\left\|\mathbf{W}\left(\tilde{\boldsymbol{\Sigma}}\tilde{\mathbf{R}}\right)\mathbf{R}^{-1}\right\|_2\sqrt{1 + \|\triangle\mathbf{F}\|_2 + \left\|\mathbf{W}\left(\tilde{\boldsymbol{\Sigma}}\tilde{\mathbf{R}}\right)\mathbf{R}^{-1}\right\|_2^2} + 2\left\|\mathbf{W}\left(\tilde{\boldsymbol{\Sigma}}\tilde{\mathbf{R}}\right)\mathbf{R}^{-1}\right\|_2^2$$

then inequalities (6.61) and (6.62) give us

$$\begin{aligned}&\leq \frac{M_{(6.61)}\varepsilon}{1 - \left\{N_{(6.61)} + K_{(6.61)}\|\mathbf{A}\|_F\left\|\mathbf{A}^\dagger\right\|_2\right\}\varepsilon}\|\mathbf{A}\|_F\left\|\mathbf{A}^\dagger\right\|_2 \\ &+ 2\frac{M_{(6.62)}\varepsilon}{1 - N_{(6.62)}\varepsilon}\|\mathbf{A}\|_F\left\|\mathbf{A}^\dagger\right\|_2 \\ &\times\sqrt{\left\{1 + \frac{M_{(6.34)}\varepsilon}{1 - N_{(6.34)}\varepsilon}\|\mathbf{A}\|_F\left\|\mathbf{A}^\dagger\right\|_2\right\}^2 + \left\{\frac{M_{(6.62)}\varepsilon}{1 - N_{(6.62)}\varepsilon}\|\mathbf{A}\|_F\left\|\mathbf{A}^\dagger\right\|_2\right\}^2}\end{aligned}$$

$$+ 2 \left\{ \frac{M_{(6.62)}\varepsilon}{1 - N_{(6.62)}\varepsilon} \|\mathbf{A}\|_F \left\|\mathbf{A}^\dagger\right\|_2 \right\}^2$$
$$\equiv \beta\varepsilon \|\mathbf{A}\|_F \left\|\mathbf{A}^\dagger\right\|_2 < 1 \; . \tag{6.79}$$

Lemma 3.6.1 now implies that $\mathbf{I} + \triangle\mathbf{G}$ is invertible, and Eq. (6.76) proves (6.74).

We would also like to estimate the norm of the inverse of $\left(\tilde{\boldsymbol{\Sigma}}\tilde{\mathbf{R}}\right)^H [\mathbf{I}+\mathbf{W}^H] \left(\tilde{\boldsymbol{\Sigma}}\tilde{\mathbf{R}}\right)$. We begin by noting that

$$(\mathbf{A} + \triangle\mathbf{A})^H(\mathbf{A} + \triangle\mathbf{A}) = \left(\tilde{\mathbf{Q}}\tilde{\mathbf{R}}\right)^H \left(\tilde{\mathbf{Q}}\tilde{\mathbf{R}}\right) = \tilde{\mathbf{R}}^H \tilde{\mathbf{Q}}^H \tilde{\mathbf{Q}}\tilde{\mathbf{R}}$$
$$= \tilde{\mathbf{R}}^H \left[\tilde{\boldsymbol{\Sigma}}^2 + \mathbf{U} + \mathbf{U}^H\right] \tilde{\mathbf{R}} = \left(\tilde{\boldsymbol{\Sigma}}\tilde{\mathbf{R}}\right)^H [\mathbf{I} + \mathbf{W}^H] \left(\tilde{\boldsymbol{\Sigma}}\tilde{\mathbf{R}}\right) + \left(\tilde{\boldsymbol{\Sigma}}\tilde{\mathbf{R}}\right)^H \mathbf{W} \left(\tilde{\boldsymbol{\Sigma}}\tilde{\mathbf{R}}\right) ,$$

so

$$\left(\tilde{\boldsymbol{\Sigma}}\tilde{\mathbf{R}}\right)^H [\mathbf{I} + \mathbf{W}^H] \left(\tilde{\boldsymbol{\Sigma}}\tilde{\mathbf{R}}\right) = (\mathbf{A} + \triangle\mathbf{A})^H(\mathbf{A} + \triangle\mathbf{A}) - \left(\tilde{\boldsymbol{\Sigma}}\tilde{\mathbf{R}}\right)^H \mathbf{W} \left(\tilde{\boldsymbol{\Sigma}}\tilde{\mathbf{R}}\right) ,$$

then Eq. (6.63) give us

$$= \mathbf{R}^H[\mathbf{I} + \triangle\mathbf{F}]\mathbf{R} - \left(\tilde{\boldsymbol{\Sigma}}\tilde{\mathbf{R}}\right)^H \mathbf{W} \left(\tilde{\boldsymbol{\Sigma}}\tilde{\mathbf{R}}\right) \equiv \mathbf{R}^H[\mathbf{I} + \triangle\mathbf{H}]\mathbf{R} , \tag{6.80}$$

where

$$\|\triangle\mathbf{H}\|_2 = \left\| \triangle\mathbf{F} - \mathbf{R}^{-H} \left(\tilde{\boldsymbol{\Sigma}}\tilde{\mathbf{R}}\right)^H \mathbf{W} \left(\tilde{\boldsymbol{\Sigma}}\tilde{\mathbf{R}}\right) \mathbf{R}^{-1} \right\|_2$$
$$\leq \|\triangle\mathbf{F}\|_2 + \left\| \left(\tilde{\boldsymbol{\Sigma}}\tilde{\mathbf{R}}\right) \mathbf{R}^{-1} \right\|_2 \left\| \mathbf{W} \left(\tilde{\boldsymbol{\Sigma}}\tilde{\mathbf{R}}\right) \mathbf{R}^{-1} \right\|_2$$

then inequality (6.77) yields

$$\leq \|\triangle\mathbf{F}\|_2 + \sqrt{1 + \|\triangle\mathbf{G}\|_2} \left\| \mathbf{W} \left(\tilde{\boldsymbol{\Sigma}}\tilde{\mathbf{R}}\right) \mathbf{R}^{-1} \right\|_2$$

and comparison with inequality (6.78) produces

$$\leq \beta\varepsilon \|\mathbf{A}\|_F \left\|\mathbf{A}^\dagger\right\|_2 < 1 \; .$$

It follows that $\mathbf{I} + \triangle\mathbf{H}$ is invertible, and Eq. (6.80) implies that

$$\left\| \left\{ \left(\tilde{\boldsymbol{\Sigma}}\tilde{\mathbf{R}}\right)^H [\mathbf{I} + \mathbf{W}^H] \left(\tilde{\boldsymbol{\Sigma}}\tilde{\mathbf{R}}\right) \right\}^{-1} \right\|_2 = \left\| \left\{ \mathbf{R}^H \left[\mathbf{I} + \triangle\mathbf{H}\right] \mathbf{R} \right\}^{-1} \right\|_2$$
$$\leq \left\| (\mathbf{I} + \triangle\mathbf{H})^{-1} \right\|_2 \left\| \mathbf{R}^{-1} \right\|_2^2 \leq \frac{1}{1 - \beta\varepsilon \|\mathbf{A}\|_F \|\mathbf{A}^\dagger\|_2} \left\| \mathbf{A}^\dagger \right\|_2^2 \; .$$

The following theorem estimates the errors in the least squares residual and solution vector, obtained by successive orthogonal projection in floating point arithmetic. It also estimates the errors in the orthonormality of the computed orthogonal projection vectors.

Theorem 6.8.5 *Assume that the hypotheses of Lemma 6.8.7 are satisfied. Given an m-vector* **b**, *let* **x** *be the n-vector that minimizes* $\|\mathbf{A}\mathbf{x}-\mathbf{b}\|_2$, *and let* $\mathbf{r}=\mathbf{b}-\mathbf{A}\mathbf{x}$ *be the residual. Define* β *by Eq.* (6.73), *and suppose that*

$$\beta\varepsilon\|\mathbf{A}\|_F\,\left\|\mathbf{A}^\dagger\right\|_2 \le \frac{\sqrt{5}-1}{2}\,.$$

Then

$$\|\tilde{\mathbf{r}}-\mathbf{r}\|_2 \le \frac{1}{\sqrt{1-\beta\varepsilon\|\mathbf{A}\|_F\,\|\mathbf{A}^\dagger\|_2}}\frac{\sqrt{n}M_{(6.34)}\varepsilon}{1-N_{(6.34)}\varepsilon}\|\mathbf{A}\|_F\,\left\|\mathbf{A}^\dagger\right\|_2\|\mathbf{r}\|_2$$
$$+\frac{M_{(6.38)}\varepsilon}{1-N_{(6.38)}\varepsilon}\|\mathbf{A}\|_F\,\left\|\mathbf{A}^\dagger\right\|_2\|\mathbf{b}\|_2+\frac{M_{(6.51)}\varepsilon}{1-N_{(6.51)}\varepsilon}\|\mathbf{b}\|_2\,, \tag{6.81}$$

$$\|\tilde{\mathbf{x}}-\mathbf{x}\|_2 \le \frac{M_{(6.34)}\|\mathbf{A}\|_F\,\left\|\mathbf{A}^\dagger\right\|_2\varepsilon}{1-\max\left\{\begin{matrix}N_{(6.34)}+[\beta+4m+4]\|\mathbf{A}\|_F\,\|\mathbf{A}^\dagger\|_2\,,\\ [4m+7]/2+2\beta\varepsilon\|\mathbf{A}\|_F\,\|\mathbf{A}^\dagger\|_2\end{matrix}\right\}\varepsilon}\left\|\mathbf{A}^\dagger\right\|_2\|\mathbf{r}\|_2$$
$$+\frac{M_{(6.41)}\varepsilon}{1-\left(m+2+N_{(6.41)}+\beta\|\mathbf{A}\|_F\,\|\mathbf{A}^\dagger\|_2\right)\varepsilon}\|\mathbf{A}\|_F\,\left\|\mathbf{A}^\dagger\right\|_2\|\mathbf{x}\|_2$$
$$+\frac{M_{(6.52)}\varepsilon}{1-\left(m+2+N_{(6.52)}+\beta\|\mathbf{A}\|_F\,\|\mathbf{A}^\dagger\|_2\right)\varepsilon}\left\|\mathbf{A}^\dagger\right\|_2\|\mathbf{b}\|_2 \text{ and} \tag{6.82}$$

$$\left\|\mathbf{I}-\left(\tilde{\mathbf{Q}}\tilde{\boldsymbol{\Sigma}}^{-1}\right)^H\left(\tilde{\mathbf{Q}}\tilde{\boldsymbol{\Sigma}}^{-1}\right)\right\|_2 \le \frac{2\left(m+2+M_{(6.62)}\|\mathbf{A}\|_F\,\left\|\mathbf{A}^\dagger\right\|_2\right)\varepsilon}{1-\left(N_{(6.62)}+\beta\|\mathbf{A}\|_F\,\|\mathbf{A}^\dagger\|_2\right)\varepsilon}\,. \tag{6.83}$$

Proof We begin by estimating the error in the computed residual. Define

$$\triangle\mathbf{r}=\tilde{\mathbf{r}}-\mathscr{P}_n{}^H\mathbf{b}\,,$$

where $\tilde{\mathbf{r}}$ is the residual determined by successive orthogonal projection in floating point arithmetic. Since

$$\mathbf{A}+\triangle\mathbf{A}\equiv\tilde{\mathbf{A}}=\tilde{\mathbf{Q}}\tilde{\mathbf{R}}=\left(\tilde{\mathbf{Q}}\tilde{\boldsymbol{\Sigma}}^{-1}\right)\left(\tilde{\boldsymbol{\Sigma}}\tilde{\mathbf{R}}\right)\,,$$

and $\mathbf{A}^H\mathbf{r}=\mathbf{0}$, we see that

$$\tilde{\boldsymbol{\Sigma}}^{-1}\tilde{\mathbf{Q}}^H\mathbf{r}=\left(\tilde{\boldsymbol{\Sigma}}\tilde{\mathbf{R}}\right)^{-H}(\mathbf{A}+\triangle\mathbf{A})^H\mathbf{r}=\left(\tilde{\boldsymbol{\Sigma}}\tilde{\mathbf{R}}\right)^{-H}\triangle\mathbf{A}^H\,\mathbf{r}\,. \tag{6.84}$$

Note that

$$\tilde{\mathbf{r}} - \mathbf{r} = \mathscr{P}_n{}^H \mathbf{b} + \triangle\mathbf{r} - \mathbf{r} = \left(\mathscr{P}_n{}^H - \mathbf{I}\right)\mathbf{r} + \mathscr{P}_n{}^H \mathbf{A}\mathbf{x} + \triangle\mathbf{r}$$

and that Eq. (6.55) gives us

$$= -\tilde{\mathbf{Q}}\tilde{\boldsymbol{\Sigma}}^{-2}\widehat{\mathbf{Q}}^H \mathbf{r} + \mathscr{P}_n{}^H \mathbf{A}\mathbf{x} + \triangle\mathbf{r}$$

then Eq. (6.56) produces

$$= -\tilde{\mathbf{Q}}\tilde{\boldsymbol{\Sigma}}^{-1}\left[\mathbf{I} + \mathbf{W}^H\right]^{-1}\left(\tilde{\mathbf{Q}}\tilde{\boldsymbol{\Sigma}}^{-1}\right)^H \mathbf{r} + \mathscr{P}_n{}^H \mathbf{A}\mathbf{x} + \triangle\mathbf{r}$$

then Eq. (6.84) leads to

$$= -\tilde{\mathbf{Q}}\tilde{\boldsymbol{\Sigma}}^{-1}\left[\mathbf{I} + \mathbf{W}^H\right]^{-1}\left(\tilde{\boldsymbol{\Sigma}}^{1}\tilde{\mathbf{R}}\right)^{-H} \triangle\mathbf{A}^H \mathbf{r} + \mathscr{P}_n{}^H \mathbf{A}\mathbf{x} + \triangle\mathbf{r}\,.$$

and finally Eq. (6.56) yields

$$= -\widehat{\mathbf{Q}}\tilde{\boldsymbol{\Sigma}}^{-1}\left(\tilde{\boldsymbol{\Sigma}}^{1}\tilde{\mathbf{R}}\right)^{-H} \triangle\mathbf{A}^H \mathbf{r} + \mathscr{P}_n{}^H \mathbf{A}\mathbf{x} + \triangle\mathbf{r}\,.$$

We can take norms to get

$$\|\tilde{\mathbf{r}} - \mathbf{r}\|_2 \le \left\|\widehat{\mathbf{Q}}\tilde{\boldsymbol{\Sigma}}^{-1}\right\|_2 \left\|\left(\tilde{\boldsymbol{\Sigma}}\tilde{\mathbf{R}}\right)^{-1}\right\|_2 \|\triangle\mathbf{A}\|_2 \|\mathbf{r}\|_2 + \left\|\mathscr{P}_n{}^H \mathbf{A}\right\|_2 \|\mathbf{x}\|_2 + \|\triangle\mathbf{r}\|_2$$

then we use inequalities (6.57), (6.74), (6.34), (6.38) and (6.51) to get

$$\le \sqrt{n}\frac{\left\|\mathbf{A}^\dagger\right\|_2}{\sqrt{1 - \beta\varepsilon\|\mathbf{A}\|_F \left\|\mathbf{A}^\dagger\right\|_2}} \frac{M_{(6.34)}\varepsilon}{1 - N_{(6.34)}\varepsilon}\|\mathbf{A}\|_F \|\mathbf{r}\|_2 + \frac{M_{(6.38)}\varepsilon}{1 - N_{(6.38)}\varepsilon}\|\mathbf{A}\|_F \left\|\mathbf{A}^\dagger\right\|_2 \|\mathbf{b}\|_2$$
$$+ \frac{M_{(6.51)}\varepsilon}{1 - N_{(6.51)}\varepsilon}\|\mathbf{b}\|_2\,.$$

This proves claim (6.81).

Next, we estimate the error in the computed least squares solution vector $\tilde{\mathbf{x}}$. We assume that $\tilde{\mathbf{x}}$ exactly solves $\tilde{\mathbf{R}}\tilde{\mathbf{x}} = \tilde{\mathbf{y}}$, and note that a more careful treatment would refer to Lemma 3.8.11 for an analysis of rounding errors in solving right-triangular linear systems. Then

$$\tilde{\mathbf{x}} - \mathbf{x} = \tilde{\mathbf{R}}^{-1}\tilde{\mathbf{y}} - \mathbf{x} = \tilde{\mathbf{R}}^{-1}\tilde{\boldsymbol{\Sigma}}\left(\widehat{\mathbf{Q}}\tilde{\boldsymbol{\Sigma}}^{-1}\right)^H \mathbf{b} + \tilde{\mathbf{R}}^{-1}\left(\tilde{\mathbf{y}} - \widehat{\mathbf{Q}}^H \mathbf{b}\right) - \mathbf{x}$$
$$= \tilde{\mathbf{R}}^{-1}\tilde{\boldsymbol{\Sigma}}\left(\widehat{\mathbf{Q}}\tilde{\boldsymbol{\Sigma}}^{-1}\right)^H (\mathbf{r} + \mathbf{A}\mathbf{x}) + \tilde{\mathbf{R}}^{-1}\left(\tilde{\mathbf{y}} - \widehat{\mathbf{Q}}^H \mathbf{b}\right) - \mathbf{x}$$

then we use Eq. (6.56):

$$= \tilde{\mathbf{R}}^{-1}\tilde{\boldsymbol{\Sigma}}\left[\mathbf{I}+\mathbf{W}^H\right]^{-1}\left(\tilde{\mathbf{Q}}\tilde{\boldsymbol{\Sigma}}^{-1}\right)^H\mathbf{r} + \tilde{\mathbf{R}}^{-1}\widehat{\mathbf{Q}}^H\mathbf{A}\mathbf{x} + \tilde{\mathbf{R}}^{-1}\left(\tilde{\mathbf{y}} - \widehat{\mathbf{Q}}^H\mathbf{b}\right) - \mathbf{x}$$

then we apply Eq. (6.84):

$$= \tilde{\mathbf{R}}^{-1}\tilde{\boldsymbol{\Sigma}}\left[\mathbf{I}+\mathbf{W}^H\right]^{-1}\left(\tilde{\boldsymbol{\Sigma}}\tilde{\mathbf{R}}\right)^{-H}\triangle\mathbf{A}\,\mathbf{r} + \tilde{\mathbf{R}}^{-1}\left(\widehat{\mathbf{Q}}^H\mathbf{A} - \tilde{\mathbf{R}}\right)\mathbf{x} + \tilde{\mathbf{R}}^{-1}\left(\tilde{\mathbf{y}} - \widehat{\mathbf{Q}}^H\mathbf{b}\right)$$

$$= \tilde{\mathbf{R}}^{-1}\tilde{\boldsymbol{\Sigma}}^2\tilde{\mathbf{R}}\left\{\left(\tilde{\boldsymbol{\Sigma}}\tilde{\mathbf{R}}\right)^H(\mathbf{I}+\mathbf{W}^H)\left(\tilde{\boldsymbol{\Sigma}}\tilde{\mathbf{R}}\right)\right\}^{-1}\triangle\mathbf{A}\,\mathbf{r} + \left(\tilde{\boldsymbol{\Sigma}}\tilde{\mathbf{R}}\right)^{-1}\tilde{\boldsymbol{\Sigma}}\left(\widehat{\mathbf{Q}}^H\mathbf{A} - \tilde{\mathbf{R}}\right)\mathbf{x}$$

$$+ \left(\tilde{\boldsymbol{\Sigma}}\tilde{\mathbf{R}}\right)^{-1}\tilde{\boldsymbol{\Sigma}}\left(\tilde{\mathbf{y}} - \widehat{\mathbf{Q}}^H\mathbf{b}\right).$$

Note that

$$\left\|\tilde{\mathbf{R}}^{-1}\tilde{\boldsymbol{\Sigma}}^2\tilde{\mathbf{R}}\right\|_2 = \left\|\mathbf{I} + \left(\tilde{\boldsymbol{\Sigma}}\tilde{\mathbf{R}}\right)^{-1}\left(\tilde{\boldsymbol{\Sigma}}^2 - \mathbf{I}\right)\left(\tilde{\boldsymbol{\Sigma}}\tilde{\mathbf{R}}\right)\mathbf{R}^{-1}\mathbf{R}\right\|_2$$

$$\le 1 + \left\|\left(\tilde{\boldsymbol{\Sigma}}\tilde{\mathbf{R}}\right)^{-1}\right\|_2\left\|\tilde{\boldsymbol{\Sigma}}^2 - \mathbf{I}\right\|_2\left\|\left(\tilde{\boldsymbol{\Sigma}}\tilde{\mathbf{R}}\right)\mathbf{R}^{-1}\right\|_2\|\mathbf{R}\|_2$$

then inequalities (6.74), (6.30), (6.77) and (6.79) imply that

$$\le 1 + \frac{\|\mathbf{A}^\dagger\|_2}{\sqrt{1-\beta\varepsilon\|\mathbf{A}\|_F\,\|\mathbf{A}^\dagger\|_2}}\frac{2(m+2)\varepsilon}{1-(4m+7)\varepsilon/2}\sqrt{1+\beta\varepsilon\|\mathbf{A}\|_F\,\|\mathbf{A}^\dagger\|_2}\|\mathbf{A}\|_2\,. \tag{6.85}$$

We can take norms of our expression for the error in the least squares solution to get

$$\|\tilde{\mathbf{x}} - \mathbf{x}\|_2 \le \left\|\tilde{\mathbf{R}}^{-1}\tilde{\boldsymbol{\Sigma}}^2\tilde{\mathbf{R}}\right\|_2\left\|\left\{\left(\tilde{\boldsymbol{\Sigma}}\tilde{\mathbf{R}}\right)^H(\mathbf{I}+\mathbf{W}^H)\left(\tilde{\boldsymbol{\Sigma}}\tilde{\mathbf{R}}\right)\right\}^{-1}\right\|_2\|\triangle\mathbf{A}\|_2\,\|\mathbf{r}\|_2$$

$$+ \left\|\left(\tilde{\boldsymbol{\Sigma}}\tilde{\mathbf{R}}\right)^{-1}\right\|_2\left\|\tilde{\boldsymbol{\Sigma}}\right\|_2\left\|\widehat{\mathbf{Q}}^H\mathbf{A} - \tilde{\mathbf{R}}\right\|_2\|\mathbf{x}\|_2 + \left\|\left(\tilde{\boldsymbol{\Sigma}}\tilde{\mathbf{R}}\right)^{-1}\right\|_2\left\|\tilde{\boldsymbol{\Sigma}}\right\|_2\left\|\tilde{\mathbf{y}} - \widehat{\mathbf{Q}}^H\mathbf{b}\right\|_2$$

then we use inequalities (6.85), (6.75), (6.34), (6.74), (6.29), (6.41), and (6.52) to get

$$\le \left\{1 + \sqrt{\frac{1+\beta\varepsilon\|\mathbf{A}\|_F\,\|\mathbf{A}^\dagger\|_2}{1-\beta\varepsilon\|\mathbf{A}\|_F\,\|\mathbf{A}^\dagger\|_2}}\frac{2(m+2)\varepsilon}{1-(4m+7)\varepsilon/2}\|\mathbf{A}\|_2\,\|\mathbf{A}^\dagger\|_2\right\}$$

$$\times\left\{\frac{\|\mathbf{A}^\dagger\|_2^2}{1-\beta\varepsilon\|\mathbf{A}\|_F\,\|\mathbf{A}^\dagger\|_2}\right\}\left\{\frac{M_{(6.34)}\varepsilon}{1-N_{(6.34)}\varepsilon}\|\mathbf{A}\|_F\right\}\|\mathbf{r}\|_2$$

$$+\left\{\frac{\|\mathbf{A}^\dagger\|_2}{\sqrt{1-\beta\varepsilon\|\mathbf{A}\|_F\,\|\mathbf{A}^\dagger\|_2}}\right\}\left\{1+\frac{(m+2)\varepsilon}{1-(3m+5)\varepsilon/2}\right\}\left\{\frac{M_{(6.41)}\varepsilon}{1-N_{(6.41)}\varepsilon}\|\mathbf{A}\|_F\right\}\|\mathbf{x}\|_2$$

$$+\left\{\frac{\|\mathbf{A}^\dagger\|_2}{\sqrt{1-\beta\varepsilon\|\mathbf{A}\|_F\,\|\mathbf{A}^\dagger\|_2}}\right\}\left\{1+\frac{(m+2)\varepsilon}{1-(3m+5)\varepsilon/2}\right\}\left\{\frac{M_{(6.52)}\varepsilon}{1-N_{(6.52)}\varepsilon}\|\mathbf{b}\|_2\right\}$$

next, we use inequalities (6.21) and (6.23) to obtain

$$\leq\left\{1+\left[1+\frac{2\beta\varepsilon\|\mathbf{A}\|_F\,\|\mathbf{A}^\dagger\|_2}{1-\beta\varepsilon\|\mathbf{A}\|_F\,\|\mathbf{A}^\dagger\|_2}\right]\frac{2(m+2)\varepsilon}{1-(4m+7)\varepsilon/2}\|\mathbf{A}\|_2\,\|\mathbf{A}^\dagger\|_2\right\}$$

$$\times\frac{M_{(6.34)}\|\mathbf{A}\|_F\,\|\mathbf{A}^\dagger\|_2\,\varepsilon}{\left(1-\beta\varepsilon\|\mathbf{A}\|_F\,\|\mathbf{A}^\dagger\|_2\right)\left(1-N_{(6.34)}\varepsilon\right)}\|\mathbf{A}^\dagger\|_2\,\|\mathbf{r}\|_2$$

$$+\left\{1+\beta\varepsilon\|\mathbf{A}\|_F\,\|\mathbf{A}^\dagger\|_2\right\}\left\{1+\frac{(m+2)\varepsilon}{1-(3m+5)\varepsilon/2}\right\}\frac{M_{(6.41)}\varepsilon}{1-N_{(6.41)}\varepsilon}\|\mathbf{A}\|_F\,\|\mathbf{A}^\dagger\|_2\,\|\mathbf{x}\|_2$$

$$+\left\{1+\beta\varepsilon\|\mathbf{A}\|_F\,\|\mathbf{A}^\dagger\|_2\right\}\left\{1+\frac{(m+2)\varepsilon}{1-(3m+5)\varepsilon/2}\right\}\frac{M_{(6.52)}\varepsilon}{1-N_{(6.52)}\varepsilon}\|\mathbf{A}^\dagger\|_2\,\|\mathbf{b}\|_2$$

then we use inequality (3.38) and the fact that $(1-x)(1-y)\geq 1-x-y$ for all nonnegative x and y to prove that

$$\leq\frac{M_{(6.34)}\|\mathbf{A}\|_F\,\|\mathbf{A}^\dagger\|_2\,\varepsilon}{1-\max\left\{\begin{array}{c}N_{(6.34)}+[\beta+4m+4]\|\mathbf{A}\|_F\,\|\mathbf{A}^\dagger\|_2\,,\\ [4m+7]/2+2\beta\varepsilon\|\mathbf{A}\|_F\,\|\mathbf{A}^\dagger\|_2\end{array}\right\}\varepsilon}\|\mathbf{A}^\dagger\|_2\,\|\mathbf{r}\|_2$$

$$+\frac{M_{(6.41)}\varepsilon}{1-\left(m+2+N_{(6.41)}+\beta\|\mathbf{A}\|_F\,\|\mathbf{A}^\dagger\|_2\right)\varepsilon}\|\mathbf{A}\|_F\,\|\mathbf{A}^\dagger\|_2\,\|\mathbf{x}\|_2$$

$$+\frac{M_{(6.52)}\varepsilon}{1-\left(m+2+N_{(6.52)}+\beta\|\mathbf{A}\|_F\,\|\mathbf{A}^\dagger\|_2\right)\varepsilon}\|\mathbf{A}^\dagger\|_2\,\|\mathbf{b}\|_2\,.$$

This proves claim (6.82).

Finally, we would like to determine the extent to which the columns of $\tilde{\mathbf{Q}}$ fail to be orthonormal. We begin by noting that

$$\left\|\mathbf{I}-\tilde{\mathbf{Q}}^H\tilde{\mathbf{Q}}\right\|_2=\left\|\mathbf{I}-\tilde{\boldsymbol{\Sigma}}^2-\mathbf{W}-\mathbf{W}^H\right\|_2$$

$$\leq\left\|\tilde{\boldsymbol{\Sigma}}^2-\mathbf{I}\right\|_2+2\|\mathbf{W}\|_2=\left\|\tilde{\boldsymbol{\Sigma}}^2-\mathbf{I}\right\|_2+2\left\|\left(\mathbf{W}\tilde{\boldsymbol{\Sigma}}\tilde{\mathbf{R}}\right)\left(\tilde{\boldsymbol{\Sigma}}\tilde{\mathbf{R}}\right)^{-1}\right\|_2$$

$$\leq\left\|\tilde{\boldsymbol{\Sigma}}^2-\mathbf{I}\right\|_2+2\left\|\mathbf{W}\left(\tilde{\boldsymbol{\Sigma}}\tilde{\mathbf{R}}\right)\right\|_2\left\|\left(\tilde{\boldsymbol{\Sigma}}\tilde{\mathbf{R}}\right)^{-1}\right\|_2$$

then inequalities (6.30), (6.62) and (6.74) produce

$$\leq \frac{2(m+2)\varepsilon}{1-(4m+7)\varepsilon/2} + \frac{2M_{(6.62)}\varepsilon}{1-N_{(6.62)}\varepsilon}\|\mathbf{A}\|_F \frac{\|\mathbf{A}^\dagger\|_2}{\sqrt{1-\beta\varepsilon\|\mathbf{A}\|_F\,\|\mathbf{A}^\dagger\|_2}}$$

then inequalities (6.23), (3.38) and (3.37) lead to

$$\leq \frac{2(m+2)\varepsilon}{1-(4m+7)\varepsilon/2} + \frac{2M_{(6.62)}\varepsilon}{1-N_{(6.62)}\varepsilon}\|\mathbf{A}\|_F\,\|\mathbf{A}^\dagger\|_2\left\{1+\beta\varepsilon\|\mathbf{A}\|_F\,\|\mathbf{A}^\dagger\|_2\right\}$$

then inequality (3.38) produces

$$\leq \frac{2(m+2)\varepsilon}{1-(4m+7)\varepsilon/2} + \frac{2M_{(6.62)}\varepsilon}{1-\left(N_{(6.62)}+\beta\|\mathbf{A}\|_F\,\|\mathbf{A}^\dagger\|_2\right)\varepsilon}\|\mathbf{A}\|_F\,\|\mathbf{A}^\dagger\|_2 \;.$$

and finally we use inequality (3.37) to get

$$\leq \frac{2\left(m+2+M_{(6.62)}\|\mathbf{A}\|_F\,\|\mathbf{A}^\dagger\|_2\right)\varepsilon}{1-\left(N_{(6.62)}+\beta\|\mathbf{A}\|_F\,\|\mathbf{A}^\dagger\|_2\right)\varepsilon} \;.$$

This is equivalent to claim (6.83).

Through the theorems in this section, Björck [8, p. 19] showed that the sensitivity of the numerical solution of least squares problems by successive orthogonal projection is similar to the sensitivity of the least squares problem itself. The computed columns of $\tilde{\mathbf{Q}}$ are not necessarily orthogonal, but this does not affect the accuracy of either the computed residual $\tilde{\mathbf{r}}$ or the computed solution $\tilde{\mathbf{x}}$ for the overdetermined least squares problem. On the other hand, inequality (6.83), which appears in Björck [8, p. 15], showed that the deviation of the normalized columns of $\tilde{\mathbf{Q}}$ from orthonormality can be on the order of the condition number of $\mathbf{A}$. Pivoting in the modified Gram-Schmidt process makes little to no improvement in this error. Thus the modified Gram-Schmidt process in Algorithm 6.8.1 or 6.8.2 should not be used *in that form* to find an orthogonal basis for the range of $\mathbf{A}$. Instead, the reader may consider using the Gram-Schmidt process combined with reorthogonalization to produce an orthonormal basis for the range of $\mathbf{A}$. We will discuss reorthogonalization in Sect. 6.8.6.

Exercise 6.8.6 For the Läuchli matrix in Examples 6.5.4 and 6.8.2, verify by hand the results of Theorems 6.8.2–6.8.5.

Exercise 6.8.7 Let $\mathbf{A}$ be the 10×5 **Hilbert matrix**

$$\mathbf{A}_{ij} = \frac{1}{1+i+j} \text{ for } 0 \leq i < 10 \text{ and } 0 \leq j < 5 \;,$$

and let $\mathbf{b}$ be the 10-vector with entries

$$\mathbf{b}_i = \frac{1}{6+i} \;.$$

Use modified Gram-Schmidt factorization and successive orthogonal projection to solve the associated least squares problem for the 5-vector **x**. Then estimate the errors in **x** by means of the rounding error estimate (6.82), and the *a posteriori* error estimate of Sect. 6.6. Discuss the relative sizes of these error estimates.

6.8.6 Reorthogonalization

As we saw in inequality (6.83), the orthonormality of the projection vectors computed by modified Gram-Schmidt factorization may deteriorate for least squares problems with large condition numbers. Fortunately, there is a way to improve the orthogonality of these vectors.

Let $\delta \in (0,1)$ be some given number; we want δ to be on the order of one, and typically take $\delta = \sqrt{2}/2 = \cos(\pi/4)$. Then reorthogonalization repeats successive orthogonal projection of the k-th column until this vector is no longer substantially shortened by successive orthogonal projection. This gives us the following

Algorithm 6.8.7 (Modified Gram-Schmidt with Reorthogonalization)

for $1 \le j \le n$
  $\omega_j^2 = \mathbf{A}_{1:m,j} \cdot \mathbf{A}_{1:m,j}$ /* LAPACK BLAS routine _dot or _dotc */
for $1 \le k \le \min\{m,n\}$
  $\sigma_k^2 = \mathbf{A}_{1:m,k} \cdot \mathbf{A}_{1:m,k}$ /* LAPACK BLAS routine _dot or _dotc */
  if $\sigma_k^2 = 0$ break /* rank deficient matrix */
  while $\sigma_k^2 \le \delta\, \omega_k^2$ /* reorthogonalization */
    for $1 \le \ell < k$
      $\varrho_{\ell k} = \mathbf{A}_{1:m,\ell} \cdot \mathbf{A}_{1:m,k} / \sigma_\ell^2$ /* LAPACK BLAS routine _dot or _dotc */
      $\mathbf{R}_{\ell k} = \mathbf{R}_{\ell k} + \varrho_{\ell k}$
      $\mathbf{A}_{1:m,k} = \mathbf{A}_{1:m,k} - \mathbf{A}_{1:m,\ell}\varrho_{\ell,k}$ /* LAPACK BLAS routine _axpy */
      $\omega_k^2 = \sigma_k^2$
      $\sigma_k^2 = \mathbf{A}_{1:m,k} \cdot \mathbf{A}_{1:m,k}$ /* LAPACK BLAS routine _dot or _dotc */
  $\sigma_k = \sqrt{\sigma_k^2}$
  $\mathbf{R}_{kk} = \sigma_k$
  $\mathbf{A}_{1:m,k} = \mathbf{A}_{1:m,k}/\sigma_k$ /* LAPACK BLAS routine _scal or _dscal */
  $\mathbf{w}_{k+1:n} = \mathbf{A}_{1:m,k+1:n}{}^H \mathbf{A}_{1:m,k}$ /* LAPACK BLAS routine _gemv */
  $\mathbf{R}_{k,k+1:n} = \overline{\mathbf{w}_{k+1:n}}$
  $\mathbf{A}_{1:m,k+1:n} = \mathbf{A}_{1:m,k+1:n} - \mathbf{A}_{1:m,k}\mathbf{R}_{k,k+1:n}$ /* LAPACK BLAS routine _ger or _geru */

For those readers who are interested in more information, we recommend Björck [10, p. 310f], who discusses reorthogonalization in more detail, and explains why only one reorthogonalization step is sufficient.

Example 6.8.4 ([72]) Suppose that the scalar $\alpha > 0$ is such that $\mathrm{fl}(1+\alpha^2) = 1$. Let

$$\mathbf{A} = \begin{bmatrix} 1 & 1 \\ \alpha & 0 \\ 0 & \alpha \end{bmatrix} \text{ and } \mathbf{b} = \begin{bmatrix} 1 \\ 0 \\ 0 \end{bmatrix} .$$

Then modified Gram-Schmidt factorization computes

$$\sigma_1 = \|\mathbf{a}_1\|_2 = \mathrm{fl}(\sqrt{1+\alpha^2}) = 1 \text{ and } \mathbf{q}_1 = \mathrm{fl}(\mathbf{a}_1/1) = \mathbf{a}_1 ,$$

$$\varrho_{12} = \mathbf{q}_1 \cdot \mathbf{a}_2 = 1 \text{ and } \mathbf{a}_2' = \mathbf{a}_2 - \mathbf{q}_1 \varrho_{12} = \begin{bmatrix} 0 \\ -\alpha \\ \alpha \end{bmatrix} ,$$

$$\sigma_2 = \|\mathbf{a}_2'\|_2 = \alpha\sqrt{2} .$$

Since the second column of **A** was substantially shortened by orthogonal projection, we apply reorthogonalization:

$$\delta\varrho_{12} = \mathbf{q}_1 \cdot \mathbf{a}_2' = \begin{bmatrix} 1 \\ \alpha \\ 0 \end{bmatrix} \cdot \begin{bmatrix} 0 \\ -\alpha \\ 0 \end{bmatrix} = -\alpha^2 \text{ and } \varrho_{12} = \mathrm{fl}(\varrho_{12} + \delta\varrho_{12}) = \mathrm{fl}(1-\alpha^2) = 1 ,$$

$$\mathbf{a}_2' = \mathbf{a}_2' - \mathbf{q}_1 \delta\varrho_{12} = \begin{bmatrix} 0 \\ -\alpha \\ \alpha \end{bmatrix} - \begin{bmatrix} 1 \\ \alpha \\ 0 \end{bmatrix} \frac{(-\alpha^2)}{1} = \begin{bmatrix} \alpha^2 \\ -\alpha \\ \alpha \end{bmatrix} ,$$

$$\sigma_2 = \|\mathbf{a}_2'\|_2 = \mathrm{fl}(\sqrt{\alpha^4 + \alpha^2 + \alpha^2}) = \alpha\sqrt{2} \text{ and } \mathbf{q}_2 = \mathbf{a}_2'/\sigma_2 = \begin{bmatrix} \alpha/\sqrt{2} \\ -1/\sqrt{2} \\ 1/\sqrt{2} \end{bmatrix} .$$

Note that reorthogonalization made no change to **R**, and no change to σ_2, but did change $\mathbf{a}_2'$ and therefore $\mathbf{q}_2$.

After Gram-Schmidt factorization with reorthogonalization, the successive orthogonal projection process computes

$$\eta_1 = \mathbf{q}_1 \cdot \mathbf{b} \begin{bmatrix} 1 \\ \alpha \\ 0 \end{bmatrix} \cdot \begin{bmatrix} 1 \\ 0 \\ 0 \end{bmatrix} = 1 \text{ and } \mathbf{b}' = \mathbf{b} - \mathbf{q}_1 \eta_1 = \begin{bmatrix} 1 \\ 0 \\ 0 \end{bmatrix} - \begin{bmatrix} 1 \\ \alpha \\ 0 \end{bmatrix} = \begin{bmatrix} 0 \\ -\alpha \\ 0 \end{bmatrix} ,$$

$$\eta_2 = \mathbf{q}_2 \cdot \mathbf{b}' = \begin{bmatrix} \alpha/\sqrt{2} \\ -1/\sqrt{2} \\ 1/\sqrt{2} \end{bmatrix} \cdot \begin{bmatrix} 0 \\ -\alpha \\ 0 \end{bmatrix} = \frac{\alpha}{\sqrt{2}} \text{ and } \mathbf{r} = \mathbf{b}' - \mathbf{q}_2 \eta_2 = \begin{bmatrix} -\alpha^2/2 \\ -\alpha/2 \\ -\alpha/2 \end{bmatrix} .$$

This is the correctly rounded value of the exact residual. Finally, we back-solve

$$\mathbf{R}\mathbf{x} = \begin{bmatrix} 1 & 1 \\ & \alpha\sqrt{2} \end{bmatrix} \begin{bmatrix} \xi_1 \\ \xi_2 \end{bmatrix} = \mathbf{y} = \begin{bmatrix} \eta_1 \\ \eta_2 \end{bmatrix} = \begin{bmatrix} 1 \\ \alpha/\sqrt{2} \end{bmatrix} \Longrightarrow \mathbf{x} = \begin{bmatrix} 1/2 \\ 1/2 \end{bmatrix} .$$

Comparison with the results in Example 6.8.2 for modified Gram-Schmidt factorization of this matrix without reorthogonalization leads to the following observations. Reorthogonalization made no change in the computed solution $\mathbf{x}$ to the least squares problem, but did produce the correctly rounded value of the exact residual. Also, reorthogonalization succeeded in producing a matrix $\tilde{\mathbf{Q}}$ with numerically mutually orthogonal columns. In contrast, modified Gram-Schmidt factorization without reorthogonalization produced off-diagonal entries of $\tilde{\mathbf{Q}}^H\tilde{\mathbf{Q}}$ with value $-\alpha/\sqrt{2}$.

Reorthogonalization is helpful in **updating** QR factorizations when columns are inserted or deleted, rows are inserted or deleted, or rank one modifications are made to the matrix. The Gram-Schmidt process can be significantly more efficient than Householder factorization for updating $m \times n$ matrices with $n \ll m$, because the latter method need to store an $m \times m$ unitary matrix $\mathbf{Q}$ during the updating. For more information regarding reorthogonalization and its importance in updating QR factorizations, see Daniel et al. [27].

Interested readers may view a C^{++} implementation of reorthogonalization in double precision, in the code for the `GramSchmidtQRFactorization` constructor.

Exercise 6.8.8 Program Gram-Schmidt QR factorization with and without reorthogonalization. Use these programs to compute the factorization of the be the 10×5 **Hilbert matrix**

$$\mathbf{A}_{ij} = \frac{1}{1+i+j} \text{ for } 0 \le i < 10 \text{ and } 0 \le j < 5 .$$

Compute $\mathbf{I} - \mathbf{Q}^H\mathbf{Q}$ for both factorizations, and discuss the results. You may want to review inequality (6.83) for this discussion.

6.8.7 *Iterative Improvement*

In Sect. 6.7.6 we presented a technique for improving the solution of least squares problems via successive reflection. The technique was useful for overcoming rounding errors that occur in the factorization of the matrix. In this section, we will present an algorithm for improving the solution of overdetermined least squares problems via successive orthogonal projection

Recall that the basic idea of least squares iterative improvement is the following. Suppose that $\tilde{\mathbf{x}}$ and $\tilde{\mathbf{r}}$ are approximate solution and residual vectors, respectively, for the overdetermined least squares problem $\min \|\mathbf{A}\mathbf{x} - \mathbf{b}\|_2$. We compute

$$\begin{bmatrix} \delta\mathbf{b} \\ \delta\mathbf{c} \end{bmatrix} = \begin{bmatrix} \mathbf{b} \\ \mathbf{0} \end{bmatrix} - \begin{bmatrix} \mathbf{I} & \mathbf{A} \\ \mathbf{A}^H & \mathbf{0} \end{bmatrix} \begin{bmatrix} \tilde{\mathbf{r}} \\ \tilde{\mathbf{x}} \end{bmatrix} = \begin{bmatrix} \mathbf{b} - \tilde{\mathbf{r}} - \mathbf{A}\tilde{\mathbf{x}} \\ -\mathbf{A}^H\tilde{\mathbf{r}} \end{bmatrix} ,$$

and then find perturbations $\delta\mathbf{r}$ and $\delta\mathbf{x}$ so that

$$\begin{bmatrix} \delta\mathbf{b} \\ \delta\mathbf{c} \end{bmatrix} = \begin{bmatrix} \mathbf{I} & \mathbf{A} \\ \mathbf{A}^H & \mathbf{0} \end{bmatrix} \begin{bmatrix} \delta\mathbf{r} \\ \delta\mathbf{x} \end{bmatrix} .$$

We expect that $\tilde{\mathbf{r}} + \delta\mathbf{r}$ should be closer to the true residual, and $\tilde{\mathbf{x}} + \delta\mathbf{x}$ should be closer to the true solution.

Suppose that we have a modified Gram-Schmidt factorization $\mathbf{A} \approx \tilde{\mathbf{Q}}\tilde{\mathbf{R}}$ where $\tilde{\mathbf{Q}}^H\tilde{\mathbf{Q}}$ is approximately the identity matrix. Then we define $\delta\mathbf{v} = \tilde{\mathbf{Q}}^H\delta\mathbf{r}$, and note that $\delta\mathbf{c} = \tilde{\mathbf{R}}^H\tilde{\mathbf{Q}}^H\delta\mathbf{r}$ implies that

$$\tilde{\mathbf{R}}^H\delta\mathbf{v} = \delta\mathbf{c} .$$

Next, we note that $\tilde{\mathbf{Q}}^H\delta\mathbf{b} = \delta\mathbf{v} + \tilde{\mathbf{R}}\delta\mathbf{x}$, so we define $\delta\mathbf{y} = \tilde{\mathbf{R}}\delta\mathbf{x}$ and see that

$$\delta\mathbf{y} = \tilde{\mathbf{Q}}^H\delta\mathbf{b} - \delta\mathbf{v} .$$

This leads to

$$\delta\mathbf{r} = \delta\mathbf{b} - \tilde{\mathbf{Q}}\delta\mathbf{y} = \left(\mathbf{I} - \tilde{\mathbf{Q}}\tilde{\mathbf{Q}}^H\right)\delta\mathbf{b} - \delta\mathbf{v} ,$$

and

$$\tilde{\mathbf{R}}\delta\mathbf{x} = \delta\mathbf{y} .$$

We can summarize these computations in the following

Algorithm 6.8.8 (Gram-Schmidt Overdetermined Iterative Improvement)

$\delta\mathbf{b} = \mathbf{b} - \tilde{\mathbf{r}} - \mathbf{A}\tilde{\mathbf{x}}$ /* LAPACK BLAS routine _gemv */
$\delta\mathbf{c} = -\mathbf{A}^H\tilde{\mathbf{r}}$ /* LAPACK BLAS routine _gemv */
solve $\tilde{\mathbf{R}}^H\delta\mathbf{v} = \delta\mathbf{c}$ /* LAPACK BLAS routine _trsv */
$\delta\mathbf{r} = \delta\mathbf{b}$
for $1 \le k \le n$ /* successive orthogonal projection */
  $\delta\mathbf{y}_k = \mathbf{q}_k \cdot \delta\mathbf{r} - \delta\mathbf{v}_k$ /* LAPACK BLAS routine _dot or _dotc */
  $\delta\mathbf{r} = \delta\mathbf{r} - \mathbf{q}_k\delta\mathbf{y}_k$ /* LAPACK BLAS routine _axpy */
solve $\tilde{\mathbf{R}}\delta\mathbf{x} = \delta\mathbf{y}$ /* LAPACK BLAS routine _trsv */
$\mathbf{x} = \tilde{\mathbf{x}} + \delta\mathbf{x}$
$\mathbf{r} = \tilde{\mathbf{r}} + \delta\mathbf{r}$

Note that it may be necessary to compute $\delta\mathbf{b}$ and $\delta\mathbf{c}$ using extended precision arithmetic.

Example 6.8.5 Suppose that we use the classical Gram-Schmidt process to solve the Läuchli problem in Example 6.5.4. That algorithm computes

$$\tilde{\mathbf{r}} = \begin{bmatrix} 0 \\ -\alpha \\ 0 \end{bmatrix} \text{ and } \tilde{\mathbf{x}} = \begin{bmatrix} 1 \\ 0 \end{bmatrix} .$$

Let us show how to use the modified Gram-Schmidt process and successive orthogonal projection to improve this solution.

Note that both the modified and classical Gram-Schmidt processes factor

$$\mathbf{A} = \tilde{\mathbf{Q}}\tilde{\mathbf{R}} = \begin{bmatrix} 1 & 0 \\ \alpha & -1/\sqrt{2} \\ 0 & 1/\sqrt{2} \end{bmatrix} \begin{bmatrix} 1 & 1 \\ & \alpha\sqrt{2} \end{bmatrix} .$$

We begin iterative improvement by computing the vectors

$$\delta\mathbf{b} = \mathbf{b} - \tilde{\mathbf{r}} - \mathbf{A}\tilde{\mathbf{x}} = \begin{bmatrix} 1 \\ 0 \\ 0 \end{bmatrix} - \begin{bmatrix} 0 \\ -\alpha \\ 0 \end{bmatrix} - \begin{bmatrix} 1 & 1 \\ \alpha & 0 \\ 0 & \alpha \end{bmatrix} \begin{bmatrix} 1 \\ 0 \end{bmatrix} = \begin{bmatrix} 0 \\ 0 \\ 0 \end{bmatrix} \text{ and}$$

$$\delta\mathbf{c} = -\mathbf{A}^H\tilde{\mathbf{r}} = -\begin{bmatrix} 1 & \alpha & 0 \\ 1 & 0 & \alpha \end{bmatrix} \begin{bmatrix} 0 \\ -\alpha \\ 0 \end{bmatrix} = \begin{bmatrix} \alpha^2 \\ 0 \end{bmatrix} .$$

Next we solve $\tilde{\mathbf{R}}^H\delta\mathbf{v} = \delta\mathbf{c}$, or

$$\begin{bmatrix} 1 & \\ 1 & \alpha\sqrt{2} \end{bmatrix} \begin{bmatrix} \delta\mathbf{v}_1 \\ \delta\mathbf{v}_2 \end{bmatrix} = \tilde{\mathbf{R}}^H\delta\mathbf{v} = \delta\mathbf{c} = \begin{bmatrix} \alpha^2 \\ 0 \end{bmatrix} \Longrightarrow \delta\mathbf{v} = \begin{bmatrix} \alpha^2 \\ \alpha/\sqrt{2} \end{bmatrix} .$$

Successive orthogonal projection computes

$$\delta\mathbf{y}_1 = \tilde{\mathbf{q}}_1 \cdot \delta\mathbf{b} - \delta\mathbf{v}_1 = -\delta\mathbf{v}_1 = -\alpha^2 ,$$

$$\delta\mathbf{r} = \delta\mathbf{b} - \tilde{\mathbf{q}}_1\delta\mathbf{y}_1 = -\begin{bmatrix} 1 \\ \alpha \\ 0 \end{bmatrix} (-\alpha^2) = \begin{bmatrix} \alpha^2 \\ \alpha^3 \\ 0 \end{bmatrix} ,$$

$$\delta\mathbf{y}_2 = \tilde{\mathbf{q}}_2 \cdot \delta\mathbf{r} - \delta\mathbf{v}_2 = \begin{bmatrix} 0 \\ -1/\sqrt{2} \\ 1/\sqrt{2} \end{bmatrix} \cdot \begin{bmatrix} \alpha^2 \\ \alpha^3 \\ 0 \end{bmatrix} - (-\alpha/\sqrt{2}) = \alpha/\sqrt{2} \text{ and}$$

$$\delta\mathbf{r} = \delta\mathbf{r} - \tilde{\mathbf{q}}_2\delta\mathbf{y}_2 = \begin{bmatrix} \alpha^2 \\ \alpha^3 \\ 0 \end{bmatrix} - \begin{bmatrix} 0 \\ -1/\sqrt{2} \\ 1/\sqrt{2} \end{bmatrix} \alpha/\sqrt{2} = \begin{bmatrix} \alpha^2 \\ \alpha/2 \\ -\alpha/2 \end{bmatrix} .$$

Next, we solve

$$\begin{bmatrix} 1 & 1 \\ & \alpha\sqrt{2} \end{bmatrix} \begin{bmatrix} \delta\mathbf{x}_1 \\ \delta\mathbf{x}_2 \end{bmatrix} = \tilde{\mathbf{R}}\delta\mathbf{x} = \delta\mathbf{y} = \begin{bmatrix} -\alpha^2 \\ \alpha/\sqrt{2} \end{bmatrix} \Longrightarrow \delta\mathbf{x} = \begin{bmatrix} -1/2 \\ 1/2 \end{bmatrix} .$$

Finally, we update

$$\mathbf{x} = \tilde{\mathbf{x}} + \delta\mathbf{x} = \begin{bmatrix} 1 \\ 0 \end{bmatrix} + \begin{bmatrix} -1/2 \\ 1/2 \end{bmatrix} = \begin{bmatrix} 1/2 \\ 1/2 \end{bmatrix} \text{ and}$$

$$\mathbf{r} = \tilde{\mathbf{r}} + \delta\mathbf{r} = \begin{bmatrix} 0 \\ -\alpha \\ 0 \end{bmatrix} + \begin{bmatrix} \alpha^2 \\ \alpha/2 \\ -\alpha/2 \end{bmatrix} = \begin{bmatrix} \alpha^2 \\ -\alpha/2 \\ -\alpha/2 \end{bmatrix} .$$

Comparison with the results in Example 6.7.1 shows that iterative improvement has produced the correctly rounded values of the least square solution vector and the residual vector.

6.8.8 Underdetermined Least Squares

We can also use modified Gram-Schmidt factorization to solve full-rank underdetermined least squares problems. Suppose that we are given an $m \times n$ matrix $\mathbf{A}$ with rank $(\mathbf{A}) = n \leq m$, and an n-vector $\mathbf{b}$. We want to find the m-vector $\mathbf{x}$ with minimum Euclidean norm so that $\mathbf{A}^H\mathbf{x} = \mathbf{b}$. We assume that we have factored $\mathbf{A} = \mathbf{Q}\mathbf{R}$ as in Sect. 6.8.1.1. Since $\mathbf{Q}$ has orthonormal columns, we see that $\mathbf{Q}^H\mathbf{Q} = \mathbf{I}$. Furthermore, $\mathbf{R}$ is right triangular. Let us see how to use this factorization to solve the underdetermined least squares problem.

Recall from Lemma 6.3.1 that the solution to the full-rank underdetermined least squares problem satisfies Eq. (6.2), which is

$$\begin{bmatrix} \mathbf{I} & \mathbf{A} \\ \mathbf{A}^H & \mathbf{0} \end{bmatrix} \begin{bmatrix} \mathbf{x} \\ -\mathbf{s} \end{bmatrix} = \begin{bmatrix} \mathbf{0} \\ \mathbf{b} \end{bmatrix} .$$

Using the factorization of $\mathbf{A}$, we can rewrite this linear system as $\mathbf{b} = \mathbf{R}^H\mathbf{Q}^H\mathbf{x}$ and $\mathbf{x} = \mathbf{Q}\mathbf{R}\mathbf{s}$. These suggest that we solve

$$\mathbf{R}^H\mathbf{y} = \mathbf{b}$$

for the m-vector $\mathbf{y}$, and then take $\mathbf{x} = \mathbf{Q}\mathbf{y}$. Then we have

$$\mathbf{Q}\mathbf{y} = \mathbf{x} = \mathbf{A}\mathbf{s} = \mathbf{Q}\mathbf{R}\mathbf{s} ,$$

so we take $\mathbf{s}$ to be the solution of the linear system

$$\mathbf{R}\mathbf{s} = \mathbf{y} .$$

Björck [10, p. 310] has developed an improved form of these computations, which can be summarized by the following

Algorithm 6.8.9 (Gram-Schmidt Underdetermined Least Squares)

solve $\mathbf{R}^H \mathbf{y} = \mathbf{b}$ /* LAPACK BLAS routine _trsv */
$\mathbf{x} = \mathbf{0}$
for $1 \leq k \leq m$
  $\omega_k = \mathbf{Q}_{1:n,k} \cdot \mathbf{x}_{1:n}$ /* LAPACK BLAS routine _dot or _dotc */
  $\mathbf{x}_{1:n} = \mathbf{x}_{1:n} - \mathbf{Q}_{1:n,k}(\omega_k - \mathbf{y}_k)$ /* LAPACK BLAS routine _axpy */
solve $\mathbf{R}\mathbf{s} = \mathbf{y} - \mathbf{w}$ /* LAPACK BLAS routine _trsv */

Note that the vector $\mathbf{s}$ is not generally needed, but could be useful for iterative improvement.

Example 6.8.6 Suppose that

$$\mathbf{A}^H = \begin{bmatrix} 1 & \alpha & 0 \\ 1 & 0 & \alpha \end{bmatrix} \text{ and } \mathbf{b} = \begin{bmatrix} 1 \\ 1 \end{bmatrix} .$$

With exact arithmetic, the solution $\mathbf{x}$ of $\mathbf{A}^H\mathbf{x} = \mathbf{b}$ with minimum Euclidean norm is

$$\mathbf{x} = \begin{bmatrix} 2 \\ \alpha \\ \alpha \end{bmatrix} \frac{1}{2 + \alpha^2} .$$

It is easy to check that $\mathbf{A}^H\mathbf{x} = \mathbf{b}$, and that $\mathbf{x} = \mathbf{A}\mathbf{s}$ where

$$\mathbf{s} = \begin{bmatrix} 1 \\ 1 \end{bmatrix} \frac{1}{2 + \alpha^2} .$$

These two facts prove that $\mathbf{x}$ is the minimum norm solution of $\mathbf{A}\mathbf{x} = \mathbf{b}$.

Next, suppose that $\mathrm{fl}(1+\alpha^2) = 1$. Then the modified Gram-Schmidt factorization in Example 6.8.2 computed $\mathbf{A} = \mathbf{Q}\mathbf{R}$ where

$$\mathbf{Q} = \begin{bmatrix} 1 & 0 \\ \alpha & -1/\sqrt{2} \\ 0 & 1/\sqrt{2} \end{bmatrix} \text{ and } \mathbf{R} = \begin{bmatrix} 1 & 1 \\ & \alpha\sqrt{2} \end{bmatrix} .$$

Next, we apply Björck's algorithm to find the minimum norm solution of $\mathbf{A}^H\mathbf{x} = \mathbf{b}$:

$$\begin{bmatrix} 1 & \\ 1 & \alpha\sqrt{2} \end{bmatrix} \begin{bmatrix} \mathbf{y}_1 \\ \mathbf{y}_2 \end{bmatrix} = \mathbf{R}^H\mathbf{y} = \mathbf{b} = \begin{bmatrix} 1 \\ 1 \end{bmatrix} \Longrightarrow \mathbf{y} = \begin{bmatrix} 1 \\ 0 \end{bmatrix}$$

$$\mathbf{x}^{(0)} = \mathbf{0}$$

$$\mathbf{w}_1 = \mathbf{q}_1 \cdot \mathbf{x}^{(0)} = \begin{bmatrix} 1 \\ \alpha \\ 0 \end{bmatrix} \cdot \begin{bmatrix} 0 \\ 0 \\ 0 \end{bmatrix} = 0$$

$$\mathbf{x}^{(1)} = \mathbf{x}^{(0)} - \mathbf{q}_1(\mathbf{w}_1 - \mathbf{y}_1) = \begin{bmatrix} 0 \\ 0 \\ 0 \end{bmatrix} - \begin{bmatrix} 1 \\ \alpha \\ 0 \end{bmatrix} (0 - 1) = \begin{bmatrix} 1 \\ \alpha \\ 0 \end{bmatrix}$$

$$\mathbf{w}_2 = \mathbf{q}_2 \cdot \mathbf{x}^{(1)} = \begin{bmatrix} 0 \\ -1/\sqrt{2} \\ 1/\sqrt{2} \end{bmatrix} \cdot \begin{bmatrix} 1 \\ \alpha \\ 0 \end{bmatrix} = -\alpha/\sqrt{2}$$

$$\mathbf{x} = \mathbf{x}^{(1)} - \mathbf{q}_2(\mathbf{w}_2 - \mathbf{y}_2) = \begin{bmatrix} 1 \\ \alpha \\ 0 \end{bmatrix} - \begin{bmatrix} 0 \\ -1/\sqrt{2} \\ 1/\sqrt{2} \end{bmatrix} (-\alpha/\sqrt{2} - 0) = \begin{bmatrix} 1 \\ \alpha/2 \\ \alpha/2 \end{bmatrix} .$$

This is the correctly rounded value of the exact solution.

If we choose, we can also solve

$$\begin{bmatrix} 1 & 1 \\ & \alpha\sqrt{2} \end{bmatrix} \begin{bmatrix} \mathbf{s}_1 \\ \mathbf{s}_2 \end{bmatrix} = \mathbf{R}\mathbf{s} = \mathbf{y} - \mathbf{w} = \begin{bmatrix} 1 \\ \alpha/\sqrt{2} \end{bmatrix} \Longrightarrow \mathbf{s} = \begin{bmatrix} 1/2 \\ 1/2 \end{bmatrix} .$$

It is interesting to note that if we had not used Björck's algorithm, and instead computed $\mathbf{x} = \mathbf{Q}\mathbf{y}$, we would have computed

$$\mathbf{x} = \begin{bmatrix} 1 & 0 \\ \alpha & -1/\sqrt{2} \\ 0 & 1/\sqrt{2} \end{bmatrix} \begin{bmatrix} 1 \\ 0 \end{bmatrix} = \begin{bmatrix} 1 \\ \alpha \\ 0 \end{bmatrix} .$$

This is *not* the correctly rounded value of the exact solution.

Next, let us discuss iterative improvement for the underdetermined least squares problem. Suppose that we have factored $\mathbf{A} \approx \tilde{\mathbf{Q}}\tilde{\mathbf{R}}$ using Algorithm 6.8.1, and we have computed an approximate solution $\tilde{\mathbf{x}}$ to the underdetermined least squares solution with $\mathbf{A}^H\mathbf{x} = \mathbf{b}$ using Algorithm 6.8.9. Assume that we have also computed an approximate value $\tilde{\mathbf{s}}$. Then we can compute the vectors $\delta\mathbf{c}$ and $\delta\mathbf{b}$ by

$$\begin{bmatrix} \delta\mathbf{c} \\ \delta\mathbf{b} \end{bmatrix} = \begin{bmatrix} \mathbf{0} \\ \mathbf{b} \end{bmatrix} - \begin{bmatrix} \mathbf{I} & \mathbf{A} \\ \mathbf{A}^H & \mathbf{0} \end{bmatrix} \begin{bmatrix} \tilde{\mathbf{x}} \\ -\tilde{\mathbf{s}} \end{bmatrix} \approx \begin{bmatrix} \mathbf{A}\tilde{\mathbf{s}} - \tilde{\mathbf{x}} \\ \mathbf{b} - \mathbf{A}^H\tilde{\mathbf{x}} \end{bmatrix} .$$

These two vectors, $\delta\mathbf{c}$ and $\delta\mathbf{b}$, may need to be computed in extended precision. We want to find perturbations $\delta\mathbf{x}$ and $\delta\mathbf{s}$ so that

$$\begin{bmatrix} \delta\mathbf{c} \\ \delta\mathbf{b} \end{bmatrix} = \begin{bmatrix} \mathbf{I} & \mathbf{A} \\ \mathbf{A}^H & \mathbf{0} \end{bmatrix} \begin{bmatrix} \delta\mathbf{x} \\ -\delta\mathbf{s} \end{bmatrix} .$$

Then we expect that $\tilde{\mathbf{x}} + \delta\mathbf{x}$ should be closer to $\mathbf{x}$.

Note that we should have $\delta\mathbf{b} = \mathbf{A}\delta\mathbf{x} = \mathbf{R}^H \mathbf{Q}^H \delta\mathbf{x} \equiv \mathbf{R}^H \delta\mathbf{y}$, so we begin by solving

$$\mathbf{R}^H \delta\mathbf{y} = \delta\mathbf{b}$$

for $\delta\mathbf{y}$. Since $\delta\mathbf{c} = \delta\mathbf{x} - \mathbf{Q}\mathbf{R}\delta\mathbf{s} \equiv \delta\mathbf{x} - \mathbf{Q}\delta\mathbf{u}$, we have

$$\mathbf{Q}^H \delta\mathbf{c} = \mathbf{Q}^H \delta\mathbf{x} - \delta\mathbf{u} = \delta\mathbf{y} - \boldsymbol{\Sigma}^2 \delta\mathbf{u} ,$$

so we can compute $\delta\mathbf{u}$ as

$$\delta\mathbf{u} = \delta\mathbf{y} - \mathbf{Q}^H \delta\mathbf{c} .$$

It follows that

$$\delta\mathbf{x} = \delta\mathbf{c} + \mathbf{Q}\mathbf{R}\delta\mathbf{s} = \delta\mathbf{c} + \mathbf{Q}\delta\mathbf{u} = \delta\mathbf{c} + \mathbf{Q}\left(\delta\mathbf{y} - \mathbf{Q}^H \delta\mathbf{c}\right) = \left(\mathbf{I} - \mathbf{Q}\mathbf{Q}^H\right)\delta\mathbf{c} + \mathbf{Q}\delta\mathbf{y} ,$$

and that

$$\mathbf{R}\delta\mathbf{s} = \delta\mathbf{y} - \mathbf{Q}^H \delta\mathbf{c} .$$

We compute these perturbations using the following algorithm:

Algorithm 6.8.10 (Gram-Schmidt Underdetermined Iterative Improvement)

$\delta\mathbf{c} = \mathbf{A}\tilde{\mathbf{s}} - \tilde{\mathbf{x}}$ /* LAPACK BLAS routine _gemv */
$\delta\mathbf{b} = \mathbf{b} - \mathbf{A}\tilde{\mathbf{x}}$ /* LAPACK BLAS routine _gemv */
solve $\mathbf{R}^H \delta\mathbf{y} = \delta\mathbf{b}$ /* LAPACK BLAS routine _trsv */
$\delta\mathbf{x} = \delta\mathbf{c}$
for $1 \le k \le m$
  $\delta\mathbf{w}_k = \mathbf{q}_k \cdot \delta\mathbf{x}$ /* LAPACK BLAS routine _dot or _dotc */
  $\delta\mathbf{x} = \delta\mathbf{x} - \mathbf{q}_k(\delta\mathbf{w}_k - \delta\mathbf{y}_k)$ /* LAPACK BLAS routine _axpy */
$\mathbf{x} = \tilde{\mathbf{x}} + \delta\mathbf{x}$
solve $\mathbf{R}\delta\mathbf{s} = \delta\mathbf{y} - \delta\mathbf{w}$ /* LAPACK BLAS routine _trsv */
$\mathbf{s} = \tilde{\mathbf{s}} + \delta\mathbf{s}$

Example 6.8.7 Suppose that

$$\mathbf{A} = \begin{bmatrix} 1 & 1 \\ \alpha & 0 \\ 0 & \alpha \end{bmatrix} \text{ and } \mathbf{b} = \begin{bmatrix} 1 \\ 1 \end{bmatrix} .$$

Also assume that $\mathrm{fl}(1 + \alpha^2) = 1$, and in floating point arithmetic we compute the approximate modified Gram-Schmidt factorization

$$\mathbf{A} = \mathbf{QR} = \begin{bmatrix} 1 & 0 \\ \alpha & -1/\sqrt{2} \\ 0 & 1/\sqrt{2} \end{bmatrix} \begin{bmatrix} 1 & 1 \\ & \alpha\sqrt{2} \end{bmatrix} .$$

Suppose that instead of using Björck's Algorithm 6.8.9, we computed

$$\tilde{\mathbf{x}} = \mathbf{Q}\mathbf{y} = \begin{bmatrix} 1 & 0 \\ \alpha & -1/\sqrt{2} \\ 0 & 1/\sqrt{2} \end{bmatrix} \begin{bmatrix} 1 \\ 0 \end{bmatrix} = \begin{bmatrix} 1 \\ \alpha \\ 0 \end{bmatrix} ,$$

where $\mathbf{y}$ is the solution of $\mathbf{R}^H\mathbf{y} = \mathbf{b}$. In order that $\tilde{\mathbf{x}} = \mathbf{A}\tilde{\mathbf{s}}$, we must have

$$\tilde{\mathbf{s}} = \begin{bmatrix} 1 \\ 0 \end{bmatrix} .$$

To apply iterative improvement, we compute

$$\delta\mathbf{c} = \mathbf{A}\tilde{\mathbf{s}} - \tilde{\mathbf{x}} = \begin{bmatrix} 0 \\ 0 \\ 0 \end{bmatrix}$$

$$\delta\mathbf{b} = \mathbf{b} - \mathbf{A}^H\tilde{\mathbf{x}} = \begin{bmatrix} -\alpha^2 \\ 0 \end{bmatrix}$$

$$\begin{bmatrix} 1 & \\ 1 & \alpha\sqrt{2} \end{bmatrix} \begin{bmatrix} \delta\mathbf{y}_1 \\ \delta\mathbf{y}_2 \end{bmatrix} = \tilde{\mathbf{R}}^H\delta\mathbf{y} = \delta\mathbf{b} = \begin{bmatrix} -\alpha^2 \\ 0 \end{bmatrix} \Longrightarrow \delta\mathbf{y} = \begin{bmatrix} -\alpha^2 \\ \alpha\sqrt{2} \end{bmatrix}$$

$$\delta\mathbf{x}^{(0)} = \delta\mathbf{c} = \begin{bmatrix} 0 \\ 0 \\ 0 \end{bmatrix}$$

$$\delta\mathbf{w}_1 = \mathbf{q}_1 \cdot \delta\mathbf{x}^{(0)} = \begin{bmatrix} 1 \\ \alpha \\ 0 \end{bmatrix} \cdot \begin{bmatrix} 0 \\ 0 \\ 0 \end{bmatrix} = 0$$

$$\delta\mathbf{x}^{(1)} = \delta\mathbf{x}^{(0)} - \mathbf{q}_1(\delta\mathbf{w}_1 - \delta\mathbf{y}_1) = \begin{bmatrix} 0 \\ 0 \\ 0 \end{bmatrix} - \begin{bmatrix} 1 \\ \alpha \\ 0 \end{bmatrix}(0 + \alpha^2) = \begin{bmatrix} -\alpha^2 \\ -\alpha^3 \\ 0 \end{bmatrix}$$

$$\delta\mathbf{w}_2 = \mathbf{q}_2 \cdot \delta\mathbf{x}^{(1)} = \begin{bmatrix} 0 \\ -1/\sqrt{2} \\ 1/\sqrt{2} \end{bmatrix} \cdot \begin{bmatrix} \alpha^2 \\ \alpha^3 \\ 0 \end{bmatrix} = \alpha^3/\sqrt{2}$$

$$\delta\mathbf{x} = \delta\mathbf{x}^{(1)} - \mathbf{q}_2(\delta\mathbf{w}_2 - \delta\mathbf{y}_2) = \begin{bmatrix} -\alpha^2 \\ -\alpha^3 \\ 0 \end{bmatrix} - \begin{bmatrix} 0 \\ -1/\sqrt{2} \\ 1/\sqrt{2} \end{bmatrix}(\alpha^3/\sqrt{2} - \alpha\sqrt{2}) = \begin{bmatrix} -\alpha^2 \\ -\alpha/2 \\ \alpha/2 \end{bmatrix}$$

$$\mathbf{x} = \tilde{\mathbf{x}} + \delta\mathbf{x} = \begin{bmatrix} 1 \\ \alpha \\ 0 \end{bmatrix} + \begin{bmatrix} -\alpha^2 \\ -\alpha/2 \\ \alpha/2 \end{bmatrix} = \begin{bmatrix} 1 \\ \alpha/2 \\ \alpha/2 \end{bmatrix}$$

Note that we had to compute $\delta\mathbf{b}$ in extended precision to get a nonzero value. The final value of $\mathbf{x}$ is the correctly rounded value of the minimum norm solution to the underdetermined least squares problem.

6.9 Givens QR Factorization

Previously in this chapter, we have described Householder factorization via successive reflection in Sect. 6.7, and Gram-Schmidt factorization via successive orthogonal projection in Sect. 6.8. The former factorization is favored by software packages such as LAPACK and MATLAB, while the latter is commonly described in introductory linear algebra courses and is used in iterative methods such as GMRES (see Sect. 2.6.2 of Chap. 2 in Volume II). In this section, we will describe yet another matrix factorization, which is based on **successive rotation**. This method is seldom used to factor a full matrix, but it is very useful to factor matrices that are nearly trapezoidal. For example, upper Hessenberg matrices (i.e., matrices with one sub-diagonal) are easily transformed to upper trapezoidal form by successive rotation. For an example of the use of successive rotation to factor such a matrix, see Sect. 1.3.7 of Chap. 1 in Volume II.

The reader should already be familiar with the following idea, which was previously developed in Sect. 3.4.8.2.

Definition 6.9.1 Let γ and σ be scalars such that $|\gamma|^2 + |\sigma|^2 = 1$. Then

$$\mathbf{G} = \begin{bmatrix} \gamma & \sigma \\ -\overline{\sigma} & \overline{\gamma} \end{bmatrix}$$

is an **elementary rotation**.

Recall that in Lemma 3.4.5 we showed that any elementary rotation satisfies

$$\mathbf{G}^H\mathbf{G} = \mathbf{I} .$$

It follows that for any 2-vector $\mathbf{x}$, we have

$$\|\mathbf{G}\mathbf{x}\|_2^2 = (\mathbf{G}\mathbf{x})^H(\mathbf{G}\mathbf{x}) = \mathbf{x}^H\mathbf{G}^H\mathbf{G}\mathbf{x} = \mathbf{x}^H\mathbf{x} = \|\mathbf{x}\|_2^2 .$$

Thus elementary rotations preserve lengths of vectors. Note, however, that an elementary rotation is not Hermitian, unless $\sigma = 0$ and γ is real.

Given two scalars ξ_1 and ξ_2, we can generate a elementary rotation $\mathbf{G} = \begin{bmatrix} \gamma & \sigma \\ -\overline{\sigma} & \gamma \end{bmatrix}$ with real γ by the following algorithm:

Algorithm 6.9.1 (Generate Plane Rotation)

$\lambda_2 = |\xi_2|$

$\gamma = 0$

$\sigma = 1$

if $\lambda_2 \leq 0$ then $\quad \varrho = \xi_1$

else

  $\lambda_1 = |\xi_1|$

  if $\lambda_1 \leq 0$ then $\varrho = \xi_2$

  else

    $\tau = \lambda_1 + \lambda_2$

    $\nu = \tau\sqrt{(\lambda_1/\tau)^2 + (\lambda_2/\tau)^2}$

    $\gamma = \lambda_1/\nu$

    $\alpha = \xi_1/\lambda_1$

    $\sigma = \alpha\overline{\xi_2}/\nu$

    $\varrho = \alpha\nu$

Division by the scalar τ serves to reduce the chance of overflow or underflow in computing ν.

The basic idea behind using elementary rotations to solve least squares problems is contained in the following lemma.

Lemma 6.9.1 *Given two scalars ξ_1 and ξ_2, define the real number γ as well as the scalars σ and ϱ by Algorithm 6.9.1. Then the elementary rotation*

$$\mathbf{G} = \begin{bmatrix} \gamma & \sigma \\ -\overline{\sigma} & \gamma \end{bmatrix}$$

is such that

$$\mathbf{G}\begin{bmatrix} \xi_1 \\ \xi_2 \end{bmatrix} = \begin{bmatrix} \varrho \\ 0 \end{bmatrix} .$$

Proof If $\xi_2 = 0$, then $\varrho = \xi_1$ and

$$\mathbf{G}\begin{bmatrix} \xi_1 \\ \xi_2 \end{bmatrix} = \begin{bmatrix} 1 & 0 \\ 0 & 1 \end{bmatrix}\begin{bmatrix} \xi_1 \\ 0 \end{bmatrix} = \begin{bmatrix} \xi_1 \\ 0 \end{bmatrix} = \begin{bmatrix} \varrho \\ 0 \end{bmatrix} .$$

On the other hand, if $\xi_2 \neq 0$ and $\xi_1 = 0$, then $\varrho = \xi_2$ and

$$\mathbf{G}\begin{bmatrix} \xi_1 \\ \xi_2 \end{bmatrix} = \begin{bmatrix} 0 & 1 \\ -1 & 0 \end{bmatrix}\begin{bmatrix} 0 \\ \xi_2 \end{bmatrix} = \begin{bmatrix} \xi_2 \\ 0 \end{bmatrix} = \begin{bmatrix} \varrho \\ 0 \end{bmatrix} .$$

Otherwise, we have $\varrho = (\xi_1/|\xi_1|)\sqrt{|\xi_1|^2 + |\xi_2|^2}$ and

$$\mathbf{G}\begin{bmatrix} \xi_1 \\ \xi_2 \end{bmatrix} = \begin{bmatrix} \frac{|\xi_1|}{\nu} & \frac{\xi_1}{|\xi_1|}\frac{\overline{\xi_2}}{\nu} \\ -\frac{\overline{\xi_1}}{|\xi_1|}\frac{\xi_2}{\nu} & \frac{|\xi_1|}{\nu} \end{bmatrix}\begin{bmatrix} \xi_1 \\ \xi_2 \end{bmatrix} = \begin{bmatrix} \frac{|\xi_1|}{\nu}\xi_1 + \frac{\xi_1}{|\xi_1|}\frac{|\xi_2|}{\nu} \\ -\frac{|\xi_1|^2}{|\xi_1|}\frac{\xi_2}{\nu} + \frac{|\xi_1|}{\nu}\xi_2 \end{bmatrix} = \begin{bmatrix} \xi_1\nu/|\xi_1| \\ 0 \end{bmatrix} = \begin{bmatrix} \varrho \\ 0 \end{bmatrix} .$$

An elementary rotation chosen to rotate from some given vector to an axis vector is called a **Givens rotation**. An algorithm similar to Algorithm 6.9.1 that generates Givens rotations has been available in LAPACK BLAS routines `_rotg`; see, for example, drotg.f or zrotg.f. These routines are now deprecated in favor of the more cautious LAPACK routines `_lartg`; see, for example, dlartg.f or zlartg.f. It should be noted that Algorithm 6.9.1 costs at most 5 multiplications and divisions, 2 moduli, 2 additions and 1 square root.

After we have generated an elementary rotation

$$\mathbf{G} = \begin{bmatrix} \gamma & \sigma \\ -\overline{\sigma} & \gamma \end{bmatrix}$$

we can replace a vector

$$\mathbf{a} = \begin{bmatrix} \alpha_1 \\ \alpha_2 \end{bmatrix}$$

by its rotation **Ga** via the following algorithm:

Algorithm 6.9.2 (Apply Plane Rotation)

$$\tau = \gamma\alpha_1 + \sigma\alpha_2$$
$$\alpha_2 = -\overline{\sigma}\alpha_1 + \gamma\alpha_2$$
$$\alpha_1 = \tau \; .$$

An algorithm similar to Algorithm 6.9.2 is available in LAPACK BLAS routines `_rot`; see, for example, drot.f or zdrot.f. This algorithm costs 4 multiplications and 2 additions. At this point, the LAPACK BLAS routines `_rot` are deprecated in favor of the LAPACK routines `_lartv`; see, for example, dlartv.f.

Next, let us show how Lemma 6.9.1 can be used to factor a $m \times n$ matrix $\mathbf{A}$ into a product $\mathbf{Q}\begin{bmatrix}\mathbf{R}\\ \mathbf{0}\end{bmatrix}$ of a unitary matrix $\mathbf{Q}$ and a right-trapezoidal matrix $\mathbf{R}$. The algorithm proceeds forwards in an outer loop from the first column to the last, and backwards in an inner loop from the last two entries in that column to the last two on or below the diagonal. For each of these pairs within a column, we construct a Givens rotation to zero the second entry, and apply that Givens rotation to the corresponding rows and all later columns. We can summarize the operations in this factorization by the following algorithm:

Algorithm 6.9.3 (Givens QR Factorization)

for $1 \leq k \leq \min\{m, n\}$
  for $m-1 \geq i \geq k$
    compute γ_{ik}, σ_{ik} and ϱ_{ik} from $\mathbf{A}_{ik}$ and $\mathbf{A}_{i+1,k}$ /* Algorithm 6.9.1; LAPACK _lartg */
    $\mathbf{A}_{ik} = \varrho_{ik}$
    for $k < j \leq \min\{i, n\}$
      $\begin{bmatrix}\mathbf{A}_{ij}\\ \mathbf{A}_{i+1,j}\end{bmatrix} = \begin{bmatrix}\gamma_{ik} & \sigma_{ik}\\ -\overline{\sigma_{ik}} & \gamma_{ik}\end{bmatrix}\begin{bmatrix}\mathbf{A}_{ij}\\ \mathbf{A}_{i+1,j}\end{bmatrix}$ /* Algorithm 6.9.2; LAPACK _lartv */

If $m \geq n$, then this algorithm costs

- $\sum_{k=1}^{n}\sum_{i=k}^{m-1} = n(2m-n-1)/2$ calls to Algorithm 6.9.1 and
- $\sum_{k=1}^{n}\sum_{i=k}^{m-1}\sum_{j=k+1}^{\min\{i,n\}} = (n-1)n(6m-n-4)/12$ calls to Algorithm 6.9.2.

The total work is on the order of $2mn^2$ multiplications and mn^2 additions, essentially all due to the work of applying the rotations to later columns. This is on the order of twice the number of arithmetic operations that are used in the Householder QR factorization. In order to store the elementary rotations, we need to save on the order of mn real scalars for the various values of γ, and mn scalars for the values of σ; if $\mathbf{A}$ is real, then this is essentially twice the storage required for the original matrix $\mathbf{A}$.

The extra work and memory required by the Givens QR factorization explain why it is not used in practice. However, in Chap. 1 of Volume II we will see that Givens rotations are very useful for certain special computations.

The following lemma describes the rounding errors that occur in computing an elementary rotation.

Lemma 6.9.2 *Assume that the relative errors in floating point addition, multiplication, division or square roots satisfy*

$$fl(\alpha + \beta) = (\alpha + \beta)(1 + \varepsilon_+) ,$$
$$fl(\alpha \times \beta) = (\alpha \times \beta)(1 + \varepsilon_\times) ,$$
$$fl(\alpha \div \beta) = (\alpha \div \beta)(1 + \varepsilon_\div) ,$$
$$fl(\sqrt{\alpha}) = \sqrt{\alpha}(1 + \varepsilon_{\sqrt{}}) \text{ and}$$
$$fl(|\alpha|) = |\alpha|(1 + \varepsilon_{|\cdot|}) ,$$

where there is a constant $\varepsilon > 1$ *so that*

$$\max\left\{|\varepsilon_+| , |\varepsilon_\times| , |\varepsilon_\div| , |\varepsilon_{\sqrt{}}| , ; |\varepsilon_{|\cdot|}|\right\} \le \varepsilon .$$

Let ξ_1 *and* ξ_2 *be two nonzero scalars, and suppose that scalars* γ, σ *and* ϱ *are evaluated as in Algorithm 6.9.1. If* $17\varepsilon < 1$, *then the errors in the floating point evaluation of these quantities satisfy*

$$|fl(\gamma) - \gamma| \le |\gamma| \frac{10\varepsilon}{1 - 15\varepsilon} . \tag{6.86a}$$

$$|fl(\sigma) - \sigma| \le |\sigma| \frac{10\varepsilon}{1 - 17\varepsilon} . \tag{6.86b}$$

$$|fl(\varrho) - \varrho| \le |\varrho| \frac{11\varepsilon}{1 - 10\varepsilon} . \tag{6.86c}$$

Proof For $i = 1, 2$, floating point computation of $|\lambda_i|$ gives

$$\tilde{\lambda}_i = \text{fl}(\lambda_i) = \lambda_i(1 + \varepsilon_{i|\cdot|}) \Longrightarrow \left|\tilde{\lambda}_i - \lambda_i\right| = \left|\lambda_i \varepsilon_{i|\cdot|}\right| \le \lambda_i \varepsilon .$$

Next, floating point computation of $\tau = \lambda_1 + \lambda_2$ gives us

$$\tilde{\tau} = \text{fl}\left(\tilde{\lambda}_1 + \tilde{\lambda}_2\right) = \left(\tilde{\lambda}_1 + \tilde{\lambda}_2\right)(1 + \varepsilon_+)$$

so

$$|\tilde{\tau} - \tau| \le \left|\left(\tilde{\lambda}_1 + \tilde{\lambda}_2\right)(1 + \varepsilon_+) - \left(\tilde{\lambda}_1 + \tilde{\lambda}_2\right)\right| + \left|\left(\tilde{\lambda}_1 + \tilde{\lambda}_2\right) - (\lambda_1 + \lambda_2)\right|$$
$$\le \left|\tilde{\lambda}_1 + \tilde{\lambda}_2\right| \varepsilon + \left|\tilde{\lambda}_1 - \lambda_1\right| + \left|\tilde{\lambda}_2 - \lambda_2\right|$$
$$\le \left(\tau + \left|\tilde{\lambda}_1 - \lambda_1\right| + \left|\tilde{\lambda}_2 - \lambda_2\right|\right) \varepsilon + \left|\tilde{\lambda}_1 - \lambda_1\right| + \left|\tilde{\lambda}_2 - \lambda_2\right|$$
$$= \tau\varepsilon + \left(\left|\tilde{\lambda}_1 - \lambda_1\right| + \left|\tilde{\lambda}_2 - \lambda_2\right|\right)(1 + \varepsilon) \le \tau\varepsilon + (\lambda_1\varepsilon + \lambda_2\varepsilon)(1 + \varepsilon) = \tau\varepsilon(2 + \varepsilon)$$

then inequality (3.38) gives us

$$\le \tau \frac{2\varepsilon}{1 - \varepsilon/2} .$$

Before proceeding further, for $i = 1, 2$ it will be helpful to estimate

$$\left|\frac{\tilde{\lambda}_i}{\tilde{\tau}} - \frac{\lambda_i}{\tau}\right| = \left|\frac{\lambda_i}{\tau\tilde{\tau}}\left(\tau(1 + \varepsilon_{i|\cdot|}) - \tilde{\tau}\right)\right| \le \frac{\lambda_i}{\tau}\frac{\tau\varepsilon + |\tilde{\tau} - \tau|}{\tau - |\tilde{\tau} - \tau|}$$
$$\le \frac{\lambda_i}{\tau}\frac{\varepsilon + 2\varepsilon/(1 - \varepsilon/2)}{1 - 2\varepsilon/(1 - \varepsilon/2)} = \frac{\lambda_i}{\tau}\varepsilon\left(1 + \frac{4\varepsilon}{1 - 5\varepsilon/2}\right)$$

and then use (3.38) to obtain

$$\le \frac{\lambda_i}{\tau}\frac{\varepsilon}{1 - 4\varepsilon} .$$

Next, we define

$$\mathbf{v} = \begin{bmatrix}\lambda_1\\ \lambda_2\end{bmatrix}\frac{1}{\tau} \text{ and } \tilde{\mathbf{v}} = \text{ fl}\left(\begin{bmatrix}\tilde{\lambda}_1\\ \tilde{\lambda}_2\end{bmatrix}\frac{1}{\tilde{\tau}}\right) = (\mathbf{I} + \mathbf{E}_\div)\begin{bmatrix}\tilde{\lambda}_1\\ \tilde{\lambda}_2\end{bmatrix}\frac{1}{\tilde{\tau}}$$

where $\mathbf{E}_\div$ is a diagonal matrix of rounding errors. Then

$$\|\tilde{\mathbf{v}} - \mathbf{v}\|_2 = \left\|(\mathbf{I} + \mathbf{E}_\div)\begin{bmatrix}\tilde{\lambda}_1\\ \tilde{\lambda}_2\end{bmatrix}\frac{1}{\tilde{\tau}} - \begin{bmatrix}\lambda_1\\ \lambda_2\end{bmatrix}\frac{1}{\tau}\right\|_2 \le \left\|(\mathbf{I} + \mathbf{E}_\div)\begin{bmatrix}\tilde{\lambda}_1/\tilde{\tau} - \lambda_1/\tau\\ \tilde{\lambda}_2/\tilde{\tau} - \lambda_2/\tau\end{bmatrix}\right\|_2 + \|\mathbf{E}_\div\mathbf{v}\|_2$$
$$\le (1 + \varepsilon)\frac{\varepsilon}{1 - 4\varepsilon}\|\mathbf{v}\|_2 + \varepsilon\|\mathbf{v}\|_2$$

then inequalities (3.38) and (3.37) produce

$$\le \frac{2\varepsilon}{1 - 5\varepsilon}\|\mathbf{v}\|_2 .$$

Since

$$\nu = \tau\|\mathbf{v}\|_2 \text{ and } \tilde{\nu} = \tilde{\tau}\,\mathrm{fl}\,(\|\tilde{\mathbf{v}}\|_2)\,(1+\varepsilon_\times)\;,$$

repeated use of the triangle inequality produces

$$\begin{aligned}
|\tilde{\nu}-\nu| &\leq |\tilde{\tau}\,\mathrm{fl}\,(\|\tilde{\mathbf{v}}\|_2)\,(1+\varepsilon_\times) - \tilde{\tau}\,\mathrm{fl}\,(\|\tilde{\mathbf{v}}\|_2)| + |\tilde{\tau}\,\mathrm{fl}\,(\|\tilde{\mathbf{v}}\|_2) - \tilde{\tau}\,\|\tilde{\mathbf{v}}\|_2| \\
&+ |\tilde{\tau}\,\|\tilde{\mathbf{v}}\|_2 - \tau\,\|\tilde{\mathbf{v}}\|_2| + |\tau\,\|\tilde{\mathbf{v}}\|_2 - \tau\,\|\mathbf{v}\|_2| \\
&\leq |\tilde{\tau}\,\mathrm{fl}\,(\|\tilde{\mathbf{v}}\|_2)|\,\varepsilon + |\tilde{\tau}|\,|\,\mathrm{fl}\,(\|\tilde{\mathbf{v}}\|_2) - \|\tilde{\mathbf{v}}\|_2| + |\tilde{\tau}-\tau|\,\|\tilde{\mathbf{v}}\|_2 + \tau\,|\|\tilde{\mathbf{v}}\|_2 - \|\mathbf{v}\|_2| \\
&\leq (\tau + |\tilde{\tau}-\tau|)\,(\|\tilde{\mathbf{v}}\|_2 + |\,\mathrm{fl}\,(\|\tilde{\mathbf{v}}\|_2) - \|\tilde{\mathbf{v}}\|_2|)\,\varepsilon + (\tau + |\tilde{\tau}-\tau|)\,|\,\mathrm{fl}\,(\|\tilde{\mathbf{v}}\|_2) - \|\tilde{\mathbf{v}}\|_2| \\
&+ |\tilde{\tau}-\tau|\,\|\tilde{\mathbf{v}}\|_2 + \tau\,\|\tilde{\mathbf{v}}-\mathbf{v}\|_2 \\
&\leq \tau\left(1+\frac{2\varepsilon}{1-\varepsilon/2}\right)\|\tilde{\mathbf{v}}\|_2\left(1+\frac{3\varepsilon}{1-3\varepsilon/2}\right)\varepsilon + \tau\left(1+\frac{2\varepsilon}{1-\varepsilon/2}\right)\|\tilde{\mathbf{v}}\|_2\,\frac{3\varepsilon}{1-3\varepsilon/2} \\
&+ \tau\frac{2\varepsilon}{1-\varepsilon/2}\,\|\tilde{\mathbf{v}}\|_2 + \tau\,\|\tilde{\mathbf{v}}-\mathbf{v}\|_2 \\
&\leq \tau\,(\|\mathbf{v}\|_2 + \|\tilde{\mathbf{v}}-\mathbf{v}\|_2)\left\{\left(1+\frac{2\varepsilon}{1-\varepsilon/2}\right)\left[\left(1+\frac{3\varepsilon}{1-3\varepsilon/2}\right)\varepsilon + \frac{3\varepsilon}{1-3\varepsilon/2}\right] + \frac{2\varepsilon}{1-\varepsilon/2}\right\} \\
&+ \tau\,\|\tilde{\mathbf{v}}-\mathbf{v}\|_2 \\
&\leq \nu\left(1+\frac{2\varepsilon}{1-5\varepsilon}\right)\left\{\left(1+\frac{2\varepsilon}{1-\varepsilon/2}\right)\left[\left(1+\frac{3\varepsilon}{1-3\varepsilon/2}\right)\varepsilon + \frac{3\varepsilon}{1-3\varepsilon/2}\right] + \frac{2\varepsilon}{1-\varepsilon/2}\right\} \\
&+ \nu\frac{2\varepsilon}{1-5\varepsilon}
\end{aligned}$$

then inequalities (3.38) and (3.37) yield

$$\leq \nu\frac{8\varepsilon}{1-7\varepsilon}\;.$$

Next, we have

$$\gamma = \frac{\lambda_1}{\nu} \text{ and } \tilde{\gamma} = \mathrm{fl}\left(\frac{\tilde{\lambda}_1}{\tilde{\nu}}\right) = \frac{\tilde{\lambda}_1}{\tilde{\nu}}(1+\varepsilon_\div)\;,$$

so the triangle inequality gives us

$$\begin{aligned}
|\tilde{\gamma}-\gamma| &\leq \left|\frac{\tilde{\lambda}_1}{\tilde{\nu}}\varepsilon_\div\right| + \left|\frac{\tilde{\lambda}_1}{\tilde{\nu}} - \frac{\tilde{\lambda}_1}{\nu}\right| + \left|\frac{\tilde{\lambda}_1}{\nu} - \frac{\lambda_1}{\nu}\right| \leq \left|\frac{\tilde{\lambda}_1}{\tilde{\nu}}\right|\varepsilon + \left|\frac{\tilde{\lambda}_1}{\tilde{\nu}\nu}(\nu-\tilde{\nu})\right| + \frac{\left|\tilde{\lambda}_1-\lambda_1\right|}{\nu} \\
&\leq \frac{\lambda_1 + \left|\tilde{\lambda}_1-\lambda_1\right|}{\nu - |\tilde{\nu}-\nu|}\varepsilon + \frac{\lambda_1 + \left|\tilde{\lambda}_1-\lambda_1\right|}{\nu\,(\nu - |\tilde{\nu}-\nu|)}|\tilde{\nu}-\nu| + \frac{\left|\tilde{\lambda}_1-\lambda_1\right|}{\nu}
\end{aligned}$$

$$\leq \frac{\lambda_1(1+\varepsilon)}{\nu\left(1-\frac{8\varepsilon}{1-7\varepsilon}\right)}\varepsilon + \frac{\lambda_1(1+\varepsilon)}{\nu^2\left(1-\frac{8\varepsilon}{1-7\varepsilon}\right)}\nu\frac{8\varepsilon}{1-7\varepsilon} + \frac{\lambda_1\varepsilon}{\nu}$$

$$= |\gamma|\left\{\frac{(1-7\varepsilon)(1+\varepsilon)}{1-15\varepsilon}\varepsilon + \frac{1+\varepsilon}{1-15\varepsilon}8\varepsilon + \varepsilon\right\} \leq |\gamma|\left\{\frac{1-6\varepsilon}{1-15\varepsilon}\varepsilon + \frac{1+\varepsilon}{1-15\varepsilon}8\varepsilon + \varepsilon\right\}$$

then inequalities (3.38) and (3.37) lead to

$$\leq |\gamma|\left\{\frac{\varepsilon}{1-15\varepsilon} + \frac{8\varepsilon}{1-15\varepsilon} + \varepsilon\right\} \leq |\gamma|\frac{10\varepsilon}{1-15\varepsilon} .$$

Note that

$$\alpha = \frac{\xi_1}{\lambda_1} \text{ and } \tilde{\alpha} = \mathrm{fl}\left(\frac{\xi_1}{\tilde{\lambda}_1}\right) = \frac{\xi_1}{\tilde{\lambda}_1}(1+\varepsilon_\div) ,$$

which imply that

$$|\tilde{\alpha}-\alpha| \leq \left|\frac{\xi_1}{\tilde{\lambda}_1}\varepsilon_\div\right| + \left|\frac{\xi_1}{\tilde{\lambda}_1} - \frac{\xi_1}{\lambda_1}\right| \leq \frac{|\xi_1|}{\lambda_1 - \left|\tilde{\lambda}_1-\lambda_1\right|}\varepsilon + \left|\frac{\xi_1}{\lambda_1\tilde{\lambda}_1}\right|\left|\lambda_1-\tilde{\lambda}_1\right|$$

$$\leq \frac{1}{1-\varepsilon}\varepsilon + \frac{1}{1-\varepsilon}\varepsilon = \frac{2\varepsilon}{1-\varepsilon} .$$

We also have

$$\sigma = \frac{\alpha\overline{\xi_2}}{\nu} \text{ and } \tilde{\sigma} = \mathrm{fl}\left(\frac{\tilde{\alpha}\overline{\xi_2}}{\tilde{\nu}}\right) ,= \frac{\tilde{\alpha}\overline{\xi_2}}{\tilde{\nu}}(1+\varepsilon_\times)(1+\varepsilon_\div) ,$$

which imply that

$$|\tilde{\sigma}-\sigma| = \left|\frac{\tilde{\alpha}\overline{\xi_2}}{\tilde{\nu}}(1+\varepsilon_\times)(1+\varepsilon_\div) - \frac{\alpha\overline{\xi_2}}{\nu}\right|$$

$$\leq \left|\frac{\tilde{\alpha}\overline{\xi_2}}{\tilde{\nu}}\right||(1+\varepsilon_\times)(1+\varepsilon_\div)-1| + \left|\frac{\tilde{\alpha}\overline{\xi_2}}{\tilde{\nu}} - \frac{\tilde{\alpha}\overline{\xi_2}}{\nu}\right| + \left|\frac{\tilde{\alpha}\overline{\xi_2}}{\nu} - \frac{\alpha\overline{\xi_2}}{\nu}\right|$$

$$\leq \frac{(|\alpha|+|\tilde{\alpha}-\alpha|)\left|\overline{\xi_2}\right|}{\nu-|\tilde{\nu}-\nu|}\frac{2\varepsilon}{1-\varepsilon/2} + \frac{(|\alpha|+|\tilde{\alpha}-\alpha|)\left|\overline{\xi_2}\right|}{\nu\left(\nu-|\tilde{\nu}-\nu|\right)}|\tilde{\nu}-\nu| + \frac{\left|\overline{\xi_2}\right|}{\nu}|\tilde{\alpha}-\alpha|$$

$$\leq \frac{\left(1+\frac{2\varepsilon}{1-\varepsilon}\right)\left|\overline{\xi_2}\right|}{\nu\left(1-\frac{8\varepsilon}{1-7\varepsilon}\right)}\frac{2\varepsilon}{1-\varepsilon/2} + \frac{\left(1+\frac{2\varepsilon}{1-\varepsilon}\right)\left|\overline{\xi_2}\right|}{\nu^2\left(1-\frac{8\varepsilon}{1-7\varepsilon}\right)}\nu\frac{8\varepsilon}{1-7\varepsilon} + \frac{\left|\overline{\xi_2}\right|}{\nu}\frac{2\varepsilon}{1-\varepsilon}$$

$$= |\sigma|\left\{\left(1+\frac{2\varepsilon}{1-\varepsilon}\right)\left(1+\frac{8\varepsilon}{1-15\varepsilon}\right)\frac{2\varepsilon}{1-\varepsilon} + \left(1+\frac{2\varepsilon}{1-\varepsilon}\right)\frac{8\varepsilon}{1-15\varepsilon} + \frac{2\varepsilon}{1-\varepsilon}\right\}$$

then inequalities (3.38) and (3.37) give us

$$\leq |\sigma| \frac{10\varepsilon}{1-17\varepsilon} \cdot$$

Finally, we have

$$\varrho = \alpha\nu \text{ and } \tilde{\varrho} = \text{ fl}\,(\tilde{\alpha}\tilde{\varrho}) \ , = \tilde{\alpha}\tilde{\varrho}(1+\varepsilon_\times) \ ,$$

so

$$|\tilde{\varrho} - \varrho| \leq |\tilde{\alpha}\tilde{\nu}\varepsilon_\times| + |\tilde{\alpha}| \ |\tilde{\nu} - \nu| + |\tilde{\alpha} - \alpha| \ |\nu|$$
$$\leq (|\alpha| + |\tilde{\alpha} - \alpha|)\,(\nu + |\tilde{\nu} - \nu|)\,\varepsilon + (|\alpha| + |\tilde{\alpha} - \alpha|)\,|\tilde{\nu} - \nu| + \nu \ |\tilde{\alpha} - \alpha|$$
$$\leq \left(1 + \frac{2\varepsilon}{1-\varepsilon}\right)\nu\left(1 + \frac{8\varepsilon}{1-7\varepsilon}\right)\varepsilon + \left(1 + \frac{2\varepsilon}{1-\varepsilon}\right)\nu\frac{8\varepsilon}{1-7\varepsilon} + \nu\frac{2\varepsilon}{1-\varepsilon}$$

then inequalities (3.38) and (3.37) produce

$$\leq |\varrho| \frac{11\varepsilon}{1-10\varepsilon} \cdot$$

Lemma 6.9.3 *Assume that the relative errors in floating point addition and multiplication satisfy*

$$fl(a+b) = (a+b)(1+\varepsilon_+) \text{ and}$$
$$fl(a \times b) = (a \times b)(1+\varepsilon_\times) \ ,$$

Suppose that

$$\tilde{\mathbf{G}} = \begin{bmatrix} \tilde{\gamma} & \tilde{\sigma} \\ -\overline{\tilde{\sigma}} & \tilde{\gamma} \end{bmatrix}$$

is an approximation to the Givens rotation

$$\mathbf{G} = \begin{bmatrix} \gamma & \sigma \\ -\overline{\sigma} & \gamma \end{bmatrix} \ ,$$

such that

$$\tilde{\gamma} = \gamma(1+\delta_\gamma) \ , \ \tilde{\sigma} = \sigma(1+\delta_\sigma) \text{ and} \max\{|\delta_\gamma| \ , \ |\delta_\sigma|\} \leq \frac{10\varepsilon}{1-17\varepsilon} \cdot$$

Given a 2-vector $\mathbf{x}$*, suppose that the matrix-vector product* $\tilde{\mathbf{G}}\mathbf{x}$ *is computed in floating point arithmetic. If* $19\varepsilon < 1$*, then*

$$\left\| fl(\tilde{\mathbf{G}}\mathbf{x}) - \mathbf{G}\mathbf{x} \right\|_2 \leq \frac{12\sqrt{2}\varepsilon}{1-19\varepsilon} \|\mathbf{x}\|_2 \,. \tag{6.87}$$

Proof We have

$$\begin{aligned}
&\left\| \mathrm{fl}\left(\tilde{\mathbf{G}}\mathbf{x}\right) - \mathbf{G}\mathbf{x} \right\|_2 \\
&= \left\| \begin{bmatrix} \{(1+\delta_\gamma)(1+\varepsilon_{\times 11})(1+\varepsilon_{+1})-1\}\gamma & \{(1+\delta_\sigma)(1+\varepsilon_{\times 12})(1+\varepsilon_{+1})-1\}\sigma \\ -\{(1+\overline{\delta_\sigma})(1+\varepsilon_{\times 21})(1+\varepsilon_{+2})-1\}\overline{\sigma} & \{(1+\delta_\gamma)(1+\varepsilon_{\times 22})(1+\varepsilon_{+2})-1\}\gamma \end{bmatrix} \mathbf{x} \right\|_2 \\
&\leq \left\| \begin{bmatrix} \{(1+\delta_\gamma)(1+\varepsilon_{\times 11})(1+\varepsilon_{+1})-1\}\gamma & \{(1+\delta_\sigma)(1+\varepsilon_{\times 12})(1+\varepsilon_{+1})-1\}\sigma \\ -\{(1+\overline{\delta_\sigma})(1+\varepsilon_{\times 21})(1+\varepsilon_{+2})-1\}\overline{\sigma} & \{(1+\delta_\gamma)(1+\varepsilon_{\times 22})(1+\varepsilon_{+2})-1\}\gamma \end{bmatrix} \right\|_F \|\mathbf{x}\|_2 \\
&\leq \max\left\{ \left|(1+\delta_\gamma)(1+\varepsilon_{\times 11})(1+\varepsilon_{+1})-1\right| , \right. \\
&\qquad \left. |(1+\delta_\sigma)(1+\varepsilon_{\times 12})(1+\varepsilon_{+1})-1| \,, \ldots \right\} \|\mathbf{G}\|_F \|\mathbf{x}\|_2 \\
&\leq \left\{ \left(1+\frac{10\varepsilon}{1-17\varepsilon}\right)\left(1+\frac{2\varepsilon}{1-\varepsilon/2}\right) - 1 \right\} \sqrt{2}\,\|\mathbf{x}\|_2
\end{aligned}$$

then inequality (3.40) gives us

$$\leq \frac{12\sqrt{2}\varepsilon}{1-19\varepsilon} \|\mathbf{x}\|_2 \,.$$

6.10 Case Study

We would like to investigate the numerical performance of successive reflection and successive orthogonal projection for the solution of large least squares problems. To this end, we have obtained a copy of the 1033×320 matrix matrix ILLC1033 from the **NIST Matrix Market**. This matrix $\mathbf{A}$ arose from a least squares problem in surveying. Its singular value decomposition estimates that $\|\mathbf{A}\|_2 = 2.14436$ and that $\|\mathbf{A}^\dagger\|_2 = 8809.13$, so its condition number is 18889.9.

In the C^{++} program LeastSquaresCaseStudy.C, we can use this matrix $\mathbf{A}$ to form a least squares problem The components of the exact solution $\mathbf{x}$ to this least squares problem are chosen to be double precision random numbers that are uniformly distributed in [0, 1]. The right-hand side for the least squares problem is chosen to be the matrix-vector product $\mathbf{b} = \mathbf{A}\mathbf{x}$, computed in double precision. Consequently, the least squares residual should be small, with value that should be bounded by the rounding errors described in Lemma 3.8.9.

We can use the C++ class SingularValueDecomposition to find the singular values of $\mathbf{A}$ in double precision. According to the results of Sect. 6.6, we can compute

$$\left\| \begin{bmatrix} \mathbf{I} & \mathbf{A} \\ \mathbf{A}^H & \mathbf{0} \end{bmatrix} \right\|_2 = 2.70188 \text{ and } \left\| \begin{bmatrix} \mathbf{I} & \mathbf{A} \\ \mathbf{A}^H & \mathbf{0} \end{bmatrix}^{-1} \right\|_2 = 7.75863 \times 10^7 \; .$$

These norms indicate that least squares problems involving $\mathbf{A}$ may have few accurate digits in single precision solutions.

Next, we use C++ class HouseholderQRFactorization to factor $\mathbf{A}$ and to solve the least-squares problem for the numerical solution $\tilde{\mathbf{x}}$. In double precision, this computation produced

$$\|\mathbf{b} - \mathbf{A}\tilde{\mathbf{x}}\|_2 = 4.39847 \times 10^{-15} \text{ and } \|\tilde{\mathbf{x}} - \mathbf{x}\|_2 / \|\mathbf{x}\|_2 = 2.66696 \times 10^{-13} \; .$$

In single precision, we obtained

$$\|\mathbf{b} - \mathbf{A}\tilde{\mathbf{x}}\|_2 = 2.06455 \times 10^{-6} \text{ and } \|\tilde{\mathbf{x}} - \mathbf{x}\|_2 / \|\mathbf{x}\|_2 = 1.05313 \times 10^{-4} \; .$$

These observed errors are much smaller than the rounding error analysis in Sect. 6.7.5 and the perturbation analysis in Sect. 6.5 would predict. This is because the *a priori* rounding error analysis estimates an upper bound on the rounding errors, not the most likely bound. The *a posteriori* error estimates give us

$$\frac{\left\| \begin{bmatrix} \tilde{\mathbf{r}} - \mathbf{r} \\ \tilde{\mathbf{x}} - \mathbf{x} \end{bmatrix} \right\|_2}{\left\| \begin{bmatrix} \mathbf{r} \\ \mathbf{x} \end{bmatrix} \right\|_2} \leq \kappa_2\left(\begin{bmatrix} \mathbf{I} & \mathbf{A} \\ \mathbf{A}^H & \mathbf{0} \end{bmatrix} \right) \frac{\left\| \begin{bmatrix} \mathbf{0} \\ \mathbf{A}^H \tilde{\mathbf{r}} \end{bmatrix} \right\|_2}{\|\mathbf{b}\|_2} = \begin{cases} 6.27349 \times 10^{-8}, & \text{double precision} \\ 32.9549, & \text{single precision} \end{cases} .$$

These *a posteriori* error estimates are much larger than the observed error in $\tilde{\mathbf{x}}$.

Afterward, we use C++ class GramSchmidtQRFactorization to factor $\mathbf{A}$ and to solve the least-squares problem for the numerical solution $\tilde{\mathbf{x}}$. In double precision, this computation produces

$$\|\tilde{\mathbf{r}}\|_2 = 2.23643 \times 10^{-15} \text{ and } \|\tilde{\mathbf{x}} - \mathbf{x}\|_2 / \|\mathbf{x}\|_2 = 3.18125 \times 10^{-14} \; .$$

In single precision, we obtain

$$\|\tilde{\mathbf{r}}\|_2 = 1.21178 \times 10^{-6} \text{ and } \|\tilde{\mathbf{x}} - \mathbf{x}\|_2 / \|\mathbf{x}\|_2 = 1.67663 \times 10^{-5} \; .$$

The *a posteriori* error estimates give us

$$\frac{\left\|\begin{bmatrix}\tilde{\mathbf{r}}-\mathbf{r}\\ \tilde{\mathbf{x}}-\mathbf{x}\end{bmatrix}\right\|_2}{\left\|\begin{bmatrix}\mathbf{r}\\ \mathbf{x}\end{bmatrix}\right\|_2} \leq \kappa_2\left(\begin{bmatrix}\mathbf{I} & \mathbf{A}\\ \mathbf{A}^H & \mathbf{0}\end{bmatrix}\right)\frac{\left\|\begin{bmatrix}\tilde{\mathbf{r}}-(\mathbf{b}-\mathbf{A}\tilde{\mathbf{x}})\\ \mathbf{A}^H\tilde{\mathbf{r}}\end{bmatrix}\right\|_2}{\|\mathbf{b}\|_2}$$

$$= \begin{cases} 4.75807\times 10^{-8}, & \text{double precision} \\ 29.0095, & \text{single precision} \end{cases}.$$

Thus for this problem, the modified Gram-Schmidt process is slightly more accurate than the Householder process. Both require essentially the same computational time.

6.11 Singular Value Decomposition

In Sects. 6.7 and 6.8, we developed reliable methods to solve least squares problems when the matrix has either full column rank (rank $(\mathbf{A}) = n$) or full row rank (rank $(\mathbf{A}) = m$). In Sect. 6.7.4.3, we saw how to use a prescribed upper bound on the condition number to select an appropriate matrix rank and solve the corresponding rank-deficient least squares problem.

In this section, we will develop an even more reliable method for determining the rank of a matrix. This approach involves measuring the distance from some given matrix to the set of matrices of some given rank. Theorem 1.5.3 of Chap. 1 in Volume II will show that this distance can be computed by using the **singular value decomposition**.

The idea behind the singular value decomposition is contained in Theorem 1.5.1 of Chap. 1 in Volume II. This theorem will prove that for every nonzero $m \times n$ matrix $\mathbf{A}$ there is an $m \times m$ unitary matrix $\mathbf{U}$, an $n \times n$ unitary matrix $\mathbf{V}$, an integer $r > 0$ and an $r \times r$ positive diagonal matrix $\boldsymbol{\Sigma}$ such that

$$\mathbf{A} = \mathbf{U}\begin{bmatrix}\boldsymbol{\Sigma} & \mathbf{0}\\ \mathbf{0} & \mathbf{0}\end{bmatrix}\mathbf{V}^H .$$

The **singular values** of $\mathbf{A}$ are the diagonal entries of $\boldsymbol{\Sigma}$. Then the distance from $\mathbf{A}$ to the nearest matrix of rank r is given by singular value of index $r+1$ in the ordering from largest to smallest.

Because the singular value decomposition involves the computation of eigenvalues and eigenvectors, we will postpone its full description until Sect. 1.5 of Chap. 1 in Volume II. In this section, we will show how to use singular value decompositions to solve least squares problems.

6.11.1 Least Squares Problems

The next lemma shows that the singular value decomposition can be very useful in solving least squares problems, even if they are rank-deficient.

Lemma 6.11.1 *Suppose that* $\mathbf{A}$ *is a nonzero* $m \times n$ *matrix with the singular value decomposition*

$$\mathbf{A} = \begin{bmatrix} \mathbf{U}_1 & \mathbf{U}_2 \end{bmatrix} \begin{bmatrix} \Sigma & 0 \\ 0 & 0 \end{bmatrix} \begin{bmatrix} \mathbf{V}_1{}^H \\ \mathbf{V}_2{}^H \end{bmatrix} .$$

Here $\begin{bmatrix} \mathbf{U}_1 & \mathbf{U}_2 \end{bmatrix}$ *is an* $m \times m$ *unitary matrix in which* $\mathbf{U}_1$ *is* $m \times r$, $\begin{bmatrix} \mathbf{V}_1 \\ \mathbf{V}_2 \end{bmatrix}$ *is an* $n \times n$ *unitary matrix in which* $\mathbf{V}_1$ *is* $n \times r$, *and* Σ *is an* $r \times r$ *diagonal matrix with positive diagonal entries. Then the pseudo-inverse of* $\mathbf{A}$ *is*

$$\mathbf{A}^\dagger = \begin{bmatrix} \mathbf{V}_1 & \mathbf{V}_2 \end{bmatrix} \begin{bmatrix} \Sigma^{-1} & 0 \\ 0 & 0 \end{bmatrix} \begin{bmatrix} \mathbf{U}_1{}^H \\ \mathbf{U}_2{}^H \end{bmatrix} . \tag{6.88}$$

Furthermore, given an m-vector $\mathbf{b}$ *let*

$$\mathbf{y} = \mathbf{U}_1{}^H \mathbf{b} \,, \; \mathbf{x} = \mathbf{V}_1 \Sigma^{-1} \mathbf{y} \text{ and } \mathbf{r} = \mathbf{b} - \mathbf{U}_1 \mathbf{y} \,.$$

Then $\mathbf{x}$ *is the n-vector with smallest Euclidean norm that minimizes* $\|\mathbf{b} - \mathbf{A}\mathbf{x}\|_2$, *and* $\mathbf{r} = \mathbf{b} - \mathbf{A}\mathbf{x}$ *is the corresponding residual vector.*

Proof We will begin by verifying the Penrose pseudo-inverse conditions (6.5), and recall that this verification was postponed from the proof of Lemma 6.4.1. We can use the formula (6.88) to compute

$$\mathbf{A}^\dagger \mathbf{A} = \begin{bmatrix} \mathbf{V}_1 & \mathbf{V}_2 \end{bmatrix} \begin{bmatrix} \mathbf{I} & \mathbf{0} \\ \mathbf{0} & \mathbf{0} \end{bmatrix} \begin{bmatrix} \mathbf{V}_1{}^H \\ \mathbf{V}_2{}^H \end{bmatrix} \text{ and}$$

$$\mathbf{A}\mathbf{A}^\dagger = \begin{bmatrix} \mathbf{U}_1 & \mathbf{U}_2 \end{bmatrix} \begin{bmatrix} \mathbf{I} & \mathbf{0} \\ \mathbf{0} & \mathbf{0} \end{bmatrix} \begin{bmatrix} \mathbf{U}_1{}^H \\ \mathbf{U}_2{}^H \end{bmatrix} .$$

Since both of these matrices are Hermitian, we have verified Penrose conditions (6.5c) and (6.5d). We can use the former of these two equations to compute

$$\left(\mathbf{A}^\dagger \mathbf{A}\right) \mathbf{A}^\dagger = \begin{bmatrix} \mathbf{V}_1 & \mathbf{V}_2 \end{bmatrix} \begin{bmatrix} \mathbf{I} & \mathbf{0} \\ \mathbf{0} & \mathbf{0} \end{bmatrix} \begin{bmatrix} \mathbf{V}_1{}^H \\ \mathbf{V}_2{}^H \end{bmatrix} \begin{bmatrix} \mathbf{V}_1 & \mathbf{V}_2 \end{bmatrix} \begin{bmatrix} \Sigma^{-1} & \mathbf{0} \\ \mathbf{0} & \mathbf{0} \end{bmatrix} \begin{bmatrix} \mathbf{U}_1{}^H \\ \mathbf{U}_2{}^H \end{bmatrix}$$

$$= \begin{bmatrix} \mathbf{V}_1 & \mathbf{V}_2 \end{bmatrix} \begin{bmatrix} \Sigma^{-1} & \mathbf{0} \\ \mathbf{0} & \mathbf{0} \end{bmatrix} \begin{bmatrix} \mathbf{U}_1{}^H \\ \mathbf{U}_2{}^H \end{bmatrix} = \mathbf{A}^\dagger \text{ and}$$

$$\mathbf{A}\left(\mathbf{A}^{\dagger}\mathbf{A}\right) = \begin{bmatrix}\mathbf{U}_1 & \mathbf{U}_2\end{bmatrix}\begin{bmatrix}\Sigma & \mathbf{0}\\ \mathbf{0} & \mathbf{0}\end{bmatrix}\begin{bmatrix}\mathbf{V}_1{}^H\\ \mathbf{V}_2{}^H\end{bmatrix}\begin{bmatrix}\mathbf{V}_1 & \mathbf{V}_2\end{bmatrix}\begin{bmatrix}\mathbf{I} & \mathbf{0}\\ \mathbf{0} & \mathbf{0}\end{bmatrix}\begin{bmatrix}\mathbf{V}_1{}^H\\ \mathbf{V}_2{}^H\end{bmatrix}$$
$$= \begin{bmatrix}\mathbf{U}_1 & \mathbf{U}_2\end{bmatrix}\begin{bmatrix}\Sigma & \mathbf{0}\\ \mathbf{0} & \mathbf{0}\end{bmatrix}\begin{bmatrix}\mathbf{V}_1{}^H\\ \mathbf{V}_2{}^H\end{bmatrix} = \mathbf{A}\,.$$

This verifies the remaining Penrose conditions (6.5a) and (6.5b).

Next, we will define

$$\mathbf{s} = \mathbf{U}_1\Sigma^{-2}\mathbf{y}$$

and show that **r**, **x** and **s** satisfy the conditions

$$\mathbf{r} + \mathbf{A}\mathbf{x} = \mathbf{b}\,,\ \mathbf{A}^H\mathbf{r} = \mathbf{0}\ \text{and}\mathbf{A}^H\mathbf{s} = \mathbf{x}\,.$$

Then Lemma 6.3.1 will verify that **x** is the minimum-norm least squares solution.

First, we compute

$$\mathbf{r} + \mathbf{A}\mathbf{x} = (\mathbf{b} - \mathbf{U}_1\mathbf{y}) + \left(\begin{bmatrix}\mathbf{U}_1 & \mathbf{U}_2\end{bmatrix}\begin{bmatrix}\Sigma & \mathbf{0}\\ \mathbf{0} & \mathbf{0}\end{bmatrix}\begin{bmatrix}\mathbf{V}_1{}^H\\ \mathbf{V}_2{}^H\end{bmatrix}\mathbf{V}_1\Sigma^{-1}\mathbf{y}\right)$$
$$= \mathbf{b} - \mathbf{U}_1\mathbf{y} + \begin{bmatrix}\mathbf{U}_1 & \mathbf{U}_2\end{bmatrix}\begin{bmatrix}\Sigma & \mathbf{0}\\ \mathbf{0} & \mathbf{0}\end{bmatrix}\begin{bmatrix}\Sigma^{-1}\mathbf{y}\\ \mathbf{0}\end{bmatrix} = \mathbf{b} - \mathbf{U}_1\mathbf{y} + \begin{bmatrix}\mathbf{U}_1 & \mathbf{U}_2\end{bmatrix}\begin{bmatrix}\mathbf{y}\\ \mathbf{0}\end{bmatrix} = \mathbf{b}\,.$$

Next, we compute

$$\mathbf{A}^H\mathbf{r} = \begin{bmatrix}\mathbf{V}_1 & \mathbf{V}_2\end{bmatrix}\begin{bmatrix}\Sigma & \mathbf{0}\\ \mathbf{0} & \mathbf{0}\end{bmatrix}\begin{bmatrix}\mathbf{U}_1{}^H\\ \mathbf{U}_2{}^H\end{bmatrix}(\mathbf{b} - \mathbf{U}_1\mathbf{y}) = \begin{bmatrix}\mathbf{V}_1 & \mathbf{V}_2\end{bmatrix}\begin{bmatrix}\Sigma & \mathbf{0}\\ \mathbf{0} & \mathbf{0}\end{bmatrix}\begin{bmatrix}\mathbf{U}_1{}^H\mathbf{b} - \mathbf{y}\\ \mathbf{0}\end{bmatrix}$$
$$= \begin{bmatrix}\mathbf{V}_1 & \mathbf{V}_2\end{bmatrix}\begin{bmatrix}\Sigma & \mathbf{0}\\ \mathbf{0} & \mathbf{0}\end{bmatrix}\begin{bmatrix}\mathbf{0}\\ \mathbf{0}\end{bmatrix} = \mathbf{0}\,.$$

Finally, we have

$$\mathbf{A}^H\mathbf{s} = \begin{bmatrix}\mathbf{V}_1 & \mathbf{V}_2\end{bmatrix}\begin{bmatrix}\Sigma & \mathbf{0}\\ \mathbf{0} & \mathbf{0}\end{bmatrix}\begin{bmatrix}\mathbf{U}_1{}^H\\ \mathbf{U}_2{}^H\end{bmatrix}\mathbf{U}_1\Sigma^{-2}\mathbf{y} = \begin{bmatrix}\mathbf{V}_1 & \mathbf{V}_2\end{bmatrix}\begin{bmatrix}\Sigma & \mathbf{0}\\ \mathbf{0} & \mathbf{0}\end{bmatrix}\begin{bmatrix}\Sigma^{-2}\mathbf{y}\\ \mathbf{0}\end{bmatrix}$$
$$= \begin{bmatrix}\mathbf{V}_1 & \mathbf{V}_2\end{bmatrix}\begin{bmatrix}\Sigma^{-1}\mathbf{y}\\ \mathbf{0}\end{bmatrix} = \mathbf{V}_1\Sigma^{-1}\mathbf{y} = \mathbf{x}\,.$$

In LAPACK the singular value decomposition is computed by routines `_gesvd`; see, for example, dgesvd.f. There are two LAPACK routines that solve least squares problems using a singular value decomposition. Routines `_gelss` use a specified condition number and the singular value decomposition to find the minimum-norm solution to a least squares problem, while routines `_gelsd` use a **divide and**

conquer approach for solving the least squares problem; see Gu and Eisenstat [49] for more information. Readers may also examine an implementation of a C++ `SingularValueDecomposition` class in files SingularValueDecomposition.H and SingularValueDecomposition.C. This class calls data type-specific LAPACK routines to perform the operations; see, for example LaDouble.C for the implementation in real double precision.

In MATLAB, the command

```
[U,S,V]=svd(A)
```

will compute the singular value decomposition of the matrix $\mathbf{A}$. Other readers may be interested in reading about the GSL Singular Value Decomposition.

The Linpack Users' Guide [34, p. 11.2] has the following practical suggestions for the use of the singular value decomposition. In floating point arithmetic, we seldom compute singular values that are exactly zero because of the influence of rounding errors. As a result, we are left with the problem of deciding when a singular value is near enough to zero to be negligible. Since the singular values of $\mathbf{A}$ change under different scalings of the rows and columns of the matrix $\mathbf{A}$, it is important that $\mathbf{A}$ be scaled properly before its singular value decomposition is computed. Suppose that we can estimate the errors in the elements of the matrix $\mathbf{A}$. (If we believe that $\mathbf{A}$ is known exactly, then we can take the error in $\mathbf{A}_{ij}$ to be a small multiple of $\mathbf{A}_{ij}\varepsilon$, where ε is machine precision.) Next, we can scale the rows and columns of $\mathbf{A}$ so that the error estimates are approximately equal to some common value δ. Then a singular value σ is considered negligible if $\sigma < \delta$. It is also important to note that row scaling may represent an impermissible change of the model in some applications.

Exercise 6.11.1 Suppose that $\mathbf{U}$ is a unitary matrix. What can you say about its singular values?

Exercise 6.11.2 Suppose that μ is a scalar. Find the matrix of rank one that is closest to

$$\mathbf{A} = \begin{bmatrix} 1 & \mu \\ 0 & 1 \end{bmatrix}$$

in the Frobenius norm.

Exercise 6.11.3 Let ε be the square-root of machine precision and

$$\mathbf{A} = \begin{bmatrix} 1 & 1 \\ \varepsilon & 0 \\ 0 & \varepsilon \end{bmatrix} , \ \mathbf{b} = \begin{bmatrix} 1 \\ 0 \\ 0 \end{bmatrix} \text{ and } \mathbf{c} = \begin{bmatrix} 1 \\ 0 \end{bmatrix} .$$

Use the MATLAB singular value decomposition to solve the following problems

1.

$$\text{minimize } \|\mathbf{A}\mathbf{x} - \mathbf{b}\|_2 \text{ and}$$

2.

$$\text{minimize } \|\mathbf{x}\|_2 \text{ subject to } A^\top \mathbf{x} = \mathbf{c} ,$$

How are your answers affected by the choice of the smallest meaningful singular value? What happens in the second problem if the singular value cutoff is significantly greater than ε?

Exercise 6.11.4 If all entries of the $n \times n$ Hilbert matrix have uncertainty on the order of machine precision times that entry, find the numerical ranks of the 10×10 and 50×50 Hilbert matrices. You may use the MATLAB singular value decomposition routine.

6.11.2 Regularization

Our next goal is to introduce the reader to the use of the singular value decomposition for **regularization**, which is also called **ridge regression** in statistics. Some applications, especially those arising from integral equations, are ill-posed in the sense that the solution is very sensitive to perturbations in the data. Regularization reduces this sensitivity by combining the original problem with some smoothing function and computing the solution of the modified problem instead.

As an example, suppose that $\mathbf{A}$ is an $m \times n$ matrix with rank $(\mathbf{A}) < \min\{m, n\}$. Let $\lambda \geq 0$ be the ridge parameter, and consider finding an n-vector $\tilde{\mathbf{x}}$ to minimize

$$\left\| \begin{bmatrix} \mathbf{A} \\ \mathbf{I}\sqrt{\lambda} \end{bmatrix} \tilde{\mathbf{x}} - \begin{bmatrix} \mathbf{b} \\ \mathbf{0} \end{bmatrix} \right\|_2^2 = \|\mathbf{A}\tilde{\mathbf{x}} - \mathbf{b}\|_2^2 + \lambda \|\tilde{\mathbf{x}}\|_2^2 .$$

For large values of λ, this least squares problem is well-conditioned, and as λ approaches zero the solution $\tilde{\mathbf{x}}$ of this modified least squares problem approaches the solution $\mathbf{x}$ of the original problem.

If $\mathbf{A} = \mathbf{U}\boldsymbol{\Sigma}\mathbf{V}^H$ is the singular value decomposition of $\mathbf{A}$, then the normal equations for $\tilde{\mathbf{x}}$ take the form

$$\mathbf{V}\left(\boldsymbol{\Sigma}^2 + \mathbf{I}\lambda\right)\mathbf{V}^H\tilde{\mathbf{x}} = \mathbf{V}\boldsymbol{\Sigma}\mathbf{U}^H\mathbf{b} .$$

As a result,

$$\tilde{\mathbf{x}} = \mathbf{V}\left(\boldsymbol{\Sigma}^2 + \mathbf{I}\lambda\right)^{-1} \boldsymbol{\Sigma}\mathbf{y} \text{ where } \mathbf{y} \equiv \mathbf{U}^H\mathbf{b} .$$

The vector $\mathbf{y}$ can be computed once, and then we can experiment with different values of λ to achieve the desired regularization in the solution.

For more information about ridge regression and regularization, see Golub and van Loan [46, p. 307ff], Hoerl [57], Press et al. [89] or Tikhonov et al. [101]

Exercise 6.11.5 Get a copy of the MATLAB Shaw test problem, which generates a matrix and right-hand side for an image restoration problem.

1. Solve the least squares problem of size 20 and $\mathbf{b} = \mathbf{A}\mathbf{e}_{10}$, without regularization. What is the condition number of the matrix? How does the computed solution differ from $\mathbf{e}_{10}$?
2. Plot the entries of the solution $\mathbf{x}$ versus the array entry number ($\mathbf{x}_i$ versus i). Does the solution $\mathbf{x}$ look smooth or wiggly?
3. Experiment with values for the ridge parameter λ to see how close you can get to retrieving the correct solution. Plot the solution versus array entry number for those values of λ, and comment on how λ affects the appearance of the plots.

6.11.3 CS Decomposition

While we are still aware of the basic properties of the singular value decomposition, we will present a useful result regarding unitary matrices. This result will involve the so-called **CS decomposition**, which will be used in Sect. 1.4.7 of Chap. 1 in Volume II to discuss invariant subspaces of general matrices. Interested readers can also find the following result in Golub and van Loan [46, p. 84].

Theorem 6.11.1 (CS Decomposition) *Suppose that*

$$\mathbf{W} = \begin{bmatrix} \mathbf{W}_1 \\ \mathbf{W}_2 \end{bmatrix}$$

where $\mathbf{W}_1$ is $m_1 \times n$, $\mathbf{W}_2$ is $m_2 \times n$, and $m_1 \geq n$. Assume that $\mathbf{W}$ has orthonormal columns:

$$\mathbf{W}^H\mathbf{W} = \mathbf{I} .$$

If $m_2 \geq n$, there exist unitary matrices $\mathbf{U}_1$, $\mathbf{U}_2$ and $\mathbf{V}$, and there exist real $n \times n$ diagonal matrices $\mathbf{C}$ and $\mathbf{S}$ so that

$$\mathbf{U}_1{}^H\mathbf{W}_1\mathbf{V} = \begin{bmatrix} \mathbf{C} \\ \mathbf{0} \end{bmatrix} , \ \mathbf{U}_2{}^H\mathbf{W}_2\mathbf{V} = \begin{bmatrix} \mathbf{S} \\ \mathbf{0} \end{bmatrix} \text{ and } \mathbf{C}^2 + \mathbf{S}^2 = \mathbf{I} .$$

On the other hand, if $m_2 < n$ then there exist unitary matrices $\mathbf{U}_1$, $\mathbf{U}_2$ and $\mathbf{V}$, and there exist real $m_2 \times m_2$ diagonal matrices $\mathbf{C}$ and $\mathbf{S}$ so that

$$\mathbf{U}_1{}^H \mathbf{W}_1 \mathbf{V} = \begin{bmatrix} \mathbf{C} & \mathbf{0} \\ \mathbf{0} & \mathbf{I} \\ \mathbf{0} & \mathbf{0} \end{bmatrix} , \; \mathbf{U}_2{}^H \mathbf{W}_2 \mathbf{V} = \begin{bmatrix} \mathbf{S} & \mathbf{0} \end{bmatrix} \text{ and } \mathbf{C}^2 + \mathbf{S}^2 = \mathbf{I} \, .$$

If $\sigma_{\min}(\mathbf{A})$ denotes the smallest singular value of $\mathbf{A}$ and $\sigma_{\max}(\mathbf{A})$ denotes the largest singular value, then we must have

$$\sigma_{\min}(\mathbf{W}_1)^2 + \sigma_{\max}(\mathbf{W}_2)^2 = 1 \, . \tag{6.89}$$

If $m_2 \geq n$, then we also have

$$\sigma_{\min}(\mathbf{W}_2)^2 + \sigma_{\max}(\mathbf{W}_1)^2 = 1 \, . \tag{6.90}$$

Proof Let the singular value decomposition of $\mathbf{W}_1$ be

$$\mathbf{W}_1 = \mathbf{U}_1 \begin{bmatrix} \Sigma \\ \mathbf{0} \end{bmatrix} \mathbf{V}^H$$

where Σ is an $n \times n$ diagonal matrix. Since $\|\Sigma\|_2 = \|\mathbf{W}_1\|_2 \leq \|\mathbf{W}\|_2 = 1$, it follows that the diagonal entries of Σ are at most one. Without loss of generality, we can assume that the singular values have been ordered so that we may partition

$$\Sigma = \begin{bmatrix} \tilde{\mathbf{C}} & \\ & \mathbf{I} \end{bmatrix} ,$$

where the $t \times t$ diagonal matrix

$$\tilde{\mathbf{C}} = \operatorname{diag}(\gamma_1, \ldots, \gamma_t)$$

has diagonal entries strictly less than one. Next, let

$$\mathbf{W}_2 \mathbf{V} = \begin{bmatrix} \mathbf{Z}_1 & \mathbf{Z}_2 \end{bmatrix} ,$$

where $\mathbf{Z}_1$ is $m_2 \times t$. Since

$$\begin{bmatrix} \mathbf{U}_1 & \\ & \mathbf{I} \end{bmatrix}^H \begin{bmatrix} \mathbf{W}_1 \\ \mathbf{W}_2 \end{bmatrix} \mathbf{V} = \begin{bmatrix} \mathbf{U}_1{}^H \mathbf{W}_1 \mathbf{V} \\ \mathbf{W}_2 \mathbf{V} \end{bmatrix} = \begin{bmatrix} \tilde{\mathbf{C}} & \mathbf{0} \\ \mathbf{0} & \mathbf{I} \\ \mathbf{0} & \mathbf{0} \\ \mathbf{Z}_1 & \mathbf{Z}_2 \end{bmatrix}$$

is a product of matrices with orthonormal columns, we have

$$\begin{bmatrix} \mathbf{I} & \\ & \mathbf{I} \end{bmatrix} = \begin{bmatrix} \tilde{C} & 0 \\ 0 & I \\ 0 & 0 \\ Z_1 & Z_2 \end{bmatrix}^H \begin{bmatrix} \tilde{C} & 0 \\ 0 & I \\ 0 & 0 \\ Z_1 & Z_2 \end{bmatrix} = \begin{bmatrix} \tilde{C}^2 + Z_1{}^H Z_1 & Z_1{}^H Z_2 \\ Z_2{}^H Z_1 & I + Z_2{}^H Z_2 \end{bmatrix} .$$

The second block diagonal entry of this equation implies that $Z_2 = \mathbf{0}$. If we define the $t \times t$ diagonal matrix $\tilde{S}$ by

$$\tilde{S} = \operatorname{diag}\left(\sqrt{1-\gamma_1^2}, \ldots, \sqrt{1-\gamma_t^2} \right) ,$$

then $\tilde{S}$ is nonsingular, and we have

$$Z_1{}^H Z_1 = \mathbf{I} - \tilde{C}^2 = \tilde{S}^2 .$$

It follows that the columns of the $m_2 \times t$ matrix $Z_1 \tilde{S}^{-1}$ are orthonormal. This implies that we must have $t \leq m_2$.

Using Corollary 3.2.1, we can extend these columns, if necessary, to an orthonormal basis for all m_2-vectors. In other words, there is an $m_2 \times m_2$ unitary matrix U_2 that can be partitioned as

$$U_2 = \left[Z_1 \tilde{S}^{-1} \; U_{22} \right] .$$

Since U_2 is unitary, we must have $U_{22}{}^H Z_1 \tilde{S}^{-1} = \mathbf{0}$, from which it follows that

$$U_{22}{}^H Z_1 = \mathbf{0} .$$

We now have

$$U_2{}^H \mathbf{W}_2 \mathbf{V} = \begin{bmatrix} \tilde{S}^{-1} Z_1{}^H \\ U_{22}{}^H \end{bmatrix} \mathbf{W}_2 \mathbf{V} = \begin{bmatrix} \tilde{S}^{-1} Z_1{}^H \\ U_{22}{}^H \end{bmatrix} [Z_1 \; \mathbf{0}] = \begin{bmatrix} \tilde{S} & \mathbf{0} \\ U_{22}{}^H Z_1 & \mathbf{0} \end{bmatrix} = \begin{bmatrix} \tilde{S} & \mathbf{0} \\ \mathbf{0} & \mathbf{0} \end{bmatrix} .$$

Recall that we assumed that $m_1 \geq n$, and found that $t \leq \min\{n, m_2\}$. In the case where $m_2 \leq n$, we define the $m_2 \times m_2$ diagonal matrices

$$C = \begin{bmatrix} \tilde{C} & \mathbf{0} \\ \mathbf{0} & \mathbf{I} \end{bmatrix} \text{ and } S = \begin{bmatrix} \tilde{S} & \mathbf{0} \\ \mathbf{0} & \mathbf{0} \end{bmatrix} .$$

Then

$$\mathrm{U}_1{}^H \mathbf{W}_1 \mathbf{V} = \begin{bmatrix} \tilde{\mathrm{C}} & \mathbf{0} \\ \mathbf{0} & \mathbf{I} \\ \mathbf{0} & \mathbf{0} \end{bmatrix} , \; \mathrm{U}_2{}^H \mathbf{W}_2 \mathbf{V} = \begin{bmatrix} \tilde{\mathrm{S}} & \mathbf{0} \end{bmatrix} \text{ and } \tilde{\mathrm{C}}^2 + \tilde{\mathrm{S}}^2 = \mathbf{I} \,.$$

Since the smallest singular value of $\mathbf{W}_1$ is either one (in which case $t = 0$) or a diagonal entry of $\tilde{\mathrm{C}}$, and the largest singular value of $\mathbf{W}_2$ is either zero (when $t = 0$) or the corresponding diagonal entry of $\tilde{\mathrm{S}}$, we find that (6.89) holds.

If $m_2 > n$, then we define the $n \times n$ diagonal matrices

$$\mathrm{C} = \begin{bmatrix} \tilde{\mathrm{C}} & \mathbf{0} \\ \mathbf{0} & \mathbf{I} \end{bmatrix} \text{ and } \mathrm{S} = \begin{bmatrix} \tilde{\mathrm{S}} & \mathbf{0} \\ \mathbf{0} & \mathbf{0} \end{bmatrix} .$$

Then

$$\mathrm{U}_1{}^H \mathbf{W}_1 \mathbf{V} = \begin{bmatrix} \mathrm{C} \\ \mathbf{0} \end{bmatrix} , \; \mathrm{U}_2{}^H \mathbf{W}_2 \mathbf{V} = \begin{bmatrix} \mathrm{S} \\ \mathbf{0} \end{bmatrix} \text{ and } \mathrm{C}^2 + \mathrm{S}^2 = \mathbf{I} \,.$$

Since the singular values of $\mathbf{W}_1$ are the diagonal entries of C, and the singular values of $\mathbf{W}_2$ are the diagonal entries of $\tilde{\mathrm{S}}$, we find that both (6.89) and (6.90) hold.

For more information about the CS decomposition, including stable numerical algorithms to compute it, and its importance in generalized singular value decompositions, see van Loan [106].

6.12 Quadratic Programming

In Lemma 6.3.1, we saw that the full-rank underdetermined least squares problem involves minimizing a simple quadratic function (the square of the norm of the solution $\mathbf{x}$) subject to a system of linear constraints. In this section, we will examine a more general formulation of the constrained least squares problem.

6.12.1 Existence and Uniqueness

The following lemma will prove the existence of a solution to the quadratic programming problem.

Lemma 6.12.1 *Suppose that n and k are positive integers. Let the $k \times n$ matrix $\mathbf{B}$ have rank k, and let the $n \times n$ matrix $\mathbf{A}$ be Hermitian and positive-definite on the nullspace of $\mathbf{B}$. Then the following hold:*

1. *The matrix* $\begin{bmatrix} \mathbf{A} & \mathbf{B}^H \\ \mathbf{B} & 0 \end{bmatrix}$ *is nonsingular.*
2. *Given an n-vector* **a** *and a k-vector* **b**, *suppose that the n-vector* **u** *solves the quadratic programming problem*

$$\min_{n\text{-vectors } \mathbf{u}} \quad P(\mathbf{u}) \equiv \mathbf{u} \cdot \mathbf{A}\mathbf{u} - \mathbf{a} \cdot \mathbf{u} - \mathbf{u} \cdot \mathbf{a} \tag{6.91a}$$

$$\text{subject to } \mathbf{B}\mathbf{u} = \mathbf{b} \; . \tag{6.91b}$$

Then $\mathbf{A}\mathbf{u} - \mathbf{a} \perp \mathscr{N}(\mathbf{B})$*; in other words,*

$$\mathbf{B}\mathbf{v} = \mathbf{0} \quad \Longrightarrow \quad \mathbf{v} \cdot (\mathbf{A}\mathbf{u} - \mathbf{a}) = 0 \; .$$

3. *If the n-vector* **u** *solves the quadratic programming problem* (6.91), *then there is a unique k-vector* **m** *so that the first-order* ***Kuhn-Tucker conditions*** *are satisfied:*

$$\begin{bmatrix} \mathbf{A} & \mathbf{B}^H \\ \mathbf{B} & 0 \end{bmatrix} \begin{bmatrix} \mathbf{u} \\ \mathbf{m} \end{bmatrix} = \begin{bmatrix} \mathbf{a} \\ \mathbf{b} \end{bmatrix} \; . \tag{6.92}$$

4. *For all n-vectors* **a** *and all k-vectors* **b** *there is a unique n-vector* **u** *that solves the quadratic programming problem* (6.91).

Proof Note that since **B** has rank k, the range of **B** is the set of all k-vectors, and the fundamental theorem of linear algebra 3.2.3 shows that the nullspace of $\mathbf{B}^H$ consists solely of the zero vector. Let the columns of the $k \times (n-k)$ matrix **Z** be a basis for the nullspace $\mathscr{N}(\mathbf{B})$. Since **A** is Hermitian and positive-definite on $\mathscr{N}(\mathbf{B})$, the $(n-k) \times (n-k)$ matrix $\mathbf{Z}^H\mathbf{A}\mathbf{Z}$ is positive-definite.

Let us prove the first claim. Suppose that

$$\begin{bmatrix} \mathbf{A} & \mathbf{B}^H \\ \mathbf{B} & \mathbf{0} \end{bmatrix} \begin{bmatrix} \mathbf{u} \\ \mathbf{m} \end{bmatrix} = \begin{bmatrix} \mathbf{0} \\ \mathbf{0} \end{bmatrix} \; .$$

Since $\mathbf{B}\mathbf{u} = \mathbf{0}$, we must have $\mathbf{u} \in \mathscr{N}(\mathbf{B})$. Thus there exists an $(n-k)$-vector **x** so that $\mathbf{u} = \mathbf{Z}\mathbf{x}$. We can multiply the first equation in the system for **u** and **m** by $\mathbf{Z}^H$ to get

$$\mathbf{0} = \mathbf{Z}^H\mathbf{A}\mathbf{u} + \mathbf{Z}^H\mathbf{B}^H\mathbf{m} = \mathbf{Z}^H\mathbf{A}\mathbf{Z}\mathbf{x} \; .$$

Since $\mathbf{Z}^H\mathbf{A}\mathbf{Z}$ is positive-definite, this implies that $\mathbf{x} = \mathbf{0}$, which in turn implies that $\mathbf{u} = \mathbf{Z}\mathbf{x} = \mathbf{0}$. Then the first equation implies that

$$\mathbf{0} = \mathbf{A}\mathbf{u} + \mathbf{B}^H\mathbf{m} = \mathbf{B}^H\mathbf{m} \; .$$

Since $\mathscr{N}(\mathbf{B}^H) = \{\mathbf{0}\}$, this implies that $\mathbf{m} = \mathbf{0}$. We have now shown that the matrix in the first claim is square, and has zero nullspace. It follows from Lemma 3.2.9 that this matrix is nonsingular, and the first claim is proved.

Let us prove the second claim. Suppose that $\mathbf{u}$ solves the quadratic programming problem (6.91). Any other feasible solution is of the form $\mathbf{u} + \mathbf{w}$ where

$$\mathbf{B}(\mathbf{u} + \mathbf{w}) = \mathbf{b} = \mathbf{B}\mathbf{u} \; .$$

Consequently, $\mathbf{B}\mathbf{w} = \mathbf{0}$, so $\mathbf{w} \in \mathscr{N}(\mathbf{B})$. Since the columns of $\mathbf{Z}$ form a basis for $\mathscr{N}(\mathbf{B})$, it follows that any feasible solution for (6.91) can be written in the form $\mathbf{u} + \mathbf{Z}\mathbf{x}$ for some $(n-k)$-vector $\mathbf{x}$. Since $\mathbf{u}$ solves (6.91) and $\mathbf{u} + \mathbf{Z}\mathbf{x}\varepsilon$ is feasible for all real scalars ε,

$$P(\mathbf{u}) \leq P(\mathbf{u}+\mathbf{Z}\mathbf{x}\varepsilon) = P(\mathbf{u}) + \varepsilon \left\{\mathbf{x}^H\mathbf{Z}^H[\mathbf{A}\mathbf{u} - \mathbf{a}] + [\mathbf{A}\mathbf{u} - \mathbf{a}]^H\mathbf{Z}\mathbf{x}\right\} + \frac{\varepsilon^2}{2}\mathbf{x}^H\mathbf{Z}^H\mathbf{A}\mathbf{Z}\mathbf{x} \; .$$

Since $\mathbf{A}$ is Hermitian and positive definite on $\mathscr{N}(\mathbf{B})$, the second variation of P is nonnegative. Since ε is arbitrary, the first variation of $P(\mathbf{u} + \mathbf{Z}\mathbf{x}\varepsilon)$ must be zero:

$$\mathbf{0} = \mathbf{x}^H\mathbf{Z}^H[\mathbf{A}\mathbf{u} - \mathbf{a}] + [\mathbf{A}\mathbf{u} - \mathbf{a}]^H\mathbf{Z}\mathbf{x} \; .$$

Since $\mathbf{x}$ is arbitrary, we can choose $\mathbf{x} = \mathbf{Z}^H[\mathbf{A}\mathbf{u}-\mathbf{a}]$ to prove that $\mathbf{A}\mathbf{u}-\mathbf{a} \in \mathscr{R}(\mathbf{Z})^\perp = \mathscr{N}(\mathbf{B})^\perp$. The second claim is now proved.

Next, we will prove the third claim. If $\mathbf{u}$ solves the quadratic programming problem (6.91), then the second claim shows that $\mathbf{A}\mathbf{u} - \mathbf{a} \perp \mathscr{N}(\mathbf{B})$. From the orthogonality of the fundamental subspaces of $\mathbf{B}$, we see that $\mathbf{A}\mathbf{u} - \mathbf{a} \in \mathscr{R}(\mathbf{B}^H)$. It follows that there exists a k-vector $\mathbf{m}$ so that $\mathbf{A}\mathbf{u} - \mathbf{a} = -\mathbf{B}^H\mathbf{m}$. This equation plus the constraint (6.91b) in the quadratic programming problem show that (6.92) is satisfied.

We turn to the fourth claim. Note that the first claim shows that the linear system

$$\begin{bmatrix} \mathbf{A} & \mathbf{B}^H \\ \mathbf{B} & \mathbf{0} \end{bmatrix} \begin{bmatrix} \mathbf{u} \\ \mathbf{m} \end{bmatrix} = \begin{bmatrix} \mathbf{a} \\ \mathbf{b} \end{bmatrix}$$

has a unique solution. The second equation in this linear system shows that the solution $\mathbf{u}$ of this linear system satisfies the constraint in the quadratic programming problem (6.91), and the first equation shows that $\mathbf{A}\mathbf{u} - \mathbf{a} \in \mathscr{R}(\mathbf{B}^H) = \mathscr{N}(\mathbf{B})^\perp$. It follows that for all $\mathbf{v} \in \mathscr{N}(\mathbf{B})$ and for all scalars ε we have $\mathbf{b} = \mathbf{B}(\mathbf{u} + \mathbf{v}\varepsilon)$ and

$$\begin{aligned} P(\mathbf{u} + \mathbf{v}\varepsilon) &= P(\mathbf{u}) + \varepsilon \left\{\mathbf{v}^H(\mathbf{A}\mathbf{u} - \mathbf{a}) + (\mathbf{A}\mathbf{u} - \mathbf{a})^H\mathbf{v}\right\} + \frac{\varepsilon^2}{2}\mathbf{v}^H\mathbf{A}\mathbf{v} \\ &= P(\mathbf{u}) + \frac{\varepsilon^2}{2}\mathbf{v}^H\mathbf{A}\mathbf{v} \geq P(\mathbf{u}) \; . \end{aligned}$$

Thus $\mathbf{u}$ solves the quadratic programming problem (6.91).

It is common to refer to the vector $\mathbf{m}$ in Eq. (6.92) as the **Lagrange multiplier** for the constraint. The function $P(\mathbf{u})$ is commonly called the **primal objective**.

6.12.2 Equality Constrained Least Squares

In Lemma 6.12.1, we proved the existence of a vector that minimizes a quadratic function subject to linear constraints. Our next lemma will specialize this result, by considering objectives that correspond to least squares problems. This lemma will construct a method for computing the solution of the constrained least squares problem.

Lemma 6.12.2 *Suppose that* $\mathbf{B}$ *is a* $k \times n$ *matrix with rank* $(\mathbf{B}) = k$. *Then there exist an* $n \times n$ *unitary matrix* $\mathbf{V}$ *and a* $k \times k$ *nonsingular right-triangular matrix* $\mathbf{R}$ *so that*

$$\mathbf{B}\mathbf{V} = \begin{bmatrix} \mathbf{0} & \mathbf{R} \end{bmatrix} . \tag{6.93}$$

If we partition

$$\mathbf{V} = \begin{bmatrix} \mathbf{V}_1 & \mathbf{V}_2 \end{bmatrix}$$

where $\mathbf{V}_1$ *is* $n \times (n-k)$, *then the columns of* $\mathbf{V}_1$ *form an orthonormal basis for the nullspace of* $\mathbf{B}$.

Next, suppose that $\mathbf{A}$ *is* $m \times n$, *and that* $\begin{bmatrix} \mathbf{A}^H & \mathbf{B}^H \end{bmatrix}$ *has rank* n. *Then there is an* $m \times m$ *unitary matrix* $\mathbf{U}$ *and a* $m \times n$ *right-trapezoidal matrix* $\mathbf{T}$ *so that*

$$\mathbf{A}\mathbf{V} = \mathbf{U}\mathbf{T} .$$

If we partition

$$\mathbf{U} = \begin{bmatrix} \mathbf{U}_1 & \mathbf{U}_2 \end{bmatrix} \text{ and } \mathbf{T} = \begin{bmatrix} \mathbf{T}_{11} & \mathbf{T}_{12} \\ & \mathbf{T}_{22} \end{bmatrix}$$

where $\mathbf{T}_{11}$ *is* $(n-k) \times (n-k)$, *then* $\mathbf{T}_{11}$ *is nonsingular.*

Finally, suppose that $\mathbf{a}$ *is an m-vector and* $\mathbf{b}$ *is a k-vector. Let*

$$\begin{bmatrix} \mathbf{c}_1 \\ \mathbf{c}_2 \end{bmatrix} = \begin{bmatrix} {\mathbf{U}_1}^H \\ {\mathbf{U}_2}^H \end{bmatrix} \mathbf{a} \, , \; \mathbf{R}\mathbf{q} = \mathbf{b} \, , \; \mathbf{T}_{11}\mathbf{p} = \mathbf{c}_1 - \mathbf{T}_{12}\mathbf{q} \text{ and } \mathbf{u} = \begin{bmatrix} \mathbf{V}_1 & \mathbf{V}_2 \end{bmatrix} \begin{bmatrix} \mathbf{p} \\ \mathbf{q} \end{bmatrix} .$$

Then $\mathbf{u}$ *minimizes* $\|\mathbf{A}\mathbf{u} - \mathbf{a}\|_2$ *subject to the constraint* $\mathbf{B}\mathbf{u} = \mathbf{b}$.

Proof We take $\mathbf{V}^H = \mathbf{H}_1 \ldots \mathbf{H}_k$ where each Householder reflector $\mathbf{H}_i$ is chosen to zero out all entries *above* entry $n + k - i$. This gives us

$$\mathbf{V}^H\mathbf{B} = \mathbf{H}_1 \ldots \mathbf{H}_k\mathbf{B} = \begin{bmatrix} \mathbf{0} \\ \mathbf{R}^H \end{bmatrix} ,$$

which is equivalent to (6.93). Next, suppose that

$$\mathbf{z} = \begin{bmatrix} \mathbf{V}_1 & \mathbf{V}_2 \end{bmatrix} \begin{bmatrix} \mathbf{y}_1 \\ \mathbf{y}_2 \end{bmatrix}$$

is in the nullspace of $\mathbf{B}$. Then

$$\mathbf{0} = \mathbf{B}\mathbf{z} = \begin{bmatrix} \mathbf{0} & \mathbf{R} \end{bmatrix} \begin{bmatrix} \mathbf{y}_1 \\ \mathbf{y}_2 \end{bmatrix} = \mathbf{R}\mathbf{y}_2 ,$$

so $\mathbf{y}_2 = \mathbf{0}$. Thus all vectors in the nullspace of $\mathbf{B}$ can be written in the form $\mathbf{z} = \mathbf{V}_1\mathbf{y}_1$; that is, the columns of $\mathbf{V}_1$ form an orthonormal basis for $\mathscr{N}(\mathbf{B})$.

Let $\mathbf{C} = \mathbf{A}\mathbf{V}$. Then we can find a product $\mathbf{U}$ of Householder reflectors in the usual way to factor

$$\mathbf{C} = \mathbf{U}\mathbf{T}$$

where $\mathbf{T}$ is right-trapezoidal. Then $\mathbf{A}\mathbf{V}_1 = \mathbf{U}_1\mathbf{T}_{11}$. If $\mathbf{z} = \mathbf{V}_1\mathbf{y}_1$ is a nonzero vector in $\mathscr{N}(\mathbf{B})$, then $\mathbf{z} \notin \mathscr{N}(\mathbf{A})$, so

$$\mathbf{0} \neq \mathbf{A}\mathbf{z} = \mathbf{A}\mathbf{V}_1\mathbf{y}_1 = \mathbf{U}_1\mathbf{T}_{11}\mathbf{y}_1 .$$

We can multiply this equation by $\mathbf{U}_1{}^H$ to see that $\mathbf{T}_{11}\mathbf{y}_1 \neq \mathbf{0}$. Thus $\mathbf{T}_{11}$ is nonsingular.

Finally, we will prove that $\mathbf{u}$ solves the constrained least squares problem. Note that

$$\mathbf{B}\mathbf{u} = \begin{bmatrix} \mathbf{0} & \mathbf{R} \end{bmatrix} \begin{bmatrix} \mathbf{p} \\ \mathbf{q} \end{bmatrix} = \mathbf{R}\mathbf{q} = \mathbf{b} ,$$

so $\mathbf{u}$ satisfies the constraint. Any other vector that satisfies the constraint is of the form $\tilde{\mathbf{u}} = \mathbf{u} + \mathbf{V}_1\mathbf{w}$, since the columns of $\mathbf{V}_1$ form a basis for the nullspace of $\mathbf{B}$. Then

$$\mathbf{A}\tilde{\mathbf{u}} - a = \mathbf{U}\left\{ \begin{bmatrix} \mathbf{T}_{11} & \mathbf{T}_{12} \\ & \mathbf{T}_{22} \end{bmatrix} \begin{bmatrix} \mathbf{p} + \mathbf{w} \\ \mathbf{q} \end{bmatrix} - \begin{bmatrix} \mathbf{c}_1 \\ \mathbf{c}_2 \end{bmatrix} \right\} = \mathbf{U} \begin{bmatrix} \mathbf{T}_{11}\mathbf{w} \\ \mathbf{T}_{22}\mathbf{q} - \mathbf{c}_2 \end{bmatrix} .$$

Since $\mathbf{T}_{11}$ is nonsingular and $\mathbf{U}$ is unitary, we can minimize the norm of this vector by choosing $\mathbf{w} = \mathbf{0}$.

Perturbation theory for this problem can be found in Eldén [37] or Cox and Higham [24].

The previous lemma constructed the following algorithm for solving the equality constrained least squares problem.

Algorithm 6.12.1 (Equality Constrained Least Squares)

factor $\mathbf{B}\mathbf{V} = \begin{bmatrix} \mathbf{0} & \mathrm{R} \end{bmatrix}$ where $\mathbf{V}$ is unitary and R is right-triangular

compute $\mathbf{C} = \mathbf{A}\mathbf{V}$

factor $\mathbf{C} = \mathrm{U}\mathbf{T}$ where U is unitary and $\mathbf{T} = \begin{bmatrix} \mathbf{T}_{11} & \mathbf{T}_{12} \\ & \mathbf{T}_{22} \end{bmatrix}$ is right-trapezoidal

solve $\mathrm{R}\mathbf{q} = \mathbf{b}$

compute $\begin{bmatrix} \mathbf{c}_1 \\ \mathbf{c}_2 \end{bmatrix} = \mathrm{U}^H \mathbf{a}$ where $\mathbf{c}_1$ is an $(n-k)$-vector

solve $\mathbf{T}_{11}\mathbf{p} = \mathbf{c}_1 - \mathbf{T}_{12}\mathbf{q}$

compute $\mathbf{u} = \mathbf{V} \begin{bmatrix} \mathbf{p} \\ \mathbf{q} \end{bmatrix}$

This algorithm is implemented in LAPACK routines `_gglse`. See, for example, dgglse.f. MATLAB users can solve constrained least squares problems via the MATLAB command lsqlin.

6.12.3 General Problems with Equality Constraints

One approach for solving the primal quadratic programming problem (6.91) is to solve the Kuhn-Tucker conditions (6.92). It should be obvious from the form of the matrix in this linear system that it is Hermitian, but cannot be positive-definite. This might suggest that we use the Hermitian indefinite factorization described in Theorem 3.13.1. However, the next lemma shows that we can avoid working with this larger matrix, and we can avoid the pivoting needed by the Hermitian indefinite factorization.

Lemma 6.12.3 *Suppose that* $\mathbf{B}$ *is a* $k \times n$ *matrix with rank* $(\mathbf{B}) = k$*. Then there exist an* $n \times n$ *unitary matrix* $\mathbf{V}$ *and a* $k \times k$ *nonsingular right-triangular matrix* R *so that*

$$\mathbf{B}\mathbf{V} = \begin{bmatrix} \mathbf{0} & \mathrm{R} \end{bmatrix} . \tag{6.94}$$

If we partition

$$\mathbf{V} = \begin{bmatrix} \mathbf{V}_1 & \mathbf{V}_2 \end{bmatrix} ,$$

then the columns of $\mathbf{V}_1$ *form an orthonormal basis for the nullspace of* $\mathbf{B}$*. Next, suppose that* $\mathbf{A}$ *is an* $n \times m$ *Hermitian matrix,and that* $\mathbf{A}$ *is positive-definite on the nullspace of* $\mathbf{B}$*. Then there is an* $(n-k) \times (n-k)$ *lower-triangular nonsingular* *matrix* $\mathbf{L}$ *so that*

$$\mathbf{V}_1{}^H\mathbf{A}\mathbf{V}_1 = \mathbf{L}\mathbf{L}^H \ .$$

Next, suppose that $\mathbf{a}$ *is an n-vector and* $\mathbf{b}$ *is a k-vector. Let*

$$\begin{bmatrix}\mathbf{c}_1 \\ \mathbf{c}_2\end{bmatrix} = \begin{bmatrix}\mathbf{V}_1{}^H \\ \mathbf{V}_2{}^H\end{bmatrix}\mathbf{a} \ , \ \mathbf{R}\mathbf{q} = \mathbf{b} \ , \ \mathbf{L}\mathbf{p} = \mathbf{c}_1 - \mathbf{T}_{12}\mathbf{q} \text{ and } \mathbf{u} = \begin{bmatrix}\mathbf{V}_1 & \mathbf{V}_2\end{bmatrix}\begin{bmatrix}\mathbf{p} \\ \mathbf{q}\end{bmatrix} \ .$$

Then $\mathbf{u}$ *minimizes* $\mathbf{u}^H\mathbf{A}\mathbf{u} - \mathbf{a}^H\mathbf{u} - \mathbf{u}^H\mathbf{a}$ *subject to* $\mathbf{B}\mathbf{u} = \mathbf{b}$.

Proof The factorization of $\mathbf{B}$ and the proof that the columns of $\mathbf{V}_1$ form an orthonormal basis for the nullspace of $\mathbf{B}$ are the same as in Lemma 6.12.2.

Let

$$\mathbf{V}^H\mathbf{A}\mathbf{V} = \begin{bmatrix}\mathbf{C}_{11} & \mathbf{C}_{12} \\ \mathbf{C}_{12}{}^H & \mathbf{C}_{22}\end{bmatrix}$$

where $\mathbf{C}_{11}$ is an $(n-k)\times(n-k)$ Hermitian matrix, and $\mathbf{C}_{22}$ is Hermitian. Since the columns of $\mathbf{V}_1$ form an orthonormal basis for the nullspace of $\mathbf{B}$, and since $\mathbf{A}$ is positive-definite on the nullspace of $\mathbf{B}$, we must have that $\mathbf{C}_{11}$ is also positive-definite. Thus we can use the Cholesky factorization, developed in Sect. 3.13.3, to compute a nonsingular left-triangular matrix $\mathbf{L}$ so that

$$\mathbf{C}_{11} = \mathbf{L}\mathbf{L}^H \ .$$

Next, let us define the vector $\mathbf{m}$ of Lagrange multipliers by

$$\mathbf{R}^H\mathbf{m} = \mathbf{c}_2 - \mathbf{C}_{12}{}^H\mathbf{p} - \mathbf{C}_{22}\mathbf{q} \ .$$

Then

$$\begin{bmatrix}\mathbf{A} & \mathbf{B}^H \\ \mathbf{B} & \mathbf{0}\end{bmatrix}\begin{bmatrix}\mathbf{u} \\ \mathbf{m}\end{bmatrix} = \begin{bmatrix}\begin{bmatrix}\mathbf{V}_1 & \mathbf{V}_2\end{bmatrix}\left\{\begin{bmatrix}\mathbf{L}\mathbf{L}^H & \mathbf{C}_{12} \\ \mathbf{C}_{12}{}^H & \mathbf{C}_{22}\end{bmatrix}\begin{bmatrix}\mathbf{p} \\ \mathbf{q}\end{bmatrix} + \begin{bmatrix}\mathbf{0} \\ \mathbf{R}^H\end{bmatrix}\mathbf{m}\right\} \\ \begin{bmatrix}\mathbf{0} & \mathbf{R}\end{bmatrix}\begin{bmatrix}\mathbf{p} \\ \mathbf{q}\end{bmatrix}\end{bmatrix}$$

$$= \begin{bmatrix}\begin{bmatrix}\mathbf{V}_1 & \mathbf{V}_2\end{bmatrix}\left\{\begin{bmatrix}(\mathbf{c}_1 - \mathbf{C}_{12}\mathbf{q}) + \mathbf{C}_{12}\mathbf{q} \\ \mathbf{C}_{12}{}^H\mathbf{p} + \mathbf{C}_{22}\mathbf{q}\end{bmatrix} + \begin{bmatrix}\mathbf{0} \\ \mathbf{R}^H\mathbf{m}\end{bmatrix}\right\} \\ \mathbf{R}\mathbf{q}\end{bmatrix} = \begin{bmatrix}\begin{bmatrix}\mathbf{V}_1 & \mathbf{V}_2\end{bmatrix}\begin{bmatrix}\mathbf{c}_1 \\ \mathbf{c}_2\end{bmatrix} \\ \mathbf{b}\end{bmatrix} = \begin{bmatrix}\mathbf{a} \\ \mathbf{b}\end{bmatrix} \ .$$

Then Lemma 6.12.1 shows that $\mathbf{u}$ and $\mathbf{m}$ solve the quadratic programming problem.

The previous lemma provides the following algorithm for solving the general quadratic programming problems.

Algorithm 6.12.2 (Quadratic Programming Problem)

factor $\mathbf{B}\mathbf{V} = [\mathbf{0}\ \mathbf{R}]$ where $\mathbf{V}$ is unitary and $\mathbf{R}$ is right-triangular

compute $\begin{bmatrix} \mathbf{C}_{11} & \mathbf{C}_{12} \\ \mathbf{C}_{12}{}^H & \mathbf{C}_{22} \end{bmatrix} = \mathbf{V}^H\mathbf{A}\mathbf{V}$ where $\mathbf{C}_{11}$ is $(n-k)\times(n-k)$

factor $\mathbf{C}_{11} = \mathbf{L}\mathbf{L}^H$

solve $\mathbf{R}\mathbf{q} = \mathbf{b}$

compute $\begin{bmatrix} \mathbf{c}_1 \\ \mathbf{c}_2 \end{bmatrix} = \mathbf{V}^H\mathbf{a}$ where $\mathbf{c}_1$ is an $(n-k)$-vector

solve $\mathbf{L}\mathbf{d} = \mathbf{c}_1 - \mathbf{C}_{12}\mathbf{q}$

solve $\mathbf{L}^H\mathbf{p} = \mathbf{d}$

compute $\mathbf{u} = \mathbf{V}\begin{bmatrix} \mathbf{p} \\ \mathbf{q} \end{bmatrix}$

For very large quadratic programming problems that arise from discretizations of partial differential equations, it is more common to use iterative methods. For more information regarding these methods, see the books by Braess [14, p. 221ff], Brenner and Scott [16, Chapter 13] and Trangenstein [102, Section 7.5].

Exercise 6.12.1 Suppose that $\mathbf{B}$ in Lemma 6.12.1 has rank less than k, and that $\mathbf{A}$ is Hermitian and positive-definite on $\mathscr{N}(\mathbf{B})$. Describe how to find a nonzero vector in the nullspace of $\begin{bmatrix} \mathbf{A} & \mathbf{B}^\top \\ \mathbf{B} & \mathbf{0} \end{bmatrix}$.

Exercise 6.12.2 Show that the second claim of Lemma 6.12.1 is valid even if the rank of $\mathbf{B}$ is less than k.

Exercise 6.12.3 Show that if $\mathbf{B}$ has rank less than k, then the Lagrange multiplier in the third claim of Lemma 6.12.1 exists, but is not unique.

Exercise 6.12.4 If $\mathbf{B}$ has rank less than k in Lemma 6.12.1, does the primal problem (6.91) have a solution? If so, is it unique?

Exercise 6.12.5 Prove that if $\mathbf{B} \in \mathbb{R}^{k\times n}$ is nonzero, then there is a constant $\sigma > 0$ so that for all $\mathbf{m} \in \mathbb{R}^k$

$$\sup_{\substack{\mathbf{u}\in\mathbb{R}^n \\ \mathbf{u}\neq\mathbf{0}}} \frac{\mathbf{m}^\top\mathbf{B}\mathbf{u}}{\|\mathbf{u}\|_2} \geq \sigma \inf_{\mathbf{y}\in\mathscr{N}(\mathbf{B}^\top)} \|\mathbf{m}-\mathbf{y}\|_2 \ .$$

Exercise 6.12.6 Suppose that $n > k$ are positive integers. Let $\mathbf{B} \in \mathbb{R}^{k\times n}$ have rank k. Assume that $\mathbf{A} \in \mathbb{R}^{n\times n}$ is Hermitian and positive-definite.

1. Given $\mathbf{a} \in \mathbb{R}^n$ and $\mathbf{b} \in \mathbb{R}^k$, suppose that $(\mathbf{u}, \mathbf{m}) \in \mathbb{R}^n \times \mathbb{R}^k$ solves the **dual quadratic programming problem**

$$\min_{\mathbf{u}\in\mathbb{R}^n,\mathbf{m}\in\mathbb{R}^k} D(\mathbf{u}, \mathbf{m}) \equiv -\frac{1}{2}\mathbf{u}^\top \mathbf{A}\mathbf{u} - \mathbf{b}^\top \mathbf{m} \tag{6.95a}$$

$$\text{subject to } \mathbf{A}\mathbf{u} + \mathbf{B}^\top \mathbf{m} = \mathbf{a} \,. \tag{6.95b}$$

Show that $\begin{bmatrix}\mathbf{A}\mathbf{u}\\ \mathbf{b}\end{bmatrix} \perp \mathscr{N}\left(\begin{bmatrix}\mathbf{A} \; \mathbf{B}^\top\end{bmatrix}\right)$; in other words,

$$\begin{bmatrix}\mathbf{A}, \mathbf{B}^\top\end{bmatrix}\begin{bmatrix}\mathbf{v}\\ \mathbf{n}\end{bmatrix} = \mathbf{0} \quad \Longrightarrow \quad \begin{bmatrix}\mathbf{v}^\top, \mathbf{n}^\top\end{bmatrix}\begin{bmatrix}\mathbf{A}\mathbf{u}\\ \mathbf{b}\end{bmatrix} = 0 \,.$$

2. Show that if $(\mathbf{u}, \mathbf{m}) \in \mathbb{R}^n \times \mathbb{R}^k$ solves the dual quadratic programming problem (6.95), then there is a unique $\mathbf{w} \in \mathbb{R}^n$ that satisfies the first-order Kuhn-Tucker conditions

$$\begin{bmatrix}\mathbf{A} & \mathbf{B}^\top\\ \mathbf{B} & 0\end{bmatrix}\begin{bmatrix}\mathbf{w}\\ \mathbf{m}\end{bmatrix} = \begin{bmatrix}\mathbf{a}\\ \mathbf{b}\end{bmatrix} \,. \tag{6.96}$$

   Also show that $\mathbf{w} = \mathbf{u}$.
3. Prove that for all $\mathbf{a} \in \mathbb{R}^n$ and $\mathbf{b} \in \mathbb{R}^k$ there is a unique $(\mathbf{u}, \mathbf{m}) \in \mathbb{R}^n \times \mathbb{R}^k$ that solves the dual quadratic programming problem (6.95).
4. If the **dual Lagrangian** is defined by

$$\mathscr{L}_D(\mathbf{u}, \mathbf{m}, \mathbf{w}) \equiv -\frac{1}{2}\mathbf{u}^\top \mathbf{A}\mathbf{u} - \mathbf{b}^\top \mathbf{m} + \mathbf{w}^\top (\mathbf{A}\mathbf{u} + \mathbf{B}^\top \mathbf{m} - \mathbf{a}) \,, \tag{6.97}$$

   show that $(\mathbf{u}, \mathbf{m}, \mathbf{w})$ is a critical point of $\mathscr{L}_D$ if and only if $\mathbf{u} = \mathbf{w}$ and $(\mathbf{w}, \mathbf{m})$ satisfies the first-order Kuhn-Tucker conditions for a minimum of D.
5. Show that for all $\mathbf{u}, \mathbf{v} \in \mathbb{R}^n$ and for all $\mathbf{m} \in \mathbb{R}^k$,

$$\mathscr{L}_D(\mathbf{v}, \mathbf{m}, \mathbf{u}) \le \mathscr{L}_D(\mathbf{u}, \mathbf{m}, \mathbf{u}) \,.$$

6. Show that if $(\mathbf{u}, \mathbf{m}) \in \mathbb{R}^n \times \mathbb{R}^k$ is feasible for the dual quadratic programming problem (6.95), then

$$\forall \mathbf{v} \in \mathbb{R}^n \,, \quad \mathscr{L}_D(\mathbf{u}, \mathbf{m}, \mathbf{v}) = D(\mathbf{u}, \mathbf{m}) \,.$$

7. Prove that if $\mathbf{u} \in \mathbb{R}^n$ satisfies $\mathbf{B}\mathbf{u} = \mathbf{b}$, then for all $\mathbf{m}, \mathbf{p} \in \mathbb{R}^k$

$$\mathscr{L}_D(\mathbf{u}, \mathbf{p}, \mathbf{u}) = \mathscr{L}_D(\mathbf{u}, \mathbf{m}, \mathbf{u}) \,.$$

8. Show that if $\mathbf{u}_P \in \mathbb{R}^n$ is feasible for the primal quadratic programming problem

$$\mathscr{L}_P(\mathbf{u}, \mathbf{m}) \equiv \frac{1}{2}\mathbf{u}^H \mathbf{A} \mathbf{u} - \mathbf{a}^H \mathbf{u} + \mathbf{m}^H (\mathbf{B}\mathbf{u} - \mathbf{b}) , \tag{6.98}$$

and $(\mathbf{u}_D, \mathbf{m}_D) \in \mathbb{R}^n \times \mathbb{R}^k$ is feasible for the dual quadratic programming problem (6.95), then

$$P(\mathbf{u}_P) \geq D(\mathbf{u}_D, \mathbf{m}_D) .$$

9. Show that if $\mathbf{u}_P \in \mathbb{R}^n$ is feasible for the primal quadratic programming problem and the gradient of the primal objective is orthogonal to the tangent plane of the constraint, i.e.,

$$\mathbf{A}\mathbf{u}_P - \mathbf{a} = \mathbf{B}\mathbf{m} \in \mathscr{R}(\mathbf{B}) = \mathscr{N}(\mathbf{B})^{\perp} ,$$

then

$$P(\mathbf{u}_P) = D(\mathbf{u}_P, \mathbf{m}) .$$

10. Show that if $(\mathbf{u}_D, \mathbf{m}_D) \in \mathbb{R}^n \times \mathbb{R}^k$ is feasible for the dual quadratic programming problem and the gradient of the dual objective is orthogonal to the tangent plane of the constraint, i.e.,

$$\begin{bmatrix} \mathbf{A}\mathbf{u}_D \\ \mathbf{b} \end{bmatrix} = \begin{bmatrix} \mathbf{A} \\ \mathbf{B} \end{bmatrix} \mathbf{m} \in \mathscr{R}\left(\begin{bmatrix} \mathbf{A} \\ \mathbf{B} \end{bmatrix} \right) = \mathscr{N}\left([\mathbf{A}, \mathbf{B}^\top] \right)^{\perp} ,$$

then

$$D(\mathbf{u}_D, \mathbf{m}_D) = P(\mathbf{u}_D) .$$

11. Show that the dual quadratic programming problem is feasible and the quadratic programming problem is bounded.
12. Show that if $\mathbf{B}$ has rank k, then the primal quadratic programming problem is feasible and the dual quadratic programming problem is bounded.

Exercise 6.12.7 Suppose that $\mathbf{B} \in \mathbb{R}^{k \times n}$ is nonzero. Show that there is a constant $\sigma > 0$ so that for all $\mathbf{u} \in \mathbb{R}^n$

$$\sup_{\substack{\mathbf{m} \in \mathbb{R}^k \\ \mathbf{m} \neq \mathbf{0}}} \frac{\mathbf{m}^\top \mathbf{B}\mathbf{u}}{\|\mathbf{m}\|_2} \geq \sigma \inf_{\mathbf{z} \in \mathscr{N}(\mathbf{B})} \|\mathbf{u} - \mathbf{z}\|_2 . \tag{6.99}$$

References

1. A.V. Aho, J.E. Hopcroft, J.D. Ullman, *The Design and Analysis of Computer Algorithms* (Addison-Wesley, Boston, 1974) 2.1, 2.2.5.3
2. E. Anderson, Z. Bai, C. Bischof, J. Demmel, J. Dongarra, J. DuCroz, A. Greenbaum, S. Hammarling, A. McKenney, S. Ostrouchov, D. Sorensen (eds.), *LAPACK Users' Guide* (SIAM, Philadelphia, 1992) 2.1, 2.10.1, 3.7.3.4
3. P.J. Asente, R.R. Swick, *X Window System Toolkit: The Complete Programmer's Guide and Specification* (Digital Press, Boston, 1991) 4.7.1.2
4. O. Axelsson, *Iterative Solution Methods* (University Press, Cambridge, 1994) 3.13.4
5. J.J. Barton, L.R. Nackman, *Scientific and Engineering C++* (Addison-Wesley, Reading, 1994) 2.1
6. E.F. Beckenbach, R.E. Bellman, *Inequalities* (Springer, Berlin, 1965) 3.5.4
7. C. Bischof, C. van Loan, The WY representation for products of householder matrices. SIAM J. Sci. Stat. Comput. **8**(1), s2–s13 (1987) 6.7.1.4
8. Å. Björck, Solving linear least squares problems by Gram-Schmidt orthogonalization. BIT Numer. Math. **7**(1), 1–21 (1967) 6.8.5, 6.8.5, 6.8.5, 6.8.5, 6.8.5
9. Å. Björck, Iterative refinement of linear least squares solutions II. BIT Numer. Math. **8**(1), 8–30 (1968) 6.7.6
10. Å. Björck, Numerics of Gram-Schmidt orthogonalization. Linear Algebra Appl. **198**, 297–316 (1994) 6.8.4, 6.8.6, 6.8.8
11. Å. Björck, *Numerical Methods for Least Squares Problems* (SIAM, Philadelphia, 1996) 6.1
12. Å. Björck, G.H. Golub, Iterative refinement of linear least squares solutions by Householder transformations. BIT Numer. Math. **7**(4), 322–337 (1967) 6.7.6.2
13. J. Blanchette, M. Summerfield (eds.), *C++ GUI Programming with Qt 4*. Prentice Hall Open Source Software Development Series, 2nd edn. (Prentice-Hall, Upper Saddle River, 2008) 4.7.1.5
14. D. Braess, *Finite Elements* (Cambridge University Press, Cambridge, 2007) 6.12.3
15. D. Brennan, D. Heller, P. Ferguson, *Motif Programming Manual*, Definitive Guides to the X Window System, vol. 6a (O'Reilly Media, Sebastopol, 1993) 4.2, 4.7.1.3
16. S.C. Brenner, L. Ridgway Scott, *The Mathematical Theory of Finite Element Methods* (Springer, New York, 2002) 6.12.3
17. R. Brent, *Algorithms for Minimization Without Derivatives* (Prentice-Hall, Englewood Cliffs, 1973) 5.6.7, 5.6.8, 5.7.6.4
18. R.C. Buck, *Advanced Calculus* (McGraw-Hill, New York, 1965) 5.2.1
19. J.R. Bunch, L. Kaufman, Some stable methods for calculating inertia and solving symmetric linear systems. Math. Comput. **137**, 163–179 (1977) 3.13.2.3

© Springer International Publishing AG, part of Springer Nature 2017
J.A. Trangenstein, *Scientific Computing*, Texts in Computational Science and Engineering 18, https://doi.org/10.1007/978-3-319-69105-3

20. J.R. Bunch, B.N. Parlett, Direct methods for solving symmetric indefinite systems of linear equations. SIAM J. Numer. Anal. **8**, 639–655 (1971) 3.13.2.2
21. G. Casella, R.L. Berger, *Statistical Inference* (Duxbur, Belmont, 1990) 6.2.1
22. S. Chapman, *Fortran 95/2003 for Scientists and Engineers* (McGraw-Hill, New York, 2007) 2.1
23. A.K. Cline, C.B. Moler, G.W. Stewart, J.H. Wilkinson, An estimate for the condition number of a matrix. SIAM J. Numer. Anal. **16**, 368–375 (1979) 3.8.7
24. A.J. Cox, N.J. Higham, Accuracy and stability of the null space method for solving the equality constrained least squares problem. BIT Numer. Math. **39**(1), 34–50 (1999) 6.12.2
25. C.W. Cryer, Pivot size in Gaussian elimination. Numer. Math. **12**, 335–345 (1968) 3
26. G. Dahlquist, Å. Björck, *Numerical Methods* (Prentice-Hall, Englewood Cliffs, 1974). Translated by N. Anderson 3, 5.1, 5.6.6
27. J. Daniel, W.B. Gragg, L. Kaufman, G.W. Stewart, Reorthogonalization and stable algorithms for updating the Gram-Schmidt factorization. Math. Comput. **30**(136), 772–795 (1976) 6.8.6
28. T.J. Dekker, Finding a zero by means of successive linear interpolation, in *Constructive Aspects of the Fundamental Theorem of Algebra*, ed. by B. Dejon, P. Henrici (Wiley-Interscience, London, 1969) 5.6.2
29. J.W. Demmel, *Applied Numerical Linear Algebra* (SIAM, Philadelphia, 1997) 3.1, 6.1
30. J.E. Dennis Jr., Toward a unified convergence theory for Newton-like methods, in *Nonlinear Functional Analysis and Applications*, ed. by L.B. Rall (Academic Press, New York, 1971), pp. 425–472 5.4.5
31. J.E. Dennis Jr., R.B. Schnabel, *Numerical Methods for Unconstrained Optimization and Nonlinear Equations* (Prentice-Hall, Englewood Cliffs, 1983) 5.1, 5.4.9.3, 5.6.8, 5.7.4.1, 5.7.4.2
32. I.C.W. Dixon, G.P. Szegö, *Towards Global Optimization*, vol. 1 (North-Holland, Amsterdam, 1975) 5.7.7
33. I.C.W. Dixon, G.P. Szegö, *Towards Global Optimization*, vol. 2 (North-Holland, Amsterdam, 1978) 5.7.7
34. J.J. Dongarra, J.R. Bunch, C.B. Moler, G.W. Stewart, *Linpack Users' Guide* (SIAM, Philadelphia, 1979) 2.10, 3.1, 3.8.7, 6.1, 6.11.1
35. M. Dowell, P. Jarratt, A modified regula falsi method for computing the root of an equation. BIT Numer. Math. **11**(2), 168–174 (1971) 5.6.4
36. N.R. Draper, H. Smith, *Applied Regression Analysis* (Wiley, New York, 1998) 6.1
37. L. Eldén, Perturbation theory for the least squares problem with linear equality constraints. SIAM J. Numer. Anal. **17**(3), 338–350 (1980) 6.12.2
38. J.E. Epperson, *An Introduction to Numerical Methods and Analysis* (Wiley-Interscience, New York, 2007) 5.6.7, 5.6.8
39. D. Flanagan, *JavaScript: The Definitive Guide* (O'Reilly, Sebastopol, 2011) 2.1, 2.9.3.4, 4.7.1.6
40. J.D. Foley, A. van Dam, S.K. Feiner, J.F. Hughes, *Computer Graphics: Principles and Practice* (Addison-Wesley Professional, Boston, 1990) 4.1, 4.2, 4.2, 4.5
41. G.E. Forsythe, C.B. Moler, *Computer Solution of Linear Algebraic Systems* (Prentice-Hall, Englewood Cliffs, 1967) 3.9.2
42. A. Freeman (ed.), *The Definitive Guide to HTML5* (Apress, New York, 2011) 4.7.1.6
43. P.E. Gill, W. Murray, Safeguarded steplength algorithms for optimization using descent methods. Technical Report NPL Report NAC 37, National Physical Laboratory, Division of Numerical Analysis and Computing, Teddington, Aug 1974 5.7.6.4
44. P.E. Gill, W. Murray, M.H. Wright, *Practical Optimization*, Tenth printing (Academic, Boston, 1993) 5.1, 5.7.6.4, 5.7.7
45. G.H. Golub, Numerical methods for solving linear least squares problems. Numer. Math. **7**, 206–216 (1965) 6.7.2
46. G.H. Golub, C.F. van Loan, *Matrix Computations*, 4th edn. (Johns Hopkins, Baltimore, 2013) 3.1, 6.1, 6.7.1.1, 6.7.2, 6.11.2, 6.11.3
47. V. Granville, M. Krivanek, J.-P. Rasson, Simulated annealing: a proof of convergence. IEEE Trans. Pattern Anal. Mach. Intell. **16**(6), 652–656 (1994) 5.7.7.2

48. A. Greenbaum, *Numerical Methods: Design, Analysis, and Computer Implementation of Algorithms* (Princeton University Press, Princeton, 2012) 5.1
49. M. Gu, S.C. Eisenstat, A divide-and-conquer algorithm for the bidiagonal SVD. SIAM J. Matrix Anal. Appl. **16**(1), 79–92 (1995) 6.11.1
50. W.W. Hager, Condition estimators. SIAM J. Sci. Stat. Comput. **5**, 311–316 (1984) 3.8.7
51. P. Halmos, *Finite-Dimensional Vector Spaces*. University Series in Higher Mathematics (van Nostrand, Princeton, 1958) 3.2.2, 3.2.2, 3.2.5, 3.2.8, 3.2.11, 3.2.12, 6.8.1
52. G.H. Hardy, *A Course of Pure Mathematics*, 10th edn. (Cambridge University Press, Cambridge, 1967) 3.2.1
53. D.D. Hearn, M.P. Baker, W. Carithers, *Computer Graphics with Open GL* (Pearson, Harlow, 2010) 4.1
54. P. Henrici, *Elements of Numerical Analysis* (Wiley, New York, 1964) 5.1
55. N.J. Higham, Fortran codes for estimating the one-norm of a real or complex matrix, with applications to condition estimation. ACM Trans. Math. Softw. **14**, 381–396 (1988) 3.8.7, 3.8.7
56. N.J. Higham, *Accuracy and Stability of Numerical Algorithms* (SIAM Publications, Philadelphia, 2002) 2.1, 3.1, 3.8.1, 3.8.1, 3.8.1, 3.8.1, 3.8.2, 3.8.3, 6.1, 6.7.1.1, 6.7.2, 6.7.5, 6.7.6
57. A.E. Hoerl, R.W. Kennard, Ridge regression: biased estimation for nonorthogonal problems. Technometrics **42**(1), 80–86 (1970) 6.11.2
58. A.S. Householder, *The Numerical Treatment of a Single Nonlinear Equation* (McGraw-Hill, New York, 1970) 5.6.5, 5.6.6
59. E.F. Johnson, K. Reichard, *X Window Applications Programming* (MIS Press, New York, 1992) 4.2
60. N.M. Josuttis, *The C++ Standard Library: A Tutorial and Reference* (Addison-Wesley, Boston, 2000) 2.1
61. W. Kahan, Personal calculator has key to solve any equation $f(x) = 0$. *Hewlett-Packard J.* **30**(12), 20–26 (1979) 5, 5.3.3, 5.6.8
62. L.V. Kantorovich, Functional analysis and applied mathematics. Uspechi Mat. Nauk. **3**, 89–185 (1948) 5.4.5, 5.4.5
63. C.T. Kelley, *Iterative Methods for Linear and Nonlinear Equations* (SIAM, Philadelphia, 1995) 5.1
64. B.W. Kernighan, D.M. Ritchie, *C Programming Language* (Prentice-Hall, Englewood Cliffs, 1988) 2.1
65. J. Kiefer, Optimal sequential search and approximation methods under minimum regularity conditions. SIAM J. Appl. Math. **5**, 105–136 (1957) 5.7.6.1
66. D. Kincaid, W. Cheney, *Numerical Analysis* (Brooks/Cole, Pacific Grove, 1991) 5.1, 5.3.1, 5.6.8
67. S. Kirkpatrick, C.D. Gelatt, M.P. Vecchi, Optimization by simulated annealing. Science **220**, 671–680 (1983) 5.7.7.2
68. D.E. Knuth, *The Art of Computer Programming, 1: Fundamental Algorithms* (Addison-Wesley, Reading, 1997) 2.1
69. E. Kreyszig, *Introductory Functional Analysis with Applications* (Wiley, New York, 1978) 3.5.1
70. D. Kruglinski, S. Wingo, G. Shepherd, *Programming Microsoft Visual C++* (Microsoft Press, Redmond, 1998) 4.2
71. F.M. Larkin, Root-finding by fitting rational functions. Math. Comput. **35**, 803–816 (1980) 5.1, 5.6.6
72. P. Läuchli, Jordan-elimination und Ausgleichung nach kleinsten Quadraten. Numer. Math. **3**, 226–240 (1961) 6.4.1, 6.5.4, 6.8.2, 6.8.3, 6.8.4
73. C.L. Lawson, R.J. Hanson, *Solving Least Squares Problems* (SIAM, Philadelphia, 1995) 6.1
74. D. Le, An efficient derivative-free method for solving nonlinear equations. ACM Trans. Math. Softw. **11**(3), 250–262 (1985) 5.6.8
75. S. Logan (ed.), *Gtk+ Programming in C* (Prentice-Hall, Upper Saddle River, 2001) 4.7.1.4
76. L.H. Loomis, S. Sternberg, *Advanced Calculus* (Addison-Wesley, Reading, 1968) 3.5.1

77. W.E. Lorensen, H.E. Cline, Marching Cubes: a high resolution 3D surface construction algorithm. Comput. Graph. **21**(4), 163–169 (1987) 4.6.3.2
78. D.G. Luenberger (ed.), *Introduction to Linear and Nonlinear Programming* (Addison-Wesley, Reading, 1973) 3.8.7
79. M. Lutz, *Learning Python: Power Object-Oriented Programming* (O'Reilly Media, Sebastopol, 2013) 2.1
80. A. Munshi, D. Ginsburg, D. Shreiner, *Open GL ES 2.0 Programming Guide* (Addison-Wesley Professional, Reading, 2008) 4.7.2.1
81. D. Musser, *STL Tutorial and Reference Guide: C++ Programming with the Standard Template Library* (Addison Wesley, Reading, 2001) 2.10.2
82. J. Neter, W. Wasserman, M.H. Kutner, *Applied Linear Statistical models: Regression, Analysis of Variance, and Experimental Designs* (R.D. Irwin, Homeword, IL, 1990) 6.1
83. T.S. Newman, H. Yi, A survey of the marching cubes algorithm. Comput. Graph. **30**, 854–879 (2006) 4.6.3.2
84. A. Nye, *Xlib Programming Manual for Version 11, Rel. 5*, Definitive Guides to the X Window System, vol. 1 (O'Reilly Media, Sebastopol, 1994) 4.7.1.1
85. J.M. Ortega, W.C. Rheinboldt, *Iterative Solution of Nonlinear Equations in Several Variables* (Academic Press, New York, 1970) 5.1, 5.7.4.1, 5.7.4.1
86. T. Parisi, *WebGL: Up and Running* (O'Reilly Media, Sebastopol, 2012) 4.7.2.3
87. D.A. Patterson, J.L. Hennessy, *Computer Organization and Design: The Hardware/Software Interface* (Morgan Kaufmann, San Francisco, 1994) 2.1
88. R. Penrose, A generalized inverse for matrices. Proc. Camb. Philos. Soc. **51**, 406–413 (1955) 6.4.1
89. W.H. Press, S.A. Teukolsky, W.T. Vetterling, B.P. Flannery, *Numerical Recipes: The Art of Scientific Computing*, 3rd edn. (Cambridge University Press, New York, 2007) 6.11.2
90. A. Ralston, P. Rabinowitz, *A First Course in Numerical Analysis* (McGraw-Hill, New York, 1978) 5.1
91. J.R. Rice, Experiments on Gram-Schmidt orthogonalization. Math. Comput. **20**, 325–328 (1965) 6.8.1
92. P. Shifrin, M. Adams, *Linear Algebra: A Geometric Approach* (W.H. Freeman, New York, 2011) 6.8.1
93. P. Shirley, M. Ashikhmin, S. Marschner, *Fundamentals of Computer Graphics* (A K Peters, Natick, 2009) 4.1
94. A. Sidi, Unified treatment of regula falsi, Newton-Raphson, secant, and Steffensen methods for nonlinear equations. J. Online Math. Appl. **6**, 1–13 (2006) 5.4.9.3
95. R.D. Skeel, Iterative refinement implies numerical stability for Gaussian elimination. Math. Comput. **35**, 817–832 (1980) 3.9.2
96. G.W. Stewart, On the continuity of the generalized inverse. SIAM J. Appl. Math. **17**, 33–45 (1969) 6.5, 6.5, 6.5
97. G.W. Stewart, *Introduction to Matrix Computations* (Academic, San Diego, 1973) 3.1, 3.2.2, 3.2.5, 3.2.7, 3.2.8, 3.2.11, 3.4.3, 3.4.5.1, 3.5.1, 3.6.1, 3.6.4, 3.9.3, 3.13.4, 3.13.8, 6.1, 6.5.2, 6.7.1.1, 6.7.6
98. G.W. Stewart, *Matrix Algorithms: Basic Decompositions* (SIAM, Philadelphia, 1998) 3.1, 6.1
99. G. Strang, *Linear Algebra and Its Applications*, 3rd edn. (Harcourt Brace and Company, San Diego, 1988) 3.2.8, 3.2.9, 3.2.11, 3.4.3, 3.7.3.3, 3.10.2, 3.13.2, 3.13.5
100. B. Stroustrup, *The C++ Programming Language*, 3rd edn. (Addison-Wesley, Boston, 1997) 2.1, 2.6.3.3
101. A.N. Tikhonov, A.V. Goncharsky, V.V. Stepanov, A.G. Yagola, *Numerical Methods for the Solution of Ill-Posed Problem* (Kluwer Academic, Boston, 1995) 6.11.2
102. J.A. Trangenstein, *Numerical Solution of Elliptic and Parabolic Partial Differential Equations* (Cambridge University Press, Cambridge, 2013) 6.12.3
103. L.N. Trefethen, D. Bau III, *Numerical Linear Algebra* (SIAM, Philadelphia, 1997) 3.1, 6.1
104. R.A. van de Geijn, E.S. Quintana-Ortí, *The Science of Programming Matrix Computations* (Lulu, Morrisville, 2008) 2.10.1

105. A. van der Sluis, Conditioning, equilibration and pivoting in linear algebraic systems. Numer. Math. **15**, 74–86 (1970) 3.9.1
106. C. Van Loan, Computing the CS and the generalized singular value decompositions. Numer. Math. **46**, 479–491 (1985) 6.11.3
107. D. Vandevoorde, N.M. Josuttis, *C++ Templates: The Complete Guide* (Addison-Wesley, Boston, 2002) 2.1
108. R.C. Whaley, A. Petitet, Minimizing development and maintenance costs in supporting persistently optimized BLAS. Softw. Pract. Exp. **35**(2), 101–121 (2005) 2.10.1
109. J.H. Wilkinson, Error analysis of direct methods of matrix inversion. J. Assoc. Comput. Mach. **8**, 281–330 (1961) 3.8.5
110. J.H. Wilkinson, *Rounding Errors in Algebraic Processes* (Prentice-Hall, Englewood Cliffs, NJ, 1963) 2.1, 3.8.2, 3.8.2
111. J.H. Wilkinson, *The Algebraic Eigenvalue Problem* (Clarendon Press, Oxford, 1965) 3.1, 3.7.5, 3.8.2, 3.8.3, 3.8.5, 6.1, 6.7.5
112. J.H. Wilkinson, C. Reinsch, *Handbook for Automatic Computation Volume II Linear Algebra*, vol. 186. Die Grundlehren der Mathematischen Wissenschaften in Einzeldarstellungen (Springer, Berlin, 1971) 2.10, 3.1, 6.1
113. P. Wolfe, Convergence conditions for ascent methods. SIAM Rev. **11**, 226–235 (1969) 5.7.4.1, 5.7.4.1
114. P. Wolfe, Convergence conditions for ascent methods. II: Some corrections. SIAM Rev. **13**, 185–188 (1971) 5.7.4.1, 5.7.4.1
115. G. Zielke, Some remarks on matrix norms, condition numbers, and error estimates for linear equations. Linear Algebra Appl. **110**, 29–41 (1988) 3.5.3

Notation Index

© Springer International Publishing AG, part of Springer Nature 2017
J.A. Trangenstein, *Scientific Computing*, Texts in Computational Science and Engineering 18, https://doi.org/10.1007/978-3-319-69105-3

Author Index

© Springer International Publishing AG, part of Springer Nature 2017
J.A. Trangenstein, *Scientific Computing*, Texts in Computational Science and Engineering 18, https://doi.org/10.1007/978-3-319-69105-3

Subject Index

© Springer International Publishing AG, part of Springer Nature 2017
J.A. Trangenstein, *Scientific Computing*, Texts in Computational Science and Engineering 18, https://doi.org/10.1007/978-3-319-69105-3

Editorial Policy

1. Textbooks on topics in the field of computational science and engineering will be considered. They should be written for courses in CSE education. Both graduate and undergraduate textbooks will be published in TCSE. Multidisciplinary topics and multidisciplinary teams of authors are especially welcome.

2. Format: Only works in English will be considered. For evaluation purposes, manuscripts may be submitted in print or electronic form, in the latter case, preferably as pdf- or zipped ps-files. Authors are requested to use the LaTeX style files available from Springer at: http://www.springer.com/authors/book+authors/helpdesk?SGWID=0-1723113-12-971304-0 (Click on ⟶ Templates ⟶ LaTeX ⟶ monographs)
Electronic material can be included if appropriate. Please contact the publisher.

3. Those considering a book which might be suitable for the series are strongly advised to contact the publisher or the series editors at an early stage.

General Remarks

Careful preparation of manuscripts will help keep production time short and ensure a satisfactory appearance of the finished book.

The following terms and conditions hold:

Regarding free copies and royalties, the standard terms for Springer mathematics textbooks hold. Please write to martin.peters@springer.com for details.

Authors are entitled to purchase further copies of their book and other Springer books for their personal use, at a discount of 33.3% directly from Springer-Verlag.

Series Editors

Timothy J. Barth
NASA Ames Research Center
NAS Division
Moffett Field, CA 94035, USA
barth@nas.nasa.gov

Michael Griebel
Institut für Numerische Simulation
der Universität Bonn
Wegelerstr. 6
53115 Bonn, Germany
griebel@ins.uni-bonn.de

David E. Keyes
Mathematical and Computer Sciences
and Engineering
King Abdullah University of Science
and Technology
P.O. Box 55455
Jeddah 21534, Saudi Arabia
david.keyes@kaust.edu.sa

and

Department of Applied Physics
and Applied Mathematics
Columbia University
500 W. 120 th Street
New York, NY 10027, USA
kd2112@columbia.edu

Risto M. Nieminen
Department of Applied Physics
Aalto University School of Science
and Technology
00076 Aalto, Finland
risto.nieminen@tkk.fi

Dirk Roose
Department of Computer Science
Katholieke Universiteit Leuven
Celestijnenlaan 200A
3001 Leuven-Heverlee, Belgium
dirk.roose@cs.kuleuven.be

Tamar Schlick
Department of Chemistry
and Courant Institute
of Mathematical Sciences
New York University
251 Mercer Street
New York, NY 10012, USA
schlick@nyu.edu

Editor for Computational Science
and Engineering at Springer:
Martin Peters
Springer-Verlag
Mathematics Editorial IV
Tiergartenstrasse 17
69121 Heidelberg, Germany
martin.peters@springer.com

Texts in Computational Science and Engineering

1. H. P. Langtangen, *Computational Partial Differential Equations*. Numerical Methods and Diffpack Programming. 2nd Edition
2. A. Quarteroni, F. Saleri, P. Gervasio, *Scientific Computing with MATLAB and Octave*. 4th Edition
3. H. P. Langtangen, *Python Scripting for Computational Science*. 3rd Edition
4. H. Gardner, G. Manduchi, *Design Patterns for e-Science*.
5. M. Griebel, S. Knapek, G. Zumbusch, *Numerical Simulation in Molecular Dynamics*.
6. H. P. Langtangen, *A Primer on Scientific Programming with Python*. 5th Edition
7. A. Tveito, H. P. Langtangen, B. F. Nielsen, X. Cai, *Elements of Scientific Computing*.
8. B. Gustafsson, *Fundamentals of Scientific Computing*.
9. M. Bader, *Space-Filling Curves*.
10. M. Larson, F. Bengzon, *The Finite Element Method: Theory, Implementation and Applications*.
11. W. Gander, M. Gander, F. Kwok, *Scientific Computing: An Introduction using Maple and MATLAB*.
12. P. Deuflhard, S. Röblitz, *A Guide to Numerical Modelling in Systems Biology*.
13. M. H. Holmes, *Introduction to Scientific Computing and Data Analysis*.
14. S. Linge, H. P. Langtangen, *Programming for Computations* - A Gentle Introduction to Numerical Simulations with MATLAB/Octave.
15. S. Linge, H. P. Langtangen, *Programming for Computations* - A Gentle Introduction to Numerical Simulations with Python.
16. H.P. Langtangen, S. Linge, *Finite Difference Computing with PDEs* - A Modern Software Approach.
17. B. Gustafsson, *Scientific Computing from a Historical Perspective*
18. J.A. Trangenstein, *Scientific Computing* - Vol. I. – Linear and Nonlinear Equations
19. J.A. Trangenstein, *Scientific Computing* - Vol. II. – Eigenvalues and Optimization
20. J.A. Trangenstein, *Scientific Computing* - Vol. III. – Approximation and Integration

For further information on these books please have a look at our mathematics catalogue at the following URL: www.springer.com/series/5151

Monographs in Computational Science and Engineering

1. J. Sundnes, G.T. Lines, X. Cai, B.F. Nielsen, K.-A. Mardal, A. Tveito, *Computing the Electrical Activity in the Heart*.

For further information on this book, please have a look at our mathematics catalogue at the following URL: www.springer.com/series/7417

Lecture Notes in Computational Science and Engineering

1. D. Funaro, *Spectral Elements for Transport-Dominated Equations.*
2. H.P. Langtangen, *Computational Partial Differential Equations.* Numerical Methods and Diffpack Programming.
3. W. Hackbusch, G. Wittum (eds.), *Multigrid Methods V.*
4. P. Deuflhard, J. Hermans, B. Leimkuhler, A.E. Mark, S. Reich, R.D. Skeel (eds.), *Computational Molecular Dynamics: Challenges, Methods, Ideas.*
5. D. Kröner, M. Ohlberger, C. Rohde (eds.), *An Introduction to Recent Developments in Theory and Numerics for Conservation Laws.*
6. S. Turek, *Efficient Solvers for Incompressible Flow Problems.* An Algorithmic and Computational Approach.
7. R. von Schwerin, ***M**ulti **B**ody **S**ystem **SIM**ulation.* Numerical Methods, Algorithms, and Software.
8. H.-J. Bungartz, F. Durst, C. Zenger (eds.), *High Performance Scientific and Engineering Computing.*
9. T.J. Barth, H. Deconinck (eds.), *High-Order Methods for Computational Physics.*
10. H.P. Langtangen, A.M. Bruaset, E. Quak (eds.), *Advances in Software Tools for Scientific Computing.*
11. B. Cockburn, G.E. Karniadakis, C.-W. Shu (eds.), *Discontinuous Galerkin Methods.* Theory, Computation and Applications.
12. U. van Rienen, *Numerical Methods in Computational Electrodynamics.* Linear Systems in Practical Applications.
13. B. Engquist, L. Johnsson, M. Hammill, F. Short (eds.), *Simulation and Visualization on the Grid.*
14. E. Dick, K. Riemslagh, J. Vierendeels (eds.), *Multigrid Methods VI.*
15. A. Frommer, T. Lippert, B. Medeke, K. Schilling (eds.), *Numerical Challenges in Lattice Quantum Chromodynamics.*
16. J. Lang, *Adaptive Multilevel Solution of Nonlinear Parabolic PDE Systems.* Theory, Algorithm, and Applications.
17. B.I. Wohlmuth, *Discretization Methods and Iterative Solvers Based on Domain Decomposition.*
18. U. van Rienen, M. Günther, D. Hecht (eds.), *Scientific Computing in Electrical Engineering.*
19. I. Babuška, P.G. Ciarlet, T. Miyoshi (eds.), *Mathematical Modeling and Numerical Simulation in Continuum Mechanics.*
20. T.J. Barth, T. Chan, R. Haimes (eds.), *Multiscale and Multiresolution Methods.* Theory and Applications.
21. M. Breuer, F. Durst, C. Zenger (eds.), *High Performance Scientific and Engineering Computing.*
22. K. Urban, *Wavelets in Numerical Simulation.* Problem Adapted Construction and Applications.
23. L.F. Pavarino, A. Toselli (eds.), *Recent Developments in Domain Decomposition Methods.*

24. T. Schlick, H.H. Gan (eds.), *Computational Methods for Macromolecules: Challenges and Applications.*
25. T.J. Barth, H. Deconinck (eds.), *Error Estimation and Adaptive Discretization Methods in Computational Fluid Dynamics.*
26. M. Griebel, M.A. Schweitzer (eds.), *Meshfree Methods for Partial Differential Equations.*
27. S. Müller, *Adaptive Multiscale Schemes for Conservation Laws.*
28. C. Carstensen, S. Funken, W. Hackbusch, R.H.W. Hoppe, P. Monk (eds.), *Computational Electromagnetics.*
29. M.A. Schweitzer, *A Parallel Multilevel Partition of Unity Method for Elliptic Partial Differential Equations.*
30. T. Biegler, O. Ghattas, M. Heinkenschloss, B. van Bloemen Waanders (eds.), *Large-Scale PDE-Constrained Optimization.*
31. M. Ainsworth, P. Davies, D. Duncan, P. Martin, B. Rynne (eds.), *Topics in Computational Wave Propagation.* Direct and Inverse Problems.
32. H. Emmerich, B. Nestler, M. Schreckenberg (eds.), *Interface and Transport Dynamics.* Computational Modelling.
33. H.P. Langtangen, A. Tveito (eds.), *Advanced Topics in Computational Partial Differential Equations.* Numerical Methods and Diffpack Programming.
34. V. John, *Large Eddy Simulation of Turbulent Incompressible Flows.* Analytical and Numerical Results for a Class of LES Models.
35. E. Bänsch (ed.), *Challenges in Scientific Computing - CISC 2002.*
36. B.N. Khoromskij, G. Wittum, *Numerical Solution of Elliptic Differential Equations by Reduction to the Interface.*
37. A. Iske, *Multiresolution Methods in Scattered Data Modelling.*
38. S.-I. Niculescu, K. Gu (eds.), *Advances in Time-Delay Systems.*
39. S. Attinger, P. Koumoutsakos (eds.), *Multiscale Modelling and Simulation.*
40. R. Kornhuber, R. Hoppe, J. Périaux, O. Pironneau, O. Wildlund, J. Xu (eds.), *Domain Decomposition Methods in Science and Engineering.*
41. T. Plewa, T. Linde, V.G. Weirs (eds.), *Adaptive Mesh Refinement – Theory and Applications.*
42. A. Schmidt, K.G. Siebert, *Design of Adaptive Finite Element Software.* The Finite Element Toolbox ALBERTA.
43. M. Griebel, M.A. Schweitzer (eds.), *Meshfree Methods for Partial Differential Equations II.*
44. B. Engquist, P. Lötstedt, O. Runborg (eds.), *Multiscale Methods in Science and Engineering.*
45. P. Benner, V. Mehrmann, D.C. Sorensen (eds.), *Dimension Reduction of Large-Scale Systems.*
46. D. Kressner, *Numerical Methods for General and Structured Eigenvalue Problems.*
47. A. Boriçi, A. Frommer, B. Joó, A. Kennedy, B. Pendleton (eds.), *QCD and Numerical Analysis III.*
48. F. Graziani (ed.), *Computational Methods in Transport.*
49. B. Leimkuhler, C. Chipot, R. Elber, A. Laaksonen, A. Mark, T. Schlick, C. Schütte, R. Skeel (eds.), *New Algorithms for Macromolecular Simulation.*

50. M. Bücker, G. Corliss, P. Hovland, U. Naumann, B. Norris (eds.), *Automatic Differentiation: Applications, Theory, and Implementations.*

51. A.M. Bruaset, A. Tveito (eds.), *Numerical Solution of Partial Differential Equations on Parallel Computers.*

52. K.H. Hoffmann, A. Meyer (eds.), *Parallel Algorithms and Cluster Computing.*

53. H.-J. Bungartz, M. Schäfer (eds.), *Fluid-Structure Interaction.*

54. J. Behrens, *Adaptive Atmospheric Modeling.*

55. O. Widlund, D. Keyes (eds.), *Domain Decomposition Methods in Science and Engineering XVI.*

56. S. Kassinos, C. Langer, G. Iaccarino, P. Moin (eds.), *Complex Effects in Large Eddy Simulations.*

57. M. Griebel, M.A Schweitzer (eds.), *Meshfree Methods for Partial Differential Equations III.*

58. A.N. Gorban, B. Kégl, D.C. Wunsch, A. Zinovyev (eds.), *Principal Manifolds for Data Visualization and Dimension Reduction.*

59. H. Ammari (ed.), *Modeling and Computations in Electromagnetics: A Volume Dedicated to Jean-Claude Nédélec.*

60. U. Langer, M. Discacciati, D. Keyes, O. Widlund, W. Zulehner (eds.), *Domain Decomposition Methods in Science and Engineering XVII.*

61. T. Mathew, *Domain Decomposition Methods for the Numerical Solution of Partial Differential Equations.*

62. F. Graziani (ed.), *Computational Methods in Transport: Verification and Validation.*

63. M. Bebendorf, *Hierarchical Matrices.* A Means to Efficiently Solve Elliptic Boundary Value Problems.

64. C.H. Bischof, H.M. Bücker, P. Hovland, U. Naumann, J. Utke (eds.), *Advances in Automatic Differentiation.*

65. M. Griebel, M.A. Schweitzer (eds.), *Meshfree Methods for Partial Differential Equations IV.*

66. B. Engquist, P. Lötstedt, O. Runborg (eds.), *Multiscale Modeling and Simulation in Science.*

67. I.H. Tuncer, Ü. Gülcat, D.R. Emerson, K. Matsuno (eds.), *Parallel Computational Fluid Dynamics 2007.*

68. S. Yip, T. Diaz de la Rubia (eds.), *Scientific Modeling and Simulations.*

69. A. Hegarty, N. Kopteva, E. O'Riordan, M. Stynes (eds.), *BAIL 2008 – Boundary and Interior Layers.*

70. M. Bercovier, M.J. Gander, R. Kornhuber, O. Widlund (eds.), *Domain Decomposition Methods in Science and Engineering XVIII.*

71. B. Koren, C. Vuik (eds.), *Advanced Computational Methods in Science and Engineering.*

72. M. Peters (ed.), *Computational Fluid Dynamics for Sport Simulation.*

73. H.-J. Bungartz, M. Mehl, M. Schäfer (eds.), *Fluid Structure Interaction II - Modelling, Simulation, Optimization.*

74. D. Tromeur-Dervout, G. Brenner, D.R. Emerson, J. Erhel (eds.), *Parallel Computational Fluid Dynamics 2008.*

75. A.N. Gorban, D. Roose (eds.), *Coping with Complexity: Model Reduction and Data Analysis.*

76. J.S. Hesthaven, E.M. Rønquist (eds.), *Spectral and High Order Methods for Partial Differential Equations.*
77. M. Holtz, *Sparse Grid Quadrature in High Dimensions with Applications in Finance and Insurance.*
78. Y. Huang, R. Kornhuber, O.Widlund, J. Xu (eds.), *Domain Decomposition Methods in Science and Engineering XIX.*
79. M. Griebel, M.A. Schweitzer (eds.), *Meshfree Methods for Partial Differential Equations V.*
80. P.H. Lauritzen, C. Jablonowski, M.A. Taylor, R.D. Nair (eds.), *Numerical Techniques for Global Atmospheric Models.*
81. C. Clavero, J.L. Gracia, F.J. Lisbona (eds.), *BAIL 2010 – Boundary and Interior Layers, Computational and Asymptotic Methods.*
82. B. Engquist, O. Runborg, Y.R. Tsai (eds.), *Numerical Analysis and Multiscale Computations.*
83. I.G. Graham, T.Y. Hou, O. Lakkis, R. Scheichl (eds.), *Numerical Analysis of Multiscale Problems.*
84. A. Logg, K.-A. Mardal, G. Wells (eds.), *Automated Solution of Differential Equations by the Finite Element Method.*
85. J. Blowey, M. Jensen (eds.), *Frontiers in Numerical Analysis - Durham 2010.*
86. O. Kolditz, U.-J. Gorke, H. Shao, W. Wang (eds.), *Thermo-Hydro-Mechanical-Chemical Processes in Fractured Porous Media - Benchmarks and Examples.*
87. S. Forth, P. Hovland, E. Phipps, J. Utke, A. Walther (eds.), *Recent Advances in Algorithmic Differentiation.*
88. J. Garcke, M. Griebel (eds.), *Sparse Grids and Applications.*
89. M. Griebel, M.A. Schweitzer (eds.), *Meshfree Methods for Partial Differential Equations VI.*
90. C. Pechstein, *Finite and Boundary Element Tearing and Interconnecting Solvers for Multiscale Problems.*
91. R. Bank, M. Holst, O. Widlund, J. Xu (eds.), *Domain Decomposition Methods in Science and Engineering XX.*
92. H. Bijl, D. Lucor, S. Mishra, C. Schwab (eds.), *Uncertainty Quantification in Computational Fluid Dynamics.*
93. M. Bader, H.-J. Bungartz, T. Weinzierl (eds.), *Advanced Computing.*
94. M. Ehrhardt, T. Koprucki (eds.), *Advanced Mathematical Models and Numerical Techniques for Multi-Band Effective Mass Approximations.*
95. M. Azaïez, H. El Fekih, J.S. Hesthaven (eds.), *Spectral and High Order Methods for Partial Differential Equations ICOSAHOM 2012.*
96. F. Graziani, M.P. Desjarlais, R. Redmer, S.B. Trickey (eds.), *Frontiers and Challenges in Warm Dense Matter.*
97. J. Garcke, D. Pflüger (eds.), *Sparse Grids and Applications – Munich 2012.*
98. J. Erhel, M. Gander, L. Halpern, G. Pichot, T. Sassi, O. Widlund (eds.), *Domain Decomposition Methods in Science and Engineering XXI.*
99. R. Abgrall, H. Beaugendre, P.M. Congedo, C. Dobrzynski, V. Perrier, M. Ricchiuto (eds.), *High Order Nonlinear Numerical Methods for Evolutionary PDEs - HONOM 2013.*
100. M. Griebel, M.A. Schweitzer (eds.), *Meshfree Methods for Partial Differential Equations VII.*

101. R. Hoppe (ed.), *Optimization with PDE Constraints - OPTPDE 2014.*

102. S. Dahlke, W. Dahmen, M. Griebel, W. Hackbusch, K. Ritter, R. Schneider, C. Schwab, H. Yserentant (eds.), *Extraction of Quantifiable Information from Complex Systems.*

103. A. Abdulle, S. Deparis, D. Kressner, F. Nobile, M. Picasso (eds.), *Numerical Mathematics and Advanced Applications - ENUMATH 2013.*

104. T. Dickopf, M.J. Gander, L. Halpern, R. Krause, L.F. Pavarino (eds.), *Domain Decomposition Methods in Science and Engineering XXII.*

105. M. Mehl, M. Bischoff, M. Schäfer (eds.), *Recent Trends in Computational Engineering - CE2014.* Optimization, Uncertainty, Parallel Algorithms, Coupled and Complex Problems.

106. R.M. Kirby, M. Berzins, J.S. Hesthaven (eds.), *Spectral and High Order Methods for Partial Differential Equations - ICOSAHOM'14.*

107. B. Jüttler, B. Simeon (eds.), *Isogeometric Analysis and Applications 2014.*

108. P. Knobloch (ed.), *Boundary and Interior Layers, Computational and Asymptotic Methods – BAIL 2014.*

109. J. Garcke, D. Pflüger (eds.), *Sparse Grids and Applications – Stuttgart 2014.*

110. H. P. Langtangen, *Finite Difference Computing with Exponential Decay Models.*

111. A. Tveito, G.T. Lines, *Computing Characterizations of Drugs for Ion Channels and Receptors Using Markov Models.*

112. B. Karazösen, M. Manguoğlu, M. Tezer-Sezgin, S. Göktepe, Ö. Uğur (eds.), *Numerical Mathematics and Advanced Applications - ENUMATH 2015.*

113. H.-J. Bungartz, P. Neumann, W.E. Nagel (eds.), *Software for Exascale Computing - SPPEXA 2013-2015.*

114. G.R. Barrenechea, F. Brezzi, A. Cangiani, E.H. Georgoulis (eds.), *Building Bridges: Connections and Challenges in Modern Approaches to Numerical Partial Differential Equations.*

115. M. Griebel, M.A. Schweitzer (eds.), *Meshfree Methods for Partial Differential Equations VIII.*

116. C.-O. Lee, X.-C. Cai, D.E. Keyes, H.H. Kim, A. Klawonn, E.-J. Park, O.B. Widlund (eds.), *Domain Decomposition Methods in Science and Engineering XXIII.*

117. T. Sakurai, S. Zhang, T. Imamura, Y. Yusaku, K. Yoshinobu, H. Takeo (eds.), *Eigenvalue Problems: Algorithms, Software and Applications, in Petascale Computing.* EPASA 2015, Tsukuba, Japan, September 2015.

118. T. Richter (ed.), *Fluid-structure Interactions.* Models, Analysis and Finite Elements.

119. M.L. Bittencourt, N.A. Dumont, J.S. Hesthaven (eds.), *Spectral and High Order Methods for Partial Differential Equations ICOSAHOM 2016.*

120. Z. Huang, M. Stynes, Z. Zhang (eds.), *Boundary and Interior Layers, Computational and Asymptotic Methods BAIL 2016.*

121. S.P.A. Bordas, E.N. Burman, M.G. Larson, M.A. Olshanskii (eds.), *Geometrically Unfitted Finite Element Methods and Applications.* Proceedings of the UCL Workshop 2016.

122. A. Gerisch, R. Penta, J. Lang (eds.), *Multiscale Models in Mechano and Tumor Biology.* Modeling, Homogenization, and Applications.

For further information on these books please have a look at our mathematics catalogue at the following URL: www.springer.com/series/3527

Printed in the United States
By Bookmasters